# Volcanoes and the Environment

*Volcanoes and the Environment* incorporates contributions from some of the foremost authorities in volcanology from around the world to form a comprehensive and accessible text. This book is an indispensable guide for all those interested in how volcanism has affected our planet's environment in the past and will continue to do so in the future. Spanning a wide variety of topics from geology to climatology and ecology, it also considers the economic and social impacts of volcanic activity on humans.

Chapters cover the role of volcanoes in shaping our planet's environment through the eons, and their effect on the geological cycle; the impacts on atmosphere and climate; impacts on the health of those living on active volcanoes; the role of volcanism on early life; effects of eruptions on modern plant and animal life and implications from these studies; links between large eruptions and mass extinctions; relationships between contrasting human societies and volcanic disasters; how volcanoes can provide heat energy and supply precious base metals, as well as raw material, for our industries; and impacts of volcanic disasters on the economy.

This book is intended for students and researchers interested in environmental change from the fields of earth and environmental science, geography, ecology, and social science. It will also interest and inform policy-makers and professionals working in natural hazard planning and mitigation.

JOAN MARTÍ is a Research Professor at the Institute of Earth Sciences of the Spanish National Research Council (CSIC) in Barcelona. He studied geology and obtained his Ph.D. at the University of Barcelona in 1986. His Ph.D. research focused on the ancient explosive volcanism of the Pyrenees. Since the beginning of his scientific career, he has conducted much research relating to the structure of calderas (giant collapse craters), and the dynamics of caldera-forming eruptions and associated hazards. Dr. Martí has studied explosive volcanism in different modern and ancient volcanic environments. In particular, he has contributed many studies on explosive volcanism in Tenerife (Canary Islands), where he continues to coordinate several research projects.

GERALD ERNST was awarded a B.Sc. in geology and mineralogy from the University of Liège, Belgium, in 1990 and a Ph.D. in Earth sciences from the University of Bristol, UK, in 1997. Following appointments as a postdoctoral fellow and lecturer in volcanology and geological fluid mechanics at the University of Bristol he joined the Belgian National Science Foundation (FWO-Vlaanderen) and the University of Ghent. as a researcher in 2003. He is currently pursuing new research focused on how volcanoes work and how they affect people, their activities, and the wider environment. Dr. Ernst is working to establish the Mercator and Ortelius Research Centre for Eruption Dynamics, and the Joseph Plateau Geological Fluid Dynamic Laboratory at the University of Ghent. He is a Honorary Research Fellow of the University of Bristol, a Scientific Collaborator of the University of Brussels, and an Adjunct Assistant Professor of Volcanology at Michigan Tech University (USA). He has been the recipient of four awards for his work including the prestigious 2002 Golden Clover Prize of the Fondation Belge de la Vocation.

# Volcanoes and the Environment

Edited by

## Joan Martí

*Institute of Earth Sciences "Jaume Almera", Consejo Superior de Investigaciones Cientificas*

and

## Gerald Ernst

*Mercaton and Ortelius Research Centre for Eruption Dynamics, and Joseph Plateau Geological Fluid Dynamics Laboratory, University of Ghent*

CAMBRIDGE UNIVERSITY PRESS

CAMBRIDGE UNIVERSITY PRESS
Cambridge, New York, Melbourne, Madrid, Cape Town, Singapore, São Paulo

Cambridge University Press
The Edinburgh Building, Cambridge CB2 2RU, UK

Published in the United States of America by Cambridge University Press, New York

www.cambridge.org
Information on this title: www.cambridge.org/9780521592543

First published 2005

Printed in the United Kingdom at the University Press, Cambridge

*A catalogue record for this book is available from the British Library*

*Library of Congress Cataloguing in Publication data*

ISBN-13 978-0-521-59254-3 hardback
ISBN-10 0-521-59254-2 hardback

# Contents

# Contributors

Peter J. Baxter
Department of Community Medicine
University of Cambridge
Addenbrooke's Hospital
Cambridge
CB2 2QQ
UK

Charlotte Benson
209 Jalan Ara
Bangsar Baru
59100 Kuala Lumpur
Malaysia

Steven N. Carey
Graduate School of Oceanography
University of Rhode Island
Narragansett
RI 02882
USA

Ray A. F. Cas
School of Geosciences
Monash University
P.O. Box 28E
Victoria, 3800
Australia

David K. Chester
Department of Geography
University of Liverpool
Liverpool
L69 3BX
UK

Virginia H. Dale
Environmental Sciences Division
Oak Ridge National Laboratory
Oak Ridge
TN 37831
USA

Johanna Delgado-Acevedo
Universidad de Puerto Rico
Departamento de Biologia
Apdo. 23360
00931 San Juan
Puerto Rico

Wendell A. Duffield
Geology Department
Northern Arizona University
Flagstaff
AZ 86011
USA

John S. Edwards
Department of Zoology
University of Washington
Seattle
WA 98195
USA

Gerald G. J. Ernst
Mercaton and Ontelius Research Centre
for Eruption Dynamics and
Joseph Plateau Geological Fluid
Dynamics Laboratory
Geological Institute
University of Ghent,
Krijgsloan 281/S8
B-9000 Ghent
Belgium

Arnau Folch
Institute of Earth Sciences "Jaume Almera"
Consejo Superior de Investigaciones Cientificas
Lluis Solé Sabaris s/n
08028 Barcelona
Spain

Harold L. Gibson
Department of Earth Sciences
Laurentian University
Sudbury
Ontario
P3E 2C6
Canada

Grant Heiken
Earth and Environmental Sciences Division
Los Alamos National Laboratory
Los Alamos
NM 87545
USA

James MacMahon
College of Science
Utah State University
5305 University Blvd
Logan
UT 84322
USA

Joan Martí
Institute of Earth Sciences "Jaume Almera"
Consejo Superior de Investigaciones Cientificas
Lluis Solé Sabaris s/n
08028 Barcelona
Spain

Stephen Self
Department of Earth Sciences
The Open University
Walton Hall
Milton Keynes
MK7 6AA
UK

Karl O. Stetter
Lehrstuhl für Mikrobiologie
Universität Regensburg
Universitätsstrasse 31
D-93053 Regensburg
Germany

Robert I. Tilling
Volcano Hazards Team
US Geological Survey
345 Middsfield Road, MS-910
Menlo Park
CA 94025
USA

Paul B. Wignall
School of Earth Sciences
University of Leeds
Leeds
LS2 9JT
UK

# Preface

Volcanic eruptions are among the most fascinating natural phenomena and can have significant impacts upon the environment. One only has to think of the 1883 eruption of Krakatau or of the 1980 eruption of Mount St. Helens to get a sense of the awesome power and wide-ranging impacts of eruptions. Some readers will remember hearing or reading about the loss of life and devastation around those volcanoes, while others will remember how even larger eruptions than these cooled the Earth's climate and affected the ozone layer. Most will recall controversial discussions about how volcanoes may have eradicated the dinosaurs some 65 million years ago or about how super-eruptions may have nearly wiped out our human ancestors some 75 000 years ago. We like to think it is fortunate that they did not, but what will happen when the next super-eruption strikes? Most recently, eruptions at ocean island volcanoes have even been proposed as triggers for catastrophic, massive volcano failure and the generation of tsunami waves of unimaginable proportion. The 26 December 2004 Banda Aceh tsunami; off Sumatra, which devastated coastal areas in SE Asia, killing 300000 and precipitating several million people into a state of absolute poverty, is a small event in comparison. Volcano-related mega-tsunamis represent a very great risk to many coastal cities around the world and to their populations, and no doubt to the world economy. Moreover, it has now been demonstrated that periods of severe cooling lasting 1000 years are unambiguously correlated with eruptions. This will no doubt fuel heated debates as to whether eruptions can indeed trigger glaciations on that timescale.

Eruptions arguably have the potential to have all these impacts and many more upon their surroundings and, in some cases, upon the global environment. This is a sobering thought, especially when considering that there are still a large number of potentially hazardous volcanoes of which we know next to nothing. For example, about 80% of active volcanoes, most of which are located in developing countries where local populations are the most vulnerable to natural hazards, remain largely unstudied. In this age of supercomputers, nanotechnology and space exploration, we know very little about these volcanoes of our own planet, most of which are not monitored to any extent, or about the impacts they could have in future. Changing this situation needs to be one of the highest priorities of volcanological efforts in coming decades.

Volcanoes also contribute positively to the environment. They have brought much to "life" in the past, and represent, at present, an important source of benefits for humanity. For example:

- Volcanoes have decisively contributed to the origin of life and Earth's atmosphere, and are often regarded as directly responsible for the existence of highly fertile soils in many parts of our planet. Uplift related to volcanism in easternmost Africa, starting millions of years ago, and associated climatic and vegetation changes, have been related to the emergence of our direct ancestors in the Rift Valley some 2–3 million years ago.
- The volcano-derived heat that may have fueled the emergence of life in the oceans is now being mined as geothermal energy on several continents. Volcanoes are also the source of ore deposits and an important source of material for industry.
- By studying the dispersal of giant clouds produced by explosive volcanic eruptions, atmospheric scientists have made great advances in understanding how the atmosphere works, and particularly in understanding its energy balance and the complexity of atmospheric circulation. Each time a large eruption happens, it is a natural experiment on the scale of our planet as a whole, enabling atmospheric scientists to put to challenging tests their models predicting our current and future climate.
- Thick volcanic ash layers from powerful eruptions have also buried and preserved the records of ancient civilizations worldwide. In Europe, this is illustrated by the late Bronze

Age towns on the Island of Thera (Santorini, Greece; arguably the legendary Atlantis or part of it), which have provided profound insights into the early development of European art, from the study of its beautiful paintings on walls and pottery. Two thousand years later, another large eruption buried the Roman town of Pompeii (near Naples, Italy), again providing invaluable clues into what made a great civilization. Of course, these buried civilizations are also a stark reminder of the extraordinary destructive power of the largest of eruptions, for which we have either no or only limited historical records.

• Studying the origin of life at hot vents on deep submarine volcanoes can also shed light on whether life may exist on other planets such as Europa, a satellite of Jupiter; and on whether life may have thrived on Mars in the past.

The list of benefits derived from volcanoes and their study extends much beyond this short list of examples as will become clear during the reading of this book.

In this textbook we have taken the broadest possible view of the environment. We have considered not just the impacts of eruptions on the atmosphere and climate, and on the flora, fauna and humans around volcanoes but also the impacts on human health, human societies, and the local and national economies. We have also considered the role of volcanism in generating precious base metal resources, the use of volcanic materials in industry and the recovery of volcano-derived thermal energy, the contribution of volcanic eruptions to past mass extinctions, and the role of volcanism and volcano-derived hydrothermal venting on the emergence of the earliest forms of life on Earth and on the development of the primitive atmosphere and ocean. We have also included a treatment of the role of volcanism in the geological cycle.

In order to enable discussions of all these interrelated impacts to be easily followed it has been necessary to introduce a basic treatment on the physical understanding of volcanic eruptions, as well as on volcanic hazards and on how eruptions can be anticipated. These topics are fascinating aspects in their own right and set the scene for the discussions that follow.

This wide coverage of topics related to the impacts of volcanism and volcanic eruptions upon the many interrelated aspects of the environment considered in the widest sense is what makes this book unique. A few years ago we came to realize that many of the aspects now discussed in this new text were covered only in part and only in scattered texts. There was not a single text attempting to present an integrated treatment. Many of the discussions appeared in specialized journal articles and books, making them inaccessible to amateur environmentalists and to most undergraduates and postgraduates. We felt that what was needed was a text that would be not only comprehensive in its treatment of the subject but also accessible to a wide audience of naturalist amateurs, undergraduate and postgraduate students in the environmental, geographical and earth sciences as well as an easy-to-carry reference text for all our research and teaching colleagues across many scientific disciplines.

Seventeen of our colleagues among the leading authorities on the subject enthusiastically shared our vision and together we started preparing what was to become this book, the very first textbook extensively discussing most aspects of the impacts of volcanic eruptions upon the environment.

We hope that like all the contributors to this book the readers will be irresistibly enthused by this exciting subject. The following paragraphs introduce the successive chapters in more detail, outlining some of the basic questions, which are discussed by the contributors.

Chapters 1–3 are foundation chapters needed before the diverse effects of volcanism can be studied. Chapter 1, by Steven Carey, a senior physical volcanologist at the University of Rhode Island, sets the scene by reviewing our understanding of volcanoes and of the physical processes associated with volcanic eruptions. Some of the questions considered include: where is volcanic activity concentrated and why? What are the main contrasting types of volcanic settings and how do they differ in the styles of the eruptions which occur there? How variable is the composition of magma and how does this come about? How are magmas generated? In what way do volcanic eruptions contribute to the development of the landscape (or the "seascape") – mountains and topographic depressions?

Volcanic hazards can affect humans and the environment. But what are the main types of hazards and how can humans cope with them? How can we assess them? Can we predict the onset of eruptions? What are the techniques used? Is it sufficient to anticipate where and when eruptions will occur to anticipate potential disasters, either environmental or human in character? These are some of the crucial questions considered, in Chapter 2, by Robert Tilling, a veteran of volcano monitoring and volcano disaster response with the US Geological Survey. Tilling discusses some of these questions by comparing the cases and responses to two classic eruptions, the eruptions in 1985 at Nevado del Ruiz, Colombia and in 1991 at Mt. Pinatubo, Philippines. He concludes by summarizing some of the key lessons that have been learnt by those who have lived through volcanic crises and volcanic disasters such as these.

Chapter 3, by Joan Martí and Arnau Folch, both experienced volcanologists and modelers at CSIC (Consejo Superior de Investigaciones Científicas, i.e., the Spanish National Research Council) in Barcelona, builds up on Chapters 1 and 2. It focuses specifically on how we can anticipate volcanic eruptions. It discusses what techniques are used in the wide variety of volcanic settings and corresponding volcano types. It also discusses how theory and laboratory work can help by providing invaluable insights and understanding. The chapter shows along its different sections that the best way to anticipate a volcanic eruption and its effects is by combining a good knowledge of the volcano's eruptive behavior (physical volcanology) and of the level of its current activity (volcano monitoring).

These three chapters are fundamental to understanding the character of volcanic perturbations, which can impact on humans and their activities or the environment at large. Chapters 4–5 explore in turn the relationships between volcanism and a specific aspect of the "environment." Chapters 4–5 are related, respectively, to the relationship of volcanism with geological time and space, and with a specific aspect of the physical environment of our planet, namely its atmosphere. Some of the key questions discussed include: what has been the contribution of volcanoes in shaping our planet and its environment through the history of the Earth? Are eruptions becoming more or less frequent now than millions of years ago? What is the role of plate tectonics and when did plate tectonics start to act as a major driving force in the evolution of our planet? What was, and what is now the contribution of volcanic degassing and volcanic eruptions to the atmosphere? – to the oceans? – to life? These questions are discussed in Chapter 4 by Ray Cas, a veteran volcanologist at Monash University who has studied volcanoes first-hand in many parts of the world.

In recent years there has been a growing realization that volcanic degassing and eruptions of a range of styles can impact upon both the chemical composition of the atmosphere and upon its radiative energy balance. This, in turn, changes the temperature distribution on the Earth and thus the weather and climate that we experience. In Chapter 5 Stephen Self, a leading volcanologist at the Open University and pioneer of research in this area, introduces the main mechanisms by which contrasting styles of eruptions can affect atmospheric composition and climate. Through a few key examples that include super-eruptions and flood basalt volcanism, he illustrates that we are increasingly able to document how large eruptions affected the atmosphere and climate in the past. This effort is crucial to help us foresee the impacts that future eruptions will have.

Chapters 6–9 explore in turn several key aspects of the relationship between volcanism and life, both in the past and at present. One of the most fascinating aspects of our relationship to volcanism is in how volcanic activity may have played a major role in our ultimate origin. That is the origin of ancestral forms of life that evolved into increasingly complex forms and eventually some 4 billion years later or so, to us *Homo sapiens*. In Chapter 6, Karl Stetter, a senior researcher at the University of California–Los Angeles and the University of Regensburg and a leading authority on the subject, considers some of the following questions: what are the most primitive forms of life still in existence today? How do they relate to volcanism? How can we describe their relationships with other species in an evolutionary tree? How diverse are these primitive forms of life and how have they survived the competition for almost 4 billion years? Professor Stetter considers these fascinating questions and more, and places them in the framework of the history of life on our planet.

An area of heated debate has been whether or not the dinosaurs were wiped out by an asteroid that fell over the Yucatan Peninsula in Mexico, or whether longer-term volcanic activity covering the largest part of India with layer upon layer of lava has anything to do with it. In Chapter 7, Paul Wignall, a geologist and paleontologist leading work in this area at the University of Leeds, seeks evidence to resolve this issue. Wignall does not limit the discussion to the great extinction of the dinosaurs and other species at the K–T boundary, but considers a variety of well-documented mass extinction events, which may be related to massive outpourings of lava on a continental scale. The chapter closes by balancing the evidence in favor and against a major role of volcanism in the mass extinction scenarios.

In Chapter 8, Virginia Dale, Johanna Delgado-Acevedo, and James MacMahon, all volcano botanists and ecologists, from the Oak Ridge National Laboratory, the University of Puerto Rico, and Utah State University, respectively, discuss how plant life is affected by volcanic eruptions. These scientists, also at the forefront of their field, consider recent eruptions of contrasting styles ranging from those emitting lava flows to those involving pyroclastic flows. What determines plant life survival? How does plant life recover after a major volcanic eruption? The chapter also summarizes the physical impacts of eruptions and their impacts on vegetation around volcanoes worldwide. They also compare the patterns of the surviving floral composition, vegetation re-establishment, and plant succession after each specific type of volcanic disturbance.

In Chapter 9, John Edwards, a leading volcano zoologist and ecologist at the University of Washington, discusses how animals may or may not survive volcanic eruptions and in the case where the eruptions are so strong that they wipe out all life forms, how and in what order the animals are observed to recolonize a barren volcanic area. The chapter not only emphasizes survival and revival of animal communities but also stresses that eruptions are a major process on evolutionary timescales of millions of years. The discussion includes the effects of ash dispersal on insects, the issue of animal survival after eruptions, the processes of recolonization by animals in six case studies, the recolonization in the zone destroyed by the "ash hurricane" at Mount St. Helens and later events in animal recolonization. One important finding is that arthropods are key players, as primary colonizers. Another key finding is that recolonization appears extremely rapid compared to evolutionary times. A largely unresolved question which deserves more attention is the "refugia question" – whether or not some animals may be able to survive extreme volcanic disasters and how this could come about? The chapter concludes by highlighting three key areas where research is much needed in order to make further advances in understanding.

Chapter 10 considers the effect of recent eruptions on human health. Over half a billion people now live in the immediate vicinity of active volcanoes, the majority of them in developing countries and highly vulnerable to volcanic emissions and eruptions. Therefore, a very important concern is to discover how volcanic degassing and eruptions can affect the health of a proportion of those people and to explore what could be done to monitor and mitigate any adverse effects on human health. What are the health effects of being exposed to volcanic gases? What are the effects of eruptions of a range of styles? Peter Baxter, a medical scientist at the University of Cambridge and the leading authority in the field of volcano medicine, has studied first-hand the effects of volcanic activity and eruptions worldwide and draws from case studies at many volcanoes to discuss these fundamental questions.

Chapter 10 also makes recommendations for good practices during volcanic crises and raises concerns for the future where appropriate.

Chapters 11–13 discuss in turn three valuable by-products of volcanism which are extensively used by humans: volcano-derived heat, precious metals, and raw materials for industry. The presence of hot magmas, below the surface in association with volcanoes, offers the prospect of harnessing a huge amount of thermal energy that can be used in our homes, both for heat or to provide electricity. This so-called geothermal energy is an important source of renewable energy and the implications for the environment are significant if we can successfully recover it. Where does geothermal energy originate? What does it take for a geothermal deposit to be economical? How do we estimate reserves? Are there any adverse effects for the environment as geothermal energy is tapped and recovered, in comparison with other types of energy? These are some of the key questions considered in Chapter 11 by Wendell Duffield, a veteran US Geological Survey scientist and volcanologist. Geothermal energy is already important and is bound to become even more so, as the reserves of fossil fuels run out and as the cost of cleaning up adverse environmental impacts can no longer be avoided in energy cost calculations.

One of the many benefits of volcanism comes from the exploitation of valuable ore deposits, which are related to the presence of volcanoes in the Earth's crust. But what do we mean when we speak of ore deposits? How do we determine whether an ore body is exploitable? What are the different types of ore bodies associated with volcanism? How can we use our understanding of volcanoes and ore deposit formation to discover the new resources that will be needed in the future. In Chapter 12, Harold Gibson, an economic geologist and volcanologist expert on this subject at Laurentian University, considers these questions and describes the volcanic environments and processes which hold the key to the formation and location of volcano-derived ore deposits.

A key area of the impact of eruptions on our environment and activities is through providing materials that can be used in industry. The exploitation of these volcanic materials in turn has a direct impact on the environment. What are the different types of materials that can be used and to what purposes? Chapter 13, by Grant Heiken, a veteran volcanologist at the Los Alamos National Laboratory, describes these different types of materials and goes on to analyze their past and current uses. In particular, pumice and scoria, which are both produced by a variety of types of explosive eruptions, have many uses ranging from raw material in the preparation of abrasives, building-blocks for cathedrals, cat litter, cement, and concrete, etc. But where can we find these materials? What problems are we faced with as we try to identify the most economical varieties of rock types, and how is this related to the processes that deposited the materials in the first place? The chapter is not limited to pumice and scoria but encompasses the whole range of volcanic materials.

Chapters 14–15 discuss additional aspects of the relationship between eruptions and humans – namely the relationship of volcanism with human societies and the impact of eruptions on the human economies. Human societies have had to respond to the many types of volcanic disasters that have occurred and continue to occur worldwide. In Chapter 14, David Chester, a geographer and volcano sociologist at the University of Liverpool and researcher leading this field, considers the relationship between contrasting human societies and volcanic disasters. The chapter starts by discussing the relationship between volcanoes and society in time and space from a historical perspective. It then moves on to review the interface between social theory and eruptions. How have societies responded to eruptions in the past? What is the modern response of contrasting societies to volcanic disasters? The chapter challenges the dominant approach that has been adopted almost up to the present and proposes more radical alternative societal responses. It concludes by highlighting areas of research that will have to be developed in the future.

It is clear that eruptions have direct and indirect effects on the economy. Do they impact only local economies, or can there be effects on a national or even global level? How big are these effects, and how is the cost of a volcanic disaster estimated? What are the key determinants of whether an eruption will or will not impact an economy? In Chapter 15, Charlotte Benson, a volcano disaster economist who has been pioneering research on this topic at the UK Overseas Development Institute, draws from documentation on the economic impacts of volcanic activity. She clearly demonstrates that this is an increasingly important and fascinating new area of research and one where much more research is needed.

We hope that you, the readers, will enjoy reading this book as much as we enjoyed putting it together. We will have succeeded in our endeavor if reading this textbook inspires some of you to pursue unanswered or poorly understood aspects of the fascinating, "volcanoes and environment" problems. More importantly, we hope that this book will enhance public awareness of our rapidly changing and evolving environment and of how volcanic eruptions contribute to the changes.

# Acknowledgments

In preparing this book, a challenging task for all involved, and extending over several years, we received tremendous help and encouragement from our families and colleagues. We are most grateful for their support and patience with us while we were preparing and editing the book.

We are indebted to many friends and scientific colleagues. Their current and past research forms the backbone around which the chapters of this book have in part been built up. Their thoughtful, incisive comments and suggestions greatly influenced the book's preparation. Particular thanks are given to Russell J. Blong, William E. Scott, Steven R. Brantley, Edward W. Wolfe, Harald Huber, Reinhard Rachel, Wendy Adams, Robert O'Neill, Patricia Parr, Roger del Moral, Frederick M. O'Hara, Linda O'Hara, Rick Sugg, Eldon Ball, Gordon Orians, Ian Thornton, Robert Fournier, Arthur Lachenbruch, Patrick Muffler, Herbert Shaw, Robert Smith, Steve Sparks, Chris German, Martin Palmer, Mike Carroll, Bill Rose, Donald White, John Sass, Michael Sorey, John Lund, Sue Priest, Kevin Rafferty, J. M. Franklin, M. D. Hannington, J. Hedenquist, C. M. Lesher, A. J. Naldrett, S. D. Scott, R. H. Sillitoe, D. H. Watkinson, R. E. S. Whitehead, R. R. Keays, and G. J. Ablay.

Permission granted to use published figures is gratefully acknowledged. In addition to the support from the respective home institutions of the authors, to which we extend our gratitude, grants from public and private institutions have also provided support for the preparation of some chapters. These generous institutions include the Deutsche Forschungsgemeinschaft; the Bundesministerium für Bildung, Wissenschaft, Forschung und Technologie; the European Union; the Fonds der Chemischen Industrie; the Minority Biomedical Research Support program; the University of Puerto Rico; the Oak Ridge Institute of Science and Education and Oak Ridge National Laboratory; The National Geographic Society; the National Science Foundation and the Mazamas; the Fondation Belge de la Vocation; the Belgian National Science Foundation (FWO-Vlaanderen) the UK Natural Environment Research Council and the UK Nuffield Foundation (NAL award to GGJE).

We wish to thank Matt Lloyd, Susan Francis, Sally Thomas, Anna Hodson and Jo Bottrill of Cambridge University Press for their expert guidance during the preparation of this book. Without their expertise and dedication to this effort, the preparation of this book might not have been possible. Special thanks are given to Silvia Zafrilla for her contribution to the editorial tasks and for the preparation of many of the figures in final form.

# Chapter 1

# Understanding the physical behavior of volcanoes

Steven N. Carey

## Introduction

Volcanism is a spectacular display of the complex way in which energy and materials are exchanged between three major components of our planet: the solid Earth, oceans, and atmosphere. Mankind has long been both fascinated and terrified by erupting volcanoes. Yet throughout history people have been drawn to their fertile slopes and have developed a unique symbiosis. In many cultures, volcanoes symbolize a source of tremendous power that must be placated by worship or sacrifice. Volcanologists, on the other hand, strive to understand how volcanoes work in order to better predict their behaviour and reduce the hazards to people who live near them. But volcanoes are not merely destructive and need to be viewed as an integral part of the dynamic Earth system. They create new land, replenish soil, and provide essential water and other gases to our oceans and atmosphere. Much of this book will focus on the relationship of volcanoes to the environment and to mankind. However, before these topics are addressed we need to begin with some fundamental concepts about the causes and processes of volcanism. In this chapter we explore how volcanoes work by examining the complex path that must be taken before an eruption takes place at the Earth's surface. This includes the generation of magma at depth, its rise, storage and evolution within the Earth's crust, and finally the factors that determine the nature of the eruption at the surface.

Each eruption represents a unique culmination of this complex series of steps, making the prediction of volcanic eruptions one of the most challenging tasks in the geosciences.

The science of volcanology draws from many different fields, including petrology, geochemistry, seismology, and sedimentology. It began with mainly qualitative observations about the distribution of volcanoes, their eruptive behavior, and the types of products they create. In AD 79, Pliny the Younger made the first written description of a volcanic eruption in two letters describing the death of his uncle, Pliny the Elder, during the eruption of Vesuvius volcano in Italy. During the past two decades the field of volcanology has undergone revolutionary changes in the understanding of volcanic activity. This has resulted from a more quantitative approach to the analysis of volcanic processes through the application of physics, chemistry, and fluid dynamics. In many instances, strides in the field were spurred on by observations of major eruptions such as Mount St. Helens, USA, in 1980, El Chichón, Mexico, in 1982, Nevado del Ruiz, Colombia, in 1985, and Mt. Pinatubo, Philippines, in 1991 (described further in Chapter 2). These eruptions often provided the opportunity to observe volcanic processes that had not previously been documented quantitatively and led to the ability to interpret such processes from deposits in the geologic record (this will be discussed further in Chapter 3). With the advent of Earth-observing satellites, the remote sensing of volcanoes from

*Volcanoes and the Environment*, eds. J. Martí and G. G. J. Ernst. Published by Cambridge University Press. © Cambridge University Press 2005.

space has led to a new appreciation for the global impact of some large-scale eruptions and provided volcanologists with new tools to assess the relationship of volcanoes to the environment (e.g. Rose *et al.*, 2000). A particularly exciting breakthrough has enabled the mapping of the ocean's bathymetry from satellites by measuring subtle changes in the elevation of the seasurface, allowing for the recognition and mapping of volcanic features on the seafloor with high resolution. As a result of land-based, marine, and space observations there is now an excellent documentation of the distribution of volcanic activity on Earth.

## Distribution of volcanic activity: the plate tectonic framework

The distribution of volcanoes on the Earth's surface provides important information about the underlying causes of volcanism, yet making a global inventory of volcanoes is not a simple task. Simkin (1993) notes that it is difficult to define precisely the number of active volcanoes on Earth. During historic times there have been some 538 with documented eruptions and over 1300 have erupted during the last 10 000 years. However, two-thirds of the planet's surface is covered by water and thus observations of volcanic activity are strongly biased towards the continents and islands. We now know that virtually all of the seafloor is composed of volcanic rock and that most volcanic activity occurs far removed from sight in the depths of the ocean.

Early workers who documented the distribution of volcanoes realized that they did not occur randomly but tended to run in belts or linear segments. This is particularly evident when one looks at the present location of volcanism on the Earth's surface (Fig. 1.1). A large majority of the Earth's active and conspicuous volcanoes can be found along a belt that rims the Pacific Ocean beginning at the southern tip of South America, proceeding north along its western coast across Mexico into the western United States (Fig. 1.1). From there it winds into Alaska, across the Aleutian islands to the western Pacific, where the line continues southward all the way to New Zealand. This remarkable clustering of volcanoes is known as the "Ring of Fire." Large earthquakes are also concentrated along this belt adding to the hazards associated with the volcanoes. Other significant belts of subaerial volcanoes occur in the Indonesian Archipelago, the Mediterranean region and in the Lesser Antilles islands of the West Indies.

It would take one of the major revolutions in the earth sciences, that of plate tectonics, to provide the unifying explanation for the distribution of volcanoes on the Earth. It has now been demonstrated that the surface of the Earth consists of numerous large plates that are constantly in motion relative to one another and relative to the deep interior of the Earth (Fig. 1.1). The theory of plate tectonics provides the fundamental linkage between our understanding of the Earth's surface and its interior (Keary and Vine, 1990; Canon-Tapia and Walker, 2003). A cross-section through the planet reveals a strongly layered internal structure (Fig. 1.2). Much of our knowledge of the Earth's interior has been derived from the study of how earthquakes are transmitted in different types of materials. By studying the arrival of different types of seismic waves, such as P- and S-types, at various positions around the globe, it is possible to infer the nature of the deep Earth. This remote-sensing approach is necessary because direct sampling of the deep Earth by drilling is usually limited to depths of about 5 km. Some deeper samples are brought to the surface naturally by volcanic eruptions. These xenoliths have been critical for determining the chemical composition and physical nature of the source areas of volcanism.

From the center of the Earth to a distance of 1228 km lies the inner core, a dense solid consisting mostly of iron with lesser amounts of nickel and sulfur. Surrounding the solid inner core out to a distance of 3500 km is a liquid outer core, also consisting of iron and sulfur. The presence of a liquid region deep in the Earth is inferred from the blocking of earthquake shear waves (S-waves), which are unable to propagate through a liquid. Within this layer, convective movement of material is responsible for the

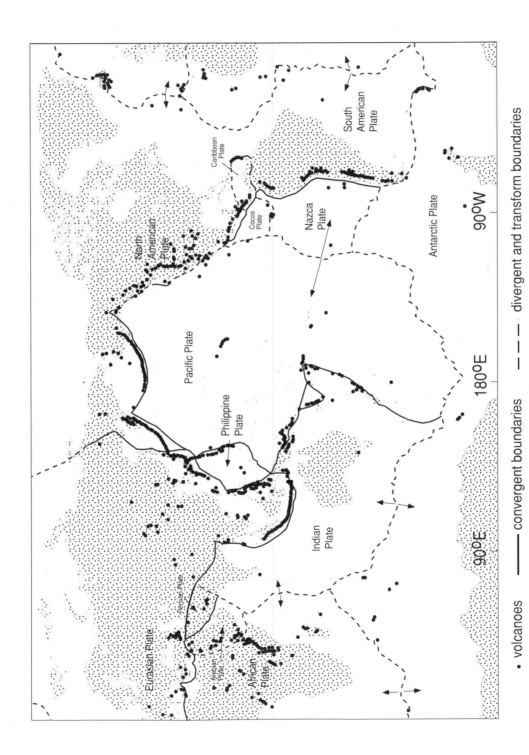

**Fig. 1.1.** Distribution of active volcanic centers (solid circles) and the boundaries between major lithospheric plates. Arrows indicate directions of relative plate motion. Modified from Sparks *et al.* (1997).

• volcanoes      —— convergent boundaries      – – – divergent and transform boundaries

Eurasian Plate

North American Plate

Persian Plate

Arabian Plate

African Plate

Caribbean Plate

Cocos Plate

Nazca Plate

South American Plate

Antarctic Plate

Pacific Plate

Philippine Plate

Indian Plate

90°E      180°E      90°W

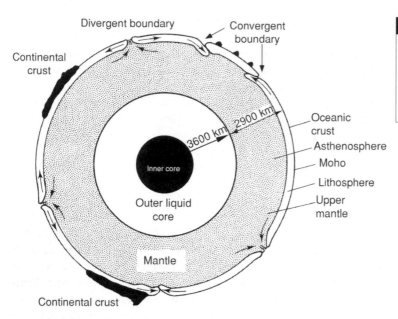

Divergent boundary

Convergent boundary

Continental crust

3600 km 2900 km

Inner core

Outer liquid core

Mantle

Continental crust

Oceanic crust

Asthenosphere

Moho

Lithosphere

Upper mantle

**Fig. 1.2.** Cross-section of the Earth's interior showing the major concentric regions. Arrows indicate the motions of the major lithospheric plates. Thickness of the lithosphere is exaggerated for viewing purposes. Modified from Frankel (1996).

generation of the Earth's magnetic field. At a distance of 3500 km from the Earth's center there is an abrupt transition to the largest volumetric component of the solid Earth, the mantle. Extending to less than 100 km from the surface, the mantle consists of silicate minerals and is the source of most magma for volcanism on the planet. As a result of gradients in pressure and temperature throughout the mantle, the assemblage of minerals changes as a function of depth. In the lower mantle, dense phases such as perovskite, magnesiowustite, and stishovite predominate, whereas above 670 km (upper mantle), minerals such as pyroxene, olivine, and an aluminum-bearing phase are stable. The mantle represents 80% of the solid Earth by volume and 67% by weight (Ringwood, 1975).

Overlying the mantle is a thin veneer, or crust, on which we live. In simple terms, the crust can divided into two principal types: oceanic and continental. Oceanic crust is relatively thin (up to 15 km), dense (3.0 g/cm$^3$), consists of rocks rich in iron and magnesium called basalt, dolerite, and gabbro, and underlies most of the ocean basins. In contrast, continental crust is thicker (up to 50 km), less dense (~2.8 g/cm$^3$), consists of rocks that are richer in silica, sodium, and potassium, such as granite, and forms most of the continental masses. The plates that are in constant motion over the Earth's surface are termed

the lithosphere, a combination of crustal rocks and upper mantle material (Fig. 1.2). They are of the order of 100 km thick and are considered to behave in a generally rigid form. Lithospheric plates ride over an underlying asthenosphere that consists of mantle rocks at higher temperatures and pressures. These conditions allow the asthenosphere material to deform under stress in a ductile fashion and to accommodate the movement of the overriding plates.

The boundaries between plates can be subdivided into three main types: divergent, convergent, and transform (Fig. 1.2). Each one represents a different sense of motion between plates. At divergent boundaries plates are moving away from one another and magma wells up from the mantle along fissures, so creating new oceanic crust and lithosphere. An example of a divergent boundary is the one between the North American and the Eurasian plates centered roughly in the middle of the Atlantic Ocean (Fig. 1.1). Convergent boundaries represent the opposite sense of motion where plates are coming together. In these areas one plate moves under another in a process called subduction (Fig. 1.2). The descending plate is typically bent downwards as it is returned to the mantle, producing a deep area, or trench, near the boundary. Earthquakes occur along the descending plate as it returns to the mantle and their location can be used to define the slope of the downgoing slab, referred to as the Benioff Zone. In the Pacific Ocean, the deepest oceanic depth of 13,000 m is associated with the trench of the Mariana subduction zone. Convergent margins can be produced by subduction of oceanic lithosphere beneath other oceanic lithosphere (island arc) or oceanic lithosphere beneath continental lithosphere (active continental margin). Finally, transform boundaries are areas where plates slide past one another, such as along the San Andreas fault system of the western United States, where the Pacific plate is moving northward relative to the North American plate (Fig. 1.1). Significant earthquake hazards are associated with this type of boundary when plate motion occurs by rock failure. At transform boundaries, there is not, however, significant volcanic activity.

Volcanoes and earthquakes are largely confined to the boundaries between lithospheric plates and are a direct consequence of plate motion. The style of volcanism, the composition of the erupted products, and the nature of volcanic features that are produced vary considerably between different boundaries and reflect the fundamental influence of tectonic forces on volcanism.

## Mid-ocean ridge volcanism (divergent boundary)

Volcanism at divergent boundaries has produced a globally encircling mountain range known as the mid-ocean ridge system. It extends throughout the major ocean basins of the world for a total length of some 70 000 km. Discovery of this undersea mountain chain and its recognition as a primary volcanic feature was one of the most important advances in the field of marine geology. Dredging of rocks from the ridge and direct observations by submersibles confirmed that fresh volcanic rock is being emplaced along this boundary, forming new seafloor that progressively moves away from the ridge axis. The rate of production has been determined by dating magnetic anomalies in the oceanic crust that are arranged in symmetrical fashion on each side of the ridge (Vine and Matthews, 1963). These anomalies record times when the Earth's magnetic field changed polarity direction by 180°.

Spreading rates are generally low by human standards, averaging only a few centimeters per year. However, over long periods of geologic time movements of plates can produce or consume entire ocean basins.

There is considerable variation in the rates of spreading along different segments of the mid-ocean ridge system. In the Atlantic Ocean the spreading along the mid-Atlantic ridge is only about 2 cm/year, whereas along the East Pacific Rise the rate is up to 18 cm/year. It should be emphasized that the spreading rates are based on calculations that average the production of new seafloor over hundreds of thousands or millions of years. There is not necessarily an addition of a few centimeters of new oceanic crust all along the mid-ocean ridge every year. Eruptions are likely to be episodic, occurring on the timescale of hundreds to thousands of years, and affecting only parts of the ridge at any given time. Only recently has there been an actual eruption detected along the mid-ocean ridge system. It occurred in 1993 when an event was recorded from a 6-km-long section of the Juan de Fuca spreading center off the northwest coast of the United States (Chadwick et al., 1995).

The crest of the ridge is generally located at a depth of several thousand meters. Topographic profiles across ridges exhibit distinctive morphologies that relate to spreading rate. The slow-spreading mid-Atlantic ridge consists of a well-defined axial valley surrounded on each side by steep, fault-bounded ridges, whereas the fast-spreading East Pacific Rise is defined by a broad elevated ridge with only a small-scale axial valley (Fig. 1.3).

Although they are primarily confined to the seafloor there are areas, such as Iceland and the African Rift Valley, where divergent boundaries are exposed on land, providing a unique opportunity to directly observe volcanic processes. Iceland is an example where an anomalous rate of magma discharge along the mid-Atlantic ridge has elevated the spreading center above sea level. The island is being enlarged through volcanic activity occurring along its neovolcanic zone that extends from the southwest to the northeast (Fig. 1.4). Along this zone is a set of prominent central volcanoes that rise to elevations up to

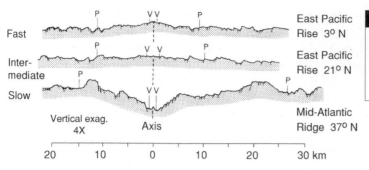

**Fig. 1.3.** Bathymetric cross-sections across the East Pacific Rise at 3° N and 21° N, and the mid-Atlantic ridge at 37° N showing the differences in morphology between fast, intermediate, and slow-spreading ridges. Modified from Brown *et al.* (1989).

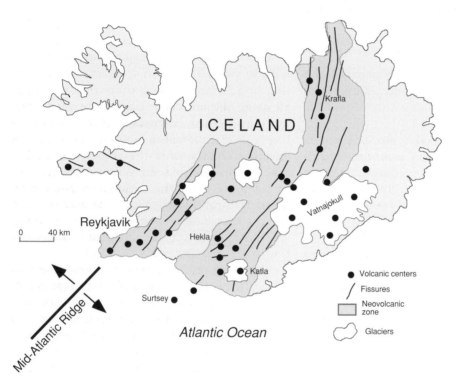

**Fig. 1.4.** Location of the neovolcanic zone (<7 Ma volcanism) in the central part of Iceland. Hotspot volcanism centered along the mid-Atlantic ridge has led to the emergence of Iceland above sea level. Modified from Gudmunsson (1996).

2000 m and are often glaciated. Between these centers are zones of fissures (rift zones) where magma can travel laterally from storage areas beneath the central volcanoes. In eastern Africa a nascent divergent boundary has created a 4000-km-long rift zone that extends from the intersection of the Red Sea and the Gulf of Aden, south to the latitude of Madagascar. This rift valley is home to some of Africa's largest volcanoes such Mt. Kilimanjaro and Mt. Kenya.

The global importance of divergent boundary volcanism can be appreciated by considering the annual budget of volcanic activity on Earth. Estimates of the total annual volcanic output from all volcanoes is about 4 km³ of magma. Of that, 3 km³, or about 75%, can be attributed to the mid-ocean system (Crisp, 1984). Thus, a large fraction of the global volcanic activity occurs at great depth in the ocean, out of view.

## Subduction zone volcanism

Most of the documented volcanic eruptions occur along convergent plate boundaries, or subduction zones, such as the ones that make up the Pacific

**Fig. 1.5.** Mount Rainier stratovolcano in Washington state, elevation 4392 m. The volcano is dominated by lava flows and breccias which are mostly andesitic in composition. Explosive eruptions have also produced pumice and ash deposits on the slopes of the volcano. Photograph by S. Carey.

"Ring of Fire" (Fig. 1.1). These eruptions are often explosive in nature, result in immediate threats to human populations, and consequently attract our attention. Many of the world's most famous volcanoes such as Krakatau, Mt. Pelée, Mount St. Helens, and Mt. Fuji are related to the subduction process. Their structure is often what we consider as a "typical" volcano; a high, steep-sided mountain with a crater at the summit (Fig. 1.5). But as we've seen, the dominant and thus more "typical" type of volcanic landform is that occurring beneath the sea in the form of mid-ocean ridges.

A striking feature of the global distribution of subduction-related volcanism is that it's highly asymmetrical, with most of it occurring around the rim of the Pacific Ocean (Fig. 1.1). The Atlantic Ocean has only two short segments of subduction volcanism; the Lesser Antilles island arc at the eastern boundary with the Caribbean Sea, and the South Sandwich island arc off the southern tip of South America. Other major subduction zones occur in the Indian Ocean (Indonesian island arc) and the Mediterranean Sea (Aeolian and Hellenic island arcs).

Just as there were significant differences in the morphology of divergent plate boundaries (Fig. 1.3) there are also marked differences in the types of convergent boundaries. Uyeda (1982) has suggested that there are at least two distinct types represented by the Mariana island arc in the western Pacific and the Andean active continental margin of South America (Fig. 1.6). A cross-section through the Marianas arc shows that subduction of oceanic crust is occurring along a steeply dipping Benioff Zone. The lack of great thrust earthquakes during historical times suggests that the mechanical coupling between the downgoing and the overriding plate is relatively weak. Volcanic activity in the Marianas occurs as a narrow, curved array of individual volcanic islands positioned about 120 km above the top of the descending plate. The curvature of the island chain is merely a reflection of the geometrical constraints of subducting a rigid plate into the surface of a sphere. In detail, however, some island arcs show distinct linear segmentation of volcanic centers along the subduction trend (e.g., Marsh, 1979; Carr, 1984). Behind the line of volcanic islands in the Marianas is a series of small oceanic basins called back-arc basins. These basins are inferred to form by spreading, in a manner similar to mid-ocean ridges, but on a smaller scale (Karig, 1971). The Marianas arc is thus considered as an example of a typical island arc involving the subduction of oceanic crust beneath other oceanic crust.

A cross-section through the west coast of South America shows a very different type of subduction process (Fig. 1.6). Here, oceanic crust is descending beneath thick continental crust at a shallow angle. The frequent occurrence of strong thrust earthquakes suggests that, unlike in the Mariana arc, there is strong coupling, or resistance, to the subduction of the plate beneath South America. Volcanism occurs as a line of individual volcanoes that typically is located at a position about 120 km above the top of the descending plate. However, because the slope of the Benioff Zone is much shallower than in the Marianas, the location of the volcanic front in the Andes occurs further back from the trench (Fig. 1.6). Another important difference is that the Andean subduction zone exhibits compression in the area behind the volcanic front and consequently this type of margin lacks the extensional back-arc basins that are characteristic of the western Pacific.

The differences in these two modes of subduction have been related to factors such as the age

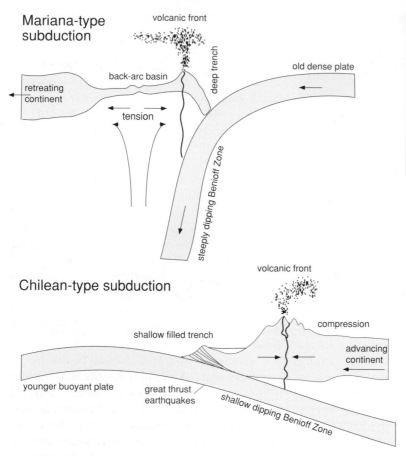

**Fig. 1.6.** Summary of the differences between the Mariana-type and Chilean-type subduction modes based on the model of Uyeda (1982). Mariana-type subduction involves steeply descending plates with few great thrust earthquakes, whereas plate subduction in the Chilean-type zones is much shallower and typically more seismically active.

of the subducting slab and the absolute motion of the overriding plate relative to the trench (Molnar and Atwater, 1978; Uyeda and Kanamori, 1978). Subduction of the Mariana type is thought to take place when old dense seafloor descends beneath oceanic crust and the overriding plate is moving away from the active trench. In the Andean type of subduction the descent of the plate is more difficult because it is younger and more buoyant. The absolute motion of the overriding plate is also towards the trench, resulting in compression behind the volcanic front.

Subduction is a process of recycling where oceanic crust is returned to the mantle. The size of the Earth has remained relatively constant and therefore the production of new seafloor at the mid-ocean ridges must be balanced by the subduction of crust at convergent margins. As discussed earlier, volcanism at mid-ocean ridges accounts for about 75% of the annual volcanic output of the Earth (Fisher and Schmincke, 1984). The volcanism associated with subduction is only 15%, and thus, even though subduction volcanism poses the largest threats to the human population and is the most conspicuous type of activity, it still represents a relatively small proportion of the annual volcanic production.

## Intraplate volcanism

There are many areas of the Earth's surface where volcanism is occurring that is unrelated to a major plate boundary. These can be found both in the ocean basins and on the continents. Focused, persistent areas of volcanic activity such as these that remain active for millions of years are known as hotspots, or mantle plume volcanism (Wilson, 1973). They are areas of anomalously high volcanic production that are related to the rise of mantle plumes from deep within

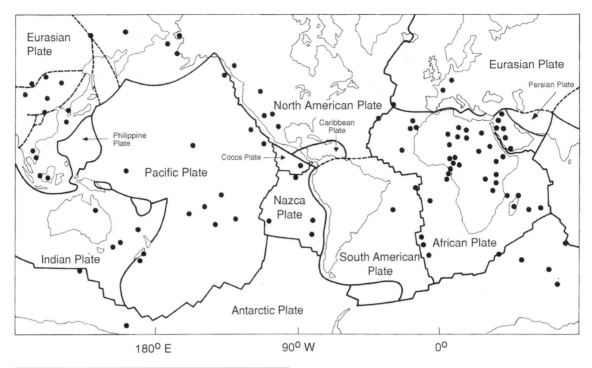

**Fig. 1.7.** Location of the major volcanic hotspots (solid circles) in the ocean basins and on the continents. Modified from Wilson (1989).

the Earth. There are currently between 50 and 100 active hotspots, depending on the factors that are used to define them (Fig. 1.7). Common features include voluminous outpourings of basaltic magma, higher than normal heat flow, thinning of the overlying crust, and development of a broad topographic high.

Because the source of magma for hotspots is derived from a long-lived source beneath the lithosphere, plate movement over millions of years results in the formation of linear volcanic features. Hotspots are thus extremely useful for inferring the absolute motions of plates relative to the deep interior of the Earth. An excellent example of this can be found in the middle of the Pacific plate where the Hawaiian islands form a distinctive linear group that trends from the southeast to the northwest. The Hawaiian islands are actually a small part of a much larger submarine linear feature that includes the Hawaiian ridge and Emperor seamount chain (Fig. 1.8).

Formation of this large-scale feature is attributed to hot spot volcanism currently located beneath the southeast flank of Hawaii. During the past 70 million years, volcanic activity from this hotspot has built a series of islands that are continually moved to the northwest by the motion of the Pacific plate. This is confirmed by the systematic increase in island age from the southeast to the northwest along the chain (Clague and Dalrymple, 1987). As the islands age, they subside and are eroded below sea level. A pronounced shift in the orientation of the Hawaiian–Emperor seamount chain occurred about 44 Ma ago when the motion of the Pacific plate changed to a more westerly direction of motion (Fig. 1.8). The most recent island associated with the Hawaiian hotspot is Loihi, now forming under water on the southeast flank of the big island of Hawaii. Other examples of linear chains of islands and seamounts includes the Tuamotu and Austral groups of the South Pacific. These chains show a similar kink associated with the major change in Pacific plate motion.

In the ocean basins some hotspots coincide with divergent plate boundaries. Iceland is an example of a hotspot that is centered along the

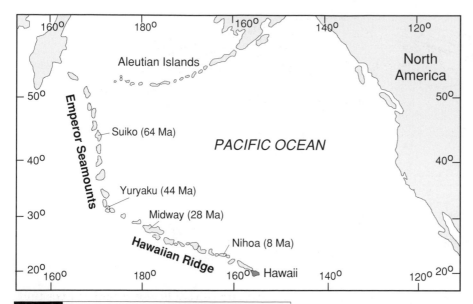

**Fig. 1.8.** Orientation of the Hawaiian ridge and Emperor seamount chain in the Pacific Ocean. Ages of various islands are shown in parentheses. Note the change in alignment of the two seamount chains at Yuryaku, corresponding to an age of 44 Ma. Modified from Clague and Dalrymple (1987).

mid-Atlantic ridge axis. The excess volcanic output of the hotspot results in the elevation of the ridge above sea level and the formation of elevated submarine ridges on each side of the Iceland Plateau. Other hotspots are located beneath the continents, especially Africa (Fig. 1.7). Plate motion of continental masses can also produce linear traces of past volcanic activity like those in the ocean basins. For example, volcanic activity beneath Yellowstone National Park in the western United States has been attributed to the location of an active hotspot. A belt of volcanic rocks ranging from Recent to 15 Ma old (Eastern Snake River Plain) can be traced from Yellowstone back across Wyoming into Idaho. This is likely to represent the record of volcanism as the North American plate migrated to the southwest (Greeley, 1982).

## Large igneous provinces (LIPs)

Much of the previous discussion has centered on describing the current distribution of volcanic activity on the Earth's surface. However, there are examples of spectacular volcanic activity that have occurred in the geologic past for which there are no modern analogues. Many continents and ocean basins contain widespread accumulations of basaltic lava flows that represent massive outpourings of magma. These are collectively referred to as large igneous provinces, or LIPs (Coffin and Eldholm, 1994). Included in this category are continental flood basalts and submarine basalt plateaus (Fig. 1.9). The Deccan Traps of India is an example of one of the largest continental flood basalt provinces on Earth. When it was formed 65 millions years ago it may have covered an area of 1.5 million km$^2$ with an average thickness of at least 1 km. Such provinces are noteworthy because of the large areas affected by volcanism and the relatively short times of formation. An estimate for the emplacement of the Deccan Traps basalts is of the order of 1 Ma (Duncan and Richards, 1991; see also Chapter 7). Such durations imply volumetric discharge rates that are up to an order of magnitude larger than those of modern volcanic provinces, such as mid-ocean ridges. As a result of the high eruption rates this type of activity does not tend to produce high individual volcanoes, but instead, eruptions from many vents leads to the accumulation of laterally extensive flows into a thick plateau. The Ontong–Java plateau, located in the western Pacific (Fig. 1.9) is an example of a large submarine flood basalt province that was formed 124 Ma ago. It is three times the area of the Deccan Traps and was produced in about 3 Ma.

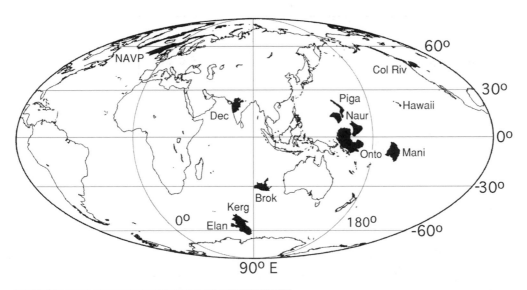

**Fig. 1.9.** Location of some of the major large igneous provinces (LIPs) NAVP, North Atlantic Volcanic Province; Dec, Deccan Traps; Kerg, Kerguelen; Elan, Elan Bank; Brok, Broken Ridge; Piga, Pigafetta Basin; Naur, Nauru Basin; Onto, Ontong–Java plateau; Mani, Manihiki; Hawaii, Hawaiian islands; Col Riv, Columbia River. From Coffin and Eldholm (1994).

The origin of LIPs remains controversial and involves models based on deep mantle plumes, shallow mantle plumes, and lithospheric rifting. Deep mantle plumes may initiate massive outpourings of lavas by actively impinging upon the base of the lithosphere, causing thinning, heating, and melt production (Richards *et al.*, 1989). The initial phase of volcanism (LIP formation) is the most voluminous because the arriving plume consists of an enlarged head followed by a long, thin conduit. Such plumes may originate near the core–mantle boundary. Alternatively, passive thermal plumes from layered mantle convection may generate large melting events when lithospheric extension allows for adiabatic decompression of hot mantle (White and McKenzie, 1989).

## Composition and physical properties of magma

### Composition of magmas
Any discussion of how volcanoes work needs first to consider the essential component of volcanism, magma. Virtually all magmas erupted on Earth are silicate liquids whose composition, including volatile content, viscosity, and temperature can vary considerably depending upon the nature of the eruption and the tectonic setting of the volcano. Magmas are complex mixtures of melt, suspended crystals, and gas bubbles on approaching the surface. Melt is usually the major component and magma consequently behaves as a fluid. Most magmas, with the exception of some rare carbonatite magmas, have $SiO_2$ contents in the range of about 45 to 77 wt.%. Other major elements include aluminum, calcium, iron, magnesium, potassium, sodium, and titanium. The compositional diversity observed in magmas is caused primarily by variations in the composition of source rocks being melted, differences in the amount of source rock melting, variations in the amount of volatiles present in the source regions, crystallization of magma during its ascent and storage in the crust, mixing with other magmas, and contamination by country rocks (Wilson, 1989).

Unraveling the details of these processes falls in the realm of igneous petrology and extensive treatment of these processes is beyond the scope of this chapter. Instead we will focus on some basic details of magma composition as they relate to the behavior of volcanic systems. As a starting point we will use a simple classification scheme for magmas based on $SiO_2$, $Na_2O$, and $K_2O$ (Fig. 1.10). This classification allows for the assignment of general names to different magma

| Oxide[a] | Mid-ocean ridge basalt (MORB) | Andesite | Dacite | Rhyolite | Peridotite (garnet lherzolite) |
|---|---|---|---|---|---|
| $SiO_2$ | 50.58 | 56.86 | 66.36 | 74.00 | 45.89 |
| $TiO_2$ | 1.49 | 0.88 | 0.58 | 0.27 | 0.09 |
| $Al_2O_3$ | 15.60 | 17.22 | 16.12 | 13.53 | 1.57 |
| $FeO$ | 9.85 | 4.26 | 2.41 | 1.16 | 6.91 |
| $Fe_2O_3$ | – | 3.29 | 2.39 | 1.47 | – |
| $MgO$ | 7.69 | 3.40 | 1.74 | 0.41 | 43.46 |
| $CaO$ | 11.44 | 6.87 | 4.29 | 1.16 | 1.16 |
| $Na_2O$ | 2.66 | 3.54 | 3.89 | 3.62 | 0.16 |
| $K_2O$ | 0.17 | 1.67 | 2.22 | 4.38 | 0.12 |

**Table 1.1** Major element composition of some common volcanic rocks and nodules

[a] Oxides are expressed as weight percentages.
*Source:* Data from Wilson (1989) and Le Maitre (1976).

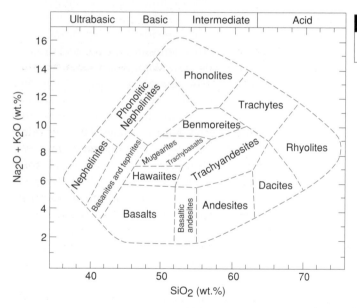

**Fig. 1.10.** Classification of volcanic rocks based on $Na_2O + K_2O$ versus $SiO_2$ content. Modified from Wilson (1989).

compositions, although it should be emphasized that there is often a complete gradation of magma compositions. Broad categories include silicic (>63% $SiO_2$), intermediate (52–63% $SiO_2$), basic (45–52% $SiO_2$), and ultrabasic (<45% $SiO_2$). The most common type of magma is basalt erupted along "normal" segments of the mid-ocean ridge system where mantle plumes are not influencing the composition of the melts. This so called mid-ocean ridge basalt, or MORB, makes up a large fraction of the seafloor (Table 1.1).

Magmas typically occur as distinct compositional suites that suggest a genetic relationship between the members. The suites consist of compositions that exhibit relatively smooth trends in the various oxide components such as $SiO_2$, FeO, MgO, CaO, and $K_2O$. Primitive members of a suite are closer to the primary magma composition derived from the source regions whereas evolved members have undergone differentiation, usually as a result of crystallization. Evolved members of a suite typically have higher contents of $SiO_2$, $K_2O$, and $Na_2O$. Different suites are associated with specific geological environments and styles of eruption. At mid-ocean ridge spreading centers, magmas commonly form part of the tholeiitic suite, a group of compositions that exhibit a strong increase in FeO at the more primitive end with a subsequent decrease towards more evolved compositions. Although the suite can include compositions that range up to >70% $SiO_2$, these solicic magma types are volumetrically minor. In contrast, subduction zone volcanism is dominated by the calcalkaline suite. This suite spans the compositional range basalt, andesite, dacite, and rhyolite, or a $SiO_2$ content of about 50–77% (Table 1.1) and is characterized by a lack of strong iron enrichment in moving from the primitive to evolved members. Unlike mid-ocean ridges, subduction zones exhibit a broader range of magma compositions with more evolved magmas, such as andesites, being volumetrically dominant.

Hotspot volcanism is dominated by basaltic magma but it differs in many respects from the voluminous magma erupted along the normal mid-ocean ridge spreading centers. In particular, the trace element and isotopic signature of plume magmas suggests derivation from source regions that are different from that feeding the normal mid-ocean ridge segments (Wilson, 1989).

Magmas typically contain small amounts of volatile components such as water, $CO_2$, sulfur, and halogens, such as Cl and F. These volatiles are dissolved in the magma and are typically more soluble at higher pressure. Their solubility is also a function of temperature, melt composition and oxygen fugacity. Figure 1.11 shows the solubility of water in magmas of different composition as a function of pressure. At a similar pressure, more evolved magmas such as rhyolites can dissolve larger amounts of water than less evolved magmas like basalts. As magmas rise towards the surface and are erupted, the dissolved gases can become supersaturated and come out of solution as bubbles. This degassing process has a fundamental influence on the style of volcanic eruptions and plays a major role in the production of explosive volcanism.

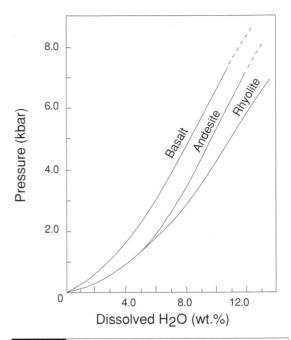

**Fig. 1.11.** Variation in dissolved water content as a function of pressure for basalt, andesite, and rhyolite magma. Modified from Burnham (1979).

Volatile contents in magmas vary considerably from one tectonic environment to another. At divergent boundaries the basalts of mid-ocean ridge systems are generally volatile-poor ($H_2O$ <1%, $CO_2$ <0.10%) and erupt quiescently. Degassing in this environment is also inhibited by the pressure of the overlying water column. At depths of 2000 m the pressure is 200 bar, or enough to retain 0.1% $H_2O$ and 0.1% $CO_2$ dissolved in the magma. In contrast, subduction zone magmas are noted for their higher volatile contents, especially magmas with $SiO_2$ contents in excess of 60%. Water is the most abundant volatile in these types of magmas and provides the driving force for the explosive volcanism that is so characteristic of the subduction zone environment. The dacite magma erupted during the May 18, 1980 Mount St. Helens eruption in Washington, USA contained about 4.5% $H_2O$ (Rutherford et al., 1985; Rutherford and Devine, 1988) and the Bishop Tuff magma of California showed $H_2O$ contents as high as 7.0% (Anderson et al., 1989). Even some basic magmas in back-arc environments have been inferred to contain up to 4% $H_2O$ (Sisson and Grove, 1993; Stolper and Newman, 1994).

Sulfur is an important volatile component in magmas because of its potential climatic effects once it is released into the atmosphere (Rampino, 1988; Sigurdsson, 1990; see also Chapter 5). Like $H_2O$ and $CO_2$ its solubility is dependent upon pressure but it is also strongly a function of the FeO content of the magma and oxygen fugacity (Carroll and Rutherford, 1988). It is dissolved as sulfide under relatively reducing conditions, and as sulfate as magmas become more oxidized. In general there is an inverse correlation between the $SiO_2$ content of magma and the amount of sulfur that can be dissolved. Basalts, for example, can contain up to to 1000 ppm (parts per million), whereas rhyolites are generally poor in sulfur, with only about 20 ppm. Some magmas, such as those erupted during the 1982 eruption of El Chichón volcano in Mexico and the 1991 eruption of Mt. Pinatubo, have sufficient sulfur and are oxidized enough to allow the mineral anhydrite ($CaSO_4$) to crystallize as a phenocryst.

## Physical properties of magmas

The physical properties of magma play a fundamental role in determining the style of a volcanic eruption. Composition is an important factor but other parameters such as temperature, proportion of crystals, amount of dissolved volatiles, and the abundance of gas bubbles all contribute to determining the rheology of the erupting magma. Temperature is a relatively simple parameter to measure and significant amounts of data are available for different magma types. There is typically an inverse correlation between eruption temperature and $SiO_2$ content. Basaltic magmas (50% $SiO_2$) erupt at temperatures of about 1200 °C, whereas silicic magmas such as rhyolites (75% $SiO_2$) are cooler and erupt in the temperature range of 700–900 °C.

One of the most important magmatic properties that influence the nature of volcanic eruptions is magma viscosity. In simple terms it is the amount of internal resistance to flow that a fluid exerts when a force is applied to it. The lower the viscosity, the more easily a fluid can flow. Specifically, viscosity is the slope of the ratio

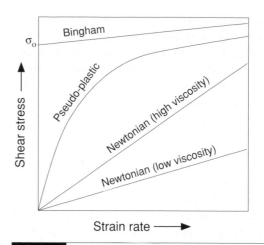

**Fig. 1.12.** Relationship between shear stress and strain rate for Bingham, pseudo-plastic, and Newtonian fluids. Modified from Wolff and Wright (1981).

of shear stress to strain rate (Fig. 1.12). Some fluids exhibit a linear relationship between shear stress and strain rate that passes through the origin on Figure 1.12. These are referred to as Newtonian fluids and they will flow even with only an infinitesimally low shear stress. In general, most magmas do not behave as Newtonian fluids unless they are at high temperatures or are relatively crystal-free. Fluids which show a non-linear relationship between stress and strain rate are described as non-Newtonian (pseudo-plastic behavior). Many magmas actually behave as Bingham fluids (Fig. 1.12). Such fluids exhibit an intercept with the stress axis at zero strain rate and a linear variation of stress and strain rate. The intercept represents a minimum shear stress that must be exceeded before a fluid will begin to flow. This parameter is called the yield strength and in magmas results from the presence of crystals, bubbles, and changes in viscosity caused by cooling during eruption. The slope of a line connecting the origin with a point on a rheological curve is called the apparent viscosity.

The viscosity of silicate melts varies as a function of temperature and composition (e.g., Bottinga and Weill, 1972; Ryan and Blevins, 1987). Silicate melts show hyperexponential dependence of viscosity on temperature, with increasing temperatures resulting in significant viscosity reduction. The effect of composition is complex but in general viscosity is inversely related to $SiO_2$ content at normal eruption temperatures. Basaltic melts have relatively low viscosities ($10^2$ to $10^3$ Pa s) compared to silicic magmas such as rhyolites ($10^6$ to $10^{12}$ Pa s). The presence of crystals and gas bubbles have important effects on magma rheology. As crystal abundance increases, viscosity increases according to a power law (Pinkerton and Stevenson, 1992). The transition from Newtonian to non-Newtonian properties can develop when the volume of suspended crystals reaches between 20% and 30%. In contrast, the effect of bubbles is more complicated and can result in either viscosity increase or decrease depending on factors such as bubble size, surface tension of the melt, and strain rate (Dingwell and Webb, 1989).

Dissolved volatiles in magmas, such as water and $CO_2$, can significantly impact magma viscosity even though their mass fraction is relatively small. The effects are due to the ways in which some of these components are held in solution. For example, water is dissolved in magma through the breaking of strong Si—O bonds and the formation of $OH^-$. This causes a reduction in the internal resistance to flow and thus a reduction in viscosity (Fig. 1.13). If 4% $H_2O$ is added to an initially dry rhyolite the viscosity will decrease by five orders of magnitude. Addition of $H_2O$ to basalt also decreases the viscosity, but the magnitude of change is considerably less (Fig. 1.13). As magmas approach the surface during an eruption they typically lose volatiles as a result of degassing induced by pressure reduction. Figure 1.13 shows that a consequence of the degassing process is a dramatic increase in the viscosity of magma, especially one with high silica content.

During eruptions, magma cools and develops mechanical strength. How magmas react to cooling plays an important role in the way they respond to stresses as the eruption proceeds. Two end member responses to cooling can be considered. Low-viscosity magmas, such as basalts, generally react to cooling by crystallization. If crystal content increases to an extent that the crystals form a touching framework, then it becomes a partially molten rock rather than a magma. This transition typically occurs at around 50% crystals

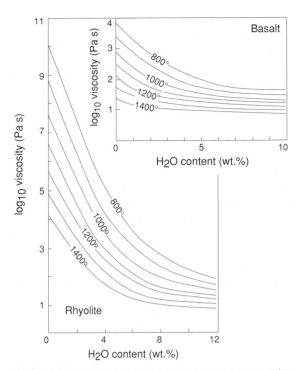

**Fig. 1.13.** Variation in magma viscosity as a function of dissolved water content for rhyolite and basalt magma at temperatures from 800 to 1400 °C. Modified from Williams and McBirney (1979).

by volume. In contrast, high-viscosity magmas react to cooling by forming supercooled melts and glasses. The glass transition temperature represents the boundary between ductile and brittle behavior. It is dependent on melt viscosity and shear rate (Dingwell and Webb, 1990). Thus, siliceous magma in a slow lava flow may be able to behave as ductile viscous fluids at much lower temperatures than in an explosive eruption where strain rates are much higher.

# Melting mechanisms for magma generation

The generation of magma occurs deep within the Earth and thus out of the range of direct observations. This process involves the melting of solid material (source rocks) to yield a melt phase that segregates and rises to the surface buoyantly. Our understanding of the melting process comes primarily from the study of the thermodynamic properties of magma and potential source rocks, together with the experimental determination of the melting behavior of geological materials under conditions appropriate for the Earth's interior. The important questions about magma generation relate to the nature of the source rocks and the conditions that are necessary to cause melting. Laboratory experiments have demonstrated that the melting of rocks consisting of more than one mineral occurs over a range of temperatures. As temperature is increased, first melting begins at the solidus and complete melting takes place above the liquidus. Within this temperature range both the amount of melt and its composition vary. For the most part, magmas erupted at the surface are generated within the Earth's mantle, although some are produced by melting of the crust.

The composition of the mantle has been inferred primarily from observed seismic velocities and the composition of nodules, thought to represent direct samples of the mantle that are brought to the surface during volcanic eruptions. Detailed geochemical and isotopic studies of magmas erupted from divergent boundaries and intraplate volcanism indicate that the source rocks in the mantle are heterogeneous on a variety of scales resulting from the recycling of material back into the mantle at subduction zones and previous melting episodes in Earth history that have depleted (i.e. impoverished) certain areas in melt. In general, the composition of the mantle is ultramafic (rocks high in magnesium and iron; see Table 1.1) and a likely source rock is known as peridotite (Yoder, 1976). The mineralogy of peridotites varies as a function of pressure (depth) as different mineral phases attain equilibrium. At the depths likely to correspond to melting zones, the peridotites consists of olivine ($(Mg,Fe)_2SiO_4$), orthopyroxene ($(Mg,Fe)SiO_3$), clinopyroxene ($(Ca,Mg)Si_2O_6$), and garnet ($MgAl_2Si_9O_{12}$). At shallower levels, the aluminum-bearing phase garnet is replaced by spinel ($MgAl_2O_4$) and then plagioclase ($NaAlSi_3O_8$–$CaAl_2Si_2O_8$).

Seismic studies indicate that except for the molten outer core, the greater part of the Earth's interior, i.e. mantle, is in the solid state. A zone of low velocity in the upper mantle may represent a small proportion of interstitial melt, but this is unlikely to represent a major source of magma for volcanism at the surface. The lack of large melted zones in the mantle indicates that the interior temperature must be below the melting temperature, or solidus, of mantle material. Magma generation must therefore involve the melting of essentially solid source rocks at depth. The temperature within the Earth increases with depth and is the result of energy accumulated during initial accretion of the planet, formation of the core by iron segregation, and radioactive heating resulting from the decay of radioactive elements (Frankel, 1996). Energy is dissipated from the Earth's interior to the surface by both conduction and convection. In the mantle, the temperature increases to close to 3000 °C near the boundary with the core and the lack of extensive melting reflects the increase in melting temperature of mantle material at high pressures.

There are essentially three ways to initiate melting of source rocks: (1) increase the temperature, (2) decrease the pressure, or (3) change the composition. It would seem most logical to suppose that melting is generally caused by an increase in temperature within a region of the Earth's interior. In fact, this is not likely to be an important way in which the majority of magmas are formed. The reason for this is the difficulty in developing enough heat locally to induce melting of source rocks. Focusing of most volcanic activity at plate boundaries indicates that the dynamics of plate motion plays a fundamental role in the melting mechanism.

## Magma generation at mid-ocean ridges

Considering that mid-ocean ridge volcanism accounts for 75% of the annual volcanic production, the generation of magma beneath the ridges represents the most important mechanism for melt generation on the Earth. The geometric configuration of a divergent boundary consists of two lithospheric plates moving apart with new seafloor created in the middle. This pattern of motion requires upwelling of material from depth in the mantle to replace material moving laterally away from the ridge. Hot mantle is brought into a lower pressure regime and

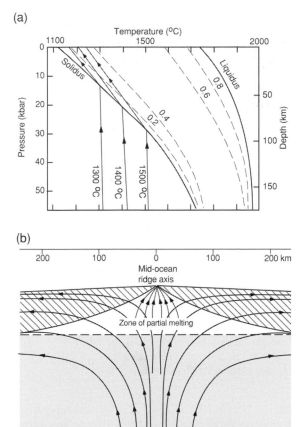

**Fig. 1.14.** (a) Paths of decompression melting for rising mantle peridotite (lines with arrowheads) at temperatures from 1300 to 1500 °C. Melting begins at the solidus in the depth range of ~50–75 km. Dashed lines indicate the fraction of melt generated at a given pressure and temperature between the solidus and liquidus. (b) Schematic representation of the partial melting zone generated beneath a mid-ocean spreading center by decompression melting of rising mantle peridotite. Modified from Brown *et al.* (1992).

can undergo melting, referred to as decompression melting (Fig. 1.14a). The solidus curve of mantle material is seen to increase in temperature with increasing depth or pressure. At all depths it is above the local temperature and melting is not predicted to take place. However, if mantle material from depth is physically transported towards the surface without losing substantial heat, i.e. adiabatically, it can intersect the solidus and begin to melt. Ascent paths are shown in Figure 1.14a for mantle material initially at 1300, 1400, and 1500 °C. Melting occurs at about

48–70 km depth, when the temperature is equal to the solidus. Continued rise of this material results in a mixture of melt and residual solid, where the fraction of melt is indicated by the dashed lines. Material initially starting at 1300 °C arrives at the surface with a temperature of about 1200 °C and the melt fraction would be about 20%. The rapid cooling of the material once it intersects the solidus (~4 °C/km) is caused by the loss of heat of fusion necessary to continue the melting process as the material decompresses.

The extent of melting and the temperature of the melt at the surface is a function of the starting temperature when the mantle material intersects the solidus. Thus, material derived from deeper depths and higher temperatures has the potential for generating more melt at higher temperatures by the time it reaches the surface. This decompression melting mechanism is shown schematically in the context of a divergent boundary (Fig. 1.14b). Below the ridge axis, rising mantle material experiences decompression melting as it rises, with the formation of a partial melting zone. The degree of melting is largest for material in the center of the upwelling, as it has the potential for the maximum decompression. Upwelling material follows flow lines that become parallel with the surface and track the lateral motion of the plates. Segregation of melt occurs within the partial melting zone, and the melt makes its way to the surface where it forms new ocean crust (Fig. 1.14b).

Experimental melting of mantle peridotites indicate that it is possible to generate a magma that resembles typical mid-ocean ridge basalt by partial melting of the order of 20%. A 6–8-km-thick oceanic crust of basaltic composition could thus be produced by upwelling and decompression melting of mantle material initially at a temperature of about 1350 °C (Sparks, 1992).

## Magma generation at subduction zones

The association of volcanism with plate subduction presents something of a paradox when considering the generation of magma. In these areas, cold, dense seafloor is being returned to the mantle by gravitational sinking and the production of magma appears counter-intuitive. Modeling of the temperature distribution in subduction

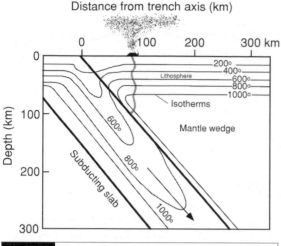

**Fig. 1.15.** Temperature distribution induced by the subduction of a lithospheric plate into the mantle. Modified from Wilson (1989).

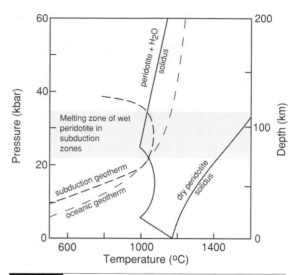

**Fig. 1.16.** Solidus curves for the melting of mantle peridotite under wet ($H_2O$-present) and dry conditions. Heavy dashed lines show the geothermal gradient in the mantle wedge beneath a subduction zone. Melting of peridotite is possible at subduction zones under wet conditions at depths of approximately 75–120 km (gray shaded zone). Modified from Wilson (1989).

zones, indeed, shows that the mantle is cooled by the descent of the slab (Fig. 1.15). A clue to solving this apparent paradox lies in the types of magmas erupted. As pointed out earlier, an important difference between the magmas erupted at divergent and convergent boundaries is that magmas in subduction zones are much richer in volatile components, such as water and carbon dioxide. Thus it is necessary to evaluate the role of these components in the melting process.

Figure 1.16 shows the solidus of mantle material with no volatiles (dry) in the context of the inferred geothermal gradient that exists beneath a subduction zone. At all depths the temperature along the geothermal gradient is less than the dry solidus for mantle material and melting is not predicted. Decompression melting could occur if material is convected upwards, but the plate motions are not conducive to the sustained large-scale upward movement of mantle material. Also shown on Figure 1.16 are the solidus curves for peridotite in the presence of water and carbon dioxide. The principal effect of these components is to dramatically reduce the melting temperature at all pressures and depths. If water is available, then the geothermal gradient beneath subduction zones can intersect the solidus and melting can take place at depths in the range 80–120 km. This prediction is in accord with the

observation that most subduction zone volcanoes are located between 100 and 150 km above the top of the descending slab.

If water is of fundamental importance to subduction zone magma genesis, what is its origin? Two possibilities are that it is present in the upper mantle in general, or it is introduced into upper mantle through the process of subduction. If the former were true then the influence of water would be evident at mid-ocean ridge spreading centers as well. However, the volatile-poor nature of MORBs suggests that high volatile contents in the mantle are a direct result of the subduction process. As new seafloor is created at mid-ocean ridges it immediately begins to cool and subside. Fracturing of the crust allows for the penetration of seawater to deep levels where chemical exchanges occur between hot rock and water. This hydrothermal circulation is an effective mechanism for enhancing heat loss from mid-ocean ridge spreading centers and sustains an exotic biological community that is based on chemosynthesis (Humphris *et al.*, 1995). One of the principal effects of this process is the hydration of the oceanic crust by the conversion of anhydrous minerals to hydrous minerals such as serpentine ($Mg_3Si_2O_5(OH)_4$) and amphibole (($Na,Ca)_2(Mg,Fe,Al)_5(Si,Al)_8O_{22}(OH)_2$) at different depths. In addition, low temperature alteration of the ocean crust takes place for millions of years as the plates move away from the spreading centers. The result is that when the plates are recycled into the mantle at subduction zones they are carrying with them water and other components that are bound in certain mineral phases. However, during the subduction process increases in pressure and temperature as the slab descends into the mantle cause the hydrous phases to become unstable and release their water. These dehydration reactions occur over a range of depths from about 60 to 130 km. Once released, the volatiles migrate upwards as a fluid phase into the overlying mantle wedge. The influx of water and other volatiles into the mantle wedge may have two effects. First, the temperature regime in the area of volatile influx may be such that melting is possible by intersection with the "wet" solidus of peridotite (Fig. 1.16). Second, if temperature conditions are not conducive for "wet" melting, then the introduction of volatiles may reduce the density of the mantle wedge sufficiently to cause diapiric uprise of material. Melting could then occur by decompression when the wet solidus is intersected during the ascent path. Whatever the ultimate mechanism, the solution to the subduction zone paradox appears to lie with the recycling of water and other components back into the mantle and their effect on the melting behavior of mantle peridotite.

### Magma generation at intraplate hotspots

Hotspot, or intraplate, volcanism appears to be largely decoupled from the motion of the lithospheric plates, implying that magma generation is not dependent upon the upper-level dynamics of plate interactions. When a hotspot is coincident with a mid-ocean ridge spreading center, such as is the case with Iceland (Fig. 1.4), the net volcanic production is high and an elevated portion of the ridge is built, yet the composition of magmas produced is broadly similar to sections of the adjacent ridge. Therefore, hotspot

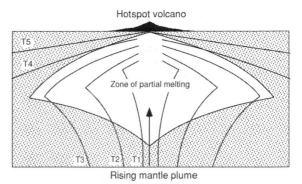

**Fig. 1.17.** Schematic representation of the partial melting of rising mantle plume material beneath a hotspot volcano. Melting occurs at greater depth in the middle of the rising plume because of the higher temperature. Modified from Brown et al. (1992).

volcanism must also involve partial melting of mantle peridotite. The higher production rate can be accounted for by applying the decompression melting model, used previously in the context of normal mid-ocean ridges. In order to yield larger amounts of melt by this mechanism it is necessary to begin with hotter mantle material (Fig. 1.14a).

Deep in the mantle, density differences caused by compositional heterogeneity or other factors trigger diapiric uprise of hot material. Based on the observation that hotspot volcanism is not uniform over time, but instead occurs as a series of pulses, it is likely that the uprise of material occurs as a stream of discrete blobs. Figure 1.17 shows the development of a partial melt zone beneath a hotspot caused by the upwelling hot mantle material. Melting is maximized in the center of the upwelling plume because this is the hottest region and consequently it begins to melt first.

## Rise and storage of magma

### Ascent of magma

Virtually all magma is generated by the partial melting of source rocks, either in the upper mantle or crust. In order for magma to be erupted it must separate from the residual source material and make its way towards the surface. The main driving force for the rise of magma is buoyancy associated with the difference in density between melt (2300–2500 kg/m$^3$) and source rock (~2700–3200 kg/m$^3$). During its rise to the surface, magma will lose heat to the surrounding rocks and begin to crystallize. If the ascent rate is too slow, the amount of crystallization may be sufficient to freeze the material en route, forming an intrusive body. Observations of erupted magmas suggest that more than 55% crystallization renders a melt "uneruptible," owing to the very high bulk viscosity as crystal content is increased (Marsh, 1981). One way for magmas to reach the surface is for ascent to take place along previous magma paths in order to take advantage of the elevated temperatures of the surrounding rocks. Thus, volcanic centers tend to focus their activity by establishing paths of minimum thermal resistance.

Ascent of magma to the surface is invariably a complex path that involves storage at different levels. Accumulation of magma within the lithosphere takes place in magma chambers, or reservoirs. Deep reservoirs (tens of kilometers in depth) may be established in the zone of partial melting where upwelling mantle material undergoes decompression (Head and Wilson, 1992). The position of such reservoirs will be controlled by the thickness and movement of the lithosphere, the transition in rheological properties of the mantle (i.e., brittle versus ductile), and existence of phase changes encountered by rising mantle material. Further ascent of magma to shallower levels may be instigated by the development of smaller diapirs, as long as the magma density is still less than the surrounding rocks and the rheological/thermal properties of the diapir allow it to rise.

Direct eruption of magma from deep reservoirs may take place occasionally but it is more common for magma to be stored in a shallow reservoir. Much more is known about the geometry and development of these storage areas and significant progress has been made during the last decade in understanding the nature of shallow chambers in different types of volcanic environments. Ryan (1987) has proposed that the creation of shallow magma chambers is controlled by the principle of neutral buoyancy, defined as the equivalency in the effective, large-scale *in situ* magma density and the density of the surrounding country rock. Under these conditions there is no longer any driving force causing the magma to rise and it will stagnate at some level beneath the surface.

For neutral buoyancy to be the controlling factor in magma chamber formation there must be some level beneath a volcano where the density of the country rock is equal to the density of the ascending magma, and additionally, that above this level the density of the country rock is lower than the magma density. At first this appears to be counter-intuitive because if a volcano is constructed out of solidified magma, shouldn't it be everywhere denser than magma? In fact, there are large variations in the density of volcanic rocks that make up the upper portions of volcanoes. This results from variations in rock composition and from the extent of fracturing. Density generally increases with depth, but in many cases exhibits a distinctive trend. From the surface to a depth corresponding to a pressure of about 200 MPa, density increases in a strongly non-linear fashion, whereas at greater depths the increase is broadly linear (Ryan, 1987). The non-linear portion of the trend is attributed to reduction in the macroscopic and megascopic pore space as pressure is increased and fractures are progressively closed and sealed. At higher pressures, the increase in density is accommodated by adjustments in the structure of the constituent minerals.

As magma rises into the upper levels of a volcano and attains neutral buoyancy, it will stagnate and begin to spread out laterally within a neutral buoyancy horizon. This horizon is expected to have limited vertical extent but a more extensive lateral expanse. Over time magma will accumulate in this zone to form a chamber from which surface eruptions may be fed. In Hawaii, *in situ* density measurements and knowledge of melt density allow predictions of the location of the neutral buoyancy horizon (Fig. 1.18). There is an excellent agreement between the predicted level of neutral buoyancy (2–4 km) and the inferred level of magma storage based on a multitude of geophysical observations (Ryan, 1987).

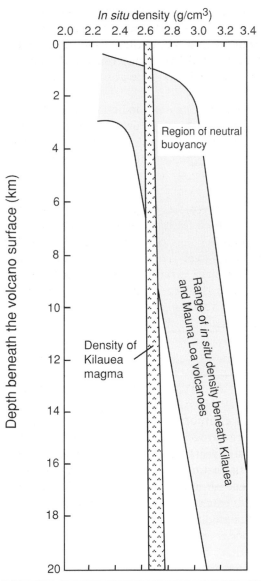

**Fig. 1.18.** Variation of *in situ* density beneath Kilauea volcano, Hawaii (shaded curve). Magma density of an olivine tholeiite composition as a function of pressure shown as stippled bar. Neutral buoyancy region occurs where magma and *in situ* density are equivalent. Modified from Ryan (1987).

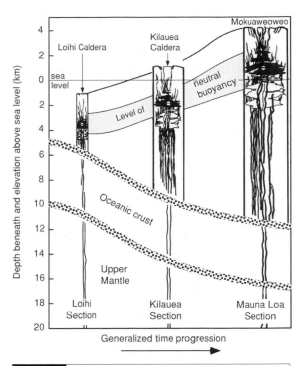

**Fig. 1.19.** Variation in the level of neutral buoyancy between Loihi, Kilauea, and Mokuaweoweo volcanic centers in the Hawaiian islands. Modified from Ryan (1987).

An important aspect of the neutral buoyancy horizon is that it is not a static level during the growth of a volcano. As eruptions occur and volcanic centers grow in elevation the density profiles will adjust to the new overlying load. The neutral buoyancy level will rise in an attempt to keep pace with the density adjustments of the edifice. In essence, a shallow magma chamber moves upwards to follow the absolute elevation of the edifice. This is portrayed in Figure 1.19 where the neutral buoyancy horizons are followed across volcanic centers of different elevation in the Hawaiian islands. The net effect is to keep the level of neutral buoyancy at a relatively constant depth beneath the top of volcano. This level will be a function of the dominant composition of the erupted products, the nature of the eruptions, and the morphological structure of the volcano.

The principle of neutral buoyancy is expected to operate in any volcanic system where magma is rising from depth and encountering a decreasing density profile towards the surface. However, because the composition of magmas and stress regimes differ considerably between tectonic environments at different plate boundaries, it is expected that the geometry and location of magma chambers should reflect these factors.

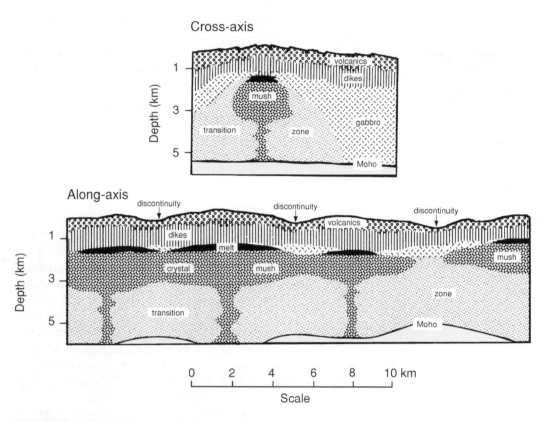

**Fig. 1.20.** Cross-axis and along-axis sections of the compound magma chamber model for a fast-spreading mid-ocean ridge system. Magma that is available for eruption is stored as a thin melt lens (black area) that is segmented along the spreading axis at axial discontinuities. From Sinton and Detrick (1992).

Magma chambers play a fundamental role in the way in which volcanoes erupt because they act as the source reservoirs for eruptions at the surface. For example, the size of an eruption is limited by how much eruptible magma is contained in a magma chamber and the style of the eruption can be related to the physicochemical changes that have taken place in the magma since its arrival in the chamber. The existence of magma chambers has been suspected for some time but it has only been recently that our view has progressed significantly from the simple "balloon and straw" model often portrayed in introductory geology textbooks. Recent advances in the use of geophysical techniques to image subsurface magma chambers has led to new insights into their shape, location, and evolution.

## Magma chambers
### Mid-ocean ridges

For many years there was considerable debate about the nature and existence of magma chambers beneath mid-ocean ridge spreading centers. Seismic experiments on the East Pacific Rise and the mid-Atlantic ridge have now led to a new model for reservoirs along divergent boundaries that takes into account differences in spreading rates (Sinton and Detrick, 1992). At fast-spreading ridges, such as the East Pacific Rise, magma derived from decompression melting of upwelling mantle material accumulates in an axial magma chamber that is located at between 1 and 2 km below the seafloor (Fig. 1.20). Geophysical evidence suggests that eruptible melt is concentrated in a relatively thin ($\sim$50 m to <1 km) lens that is about 1–2 km in width. This melt lens is located directly beneath the spreading axis and overlies a more extensive zone of crystal mush (small amounts of interstitial melt and crystals). The mush has a vertical extent of several kilometers and merges with a transition zone consisting of hot, but essentially solid rock.

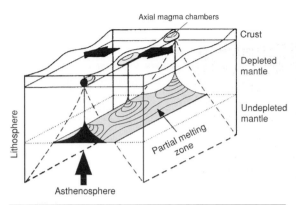

**Fig. 1.21.** Model for the development of axial magma chambers by Rayleigh–Taylor instabilities in the zone of partial melting beneath a mid-ocean ridge spreading center. Instabilities within the partial melt zone feed overlying magma chambers at regularly spaced intervals. Modified from Schouten et al. (1985).

At fast-spreading centers, the melt lens and mush zone are considered to be in a quasi-steady state, and can easily feed new eruptions to form ocean crust. Along the axis there are variations in the relative volumes of melt mush that appear to be coincident with morphological discontinuities of the spreading boundary (Fig. 1.20). These differences may be related to along-axis variations in the supply of upwelling magma diapirs beneath spreading centers (e.g., Whitehead et al., 1984; Schouten et al., 1985). As upwelling mantle encounters overlying mantle of greater density and viscosity, gravitational instabilities with regular spacing are predicted to form (Fig. 1.21). These are analogous to Rayleigh–Taylor instabilities produced whenever a fluid of low density is placed beneath a fluid of higher density. The spacing of the resulting diapirs is a function of the thickness and rheological properties of the two fluids. This process can therefore lead to the regular focusing of magma supply to the shallow chamber beneath fast-spreading ridges and have a fundamental influence on the morphological signature of the accretion process.

Geophysical observations at slow-spreading ridges, such as the mid-Atlantic ridge, indicate a major difference in the nature of the subsurface reservoir system. There is currently no geophysical evidence for the existence of a steady-state magma lens, like that imaged on the East Pacific Rise. Instead, Sinton and Detrick (1992) propose that a mush and hot rock zone is located several kilometers below the axial rift valley (Fig. 1.22). The supply rate of magma from depth is insufficient to produce accumulations of melt that can be drawn upon to feed surface eruptions. It is more likely that surface eruptions coincide with injection events of new magma into the mush zone.

## Subduction zones

In contrast to the elongated nature of mid-ocean ridges, volcanism at subduction zones is focused at individual volcanoes and thus the geometry of the underlying magma chambers is expected to be more analogous to the traditional view of a spherical reservoir connected to the surface by a conduit. Considerable effort has been made to infer the nature of magma chambers beneath subduction zone volcanoes by a variety of geophysical techniques such as seismic, gravity, magnetic, and electrical techniques (e.g., Ryan, 1988; Iyer et al., 1990; Barker and Malone, 1991). A particularly useful technique is seismic tomography which relies on the attenuation of seismic waves (natural or artificial) by the presence of magma. The technique enables the generation of a three-dimensional image of magma chambers, although it is not possible to quantitatively infer the amount of melt, or its composition. In the Cascade subduction zone of the western United States three types of situation have been found using this technique (Iyer et al., 1990): (1) no discernable magma chambers, (2) magma chambers with volumes between 200 and 1000 km$^3$, and (3) small chambers (few cubic kilometers) embedded in intrusions within the upper 5 km of crust.

A good example of a well-studied subduction zone magma reservoir is the one beneath Mount St. Helens volcano in the western United States (Pallister et al., 1992). As a result of the 1980 explosive eruption a large variety of geophysical, petrological, and experimental data has contributed to the definition of the system's location and geometry. Its shape has been largely inferred from the location of an earthquake-free zone in the depth

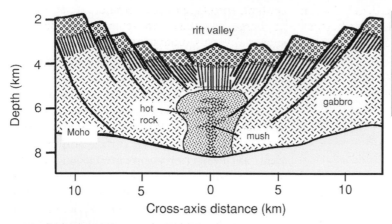

**Fig. 1.22.** Model for the magma chamber configuration beneath a slow-spreading mid-ocean ridge center. In contrast to fast-spreading ridges there is no evidence for a long-lived lens of melt beneath the axial valley. From Sinton and Detrick (1992).

range of 7–15 km, where the presence of magma precludes brittle failure (Fig. 1.23). The depth to the top of the chamber based on seismic evidence coincides well with petrologic determinations of the pressure/temperature conditions of the Mount St. Helens magma just prior to the 1980 eruption (Rutherford *et al.*, 1985). Ascent of magma during the 1980 eruption occurred through a conduit that was of the order of 50 m diameter (Carey and Sigurdsson, 1985; Scandone and Malone, 1985).

An important point about subduction zone magma chambers is their size relative to those beneath the mid-ocean ridge system. At spreading centers the chambers are relatively small and the amount of eruptible magma per unit length of ridge is only of the order of <0.5 km$^3$. In contrast, the chambers beneath subduction zone volcanoes can vary in size by several orders of magnitude. At Mount St. Helens the inferred volume of the reservoir system is about 10 km$^3$. During the 1980 eruption only 0.5 km$^3$ was erupted from the chamber and thus the majority remained at depth. Other eruptions of subduction zone volcanoes have discharged hundreds or even thousands of cubic kilometers of magma during single eruptions so the volume of their chambers may be $10^3$–$10^4$ km$^3$ in size.

The establishment of shallow magma chambers beneath subduction zone volcanoes is also likely to be governed by the principle of neutral buoyancy (Ryan, 1987). However, the dominance of more evolved (higher silica), lower density rocks in the crust of subductions zone and the subaerial expression of many of the volcanic

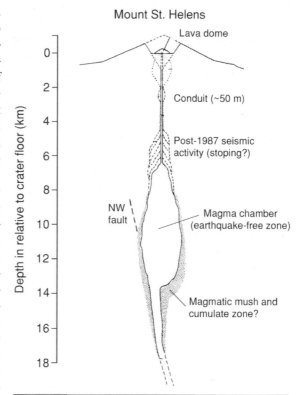

**Fig. 1.23.** Inferred configuration of the magma chamber and conduit system beneath Mount St. Helens prior to the 1980 eruptions, based on geophysical and petrologic data. Modified from Pallister *et al.* (1992).

centers may favor stabilization at deeper levels beneath the surface.

### Intraplate hotspots

Perhaps in no other environment has the complexity of magma chambers and their plumbing systems been revealed with more detail than at Hawaii, a prime example of an intraplate hotspot. Decades of observations on eruptive activity and seismic events at Kilauea volcano have been synthesized to reveal the internal three-dimensional system of magma storage (Ryan, 1988). In particular, specific types of earthquake have been used to infer the location and direction of magma movement beneath the volcano. Magma rises beneath Kilauea through a primary conduit that extends from the upper mantle to a shallow reservoir. This conduit is an integrated zone of magma ascent that is about 3 km in diameter.

Magma is stored within a complex summit reservoir that extends from 2 to 7 km depth and whose formation is governed by neutral buoyancy. Based on inflation of the summit area, the reservoir is envisioned to be a network of magma veins separated by a boxwork of country-rock blocks. Eruptions at Kilauea occur in the Kilauea caldera and along the East Rift zone. These eruptions are fed from the summit reservoir crown and from the summit reservoir base via the upper East Rift zone pipe. It should be emphasized that this picture of the Kilauea system is an integrated one built up from data collected during many periods of activity and thus may not represent the configuration of the system at any one time. Nevertheless, it provides important insights into the complexities of magma ascent, storage, and eruption in a large hotspot volcano.

# Styles of volcanic activity

When magma erupts at the surface the discharge can take two different forms. If the magma is volatile-poor then it will remain intact and flow as a mixture of melt and crystals, eventually solidifying as it cools to ambient temperature. This type of activity is referred to as effusive, or quiescent. However, if the magma contains significant amounts of dissolved volatiles, or if it comes into contact with water near the surface, it can be catastrophically disrupted into small pieces, or pyroclasts, by the rapid expansion of gases. Such events are termed explosive eruptions. The style of an eruption depends on many factors including the composition of the magma, gas content, structure of the volcano, rate of magma discharge, and the environment of eruption, i.e., subaerial versus submarine.

## Effusive volcanism

One of the most common forms of volcanic activity on Earth is the eruption of basaltic magma as lava flows. Lava flows are produced where magma issues from a vent quiescently or from high magma discharge lava-fountains and flows away from the source. These vents may be located at the summit of a volcano, on the flanks, or adjacent to the base. The geometry of the vent can take two different forms: a central vent or a linear fissure. Lava flow behavior can be quite complex and is related to factors such as magma composition, volatile content, crystal content, viscosity, local slope of the terrain, and eruption rate.

The principal driving force for the movement of lava flows is gravity and thus movement will be constrained by the local topography. In order for flow to occur the internal viscous forces of the magma must be overcome. If lavas were truly Newtonian fluids they would flow on any slope, even though it might be very slow. However, most lavas have a finite yield strength (Fig. 1.12) and behave as Bingham fluids. Consequently, there is a minimum amount of shear stress that needs to be exceeded before flow will proceed. Yield strength is a result of the partial solidification of magma as it migrates to the surface and is erupted. It is a function of magma composition and the extent to which crystallization has occurred.

For a given slope, the lava must be a certain thickness in order for the weight to overcome the finite yield strength. The thickness of the flow, $t$, is given by

$$t = \frac{\tau}{\rho g \tan \alpha} \tag{1.1}$$

where $\tau$ is the yield strength, $\rho$ is the lava density, $g$ the acceleration of gravity, and $\alpha$ is the slope in degrees (Hulme, 1974). It can be seen from this relationship that the higher the yield strength, the thicker the lava flow must be in order to move.

Once a flow is moving the velocity of the flow will be affected by many factors. In general, the

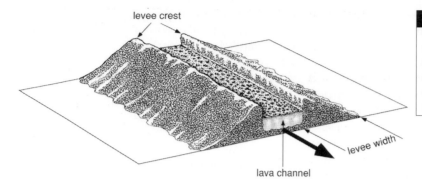

levee crest

**Fig. 1.24.** Levee development in a basaltic lava flow. During movement solidified pieces of the lava flow build up along the sides to form a confined channel within which lava can continue to flow. Modified from Francis (1993).

levee width

lava channel

speed of the flow, $V$, can be approximated by this relationship:

$$V = \frac{\rho\, gt^2}{B\eta} \sin\alpha \qquad (1.2)$$

where $\eta$ is the lava viscosity, $\rho$ is its density, $g$ is the acceleration of gravity, $t$ is flow thickness, $B$ is a constant, and $\alpha$ is the slope. An important result of this equation is that the speed is critically dependent upon the lava viscosity. As viscosity increases, the speed will decrease. Hawaiian lava flows have been measured at speeds of up to 60 km/hr on slopes of 10–28°. This represents the upper range of lava flow speeds, with most moving considerably more slowly.

The flow regime of moving lava can be interpreted using the dimensionless Reynolds number given by

$$R_e = 2\rho HV/\eta \qquad (1.3)$$

where $\rho$ is the magma density, $H$ is the depth of the flow, $V$ is the velocity, and $\eta$ is the magma viscosity. A Reynolds number value of less than 2000 corresponds to laminar flow where movement is smooth and flow lines do no cross. If the $R_e$ is greater than 2000, the flow becomes turbulent with eddies and crossing flow lines. For the majority of basaltic eruptions the viscosity of the magmas is such that laminar flow conditions prevail. Only in the case of extremely low viscosities or high velocities, such as might occur over very steep topography, would flow conditions become turbulent.

Discharge of magma at the vent may be passive, in the form of slow outpourings of magma, or it may be more spectacular with fountains of magma (lava-fountaining) that can attain heights

of over 1000 m. Fallback of material from lava-fountaining may form quenched scoria fragments or, if the fragments are still very hot, spatter develops around the vent, fuses together, and may begin to move away as a lava flow.

## Subaerial conditions

As magma is discharged from a vent under subaerial conditions cooling takes place by conduction, convection and radiation. At the high temperature of eruption, radiation is initially the most important mechanism of heat loss and can be approximated by

$$Q \approx \sigma T^4 \qquad (1.4)$$

where $Q$ is the rate of heat loss, $\sigma$ is the Stefan–Boltzman constant ($5.67 \times 10^{-12}$ J s$^{-1}$ cm$^2$ °C$^{-4}$), and $T$ is the absolute temperature. Because $Q$ is proportional to the fourth power of temperature there is a dramatic decrease in the rate of heat loss associated with only slight decreases in temperature. Loss of heat by radiation immediately lowers the surface temperature and changes the color of a flow. This results in an increase in viscosity and a decrease in flow speed as pieces of chilled lava appear on the surface. These are carried along on the lava surface and only parts of the incandescent interior remain visible. As pieces of quenched lava fall off the sides of a lava flow they build up a natural levee that confines the flow and insulates it from cooling (Fig. 1.24). This process results in lava being funneled down the slope of a volcano along a similar path for long periods of time. The width of the levees is a function of the yield strength of the magma and of the local slope. Hulme (1974) has developed a relationship for quantitatively estimating yield strength, $\tau$, from levee dimensions as follows

$$\tau = 2w_b g \rho \alpha^2 \qquad (1.5)$$

where $w_b$ is levee width, $g$ is local gravity, $\rho$ is magma density, and $\alpha$ is the slope.

Continued development of a quenched crust on the lava flow surface may eventually lead to isolation of the flowing interior from the atmosphere as a roof of lava crust forms over an interior lava tube. Because lava and quenched magma are poor conductors of heat, but have high heat capacities, lavas may continue to flow for considerable distances from source without suffering significant heat loss. The confinement of lava to these subsurface tubes is thus an effective mechanism for flows to increase their maximum runout from source. When an eruption ceases, downslope drainage of the lava may result in the formation of lava tunnels that can extend for many kilometers away from the vent. In the distal parts of lava flows, the pressure build-up within the interior of the flow may causing upwelling of the surface and occasional breakout of magma. The uplifted breaks in the surface are referred to as tumuli.

An important question about lava flows is: how far will they travel? Most historic lava flows have traveled several to tens of kilometers from the source vent. However, there is evidence in the geologic record of basaltic lava flows that have traveled up to 300 km from source. The 14 Ma Pomona flow in the western United States was erupted in Idaho and can be traced almost to the Pacific coast in Oregon (Fig. 1.25). Walker (1974) proposed that the discharge rate is the most important factor in determining the ultimate length of lava flows. Discharge rates for historic lava flow have varied by several orders of magnitude (0.5 to $5 \times 10^3$ m$^3$/s), and there is evidence for truly enormous rates of discharge ($\sim 1 \times 10^4$ m$^3$/s) during some flood basalt episodes (Swanson et al., 1975). In addition, flow volume and cross-sectional area appear to correlate with flow length (Malin, 1980), although this is to be expected if discharge rate plays a major role in flow runout (Woods, 1988). Another factor that influences how far lava flows travel is whether a flow is channel or tube-fed. The latter will favor

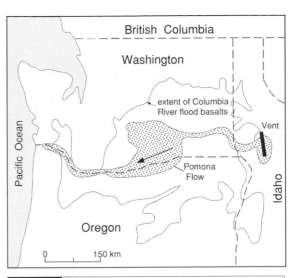

**Fig. 1.25.** Distribution of the 14 Ma Pomona basalt lava flow in Washington state. Modified from Francis (1993).

**Fig. 1.26.** Surface morphologies of an aa lava flow (background) and a pahoehoe lava flow (foreground) from Kilauea volcano, Hawaii. Differences in the texture are attributed to variations in shear strain and magma viscosity. Photograph by S. Carey.

longer runouts because heat loss, and thus viscosity increase, is reduced for flows in subsurface tubes.

Quiescent subaerial discharge of basaltic magma generally leads to the production of two distinct type of lava flows: aa and pahoehoe. Aa is the most common type and consists of flows that have a rubbly top with sharp angular blocks and clasts (Fig. 1.26). These flows tend to form units 10–100 m thick that cover areas of between

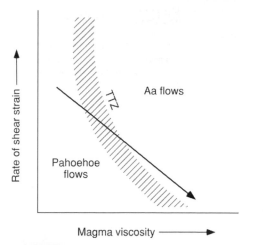

**Fig. 1.27.** Configuration of the proposed transition threshold zone (TTZ) defining the behavior of basaltic lava flows. An increase in rate of shear strain or magma viscosity favors the development of aa type flows over pahoehoe. Modified from Peterson and Tilling (1980).

1 and 100 km². In cross-section the flow typically consists of an upper rubble zone overlying a massive interior. The interior often exhibits abundant vesicles formed by the exsolution of dissolved gases. At the base of the flow there is also a thin rubble zone separating the massive flow from the ground surface. In contrast, pahoehoe flows are characterized by a much smoother surface with ropey and entrail-like morphology (Fig. 1.26). These flows are generally thinner (<15 m thick) and can cover areas of 1–1000 km².

In many cases the composition of aa and pahoehoe flows is identical and thus composition alone can not be called upon to explain the origin of the two different flow morphologies. Most Hawaiian basaltic eruptions begin as pahoehoe but some undergo a transition to aa. The reverse, however, has never been observed. Peterson and Tilling (1980) suggest that the transition from pahoehoe to aa is determined by the relationship between magma viscosity and rate of shear strain. The latter is considered to be a measure of how fast differential motion occurs between adjacent parts of the flowing lava. They proposed the existence of a transition threshold zone (TTZ) that separates lava flow behavior into pahoehoe and aa regimes (Fig. 1.27). The rate of shear strain necessary to initiate transition to aa behavior is

inversely related to viscosity. Thus, high-viscosity magmas require lower rates of shear strain to make the transition to aa behavior. Several studies have proposed that discharge rate plays a major role in determining the production of aa versus pahoehoe (e.g., Pinkerton and Sparks, 1976; Rowland and Walker, 1990). Field observations suggest that the formation of aa flows is favored by higher discharge rates (>5–10 m³/s).

Magmas of more evolved composition are also erupted as lava flows but their resulting morphologies reflect fundamental differences in rheology. In general as silica content increases, viscosity increases (Fig. 1.13) and thus flow mobility is limited. Flows of andesitic or dacitic composition tend to be thicker and less widespread compared to basaltic flow. Aspect ratio is a convenient parameter to characterize the general shapes of lava flows. In this case it is defined as the ratio of the thickness of a flow to the area that it covers. Because of their low viscosity, basaltic magma generally form thin and widespread flows, whereas more viscous rhyolite flows are fat and compact (Fig. 1.28).

Some evolved magmas have such high viscosities that they are unable to flow any substantial distance away from the vent, and instead build up domes. Extrusion of viscous magma can lead to four principal types of domes: Peléean type, low lava dome, upheaved plug, or coulee (Blake, 1989). Peléean domes are named after the famous structure that grew in the crater of Mt. Pelée following the explosive eruption of 1902 (Lacroix, 1904). These structures are characterized by very steep sides and a sharp, craggy spine at the top (Fig. 1.29). Surrounding their base is an apron of debris generated by repeated collapse of the dome's sides. A low lava dome has a more subdued profile and symmetrical distribution around the source vent (Fig. 1.29). It grows by outward displacement of previously erupted material as new magma is introduced internally. Coulees represent something of a transition between lava domes and lava flows. They are generated when viscous magma is erupted on a steep slope and there is sufficient shear stress for some downslope movement. This flow results in an asymmetrical distribution about the vent, but near source they may still be as thick as low

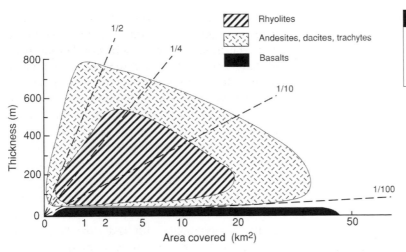

**Fig. 1.28.** Fields of thickness versus area covered for lava flows of different composition. Fractions indicate the ratio of thickness to area. Modified from Walker (1973).

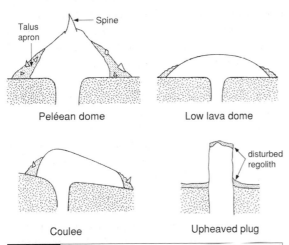

**Fig. 1.29.** Different types of domes produced by the extrusion of viscous magma. Variations in dome morphology are related to differences in magma viscosity and extrusion rate. Modified from Blake (1989).

lava domes (Fig. 1.29). Finally, upheaved plugs are pushed up sections of viscous magma that rise almost vertically and lift sections of the surface stratigraphy (Fig. 1.29).

### Submarine conditions

The majority of effusive volcanism on Earth is basaltic in composition and occurs underwater either along the mid-ocean ridge system or at submarine hotspots. Discharge of magma into water produces distinctive types of flows that reflect the more efficient cooling compared to the subaerial environment. Submersible observations and bottom photography have revealed two principal types of basaltic submarine lava flows along mid-ocean ridge spreading centers: pillows and sheet flows. Pillow lavas consist of masses of interconnected flow lobes with bulbous, spherical, or elongated morphologies (Fig. 1.30). Individual pillows range from about 10 cm to 1 m in diameter. The surface of pillows consists of a thin glassy crust (<1 cm) formed by the rapid quenching of magma by seawater. Inside, the pillows exhibit radial fractures, vesicles, and a more crystalline groundmass resulting from slower cooling. The outside of some elongated pillows is decorated with corrugations and contraction joints formed as pillows grow and subsequently cool. On the Juan de Fuca ridge two types of pillow morphologies have been observed (Chadwick and Embley, 1994). One type was characterized by large elongated pillows with striated surfaces, while the other consisted of smoothed surface pillow lobes that resembled subaerial pahoehoe toes.

Sheet flows, as their name imply, are flows with a large horizontal extent relative to their thickness (Ballard et al., 1979). The surfaces of the flows can be relatively smooth or exhibit different types of deformation features, such as ropey, folded, or coiled morphology. Some, referred to as hackly flows, have irregular surfaces that are made up of jagged blocks about 1 m in size. These bear some resemblance to the rough surfaces of some subaerial aa flows. The glassy rinds of sheet flows are commonly thicker than those of pillow basalt, measuring about 5 to 15 cm (Hekinian

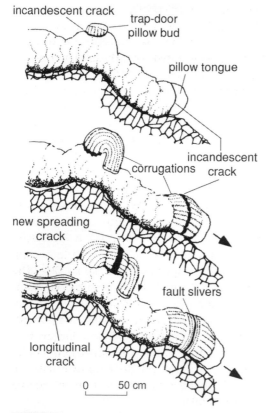

incandescent crack

trap-door
pillow bud

pillow tongue

incandescent
corrugations crack

new spreading
crack

fault slivers

longitudinal
crack

0      50 cm

**Fig. 1.30.** Formation of pillow basalt in the submarine environment. Rapid cooling of lava against seawater generates a solid crust. Continued growth occurs by budding and elongation of pillow lobes as magmatic pressure exceeds the strength of the pillow walls. Modified from Brown *et al.* (1989).

As magma is erupted slowly it is rapidly chilled to form a glassy rind that inflates due to the internal magma pressure. The flow advances intermittently in a series of pulses, usually moving a few meters at a time. Formation of the quenched crust temporarily stops the flow advancement until the internal pressure is sufficient to rupture the skin and a new pillow is budded out. This new growth may occur from the front of the pre-existing pillow or, more commonly, through the side. The directional changes caused by laterally advancement leads to the complex interconnected appearance of pillow lava fields. Changes in the rate of magma discharge appears to control the fine-scale structure of pillows (Ballard and Moore, 1977). Low magma discharge produces elongated pillows with cracks and striations because a thick brittle crust is developed and budding to form new pillows occurs through irregular fractures. At higher discharge rates the quenched skin is thinner and thus can stretch without fracturing, producing a smoother surface.

The formation of submarine sheet flows has never been directly observed but their morphological features provide important clues to their origin. Using analogies with subaerial lava flows Ballard *et al.* (1979) proposed a model for the formation of pillow and sheet flows. Sheet flows are thought to form from high-discharge events of fluid lava from fissures. They are thus analogous to high-discharge-rate subaerial surface-fed pahoehoe in which the flow is not tightly constrained by channels. The rapid discharge of lava can produce ponding in topographic depressions. As breakout of magma occurs at the margins of the pond, or drainback to the source vents takes place, collapse of the quenched surface produces pits. If the drainage is discontinuous, a record of the changes will be preserved in the form of horizontal terraces along the pit margin. The pillars that are often found in association with sheet flows most likely represent places where water that was trapped below the flow has been ejected upwards through the flow creating a cylindrical conduit from quenched lava.

The important role of discharge rate on flow morphology of submarine lava flows has been demonstrated by laboratory experiments using

*et al.*, 1989). Sheet flows are often found in association with collapse pits in summit calderas along spreading ridges. These pits are up to several hundred meters in diameter and can be 25 m deep (Francheteau *et al.*, 1979). Inside there is commonly basaltic rubble scattered across the floor, horizontal ledges at various levels along the sides, and unusual circular pillars with distinctive ribbed surfaces.

In many submarine volcanic fields the pillow and sheet lavas have the same composition and thus the origin of the contrasting morphologies cannot be dependent on magma composition alone. The formation of pillow lavas has been well documented from observations of lava flows entering the sea off Hawaii (Moore, 1975).

polyethylene glycol wax (Griffiths and Fink, 1992; Gregg and Fink, 1995). Different flow morphologies can be characterized by a dimensionless parameter, $\psi$, defined by

$$\psi = t_s/t_a \qquad (1.6)$$

where $t_s$ is the time required for the formation of a solid crust on the surface of the flow and $t_a$ is some characteristic timescale for horizontal advection of the flow. Pillow lavas form when the value of $\psi$ is less than 3, whereas various types of sheet flows are produced if $\psi$ is between 3 and 25.

The distribution of different submarine lava flow types has been studied in a number of different mid-ocean spreading areas such as the mid-Atlantic ridge (slow spreading rate), the Juan de Fuca ridge and Galapagos Rift (intermediate spreading rate) and the East Pacific Rise (fast-spreading rate). At slow-spreading ridges, pillow basalts are dominant, whereas sheet flows are most common along intermediate and fast-spreading ridges. This suggests that magma discharge rates are likely to vary between these different spreading environments. In addition, there are variations in the relative abundances of different flow types along the axis of ridge segments. Francheteau and Ballard (1983) found that sheet flows were more abundant at the topographic highs of first order ridge segments on the East Pacific Rise. These areas correspond to zones of high magma supply and thus eruptions can tap into the subaxial magma chambers for voluminous discharges of fluid magma (Fig. 1.20). With increasing distance from the topographic high, the magma supply diminishes and eruptions of lower discharge rate produce increasing amounts of pillowed flows. In contrast, the lack of a well-developed melt lens at slow-spreading ridges (Fig. 1.22) limits the amount of available magma and leads to common eruptions of low discharge rate and production of large amounts of pillowed lavas on the rift axis floor.

## Explosive volcanism
### Fragmentation of magma by volatile degassing
One of the most important mechanisms of magma fragmentation is degassing of dissolved volatile components such as water and carbon dioxide (e.g., Verhoogen, 1951; McBirney, 1973). At depth, magma may be undersaturated, saturated, or supersaturated with these components. The degassing process has been largely inferred from experimental and theoretical work on bubble growth in gas-saturated liquids, coupled with observations of erupted products (e.g. Sparks, 1978; Cashman and Mangan, 1994; Sparks et al., 1994). In order for bubbles to begin to grow, supersaturation of only a few tens of bars is required for heterogeneous nucleation (Hurwitz and Navon, 1994). During ascent, bubbles increase in size by a combination of diffusion (mass transfer) and depressurization (Sparks, 1978). Diffusional growth dominates at depth while decompression is more important near the surface. The ultimate size of the bubbles is a function of magma ascent rate, initial volatile content, and diffusion coefficient of the volatile component. Bubbles in basaltic magma tend to be larger than in silicic magma because the diffusion coefficients of most volatile components in basalt are high and the viscous forces that impede bubble growth are smaller.

When bubbles form in ascending magma they are less dense than the surrounding liquid and will also begin to rise. The rate at which they rise will be determined by their size, density contrast with the magma, and magma viscosity. If the rise rate is similar to or much less than the rise rate of the magma, then the bubbles are essentially locked into the volume of magma from which they grew. This situation typically occurs for silicic magmas whose viscosity is so high that bubbles cannot rise relative to the magma even for low rates of magma rise. In contrast, the viscosities of basaltic magmas can be low enough such that bubbles can rise faster than the magma. This can result in the accumulation of gases in parts of the magma chamber or conduit system.

As magma approaches the surface the volume fraction of gas continues to increase as bubbles grow larger. An important aspect of the degassing process is the rheological changes associated with the decrease in dissolved volatiles during ascent. Figure 1.13 shows that for a rhyolite magma, the loss of only 1.0% water causes an increase in viscosity of about an order of magnitude. Consequently, the magma is essentially transformed into a foam that becomes increasingly rigid. Crowding of bubbles and the increase in strength of the foam makes it difficult for further bubble growth and excess pressure builds within bubbles. Eventually the internal pressure in the bubbles is high enough to cause bubble bursting and fragmentation of the magma into a mixture of exsolved gases and hot, liquid particles. This transition from a state in which the magma was the continuous phase to one where the gas is the continuous phase is referred to as the fragmentation level. The position of this level will be a function of the initial gas content of the magma, the magma ascent rate, and the magma composition. In general, fragmentation during explosive eruptions occurs at depths of less than 1 km below the vent, although in some cases it may migrate to deeper levels. Above this level the mixture is rapidly accelerated due the expansion of the gases under high pressure. It is this expansion that is the driving force for the high-speed ejection of material from a vent during an explosive eruption. The degassing process can lead to a variety of eruption styles depending on the amount of volatiles in the magma, the composition of the magma, and the rate at which it is supplied from depth.

## LAVA-FOUNTAINING AND STROMBOLIAN ERUPTIONS

Lava-fountaining is the disruption of basaltic magma into a spray that is ejected up to over 1000 m above a vent, although typical heights are usually tens to hundreds of meters. An important aspect of this style of eruption is that it produces activity that may be sustained for several hours or days. It is a common type of activity for Hawaiian volcanoes and can occur from both central vents or fissures. The degree of fragmentation is typically poor and thus relatively large clots up to several tens of centimeters are formed. Because the clasts can be large and the ejection heights modest, the pyroclasts from lava-fountaining may remain molten during their transport and coalesce after fallout to form a lava flow. Finer pyroclasts may cool sufficiently during fallout to produce a deposit of solid tephra particles that is controlled by the direction and speed of the local

winds. The heights of the fountains are a function of magma volume flux and volatile content (Head and Wilson, 1987).

In contrast to the continuous fountain of basaltic magma produced by lava-fountaining, Strombolian eruptions are characterized by a series of discrete explosions separated by <0.1 s to several hours. Each explosion represents the bursting of one or more very large bubbles (up to meter-size diameter) near the surface, usually within a standing lava lake (Blackburn et al., 1976). Fragmentation is also relatively poor during these types of events and incandescent clasts are typically ejected along well-defined ballistic trajectories through the atmosphere (e.g., Chouet et al., 1974). Accumulation of coarse ejecta from Strombolian eruptions can construct steep-sided cones such as the one produced during the 1943–52 eruption of Parícutin volcano in Mexico (e.g., Riedel et al., 2003).

The mechanism of lava-fountaining and Strombolian eruptions involve fragmentation of low-viscosity basaltic magma by volatile degassing. Parfitt and Wilson (1995) have suggested that a fundamental factor that differentiates lava-fountaining from Strombolian eruptions is the extent to which bubble coalescence occurs during magma ascent. In basaltic magma the viscosity is sufficiently low lava that bubble rise due to buoyancy can approach or exceed the rise rate of the magma. When bubbles are able to rise relative to the magma the probability of coalescence increases dramatically as portions of the magma become enriched in bubbles. As magma rise rate increases bubbles can no longer move relative to the magma and are thus fixed to the portion of melt from which they exsolved, reducing the potential for coalescence.

Lava-fountaining is favored by high rates of magma rise. Under these conditions bubble growth occurs more homogeneously throughout the rising magma and fragmentation can be attained at deep levels. This leads to a continuous discharge of fragmented spray above the vent. Modeling by Parfitt and Wilson (1995) suggests that for volatile contents from 0.1% to 1.0%, lava-fountaining will occur if magma rise speeds are in excess of 0.1 m/s (Fig. 1.31). As magma rise rate is decreased, bubbles are able to move and

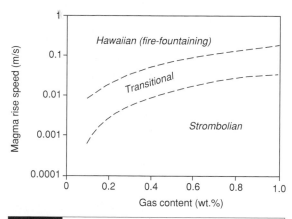

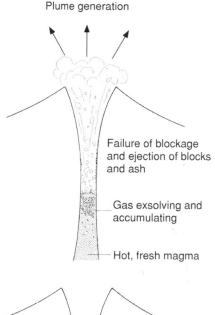

**Fig. 1.31.** Predicted conditions for the production of lava-fountaining versus Strombolian eruptions during the discharge of basaltic magma. Based on the modeling of Parfitt and Wilson (1995).

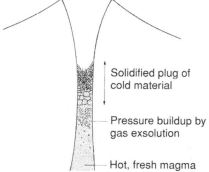

**Fig. 1.32.** Cross-section showing the development of a Vulcanian eruption (bottom to top). Modified from Wilson (1980).

concentrate relative to the rising magma. Some parts of the magma become volatile-depleted and can thus rise to the surface without fragmentation. Other portions, where volatiles accumulate and bubble coalescence takes place, will contain large bubbles that burst intermittently as they arrive at the surface. Low magma rise speeds thus favor Strombolian eruptions (Fig. 1.31) and transitions in eruption style are more likely to be determined by changes in ascent rate as opposed to variations in volatile content.

VULCANIAN ERUPTIONS

Another type of explosive eruption that involves a series of discrete events, but which discharges more evolved magma, such as basaltic andesite or andesite, is called Vulcanian. These eruptions occur at intervals of a few minutes to several hours and eject pyroclastic material at velocities up to 400 m/s (Self et al., 1979). Fragmentation of magma and country rock is more efficient than in Strombolian eruptions and the ejected mixture typically forms a turbulent plume of particles and gases that rises by thermal convection above the vent to altitudes up to several thousand meters. The plumes generated by such eruption may reach as high as 20 km into the atmosphere and result in much wider dispersal of material compared with Strombolian eruptions. It should be emphasized, however, that the division bet-

ween these types of explosive eruptions is likely to be transitional and complicated by the different mechanism by which they are generated.

Vulcanian eruptions are generally attributed to the sudden release of pressure within a conduit or dome as a result of the failure of a partially cooled overlying carapace (Fig. 1.32). As magma rises into the edifice of a volcano, crystallization and volatile exsolution can build up high pressures. If the magma is blocked by a plug of solidified magma or country rock, the pressure may continue to build until it exceeds the strength of the cap. Sudden pressure release allows the gases to rapidly expand and fragment both magma and the solidified cap. Observed

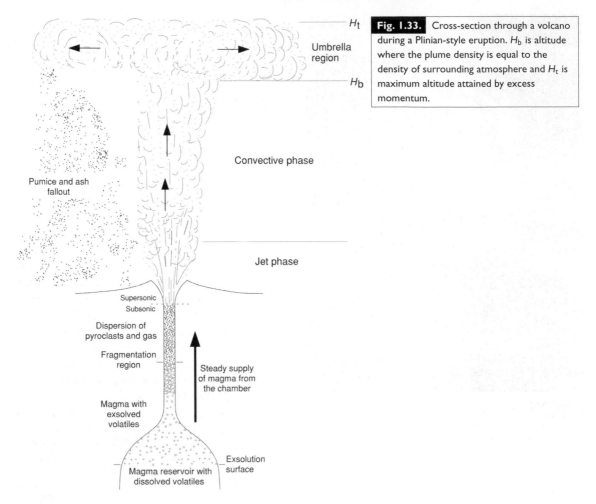

$H_t$

Umbrella region

$H_b$

**Fig. 1.33.** Cross-section through a volcano during a Plinian-style eruption. $H_b$ is altitude where the plume density is equal to the density of surrounding atmosphere and $H_t$ is maximum altitude attained by excess momentum.

Convective phase

Pumice and ash fallout

Jet phase

Supersonic
Subsonic

Dispersion of pyroclasts and gas

Fragmentation region

Steady supply of magma from the chamber

Magma with exsolved volatiles

Exsolution surface

Magma reservoir with dissolved volatiles

ejection velocities of up to 200 m/s are consistent with the shallow level degassing of several weight percent of dissolved volatile components from magmas (Wilson, 1980). In some cases, Vulcanian eruptions may be driven by the interaction of magma with groundwater, whereby ejection velocities up to 400 m/s may be developed.

PLINIAN ERUPTIONS

The most intense type of explosive eruption involving fragmentation of magma by volatile degassing occurs when large volumes of silicic magma are erupted in a quasi-steady-state fashion for periods of hours to days. These events are referred to as Plinian eruptions in honor of Pliny the Younger who described the classic AD 79 explosive eruption of Vesuvius which buried the cities of Pompeii and Herculaneum in Italy.

During these eruptions magma is withdrawn from a crustal magma chamber and undergoes gas exsolution as it moves toward the surface. Initially, bubbles grow and move with the magma because of its high viscosity (Fig. 1.33). Recent experimental and theoretical work suggests that a large part of the degassing may actually occur over a limited region of the conduit (Mader *et al.*, 1994; Sparks *et al.*, 1994). Large degrees of supersaturation caused by magma ascent may result in homogeneous nucleation of bubbles and catastrophic gas release. Rapid acceleration of a stiff foam and breakage of bubble membranes results in disintegration into a gas/pyroclastic mixture within a fragmentation zone. Above this zone the mixture moves in a turbulent, high-speed fashion before being ejected from the vent.

A key feature of Plinian eruptions is the sustained discharge of a highly fragmented

**Fig. 1.34.** Plinian-style plume from an explosive eruption of Mount St. Helens in 1980. Photograph by M. Doukas (US Geological Survey).

gas/pyroclast mixture for long periods of time. Exsolution of only 1% to 5% water is sufficient to produce exit velocities of 100 to 500 m/s at the vent (Wilson *et al.*, 1980). The hallmark of Plinian eruptions is the production of a spectacular mushroom-shaped plume (Figs. 1.33, 1.34), such as described by Pliny the Younger during the AD 79 eruption of Vesuvius. As the gas/pyroclast mixture is ejected from the vent its bulk density is greater than the atmosphere density and it will initially rise to several hundreds or thousands of meters owing to its momentum (Sparks, 1986). During this jet phase the density of the mixture must decrease in order for continued rise to take place above the volcano. This is accomplished by the entrainment and heating of ambient air at the sides of the high-velocity jet. A transition occurs where the motion of the material is now determined by buoyant thermal convection (Fig. 1.33). Essentially the plume rises like a hot-air balloon. This convective region typically

constitutes a large part of the rising plume that may eventually reach several tens of kilometers in height. However, because the atmosphere is stratified, the plume will eventually reach an altitude, $H_b$, where its bulk density is equal to that of the surrounding air and there is no longer sufficient thermal energy to heat entrained air. At that point the plume will begin to spread out laterally into an umbrella region (Fig. 1.33). The top of the umbrella region, $H_t$, represents the maximum height attained by the rising mixture caused by its momentum when it arrives at the level of neutral bouyancy. Away from the convective portion of the plume the umbrella region spreads like a giant gravity current in the atmosphere (Bursik *et al.*, 1992a, 1992b). The lateral velocities can be very large near source and overwhelm the strength of the local winds allowing an umbrella current to spread radially for great distances (e.g., Sparks *et al.*, 1986). Four hours after the beginning of the 1991 explosive eruption of Mt. Pinatubo the umbrella region had grown to a diameter in excess of 400 km.

The height of a Plinian plume is mainly determined by the thermal flux at the vent and to a first approximation behaves in a manner similar to turbulent convective plumes in a stratified environment where the height, $h$, is given by

$$h = bQ^{0.25} \tag{1.7}$$

and $Q$ is the thermal flux (directly related to magma discharge rate), and $b$ is a parameter related to the density stratification of the atmosphere (Sparks, 1986). Most Plinian eruptions generate sufficient thermal output to form plumes that extend well into the Earth's stratosphere. For example, the 1991 eruption of Mt. Pinatubo in the Philippines produced a giant mushroom-shaped plume that reached between 35 and 40 km altitude (Koyaguchi and Tokuno, 1993). More sophisticated models of Plinian plumes allow for the specific calculation of plume density, temperature, and velocity as a function of altitude (e.g., Woods, 1988).

Fallout of pumice and ash occurs from the sides of Plinian eruption columns and the base of the umbrella region (Carey and Sparks, 1986; Bursik *et al.*, 1992a; Bonadonna *et al.*, 1998). Because of the great height of the plumes and

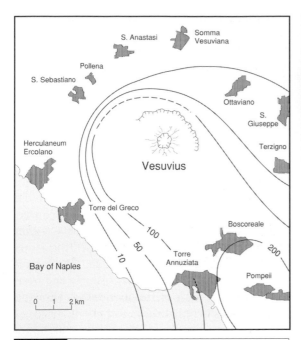

**Fig. 1.35.** Isopachs of the AD 79 Plinian pumice fall deposit from Vesuvius volcano. Thicknesses are in centimeters. Modified from Sigurdsson *et al.* (1985).

**Fig. 1.36.** Pyroclastic flow descending the slopes of Mount St. Helens during the August 7, 1980 eruption. Photograph by P. Lipman (US Geological Survey).

continuous nature of Plinian eruptions, large areas can be impacted by thick accumulation of tephra fall. The AD 79 eruption of Vesuvius lasted for approximately 19 hours and resulted in the burial of Pompeii under about 2 m of pumice fallout (Fig. 1.35).

### Generation of pyroclastic flows and surges

One of the deadliest types of behavior during explosive eruptions is the generation of mixtures of hot gases and particles that descend down the slopes of a volcano at high speeds (Fig. 1.36). They can travel at velocities in excess of 100 km/hr and maintain temperatures of >400 °C at great distances from source. Their speed and high temperature make them particularly lethal to human populations around volcanoes and have resulted in tens of thousands of fatalities during historic times (Tilling, 1989). Pyroclastic flows are high-particle concentration types of flows that tend to be strongly controlled by the local topography. They can produce widespread deposits of coarse pumice and ash known as ignimbrites. Pyroclastic surges are more dilute flows that move in a

highly turbulent fashion and can overrun many topographic barriers (Fisher *et al.*, 1980; Carey, 1991). They tend to produce a thinner, stratified deposit of fine-grained pumice and ash.

An important mechanism for the generation of flows involves the behavior of the high-speed jet of gas and pyroclasts after it leaves the vent. The gas/pyroclast mixture that is ejected from vents during explosive eruptions is in most cases denser than the surrounding atmosphere and can only continue to rise if it entrains and heats enough ambient air to reduce its bulk density to less than that of the atmosphere (Sparks and Wilson, 1976). When that occurs, a high-altitude convective plume is formed, as described in the previous section. However, if the mixture is unable to become buoyant, then it will collapse to form pyroclastic flows or surges. The collapse may affect the entire jet, in which case a low collapsing fountain may occur over the vent, or

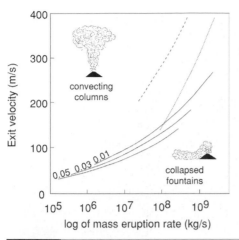

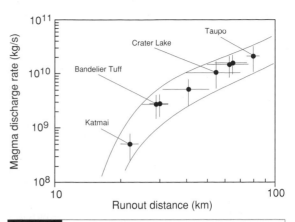

**Fig. 1.37.** Predicted conditions for the generation of convecting columns versus collapsing fountains (pyroclastic flow generation). Values next to the boundary curves indicate the volatile mass fraction in the magma. Modified from Sparks et al. (1997).

**Fig. 1.38.** Inferred magma discharge rates versus runout distance for some large-scale pyroclastic-flow-generating eruptions. Modified from Bursik and Woods (1996).

it can take place only along the column margin, with some parts of the column continuing to rise convectively (Carey et al., 1988).

The main factors that determine whether an erupting mixture will form a rising plume or a collapsing fountain are the magmatic volatile content, exit velocity, vent radius, and magma discharge rate. Figure 1.37 shows the configuration of the two different regimes based on theoretical modeling of eruption column behavior (Sparks et al., 1997). Exit velocity is strongly dependent on magmatic volatile content (Wilson, 1980) and magma discharge rate is controlled in part by the vent radius (Wilson et al., 1980). The modeling results indicate that the generation of pyroclastic flows and surges is favored by high magma discharge rates and low exit velocities. A common feature of many Plinian eruptions is for the event to begin with a high-altitude convective plume that forms widespread fallout followed by a transition to the generation of pyroclastic flows. This transition may be the result of decreasing volatile content of the magma or increases in the magma discharge rate during the course of the eruption. The AD 79 Plinian eruption of Vesuvius exhibited this type of evolution and calculation of eruption parameters, such as magma discharge rate and exit velocity, are in

accord with theoretical predictions of eruption column behavior (Sigurdsson et al., 1985; Carey and Sigurdsson, 1987).

Pyroclastic flows and surges may also be generated by the collapse of growing lava domes by explosive or gravitational failure of the sides or explosive eruption. A lateral blast from a growing dome on the summit of Mt. Pelée in Martinique on May 8, 1902 generated a devastating pyroclastic surge that killed more than 28 000 people in the city of St. Pierre (Fisher et al., 1980). More recently, pyroclastic flows and surges have been generated frequently during dome growth at Unzen volcano in Japan and the Soufrière Hills of Montserrat (Young et al., 1997).

Once generated, pyroclastic flows and surges travel downslope under the influence of gravity and entrain air as they move. Deposition of material from the base and heating of entrained air results in the generation of a buoyant co-ignimbrite plume that rises off the top of the flow (Fig. 1.36) and may match or exceed the heights of some Plinian plumes (Woods and Wohletz, 1991). An important aspect of pyroclastic flows, as with lava flows, is how far they travel from source. Bursik and Woods (1996) recently developed a model for the dynamics and thermodynamics of large pyroclastic flows. They found that model-derived calculations of magma discharge rate for a number of large pyroclastic-flow-generating eruptions were well correlated with the runout distance (Fig. 1.38). Furthermore, the inferred magma

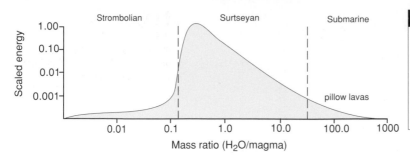

**Fig. 1.39.** Variation in the scaled energy (efficiency of conversion of thermal energy to explosive magma disruption) versus the mass ratio of water to magma. Maximum efficiency occurs at a mass ratio of about 0.3. Modified from Wohletz and McQueen (1984).

discharge rates of the associated Plinian phases of these eruptions were always less than the pyroclastic flow-phase rates, in support of predicted eruption column behavior shown in Fig. 1.37.

Generation of pyroclastic flows during large-volume silicic eruptions represents one of the most spectacular forms of volcanism known on Earth. Enormous volumes of magma can be expelled during single events to form widespread deposits known as ignimbrites, or ash flow tuffs (Ross and Smith, 1961). Such deposits consist of a poorly sorted mixture of pumice, ash, crystals, and lithics. Individual flow units may reach thicknesses of tens of meters and can extend in excess of 100 km from source. In some cases the deposits are emplaced at such high temperatures and with sufficient thickness that individual glassy particles fuse together to form a dense welded tuff. Historic eruptions have produced pyroclastic flow deposits with volumes up to several tens of cubic kilometers, but there are many examples in the geological record of much larger deposits (Smith, 1979). For example, the volume of the Fish Canyon Tuff on the La Garita caldera in the San Juan Mountains of Colorado, USA, has been estimated as >3000 km³ (Steven and Lipman, 1976).

### Fragmentation of magma by interaction with external water

The second major mechanism for the fragmentation of magma to generate explosive volcanism is the interaction with some external source of water. This may include groundwater, rivers, lakes, seawater, or ice and snow (e.g., Colgate and Sigurgeirsson, 1973). Magma is fragmented by the rapid conversion of water to steam or by rapid quenching. Kokelaar (1986) has suggested three principal types of fragmentation mecha-

nisms involving external water. The first, contact-steam explosivity, involves the generation of a thin film of vapor at the interface between hot magma and water. The film subsequently collapses as a result of cooling by the remaining reservoir of water and disrupts the surface into small droplets. This process exposes new hot magma below and the process repeats itself. Each cycle occurs on a timescale of microseconds and evolves into an explosive process by high-frequency repetition (Wohletz, 1986). However, explosive disruption of magma does not always occur when it comes in contact with water. A major factor in determining the nature of the interaction is the mass ratio of water to magma. Explosive fragmentation is favored by a ratio of about 0.3 (Fig. 1.39). Contact-steam explosivity can occur during underwater eruptions although the maximum depth at which the process can take place is poorly known.

The second mechanism for magma fragmentation by external water, called bulk interaction steam explosivity, involves the local trapping of water and conversion to vapor. For example, a lava flow erupted underwater may trap water-rich sediments beneath the flow. Heat from the flow leads to superheating of the trapped water and the build-up of high pressure as vapor is produced. If the pressure exceeds the strength of the overlying flow, then fragmentation occurs as the cap fails catastrophically. This process is likely to occur in water depths up to 2 to 3 km (Kokelaar, 1986), being limited by the pressure at which the volume change of water from liquid to vapor is no longer significant enough to cause high internal pressure. As mentioned previously, some Vulcanian-style eruptions may be triggered by the heating of groundwater by magma that is blocked by a solidified plug. As pressure increases the plug eventually fails

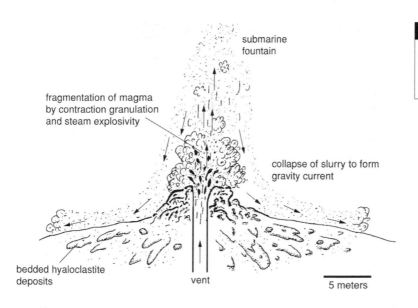

submarine fountain

fragmentation of magma by contraction granulation and steam explosivity

collapse of slurry to form gravity current

bedded hyaloclastite deposits

vent

5 meters

**Fig. 1.40.** Structure of an underwater lava fountain that generates hyaloclastite gravity currents by collapse. Modified from Smith and Batiza (1989).

allowing the vapor to expand and the magma to undergo further degassing of dissolved volatile components.

A third mechanism of magma fragmentation is significantly less explosive than the other two, but nevertheless can result in significant production of fragmental material during volcanic eruptions. As magma comes into contact with water, rapid quenching transforms the melt to a glass. Significant contraction of the material takes place as a result, leading to the fracturing of glass into small pieces, or hyaloclastites. This cooling contraction granulation is an important process associated with the emplacement of submarine lava flows. Unlike the other fragmentation mechanisms, there is no depth limit to contraction granulation, as steam expansion is not a consideration.

### SUBMARINE LAVA-FOUNTAINING

Lava-fountaining is a common style of volcanic activity for many subaerial basaltic eruptions and there is now evidence that an analogous process may occur in relatively deep water as a result of magma–water interactions. Submersible work on seamounts near the East Pacific Rise has discovered the common occurrence of bedded hyaloclastite deposits at depths from 1240 to 2500 m (Smith and Batiza, 1989). The deposits are closely associated with pahoehoe-like flows suggesting that high eruption rates are necessary for their formation. Smith and Batiza (1989) propose that

rapid discharge of magma generates a fountain of magma above the submarine vent (Fig. 1.40). Fragmentation occurs by cooling contraction granulation and steam explosivity as seawater mixes with magma to produce a slurry of basaltic glass shards, hot water, and possibly steam. Collapse of the slurry from the top of the fountain produces gravity currents that move away from the vent and eventually deposit the bedded hyaloclastites (Fig. 1.40).

### SURTSEYAN ERUPTIONS

The construction of many oceanic volcanoes begins in deep water but progresses towards the surface as new volcanic material is added. A Surtseyan eruption is an explosive event associated with discharge of magma as a volcano approaches the surface and becomes emergent. Vigorous explosions send steam-rich cocks-tail style plumes along parabolic trajectories (Fig. 1.41). The tephra-laden plumes are produced as a series of discrete events that may take place on a variety of timescales ranging from minutes to hours. In 1963, the island of Surtsey emerged from the sea south of Iceland in a series of explosive eruptions that are considered the type example of this activity. A cone was built up from the accumulation of basaltic ash and scoria ejected by the eruptions. Kokelaar (1983) has proposed that in the vent area of a Surtseyan eruption a slurry of hot water and tephra is mixed with new magma

**Fig. 1.41.** Eruption of Surtsey volcano, Iceland in 1963. Photograph by I. Einarsson.

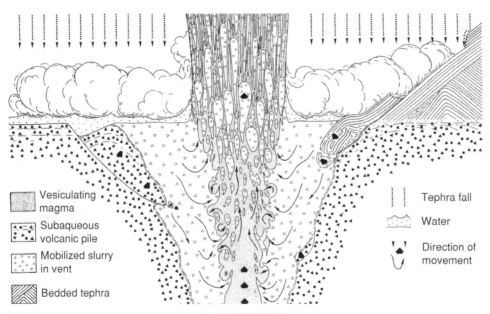

Vesiculating magma

Subaqueous volcanic pile

Mobilized slurry in vent

Bedded tephra

Tephra fall

Water

Direction of movement

**Fig. 1.42.** Internal structure of a Surtseyan eruption column. Rising, vesiculating magma is mixed with a slurry of water and volcaniclastic particles near the surface and then rapidly ejected as the liquid is converted to steam. Modified from Kokelaar (1983).

(Fig. 1.42). Upward migration of the mixture and heat transfer cause the mixture to flash to steam and be rapidly accelerated out of the vent as a jet. The primary mechanisms of magma fragmentation are thus contact-surface explosivity coupled with degassing of juvenile volatile components dissolved in the magma. Explosive activity at Surtsey ceased once an island was formed with sufficient relief to isolate interaction of magma with seawater, indicating that the dominant fragmentation mechanism was related to interaction of magma with seawater.

## Construction of volcanic edifices

The great variety in magma compositions, styles of eruption, and rates of discharge combine to

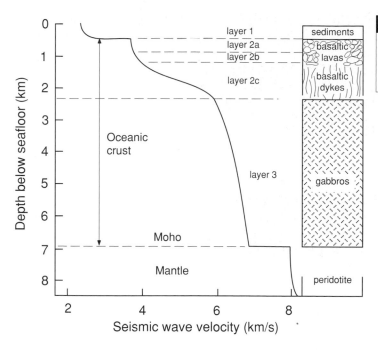

**Fig. 1.43.** Idealized stratigraphic section through normal oceanic crust showing lithologic subdivisions and their associated seismic velocities. Modified from Brown *et al.* (1989).

produce the myriad of volcanic structures that have been recognized throughout the world. Understanding the physical mechanisms of volcanic eruptions and the types of volcanic deposits that they produce has proved critical to reconstructing the evolution of volcanic landforms. In many cases the growth and development of volcanoes has been directly witnessed but in other cases, such as in the marine environment, the development of models for the evolution of volcanoes has had to rely largely on the interpretation of the erupted products and indirect measurements of their structure.

## Submarine and subaerial rift structures

Most volcanism on Earth occurs along linear rift systems where the surface plates are in extension and volcanism produces new ocean crust. The mid-ocean ridge system is a globally encircling, elevated volcanic feature that extends for a total length of about 70 000 km (Fig. 1.1). Its structure varies considerably and reflects the balance between tectonism and magmatism. Where the mid-ocean ridge is spreading slowly a central rift valley bounded on each side by faulted ridges is usually developed (Fig. 1.3). The central valley is less than 3 km wide and approximately 400 m deep. At high rates of spreading the

across-axis profile is typically smoother, with a central axial high. The height of the mid-ocean ridges is the result of the elevated temperatures of the upwelling mantle material. As new ocean crust is formed and spreads away from the ridge it cools and becomes denser. The increase in density results in subsidence of the plate according to a systematic relationship where the depth is related to the square root of the age.

Volcanism at the ridge is dominated by the quiescent extrusion of basaltic magma as pillow lavas and sheet flows. The magma for these flows is derived from a subaxial melt lens in the case of fast-spreading ridges and from isolated melt bodies beneath slow-spreading ridges (Figs. 1.20 and 1.22). This style of volcanism produces a distinctive layered structure characteristic of oceanic crust as inferred from geophysical studies and the examination of uplifted submarine sequences known as ophiolites (Cas, Chapter 4, this volume) (Fig. 1.43). Layer 1 consists of a sequence of deep-sea sediments that accumulates on top of the oceanic crust as it migrates away from the ridge. Layer 2 is subdivided into three units based on seismic velocity and morphology of erupted products. Layers 2a and 2b are composed of a sequence of overlapping pillow basalts and sheet flows formed by the extrusion of magma onto the seafloor. Layer 2c lies beneath these extrusives and is made up of vertically oriented dykes of similar composition to the pillow basalts and sheet flows. These dykes represent the feeder system whereby magma is transferred from the subaxial magma chamber to the surface. Magma that remains in these conduits solidifies to form the dykes. An abrupt change in lithology occurs between the base of layers 2 and 3 where gabbros and metagabbros are encountered (Fig. 1.43). These coarse-grained rocks are compositionally equivalent to basalt but have crystallized over much longer periods of time. They represent magma that was unable to be erupted at the surface and has frozen in place as the lithospheric plate migrated away from the ridge. Below the gabbros is another sharp contact that marks the transition to periodotites of ultra-mafic composition (Fig. 1.43). These mark the base of the oceanic crust sequence and represent the residual mantle material after magma has been extracted by pressure-release melting.

The layered structured of the oceanic crust is a general feature that persists in many types of spreading environments. Other important structural features of the ridge system include along-axis segmentation (e.g., Macdonald *et al.*, 1991). The nature of segmentation varies considerably in scale along the ridge and again reflects competing processes of tectonism and volcanism. First-order segments are long sections of the ridge (up to hundreds of kilometers) that are bounded on each end by a transform fault. Their production is tied to the tectonics and rate of plate spreading. Prime segments are sections of the ridge that are bounded by smaller-scale ridge axis discontinuities such as overlapping spreading centers or axial offsets of >0.5 km and 20 m elevation (Tighe, 1997). They are of the order of 100 km in length along the East Pacific Rise and coincide closely with magma that can be related to a single parental magma by fractional crystallization. Within each of the prime segments there is a single or series of axial volcanoes defined by <0.5 km axial offsets and single bathymetric highs. These are interpreted to represent the locus of melt uprise and eruption. Lateral flow of magma along the ridge is likely to take place away from the

bathymetric highs leading to flanking fissure eruptions. Little is known about the frequency of eruptions that take place along the mid-ocean ridge system although morphological constraints suggest that individual axial volcanoes may have a lifetime of the order of $10^2$ to $10^3$ years.

Rift-dominated volcanism is also developed on some of the continents and although some represent extensions of mid-ocean spreading ridges, others are more complex. In general, many rifts occur within a broad topographic elevation that is attributed to hot, buoyant mantle rising from below. A fundamental question for many rifts is the relationship between the rifting and volcanism. Two alternatives are that (1) rifting and volcanism are the product of the active rise of mantle material from depth, such as a hotspot, or (2) volcanism is a passive response to rifting of continental crust that is driven by a wider regional tectonism. The East African rift is often cited as a classic continental rift where extension is accompanied by active volcanism. This rift is an extension of a triple junction occurring at the intersection of the Red Sea and the Gulf of Aden. It extends to the southwest for about 4000 km with a width of 100 km. Within the rift are some of Africa's largest volcanoes, such as Kilimanjaro, Nyiragongo, and Longonot. Formation of the East African rift is attributed to the formation of a hotspot beneath Africa that is progressively forming a series of new small ocean basins by thinning and extension of the continental crust.

The association of some continental flood basalt provinces with the location of ancient rift zones has led to the suggestion that hotspot development beneath continents will necessarily lead to rifting. However, areas such as the Columbia River flood basalts and the Siberian province do not show evidence of rifting (Francis, 1993). Clearly the relationship between hotspots, flood basalt provinces, and continental rifting has yet to be precisely understood.

### Shield volcanoes

Large-volume discharge of basaltic magma as lava flows is responsible for the construction of the world's largest type of individual volcano, known as shields. They are characterized by gentle slopes

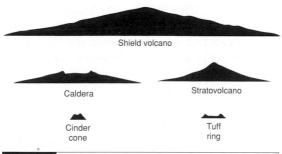

**Fig. 1.44.** Comparison of different volcano sizes and morphologies. Modified from Decker and Decker (1997).

of usually less than 10° and broad extent giving the overall appearance of an overturned shield (Fig. 1.44). Mauna Loa on the island of Hawaii is the world's largest shield with an elevation of 10 km as measured from the seafloor to the summit and basal diameter of 190 km. The volume of Mauna Loa has been estimated as at least 40,000 km³, or an order of magnitude higher than most other volcanoes. It has been built by the repeated eruption of low-viscosity tholeiitic magma beginning at water depths of several thousand meters and eventually building up above sea level for 4 km. The morphology of shield volcanoes reflects the ability of lava flows to travel far from the vent and produce gentle slopes. In many instances the enhanced transport of flows is facilitated by movement through lava tubes which insulate the flows from cooling and greatly extend their runout. At the summit of many shield volcanoes are calderas where lava lakes may be present during eruptions. The formation of these structures is due to the draining of magma from the summit regions into fissures feeding flank eruptions or lateral intrusions.

There are differences in the morphology and size of shields in different geologic settings. For example, some shield volcanoes of the Galapagos Islands exhibit a distinctive profile with a relatively flat top, moderately steep upper slopes (>10°), and a relatively abrupt change to more gentle distal slopes (Francis, 1993). In addition, many of the Galapagos shields have much deeper summit calderas (up to 800 m deep) than their Hawaiian counterparts.

Many of the large shield volcanoes form oceanic islands and thus have evolved from submarine to subaerial conditions. The internal structure of such volcanoes reflects the influence of these different environments on the nature of the eruptive style and morphology of the resulting deposits. A well-studied example is the island of Gran Canaria in the Canary Islands of the Atlantic. The evolution of the island has been reconstructed by a combination of land-based and marine geological studies (Schmincke, 1994; Schmincke et al., 1995). Submarine growth of the island produced a series of distinct overlapping facies (Fig. 1.45). The core of the volcano is built up of intrusions, pillow lavas, pillow breccias, debris flows, and hyaloclastites. With increasing height above the seafloor the pillow lavas in the core become more vesicular as gases are able to exsolve under lower pressure and the abundance of hyaloclastites is higher. Surrounding the core facies is a seismically chaotic flank facies composed of pillow breccias, hyaloclastites, and debris flow deposits (Fig. 1.45). This facies is constructed by shallow submarine eruptions and the eventual emergence of the island above sea level. Overlying the flank facies are the slope and basin facies that define the submarine morphology of the volcano. The slope facies is more proximal to the source and represents a series of slumps, debris flows, and massive units that were produced by subaerial eruptions discharging clastic material into the sea and the redeposition of epiclastic materials. At Gran Canaria this facies extends from the shoreline to about 45 km offshore. With increasing distance away from the volcano the slope facies interfingers with a more well-bedded basin facies. This consists of interbedded pelagic sediment, fine-grained turbidites, and tephra fall layers. Its extent is up to hundreds of kilometers from source and marks the edge of the island's volcanic apron (Fig. 1.45).

An important discovery related to the growth of oceanic shield volcanoes is that they are susceptible to major mass-wasting processes that can drastically modify their flanks. In the Hawaiian islands seafloor mapping has revealed the presence of enormous slumps and debris avalanches on the flanks and adjacent to many of the islands (e.g., Moore and Normark, 1994). Slumps are slow displacements of large portions

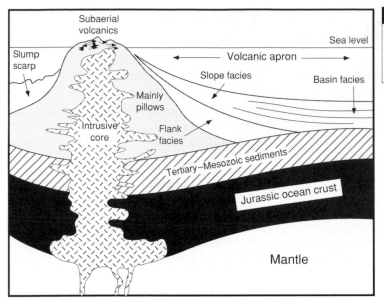

**Fig. 1.45.** Cross-section through Gran Canaria island based on seismic stratigraphy and inferences about submarine volcanic process. Modified from Schmincke *et al.* (1995).

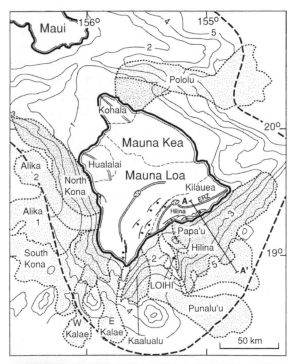

**Fig. 1.46.** Distribution of major submarine slumps and debris avalanches (stippled patterns) around the island of Hawaii. Modified from Moore *et al.* (1995).

from the volcanic rift zone seaward to the base of the volcanic pile. The slumps can be as much as 110 km wide and 10 km thick, with gradients of 3°. The Hilina slump on the south coast of the island of Hawaii serves as a good example of an active slump on the flank of an oceanic island (Fig. 1.46). On land the slump shows up as a series of normal faults that traverse the south coast of the island. In 1975 during a 7.5-magnitude earthquake, a 60-km-long section of Kilauea's coast subsided by as much as 3.5 m and moved seaward by 8 m.

Debris avalanches tend to be longer, thinner, and less steep than the slumps, and often can be traced back upslope to a well-defined amphitheater. A characteristic feature of debris avalanches that allows for their recognition in the deep sea is the presence of abundant hummocks, similar to debris avalanche deposits on land. On sonar returns they show up as a speckled pattern. The size distribution of hummocks can vary significantly from one slide to another. Some are 1 km in diameter whereas others are up to 10 km. High-resolution bottom photography indicates that the areas between large hummocks are filled with much smaller fragments that do not show up on the large-scale mapping images. The Alika-2 debris avalanche exhibits hummock sizes and spacing that are similar to the 1980 Mount St. Helens debris avalanche deposit. This suggests

of an island's flank that occur over an extended period of time, keeping pace with the production of new volcanic material. They are thought to be deeply rooted in the volcanic edifice, extending

**Fig. 1.47.** Mount St. Helens volcano prior to the catastrophic eruption of May 18, 1980. Spirit Lake is in the foreground.

that the debris avalanches are rapidly moving events, in contrast to the slow-moving slumps.

There are two major hazards associated with the generation of major submarine debris avalanches on the flanks of oceanic shield volcanoes. First, if the debris avalanche begins on land then significant portions of an island's coast may rapidly slide into the sea, destroying property and lives. The second principal hazard is related to the displacement of water during the generation of a debris avalanche and the formation of large tsunamis. Although this has not been witnessed directly there is evidence in the geological record that such events have occurred and may be common in the evolution of island volcanoes (Moore et al., 1994).

## Volcanic cones and composite volcanoes

On a small scale, localized volcanism often builds volcanic cones that represent a single eruptive episode (Fig. 1.44) (Riedel et al., 2003). The duration of such episodes may vary from a few days to several years. These volcanoes are known as monogenetic because once the eruptive episode has ceased there is no more activity at the site. Accumulation of scoria and lava flows typically builds a steep-sided cone (~33°) several hundred meters in height, with a relatively flat top.

The scoria cone of Parícutin in Mexico was constructed during an eruptive episode from 1943 to 1952 and is just one of many such cones on the flank of Pico de Tancitaro in the Trans-Mexican volcanic belt. In some cases the eruption of basalt during monogenetic activity will interact with water and become explosive. Maars are low craters formed by explosive interaction of groundwater and magma. The explosions excavate the subsurface rocks to form a crater surrounded by a rim of ejected debris. If the explosive discharge leads predominantly to accumulation of material over the subsurface rocks by Surtseyan-type explosions then the feature is generally referred to as a tuff ring.

Most volcanoes have a more complex evolution and are built up from both effusive and explosive volcanism. Their internal structures thus consist of a combination of lava flows and pyroclastic deposits. These composite volcanoes, or stratovolcanoes, differ significantly in morphology from shield volcanoes, having much steeper slopes and flank profiles that are concave downward. They are most common in subduction zone environments (island arcs and continental margins) where the magma compositions are more evolved and generally carry more dissolved volatiles. Mount St. Helens in the northwest United States is a good example of a stratovolcano (Fig. 1.47).

Composite volcanoes are built up over extended periods of time by a large number of

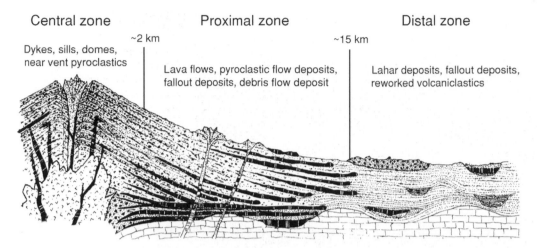

**Central zone**

Dykes, sills, domes,
near vent pyroclastics

~2 km

**Proximal zone**

Lava flows, pyroclastic flow deposits,
fallout deposits, debris flow deposit

~15 km

**Distal zone**

Lahar deposits, fallout deposits,
reworked volcaniclastics

**Fig. 1.48.** Cross-section through a composite volcano showing the changes in structure as a function of distance from the peak. Modified from Williams and McBirney (1979).

eruptive events and are referred to as polygenetic. The location of volcanism remains relatively fixed, but the nature of the eruptions and the development of the peak may change dramatically. The core of the volcano consists of intrusive bodies and dykes that represent the pathways for magmas to the surface and locations where magma was stored without eruption. Individual eruptive vents may migrate around a central location but the integrated products still produce a structure with conical symmetry. At many composite volcanoes, the summit area consists of a complex of domes formed by effusive activity. Many have some type of summit crater that formed as a result of the ejection of material by explosive eruptions or collapse associated with magma withdrawal. The flanks are built up of multiple layers of lava flows, primary pyroclastic deposits, and the reworked products of the volcano (Fig. 1.48). Slopes on the flanks may vary from about 10° on the lower parts to up to 35° near the summit.

The accumulation of alternating layers of lava flows and pyroclastic deposits at relatively steep angles on the slopes of composite volcanoes often leads to unstable conditions. Catastrophic collapse of the volcano's flank may lead to the generation of landslides, debris avalanches, and debris flows, such as occurred at the onset of the May 18, 1980 eruption of Mount St. Helens (Lipman and Mullineaux, 1981). It is now apparent that catastrophic sector collapse of volcanoes is a relatively common phenomenon and a natural process in the evolution of composite volcanoes (Siebert, 1984). Volcanic debris avalanches are quite different from normal dry rock avalanches in the amount of material that is involved. Non-volcanic dry rock avalanches tend to form along bedding or joint planes and are shallow relative to their length. In contrast, volcanic debris avalanches are much deeper relative to their length and have a steeply sloping back wall. The scar that remains on the flank of the volcano often appears as a horseshoe-shaped crater (Fig. 1.49).

One of the most distinctive features of debris avalanche deposits is the occurrence of numerous hills and depressions on their surfaces, giving an overall hummocky topography. In addition, there are also longitudinal and transverse ridges. The deposits usually consist of a very poorly sorted mixture of brecciated debris. Most of the material is lithic clasts and blocks from the volcanic edifice. However, there can be some juvenile material. The majority of debris avalanches travel in excess of 10 km from the volcano, with several events reaching 50–60 km. These are thus similar to the runout of some pyroclastic flows. Velocities can also be high, with many in excess of 150 km/hr.

An important aspect of debris avalanches is the sudden release of pressure on the volcano as a result of slope failure. This can lead to triggering of explosive activity. There are two types of eruptions that generally occur (Siebert *et al.*, 1987). The Bezimianny-type involves a magmatic component

**Fig. 1.49.** Summit area of Mount St. Helens showing the horseshoe-shaped crater that formed as a result of the May 18, 1980 explosive eruption. Photograph from Lipman and Mullineaux (1981).

and is driven by degassing. It may be preceded by several days to months of precursory seismic and volcanic activity. In contrast, the Bandai-type is phreatic with no juvenile magma discharge and is driven by expansion of the hydrothermal system beneath the volcano. These types of events may occur with or without preceding seismic activity. The horseshoe-like craters formed by flank collapse tend to "heal" by the growth of lava domes, although the rates can vary widely. At Bezimianny the 1956 amphitheater has been almost completely filled in. At Mount St. Helens a new lava dome has been building since the 1980 eruption and the crater is about one-third filled. The healing process makes it difficult to determine whether a volcano has experienced a collapse in the recent past.

## Calderas

The largest volcanic structures that are formed during the course of single eruptions are the great circular depressions known as calderas. These spectacular features are typically associated with voluminous silicic volcanism that generates thick pyroclastic flow deposits, or ignimbrites (Smith, 1979). The size of calderas can vary significantly, with the largest attaining dimensions of tens of kilometers in diameter. Irrespective of size, the fundamental mechanism for their origin is similar. As magma is withdrawn from shallow magma chambers during eruption, the overlying volcanic center must adjust its structure in order to accommodate the reduction in volume. If the amount and rate of withdrawal are low, the reduction in volume may be adjusted by general subsidence without significant surface deformation. However, large volume discharges at high rates may reduce the pressure of the magma chamber to such an extent that the overlying rocks are left unsupported and collapse of the roof occurs (Fig. 1.50). Such collapse forms calderas that can vary in size from a few kilometers to a hundred kilometers in diameter. Foundering of the magma chamber roof can produce a dynamic pressure on the remaining magma causing high rates of magma discharge through radial fissures and favoring the production of additional pyroclastic flows (Druitt and Sparks, 1984).

Smaller calderas, a few kilometers in diameter, typically form within a pre-existing edifice and thus mostly involve the collapse of volcanic rocks. A recent example is the formation of a ~6-km-diameter caldera on Tambora volcano

(a)

(b)

(c)

(d)

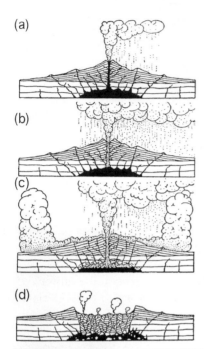

**Fig. 1.50.** Schematic representation of the formation of a caldera during a large explosive eruption of a stratovolcano. An intitial phase of pumice fallout (a) and (b) is followed by the generation of pyroclastic flows (c). Rapid evacuation of the underlying magma chamber eventually results in the collapse of the volcanic edifice to form a caldera (d). Modified from Williams and McBirney (1979).

**Fig. 1.51.** View of the caldera of Tambora volcano in Indonesia. The caldera was formed during the great 1815 explosive eruption and is approximately 6 km in diameter and 1.2 km deep. Photograph by S. Carey.

in Indonesia (Fig. 1.51). Prior to 1815, Tambora was a broad stratovolcano with an elevation of about 4000 m. In 1815 the volcano produced the largest explosive eruption of historic times, ejecting about 50 km³ of magma during a 2-day span. Rapid evacuation of a shallow magma chamber resulted in the generation of voluminous pyroclastic flows and eventual collapse of the upper 1200 m of the summit. The resulting caldera is about 1250 m deep from its present rim and about 6 km across. Studies of the erupted products indicate that the caldera volume is similar to that of the widely dispersed pyroclastic material.

Tambora caldera is an impressive sight from the top of the volcano (Fig. 1.51) but its size is dwarfed by the largest of calderas found on Earth. These giant structures can be close to 100 km in diameter and in some cases are filled with water. One of the largest calderas in the world is Lake Toba on the island of Sumatra in Indonesia (Fig. 1.52). It was produced by a series of very large explosive eruptions, the last of which occurred about 74,000 years ago (Chesner and Rose, 1991). Approximately 2000 km³ of silicic magma was ejected during the eruption, the greater part by pyroclastic flows that formed extensive ignimbrite deposits inside and outside the caldera. The last Toba eruption occurred at about the same time that the Earth's climate entered an ice age and it has been speculated that the injection of ash and aerosols into the atmosphere might have been a triggering mechanism for catastrophic climatic change (Rampino and Self, 1992). Other examples of giant calderas include many of the well-studied centers in the western United States such as Yellowstone in Wyoming, Valles in New Mexico, and Long Valley in California.

The size of the largest calderas, such as the ones in the western United States, is so great that the collapse that formed them includes not only pre-existing volcanic rocks but large pieces of the upper crust. It is difficult to imagine the scale of an eruption that is associated with this type of structure. Observations of historic eruptions are limited to the formation of relatively small calderas on existing volcanoes where only a few tens of cubic kilometers of magma were erupted. In contrast, the formation of the large calderas likely involves discharge of thousands of cubic kilometers of magma (Smith, 1979). Massive

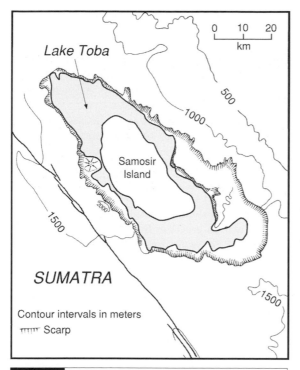

**Fig. 1.52.** Lake Toba caldera on the island of Sumatra in Indonesia. The caldera has been produced by multiple large-volume silicic explosive eruptions. Modified from Francis (1993).

the existence of a topographic high in the central part of the caldera. This center is attributed to uplift after the formation of the caldera by a process known as resurgence (Smith and Bailey, 1968). In most cases this uplift is not associated with the eruption of magma. The ultimate cause of resurgence is still being debated but may be the result of upward pressure exerted by magma remaining in the chamber as regional subsidence occurs over an area larger than the caldera or expansion of magma remaining the chamber as a result of exsolution of dissolved volatile components.

Calderas are also formed on basaltic shield volcanoes, although their sizes are usually smaller than those formed by explosive eruption of evolved magmas. The caldera on the summit of Kilauea volcano in Hawaii is about 4 km in diameter with steep, almost vertical sides. Caldera formation in basaltic systems typically involves lateral transfer of magma away from a summit magma chamber through a system of dykes and fissures. For example, magma supplied to the Kilauea summit is distributed through eruptions in the caldera and along the upper and middle east rift zones that extend more than 20 km from the summit.

deposits of pyroclastic flows (ignimbrites) are generated during this type of event and may extend up to a hundred kilometers from source. In addition, because caldera collapse occurs over such a large area magma can be discharged along many of the ring faults that define the subsiding block.

Despite, or perhaps more accurately, because of their great size, some calderas have eluded detection. A good example is the Cerro Galán caldera in northwest Argentina which was only recognized as a caldera from satellite images (Francis *et al.*, 1978). Its structure exhibits many features of large calderas from around the world. The boundaries are defined by a series of ring fractures that have acted as passageways for magma to escape from a large shallow chamber. Exposures of Paleozoic basement occur along the caldera margin indicating the involvement of large crustal blocks during the collapse. Another prominent feature of the Cerro Galán caldera and many other large calderas around the world is

# References

Anderson, A. T., Jr., Newman, S., Williams, S., *et al.* 1989. H$_2$O, CO$_2$, Cl, and gas in Plinian and ash-flow Bishop rhyolite. *Geology*, **17**, 221–225.

Ballard, R. and Moore, J. 1977. *Photographic Atlas of the Mid-Atlantic Ridge Rift Valley*. Berlin, Springer-Verlag.

Ballard, R., Holcomb, R., and van Andel, T. 1979. The Galapagos Rift at 86° W. III. Sheet flows, collapse pits, and lava lakes of the rift valley. *Journal of Geophysical Research*, **84**, 5407–5422.

Barker, S. and Malone, S. 1991. Magmatic system geometry at Mount St. Helens modeled from the stress field associated with posteruptive earthquakes. *Journal of Geophysical Research*, **96**, 11883–11894.

Blackburn, E., Wilson, L., and Sparks, R. S. J. 1976. Mechanisms and dynamics of strombolian activity. *Journal of the Geological Society of London*, **132**, 429–440.

Blake, S. 1989. Viscoplastic models of lava domes. *IAVCEI Proceedings in Volcanology, Lava Flows and Domes*, **2**, 88–126.

Bonadonna, C., Ernst, G. G. J., and Sparks, R. S. J. 1998. Thickness variations and volume estimates of tephra fall deposits: the importance of particle Reynolds number. *Journal of Volcanology and Geothermal Research*, **81**, 173–187.

Bottinga, Y. and Weill, D. 1972. The viscosity of magmatic silicate liquids: a model for calculation. *American Journal of Science*, **272**, 438–475.

Brown, G. C., Hawkesworth, C., and Wilson, M. 1992. *Understanding the Earth*. Cambridge, UK, Cambridge University Press.

Brown, J., Colling, A., Park, D., *et al.* 1989. *The Ocean Basins: Their Structure and Evolution*. New York, Pergamon Press.

Burnham, C. W. 1979. The importance of volatile constituents. In H. Yoder (ed.) *The Evolution of Igneous Rocks*. Princeton, NJ, Princeton University Press, pp. 439–478.

Bursik, M. and Woods, A. 1996. The dynamics and thermodynamics of large ash flows. *Bulletin of Volcanology*, **58**, 175–193.

Bursik, M. I., Sparks, R. S. J., Gilbert, J. S., *et al.* 1992a. Sedimentation of tephra by volcanic plumes. I. Theory and its comparison with a study of the Fogo A plinian deposit, São Miguel (Azores). *Bulletin of Volcanology*, **54**, 329–344.

Bursik, M., Carey, S., and Sparks, R. S. J. 1992b. A gravity current model for the May 18, 1980 Mount St. Helens plume. *Geophysical Research Letters*, **19**, 1663–1666.

Carey, S. N. 1991. Transport and deposition of tephra by pyroclastic flows and surges. In R. V. Fisher and G. Smith (eds.) *Sedimentation in Volcanic Settings*, Special Publication no. 45. Tulsa, OK, Society for Sedimentary Geology, pp. 39–57.

Carey, S. N. and Sigurdsson, H. 1985. The May 18, 1980 eruption of Mount St. Helens. II. Modeling of dynamics of the plinian phase. *Journal of Geophysical Research*, **90**, 2948–2958.

1987. Temporal variations in column height and magma discharge rate during the 79 AD eruption of Vesuvius. *Geological Society of America Bulletin*, **99**, 303–314.

Carey, S. N. and Sparks, R. S. J. 1986. Quantitative models of the fallout and dispersal of tephra from volcanic eruption columns. *Bulletin of Volcanology*, **48**, 109–125.

Carey, S. N., Sigurdsson, H., and Sparks, R. S. J. 1988. Experimental studies of particle-laden plumes. *Journal of Geophysical Research*, **93**, 15314–15328.

Caroll, M. and Rutherford, M. 1988. Sulfur speciation in hydrous experimental glasses of varying oxidation state: results from measured wavelength shifts of sulfur x-rays. *American Mineralogist*, **73**, 845–849.

Carr, M. J. 1984. Symmetrical and segmented variation of physical and geochemical characteristics of the Central American volcanic front. *Journal of Volcanology and Geothermal Research*, **20**, 231–252.

Cashman, K. and Mangan, M. 1994. Physical aspects of magma degassing. II. Constraints on vesiculation processes from textural studies of eruptive processes. *Reviews in Mineralogy*, **30**, 447–474.

Chadwick, W. and Embley, R. 1994. Lava flows from a mid-1980s submarine eruption on the Cleft segment, Juan de Fuca Ridge. *Journal of Geophysical Research*, **99**, 4761–4776.

Chadwick, W., Embley, R., and Fox, C. 1995. SeaBeam depth changes associated with recent lava flows, CoAxial segment, Juan de Fuca Ridge: evidence for multiple eruptions between 1981–1993. *Geophysical Research Letters*, **22**, 167–170.

Chesner, C. and Rose, W. I. 1991. Stratigraphy of the Toba Tuffs and evolution of the Toba caldera complex, Sumatra, Indonesia. *Bulletin of Volcanology*, **53**, 343–356.

Chouet, B., Hamisevicz, N., and McGuetchin, T. R. 1974. Photoballistics of volcanic jet activity at Stromboli, Italy. *Journal of Geophysical Research*, **79**, 4961–4976.

Clague, D. and Dalrymple, G. B. 1987. The Hawaiian–Emperor volcanic chain. I. Geologic evolution. In *Volcanism in Hawaii*, US Geological Survey Professional Paper no. 1350. Washington, DC, US Government Printing Office, pp. 5–54.

Coffin, M. and Eldholm, O. 1994. Large igneous provinces: crustal structure, dimensions, and environmental consequences. *Reviews of Geophysics*, **32**, 1–36.

Colgate, S. and Sigurgeirsson, T. 1973. Dynamic mixing of water and lava. *Nature*, **244**, 552–555.

Crisp, J. A. 1984. Rates of magma emplacement and volcanic output. *Journal of Volcanology and Geothermal Research*, **20**, 177–211.

Decker, R. and Decker, B. 1997. *Volcanoes*. New York, W. H. Freeman.

Dingwell, D. B. and Webb, S. L. 1989. Structural relaxation in silicate melts and non-Newtonian melt rheology in geologic processes. *Physics and Chemistry of Minerals*, **16**, 508–516.

1990. Relaxation in silicate melts. *European Journal of Mineralogy*, **2**, 427–449.

Druitt, T. and Sparks, R. S. J. 1984. On the formation of calderas during ignimbrite eruptions. *Nature*, **310**, 679–681.

Duncan, R. and Richards, M. 1991. Hotspots, mantle plumes, flood basalts and true polar wanderings. *Reviews of Geophysics*, **29**, 31–50.

Fisher, R. V. and Schmincke, H. U. 1984. *Pyroclastic Rocks*. New York, Springer-Verlag.

Fisher, R. V., Smith, A. L., and Roobol, M. J. 1980. Destruction of St. Pierre, Martinique, by ash cloud surges. *Geology*, **8**, 472–476.

Francheteau, J. and Ballard, R. 1983. The East Pacific Rise near 21° N, 13° N, and 20° N: inferences for along-strike variability of axial processes of the mid-ocean ridge. *Earth and Planetary Science Letters*, **64**, 93–116.

Francheteau, J., Juteau, T., and Rangin, C. 1979. Basaltic pillars in collapsed lava-pools on the deep sea floor. *Nature*, **281**, 209–211.

Francis, P. 1993. *Volcanoes: A Planetary Perspective*. Oxford, UK, Oxford University Press.

Francis, P. W., Hammill, M., Kretzschmar, G., et al. 1978. The Cerro Galán Caldera, Northwest Argentina. *Nature*, **274**, 749–751.

Frankel, C. 1996. *Volcanoes of the Solar System*. Cambridge, UK, Cambridge University Press.

Greeley, R. 1982. The Snake River Plain, Idaho: representative of a new category of volcanism. *Journal of Geophysical Research*, **87**, 2705–2712.

Gregg, T. and Fink, J. 1995. Quantification of submarine lava-flow morphology through analog experiments. *Geology*, **23**, 73–76.

Griffiths, R. and Fink, J. 1992. Solidification and morphology of submarine lavas: a dependence in extrusion rate. *Journal of Geophysical Research*, **97**, 19729–19737.

Gudmunsson, A. 1996. *Volcanoes in Iceland*. Reykjavik, Vaka-Helgafell.

Head, J. and Wilson, L. 1987. Lava fountain heights at Pu'u 'O'o, Kilauea, Hawaii: indicators of amount and variations of exsolved magma volatiles. *Journal of Geophysical Research*, **92**, 13715–13719.

1992. Magma reservoirs and neutral buoyancy zones on Venus: implications for the formation and evoution of volcanic landforms. *Journal of Geophysical Research*, **97**, 3877–3903.

Hekinian, R., Thompson, G., and Bidcau, D. 1989. Axial and off-axial heterogeneity of basaltic rocks from the East Pacific Rise at 12° 35′ N–12° 51′ N and 11° 26′ N–11° 30′ N. *Journal of Geophysical Research*, **94**, 17437–17463.

Hulme, G. 1974. Interpretation of lava flow morphology. *Royal Astronomical Society Geophysical Journal*, **39**, 361–383.

Humphris, S., Zierenberg, R., Mullineaux, L., et al. 1995. *Seafloor Hydrothermal Systems: Physical, Chemical, Biological and Geological Interactions*, Geophysical Monograph no. 91. Washington, DC, American Geophysical Union.

Hurwitz, S. and Navon, O. 1994. Bubble nucleation in rhyolitic melts: experiments at high pressure, temperature and water content. *Earth and Planetary Science Letters*, **122**, 267–280.

Iyer, H., Evans, J., Dawson, P., et al. 1990. Differences in magma storage in different volcanic environments as revealed by seismic tomography: silicic volcanic centers. In M. Ryan (ed.) *Magma Transport and Storage*. Chichester, UK, John Wiley, pp. 293–316.

Karig, D. 1971. Origin and development of marginal basins in the western Pacific. *Journal of Geophysical Research*, **76**, 2542–2561.

Keary, P. and Vine, F. 1990. *Global Tectonics*. Oxford, UK, Blackwell Scientific Publications.

Kokelaar, P. 1983. The mechanism of Surtseyan volcanism. *Journal of the Geological Society of London*, **140**, 939–944.

1986. Magma–water interactions in subaqueous and emergent basaltic volcanism. *Bulletin of Volcanology*, **48**, 275–289.

Koyaguchi, T. and Tokuno, M. 1993. Origin of the giant eruption cloud of Pinatubo, June 15, 1991. *Journal of Volcanology and Geothermal Research*, **55**, 85–96.

Lacroix, A. 1904. *La Montagne Pelée et ses eruptions*. Paris, Masson.

Le Maitre, R. 1976. The chemical variability of some common igneous rocks. *Journal of Petrology*, **17**, 589–637.

Lipman, P. and Mullineaux, D. 1981. *The 1980 Eruptions of Mount St. Helens, Washington*, US Geological Survey Professional Paper no. 1250. Washington, DC, US Government Printing Office.

Macdonald, K. C., Scheirer, C., and Carbotte, S. 1991. Mid-ocean ridges: discontinuities, segments, and giant cracks. *Science*, **253**, 986–994.

Mader, H., Zhang, Y., Phillips, J., et al. 1994. Experimental simulations of explosive degassing of magma. *Nature*, **372**, 85–88.

Malin, M. 1980. The lengths of Hawaiian lava flows. *Geology*, **8**, 306–308.

Marsh, B. 1979. Island-arc volcanism. *American Scientist*, **67**, 161–172.

1981. On the crystallinity, probability of occurrence, and rheology of lava and magma. *Contributions to Mineralogy and Petrology*, **78**, 85–98.

McBirney, A. R. 1973. Factors governing the intensity of explosive andesitic eruptions. *Bulletin of Volcanology*, **36**, 443–453.

Molnar, P. and Atwater, T. 1978. Interarc spreading and Cordilleran tectonics as alternates related to the age of subducted oceanic lithosphere. *Earth and Planetary Science Letters*, **41**, 330–340.

Moore, J. 1975. Mechanism of formation of pillow lava. *American Journal of Science*, **63**, 269–277.

Moore, J. G. and Normark, W. 1994. Giant Hawaiian landslides. *Annual Review of Earth and Planetary Sciences*, **22**, 119–144.

Moore, J. G., Bryan, W. B., and Ludwig, K. 1994. Chaotic deposition by a giant wave, Molokai, Hawaii. *Geological Society of America Bulletin*, **106**, 962–967.

Moore, J. G., Bryan, W., Beeson, M., et al. 1995. Giant blocks in the South Kona landslide, Hawaii. *Geology*, **23**, 125–128.

Pallister, J., Hoblitt, R., Crandell, D., et al. 1992. Mount St. Helens a decade after the 1980 eruptions: magmatic models, chemical cycles, and a revisted hazards assessment. *Bulletin of Volcanology*, **54**, 126–146.

Parfitt, E. and Wilson, L. 1995. Explosive volcanic eruptions. IX. The transition between Hawaiian-style lava fountaining and strombolian explosive activity. *Geophysical Journal International*, **121**, 226–232.

Peacock, S. M. 1996. Thermal and petrologic structure of subduction zones. In G. Bebout, D. Scholl, S. Kirby, et al. (eds.) *Subduction Top to Bottom*. Geophysical Monograph no. 96. Washington, DC, American Geophysical Union, pp. 119–134.

Peterson, D. and Tilling, R. 1980. Transition of basaltic lava from pahoehoe to aa, Kilauea volcano, Hawaii: field observations and key factors. *Journal of Volcanology and Geothermal Research*, **7**, 271–293.

Pinkerton, H. and Sparks, R. S. J. 1976. The 1975 subterminal lavas, Mount Etna: a case history of the formation of a compound lava field. *Journal of Volcanology and Geothermal Research*, **1**, 167–182.

Pinkerton, H. and Stevenson, R. 1992. Methods of determining the rheological properties of magmas at sub-liquidus temperatures. *Journal of Volcanology and Geothermal Research*, **53**, 47–66.

Rampino, M. 1988. Introduction: the volcano/climate connection. *Annual Reviews of Earth and Planetary Sciences*, **16**, 73–99.

Rampino, M. and Self, S. 1992. Volcanic winters and accelerated glaciations following the Toba super-eruption. *Nature*, **359**, 50–52.

Richards, M., Duncan, R., and Courtillot, V. 1989. Flood basalts and hot spot tracks: plume heads and tails. *Science*, **246**, 103–107.

Riedel, C., Ernst G. G. J., and Riley, M. 2003. Controls on the growth and geometry of pyroclastic constructs. *Journal of Volcanology and Geothermal Research*, **127**, 121–152.

Ringwood, A. E. 1975. *Composition and Petrology of the Earth's Mantle*. New York, McGraw-Hill.

Ross, C. and Smith, R. L. 1961. *Ash-Flow Tuffs: Their Origin, Geological Reactions and Identification*. US Geological Survey Professional Paper no. 366. Washington, DC, US Government Printing Office.

Rowland, S. and Walker, G. P. L. 1990. Pahoehoe and aa in Hawaii: volumetric flow rate controls the lava structure. *Bulletin of Volcanology*, **52**, 615–628.

Rutherford, M. J. and Devine, J. 1988. The May 18, 1980 eruption of Mount St. Helens. III. Stability and chemistry amphibole in the magma chamber. *Journal of Geophysical Research*, **93**, 1310, 11949–11959.

Rutherford, M., Sigurdsson, H., Carey, S., et al. 1985. The May 18, 1980 eruption of Mount St. Helens. I. Melt composition and experimental phase equilibria. *Journal of Geophysical Research*, **90**, 2929–2947.

Ryan, M. P. 1987. Neutral buoyancy and the mechanical evolution of magmatic systems. In B. O. Mysen (ed.) *Magmatic Processes: Physiochemical Principles*. Special Publication no. 1. Chichester, UK, Geochemical Society, pp. 259–287.

1988. The mechanics and 3-d internal structure of active magma systems: Kilauea volcano, Hawaii. *Journal of Geophysical Research*, **93**, 4213–4248.

Ryan, M. P. and Blevins, J. Y. K. 1987. The viscosity of synthetic and natural silicate melts and glasses at high temperature and 1 bar ($10^5$ Pascals) pressure and at higher pressure. *US Geological Survey Bulletin*, **1764**, 1–563.

Scandone, R. and Malone, S. 1985. Magma supply, magma discharge and readjustment of the feeding system of Mount St. Helens during 1980. *Journal of Volcanology and Geothermal Research*, **23**, 239–262.

Schmincke, H. U. 1994. *Geological Field Guide of Gran Canaria*, 6th edn. Kiel, Pluto Press.

Schmincke, H. U., Weaver, P. P. E., Firth, J. V., *et al.* 1995. *Proceeding of the ODP, Initial reports* no. 157. College Station, TX, Ocean Drilling Program.

Schouten, H., Klitgord, K., and Whitehead, J. 1985. Segmentation of mid-ocean ridges. *Nature*, **317**, 225–229.

Self, S., Wilson, L., and Nairn, L. 1979. Vulcanian eruption mechanisms. *Nature*, **277**, 440–443.

Siebert, L. 1984. Large volcanic debris avalanches: characteristics of source areas, deposits and associated eruptions. *Journal of Volcanology and Geothermal Research*, **22**, 163–197.

Siebert, L., Glicken, H. X., and Ui, T. 1987. Volcanic hazards from Bezymianny- and Bandai-type eruptions. *Bulletin of Volcanology*, **49**, 435–459.

Sigurdsson, H. 1990. Evidence of volcanic loading of the atmosphere and climate response. *Paleogeography, Paleoclimatology, Paleoecology*, **89**, 277–289.

Sigurdsson, H., Carey, S., Cornell, W., *et al.* 1985. The eruption of Vesuvius in AD 79. *National Geographic Research*, **1**(3), 332–387.

Simpkin, T. 1993. Terrestrial volcanism in space and time. *Annual Reviews of Earth and Planetary Sciences*, **21**, 427–452.

Sinton, J. M. and Detrick, R. S. 1992. Mid-ocean ridge magma chambers. *Journal of Geophysical Research*, **97**, 197–216.

Sisson, T. and Grove, T. 1993. Experimental investigations of the role of $H_2O$ in calc–alkaline differentiation and subduction zone magmatism. *Contributions to Mineralogy and Petrology*, **113**, 143–166.

Smith, R. L. 1979. *Ash flow magmatism*. Special Paper no. 180. Boulder, CO, Geological Society of America, pp. 5–27.

Smith, R. L. and Bailey, R. A. 1968. Resurgent cauldrons. In R. L. Hay and C. A. Anderson (eds.) *Studies in Volcanology*, Geological Society of America Memoir no. 116. Boulder, CO, Geological Society of America, pp. 153–210.

Smith, T. and Batiza, R. 1989. New field and laboratory evidence for the origin of hyaloclastite flows on seamount summits. *Bulletin of Volcanology*, **51**, 96–114.

Sparks, R. S. J. 1978. The dynamics of bubble formation and growth in magmas. *Journal of Volcanology and Geothermal Research*, **3**, 1–37.

1986. The dimensions and dynamics of volcanic eruption columns. *Bulletin of Volcanology*, **48**, 3–15.

1992. Magma generation in the Earth. In *Understanding the Earth*. Cambridge, UK, Cambridge University Press, pp. 91–114.

Sparks, R. S. J. and Wilson, L. 1976. Model for the formation of ignimbrite by gravitational column collapse. *Journal of the Geological Society of London*, **132**, 441–452.

Sparks, R. S. J., Moore, J. G., and Rice, C. J. 1986. The initial giant umbrella cloud of the May 18, 1980 explosive eruption of Mount St. Helens. *Journal of Volcanology and Geothermal Research*, **28**, 257–274.

Sparks, R. S. J., Barclay, J., Jaupart, C., *et al.* 1994. Physical aspects of magma degassing. I. Experimental and theoretical constraints on vesiculation. *Reviews in Mineralogy*, **30**, 413–443.

Sparks, R. S. J., Bursik, M., Carey, S., *et al.* 1997. *Volcanic Plumes*. Chichester, UK, John Wiley.

Steven, T. A. and Lipman, P. W. 1976. *Calderas of the San Juan Volcanic Field, Southwestern Colorado*, US Geological Survey Professional Paper no. 958. Washington, DC, US Government Printing Office.

Stolper, E. and Newman, S. 1994. The roll of water in the petrogenesis of Mariana Trough magmas. *Earth and Planetary Science Letters*, **121**, 293–325.

Swanson, D., Wright, T., and Helz, R. 1975. Linear vent systems and estimated rates of eruption for the Yakima basalt on the Columbia Plateau. *American Journal of Science*, **275**, 877–905.

Tighe, S. 1997. The morphological characteristics and geological implications of prime segments and axial volcanoes along the East Pacific Rise. Unpublished Ph.D. thesis, University of Rhode Island.

Tilling, R. I. 1989. Introduction and overview. In R. I. Tilling (ed.) *Short Course in Geology*, vol. 1, *Volcanic Hazards*. Washington, DC, American Geophysical Union, pp. 1–8.

Uyeda, S. 1982. Subduction zones: an introduction to comparative subductology. *Tectonophysics*, **81**, 133–159.

Uyeda, S. and Kanamori, H. 1978. Back-arc opening and the mode of subduction. *Journal of Geophysical Research*, **84**, 1049–1061.

Verhoogen, J. 1951. Mechanics of ash formation. *American Journal of Science*, **249**, 239–246.

Vine, F. J. and Matthews, D. H. 1963. Magnetic anomalies over oceanic ridges. *Nature*, **199**, 947–949.

Walker, G. P. L. 1974. The lengths of lava flows. *Philosophical Transactions of the Royal Society of London, Series A*, **274**, 107–118.

White, R. and McKenzie, D. 1989. Magmatism at rift zones: the generation of volcanic continental margins and flood basalts. *Journal of Geophysical Research*, **94**, 7685–7729.

Whitehead, J. A., Dick, H. J. B., and Schouten, H. 1984. A mechanism for magmatic accretion under spreading centers. *Nature*, **312**, 146–148.

Williams, H. and McBirney, A. R. 1979. *Volcanology*. San Francisco, CA, Freeman, Cooper and Co.

Wilson, J. T. 1973. Mantle plumes and plate motions. *Tectonophysics*, **19**, 149–164.

Wilson, L. 1980. Relationships between pressure, volatile content and ejecta velocity in three types of volcanic explosion. *Journal of Volcanology and Geothermal Research*, **8**, 297–313.

Wilson, L., Sparks, R. S. J., and Walker, G. P. L. 1980. Explosive volcanic eruptions. IV. The control of magma properties and conduit geometry on eruption column behaviour. *Geophysical Journal of the Royal Astronomy Society*, **63**, 117–148.

Wilson, M. 1989. *Igneous Petrogenesis: A Global Tectonic Approach*. London, Unwin Hyman.

Wohletz, K. 1986. Explosive magma–water interactions: thermodynamics, explosion mechanisms, and field studies. *Bulletin of Volcanology*, **48**, 245–264.

Wohletz, K. and McQueen, R. 1984. Experimental studies of hydromagmatic volcanism. In *Explosive Volcanism: Inception, Evolution, and Hazards*. Washington, DC, National Academy Press, pp. 158–169.

Wolff, J. and Wright, J. 1981. Rheomorphism of welded tuffs. *Journal of Volcanology and Geothermal Research*, **10**, 13–34.

Woods, A. W. 1988. The dynamics and thermodynamics of eruption columns. *Bulletin of Volcanology*, **50**, 169–191.

Woods, A. and Wohletz, K. 1991. Dimensions and dynamics of co-ignimbrite eruption columns. *Nature*, **350**, 225–227.

Yoder, H. S. 1976. *Generation of Basaltic Magma*. Washington, DC, National Academy of Sciences.

Young, S., Sparks, R. S. J., Robertson, R., *et al.* 1997. Eruption of Soufrière Hills volcano in Montserrat continues. *Eos*, **78**, 401–409.

# Chapter 2

# Volcano hazards

Robert I. Tilling

## Introduction

Not only do "Volcanoes assail the senses . . ." (Decker and Decker, 1997, p. vii), but they also assail the environment when they erupt, terrifying and fascinating humankind for countless millennia. Volcanic processes and products – beneficial and hazardous – have profoundly impacted and continue to impact society (Chester, 1993, Chapter 14, this volume).

It is estimated that about 10% of the world's population live within proximity of active and potentially active volcanoes. With projected population growth, by the twenty-first century more than 500 million people could be at risk from volcano hazards (Peterson, 1986; Tilling and Lipman, 1993). Of these half billion people at risk, roughly equal to the estimated entire world population at the beginning of the seventeenth century, about 90% live in the circum-Pacific region. Also in the twenty-first century, with continued increasing urbanization there will be more than 100 cities with greater than 2 million population, and many of these will be located within 200 km of volcanically active subduction zones (McGuire, 1995a, Fig. 15.1).

Over 1500 subaerial volcanoes have been active during the Holocene (i.e., the past 10 000 years), and more than a third of these have erupted one or more times during recorded history (Simkin and Siebert, 1994). Active or potentially active volcanoes occur in narrow belts that collectively comprise less than 1% of the Earth's total surface area, dotting the divergent and convergent boundaries between the tectonic plates and above well-documented intraplate "hot spots," such as Hawaii and the Galapagos (Simkin *et al.*, 1994). Even though convergent-plate (subduction-zone) volcanism accounts for only about 15% of the averaged global volcanic output (Simkin, 1993), it produces more than 80% of documented historical eruptions (Simkin and Siebert, 1994; Tilling, 1996) (Fig. 2.1a). This reflects the fact that the overwhelming majority of the Earth's eruptive activity takes places on the deep ocean floor along the global oceanic ridge systems – sight unseen and posing no volcanic hazards.

In a previous review of the topic of volcanic hazards, British volcanologist George P. L. Walker commented (wryly?): "In principle, volcanic risk could be eliminated by the total abandonment of all volcanic areas." (Walker, 1982, p. 156). However, he also stated clearly that such an option "is not realistic" because these areas are already inhabited – some for many centuries or millennia. Indeed, the combined pressures of population rise and urbanization, human settlement, agricultural cultivation, and industrial development increasingly encroach upon volcanoes heretofore considered "remote" and, hence, presumed to pose minimal risk to people and property. Moreover, explosive eruptions of still-remote volcanoes (e.g., many along the Aleutian, Kamchatka, and Kurile volcanic arcs) constitute a substantial and growing risk to aircraft flying over them along the heavily traveled North Pacific air routes (Casadevall, 1994a, 1994b; Casadevall

*Volcanoes and the Environment*, eds. J. Martí and G. G. J. Ernst. Published by Cambridge University Press. © Cambridge University Press 2005.

**Fig. 2.1.** (a) A powerful explosive eruption of Mt. Pinatubo, Philippines, on June 12, 1991; the climactic eruption on June 15 was more voluminous ($\sim$5 km$^3$), perhaps the second largest eruption in the twentieth century. About 250 times larger than the 1985 Ruiz eruption, the Pinatubo eruption had, and continues to have, tremendous impact on the environment, yet it caused only a few hundred deaths (for reasons, see discussion on the case histories of two eruptions). Photograph by David H. Harlow (US Geological Survey). (b) Aerial view of the remains of the city of Armero, which was devastated by lahars triggered by a small-volume (0.02 km$^3$) magmatic eruption of Nevado del Ruiz, Colombia, on November 13, 1985; the lahars poured from the canyon mouth of Río Lagunillas (upper right corner), killing more than 22 000 people. The Ruiz catastrophe is the worst volcanic disaster in the world since the 1902 eruption of Mt. Pelée (Martinique). (Photograph by Darrell G. Herd, US Geological Survey.)

the more severe will be the losses once an eruption occurs.

Thus, a realistic approach to reducing volcano risk must necessarily embody the main theme of the 1988 Kagoshima International Conference on Volcanoes: "Towards better coexistence between human beings and volcanoes" (Kagoshima Conference, 1988). We must become more resourceful in enjoying the long-term benefits of volcanoes during their repose, while planning for, and coping with, the short-term hazardous consequences when they erupt.

This chapter is not intended as a comprehensive treatment of the wide-ranging topic of volcano hazards, but instead it highlights some successes and failures in volcano-hazards mitigation as well as some challenging issues confronting volcanologists in this century. In this selective review, I draw liberally, in places verbatim, from several relatively recent summary papers and books on the topic: Blong (1984); Crandell *et al.* (1984); Latter (1988); Tilling (1989a, 1989b); Scott (1989a, 1989b); Ewert and Swanson (1992); Chester (1993); Peterson and Tilling (1993); Tilling and

*et al.*, 1999). David Chester (1993, p. 232) cogently summarizes the crux of the problem:

Without people there can be no natural hazards; they are an artifice of the interaction between people and nature. In general, the greater the population density and the higher the level of economic development,

**Fig. 2.1.** *(cont.)*

Lipman (1993); Simkin and Siebert (1994); Mc-Guire *et al.* (1995); and Scarpa and Tilling (1996). For more detailed discussion of any particular aspect of volcano hazards, the interested reader is referred to these works and the references contained therein.

## Volcano hazards

Hazardous processes associated with volcanic eruptions largely were little appreciated perhaps until the 1883 eruption of Krakatau (Sunda Strait, between Java and Sumatra, Indonesia), when more than 36 000 people were killed, mostly by volcanogenic tsunamis (Simkin and Fiske, 1983, 1984). The news of this destructive eruption quickly became public knowledge – via the newly invented telegraph communications system – following observations by the Dutch and British scientific expeditions dispatched quickly to the scene (Verbeek, 1885; Symons, 1888).

Ironically, earlier in the nineteenth century, the 1815 eruption of Tambora (Sumbawa Island, Indonesia), the largest and deadliest eruption (92 000 deaths) in recorded history, received scant notice worldwide because of the volcano's remote location and the poor global communications and meager scientific understanding of volcanic phenomena prevailing at the time. It was not until the 1902 eruption of Mt. Pelée (island of Martinique, West Indies), whose pyroclastic flows obliterated the city of St. Pierre and swiftly killed all but two of its 30 000 inhabitants, that the devastating impacts of volcano hazards captured widespread public and scientific attention, spurring the establishment of instrumented volcano observatories around the world and fostering the development of the modern science of volcanology.

The powerful impact of hazardous volcanic processes on the environment and society has been vividly demonstrated by several eruptions in recent decades, especially the voluminous explosive eruptions of Mt. Pinatubo (Luzon, Philippines) in 1991 (Newhall and Punongbayan,

**Fig. 2.2.** Looking eastward from the site of the former capital city of Plymouth, near Sagar Bay on the southwestern side of Montserrat. Structures have been damaged and destroyed by a combination of pyroclastic flows and lahars. Soufrière Hills Volcano is visible in the background; the pyroclastic flows originated from collapse of the unstable lava some located between the two prominent peaks. (Photograph by Richard P. Hoblitt, US Geological Survey.)

1996b) (Fig. 2.1a). However, eruptions need not be large to be deadly and (or) cause massive socio-economic disruption. For example, the 1985 eruption of Nevado del Ruiz (Colombia) was quite small, erupting only ~0.02 km³ of magma compared to the ~5 km³ for the climactic eruption of Pinatubo, but it resulted in more than 22 000 fatalities (Williams, 1990a, 1990b) (Fig. 2.1b). Moreover, the hazardous juxtaposition of society and active volcanoes also is emphasized by the recent eruption of Soufrière Hills Volcano, on the island of Montserrat (British West Indies), located approximately 250 km north-northwest of Mt. Pelée along the Caribbean volcanic arc. This eruption, which began in mid July 1995, was the first on Montserrat since European settlement in

the Caribbean region and has resulted in substantial socio-economic impact and human misery for the approximately 12 000 inhabitants of this tiny island (<160 km² in area). For a detailed account of the eruptive activity through 1997 and related topical studies, see Aspinall *et al.* (1998), Young *et al.* (1998b), and Druitt and Kokelaar (2002), but a brief summary is given below.

The initial phase of the Soufrière Hills eruption (July to November 1995) was characterized by phreatic explosions producing steam and ash columns, the most energetic of which generated cold base-surge ash clouds and light ash-falls of non-juvenile materials blasted from the existing domes (Young *et al.*, 1998a). The eruptive activity became magmatic in mid November 1995 and has involved pyroclastic flows, surges and co-ignimbrite ash clouds, generated by a series of collapses of actively growing, gravitationally unstable lava domes. Tephra falls and pyroclastic flows from an explosive eruption in mid September 1996 devastated all flanks of the volcano (Fig. 2.2), which make up the entire southern half of the island (Young *et al.*, 1998a). Pyroclastic flows and associated ash-cloud surges triggered by a large dome collapse on June 25, 1997 destroyed hundreds of houses and killed 19 people (Aspinall *et al.*, 1998). Rapid magma extrusion immediately ensued and, by December 25 (Christmas Day) 1997, the lava dome had attained a volume of about $115 \times 10^6$ m$^3$. Then on December 26 (Boxing Day) 1997, a massive collapse of the dome occurred, involving about 40% of its volume and producing debris avalanches, pyroclastic flows and surges, co-ignimbrite ash clouds, lateral blasts, and a small tsunami. The Boxing Day processes and deposits, which perhaps constituted the volcano's most intense outburst to date, produced no human fatalities but severely damaged or completely buried several (already evacuated) settlements 2–3 km downslope from the summit. After the collapse, magma extrusion and dome growth quickly resumed and continued at varying rates through mid March 1998, accompanied by small collapses, explosions, and weak to moderate ash venting. From April 1998 through mid November 1999, even though no new magma was added to the dome complex, numerous so-called "passive" gravitational collapses took place. These events typically only produced small to moderate rockfalls and localized associated pyroclastic flows and ash clouds, but some resulted in substantial changes in dome morphology. On November 29, however, another new dome emerged, and the first new magma erupted since March 1998. Dome growth then continued episodically through 2001, and the dome reached its largest size ($>120 \times 10^6$ m$^3$) and highest elevation ($>1000$ m) since the eruption began in 1995.

Because of the intensive volcano monitoring and effective communications of hazard warnings to government officials by the newly and quickly established Montserrat Volcano Observatory, eruption-caused fatalities were relatively few. However, nearly all of the island's means of livelihood and infrastructure, including the principal city of Plymouth and the airport, were lost during the eruption, necessitating the evacuation of most of the population to the northern (less hazardous) end of the island and to other locales, including Great Britain. In late 1997, only about 3300 people remained on the island; at present, about 6500 people, or about half of island's pre-crisis population, are still living elsewhere (Rozdilsky, 2001). The Montserrat and British governments have developed long-range plans (e.g., Government of Montserrat and Her Majesty's Government, 1998) for rehabilitation of the island and restoration of its infrastructure, to allow the eventual return of the people evacuated. Aspinall *et al.* (1998, p. 3387) remarked: "Montserrat is a small island, and the hazards there are hard to avoid . . ." According to media reports and observers, toward the end of 2001 infrastructure rebuilding is under way and the general quality of life is beginning to improve, but housing remains in short supply. However, entering the year 2004, Soufrière Hills Volcano remains intermittently active with dome growth and collapses, rockfalls, and pyroclastic flows. Because of the continuing activity and associated volcanic hazards, the southern two-thirds of the island, with exception of a small daytime entry zone on the western coast, remains an Exclusion Zone (Montserrat Volcano Observatory, 2001), which is off-limits to ordinary human activities and only accessible for purposes of scientific monitoring and national security. Thus, the volcanic crisis that began at Montserrat in 1995 is persisting into the twenty-first century.

Hazardous volcanic processes (i.e., volcano hazards) have been well described elsewhere (e.g., Macdonald, 1972; Blong, 1984; Crandell *et al.*, 1984a; Scott, 1989a; Myers *et al.*, 1997; Sigurdsson *et al.*, 2000; US Geological Survey, 2005). Therefore, only a short, selective summary is given herein, drawing heavily from Blong (1984) and, especially, Scott (1989a) and emphasizing the hazards that have occurred frequently in historical time. Volcano hazards can be grouped conveniently into two broad categories – *direct* and *indirect* (Table 2.1) – although the distinction between them can be arbitrary and subjective in some instances, depending on the time elapsed between the eruptive activity and the occurrence of the hazardous event(s).

## Direct volcano hazards

Hazardous events that are produced during or shortly following the eruption (i.e., within minutes to several days) are considered to be

**Table 2.1** | Principal types of volcanic hazards and selected examples

| Type | Selected examples | References |
|---|---|---|
| *Direct volcano hazards* | | |
| Fall processes | | |
|   Tephra falls | Rabaul, 1994 | Blong and McKee (1995) |
|   Ballistic projectiles | Soufrière (St. Vincent), 1812 | Anderson and Flett (1903) |
| Flowage processes | | |
|   Pyroclastic flows, surges | Mt. Pelée, 1902 | Fisher *et al.* (1980) |
|   Laterally directed blasts | Bezimianny, 1956 | Gorshkov (1959); Belousov (1996) |
| | Mount St. Helens, 1980 | Hoblitt *et al.* (1981); Kieffer (1981) |
|   Debris avalanches | Mount St. Helens, 1980 | Voight *et al.* (1981); Glicken (1998) |
|   Primary debris flows | Nevado del Ruiz, 1985 | Pierson and Janda (1990) |
|    (lahars) | | |
|   Floods (jökulhlaups) | Katla, 1918 | Thorarinsson (1957) |
| | Grímsvötn (Vatnajökull), 1996 | Gudmundsson *et al.* (1997); Jónsson *et al.* (1998) |
|   Lava flows | Kilauea, 1959–60 | Macdonald (1962) |
| | Kilauea, 1983–present | Wolfe *et al.* (1988); Heliker *et al.* (1998) |
| Other processes | | |
|   Phreatic explosions | Soufrière (Guadeloupe), 1976 | Feuillard *et al.* (1983) |
|   Volcanic gases and acid | Dieng Plateau (Indonesia), 1979 | Le Guern *et al.* (1982) |
|    rains | Lake Nyos (Cameroon), 1986 | Kling *et al.* (1987) |
| | Kilauea, 1983–present | Sutton *et al.* (1997) |
| | Long Valley Caldera, 1989–present | Farrar *et al.* (1995); Sorey *et al.* (1998) |
| *Indirect volcano hazards* | | |
| Earthquakes and ground movements | Sakurajima, 1914 | Shimozuru (1972) |
| Tsunami (seismic seawave) | Krakatau, 1883 | Simkin and Fiske (1983) |
| Secondary debris flows (lahars) | Mt. Pinatubo, 1991–2 | Rodolfo *et al.* (1996) |
| Secondary pyroclastic flows | Mt. Pinatubo, 1991–3 | Torres *et al.* (1996) |
| Post-eruption erosion and sedimentation | Mt. Pinatubo, 1991–4 | Punongbayan *et al.* (1996a) |
| Atmospheric effects | Mayon, 1814 | COMVOL (1975) |
| Climate change | Tambora, 1815–16 | Stommel and Stommel (1983) |
| | Mt. Pinatubo, 1991–3 | Self *et al.* (1996) |
| Post-eruption famine and disease | Lakagígar (Laki), 1783 | Thorarinsson (1979) |
| Aircraft encounters with volcanic ash | Redoubt, 1989–90 | Casadevall (1994a) |
| | Mt. Pinatubo, 1991 | Casadevall *et al.* (1996) |

*Source:* Updated and expanded from Tilling (1989a, Table 2).

*direct* volcano hazards. Direct hazards include (see Table 2.1): *fall* processes, involving the fall and accumulation of air-borne volcanic ejecta; *flowage* processes, involving eruption-triggered, ground-hugging movement and deposition of primary volcanic materials and (or) of mixtures of volcanic debris with eruption-induced meltwater or other runoff (e.g., co-eruption rainfall or evacuation of crater lakes); and *other* processes, involving phreatic explosions, emission of volcanic gases, and eruption-related acid rains or volcanic air pollution. Some common fall and flow processes are briefly discussed below to provide context for the discussion to follow. However, because the hazards associated with phreatic explosions, volcanic gases, and acid rains are comparatively minor, they will not be considered in this chapter, and the interested reader is referred to the excellent summaries in Blong (1984) and Scott (1989a).

## Fall processes

The ejection and deposition of air-borne fragmental volcanic materials from explosive eruptions perhaps constitute the most common, if not most severe, volcano hazard. The origin of the fragmental ejecta may be *juvenile* (formed of magma involved in the eruption), *accidental* (derived from pre-existing rocks), or, most commonly, a mixture of both.

### TEPHRA FALLS AND BALLISTIC PROJECTILES

Falls of *tephra* – air-borne fragments of rock and lava of any size or shape expelled during explosive eruptions – constitute the commonest and most hazardous fall process. The temperature of the erupted material and mass eruption rate determine the height of an eruption column, which, along with wind strength and direction, exert the principal controls on the long-distance transport and, ultimately, deposition of tephra. Tephra typically becomes finer-grained and forms thinner deposits, with increasing distance downwind from the eruptive vent.

Depending on the volume and duration of the eruption, tephra falls can blanket many tens of thousands of square kilometers of areas around the volcano. For example, the 1.1 $km^3$ (bulk volume) of tephra ejected during the 9-hour-long climactic eruption (May 18, 1980) of Mount St. Helens covered more than 57 000 $km^2$, with accumulated thickness of many meters in the proximal areas and measurable thicknesses ($\geq 0.5$ mm) at distances of more than 700 km from the vent (Sarna-Wojcicki *et al.*, 1981). While it had a devastating impact on surrounding areas, this eruption was a relatively modest-size event, when compared to some voluminous explosive eruptions in the geologic past, such as those forming Yellowstone Caldera (northwestern Wyoming) and Long Valley Caldera (east-central California) in the United States. These huge eruptions erupted several orders of magnitude more magma ($10^2$–$10^3$ $km^3$) (Christiansen, 1984) than did Mount St. Helens in 1980, and the impact of the tephra falls from such eruptions covered areas on a continental scale.

The high-velocity ejection and fall of large *ballistic projectiles* (volcanic *bombs* and *blocks*) during energetic explosive eruptions constitute another common, but more localized, fall hazard. Because such projectiles exit the vent at speeds of tens to hundreds of meters per second on trajectories, they are only minimally affected by eruption column dynamics or the wind; their areas of impact are typically restricted to within 5 km of vents (Blong, 1984).

Tephra fall and ballistic projectiles affect the environment by: (1) the force of impact of falling fragments, (2) burial, (3) production of a suspension of fine-grained particles in air and water, and (4) venting of noxious gases, acids, salts, and, close to the vent, heat. The most-damaging impacts, however, are the collapse of roofs of buildings, interruption of power, disruption of societal infrastructures (e.g., water, waste-treatment, power, transportation, and communications systems), and the damage or killing of vegetation, including agricultural crops.

## Flowage processes

Flowage processes are among the most deadly of the direct volcano hazards, even though their impacts are much more restricted to the immediate areas around the volcano than are those of fall processes.

## PYROCLASTIC FLOWS, PYROCLASTIC SURGES, AND LATERALLY DIRECTED BLASTS

These phenomena all involve rapidly moving, ground-hugging mixtures of rock fragments and gases. Pyroclastic flows and surges are masses of hot (300 – >800 °C), dry, pyroclastic debris and gases that sweep along the ground surface at extremely high velocities, ranging from ten to several hundred meters per second. Such processes are initiated by discontinuous or continuous collapse of an eruption column, buoyant upwelling at the vent, and gravitational or explosive collapse of a growing lava dome (Scott, 1989a) (Fig. 2.2).

Pyroclastic surges are distinguished from pyroclastic flows by having a relatively low ratio of solid materials to gases and, hence, lower density, even though a continuum exists between them. Because of the density difference, pyroclastic flows tend to be more controlled by topography, mostly restricted to valley floors, whereas less dense, more mobile surges can affect areas high on valley walls and even overtop ridges to enter adjacent valleys. Pyroclastic surges can either be hot or cold. Hot pyroclastic surges are closely associated with pyroclastic flows, both generated by the same processes; cold pyroclastic surges, however, are generated by hydromagmatic or hydrothermal explosions (Scott, 1989a). Convecting clouds of finer ash commonly accompany pyroclastic flows and surges and form one type of tephra-fall deposit.

Pyroclastic flows and surges are common at many andesitic and dacitic composite volcanoes and at silicic calderas. Pyroclastic flows containing abundant dense to slightly vesicular lithic fragments entrained in an ash matrix ("block-and-ash" flows) are generally of smaller volume and typically restricted to within a few tens of kilometers of vents. In contrast, large pumiceous pyroclastic flows, composed mostly of lapilli and ash, can extend up to 200 km from vents and can cover tens of thousands of square kilometers.

Laterally directed blasts (also called *lateral blasts*) share characteristics of pyroclastic flows and surges but have in addition an initial low-angle component of explosive energy release that can fan out and affect large sectors (up to 180°) of the volcano out to distances of tens of kilometers. A laterally directed blast is little affected by topographic features. Large directed blasts result from the sudden depressurization of a magmatic and (or) hydrothermal system within a volcano, commonly by flank failure and associated landsliding, as occurred at Mount St. Helens in 1980 (Christiansen and Peterson, 1981), affecting an area of about 600 km². Well-documented notable catastrophic directed blasts during historical time took place at Bezimianny Volcano, Kamchatka, in 1956 (Gorshkov, 1959; Belousov, 1996).

Owing to their mass, high temperature and gas content, high velocity, and great mobility, pyroclastic flowage processes are among the most deadly of volcano hazards, causing some or all of the following consequences: asphyxiation, burial, incineration, and physical impact. In addition to these direct effects, pyroclastic flows can mix with surface water or water melted from snow and ice to form destructive lahars and floods that can affect valleys farther downstream. Volcanic catastrophes (each causing 2000 or more fatalities) involving pyroclastic flows and (or) surges include: Vesuvius, Italy, in AD 79; Mt. Pelée, Martinique, in 1902; Mt. Lamington, Papua New Guinea, in 1951; and El Chichón, Mexico, in 1982.

## DEBRIS AVALANCHES

Active and potentially active volcanoes are inherently unstable structures, because they are built of weak or unconsolidated rock, have steep slopes, are highly fractured and faulted, and may be undergoing deformation related to magma movement and (or) hydrothermal pressurization. Thus, structural collapse at restless volcanoes can precede, accompany, or follow eruptive activity, triggering rockfalls, rockslides, and debris avalanches, which can move rapidly downslope and may pose significant hazards. Most volcanic debris avalanches have followed days to months of precursory activity (e.g., seismicity, ground deformation, phreatic explosions), but some apparently have occurred with little detectable precursors. In addition to the well-documented 1980 event at Mount St. Helens (Voight *et al.*, 1981, 1983; Glicken, 1998), debris avalanches have occurred at numerous composite volcanoes in historical time (e.g., Schuster and Crandell, 1984; Siebert, 1984, 1996).

Volcanic debris avalanches apparently are more mobile than their non-volcanic counterparts, and, for a given volume and vertical drop, volcanic debris avalanches travel farther (Scott, 1989a, Fig. 2.5). Large-volume avalanches can extend as far as 85 km beyond their sources and can deposit volcanic debris tens of meters thick over areas $10^2-10^3$ km$^2$ (Siebert, 1996). With sufficient momentum, debris avalanches can run up slopes and cross topographic barriers up to several hundred meters high. The Mount St. Helens debris avalanche (2.5 km$^3$ volume) was the largest such event, volcanic or non-volcanic in origin, in historical time (Glicken, 1998). However, a prehistoric debris avalanche (~300 ka) at Mt. Shasta Volcano, California covered nearly 700 km$^2$ and involved a volume of 45 km$^3$ (Crandell et al., 1984b; Crandell, 1989).

Debris avalanches are highly destructive, burying and destroying everything in their paths. Moreover, the "dewatering" of a debris avalanche can generate lahars and floods downvalley, as observed at Mount St. Helens in 1980 (Janda et al., 1981). Lahars and floods can also be triggered by the catastrophic failure and draining of lakes formed by damming of streams by avalanches (Costa and Schuster, 1988; Scott, 1988). Avalanches may also cause indirect hazards, when the volcanic debris enters lakes or bays and suddenly displaces large volumes of water to produce high waves, or enters the sea to generate volcanogenic tsunamis (Scott 1989a; Siebert, 1996).

PRIMARY DEBRIS FLOWS (LAHARS)
    AND FLOODS

A lahar (an Indonesian term for volcanic mudflow) is a slurry of rock debris and water that originates on the slopes of volcanoes. Such flows are called primary if they occur during eruptive activity, and secondary if they are post-eruption. The source(s) of the water to mobilize lahars can be one or more of the following: melting of ice and snow by hot volcanic ejecta; crater lakes and other surface waters; water in the groundwater and geothermal systems; and torrential rains. Lahars and floods are end members of a con-

tinuum of processes ranging from dense lahars (with consistency of wet concrete) dominated by laminar flow to turbulent watery floods. The transition downstream from a lahar, to a lahar-runout flow, and then to normal stream flow (i.e., flood) is primarily a function of decreasing sediment concentration; see Scott (1988, 1989) for details of the transition and the controlling factors (grain size, water content, yield strength, bulk density, etc.). As mentioned earlier, water-saturated debris avalanches can transform into lahars.

Measured velocities of historical lahars have varied greatly depending on volume, grain-size distribution, channel dimensions and configuration, and slope gradient, ranging from about 1 to 40 m/s, with mean values on the order of 10–20 m/s (Macdonald, 1972; Janda et al., 1981; Blong, 1984). High-velocity lahars have sufficient momentum to rise on the outside of bends of channels and to surmount topographic barriers. Large-volume lahars, rich in clay and confined to narrow channels, can travel hundreds of kilometers down valleys.

Because of their high bulk density and velocity, lahars pose a major hazard along valleys draining the volcano, even for communities at great distances from the volcano. Lahars are highly destructive, destroying and burying everything in their paths. They can also fill and modify stream channels, thus decreasing the channels' normal carrying capacity and increasing the potential for water floods; also, increased sedimentation in channels by lahars can affect their navigation.

Lahars and pyroclastic flows and surges have been the deadliest volcano hazards in recorded history (Blong, 1984; Tilling, 1989a, 1989b). Compared to pyroclastic flow and surge hazards, lahars and flood hazards – in theory at least – can be more easily mitigated, because they have relatively well-defined upper limits of potential impact along valleys. With early detection of a lahar close to its source, and if such detection is communicated effectively to the settlements downvalley, people at risk can quickly climb to safety if high-ground safe areas are identified beforehand. Had such a detection and communication system been established at Volcán Nevado del Ruiz (Colombia) in 1985, many lives probably could have been saved (see p. 74). Also, "sabo" and other engineering works along valleys (e.g., check dams, sediment-retention and flood-control structures, levees) can mitigate the hazards of small-volume lahars but not large ones (Scott, 1989a).

LAVA FLOWS

Lava flows constitute the most common volcano hazards from non-explosive eruptions, especially at basaltic systems. How far a lava flow can travel from the vent depends on its viscosity (itself determined by temperature, chemical composition, gas content, and crystallinity), effusion rate, and ground slope. At low effusion rates ($<10$ m$^3$/s), basaltic lava tends to produce many small flows that puddle and pile up near the vent, whereas at higher rates ($10^1$–$10^3$ m$^3$/s), the flows produced can travel tens of kilometers and can cover hundreds of square kilometers. For example, the 1783 Lakagígar (Laki) fissure eruption in Iceland, with an effusion rate of 5000 m$^3$/s, generated lava flows that covered more than 500 km$^2$ (Thorarinsson, 1969). During episodes of flood-basalt volcanism in the geologic past, such as those producing the Columbia River Basalts in the United States, single flows discharged at estimated rates of about $1 \times 10^6$ m$^3$/s covered tens of thousands of square kilometers (Swanson *et al.*, 1975).

Unlike the sheet-like flows formed of basaltic and other mafic lavas, more viscous lavas (e.g., andesite, dacite, and rhyolite) are typically erupted at low rates, forming short, stubby lava flows or steep-sided domes that cover only a few square kilometers. Rates of movement of lava flows vary considerably, from a few to hundreds of meters per hour for silicic lava flows to several kilometers per hour for basaltic lava flows on steep slopes. Because of their slow rates of movement, lava flows seldom threaten human life, but they can be highly destructive, burying, crushing, or burning everything in their paths. In addition, fires started by lava flows can affect areas far beyond their borders. Lava flows also melt snow and ice that can produce small debris flows and floods; however, catastrophic floods can occur if the meltwater can be stored and later released suddenly in large quantities, as during Icelandic jökulhlaups.

Because the movement of lava flows is primarily controlled by gravity, once potential or actual vents of lava flows are identified, their paths can be predicted based on the surrounding topography. With sufficient data, accurate hazard-zonation maps for lava flows (e.g., Wright and others, 1992) provide a basis for emergency-response plans. The most cost-effective mitigation of lava-flow hazards is prudent land-use planning, but several attempts have been made in recent decades to divert or control the paths of lava flows after the start of eruption (for examples, see Macdonald, 1972; Scott, 1989a; Williams, 1997).

## Indirect volcano hazards

As the term implies, indirect volcano hazards are destructive processes that are incidental to the eruptive phenomena themselves and include (see Table 2.1): ground-shaking and movements caused by volcanogenic earthquakes; tsunamis generated by eruption-induced collapse, debris avalanche or slump of a volcanic edifice; "secondary" debris flows and floods triggered by heavy rainfall and other post-eruption causes; post-eruption erosion and sedimentation; atmospheric effects (electrical discharges, shock waves); climate change; post-eruption famine and disease; and, only within recent decades, damage to aircraft encountering volcanic ash clouds. Only the most severe of these indirect hazards will be reviewed briefly below; for the others, the interested reader is referred to Blong (1984), Scott (1989a), and the relevant references given in Table 2.1.

### Volcanogenic tsunamis

A *tsunami* (a Japanese word meaning "harbor wave") is a long-period seawave that is generated by the sudden displacement of water, most commonly by fault displacements of the seafloor by tectonic earthquakes (e.g., along subduction zones). Volcanogenic tsunamis, much less common, can be produced by one or more of the following mechanisms: volcanic or volcano-tectonic earthquakes; explosions; collapse or subsidence of volcanic edifices; debris avalanches, lahars, or pyroclastic flows entering water bodies; and, atmospheric shock waves that couple with the sea (Latter, 1981).

Tsunamis travel at high speeds (>800 km/hr) as imperceptible low broad waves in the open ocean but build to great heights as they approach shore and "touch bottom." Because of the tremendous release of wave energy upon slamming onto the shore, tsunamis are highly hazardous for life and property on low-lying shore areas of lakes and oceans, even at great distances ($10^3$ km) from the tsunami source. For example, the 1883 Krakatau eruption produced tsunamis that ran up coasts to heights of 35 m, devastating nearly 300 coastal villages and claiming more than 30 000 victims (see numerous references in Simkin and Fiske, 1983). Another deadly volcanogenic tsunami occurred in 1792, when a massive debris avalanche produced by the collapse of the Mayuyama dome (Unzen Volcano, Japan) entered the Ariake Sea, generating a tsunami that killed about 5000 people (Tanguy et al., 1998).

The international Pacific Tsunami Warning Center (Honolulu, Hawaii), created in 1965, employs a circum-Pacific network of seismic and tide-gauge stations to provide timely warnings of approaching tsunamis to areas hundreds to thousands of kilometers from sources. Such early warnings are very useful for tsunamis generated by distant sources, allowing several hours for officials and people to take safety measures. However, because of the high velocity of tsunamis, it is impossible to provide similar early warnings to people living close to a locally generated tsunami.

### Secondary debris flows (lahars)

Lahars and floods generated after or between eruptions occur commonly at volcanoes. This comes as no surprise because most active or geologically young volcanoes are constructed of poorly consolidated volcanic products and have steep slopes. In addition, valleys draining volcanoes typically contain abundant unconsolidated volcanic debris from previous eruptive activity, and such deposits act to disrupt the established drainages. Given such conditions, all that is needed to produce secondary lahars is a sudden massive infusion of water, which is most frequently supplied by torrential rainfall, quite common in many volcanic regions during the rainy or monsoon season. A less common mechanism for generating secondary lahars is the sudden release of water impounded by natural dams – formed by lava flows, lahars, debris avalanches, pyroclastic flows, or crater rims – when such dams fail or are overtopped (e.g., Houghton et al., 1987; Costa and Schuster, 1988; Lockwood et al., 1988).

That hazards associated with secondary lahars are not inconsequential is well illustrated at Mt. Pinatubo (Luzon, Philippines). Since Pinatubo's climactic eruption in 1991, monsoon rains have caused massive redistribution of the 1991 eruptive and lahar deposits, generating numerous secondary lahars that have buried villages and valuable agricultural lands and, to date, have killed about an additional 100 people since the eruption, during which 250–300 people perished. Secondary lahars, involving progressively declining volumes, are expected to continue well into the next decade, as Pinatubo's drainages gradually become re-established (Wolfe and Hoblitt, 1996).

### Post-eruption famine and disease

In addition to the direct threat to human safety, volcano hazards also can adversely affect the daily lives of people by destruction of dwellings and societal infrastructure but, more importantly, by the disruption of food supply by the immediate loss of livestock and crops, and by the longer-term (years to decades) loss of agricultural productivity of farm lands buried by eruptive materials. Before the twentieth century, post-eruption famine and epidemic disease – in the aftermath of the 1783 eruption of Laki (Iceland) and the 1815 Tambora eruption – constituted the deadliest volcano hazard (Fig. 2.3). Previous estimates of the deaths caused by post-eruption starvation after the Tambora eruption were on the order of 80 000 (e.g., Blong, 1984; Tilling, 1989a), but a recent estimate is ~49 000 (Tanguy et al., 1998). The estimate of the number of starvation deaths (~10 000) caused by the Laki eruption is more reliable; the deaths represented about a quarter of Iceland's entire population at the time. In the twentieth century, fatalities from post-eruption famine and epidemic disease have been greatly reduced (Fig. 2.3).

### Aircraft encounters with volcanic ash

A significant and growing indirect volcano hazard is the encounter between aircraft and

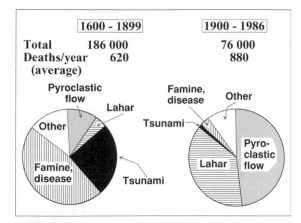

**Fig. 2.3.** Eruption-related fatalities according to principal cause for the periods 1600–1899 and 1900–1986. Modified from Tilling (1996, Fig. 2.2); data sources: Blong (1984); Tilling (1989a).

volcanic ash clouds, which are not detectable by the aircraft's onboard radar instrumentation. This problem only has emerged in recent decades, with the advent of high-performance jet engines. The hazard stems from the ingestion of silicate ash into the aircraft's jet engines when operating in volcanic ash clouds (for details, see Casadevall, 1992; Dunn and Wade, 1994). The ash particles can abrade, and (or) accumulate in, nozzle guide vane, ducts, compressor and turbine blades, and other engine parts; also, the high operating engine temperatures are sufficient to partially melt the ash, which then adheres to or clogs critical engine parts and openings. Ash ingestion degrades engine performance and, in the worst case, causes engine flame-out and loss of power. The ash can also cause exterior damage, such as abrasion of the windshield and windows, erosion of leading edges, etc.

Since the early 1970s, there have been more than 60 volcanic ash–aircraft encounters, with several of them experiencing total power loss, necessitating emergency landings. Most encounters occur within about 250 km of the erupting volcano, but many have happened at greater distances (>900 km) from the volcano. Drifting volcanic ash clouds, following stratospheric winds, can be tracked for thousands of kilometers before dissipation. A very costly encounter was during the eruption of Redoubt Volcano (Alaska, USA) in

December 1989. A new Boeing 747–400 encountered Redoubt's volcanic ash cloud and temporarily lost all four engines, which (luckily) were able to be restarted in ash-free atmosphere after a powerless descent of nearly 4000 meters, and the plane landed safely at Anchorage airport. Fortunately, no passengers were injured, but the costs to replace all four engines and repair other damage exceeded US$80 million (Casadevall, 1992).

## Spatial and temporal scales of volcano hazards

From the previous discussion, it is obvious that the various volcano hazards differ markedly, in spatial and temporal scale, in their impact on the physical environment and on people. Of the direct hazards, flowage processes (pyroclastic flows and surges, lateral blasts, debris avalanches and flows, and lava flows) have an immediate impact on the environment. With the exception of basaltic lava flows, most other direct flowage hazards typically occur over a timescale of seconds to days. As exemplified by the current eruption at Kilauea Volcano (Hawaii, USA), which began in January 1983 and has continued to date (as of February 2000) virtually non-stop in a quasi-steady-state manner (Wolfe *et al.*, 1988; Mattox *et al.*, 1993; Heliker *et al.*, 1998), basaltic lava flows, on the surface and below (in lava tubes), can occur on a timescale of hours to years. In general, the environmental impacts of all flowage hazards are restricted to areas around the volcano and its principal drainages out to distances on the order of several hundred kilometers maximum.

Tephra fall poses the widest-ranging direct hazard from volcanic eruptions. For instance, areas of $10^3$ to $10^4$ km$^2$ may be covered with >10 cm of tephra during some large eruptions, and fine ash can be carried over areas of continental size or larger. For this reason, tephra layers produced by eruptions of known ages serve as important regional time markers in stratigraphic and geologic studies over extensive areas. For example, the Bishop tuff deposit (~600 km$^3$ volume) produced by a cataclysmic eruption that formed Long Valley Caldera (eastern California, USA) 760 000 years ago can be traced as a stratigraphic marker in the mid-continent of the United States, thousands of kilometers distant from the caldera (Izett *et al.*, 1988).

Two indirect volcano hazards also cause wide-ranging impact on the environment, but on different temporal scales: volcanogenic tsunamis and eruption-influenced global climate change. The series of powerful tsunamis generated by the 1883 Krakatau eruption not only wreaked havoc locally, but the larger waves were recorded within about 12 hours by tide gauges as far away as the Arabian Peninsula, more than 7000 km from the eruption site. Recent studies demonstrate that the largest-scale, but lowest-frequency, debris avalanches are of submarine rather than sub-aerial origin, from collapses of flanks of volcanic islands in the Pacific, Indian, and Atlantic Oceans (e.g., Moore *et al.*, 1989, 1994; Holcomb and Searle, 1991; Barsczus *et al.*, 1992). Some studies have suggested that these prodigious avalanches have produced tsunamis orders of magnitude greater than any observed in historical time, leaving deposits of their occurrence >300 m above current sea level (Moore and Moore, 1984, 1988; Young and Bryant, 1992).

Perhaps the widest-ranging volcano hazard is global climate change caused by ejection of volcanic gases (principally $SO_2$) into the stratosphere during powerful explosive eruptions. The $SO_2$ forms an aerosol layer of sulfuric acid droplets, which tends to cool the troposphere by reflecting solar radiation, and to warm the stratosphere by absorbing radiated Earth heat. Such global climate change is best illustrated by the 1815 Tambora eruption. The stratospheric volcanic cloud associated with this huge eruption – the largest in recorded history – resulted in the well-documented "Year without Summer" the following year (1816), when usually cold weather prevailed across the northern hemisphere, marked by severe frosts in July and crop failures (Stommel and Stommel, 1983). For example, the annual summer temperature for 1816 for New Haven (Connecticut, USA) dropped 3 °C below the 145-year average. The more recent eruptions of El Chichón Volcano in 1982 and of Mt. Pinatubo in 1991 also produced measurable temperature decreases for the northern hemisphere, by 0.2 to 0.5 °C respectively (Simarski, 1992; Self *et al.*, 1996). However, such perturbations

of the atmosphere by volcanic aerosols and associated changes in global climate are short-lived, lasting only for a few years.

It also should be emphasized that more than one hazardous process can occur during the same eruption, each process with its own temporal and spatial scales. For example, climactic eruption of Mount St. Helens on May 18, 1980, within the span of a few hours, involved flank collapse, laterally directed blast, debris avalanches and flows, pyroclastic flows, and areally extensive tephra fall (Fig. 2.4; see also Lipman and Mullineaux, 1981; Tilling *et al.*, 1990). The lateral blast initiated within 2 s after the flank collapse and onset of the debris avalanche (estimated at 08:32:21.0 local time). The blast itself probably lasted no more than about 30 s at the vent, but the associated expanding and radiating blast cloud persisted for about another minute; during this short time, the blast affected a large sector (600 km$^2$) north of the volcano. A few seconds following the blast, the vertically directed tephra column reached an altitude of 26 km in less than 15 min. Eruption of tephra continued for the next 9 hr, and the fallout area covered more than 57 000 km$^2$ downwind. The finer ash and volcanic gases remained air-borne, spreading across the USA in 3 days and circling the world in 15 days. Several small pyroclastic flows, though first directly observed shortly after noon, probably were erupted shortly after the lateral blast, pyroclastic flow activity continued intermittently for about 5 hr. Small lahars in the South Fork of the Toutle River were observed as early as 08:50, and the larger and more destructive lahars in the North Fork of the Toutle River developed several hours later (Fig. 2.4).

## Global and historical perspective

Compared to other hazards, natural or man-made, volcano hazards have caused far fewer fatalities, largely because they occur infrequently; floods, storms, earthquakes, and droughts are the most frequently occurring natural hazards and affect far more people (Tilling, 1989b, Figs. 1.1 and 1.2). Since AD 1000, direct and indirect volcano hazards have killed about 300 000 people, which is much less than the fatalities for a single worst earthquake or

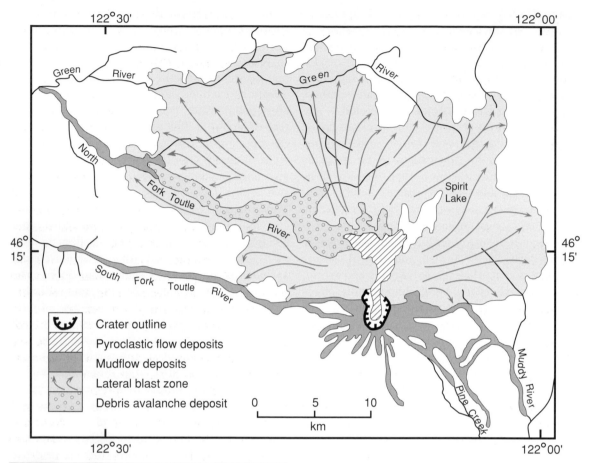

**Fig. 2.4.** Generalized geologic map showing the impacts and proximal deposits of the climactic eruption of Mount St. Helens Volcano on May 18, 1980; not shown are the tephra-fall deposits. The hazardous processes that produced these deposits all occurred within about one hour (see text). Modified from Tilling et al. (1990, p. 22); original data source: Kieffer (1981, Fig. 219).

hurricane disaster, such as the 1556 Huahsien (Shensi, China) earthquake which killed more than 820 000 or the 1970 Ganges Delta (Bangladesh) hurricane in which about 500 000 perished. From a global and historical standpoint, the adverse consequences of volcano hazards on people are comparatively quite small, but volcanic disasters still can have significant short-term human and economic impact (Chester, 1993; Chapter 14, this volume).

During the seventeenth to nineteenth centuries, most eruption-related casualties were caused by indirect volcano hazards, most notably tsunamis and post-eruption famine and disease; in contrast, the principal causes of eruption-related deaths in the twentieth century have been the direct volcano hazards – pyroclastic flows (including surges) and lahars (Fig. 2.3). As noted previously by Tilling (1989a, p. 260), the reduction in the incidence of deaths resulting from post-eruption starvation in the twentieth century "is real and reflects the existence of modern, rapid communications and disaster relief-delivery systems," which were lacking in centuries past. However, the decrease in the number of volcano-related tsunami deaths in the twentieth century (Fig. 2.3), while real, cannot simply be credited to any technological or societal advance. The observed decrease could merely reflect the fortunate circumstance that no significant volcanogenic tsunami has occurred since 1883.

With the international Pacific Tsunami Warning Center (Honolulu, Hawaii) in operation since the mid-1960s, it is now possible to provide early warnings for future, remote volcanogenic tsunamis that are expected to arrive onshore an hour or more after the triggering event (Lander and Lockridge, 1989). However, such expectation can be realized only if the volcanogenic source of the tsunami is centered at least 1000 km distant from the threatened coastal areas. Given the high velocity at which tsunamis travel in the open ocean (~800 km/hr), it would be nearly impossible for timely warnings to be given to people at risk from locally generated tsunamis, with triggering sources within 1000 km or less.

Most of the countries in the developing regions facing severe risk from volcano hazards lack the economic and technical resources to adequately study their dangerous volcanoes and to develop effective volcanic-emergency management systems (UNDRO/UNESCO, 1985). Not surprisingly, more than 99% of the eruption-related deaths in the twentieth century have occurred in the developing regions, reflecting one or more of the following factors: (1) high population density and growth, (2) insufficient scientific and economic resources, and (3) the fact that no major eruptions took place in any densely populated volcanic area in an industrialized region (e.g., Japan, Italy) during the twentieth century.

The last two decades of the twentieth century saw the emergence of a new indirect volcano hazard: volcanic ash in eruption clouds, which severely affects jet aircraft flying into them (Casadevall, 1994b). To date, while no fatal accidents have happened from in-flight encounters between aircraft and volcanic ash, many hundreds of millions of dollars have been incurred from damage to aircraft (including total power loss), unscheduled emergency landings, and increased flight time and fuel consumption in rerouting the planes from known or suspected volcanic ash plumes. With continued rapidly escalating air travel across the North Pacific and through other air routes over active volcanic belts (Casadevall *et al.*, 1999), the risk posed by airborne ash will increase proportionately and could well become the prime indirect volcano hazard in this century. Volcanologists are now working closely with aircraft controllers, the International Civil Aviation Organization, the aviation industry, and pilots to develop more effective systems to provide timely warnings of eruptions from volcanoes along air routes and to predict and track the possible paths of the volcanic ash clouds once formed (Casadevall, 1994b).

The average number of eruption-caused deaths per year for the twentieth century (880) is greater than that for the seventeenth to nineteenth centuries (620) (Fig. 2.3). Inasmuch as eruption frequency and severity have not increased in recent centuries (Simkin and Siebert, 1994), the higher toll of casualties in the twentieth century could be interpreted as reflecting world population growth. Such interpretation, however, is inconsistent with the approximately ten-fold increase in world population during the seventeenth to twentieth centuries; thus the frequency of eruption-caused deaths per capita worldwide actually decreased in the twentieth century. Lacking an analysis of relevant population and eruption fatality data specific to the volcanic regions, a plausible inference would be that progress in volcano-hazards mitigation has more than kept pace with population growth.

Perhaps the data for eruption-caused fatalities in the twentieth century (Fig. 2.5) may support this inference, if the 1902 Mt. Pelée and 1985 Nevado del Ruiz disasters are not taken into account. The rationale for such exclusion is that, for the 1902 disaster, the inhabitants of city of St. Pierre were not ordered to evacuate and all but two perished during the eruption. Even though political motivation was a factor (Heilprin, 1903), the primary reason for non-evacuation of the city was that its inhabitants and officials had no scientific understanding of eruptive phenomena and were totally ignorant of the developing unrest of Mt. Pelée and its ominous portent of the calamity to come. Awareness of the destructive power of pyroclastic flows and surges and the emergence of volcanology as a modern science came after the obliteration of the city. As discussed in detail later (p. 74), in the case of the Ruiz catastrophe, the high death tolls resulted from a failure in communication and human error, despite the availability of monitoring data and a hazards assessment; government officials simply failed to order evacuations even though the volcano and the scientists had given sufficient warning.

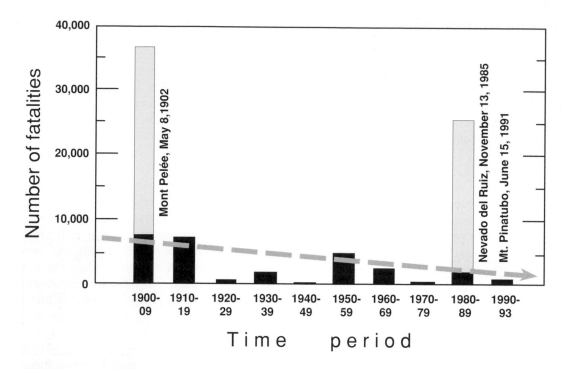

**Fig. 2.5.** Eruption-related fatalities by decade for the twentieth century through 1993. The shaded bars show the fatalities for the two deadliest eruptions of the century: the 1902 Mt. Pelée eruption (deaths from pyroclastic flows), and the 1985 Nevado del Ruiz eruption (deaths from mudflows). Dashed arrow hints a possible decline in fatalities with time if the fatalities from these two events are not considered (see text). Modified from Tilling and Lipman (1993, Figure 1).

A comparison of the 1902 Mt. Pelée eruption with the 1991 Mt. Pinatubo eruption also bears on the notion that a possible decreasing trend of eruption-related deaths for the twentieth century (Fig. 2.5) reflects advances in scientific understanding of eruptive phenomena and in volcano-hazards mitigation. "If Pinatubo (with its current population) had erupted in 1902, the deaths would have been at least comparable in number to those in St. Pierre. In these two cases – in both of which devastating pyroclastic flows impacted some densely populated areas – the decrease in casualties truly reflects a century of advance in hazards assessment, monitoring, and communication." (Edward W. Wolfe, 1998, personal communication.)

## Coping with volcano hazards

Given the projected growth in world population and the observation that on average 50–60 volcanoes are active each year (Simkin and Siebert, 1994), the conclusion seems inescapable that the number of people and the quantity of economic infrastructure at risk from volcano hazards will increase (Fig. 2.6). With this daunting demographic scenario, which is unlikely to change much during the next several decades, what can be done to reduce risk from volcanoes? For active or potentially active volcanoes, the three most important elements of an effective approach to volcano-hazards mitigation are:

(1) Acquisition of a good understanding of a volcano's *past behavior* by deciphering its eruptive history from basic geologic, geochronologic, and other geoscience studies. Such an understanding is prerequisite for the preparation of a hazards assessment and associated hazards-zonation map, as well as for long-term forecasts of potential future activity.

(2) Initiation or expansion of volcano monitoring to determine a volcano's *current behavior*, to characterize its "normal" baseline level of

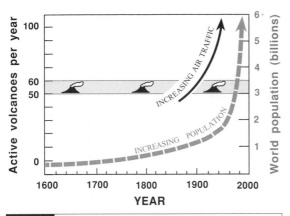

**Fig. 2.6.** Schematic diagram showing that, given a fixed global eruption frequency of 50–60 volcanoes active each year on average (shaded bar with erupting volcano symbols) and an increasing world population (heavy dashed curve), more and more people inevitably will be at risk from volcanoes into the foreseeable future.

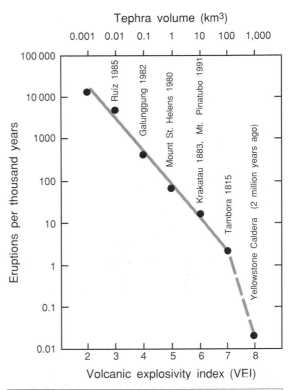

**Fig. 2.7.** Frequency and eruption size expressed in terms of the volcanic explosivity index (VEI), with some representative examples; extrapolation to the huge Yellowstone caldera-forming eruption 2 million years ago is unconstrained and only shown for relative comparison. The VEI is a semiquantitative estimate of the "magnitude" of explosive eruptions in historical time; a non-explosive eruption is assigned a VEI = 0 (see Newhall and Self, 1982). As observed also for earthquakes, small events (VEI ≤ 2) are much more frequent than larger events (VEI ≥ 5). Eruptions of VEI 6 (e.g., Krakatau 1883 and Mt. Pinatubo 1991) are very infrequent, occurring on average about once a century, and the *only* VEI 7 is given to the world's largest historical eruption (Tambora 1815). Modified from Simkin and Siebert (1994, Fig. 10).

activity, and to detect, reliably and quickly, any significant departures from it. Volcano-monitoring data provide the scientific basis for short-term forecasts of possible *future behavior*.

(3) Establishment or maintenance of credible, effective communications with civil authorities, the media, and the general public, while striving to accomplish the other two elements. Equally important is the creation of emergency-response or "preparedness" plans; effective communications and mutual trust will promote the development and testing of such plans. *Non-achievement of element (3) renders elements (1) and (2) useless.*

## Hazards assessment, hazards-zonation maps, and long-term forecasts

An assessment of a volcano's natural hazards is only as good as the abundance and quality of geologic, geochronologic, petrologic, geochemical and other pertinent data collected to reconstruct its eruptive history in the most comprehensive manner. The key questions to be answered are straightforward but crucial: what is the typical or characteristic style and frequency of its past activity? Explosive or non-explosive eruptions? Eruption sizes always similar? Frequent small

eruptions (volcanic explosivity index (VEI) ≥ 2) or rare huge events (VEI ≥ 7) (Fig. 2.7)? Which sectors of the volcano have been repeatedly affected by hazardous processes? Definitive answers to these and related questions provide the key constraints on the types, scales, duration, recurrence, and other characteristics of hazardous processes that have dominated the volcano's eruptive history. An in-depth understanding of eruptive history and characteristic behavior in turn provides the best basis for *long-term* forecasts of future eruptive activity and associated hazards.

A *long-term* forecast is generally taken to mean a forecast made one year or longer in advance of the anticipated activity. An excellent example of a successful long-term forecast was that made by US Geological Survey scientists in February 1975 (Crandell *et al.*, 1975) that Mount St. Helens in the Cascade Range was the one volcano in the conterminous United States most likely to reawaken and to erupt "perhaps before the end of this century" (i.e., the twentieth century). Five years later (March 1980), Mount St. Helens erupted. For more information on the related broad topics of hazards assessment, hazard-zonation maps, and long-term forecasts, the interested reader is referred to the review papers of Crandell *et al.* (1984a), Scott (1989b), and Tilling (1989a). Also, the volumes of Latter (1988) and Scarpa and Tilling (1996) each contain several papers that treat more specialized aspects (including probabilistic analysis) of hazards assessment.

For volcanically active areas that still are *sparsely populated* and relatively undeveloped, we must accelerate basic geoscience studies of volcanoes to make the best possible hazards-zonation maps to guide long-term land-use planning by government authorities, developers, and the insurance industry. High-density development in high-risk zones should be avoided or minimized to the extent practical economically. In theory at least, scientifically based land-use planning would allow land managers to follow George Walker's (1982) ideal dictum to mitigate volcano hazards: simply do not live or work near them. However, for the much more common situation where the volcanic area is already *densely populated* and developed and (or) where land-use patterns have long been fixed by tradition and culture, the option of prudent land-use planning to mitigate potential hazards cannot be exercised. Nonetheless, even in these situations, accurately prepared hazards-zonation maps provide essential information needed by civil authorities to develop contingency plans for an acute volcanic crisis that might culminate in eruption. Specifically, the hazards-zonation maps should help the officials to determine – *ideally*, before a crisis strikes – the least-risk zones for staging areas, escape routes, resettlement shelters, etc., should evacuation or other public-safety measures be required. Also, citizens should have sufficient information about the hazards to make knowledgeable decisions about where to live and work within the context of level of acceptable risk.

## Volcano monitoring and short-term forecasts

Volcano monitoring refers to the systematic visual observation and instrumental measurement of structural, geochemical, and seismic and other geophysical changes in the state of a restless volcanic system. Although the regular monitoring of active volcanoes – during and between eruptions – was the *raison d'être* for establishing modern volcano observatories early in the twentieth century, it also was the catalyst that quickened the transformation of volcanology into the multidisciplinary science that it is today. *Only* volcano-monitoring data can provide a diagnosis of a volcano's current behavior, which in turn constitutes the *only* scientific basis for short-term (hours to months in advance) forecasts or "predictions" of impending eruption, or of midcourse changes of an ongoing eruption.

Since the 1972 UNESCO review of volcano-surveillance techniques (UNESCO, 1972), there have been many hundreds (thousands?) of papers on the general theme of volcano monitoring and its applications at diverse volcanic systems around the world. This remarkable surge in volcano-monitoring studies clearly reflects the greatly enhanced general awareness of eruptions and their impacts that began with the May 18, 1980 eruption of Mount St. Helens, which caused the worst volcanic disaster in the history of the United States. The increased awareness whetted the public's appetite for information on volcanoes and eruptions, encouraged (justified?) public officials to increase funding for volcanologic investigations, and spurred scientists to pursue an improved understanding of "how volcanoes work."

It is beyond the scope of this chapter to treat volcano monitoring in detail; the interested reader is referred to the following works and the references contained therein: Civetta *et al.*

(1974), Fournier d'Albe (1979), Lipman *et al.* (1981), Swanson *et al.* (1983, 1985), Tazieff and Sabroux (1983), Hill (1984), Latter (1988), Banks *et al.* (1989), Ewert and Swanson (1992), Swanson (1992), Andres and Rose (1995), McGuire (1995a, 1995b), McGuire *et al.* (1995), Murray *et al.* (1995, 1996), Tedesco (1995), Tilling (1995), Chouet (1996a, 1996b), Francis *et al.* (1996b), Giggenbach (1996), McNutt (1996), Rymer (1996), Scarpa and Gasparini (1996), Scarpa and Tilling (1996), Dvorak and Dzurisin (1997), and Sigurdsson *et al.* (2000). Instead, the discussion below is selective, focusing on some notable recent advances and developments in volcano monitoring.

Seismic monitoring and geodetic monitoring remain the primary means of volcano surveillance (Scarpa and Gasparini, 1996). Singly or combined, "these two techniques measure the direct responses of the volcanic system – brittle fracture, inflation, and deflation – in adjusting to subsurface magma movement and accompanying stresses and (or) hydrothermal-pressurization effects" (Tilling, 1995, p. 371). Thus, they reliably detect the earliest signs of departure from a volcano's "normal" (i.e., baseline) behavior. In recent decades, with improvements in instrumentation, electronics, digital-telemetry technology, and computerized data collection and processing, modern seismic networks can monitor the seismicity of a volcanic system in "real time" or "near real time" in different ways. Thus, an overall improvement in seismic monitoring networks has been achieved at many, perhaps most, permanent volcano observatories. In addition, the recent development of robust, inexpensive, and easily portable personal computer (PC)-based systems for various "real-time" seismic-data analysis software, including "real-time seismic amplitude measurement" (RSAM), real-time "seismic spectral amplitude measurement" (SSAM), and automatic earthquake locations (Lee, 1989), permits rapid deployment at restless volcanoes not monitored by any permanent observatories. The RSAM system (Endo and Murray, 1991) enables the continuous monitoring of total seismic energy release even when conventional analogue systems are saturated, and the SSAM (Stephens *et al.*, 1994) aids in the rapid identification of precursory long-period (LP) volcanic seismicity.

Both the RSAM and SSAM systems – together with associated software for collecting, processing, and displaying in near-time their data along with non-seismic monitoring data – form the core of an integrated "mobile volcano observatory" successfully used at Mount St. Helens, Redoubt Volcano, and Mt. Pinatubo (Murray *et al.*, 1996).

The increasing deployment of networks of broadband seismometers (e.g., Dawson *et al.*, 1998) at some volcanoes is now enabling the complete recording and sophisticated analysis of LP and VLP (very long-period, $>10$ s) seismic signals. Such research is expected to quantify our fundamental knowledge of magma transport, source mechanism of volcanic tremor, magma or fluid pathways, and other attributes of magmatic systems. A refined understanding of the source mechanisms producing long-period volcanic seismicity, which almost invariably precedes eruptive activity, is critical to improved short-term forecasts (Chouet *et al.*, 1994; Chouet, 1996a, 1996b; McNutt, 1996).

It has long been recognized that optimum volcano monitoring is achieved by a combination of techniques rather than sole reliance on seismic monitoring, or some other single technique. In this regard, in recent decades geodetic monitoring methods and networks have also improved significantly (Murray *et al.*, 1995), with increasing use of satellite-based technology such as the global positioning system (GPS) and synthetic aperture radar interferometry (InSAR) (Massonnet *et al.*, 1995; Nunnari and Puglisi, 1995, Sigmundsson *et al.*, 1995; Francis *et al.*, 1996b; Lanari *et al.*, 1998; Wicks *et al.*, 1998). While GPS and InSAR monitoring of ground deformation has several major advantages over most ground-based systems (e.g., immense ground coverage, an areal vs. point perspective, "all weather"), at present these techniques are still not cost-effective, require much computer data manipulation, and generally cannot acquire and process data continuously or quickly enough to be useful during a rapidly evolving volcanic crisis. However, relatively low-cost, prototype continuous GPS monitoring methods are being developed and tested in the United States, Italy, and elsewhere.

Monitoring of volcanic gases has also advanced significantly, especially with improvements in the remote techniques for measurement of sulfur dioxide ($SO_2$) and carbon dioxide ($CO_2$) – the two most abundant gases after water vapor – that can be applied in the field rather than in the laboratory. The correlation spectrometer (COSPEC) is being increasingly used, in a ground-based or air-borne mode, at volcanoes for measurement of $SO_2$ emission (Andres and Rose, 1995). Another recently developed remote technique showing promise involves Fourier transform infrared (FTIR) spectroscopy for the measurement of $SO_2$, $SO_2/HCl$ ratio, and $SiF_4$ in volcanic plumes and has been successfully used at Kilauea, Etna, Vulcano, Asama, and Unzen volcanoes (see Francis *et al.*, 1996a; McGee and Gerlach, 1998; and references therein). A highly useful satellite-based method for the measurement of $SO_2$ release of recent eruptions utilizes the total ozone mapping spectrometer (TOMS), an instrument originally designed for ozone measurement but subsequently discovered to be able also to measure $SO_2$ in volcanic plumes that reach the upper atmosphere (Krueger *et al.*, 1990; Bluth *et al.*, 1993). Particularly important have been the advances in the measurement of, and acquisition of time-series data on, $SO_2$ and $CO_2$ flux (i.e., volcanic degassing) at many volcanoes in repose or activity (e.g., Mt. Etna, Strómboli, Vulcano, Popocatépetl, Mount St. Helens, Kilauea, Long Valley Caldera, Grimsvötn), using the COSPEC, TOMS, and infrared $CO_2$ analyzers (e.g., LI-COR) (Gerlach *et al.*, 1997). Such studies furnish additional insights into and constraints on intrusion of magma to shallow levels and other dynamic processes within volcanic systems. Measurement of helium emission from fumaroles also provide a means of determining the magmatic component in volcanic gas (e.g., Sorey *et al.*, 1998).

As mentioned previously, socio-economic and demographic realities preclude the utilization of prudent land-use planning and zoning to mitigate volcano hazards as a viable strategy for most active volcanic areas of the world. Thus, the regular monitoring of high-risk volcanoes in already densely populated regions – to enable early detection of volcanic unrest and, if the data warrant,

to alert officials of possible activity – constitutes the only practical strategy to reduce volcano risk. In such cases, the primary focus should be to save human lives, by the timely issuance of monitoring-based warnings.

## Effective communication and emergency-preparedness plans

The need for effective communications for the successful mitigation of volcano hazards has been emphasized in many previous studies (e.g., Fiske, 1984, 1988; Peterson, 1986, 1988, 1996; Tilling, 1989a, 1989b; Voight, 1990, 1996; Peterson and Tilling, 1993; Tilling, 1995). The critical importance of effective communications, or lack thereof, will become apparent in the two case histories discussed below, but the essence is:

> Monitoring data or eruption forecasts, no matter how timely or precise, have little practical consequence unless they can be communicated effectively to, and acted upon in a timely manner by, emergency-management authorities.
>
> (Tilling, 1995, p. 396a)

## Two instructive recent case histories

The case histories selected for review here were chosen because they illustrate two very different outcomes of recent scientific and public response to a volcanic crisis: the disastrous outcome at Nevado del Ruiz represents a tragic failure in hazards mitigation, whereas the outcome at Mt. Pinatubo in 1991 is widely hailed as a genuine volcanologic success story, even if luck played a key role in it (Punongbayan and Newhall, 1995; Newhall and Punongbayan, 1996a). The discussion below draws liberally, in places verbatim, from a review by Tilling (1995), amended and updated as appropriate in light of the recent comprehensive studies of Voight (1996) for Ruiz, and of Newhall and Punongbayan (1996a, 1996b) for Pinatubo and works cited therein.

### Nevado del Ruiz, Colombia, 1985: A tragic failure in hazards mitigation

At 15:06 h of November 13, 1985, a brief phreatic burst, lasting about a quarter hour, occurred at the summit of Volcán Nevado del Ruiz, a 5389-m-high, glacier-capped mountain that is the northernmost active volcano in the Andes (Williams, 1990a, 1990b). Volcanic tremor then persisted through 21:08 h, when a small-volume ($\sim$0.02 km$^3$) magmatic eruption began and several pyroclastic flows were emplaced in rapid succession onto the glacier surface. The hot ejecta melted and mixed with snow and ice to form several mudflows that swept down the steep, narrow valleys draining the volcano. These mudflows destroyed or buried everything in their paths, killing about 23 000 people, the vast majority in Armero, an agricultural community in the valley of Río Lagunillas. The mudflows also injured another 5000 people, left about 7700 homeless, and caused over US$7.7 billion in economic loss, representing about 20% of Colombia's GNP in 1985 (Voight, 1996) (Table 2.1). The 1985 Ruiz catastrophe was the worst volcanic disaster in the recorded history of Colombia and the most deadly in the world since the 1902 eruption of Mt. Pelée on the Caribbean island of Martinique.

The 1985 eruption and associated mudflows at Ruiz should not have come as a surprise, because similarly destructive events have occurred in the past (AD 1595, AD 1845) and inundated the same area upon which Armero later developed (Voight, 1990, 1996, Fig. 13). In addition, efforts were made to conduct monitoring studies in response to the precursory signals at Ruiz (felt earthquakes, increased fumarolic activity, phreatic explosions, bursts of volcanic tremor) that were noticed in late November 1984. Due to administrative inertia and logistical problems, it was not until the summer of 1985 that rudimentary seismic monitoring began. However, when a strong phreatic eruption on September 11 produced measurable ash fall at Manizales, the capital of Caldas Province (population 230 000), and several sizeable debris flows, governmental concern was heightened and monitoring increased. By mid October, a monitoring network of five smoke-drum seismometers and four precise levelling ("dry tilt") arrays was established; two electronic tiltmeters operated briefly in early November (Banks et al., 1990). A preliminary volcanic hazard map was issued on October 7, clearly showing Armero to be located in a high hazard zone for mudflows (lahars). A version of this map containing notable errors was published on October 9 in the Bogotá newspaper El Espectador (Voight, 1996, Fig. 5a, c).

Analysis of fumarolic gases in late October indicated substantial magmatic contributions of $CO_2$ and $SO_2$ and the need "to seriously consider the possibility of an impending magmatic eruption." (Barberi et al., 1990, p. 5). Although expanded in the month prior to the November 13, 1985 eruption and catastrophe, monitoring studies still were "too little–too late" to allow any precise forecast of impending eruption. As Hall (1990, p. 114) concluded: "Limited scientific data from a marginal monitoring program, *no baseline data* (italics added), and greatly delayed processing precluded a realistic attempt to understand or predict an eruptive event."

In hindsight, however, it is arguable that, even if it had been possible to make a precise short-term forecast based on more extensive monitoring, the Ruiz disaster could have been averted. Well-documented post-mortems of the Ruiz tragedy (e.g., CERESIS, 1990; Hall, 1990, 1992; Voight, 1990, 1996) have clearly demonstrated that, whatever might have been the inadequacies of the scientific information available prior to the November 13 eruption, the root cause of the disaster was the lack of effective and timely warnings, which in turn stemmed from the lack of effective communications between the scientists, the responsible authorities, the media, and the populations affected. As first emphasized by Voight (1990) and later reiterated (Voight, 1996, p. 764):

The Armero catastrophe was not produced by technological ineffectiveness, nor by an overwhelming eruption, nor by an improbable run of bad luck. The disaster happened because of cumulative human error – by misjudgment, indecision and bureaucratic shortsightedness . . . The national and provincial authorities were unwilling to bear the economic or political costs of early evacuation or a false alarm, and they delayed action to the last possible minute.

Tragically, the authorities' delaying "action to the last possible minute" was too late for 23 000 people.

## Mount Pinatubo, Philippines, 1991: a volcanologic success story

Mount Pinatubo is a dacitic volcanic complex located about 100 km northwest of Manila, on the island of Luzon in the Philippines. Reconnaissance geologic and radiocarbon dating studies in the early 1980s, in part to assess geothermal potential, indicated that its summit dome complex was flanked by extensive fans of young (600–8000 BP (years before present)) pyroclastic and lahar deposits (Pinatubo Volcano Observatory Team, 1991). Because of three radiocarbon ages available in the 1970s (635 ± 80 BP dating the most recent explosive eruption), Pinatubo was reclassified in 1987 from "inactive" to "active," even though it had not erupted in the recorded history of the Philippines (Punongbayan, 1987; PHIVOLCS, 1988).

On April 2, 1991, phreatic explosions began about 1.5 km northwest of the summit, accompanied by earthquake activity and increased steaming at the geothermal vents. A few days later, scientists from the Philippine Institute of Volcanology and Seismology (PHIVOLCS) began to monitor the continuing high seismicity with portable seismographs. Continuing activity prompted on April 7 the evacuation of more than 5000 people living within a 10-km radius of the volcano. A team of scientists from the US Geological Survey's (USGS) Volcano Disaster Assistance Program (VDAP) arrived in late April in response to a request from the Philippine government for assistance; the VDAP is sponsored jointly by the USGS and the Office of Foreign Disaster Assistance of the US Department of State. With increasing concern about the possibility of a large magmatic eruption, monitoring was intensified and included the use of a correlation spectrometer (COSPEC) for measuring $SO_2$ emission and two electronic tiltmeters for measuring ground deformation. A radio-telemetered network of seven seismic stations was also installed by a PHIVOLCS–USGS team, making possible the near real-time monitoring of earthquake location, total seismic energy, RSAM, and SSAM.

At the same time, PHIVOLCS implemented a five-level alert scheme (Table 2.2) that could be easily communicated to, and understood by, emergency-management officials. This scheme, based on qualitative criteria, was not intended to "make predictions," but rather only to reflect "increasing levels of unrest" and "what might occur," to "provide a simple set of steps" for emergency-management officials to "design corresponding response plans" (Punongbayan et al., 1996b, p. 73). On May 23, a hazard-zonation map, prepared from rapid geologic reconnaissance by PHIVOLCS–USGS scientists, was issued to provincial authorities and to the officials of the US Air Force at Clark Air Base, located on the eastern foot of Mt. Pinatubo. During the course of the eruption, the map was used to determine hazard zones, expressed as increasing radii from the volcano, recommended by PHIVOLCS for evacuation (Punongbayan et al., 1996b).

$SO_2$ flux increased ten-fold during the latter part of May, from 500 tons/day to 5000 tons/day and then decreased sharply to only 280 tons/day by June 4 (Pinatubo Volcano Observatory Team, 1991; Daag et al., 1996). This sudden decrease in $SO_2$ emission, together with accompanying increased shallow seismicity, occurrence of volcanic tremor, and recording of deep long-period events beginning in late May (Harlow et al., 1996; White, 1996), was interpreted to indicate upward movement of magma and, ultimately, blockage of escaping gas within the conduit. This development led PHIVOLCS to declare on June 5 an Alert Level 3, which states: "If trend of increasing unrest continues, eruption possible within 2 weeks." (Table 2.2). In essence, an Alert Level 3 constituted a short-term forecast. Intense shallow seismicity and a sharp inflationary tilt during the following two days culminated in the extrusion of a lava dome on June 7, marking the beginning of magmatic activity. The emergence and growth of the dome, combined with a shift of seismic foci to directly beneath the dome and an irregular increase in the amplitude and frequency of occurrence of volcanic tremor, collectively provided the basis for raising the alert level to 4 on June 7, and then to 5 on June 9 (Table 2.2). The increasing activity and attendant elevations of alert level led to a 20-km radius of evacuation, involving an additional 20 000 evacuees from Zambales, Tarlac, and Pampanga Provinces; all aircraft and supporting facilities at Clark Air Base were moved to safer locations elsewhere. On

**Table 2.2** The five-level alert scheme for the 1991 eruption of Mt. Pinatubo, Luzon, Philippines, implemented May 13, 1991

| Alert level[a] | Criteria | Interpretation | Date declared |
|---|---|---|---|
| No alert | Background, quiet | No eruption in foreseeable future | n/a |
| 1 | Low-level seismic, fumarolic, other unrest | Magmatic, tectonic, or hydrothermal disturbance; no eruption imminent | n/a |
| 2 | Moderate level of seismic, other unrest, with positive evidence for involvement of magma | Probable magmatic intrusion; could eventually lead to an eruption | May 13, 1991 |
| 3 | Relatively high and increasing unrest including numerous b-type earthquakes, accelerating ground deformation, increased vigor of fumaroles, gas emission | If trend of increasing unrest continues, eruption possible within 2 weeks | June 5, 1991 |
| 4 | Intense unrest, including harmonic tremor and/or many "long-period" (= low-frequency) earthquakes | Eruption possible within 24 hours | June 7, 1991 17:00 h |
| 5 | Eruption in progress | Eruption in progress | June 9, 1991 17:15 h |

[a] Stand-down procedures: in order to protect against "lull before the storm" phenomena, alert levels will be maintained for the following periods after activity decreases to the next lower level. From level 4 to level 3, wait 1 week; from level 3 to level 2, wait 72 hours.
*Source:* Pinatubo Volcano Observatory Team (1991, Table 1).

June 10, more than 14 500 military personnel and dependents were evacuated.

At 08:51 h on the morning of June 12, the first of several vigorous Plinian eruptions began, producing a 19-km-high ash column. This escalation in activity prompted PHIVOLCS to declare a 40-km radius "danger zone" for the next three days (Punongbayan and Newhall, 1995). A series of strong explosions ensued the next several days, culminating in the climactic eruption that began at 13:42 h (local time) on 15 June and lasted about 9 hours (Wolfe and Hoblitt, 1996). The climactic event was one of largest volcanic eruptions in the twentieth century (Dartevelle *et al.*, 2002). The combined dense-rock-equivalent volume of pyroclastic-flow and tephra deposits was about 3.7–5.3 km$^3$ (Scott *et al.*, 1996; Wolfe and Hoblitt, 1996), and about 17 Mt of $SO_2$ were released into the stratosphere (Gerlach *et al.*, 1996). A new 2.5-km-diameter caldera was formed, centered about 1 km north of the former summit, which was destroyed; the new summit is about 200 m lower. Pyroclastic flows and tephra deposits blanketed an area of more than 100 km$^2$ around the volcano and greatly altered local drainage patterns; the loose debris making up these new deposits provided ample additional solid materials for subsequent lahars. Heavy rains associated with Typhoon Yunya, which passed only 50 km north of Pinatubo on the same day as the climactic eruption, triggered destructive debris flows and also wetted the ash that accumulated on roofs. The thick accumulation of wet ash, together with intense seismicity accompanying the climactic eruption, resulted in extensive roof collapse, the principal cause of eruption-related deaths (~250).

After the climactic eruption, seismicity and tephra emission gradually diminished over the next several months, and the alert level was lowered from 5 to 2 by December 1991. A 10-km radius danger zone was maintained, however. Following the emplacement of a lava dome within the new summit caldera during July–October 1992, eruptive and seismic activity at Pinatubo continued to declined gradually through mid 1994 (Punongbayan et al., 1996b; Wolfe and Hoblitt, 1996). However, during the 1991 and 1992 rainy seasons (typically June–November), the long-term, post-eruption hazards involving secondary lahars became increasingly apparent. Numerous secondary lahars, triggered by rainfall as well as by breakouts of temporary lakes impounded by natural dams of volcanic debris, resulted in widespread sedimentation in towns and covered valuable farmlands. Lahar-related erosion of still-hot volcanic debris caused numerous rootless, secondary explosions and large secondary pyroclastic flows that traveled several kilometers from source. From June 1991 through November 1992, over 8000 dwellings were totally destroyed and about 73 000 were partially damaged, affecting 329 000 families (~2.1 million people), or about one-third of the region's population; the overriding cause of such massive damage and human suffering was the destructive lahar activity during the 1991 and subsequent rainy seasons (Mercado et al., 1996; Newhall and Punongbayan, 1996b).

In retrospect, considering the huge size of the June 15, 1991 Pinatubo eruption and the large number of people affected, the death toll directly related to the climactic eruption was comparatively small, only between 250 and 300 (Punongbayan and Newhall, 1995). The effectiveness of the scientific and emergency-management response to the reawakening of Pinatubo after a 600-year dormancy can be attributed in large part to:

(1) Provision of international assistance to supplement the scientific resources of PHIVOLCS, especially the rapid deployment of a USGS–VDAP mobile observatory (Murray et al., 1996) to initiate near-real-time volcano monitoring. The monitoring data, plus geologic observations, allowed the detection and evaluation of critical shifts in activity, thereby partly compensating for the lack of any previous baseline monitoring data, and made possible reliable forecasts during the course of the eruption.

(2) Preparation of a hazards map – based on reconnaissance fieldwork in a short time – accurately identified the zones around the volcano most vulnerable to pyroclastic flow and lahar hazards. Geologic studies of prehistoric volcanic deposits also indicated that Pinatubo had produced numerous large-volume magmatic eruptions in its history, providing diagnostic evidence in assessing volcano hazards.

(3) PHIVOLCS' very effective use of a simple alert-level scheme (Table 2.2), together with near-real-time analysis and interpretation of geologic and monitoring data, was critical in prompting civil authorities to order timely evacuations of areas at risk and in serving as an understandable framework for communicating volcano information to the officials, media, and the general public.

(4) The doggedness and aggressive campaigning of the PHIVOLCS–USGS scientists to educate the local authorities and the people at risk, and to persist in the face of indifference, denial, skepticism, and (or) rebuffs (see Punongbayan and Newhall (1995) and Newhall and Punongbayan (1996a) for discussion). A major tool in this aggressive campaign was the use, during the weeks building up to the climactic eruption, of a still-unfinished version of Understanding Volcanic Hazards, a 30-minute videotape (IAVCEI, 1995) that vividly illustrates the devastating impacts of volcanic hazards. It was produced for the International Association of Volcanology and Chemistry of the Earth's Interior (IAVCEI) by the late volcanologist Maurice Krafft. Sadly, both Maurice and his wife (Katia Krafft) were killed while filming the eruption at Unzen Volcano, Japan, on June 3, 1991, while his work was being used so effectively at Pinatubo. The video was instrumental in convincing the government authorities and the populace that Pinatubo posed a very real danger, and, consequently, official evacuation orders were heeded with little or no reluctance.

Finally, as cautioned by Newhall and Punong-bayan (1996a, p. 807), despite the positive outcome at Pinatubo, "a suite of problems – including skepticism, bureaucratic and logistic difficulties, and short warning from the volcano – practically eliminated the margin of safety. Ultimately, luck and last-minute efforts played key roles, but *warnings and response could easily have been too late*." (Italics added).

## Lessons and trends

From this brief review and numerous other publications addressing the broad topic of monitoring and mitigation of volcano hazards, many specific "lessons," observations, and recommendations have been emphasized. Not surprisingly, many of these statements are variations on several similar but paramount themes. As we enter the twenty-first century, it seems appropriate to reiterate some of the main lessons learned from recent volcanic crises and disasters of the previous century, in the hope that the volcanologic community and civil authorities can better apply these lessons in responding to volcanic emergencies. Outlined below are some prime points of general agreement (not in any order of priority) that emerge from the collective experience in confronting volcano hazards to date:

• Because the eruption frequency (50 to 60 volcanoes active each year on average) is unlikely to decrease, the global problem of reducing volcano risk can only worsen with time, with continued growth in world population, commerce and economic development, urbanization, and air traffic. While the industrialized countries with active volcanoes (e.g., Japan, the United States, Italy, France, Iceland, and New Zealand) must also confront this reality, this problem is more acute for developing countries, which host most of the world's high-risk volcanoes but possess inadequate resources for their study and monitoring.

• Fundamental geologic and geochronologic data for reconstruction of eruptive histories are lacking for most of the world's active and potentially active volcanoes. Because such information constitutes the basis for hazards assessments, hazards-zonation maps, and long-term forecasts of potential activity we must promote and accelerate such basic geoscience studies.

• "Too few active and potentially active volcanoes of the world are being monitored adequately, and many are not monitored at all or only minimally." (Tilling, 1995, p. 395; see also Scarpa and Gasparini, 1996, Table 1.) As discussed previously, because the option of hazards-assessment-based land-use planning to reduce volcano risk is impractical for many volcanic areas, volcano monitoring and the capability to make short-term forecasts provide the only means to mitigate hazards by providing the essential data to officials to make well-informed decisions about evacuations and other countermeasures.

• Therefore, we must monitor more of the world's high-risk volcanoes in densely populated areas, to begin acquiring monitoring data, ideally *before* a volcanic crisis strikes. A long-term approach to increase monitoring worldwide would be to establish well-equipped and well-staffed permanent observatories at many more high-risk volcanoes, but such an endeavor would be costly and is unlikely to be supported any time soon by national or international funding agencies, given the current world economy. However, as an interim solution and as demonstrated by the Pinatubo case history, under favorable circumstances, the rapid deployment of a mobile observatory (Murray *et al.*, 1996) can be highly effective and achieve successful mitigation.

• The longer the pre-crisis monitoring period, the earlier the reliable detection of departure of a dormant volcanic system from its baseline or "normal" behavior. Early detection allows more lead time for augmented monitoring and for scientists to work with – and to establish credibility and rapport with – public officials and the public in the development of emergency-response plans, hopefully under non-crisis conditions.

• Despite recent advances in volcano monitoring, successful forecasting of explosive eruptions, with rare exception (e.g., Mt. Pinatubo in

1991), still frustrate and elude scientists. With present state-of-the-art in volcanology, "Monitoring data, no matter how good or complete, do not guarantee 'successful' eruption forecasts" (Tilling, 1995, p. 396). There is an urgent need in this century to conceptualize and develop new approaches to eruption prediction, apart from the current empirical approach largely based on pattern recognition in monitoring data. In this regard, the recent research (e.g., Chouet, 1996a, 1996b) on long-period seismicity produced by magma movement or hydrothermal-pressurization phenomena, coupled with a refined petrological and geochemical understanding of the behavior of magmatic and hydrothermal fluids before and during eruption, may hold promise.

- As emphasized and recommended in numerous works (e.g., Tilling, 1989a; Tilling and Lipman, 1993; Punongbayan and Newhall, 1995; Newhall and Punongbayan, 1996a, 1996b; Voight, 1996), there needs to be more effective international cooperation in responding to volcanic crises. The Pinatubo case history clearly demonstrates the effectiveness of such collaboration. There are simply too many high-risk volcanoes and too few trained scientists with first-hand experience or knowledge of eruptions in each nation. Thus, it is vital to pool resources and to share and exchange data and knowledge; this should become increasingly more convenient with advances in telecommunications (e.g., e-mail, Internet). In particular, scientists in the developed countries must become more involved in bilateral or international programs to work with their counterparts in the developing countries, many of which lack self-sufficiency in volcanology and (or) the necessary economic resources to conduct monitoring and other hazards-mitigation studies.

- Continuing the trend already in progress, volcano-monitoring systems in this century will increasingly employ remote techniques that will acquire, process, and display the data in real-time or near real-time. Satellite-based systems for monitoring ground deformation (e.g., GPS, InSAR) are likely to increasingly complement and, in some cases, supplant the conventional ground-based techniques currently

used. Continuous monitoring of $SO_2$, $CO_2$, and possibly other volcanic gases will almost certainly also become routine. However, the concept of "keeping monitoring as simple as practical" (Swanson, 1992) should not be dismissed, because in some situations simple methods might be most cost-effective and applicable. Certainly, simple monitoring approaches are better than no monitoring at all.

- Scientists must redouble efforts to establish and maintain effective communications with the emergency-management authorities, the media, and the populations at risk from volcano hazards. A prerequisite of effective communications, credibility, and mutual trust between all involved parties is absolutely critical in the development of emergency-preparedness plans, based on the data obtained from hazards assessments and a volcano-monitoring program as well as the consideration of logistical, socio-economic, political, cultural, and other human factors of the jurisdiction in which the volcano(es) is (are) situated. In working with emergency-response officials, the media, and the public, scientists must learn patience and be diligent and persistent in overcoming indifference and skepticism of many community leaders and citizens.

- As was done successfully at Pinatubo, scientists need to work actively to launch broadly based public-education campaigns focused on people at risk in hazardous areas, the civil authorities, and the emergency-response communities. The themes of "Prevention begins with information" and "Building a culture of prevention" are resoundingly sounded in the press kit for the 1998 World Disaster Reduction Campaign of the United Nations' Secretariat for the International Decade for Natural Disaster Reduction (1990–2000).

We must never forget the many victims of the 1985 Ruiz catastrophe, who perished needlessly because of failed communications and the lack of an emergency-preparedness plan. By the same token, we also must remember that many thousands of lives at Pinatubo were saved because the scientists effectively communicated the findings and implications of their studies and convinced the officials to order timely evacuations, guided by a simple but effective alert-level scheme. There is no assurance that the successful outcome at Pinatubo in 1991 can be routinely duplicated for future volcanic crises, but it certainly defines a worthy goal.

## Note added in proof

By May 2003, the growing lava dome at Soufrière Hills Volcano, Montserrat, attained its maximum size. On July 12–13, the dome failed catastrophically and resulted in the largest dome-collapse event (>12 million m³) since the onset of eruptive activity in July 1995. Moreover, large explosions accompanied the collapse, producing voluminous pyroclastic flows and ash clouds as high as 15 km. Yet, despite the huge eruption size and associated hazardous processes, there were no deaths or reported injuries. This is because scientists at the Montserrat Volcano Observatory – anticipating that such a large collapse was the most likely scenario based on volcano-monitoring data – weeks before had advised government authorities to enforce rigorously the Exclusion Zone around the volcano. Intermittent activity was still continuing at the volcano as this book went to press.

## Acknowledgments

This brief article builds on the large and growing scientific literature on volcanic phenomena and, more recently, on volcano hazards and their mitigation. Many of the papers cited herein are works published by many good friends and colleagues, within and outside the United States, with whom I have worked or discussed topics of mutual interest. Because my paper draws so heavily on their comprehensive reviews on volcano hazards, I would to express my appreciation to Russell J. Blong (Natural Hazards Research Centre, Macquarie University, North Ryde, NSW, Australia) and to William E. Scott (US Geological Survey, Cascades Volcano Observatory, Vancouver, Washington) for allowing me to freely excerpt, paraphrase, or distill their texts. An earlier draft of this paper was critically reviewed by Steven R. Brantley (US Geological Survey, Hawaiian Volcano Observatory), Edward W. Wolfe (US Geological Survey, Cascades Volcano Observatory), and co-editor Joan Martí. Their thoughtful, incisive comments and suggestions greatly influenced my preparation of the final manuscript. I owe them hearty thanks for their help, but I remain solely responsible for any errors, illogic, or misplaced emphasis in the paper.

## References

Anderson, T. and Flett, J. S. 1903. Report on the eruptions of the Soufrière in St. Vincent, and on a visit to Montagne Pelée in Martinique. *Philosophical Transactions of the Royal Society of London, Series A*, **200**, 353–553.

Andres, R. J. and Rose, W. I. 1995. Remote sensing spectroscopy of volcanic plumes and clouds. In W. McGuire, C. Kilburn, and J. Murray (eds.) *Monitoring Active Volcanoes: Strategies, Procedures and Techniques*. London, UCL Press, pp. 301–314.

Aspinall, W. P., Lynch, L. L., Robertson, R. E. A., *et al.* (eds.) 1998. Special section on the Soufrière Hills eruption, Montserrat, British West Indies, Part 1. *Geophysical Research Letters*, **25**, 3387–3440.

Banks, N. G., Carvajal, C., Mora, H., *et al.* 1990. Deformation monitoring at Nevado del Ruiz, Colombia, October 1985–March 1988. *Journal of Volcanology and Geothermal Research*, **41**, 269–295.

Banks, N. G., Tilling, R. I., Harlow, D. H., *et al.* 1989. Volcano monitoring and short-term forecasts. In R. I. Tilling (ed.) *A Short Course in Geology*, vol. 1, *Volcanic Hazards*. Washington, DC, American Geophysical Union, pp. 51–80.

Barberi, F., Martini, M., and Rosi, M. 1990. Nevado del Ruiz volcano (Colombia): pre-eruption observations and the November 13, 1985 catastrophe. *Journal of Volcanology and Geothermal Research*, **42**, 1–12.

Barsczus, H. G., Filmer, P. M., and Desonie, D. 1992. Cataclysmic collapse and mass wasting processes in the Marquesas. *Eos, Transactions of the American Geophysical Union*, **73** (14), 313.

Belousov, A. 1996. Deposits of the 30 March 1956 directed blast at Bezymianny volcano, Kamchatka, Russia. *Bulletin of Volcanology*, **57**, 649–662.

Blong, R. J. 1984. *Volcanic Hazards: A Sourcebook on the Effects of Eruptions*. San Diego, CA, Academic Press.

Blong, R. J. and McKee, C. 1995. *The Rabaul Eruption 1994: Destruction of a Town*. North Ryde, NSW, Natural Hazards Research Centre, Macquarie University.

Bluth, G. J. S., Schnetzler, C. C., Kreuger, A. J., *et al.* 1993. The contribution of explosive volcanism to global atmospheric sulfur dioxide concentrations. *Nature*, **366**, 327–329.

Casadevall, T. J. 1992. Volcanic hazards and aviation safety: lessons of the past decade. *Federal Aviation Administration Aviation Safety Journal*, **2** (3), 9–17.

1994a. The 1989–1990 eruptions of Redoubt Volcano, Alaska: impacts on aircraft operations. *Journal of Volcanology and Geothermal Research*, **62**, 301–316.

(ed.) 1994b. *Volcanic Ash and Aviation Safety*, Proceedings of the 1st International Symposium on Volcanic Ash and Aviation Safety, US Geological Survey Bulletin no. 2047. Washington, DC, US Government Printing Office.

Casadevall, T. J., Delos Reyes, P. J., and Schneider, D. J. 1996. The 1991 Pinatubo eruptions and their effects on aircraft operations. In C. G. Newhall and R. S. Punongbayan (eds.) *Fire and Mud: Eruption and Lahars of Mount Pinatubo, Philippines*. Quezon City, Philippines, Philippine Institute of Volcanology and Seismology, pp. 1071–1088.

Casadevall, T. J., Thompson, T. B., and Fox, T. 1999. *World Map of Volcanoes and Principal Aeronautical Features*, US Geological Survey Investigations Series Map no. I-2700. Washington, DC, US Government Printing Office.

CERESIS 1990. Riesgo volcánico: evaluación y mitigación en América Latina: aspectos sociales, institucionales y científicos. Lima, Perú, Centro Regional de Sismología para América del Sur (CERESIS).

Chester, D. 1993. *Volcanoes and Society*. London, Edward Arnold.

Chouet, B. A. 1996a. Long-period volcano seismicity: its source and use in eruption forecasting. *Nature*, **380**, 309–316.

1996b. New methods and future trends in seismological volcano monitoring. In R. Scarpa and R. I. Tilling (eds.) *Monitoring and Mitigation of Volcano Hazards*. Heidelberg, Germany, Springer-Verlag, pp. 23–97.

Chouet, B. A., Page, R. A., Stephens, C. D., *et al.* 1994. Precursory swarms of long-period events at Redoubt Volcano (1989–1990), Alaska: their origin and use as a forecasting tool. *Journal of Volcanology and Geothermal Research*, **62**, 95–135.

Christiansen, R. L. 1984. Yellowstone magmatic evolution: its bearing on understanding large-volume explosive volcanism. In Geophysics Study Committee (National Research Council) *Explosive Volcanism: Inception, Evolution and Hazards*. Washington, DC, National Academic Press, pp. 84–95.

Christiansen, R. L. and Peterson, D. W. 1981. Chronology of the 1980 eruptive activity. In P. W. Lipman and D. R. Mullineaux (eds.) *The 1980 Eruptions of Mount St. Helens, Washington*, US Geological Survey Professional Paper no. 1250. Washington, DC, US Government Printing Office, pp. 17–67.

Civetta, L., Gasparini, P., Luongo, G., *et al.* (eds.) 1974. *Developments in Solid Earth Geophysics*, vol. 6, *Physical Volcanology*. Amsterdam, The Netherlands, Elsevier.

COMVOL 1975. *Mayon Volcano*. Manila, Philippines, Commission on Volcanology.

Costa, J. E. and Schuster, R. L. 1988. The formation and failure of natural dams. *Geological Society of America Bulletin*, **100**, 1054–1068.

Crandell, D. R. 1989. *Gigantic Debris Avalanche of Pleistocene Age from Ancestral Mount Shasta Volcano, California, and Debris-Avalanche Hazard Zonation*, US Geological Survey Bulletin no. 1861. Washington, DC, US Government Printing Office.

Crandell, D. R., Mullineaux, D. R., and Rubin, M. 1975. Mount St. Helens: recent and future behavior. *Science*, **187** (4175), 438–441.

Crandell, D. R., Booth, B., Kusumadinata, K., *et al.* 1984a. *Source-Book for Volcanic-Hazard Zonation*. Paris, France, UNESCO.

Crandell, D. R., Miller, C. D., Glicken, H. X., *et al.* 1984b. Catastrophic debris avalanche from ancestral Mount Shasta volcano, California. *Geology*, **12**, 143–146.

Daag, A. S., Tubianosa, B. S., Newhall, C. G., *et al.* 1996. Monitoring sulfur dioxide emission at Mount Pinatubo. In C. G. Newhall and R. S. Punongbayan (eds.) *Fire and Mud: Eruption and Lahars of Mount Pinatubo, Philippines*. Quezon City, Philippines, Philippine Institute of Volcanology and Seismology, pp. 409–414.

Dartevelle, S., Ernst, G. G. J., Stix, J., *et al.* 2002. Origin of the Mount Pinatubo climactic eruption cloud: implications for volcanic hazards and atmospheric impacts. *Geology*, **30**, 663–666.

Dawson, P. B., Dietel, C., Chouet, B. A., *et al.* 1998. *A Digitally Telemetered Broadband Seismic Network at Kilauea Volcano, Hawaii*, US Geological Survey

Open-File Report no. 98–108. Washington, DC, US Government Printing Office.

Decker, R. and Decker, B. 1997. *Volcanoes*, 3rd edn. New York, W. H. Freeman.

Druitt, T. H. and Kokelaar, B. P. (eds.) 2002. *The Eruption of Soufrière Hills Volcano Montserrat from 1995 to 1999.* Geological Society Memoir no. 21.

Dunn, M. G. and Wade, D. P. 1994. Influence of volcanic ash clouds on gas turbine engines. In T. J. Casadevall (ed.) *Volcanic Ash and Aviation Safety*, Proceedings of the 1st International Symposium on Volcanic Ash and Aviation Safety, US Geological Survey Bulletin no. 2047. Washington, DC, US Government Printing Office, pp. 107–117.

Dvorak, J. J. and Dzurisin, D. 1997. Volcano geodesy: the search for magma reservoirs and the formation of eruptive vents. *Reviews of Geophysics*, **35**, 343–384.

Endo, E. T. and Murray, T. L. 1991. Real-time seismic amplitude measurement (RSAM): a volcano monitoring and prediction tool. *Bulletin of Volcanology*, **53**, 533–545.

Ewert, J. W. and Swanson, D. A. (eds.) 1992. *Monitoring Volcanoes: Techniques and Strategies Used by the Staff of the Cascades Volcano Observatory, 1980–90*, US Geological Survey Bulletin no. 1966. Washington, DC, US Government Printing Office. [A Spanish translation of this volume, *Vigilando volcanes: Técnicas y estrategias empleadas por el personal del Observatorio Vulcanológico Cascades, 1980–90*, was published in 1993 in cooperation with the Office of Foreign Disaster Assistance (OFDA), US Agency for International Development; it is available upon request from the Scientist-in-Charge, Cascades Volcano Observatory, US Geological Survey, 5400 MacArthur Blvd., Vancouver, WA 98661, USA.]

Farrar, C. D., Sorey, M. L., Evans, W. C., *et al.* 1995. Forest-killing diffuse $CO_2$ emissions at Mammoth Mountain as a sign of magmatic unrest. *Nature*, **376**, 675–678.

Feuillard, M., Allegre, C. J., Brandeis, G., *et al.* 1983. The 1975–1977 crisis of la Soufrière de Guadeloupe (F.W.I.): a still-born magmatic eruption. *Journal of Volcanology and Geothermal Research*, **16**, 317–334.

Fisher, R. V., Smith, A. L., and Roobol, M. J. 1980. Destruction of St. Pierre, Martinique by ash cloud surges, May 8 and 20, 1902. *Geology*, **8**, 472–476.

Fiske, R. S. 1984. Volcanologists, journalists, and the concerned local public: a tale of two crises in the eastern Caribbean. In Geophysics Study Committee (National Research Council) *Explosive Volcanism: Inception, Evolution, and Hazards.* Washington, DC, National Academy Press, pp. 170–176.

1988. Volcanoes and society: challenges of coexistence. *Proceedings of the Kagoshima International Conference on Volcanoes*, Kagoshima Prefectural Government and the National Institute for Research Advancement (NIRA), Japan, pp. 14–21.

Fournier d'Albe, E. M. 1979. Objectives of volcanic monitoring and prediction, *Journal of Geological Society of London*, **136**, 321–326.

Francis, P. W., Chaffin, C., Maciejewski, A., *et al.* 1996a. Remote determination of $SiF_4$ in volcanic plumes: a new tool for volcano monitoring. *Geophysical Research Letters*, **23**, 249–252.

Francis, P. W., Wadge, G., and Mouginis-Mark, P. J. 1996b. Satellite monitoring of volcanoes. In R. Scarpa and R. I. Tilling (eds.) *Monitoring and Mitigation of Volcano Hazards*. Heidelberg, Germany, Springer-Verlag, pp. 257–298.

Gerlach, T. J., Westrich, H. R., and Symonds, R. B. 1996. Preeruption vapor in magma of the climactic Mount Pinatubo eruption: source of the giant stratospheric sulfur dioxide cloud. In C. G. Newhall and R. S. Punongbayan (eds.) *Fire and Mud: Eruption and Lahars of Mount Pinatubo, Philippines*. Quezon City, Philippines, Philippine Institute of Volcanology and Seismology, pp. 415–433.

Gerlach, T. J., Delgado, H., McGee, K. A., *et al.* 1997. Application of the LI-COR $CO_2$ analyzer to volcanic plumes: a case study, volcán Popocatépetl, Mexico, June 7 and 10, 1995. *Journal of Geophysical Research*, **102**, 8005–8019.

Giggenbach, W. F. 1996. Chemical composition of volcanic gases. In R. Scarpa and R. I. Tilling (eds.) *Monitoring and Mitigation of Volcano Hazards*. Heidelberg, Germany, Springer-Verlag, pp. 221–256.

Glicken, H. 1998. Rockslide-debris avalanche of May 18, 1980, Mount St. Helens Volcano, Washington. *Bulletin of Geological Society of Japan*, **49**, 55–106.

Gorshkov, G. S. 1959. Giant eruption of the volcano Bezymianny. *Bulletin Volcanologique*, **20**, 76–109.

Government of Montserrat and Her Majesty's Government 1998. *Sustainable Development Plan: Montserrat Social and Economic Recovery Programme – A Path to Sustainable Development.* HGM Office of the Governor, Olveston, Montserrat, Government of Montserrat and Her Majesty's Government.

Gudmundsson, M. T., Sigmundsson, R., and Björnsson, B. 1997. Ice-volcano interaction of the 1996 Gjálp subglacial eruption, Vatnajökull, Iceland. *Nature*, **389**, 954–957.

Hall, M. L. 1990. Chronology of the principal scientific and governmental actions leading up to the November 13, 1985 eruption of Nevado del Ruiz, Colombia. *Journal of Volcanology and Geothermal Research*, **42**, 101–115.

—— 1992. The 1985 Nevado del Ruiz eruption: scientific, social, and governmental responses and interaction before the event. In G. J. H. McCall, D. J. C. Laming, and S. C. Scott, *Geohazards: Natural and Man-Made*. London, Chapman and Hall, pp. 43–52.

Harlow, D. H., Power, J. A., Laguerta, E. P., *et al.* 1996. Precursory seismicity and forecasting of the June 15, 1991, eruption of Mount Pinatubo. In C. G. Newhall and R. S. Punongbayan (eds.) *Fire and Mud: Eruption and Lahars of Mount Pinatubo, Philippines*. Quezon City, Philippines, Philippine Institute of Volcanology and Seismology, pp. 285–305.

Heilprin, A. 1903. *Mount Pelée and the Tragedy of Martinique*. Philadelphia, PA, J. B. Lippincott.

Heliker, C. C., Mangan, M. T., Mattox, T. N., *et al.* 1998. The character of long-term eruptions: inferences from episodes 50–53 of the Pu'o 'O'o – Kupaianaha eruption of Kilauea Volcano. *Bulletin of Volcanology*, **59**, 381–393.

Hill, D. P. 1984. Monitoring unrest in a large silicic caldera, the Long Valley–Inyo Craters Volcanic Complex in east-central California. *Bulletin Volcanologique*, **47**, 371–396.

Hoblitt, R. P., Miller, C. D., and Vallance, J. W., 1981. Origin and stratigraphy of the deposit produced by the May 18 directed blast. In P. W. Lipman and D. R. Mullineaux (eds.) *The 1980 Eruptions of Mount St. Helens, Washington*, US Geological Survey Professional Paper no. 1250. Washington, DC, US Government Printing Office, pp. 401–419.

Holcomb, R. T. and Searle, R. C. 1991. Large landslides from oceanic volcanoes. *Marine Geotechnology*, **10**, 19–32.

Houghton, B. F., Latter, J. H., and Hackett, W. R. 1987. Volcanic hazard assessment for Ruapehu composite volcano, Taupo volcanic zone, New Zealand. *Bulletin of Volcanology*, **49**, 737–751.

IAVCEI 1995. *Understanding Volcanic Hazards*: a 30-minute videotape (copyrighted), International Association of Volcanology and Chemistry of the Earth's Interior (IAVCEI), produced by the (late) Maurice Krafft. Available from: Northwest Interpretive Association, 3029 Spirit Lake Highway, Castle Rock, Washington 98611, USA.

Izett, G. A., Obradovich, J. D., and Mehnert, H. H. 1988. *The Bishop Ash Bed (middle Pleistocene) and some Older (Pliocene and Pleistocene) Chemically and Mineralogically Similar Ash Beds in California, Nevada, and Utah*, US Geological Survey Bulletin no. 1675. Washington, DC, US Government Printing Office.

Jackson, D. B., Kauahikaua, J., and Zablocki, C. J. 1985. Resistivity monitoring of an active volcano using the controlled-source electromagnetic technique, Kilauea, Hawaii. *Journal of Volcanology and Geothermal Research*, **90**, 545–555.

Janda, R. J., Scott, K. M., Nolan, K. M., *et al.* 1981. Lahar movement, effects, and deposits. In P. W. Lipman and D. R. Mullineaux (eds.) *The 1980 Eruptions of Mount St. Helens, Washington*, US Geological Survey Professional Paper no. 1250. Washington, DC, US Government Printing Office, pp. 461–478.

Jónsson, P., Sigurdsson, O., Snorrason, A., *et al.* 1998. Course of events of the jökulhlaup on Skeidarársandur, Iceland, in November 1996. Abstracts Volume, *15th International Sedimentological Congress*, Alicante, Spain, 12–17 April, pp. 456–457.

Kagoshima Conference 1988. *Proceedings of the Kagoshima International Conference on Volcanoes, 19-23 July 1988*. Kagoshima, Japan, Kagoshima Prefectural Government and the National Institute for Research Advancement (NIRA).

Kieffer, S. W. 1981. Fluid dynamics of the May 18 blast at Mount St. Helens. In P. W. Lipman and D. R. Mullineaux (eds.) *The 1980 Eruptions of Mount St. Helens, Washington*, US Geological Survey Professional Paper no. 1250. Washington, DC, US Government Printing Office, pp. 379–400.

Kling, G. W., Clark, M. A., Compton, H. R., *et al.* 1987. The 1986 Lake Nyos gas disaster in Cameroon, west Africa. *Science*, **236**, 169–175.

Krueger, A. J., Walter, L. S., Schnetzler, C. C., *et al.* 1990. TOMS measurement of the sulphur dioxide emitted during the 1985 Nevado del Ruiz eruptions. *Journal of Volcanology and Geothermal Research*, **41**, 7–15.

Lanari, R., Lundgren, P., and Sansosti, E. 1998. Dynamic deformation of Etna volcano observed by satellite radar interferometry. *Geophysical Research Letters*, **25**, 1541–1548.

Lander, J. F. and Lockridge, P. A. 1989. *United States Tsunamis (including United States Possessions)*

*1690–1988*, National Oceanic and Atmospheric Administration Publication no. 41–2. Washington, DC, US Government Printing Office.

Latter, J. H. 1981. Tsunamis of volcanic origin. *Bulletin Volcanologique*, **44**, 467–490.

(ed.) 1988. *Volcanic Hazards: Assessment and Monitoring*, IAVCEI Proceedings in Volcanology vol. 1. Heidelberg, Germany, Springer-Verlag.

Lee, W. H. K. (ed.) 1989. *Toolbox for Seismic Data Acquisition, Processing, and Analysis*, International Association of Seismology and Physics of the Earth's Interior (IASPEI) Software Library. El Cerrito, CA, Seismological Society of America, **1**, 21–46.

Le Guern, F., Tazieff, H., and Faivre-Pierret, R. 1982. An example of health hazards, people killed by gas during a phreatic eruption: Dieng Plateau (Java, Indonesia), February 19, 1979. *Bulletin Volcanologique*, **45**, 153–156.

Lipman, P. W. and Mullineaux, D. R. (eds.) 1981. *The 1980 Eruptions of Mount St. Helens, Washington*, US Geological Survey Professional Paper no. 1250. Washington, DC, US Government Printing Office.

Lipman, P. W., Moore, J. G., and Swanson, D. A. 1981. Bulging of the north flank before the May 18 eruption: geodetic data. In P. W. Lipman and D. R. Mullineaux (eds.) *The 1980 Eruptions of Mount St. Helens, Washington*, US Geological Survey Professional Paper no. 1250. Washington, DC, US Government Printing Office, pp. 143–155.

Lockwood, J. P., Costa, J. E., Tuttle, M. L., *et al.* 1988. The potential for catastrophic dam failure at Lake Nyos maar, Cameroon. *Bulletin of Volcanology*, **50**, 340–349.

Macdonald, G. A. 1962. The 1959 and 1960 eruptions of Kilauea Volcano, Hawaii, and the construction of walls to restrict the spread of the lava flows. *Bulletin Volcanologique*, **24**, 249–294.

1972. *Volcanoes*. Englewood Cliffs, NJ, Prentice-Hall.

Massonnet, D., Briole, P., and Arnaud, A. 1995. Deflation of Mount Etna monitored by spaceborne radar interferometry. *Nature*, **75**, 567–570.

Mattox, T. N., Heliker, C., Kauahikaua, J., *et al.* 1993. Development of the 1990 Kalapana Flow Field, Kilauea Volcano, Hawaii. *Bulletin of Volcanology*, **55**, 407–413.

McGee, K. A. and Gerlach, T. M. 1998. Airborne volcanic plume measurements using a FTIR spectrometer, Kilauea volcano, Hawaii. *Geophysical Research Letters*, **25**, 615–618.

McGuire, W. J. 1995a. Monitoring active volcanoes: an introduction. In W. J. McGuire, C. Kilburn, and J. Murray (eds.) *Monitoring Active Volcanoes: Strategies, Procedures and Techniques*. London, UCL Press, pp. 1–31.

1995b. Prospects for volcano surveillance. In W. J. McGuire, C. Kilburn, and J. Murray (eds.) *Monitoring Active Volcanoes: Strategies, Procedures and Techniques*. London, UCL Press, pp. 403–410.

McGuire, W. J., Kilburn, C., and Murray, J. (eds.) 1995. *Monitoring Active Volcanoes: Strategies, Procedures and Techniques*. London, UCL Press.

McNutt, S. R. 1996. Seismic monitoring and eruption forecasting of volcanoes: a review of the state-of-the-art and case histories. In R. Scarpa and R. I. Tilling (eds.) *Monitoring and Mitigation of Volcano Hazards*. Heidelberg, Germany, Springer-Verlag, pp. 99–146.

Mercado, R. A., Lacsamana, J. B. T., and Pineda, G. L. 1996. Socioeconomic impacts of the Mount Pinatubo eruption. In C. G. Newhall and R. S. Punongbayan (eds.) *Fire and Mud: Eruption and Lahars of Mount Pinatubo, Philippines*. Quezon City, Philippines, Philippine Institute of Volcanology and Seismology, pp. 1063–1069.

Montserrat Volcano Observatory 2001. Scientific and hazards assessment of the Soufrière Hills Volcano, Montserrat: The Montserrat Volcano Observatory (MVO), 6–8 September 2001. Available online at: http://www.mvo.ms/shvha.htm

Moore, J. G. and Moore, G. W. 1984. Deposit from a giant wave on the Island of Lanai, Hawaii. *Science*, **226**, 1312–1315.

1988. Large-scale bedforms in boulder gravel produced by giant waves in Hawaii. *Geological Society of America Special Paper*, **229**, 101–109.

Moore, J. G., Clague, D. A., Holcomb, R. T., *et al.* 1989. Prodigious submarine landslides on the Hawaiian Ridge. *Journal of Geophysical Research*, **94**, 17465–17484.

Moore, J. G., Normark, W. R., and Holcomb, R. T. 1994. Giant Hawaiian landslides. *Annual Reviews of Earth and Planetary Science*, **22**, 119–144.

Murray, J. B., Pullen, A. D., and Saunders, S. 1995. Ground deformation surveying of active volcanoes. In W. J. McGuire, C. Kilburn, and J. Murray (eds.) *Monitoring Active Volcanoes: Strategies, Procedures and Techniques*. London, UCL Press, pp. 113–150.

Murray, T. L., Ewert, J. W., Lockhart, A. B., *et al.* 1996. The integrated mobile volcano-monitoring system used by the Volcano Disaster Assistance Program

(VDAP). In R. Scarpa and R. I. Tilling (eds.) *Monitoring and Mitigation of Volcano Hazards*. Heidelberg, Germany, Springer-Verlag, pp. 315–362.

Myers, B., Brantley, S. R., Stauffer, P., *et al.* 1997. *What Are Volcano hazards?* US Geological Survey Fact Sheet no. 002–97. Washington, DC, US Government Printing Office.

Newhall, C. G. and Punongbayan, R. S. 1996a. The narrow margin of successful volcanic-risk mitigation. In R. Scarpa and R. I. Tilling (eds.) *Monitoring and Mitigation of Volcano Hazards*. Heidelberg, Germany, Springer-Verlag, pp. 807–838.

(eds.) 1996b. *Fire and Mud: Eruption and Lahars of Mount Pinatubo, Philippines*. Quezon City, Philippines, Philippine Institute of Volcanology and Seismology.

Newhall, C. G. and Self, S. 1982. The volcanic explosivity index (VEI): an estimate of explosive magnitude for historical volcanism. *Journal of Geophysical Research* (Oceans and Atmospheres), **87**, 1231–1238.

Nunnari, G. and Puglisi, G. 1995. GPS: monitoring volcanic deformation from space. In W. J. McGuire, C. Kilburn, and J. Murray (eds.) *Monitoring Active Volcanoes: Strategies, Procedures and Techniques*. London, UCL Press, pp. 151–183.

Peterson, D. W. 1986. Volcanoes: tectonic setting and impact on society. In Geophysics Study Committee (National Research Council) *Active Tectonics*. Washington, DC, National Academy Press, pp. 231–246.

1988. Volcanic hazards and public response. *Journal of Geophysical Research*, **93**, 4161–4170.

1996. Mitigation measures and preparedness plans for volcanic emergencies. In R. Scarpa and R. I. Tilling (eds.) *Monitoring and Mitigation of Volcano Hazards*. Heidelberg, Germany, Springer-Verlag, pp. 701–718.

Peterson, D. W. and Tilling, R. I. 1993. Interactions between scientists, civil authorities and the public at hazardous volcanoes. In C. R. J. Kilburn and G. Luongo (eds.), *Active Lavas: Monitoring and Modelling*. London, UCL Press, pp. 339–365.

PHIVOLCS 1988. *Distribution of active and inactive volcanoes in the Philippines* (educational poster, 1 sheet). Manila, Philippines, Philippine Institute of Volcanology and Seismology (PHIVOLCS).

Pierson, T. C. and Janda, R. J. 1990. Perturbation and melting of snow and ice by the 13 November 1985 eruption of Nevado del Ruiz, Colombia, and consequent mobilization, flow and deposition and lahars. *Journal of Volcanology and Geothermal Research*, **41**, 17–66.

Pinatubo Volcano Observatory Team 1991. Lessons from a major eruption: Mt. Pinatubo, Philippines. *Eos, Transactions of the American Geophysical Union*, **72**, 545, 552–553, 555.

Punongbayan, R. S. 1987. Disaster preparedness systems for natural hazards in the Philippines: an assessment. In *Geologic Hazards and Disaster Preparedness Systems*. Quezon City, Philippines, Philippine Institute of Volcanology and Seismology, pp. 77–101.

Punongbayan, R. S. and Newhall, C. G. 1995. Warning the public about Mount Pinatubo: a story worth repeating. *Stop Disasters*, 25(III), 11–14.

Punongbayan, R. S., Newhall, C. G., and Hoblitt, R. P. 1996a. Photographic record of rapid geomorphic change at Mount Pinatubo, 1991–94. In C. G. Newhall and R. S. Punongbayan (eds.) *Fire and Mud: Eruption and Lahars of Mount Pinatubo, Philippines*. Quezon City, Philippines, Philippine Institute of Volcanology and Seismology, pp. 21–66.

Punongbayan, R. S., Newhall, C. G., Bautista, M. A., *et al.* 1996b. Eruption hazard assessments and warnings. In C. G. Newhall and R. S. Punongbayan (eds.) *Fire and Mud: Eruption and Lahars of Mount Pinatubo, Philippines*. Quezon City, Philippines, Philippine Institute of Volcanology and Seismology, pp. 67–85.

Rodolfo, K. S., Umbal, J. V., Alonso, R. A., *et al.* 1996. Two years of lahars on the western flank of Mount Pinatubo: initiation, flow processes, deposits, and attendant geomorphic and hydraulic changes. In C. G. Newhall and R. S. Punongbayan (eds.) *Fire and Mud: Eruption and Lahars of Mount Pinatubo, Philippines*. Quezon City, Philippines, Philippine Institute of Volcanology and Seismology, pp. 989–1014.

Rozdilsky, J. L. 2001. Second hazards assessment and sustainable hazards mitigation: disaster recovery on Montserrat. *Natural Hazards Review*, **2**, 64–71.

Rymer, H. 1996. Microgravity monitoring. In R. Scarpa and R. I. Tilling (eds.) *Monitoring and Mitigation of Volcano Hazards*. Heidelberg, Germany, Springer-Verlag, pp. 169–197.

Sarna-Wojcicki, A. M., Shipley, S., Waitt, R. B., Jr., *et al.* 1981. Areal distribution, thickness, mass, volume, and grain size of air-fall ash from the six major eruptions of 1980. In P. W. Lipman and D. R. Mullineaux (eds.) *The 1980 Eruptions of Mount St. Helens, Washington*, US Geological Survey

Professional Paper no. 1250. Washington, DC, US Government Printing Office, pp. 577–600.

Scarpa, R. and Gasparini, P. 1996. A review of volcano geophysics and volcano-monitoring methods. In R. Scarpa and R. I. Tilling (eds.) *Monitoring and Mitigation of Volcano Hazards*. Heidelberg, Germany, Springer-Verlag, pp. 3–22.

Scarpa, R. and Tilling, R. I. (eds.) 1996. *Monitoring and Mitigation of Volcano Hazards*. Heidelberg, Germany, Springer-Verlag.

Schuster, R. L. and Crandell, D. R. 1984. Catastrophic debris avalanches from volcanoes. *Proceedings of the 4th International Symposium on Landslides*, Toronto, Canada, vol. 1, pp. 567–572.

Scott, K. M. 1988. *Origins, Behavior, and Sedimentology of Lahars and Lahar-Runout Flows in the Toutle–Cowlitz River System*, US Geological Survey Professional Paper no. 1447-A. Washington, DC, US Government Printing Office.

1989. *Magnitude and Frequency of Lahars and Lahar-Runout Flows in the Toutle-Cowlitz River System*, US Geological Survey Professional Paper no. 1447-B. Washington, DC, US Government Printing Office.

Scott, W. E. 1989a. Volcanic and related hazards. In R. I. Tilling (ed.) *A Short Course in Geology*, vol. 1, *Volcanic Hazards*. Washington, DC, American Geophysical Union, pp. 9–23.

1989b. Volcanic-hazards zonation and long-term forecasts. In R. I. Tilling (ed.) *A Short Course in Geology*, vol. 1, *Volcanic Hazards*. Washington, DC, American Geophysical Union, pp. 25–49.

Scott, W. E., Hoblitt, R. P., Torres, R. C., *et al.* 1996. Pyroclastic flows of the June 15, 1991, climactic eruption of Mount Pinatubo. In C. G. Newhall and R. S. Punongbayan (eds.) *Fire and Mud: Eruption and Lahars of Mount Pinatubo, Philippines*. Quezon City, Philippines, Philippine Institute of Volcanology and Seismology, pp. 545–570.

Self, S., Zhao, J.-X., Holasek, R. E., *et al.* 1996. The atmospheric impact of the 1991 Mount Pinatubo eruption. In C. G. Newhall and R. S. Punongbayan (eds.) *Fire and Mud: Eruption and Lahars of Mount Pinatubo, Philippines*. Quezon City, Philippines, Philippine Institute of Volcanology and Seismology, pp. 1089–1115.

Shimozuru, D. 1972. A seismological approach to the prediction of volcanic eruptions. In *The Surveillance and Prediction of Volcanic Activity*. Paris, France, UNESCO, pp. 19–45.

Siebert, L. 1984. Large volcanic debris avalanches: characteristics and source areas, deposits, and associated eruptions. *Journal of Volcanology and Geothermal Research*, **22**, 163–197.

1996. Hazards of large volcanic debris avalanches and associated eruptive phenomena. In R. Scarpa and R. I. Tilling (eds.) *Monitoring and Mitigation of Volcano Hazards*. Heidelberg, Germany, Springer-Verlag, pp. 541–572.

Sigmundsson, F., Vadon, H., and Massonnet, D. 1995. Readjustment of the Krafla spreading segment to crustal rifting measured by Satellite Radar Interferometry. *Geophysical Research Letters*, **24**, 1843–1846.

Sigurdsson, H., Houghton, B., McNutt, S. R., *et al.* (eds.) 2000. *Encyclopedia of Volcanoes*. San Diego, CA, Academic Press.

Simarski, L. T. 1992. *Volcanism and Climate Changes: Special Report*. Washington, DC, American Geophysical Union.

Simkin, T. 1993. Terrestrial volcanism in space and time. *Annual Reviews Earth and Planetary Science*, **21**, 427–452.

Simkin, T. and Fiske R. S. 1983. *Krakatau 1883: The Volcanic Eruption and its Effects*. Washington, DC, Smithsonian Institution Press.

1984. Krakatau 1883: a classic geophysical event. In C. S. Gillmor (ed.) *History of Geophysics*, vol. 1. Washington, DC, American Geophysical Union, pp. 46–48.

Simkin, T. and Siebert, L. 1994. *Volcanoes of the World: A Regional Directory, Gazetteer, and Chronology of Volcanism during the Last 10,000 Years*. Tucson, AZ, Geoscience Press.

Simkin, T., Unger, J. D., Tilling, R. I. *et al.* (compilers) 1994. *This Dynamic Planet: World Map of Volcanoes, Earthquakes, Impact Craters, and Plate Tectonics*, US Geological Survey Special Map, 2nd edn. Washington, DC, US Government Printing Office.

Sorey, M. L., Evans, W. C., Kennedy, B. M., *et al.* 1998. Carbon dioxide and helium emissions from a reservoir of magmatic gas beneath Mammoth Mountain, California. *Journal of Geophysical Research*, **103**, 303–315, 323.

Stephens, C. D., Chouet, B. A., Page, R. A., *et al.* 1994. Seismological aspects of the 1989–1990 eruptions at Redoubt Volcano, Alaska: the SSAM perspective. *Journal of Volcanology and Geothermal Research*, **62**, 153–182.

Stommel, H. and Stommel, E. 1983. *Volcano Weather: The Story of 1816, The Year without a Summer*. Newport, RI, Seven Seas Press.

Sutton, J., Elias, T., Hendley, J. W., II, *et al.* 1997. *Volcanic Air Pollution: A Hazard in Hawaii*, US Geological Survey Fact Sheet no. 169–97. Washington, DC, US Government Printing Office.

Swanson, D. A. 1992. The importance of field observations for monitoring volcanoes, and the approach of "keeping monitoring as simple as practical." In J. W. Ewert and D. A. Swanson (eds.) *Monitoring Volcanoes: Techniques and Strategies used by the Staff of the Cascades Volcano Observatory, 1980–90*, US Geological Survey Bulletin no. 1966. Washington, DC, US Government Printing Office, pp. 219–223.

Swanson, D. A., Wright, T. L., and Helz, R. T. 1975. Linear vent systems and estimated rates of magma production and eruption for the Yakima Basalt on the Columbia Plateau. *American Journal of Science*, **275**, 877–905.

Swanson, D. A., Casadevall, T. J., Dzurisin, D., *et al.* 1983. Predicting eruptions at Mount St. Helens, June 1980 through December 1982. *Science*, **221**, 1369–1376.

1985. Forecasts and predictions of eruptive activity at Mount St. Helens, USA: 1975–1984. *Journal of Geodynamics*, **3**, 397–423.

Symons, G. J. (ed.) 1888. *The Eruption of Krakatoa, and Subsequent Phenomena*, Report of the Krakatoa Committee of the Royal Society of London. London, Trübner & Co.

Tanguy, J.-C., Ribière, C., Scarth, A., *et al.* 1998. Victims from volcanic eruptions: a revised database. *Bulletin of Volcanology*, **60**, 137–144.

Tazieff, H. and Sabroux, J.-C. 1983. *Developments in Volcanology*, vol. 1, *Forecasting Volcanic Events*. Amsterdam, The Netherlands, Elsevier.

Tedesco, D. 1995. Monitoring fluids and gases at active volcanoes. In W. J. McGuire, C. Kilburn, and J. Murray (eds.) *Monitoring Active Volcanoes: Strategies, Procedures and Techniques*. London, UCL Press, pp. 315–345.

Thorarinsson, S. 1957. The jökulhlaup from the Katla area in 1955 compared with other jökulhlaups in Iceland. *Jökull*, **7**, 21–25.

1969. The Lakagigar eruption of 1783. *Bulletin Volcanologique*, **33**, 919–929.

1979. On the damage caused by volcanic eruptions with special reference to tephra and gases. In P. D. Sheets and D. K. Grayson (eds.) *Volcanic Activity and Human Ecology*. San Diego, CA, Academic Press, pp. 125–159.

Tilling, R. I. 1989a. Volcanic hazards and their mitigation: progress and problems. *Reviews of Geophysics*, **27**, 237–269.

(ed.) 1989b. *A Short Course in Geology*, vol. 1, *Volcanic Hazards*. Washington, DC, American Geophysical Union. [A Spanish translation of this volume, *Los Peligros Volcánicos*, was published in 1993 by the World Organization of Volcano Observatories (WOVO), a Commission of the International Association of Volcanology and Chemistry of the Earth's Interior (IAVCEI); it is available through the editor.]

1995. The role of monitoring in forecasting volcanic events. In W. J. McGuire, C. Kilburn, and J. Murray (eds.) *Monitoring Active Volcanoes: Strategies, Procedures and Techniques*. London, UCL Press, pp. 369–402.

1996. Hazards and climatic impact of subduction-zone volcanism: a global and historical perspective. In G. E. Bebout, D. W. Scholl, S. H. Kirby, *et al.* (eds.) *Subduction: Top to Bottom*, Geophysical Monograph no. 96. Washington, DC, American Geophysical Union, pp. 331–335.

Tilling, R. I. and Lipman, P. W. 1993. Lessons in reducing volcano risk. *Nature*, **364**, 277–280.

Tilling, R. I., Topinka, L., and Swanson, D. A. 1990. *Eruptions of Mount St. Helens: Past, Present, and Future*, US Geological Survey General-Interest Publications Series. Washington, DC, US Government Printing Office.

Torres, R. C., Self, S., and Martinez, M. M. L. 1996. Secondary pyroclastic flows from the June 15, 1991, ignimbrite of Mount Pinatubo. In C. G. Newhall and R. S. Punongbayan (eds.) *Fire and Mud: Eruption and Lahars of Mount Pinatubo, Philippines*. Quezon City, Philippines, Philippine Institute of Volcanology and Seismology, pp. 665–678.

UNDRO/UNESCO 1985. *Volcanic Emergency Management*. New York, Office of the United Nations Disaster Relief Co-ordinator (UNDRO), United Nations Educational, Scientific, and Cultural Organization (UNESCO).

UNESCO 1972. *The Surveillance and Prediction of Volcanic Activity: A Review of Methods and Techniques*. Paris, France, United Nations Educational, Scientific, and Cultural Organization (UNESCO).

US Geological Survey 2005. *Photoglossary*. Available online at: http://volcanoes.vsgs.gov

Verbeek, R. D. M. 1885. *Krakatau*. Batavia, Indonesia, Imprimerie de l'Etat.

Voight, B. 1990. The 1985 Nevado del Ruiz volcano catastrophe: anatomy and retrospection. *Journal of Volcanology and Geothermal Research*, **42**, 151–188.

1996. The management of volcano emergencies: Nevado del Ruiz. In R. Scarpa and R. I. Tilling (eds.) *Monitoring and Mitigation of Volcano Hazards.* Heidelberg, Germany, Springer-Verlag, pp. 719–769.

Voight, B., Glicken, H., Janda, R. J., *et al.* 1981. Catastrophic rockslide avalanche of May 18. In P. W. Lipman and D. R. Mullineaux (eds.) *The 1980 Eruptions of Mount St. Helens, Washington,* US Geological Survey Professional Paper no. 1250. Washington, DC, US Government Printing Office, pp. 347–377.

Voight, B., Janda, R. J., Glicken, H., *et al.* 1983. Nature and mechanism of the Mount St. Helens rockslide avalanche. *Geotechnique,* **33**, 224–273.

Walker, G. P. L. 1982. Volcanic risk. *Interdisciplinary Science Reviews,* **7**, 148–157.

White, R. A. 1996. Precursory deep long-period earthquakes at Mount Pinatubo: spatio-temporal link to a basalt trigger. In C. G. Newhall and R. S. Punongbayan (eds.) *Fire and Mud: Eruption and Lahars of Mount Pinatubo, Philippines.* Quezon City, Philippines, Philippine Institute of Volcanology and Seismology, pp. 307–327.

Wicks, C., Jr., Thatcher, W., and Dzurisin, D. 1998. Migration of fluids beneath Yellowstone Caldera inferred from satellite radar interferometry. *Science,* **282** (5388), 458–462.

Williams, R. S., Jr. (ed.) 1997. *Lava-Cooling Operations during the 1973 Eruption of Eldfell Volcano, Heimaey, Vestmannaeyjar, Iceland,* US Geological Survey Open-File Report no. 97–724. Washington, DC, US Government Printing Office.

Williams, S. N. (ed.) 1990a. Special Issue on Nevado del Ruiz, Colombia, I. *Journal of Volcanology and Geothermal Research,* **41**, 1–377.

(ed.) 1990b. Special Issue on Nevado del Ruiz, Colombia, II. *Journal of Volcanology and Geothermal Research,* **42**, 1–224.

Wolfe, E. W. and Hoblitt, R. P. 1996. Overview of the eruptions. In C. G. Newhall and R. S. Punongbayan (eds.) *Fire and Mud: Eruption and Lahars of Mount Pinatubo, Philippines.* Quezon City, Philippines, Philippine Institute of Volcanology and Seismology, pp. 3–20.

Wolfe, E. W., Neal, C. A., Banks, N. G., *et al.* 1988. Geologic observations and chronology of eruptive events. In E. W. Wolfe (ed.) *The Puu Oo eruption of Kilauea Volcano, Hawaii: Episodes 1 through 20, January 3, 1983, through June 8, 1984,* US Geological Survey Professional Paper no. 1463. Washington, DC, US Government Printing Office, pp. 1–97.

Wright, T. L., Chun, J. Y. F., Esposo, J., *et al.* 1992. *Map Showing Lava Flow Hazard Zones, Island of Hawaii,* US Geological Survey Miscellaneous Field Studies Map no. MF-2193. Washington, DC, US Government Printing Office.

Young, R. W. and Bryant, E. A. 1992. Catastrophic wave erosion on the southeastern coast of Australia: impact of the Lanai tsunami ca. 105 ka? *Geology,* **20**, 199–202.

Young, S. R., Sparks, R. S. J., Aspinall, W. P., *et al.* 1998a. Overview of the eruption of Soufrière Hills volcano, Montserrat, 18 July 1995 to December 1997. *Geophysical Research Letters,* **25**, 3389–3392.

Young, S. R., Voight, B., Sparks, R. S. J., *et al.* (eds.) 1998b. Special section on the Soufrière Hills eruption, Montserrat, British West Indies, Part 2. *Geophysical Research Letters,* **25**, 3651–3700.

# Chapter 3

# Anticipating volcanic eruptions

## Joan Martí and Arnau Folch

## Introduction

Volcanic activity involves movements of fluids, magma, and vapor, inside the volcano and its feeding systems. These movements will cause external signals, eruption precursors, which may alert us to the proximity of a volcanic event. Such precursory signals can be detected by an adequate volcano monitoring program. Monitoring techniques include a range of geophysical and geochemical techniques, encompassing seismic, ground-deformation, gravity, and magnetic observations, gas monitoring, and remote sensing. Successful forecasting of volcanic events depends on the precision of the surveillance network in detecting any changes in the volcano's current behavior. To interpret the geochemical and geophysical precursors correctly, however, it is also important to understand the physics of the volcanic processes involved in volcanic eruptions. Detailed knowledge of the volcano, its internal structure and style, and potential triggering mechanisms of past eruptions must be combined with adequate monitoring if future volcanic eruptions are to be anticipated and their effects mitigated (Fig. 3.1). In summary, prediction of volcanic activity has the aims of determining *when* and *where* a future eruption will occur, and *how* it will proceed. However, it is also important to understand *why* the next eruption will occur.

Volcanic eruptions are caused mainly by processes occurring in magma chambers at depth.

During the lifetime of the volcano the corresponding magma chamber may change in size and shape. A chamber's shape and size strongly influence the stress field associated with the magma-filled reservoir, which controls the geometry and frequency of dyke injection and extrusion from the volcano. Eruptions are typically triggered by the injection of new magma into the chamber from deeper levels, leading to chamber overpressure either directly through the added magma volume, or indirectly by sudden cooling and crystallization of the new magma leading to wholesale volatile exsolution. Improved understanding of magma chamber evolution, host-rock mechanics, volatile solubilities, and degassing is needed if eruptive triggers are to be understood, and quantitative models of eruption probability developed.

Few large volcanic eruptions have been observed directly, which imposes important restrictions on the study of eruptive phenomena. Our knowledge of eruption dynamics is based mostly on studies of the solid products of past eruptions, and the development of appropriate theoretical and experimental models. Significant advances have been made in these fields over recent years, though there remains much to learn from this combined field, experimental, and theoretical approach. Causes, dynamics, and effects of volcanic eruptions have traditionally been studied separately, due to the complexity of the problems involved and due to difficulties inherent in a multidisciplinary approach. However, that a volcanic eruption, its causes and effects,

*Volcanoes and the Environment*, eds. J. Martí and G. G. J. Ernst. Published by Cambridge University Press. © Cambridge University Press 2005.

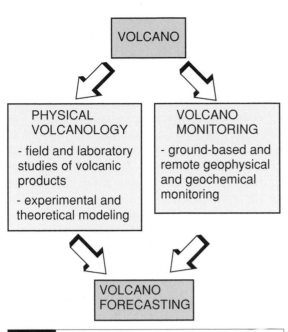

VOLCANO

PHYSICAL VOLCANOLOGY
- field and laboratory studies of volcanic products
- experimental and theoretical modeling

VOLCANO MONITORING
- ground-based and remote geophysical and geochemical monitoring

VOLCANO FORECASTING

**Fig. 3.1.** Integration of studies of physical volcanology and volcano monitoring is necessary to understand a volcano's past and present activity and to predict its future behavior.

should be regarded as a continuum of related phenomena is increasingly being accepted by volcanologists.

Understanding the physical principles that control pre-eruptive and eruptive processes is not only crucial in reducing the risk to human lives and properties in volcanic areas but also in assessing the effects of volcanic eruptions on the environment. Though the precise prediction of future volcanic activity remains elusive, significant progress has been made in understanding the processes involved, aiding the assessment of precursors of eruptions. In this chapter, we will review the processes that accompany volcanic eruptions and the methods typically used in their characterization. We will describe in some detail the theoretical modeling approach for a general audience in recognition of its increasing success in anticipating volcanic eruptions and their effects. We refer mainly to explosive eruptions of large silicic volcanoes, as these represent perhaps the most important of all volcanic hazards. However, the concepts covered in this chapter are also applicable to quieter, less dangerous eruptions of basaltic volcanoes. Volcano monitoring

has been treated in considerable detail elsewhere (e.g., McGuire *et al.*, 1995; Scarpa and Tilling, 1996 and references therein) and this chapter also gives a brief review of the main geophysical and geochemical monitoring methods, outlining the fundamental concepts and current state of the art in eruption forecasting.

## Magmas and magma chambers

The causes of volcanic activity are ultimately explained within the framework of the plate tectonics theory (see Chapters 1 and 4, this volume) relating global tectonics and magmatism. Most volcanic activity is driven by basaltic magma that forms in the mantle and rises to lithospheric levels when significant volume is available. Magma ascent is controlled by density differences with the host rocks and by the rheology of the magma and host rocks. At deeper levels magma generated by fusion of the mantle will rise diapirically while surrounding host rocks deform plastically, while at shallower levels in the brittle lithosphere it will ascend though dyke-like fractures. Although eruptions directly from the source region are thought to occur sometimes, magma will generally accumulate at levels above the source region, forming reservoirs or magma chambers where cooling will cause physical and chemical changes. The intrusion of hot, mantle-derived basaltic magma into the crust can also cause extensive melting of shallow host rocks, generating crustal-derived, silicic magmas (Huppert and Sparks, 1988).

Many factors will influence whether or not a magma will erupt at the Earth's surface, and how and when this will occur. These include tectonic setting, source rock composition and melting process, magma chamber characteristics, and magmatic evolution. The chemistry and physical properties of magmas, as well as their ascent and magma chamber formation, are covered in Chapter 1. Here, we use these concepts to explain how magmatic evolution can cause a volcanic eruption.

Volcanic eruptions range from quiet lava outpourings to explosive events that can inject material into the stratosphere and cause damage over

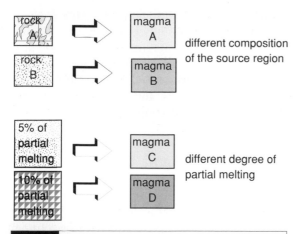

different composition of the source region

different degree of partial melting

**Fig. 3.2.** The influence of source rock composition and degree of partial melting on the composition of primary (first-generated) magmas.

thousands of square kilometers. The reasons for these differences are likely to be related to storage and differentiation of magma, its movement in dykes connected to the surface, and the near-surface degassing of volatiles. The chemical composition of a magma, including the presence of volatiles, will strongly influence its physical properties, specifically its rheology, and the manner in which it will erupt. Useful additional information on the genesis and evolution of magmas may be found in several petrological textbooks (e.g., Carmichael *et al.*, 1974; McBirney, 1984; Wilson, 1989; Hall, 1996).

Magma is a silicate liquid melt that may contain solid (mineral) and volatile (dissolved and exsolved fluid) components. It originates by partial melting of mantle or crustal rocks. During partial melting, under given conditions of pressure and temperature, minerals with low melting points will melt, leaving part of the crustal or mantle source rock unmelted. Mantle and crustal source rocks are composed mainly of silicate minerals, whose chemical components, and the proportion in which these are added to the melt, define the resulting primary magma composition (Fig. 3.2).

Mantle-derived magmas erupt in all tectonic environments, notably at mid-ocean ridges, oceanic islands, continental and oceanic arcs, and continental rifts (see Chapters 1 and 4). Partial melting of mantle, whatever the tectonic

setting, gives rise to primary magmas generically termed basalts, characterized by low volatile contents, high density, high temperature, and low viscosity. In contrast, when the crust melts the resulting magma is more silicic, cooler and lower in density, and more viscous and richer in volatile species, particularly water. The differences between mantle and crustal-derived primary magmas are mainly due to different source rocks compositions. Primary basaltic magmas also vary in composition depending on the nature of the source mantle which varies with tectonic setting. The conditions of pressure and temperature at which partial melting occurs, also dependent on tectonic setting, are different. They also influence the composition of the resulting primary melts, as the composition of the source rocks and the conditions of melting are different in each case.

The wide diversity of primary magma compositions shows consequently significant variations in rheology and other physical properties. As magma is of lower density than its solid host rocks, it will tend to move away from the source region by ascending to shallower levels. As it rises it will cool and the primary magma will begin to crystallize and experience magmatic differentiation. It may accumulate in shallower reservoirs, interact with wall rocks, and mix with other magmas (Fig. 3.3). All of these influences on evolving magma composition will cause variations in rheology, eventually influencing the characteristics of any eruption. Any combination of magmatic differentiation, mixing, and contamination might occur during the evolution of a magma from source to eruption and solidification at the Earth's surface. The complexity of magmatic evolution has been revealed by studies of magmatic rocks solidified at the Earth's surface (volcanic rocks) or beneath the surface (plutonic rocks and hypabyssal rocks). Studies of the chemistry and mineralogy of volcanic rocks are a key component in determining the magma's evolution beneath the surface. Petrological studies are particularly important in determining the pre-eruptive conditions of the magma and in predicting a volcano's future behavior. Eruption style is a direct consequence of a magma's chemical and physical properties, so that petrological

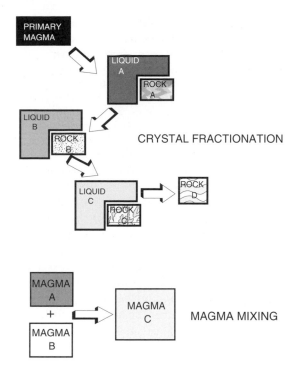

**Fig. 3.3.** The effect of fractional crystallization, magma mixing and crustal assimilation on magma composition. Magmatic differentiation describes mechanisms by which a magma changes its physicochemical (due to phase changes) composition during cooling. Fractional crystallization, the most important mechanism of magmatic differentiation, involves the separation of crystals from the magmatic liquid. Fractional crystallization of magma by gravitational separation of crystals that progressively form during cooling may occur over an extended period and by the accumulation of magma in reservoirs or magma chambers. Removal of crystals induces the composition of the remaining liquid magma to change. In addition to changing its composition, a magma may experience mixing with other magmas and host rock contamination. In both cases the physicochemical composition of the magma will also change.

magma chambers have been inferred convincingly from geophysical studies able to detect the presence of molten rock within the solid lithosphere, but also following from the existence of plutons, igneous bodies representing ancient magma chambers at the Earth's surface. The chemical and mineralogical features of volcanic rocks also typically indicate the temporary arrest and differentiation of magma at a certain depth, typically between 2 and 8 km, before eruption. Magma chambers apparently range in size from less than 1 km$^3$ to more than 1000 km$^3$. The size of chamber attained, and the volume of magma that may be accumulated locally within the lithosphere, will depend mainly on the stress field that dominates at the site, on the composition and rheology of the magma, and on the rate of injection of new magma into the chamber. Over its lifetime, a magma chamber may undergo episodes of inflation, when new magma is injected into it, and episodes of deflation when magma leaves the chamber and is intruded into the host rocks or erupted at the surface. Crystallization and subsequent differentiation of magma will tend to lead to a progressive change in composition and physical properties. This may eventually cause the magma to become oversaturated in volatiles, which will vaporize and cause the chamber to expand until magma or gas is voided from the chamber.

Chapter 1 describes how magma chambers form at neutral buoyancy levels, where the bulk density of magma is equal to that of the host rocks. Levels of neutral buoyancy are associated with mechanical discontinuities in the Earth, such as the asthenosphere–lithosphere boundary, the mantle–crust or Moho discontinuity, the base of a volcano edifice, or stratigraphic discontinuities in the interior of a volcanic pile. Depending on its location, the regional stress field influencing the chamber's development will vary.

The magma chamber feeding a volcano will change shape during the volcano's lifetime. A chamber's shape and size will strongly influence the local stress field, associated with the presence of the chamber, controlling magma intrusion and eruption (Gudmundsson, 1988). Understanding the geometric evolution of a magma chamber is thus a prerequisite for understanding

studies of products of past eruptions are crucial in explaining eruptive phenomena.

Magma chambers are reservoirs of molten silicate liquid plus any crystals it may contain (Marsh, 1989). The existence and dimensions of

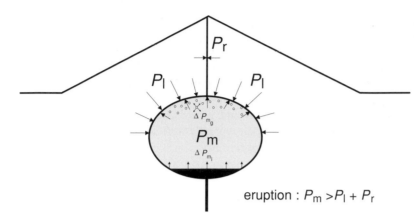

eruption : $P_m > P_l + P_r$

**Fig. 3.4.** Triggering mechanisms of explosive eruptions. Pressure inside the magma chamber ($P_m$) may increase by an amount ($\Delta P_{m_g}$) due to volatile oversaturation or to intrusion of fresh magma ($\Delta P_{m_i}$). If the chamber pressure exceeds lithostatic pressure ($P_l$) plus the tensile strength of the surrounding rocks ($P_r$), tensile vertical fractures may open allowing magma to escape from the chamber, sometimes reaching the surface to cause a volcanic eruption.

the eruption patterns of volcanoes. Moreover, the behavior of volcanic systems depends greatly on how magma accumulates at shallow levels in the crust, and at what depth and how the resulting magma chamber connects to the surface. The distribution of volcanic vents at the surface will depend on the combination of regional tectonic structure and that of the active magmatic system, including the strength of host rocks, the intensive parameters of the magma chamber, and its deformation history of inflation and deflation episodes, and by the stress distribution within the volcanic edifice. Understanding of vent distribution and vent development during explosive and effusive volcanic eruptions constitutes one of the most important elements of volcanic hazard assessment.

## Eruptive processes

As we have seen, a volcanic cycle typically starts with the intrusion and accumulation of deeply sourced magma into the crust and its accumulation to a magma chamber. Once magma is accumulated in a shallow chamber, heat is slowly lost through the chamber walls and the magma gradually changes its composition and physical properties as it differentiates. If the conditions for magma to exit the chamber are never achieved, the chamber will eventually completely crystallize to form plutonic rocks. However, during its lifetime the chamber may become overpressurized occasionally leading to eruption.

The pressure increase can result from two main causes discussed above: volatile oversaturation driven by differentiation or the injection of fresh magma into the chamber (Blake, 1981, 1984; Tait *et al.*, 1989; Folch and Martí, 1998) (Fig. 3.4). Other, less common, mechanisms such as seismic excitation of entrapped bubbles (Sturtevant *et al.*, 1996) and tectonic triggers (Linde and Sacks, 1998) have also been recognized. Whatever the case, the magma chamber responds to restore mechanical equilibrium with its surroundings by injecting dykes or deforming the wall rocks. If the excess pressure cannot be released, a magma-filled fracture may propagate from the chamber to the surface and trigger an eruption. A volcanic eruption typically ends once mechanical equilibrium has been restored by the withdrawal of a few percent of the stored magma. When the eruption ceases, the remaining magma can continue to cool and evolve, and a new eruptive event can occur if conditions for overpressure are again reached. The period between consecutive eruptions is known as the repose period, and is typically of the order of a few years to a few thousands of years. A silicic magma chamber will commonly produce a sequence of eruptions rather than a single event; the products extruded contributing to the construction of a volcanic edifice.

A correlation between the length of volcanic repose periods and the volatile content and explosivity of the subsequent eruption suggested to early workers that the exsolution of magmatic volatiles within the chamber could produce overpressures sufficiently high to trigger an explosive eruptive event (Morey, 1922; Smith and Bailey, 1968; Smith, 1979; Blake, 1984; Tait *et al.*, 1989). This hypothesis was supported by petrological evidence that magmatic water content progressively increased during the repose periods. A further section describes mathematical models that have allowed quantification of the effect of volatile oversaturation on the increase of magmatic pressure.

The second, and probably the most common, mechanism triggering volcanic eruptions is the injection of new magma into the chamber (Blake, 1981). Essentially the volume of material able to be stored within the chamber is limited by the compressibility of the resident magma and the expansivity of the chamber walls. This critical volume cannot be exceeded, and excess magma is forced to exit the chamber through dykes. Evidence that magma chambers are open systems periodically experiencing inputs of magma is compelling. For example, volcano inflation and ground deformation (see p. 113), often accompanying seismic activity, can be explained by an influx of fresh magma into the chamber. Active eruptive periods separated by repose periods of a few years are suggested to be related to the uprise of magma at rates of several cubic meters per second (Blake, 1981). A further section gives a detailed description of theoretical models developed to quantify the effect of magma injection on triggering volcanic eruptions.

The physics of explosive volcanism began to be understood during the late 1970s (e.g., Walker, 1973; Wilson, 1976; Wilson *et al.*, 1980). Processes related to explosive volcanism have become a subject of major interest to scientists in the last two decades, due to the potential hazards for human life and environment.

Explosive volcanic activity is strongly related to the presence of volatiles originally dissolved in the magma. The solubilities of the main volatile species ($H_2O$, $CO_2$, $SO_2$) are highly dependent on pressure. Therefore, during the ascent of magma, a level is reached where magma pressure equals the exsolution pressure of the dissolved gas species and vapor is exsolved to form small bubbles. The level at which this occurs is known as the exsolution level (Fig. 3.5). The exsolution level can be located either in the chamber, or

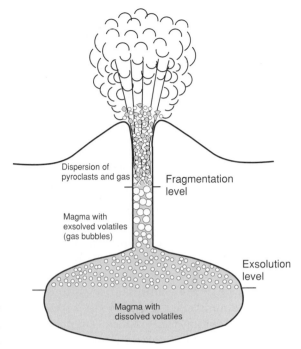

**Fig. 3.5.** Volcano undergoing an explosive eruption. Exsolution of volatiles may occur by decompression when magma approaches the surface, or by oversaturation in the residual liquid as magma cools down and crystallizes. Whichever the case, volatile exsolution causes small bubbles to form above the exsolution level, where magma becomes a two-phase mixture of liquid and gas. Once nucleated, gas bubbles grow by a combination of mass transfer of volatile components from magma to bubbles and by depressurization (Sparks, 1978). This occurs until magma disrupts at the fragmentation level due to the bursting of bubbles. The magma thus becomes a gas continuum with dispersed magmatic particles called pyroclasts, which are ejected from the vent at high velocity. After Fisher and Schmincke (1984).

within the conduit leading to the surface. Above the exsolution level magma becomes a two-phase mixture composed of a gas-saturated liquid with gas bubbles dispersed in it. Gas bubbles, once nucleated, grow by several mechanisms. These include diffusion as more exsolved gas migrates into the bubbles, decompression as ascent to shallow levels lowers ambient pressure and expands the bubbles, and coalescence when small gas bubbles join to form larger ones (Sparks, 1978; Prousevitch *et al.*, 1993; Toramaru, 1995). Bubble size, however, cannot be arbitrarily large and a level is reached where magma fragments become a gas continuum containing liquid magma fragments, called pyroclasts, dispersed within. This level is known as the fragmentation level, above which the gas–pyroclast mixture is accelerated to supersonic velocities.

Magma fragmentation is a defining feature of explosive volcanism but is still not well understood and controversial (Dingwell, 1998). The debate over fragmentation encompasses two major models. The first to be developed (Sparks, 1978) concentrates on strain criteria, where magma disrupts and fragments when its vesicularity attains a critical value. During ascent the exsolution of volatiles from the magma causes viscosity to increase by several orders of magnitude. The viscosity increase of the bubbly mixture prevents further growth of bubbles, allowing them to generate sufficient stress to disrupt the surrounding magma. This model is supported by the observation that pumice clasts produced in explosive eruptions typically exhibit vesicularities in the range 70–75%, the value for maximum spherical bubble-packing. The second model concentrates on strain rate criteria. The vigorous kinematics of the mixture of gas and pyroclasts in the uppermost parts of the conduit may produce strain rates sufficient to induce a viscoelastic response of the liquid melt and subsequent fragmentation when stresses exceed the magma tensile strength (Papale, 1998; Martí et al., 1999).

In the foregoing paragraphs of this section we have briefly summarized the main features of explosive volcanism. Characteristics of explosive eruptions and their products have already been described in Chapter 1. The reader will find additional information on the subject in several recent textbooks and specialized volumes and reviews (Willams and McBirney, 1979; Fisher and Schmincke, 1984; Cas and Wright, 1987; Wohletz and Heiken, 1992; Martí and Araña, 1993; Papale, 1996; Sparks et al., 1997; Freund and Rosi, 1998; Gilbert and Sparks, 1998; Sigurdsson, 2000).

## Field and laboratory studies: the reconstruction of past eruptions

Products derived from a volcanic eruption are incorporated into the geological record and form a basis for reconstructing previous eruptions. Such studies might also reveal to us the most probable cause of an eruption. A detailed field study of products of previous eruptions and their comprehensive petrological and geochemical analysis can, therefore, provide information necessary to constrain the pre-eruptive conditions of magma and indicate conditions likely to trigger future eruptions. Field and laboratory data also include the measurements of the physicochemical properties of volcanic materials necessary to constrain physical and numerical models of volcano behavior and eruption dynamics used to anticipate future volcanic activity.

The reconstruction of past eruptions and their comparison with observed examples constitutes a key to understanding the eruptive behavior of volcanoes (e.g., Bonadonna et al., 1998). While direct observations of volcanic eruptions are restricted in number, improved observational techniques have permitted detailed records of the different phases of these eruptions to be compiled allowing correlation between volcanic products and their corresponding phases of the eruption. Such observations also improve our knowledge of the emplacement dynamics of the eruption products. The direct observation of recent eruptions, such as that of Mt. Pinatubo in the Philippines in 1991, has demonstrated how complex the interaction may be between the deposition of volcanic products and later erosion and secondary processes that may rapidly transform their original appearance. There are many volcanoes, however, from which no eruption has been recorded in historic times, but which may become active in the near future. Again Pinatubo is a good example of such a volcano. In such cases, the only way to anticipate the volcano's future behavior is the study of the products of past eruptions and comparison with observed eruptions, and inferences from theoretical and experimental models.

### Field studies

Field studies of the products of an eruption are performed to characterize the sequence of volcanic deposits and to reconstruct the sequence of eruptive and depositional events (Fig. 3.6). Volcanological field studies aim to determine the relative stratigraphy and distribution of the different units forming a particular eruption sequence

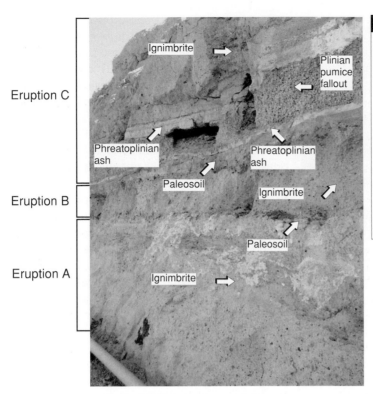

Eruption C

Eruption B

Eruption A

Ignimbrite

Plinian pumice fallout

Phreatoplinian ash

Phreatoplinian ash

Paleosoil

Ignimbrite

Paleosoil

Ignimbrite

**Fig. 3.6.** Interpretation of the dynamics of past eruptions requires detailed field studies of the stratigraphy, distribution and nature of their products. The photograph shows a detail of the stratigraphic sequence of youngest (0.3–0.2 Ma) eruptive episodes of the Las Cañadas volcano in Tenerife (Canary Islands). The products of three different eruptions are visible clearly separated by paleosoils. The first two eruptions are represented by ignimbrite deposits, while the youngest shows a more complex sequence of products including two phreatoplinian ash deposits separated by a Plinian pumice fallout deposit at the base, and an ignimbrite on top.

(Rosi, 1996). A volcanic eruption may involve several phases, each giving rise to different products. For example, the eruption may include the formation of a Plinian column that generates units of fallout deposits, after which several collapses of the Plinian column may produce pyroclastic flows and surges. Each of the deposits produced by such an eruption will exhibit different lithological, sedimentological, and stratigraphic characteristics, as well as having different distributions around the volcano. Plinian fallout deposits will be widely distributed and will tend to mantle the surrounding topography. Deposits from pyroclastic flows and surges tend to accumulate in low-lying areas (see Chapter 1). Following deposition eruptive products may be affected by other geological processes producing secondary volcanic deposits. Primary and secondary deposits from a particular eruption appearing in the geological record show complex stratigraphic relationships depending on the characteristics of the eruption, the topography of the surrounding, and the environment in which the eruption occurs.

Field studies should provide the information necessary to identify the different products and phases of a particular eruption, and to discriminate them from others related to different eruptions or to non-volcanic processes. Geological mapping to determine the distribution of volcanic units, may be carried out at scales from regional to local, depending on the size of the eruption. This will determine what combination of photogeological or remote sensing and direct field reconnaissance is required. Stratigraphic correlation is necessary in order to confirm the distribution of deposits but also to characterize lateral variations in thickness, geometry, and lithology, which may occur in each unit or facies. Stratigraphic studies are important also in distinguishing groups of deposits from different eruptions or volcanic centers. Field lithological studies of volcanic deposits are needed to identify an appropriate sampling policy for accompanying mineralogical and geochemical studies. Moreover, identifying the sedimentological characteristics, such as grain-size distribution and sedimentary structures, is crucial in determining emplacement mechanisms of the deposits.

Each volcanic unit should be described in detail, its stratigraphic relationships noted, and outcrops spatially correlated. In a sequence of volcanic deposits the different units should be

**Fig. 3.7.** Interpreting the products of volcanic eruptions exposed in the geological record becomes more difficult when products of different vents are interbedded in the same stratigraphic sequence. In this example from Tenerife, a distal Plinian fallout deposit, light in color, is interbedded with the dark-colored proximal products of several small basaltic cones.

widespread paleosoils or erosion and alteration surfaces allowing unequivocal recognition of different eruption sequences, are thus desirable. Such criteria can also be useful in reconstructing the long-term evolution of volcanic systems, identifying volcanic cyclicity and making future volcanic activity more predictable. Stratigraphic correlations between outcrops are also necessary to identify the provenance and thickness variations of the different volcanic deposits. Here, the use of isopach and isoplet maps (Fig. 3.8) will constrain the source vent for each deposit. However, the exact location of eruptive vents is not always possible to identify, particularly in complex volcanic fields where the activity of several volcanoes may coincide in time and space.

While stratigraphic studies are vital, determinations of the absolute age of a volcano's deposits will help greatly in reconstructing its eruptive history. Absolute age determinations, typically estimated from isotopic compositions, will allow different eruptions and possible cycles of eruptions to be recognized. However, recovering an absolute age of a deposit is not always possible, depending on the dating technique and the type and alteration of the sample. The establishment of the relative age scheme for a group of volcanic products using stratigraphic methods should therefore always be a step before determining the chronology of a sequence using radiometric methods.

## Laboratory studies of volcanic rocks

Studies of the samples of volcanic units collected from the field continue in the laboratory with mineralogical and geochemical analysis. This will enable the magma composition and possibly also its pre-eruptive conditions to be determined. When a magma erupts at the Earth's surface it cools rapidly and its liquid part quenches to a glass. The crystals observed in the lavas and pyroclasts forming the magmatic products comprise the mineral assemblage stable during cooling of the magma in the chamber (Fig. 3.9). Some reactions of the crystals with the liquid may have occurred during the eruptive processes. However, it is generally assumed that the observed compositions and assemblage of large crystals (phenocrysts), and the quenched glassy groundmass,

grouped according to a hierarchy allowing temporal succession of volcanic events to be recognized (Fisher and Schmincke, 1984; Martí, 1993). Each eruption may last days to months, or several years, and may include several phases with durations of hours to days. Moreover, each phase may include several pulses of activity of seconds to minutes duration. Different deposits may be generated by each eruptive phase or pulse, so that the resulting sequence of deposits will be as complex as the eruption. Where direct observation of an eruption is lacking, a good interpretation of the lithological nature and stratigraphic position of each volcanic deposit will provide a basis for understanding a particular eruption and predicting a volcano's potential future behavior.

Understanding a sequence of deposits from a volcanic eruption becomes more difficult as the age of the deposits increases. This is because of post-eruptive processes that may remove or transform the deposits and the incorporation into the stratigraphic sequence of products from other eruptions (Fig. 3.7). Stratigraphic criteria, such as

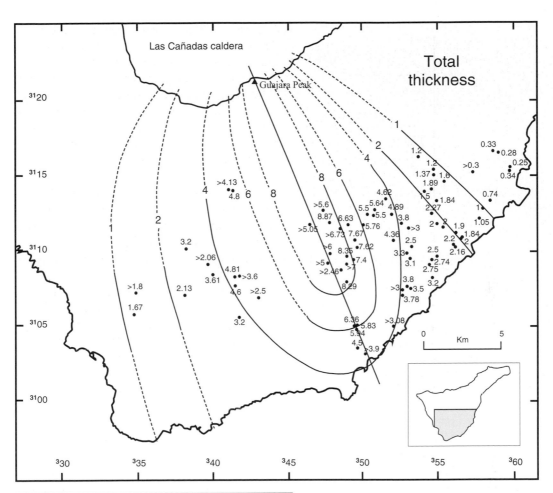

**Fig. 3.8.** Isopach maps of the Granadilla pumice fallout deposit, Tenerife. Isopach maps show thickness variations of the deposit. Each contour joins points where the deposit has a particular thickness value. Isopach maps, together with isoplet maps showing grain-size variations within the deposit, are useful for discriminating fallout deposits from other kinds of pyroclastic deposits formed by flow processes, and help to identify the vent area and distribution of the deposit. Simplified from Bryan et al. (2000).

will be representative of the magma composition immediately before the eruption. Petrological investigation of the mineralogy and geochemistry of the minerals and glass can also constrain physicochemical conditions within the magma chamber prior to eruption.

A magma's mineralogy and chemical composition reflect its thermodynamic history. For example, mineral assemblages and compositions vary in response to pressure, temperature, and liquid composition. If minerals and liquid were in equilibrium with local conditions then thermodynamic techniques such as geothermometry and geobarometry can be used to determine the temperature and pressure at which a mineral assemblage formed (Wood and Fraser, 1976). Experimental petrology also provides tools to determine the pre-eruptive conditions of magmas. Natural volcanic samples can be melted and recrystallized under a variety of physical conditions until the mineral assemblage contained in the natural sample is reproduced and natural conditions deduced. Of particular interest in experimental petrological studies of volcanic rocks are those focused on the stability relations of hydrous minerals, i.e., minerals that contain water or other volatiles in their structures. The stability of hydrous minerals is closely linked to the activity of water in the magma, which provides a major driving force for explosive eruptions.

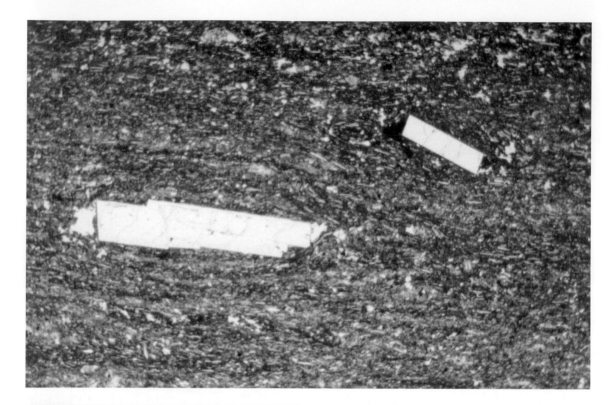

**Fig. 3.9.** Thin section of a phonolitic lava flow from Tenerife showing large phenocrysts of alkali feldspar, grown in the magma chamber before the eruption, embedded in a glassy groundmass formed by chilling of magmatic liquid at the Earth's surface.

One of the key parameters controlling fragmentation and eruption dynamics is pre-eruptive magmatic gas content. Estimating the pre-eruptive gas content of a magma from its eruptive products is thus desirable in order to characterize the hazard potential of related future volcanic activity. Analysis of hydrous mineral stability relations is one possible way to estimate pre-eruptive gas content, but this will give only the gas content at the time prior to that at which the minerals equilibrated. The gas content at the time of fragmentation might be estimated from either analysis of bubble size distributions or water contents of pumice produced by an eruption. However, this would require a firm understanding of post-eruptive degassing processes, perhaps achievable through theoretical calculations. Laboratory studies aimed at determining the pre-eruptive gas content of magmas

also include the analysis of samples of melt included in crystals. Such melt inclusions represent trapped samples of magmatic liquid preserving the original volatile content of the magma, unmodified by degassing prior to or during the eruption. Melt inclusion data from a sequence of eruptions from the same volcano may show temporal variations in the content and composition of volatiles. Such variations can be interpreted in light of eruptive behavior, making future trends and activity more predictable.

Concentrations of many stable and radioactive isotopes, such as U, Th, Ra, Sr, Nd, O, H, and S, are measured in the laboratory, in order to characterize the origin and evolution of magmas. The source region of a particular magma has an isotopic signature that is transferred to the magma it generates, allowing us to distinguish between different melting zones and sources. If magma assimilates host rocks in its way to the surface the isotopic signature of the host rock will be imported to the magma, allowing isotopes to detect contamination processes during ascent. In addition, isotopes are important in constraining processes occurring in the shallow magmatic plumbing systems. From isotopic petrogenetic studies it is possible to identify the presence of a shallow reservoir and constrain its volume, the average residence time of the magma, and the rates of magma replenishment, fractional crystallization, and magmatic degassing.

Radioactive disequilibrium between short-lived nuclides of the U and Th decay series can provide abundant information on magmatic fractionation processes and their timescales and has become a subject of increasing interest in modern volcanology (Lambert *et al.*, 1986; Condomines *et al.*, 1988, 1995). Increasingly precise measurements and theoretical modeling of short-lived daughter nuclides of the U and Th decay series can be used to constrain timescales of magmatic processes ranging from days to thousands of years. Investigation and modeling of the systematics of the daughter isotopes of radon ($^{210}$Pb, $^{210}$Po, $^{210}$Bi), which can occur in volatile S and Cl compounds, also contribute to understanding degassing timescales and mechanisms and open up new possibilities for the use of volcanic gas monitoring in eruption prediction (Lambert *et al.*, 1986; Le Cloarec and Marty, 1991; Le Cloarec *et al.*, 1992; Pennisi and Le Cloarec, 1998).

## Scale and analogue modeling

Scale and analogue modeling has, over the last few decades, become a useful tool in the study of volcanic processes, particularly applied to the dynamics of shallow magma chambers, eruption mechanisms, volcanic plumes, pyroclastic flows, and volcano stability. Scale experiments in the laboratory can simulate volcanic processes under conditions analogous to those occurring in nature, allowing visualization of phenomena that cannot normally be directly observed. Experimental work, together with field and laboratory data, is also important in validating numerical models.

What are the differences between analogue and scale models? An analogue model is one that reproduces some fundamental aspect of the natural phenomenon without direct physical scaling between the two. For example, opening a bottle of champagne is a simple analogue of the tapping of a volatile-rich magma chamber. Such models provide no real quantification of the factors critical in nature, as their scales and proportions are different. Scale experiments, on the other hand, allow quantitative comparison between the experimental model and the natural system, by reproducing the geometry and relative proportions of forces pertaining in the natural case. Scale experiments require not only dimensional or scale equivalence (geometrical similarity) with the natural model, but also the equivalence between body and surface forces acting on the system (dynamical similarity). The major difficulty with such experiments is the scaling of gravity forces, which requires a centrifuge, providing for unworkable complexity in experimental design. This restriction has significantly limited the application of true scale experiments in volcanology and has favored analogue and dimensionally scaled approaches.

Analogue experiments have become useful in visualizing and understanding hidden processes occurring in magma chambers, which can be identified from the study of volcanic products but cannot be observed directly. These processes, including fractional crystallization, mixing, convection, and zonation, help to determine the characteristics of the eruptive processes and should be considered a tool for forecasting volcano behavior. Certain models involve the use of saline solutions with contrasting physical properties, to simulate interactions between magmas, such as mixing or density stratification. In some experiments a hot, dense saline solution is injected into the bottom of a laboratory tank containing a cooler, lighter solution, to simulate replenishment of a shallow magma chamber by fresh magma (Huppert and Sparks, 1980; Huppert *et al.*, 1983). Convection in homogeneous magma chambers has been simulated by an experimental tank containing a homogeneous saline solution heated from below (Turner and Campbell, 1986). Analogue experiments with fluids of different physical properties have also been used to investigate magma mixing during flow in volcanic conduits during explosive eruptions (Freund and Tait, 1986; Blake and Fink, 1987).

Experimental models have been specifically developed to understand eruption dynamics, from bubble growth to the emplacement of pyroclastic flows. Bubble nucleation and growth in magmas is a crucial aspect of explosive volcanism mainly understood through theoretical modeling. Experiments performed in order to improve our understanding of magma vesiculation (Hurwitz and Navon, 1994; Lyakhovsky et al., 1996) mainly use natural rhyolites heated and hydrated at high pressures to form volatile saturated melts, and then decompressed instantaneously allowing bubbles to nucleate and grow. Quenching of samples after different times at the new pressure permits examination of the number, density, and spatial distribution of the bubbles formed. In other experiments, natural samples are heated and the growth of individual, pre-existing bubbles observed and compared with the viscosity and temperature variation of the sample (Bagdassarov et al., 1996).

An important group of experimental models of eruption dynamics, which has received great attention over the last few years, refers to magma fragmentation during explosive eruptions. Fragmentation experiments are basically designed to produce controlled explosions in shock tubes, and include different techniques (Mader, 1998). Some experiments are an extension of the vesiculation experiments described above and involve explosive vaporization of a volatile saturated liquid by sudden decompression, producing explosive boiling. Other fragmentation experiments suddenly decompress a sample of highly viscous vesicular magma, producing experimental pyroclasts very similar to natural samples from Peléan and Vulcanian eruptions (Alidibirov and Dingwell, 1996). These experiments show that brittle fragmentation of magma may occur by expansion of gases in nearly cooled magmatic materials. To simulate the triggering effect of exsolution of volatiles from silicic magmas, experiments have used chemical reactions to generate large volumes of $CO_2$ within non-magmatic liquids, which are then subjected to sudden decompression at room temperature (Mader et al., 1994, 1996). These experiments cause vesiculation and disruption of the magma analogue and suggest that magma fragmentation results from acceleration of the gas–liquid mixture by gas expansion during decompression. Such experiments

provide good analogues of natural process and assist understanding of the physics of magma fragmentation during explosive eruptions. However, experiments involving analogue magmas and shock-tubes cannot be properly scaled to nature restricting the quantitative interpretation of experimental results derived from them.

Phreatomagmatic eruptions, involving the interaction of magma and surface or groundwater, have been successfully investigated using experimental models (Zimanowski, 1998). In such eruptions, the thermal energy of magma is transformed into mechanical energy when water is heated and suddenly vaporized. Experiments designed specifically to reproduce this particular type of explosive interaction have been performed by several researchers (Wohletz and Sheridan, 1983; Wohletz, 1986; Zimanowski et al., 1991, 1997). In pioneering work, Wohletz and Sheridan (1983) used metallic melt, in some cases mixed with molten quartz sand, to approximate the silicate magma, to provide crucial information concerning the control of water/melt ratios on the resulting pyroclasts formation. Further experimental work by Zimanowski and co-workers used remelted volcanic rocks to provide new insights into the process.

The dynamics of eruption columns and pyroclastic flows has been also investigated through experimental models. Experiments on eruption columns (Carey et al., 1988; Sparks et al., 1991; Ernst et al., 1994, 1996) simulate the physical processes that control column behavior and particle sedimentation, allowing one to establish predictive models of ejecta dispersal. The experiments include a constant source of fresh, particle-laden water injected at the base of a tank filled with aqueous saline solution (Fig. 3.10). The experimental design allows variation of the injection rate and the particle loading, two of the most important variables controlling eruption columns.

In a similar way, the emplacement of pyroclastic flows has been investigated through analogue models. Pyroclastic flows represent one of the most important volcanic hazards and a good knowledge of their physics is crucial for elaborating accurate predictive models. Unable to observe the interior of a moving pyroclastic flow,

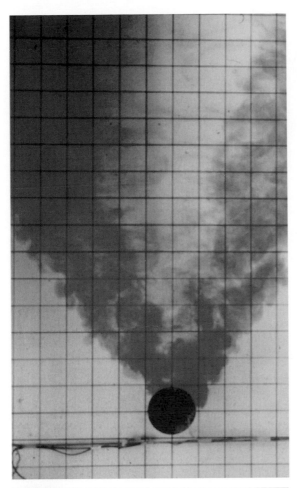

**Fig. 3.10.** Plan-view photograph of an experiment to study the dynamics of eruption plumes. The experiment was designed to study plume bifurcation (Ernst et al., 1994), a phenomenon observed in laboratory experiments and natural eruptions, including that of Rabaul (Papua New Guinea) in 1994. The experiment consists of a low-speed water channel where dyed (cold or warm) water can be injected vertically through a nozzle on the channel floor into cold flowing water, generating turbulent buoyant plumes in a cross-current. Photograph by G. G. J. Ernst.

structures that are observed in natural deposits, helping in the interpretation of their transport physics. Other experiments have investigated the effects of topography on the emplacement of pyroclastic flows (Woods and Bursik, 1994; Woods et al., 1998). In this case, the experiments use a laboratory tank, 1–2 m long, which contains an ambient salt-water medium. An inclined plane simulating a ridged topography is placed in the tank and a finite volume of dense fluid consisting of a mixture of fresh water and solid particles is released into the tank over the inclined plane. The experiments contribute to the understanding of the relative importance of sedimentation and entrainment in generating buoyancy in the pyroclastic flow and formation of co-ignimbrite clouds (see Chapter 1).

Scale and analogue models have been developed to investigate both the geometry of relatively small volcanic cones (e.g., Riedel et al., 2003) on the internal structure and stability of large volcanoes. Formation of collapse calderas and failure of volcano slopes have been successfully reproduced with simple analogue experiments, which allow the reproduction of permanent deformation structures such as fractures and faults. The experiments simulate the rigid crust using cohesive, dry, powder mixtures (sand, fused alumina, flour, etc.), while silicone is used to simulate magma or a mantle plastic layer. For example, three different experimental models have been designed to study the formation of collapse calderas, volcanic depressions originated by the collapse of the magma chamber roof during large volume eruptions. In all cases, a laboratory tank was filled with powder material and the main difference between experiments is the nature of the experimental magma chamber: an ice ball (Komuro et al., 1984), a inflatable balloon (Martí et al., 1994), and silicone (Roche et al., 2000), respectively. The results obtained are similar and allow one to reproduce the formation of tectonic structures (inward dipping faults, ring faults, radial faults) commonly associated with the formation of collapse calderas (Fig. 3.11). Analogue experiments using a cohesive mixture of powder materials simulating a volcanic cone and silicone simulating intrusion of magma or a plastic mantle layer that deforms due to the load of the

volcanologists have designed experimental models addressed to investigate their emplacement mechanism. Some experiments have focused on the effect of fluidization on the emplacement of pyroclastic flows (Wilson, 1980, 1984). This has been achieved in the laboratory by passing gas upwards through a natural pyroclastic deposit, in a similar way to what is used in some industrial applications. These experiments have reproduced

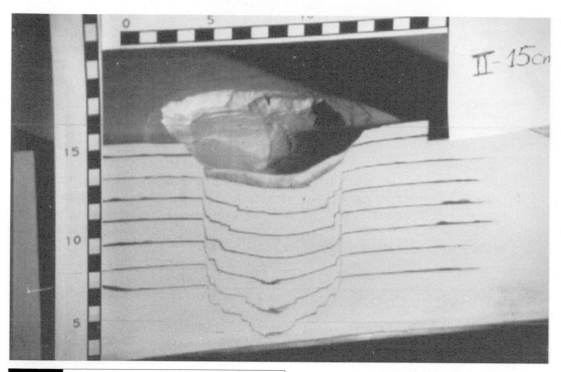

**Fig. 3.11.** Analogue experiments on the formation of a collapse caldera. The objective was to investigate the influence of the geometry and location of the magma chamber on the geometry and location of the resulting caldera. An elastic-skinned balloon was inflated and emplaced within a cohesive powdered material (fused alumina), and deflated to simulate eruption and decompression of a shallow silicic magma chamber (Martí *et al.*, 1994). Scaling of material properties ensures geometrical similarity of the model with nature. The experimental model provides insights into the geometry and motion of faults that control the processes of caldera collapse.

volcanic edifice have been performed to investigate the effect of cryptodomes (Donadieu and Merle, 1998) and spreading (Merle and Borgia, 1996), respectively, on the stability of volcanic edifices.

## Theoretical modeling in volcanology

Theoretical models based on the physical disciplines of thermodynamics, solid and fluid mechanics, etc., have been applied to many different volcanic phenomena and have progressively become an indispensable tool. In several cases, the complexity of the mathematics needed to describe theoretical models has led to the application of computer-based numerical simulations to volcanic processes.

A theoretical model is a simplified abstraction of a given observed phenomenon. Such a model is acceptable when it can not only reproduce experimental data but can also make predictions confirmable by further measurements. Any such theoretical model is characterized by a set of equations, termed the governing equations of the model, which describe the physics of the problem in mathematical notation. Typically, solution of the governing equations cannot be achieved analytically with the use of simple algebra, and numerical techniques must be employed. A numerical simulation is simply the solution of a particular theoretical model under a given (or assumed) set of conditions. Such conditions might include the boundary and initial conditions for the commonest type of governing equations, which are time-dependent differential equations that explain and predict the phenomena. A numerical simulation performed with a certain technique can be checked against other numerical procedures to see whether they yield similar results when applied to the same problem.

## Pre-eruptive models

Modeling pre-eruptive phenomena is important in understanding processes that potentially lead to volcanic eruptions. Correct comprehension of the problem is a crucial aspect in interpreting the meaning of some of the geophysical and geochemical precursors signaling an eruptive event. For example, seismic signals may be recorded months or years before an eruption begins. On the other hand, eruptive products sometimes show clear evidences that fresh magma was intruded into the magma chamber shortly, perhaps weeks, months, or a few years, prior to eruption. Based on previous sections, this fact may be interpreted to indicate that an eruption was triggered by a pressure increase inside the magma chamber related to the intrusion. Consequently, preceding seismic signals may be interpreted to record such intrusions of new magma into the chamber that have caused a critical pressure increase and a subsequent rupture of the chamber walls. The eruption of Mt. Pinatubo, Philippines, in 1991 is a good example of this sequence of events (Newhall and Punongbayan, 1996). Clearly, a good knowledge of the processes occurring during replenishment of a shallow magma chamber might help to predict the time lag between the seismic precursors and the eventual eruption. As shown previously, a first approach to the understanding of magma chamber processes may be obtained from experimental modeling and petrological studies. However, if the magnitude and duration of these thermodynamic and mechanical changes affecting the magma chamber are to be quantified, a theoretical model is required. This is just an example to illustrate how theoretical modeling of pre-eruptive processes can help us to anticipate volcanic eruptions.

Several groups of theoretical models have been developed to understand relevant pre-eruptive processes occurring in shallow magma chambers. Important aspects include cooling, differentiation, and mixing of magma, magma chamber degassing, and the achievement of critical conditions leading to rupture of the magma chamber walls. Magma chamber pressure may increase owing to oversaturation of volatiles when magma cools and differentiates, or by the addition of new magma into the chamber. Two types of theoretical model have been developed to determine the main physical processes that occur in each case. The first group may be called "closed system models," because they investigate the effect of volatile oversaturation on magma pressure, assuming that no new magma is added to or removed from the chamber. The first model of this type to be developed was that of Blake (1984). Although the model was limited to the effect of water in rhyolitic magmas, its general approach can be applied to other magma-volatile systems. This model allowed quantification of the water content required to generate a critical overpressure equal to the tensile strength of the country rock, in terms of parameters such as chamber depth, magma compressibility, and strength of the chamber walls. According to Blake's (1984) model, water contents in the range of 3–6 wt.% are in most cases sufficient to rupture the chamber. Another important conclusion of the study is that fractionated magmas evolving at shallow depth cannot attain water contents much above 6 or 7 wt.%, at which point the chamber becomes critically overpressurized and a volcanic eruption is triggered. More elaborate models aim to calculate chamber overpressure as a function of the amount of crystallization and the solubility law of the volatile species present, assuming some initial mass of gas (Tait et al., 1989; Bower and Woods, 1997; Folch et al., 1998) (Fig. 3.12). These models have shown that the more soluble the volatile species, the more significant is the development of chamber overpressure.

More complex models consider a magma chamber as an open system, where pressure increase is caused by addition of new magma into the chamber. The first quantitative model explaining magma chamber replenishment as a first-order mechanism to trigger volcanic eruptions was that of Blake (1981). According to his model a volume of injected magma of about 0.1% of the chamber volume or approximately 1% when a gas phase is present (Bower and Woods, 1997) is able to produce sufficient overpressure to trigger an eruption.

The period of repose occurring between injection of new magma and any subsequent eruption suggests that the injection of a volume of

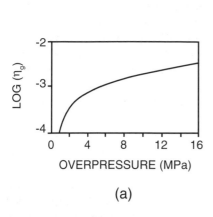

(a)

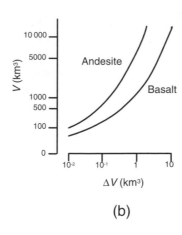

(b)

**Fig. 3.12.** Theoretically calculated magma chamber pressure increases due to volatiles oversaturation during crystallization (a) or replenishment (b). (a) The relationship between the volume fraction of gas ($\eta_g$) and the overpressure as crystallization proceeds, assuming that no gas is present at $t = 0$. Simplified from Tait et al. (1989). (b) The volume of magma, $\Delta V$, that must be introduced into a chamber of constant volume, $V_c$, to trigger eruption, for different magma compositions. Simplified from Blake (1981).

new magma is not, in itself, sufficient to provoke an eruption immediately, and that some secondary effect(s) of the injection causes the critical overpressure. This observation has led to an important group of pre-eruptive models focused on the effects of magma mixing as a volcanic eruption trigger. This idea was first introduced to explain the 1875 Plinian eruption of Askja Volcano, Iceland (Sparks et al., 1977) and has since then been invoked to explain other eruptions including the 1991 eruption of Mt. Pinatubo, Philippines (Pallister et al., 1992). When mafic magma is injected into the base of a chamber containing lighter and cooler silicic magma, a complex sequence of processes occurs (Sparks et al., 1977; Huppert et al., 1982a, 1982b; Campbell and Turner, 1986; Sparks and Marshall, 1986; Snyder and Tait, 1995, 1996). These include cooling and crystallization of the mafic magma, heating and convection of the silicic magma, volatile-oversaturation and subsequent gas release from both magmas, convective overturning, and large-scale mixing between the two magmas. Eruption may occur at any time during this sequence of mixing events if the critical overpressure is attained. To summarize, exsolution of volatiles from the felsic magma due to a increase in temperature or upward convection, or exsolution of volatiles from the mafic magma during cooling and crystallization, have both been cited as mechanisms by which the injection of new magma and subsequent mixing events could contribute to overpressurizing the magma chamber. Folch

and Martí (1998) give a quantification of the relative importance of these processes in causing chamber overpressure.

Volcanic eruptions are mostly supplied with magma through dykes, or other types of sheet intrusions. However, commonly, the eruption frequency of a volcano is only a fraction of the inferred sheet-injection frequency of its source magma chamber, meaning that many sheets must stop within the crust, never reaching the surface to erupt. Thus in order to be able to predict volcanic eruptions and to estimate volcanic risk, not only must the condition for injection of sheets from the chamber be known, but also the condition for sheet arrest (Gudmundsson et al., 1999 and references therein). The types of models mentioned above provide valuable insights into the causes of magma overpressure, but do not treat the resistance limits and rupture mechanisms of the country rock. Another set of theoretical models based mostly on principles of rock mechanics have been developed to explain rock failure and dyke injection under conditions of magmatic overpressure. These models are significant, allowing interpretation of the seismic signals that precede volcanic eruptions, as well as geodetic and gravimetric data from volcano monitoring. They treat the physical principles controlling rupture of dykes and dyke injection from a magma chamber and the general mechanics of dyke propagation and arrest, as well as the proportion of dykes that reach the surface to erupt (Roberts, 1970; Gudmundsson, 1988; Rubin, 1995).

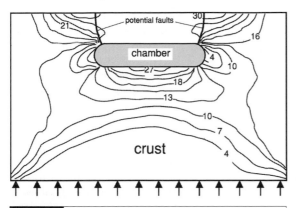

**Fig. 3.13.** Contours of the maximum tensile stress, in MPa, around a sill-like magma chamber at the base of the volcanic area subjected to 10 MPa magmatic excess pressure. The results allow deduction of the position of ring faults allowing a caldera collapse event to occur, and show that the depression would be very similar in dimension to the cross-sectional area of the source chamber. In these boundary-element calculations the static Young's modulus of the crustal plate is 40 GPa and Poisson's ratio is 0.25. From Gudmundsson *et al.* (1997).

The stress field associated with the source magma chamber is the main parameter controlling whether a sheet is intruded into surrounding rocks from the chamber. This stress field requires detailed analysis, particularly regarding the three-dimensional shape of the chamber and the location of stress concentrations where sheet injection is most likely. Traditional models of rupture of magma chambers and dyke propagation, however, make drastic simplifying assumptions, not taking into account the complexities of magma dynamics and the geometric evolution of dykes during emplacement. Such complexities must be analyzed numerically in combination with analytical work to advance our understanding. Numerical modeling is now widely used to simulate the mechanics of magma chambers and to study the conditions of stress distribution for magma chamber rupture for different chamber depths, shapes, volumes, rock strengths, and loading conditions (Gudmundsson, 1988, 1998; Gudmundsson *et al.*, 1997). These numerical results provide, for example, good insights on the most favorable stress conditions that in each case can determine the chamber rupture and eventually a volcanic eruption (Fig. 3.13).

The conditions leading to sheet arrest form a subject that has received little attention and is poorly understood. One possibility is that sheet propagation is arrested when magma flow into the sheet tip is blocked by solidification when the propagation velocity decreases below a critical level. Alternatively, arrest may occur when the sheet enters crustal layers where sheet-normal compressive stresses exceed the magmatic overpressure driving the sheet. These and others possibilities might be tested in detail by combining extensive field studies of exposed sheet tips with theoretical modeling (Gudmundsson *et al.*, 1999).

Magma degassing has also been modeled theoretically. Differences in eruptive behavior broadly correlate with differences in magma type, with gas-poor basaltic magma more likely to erupt as lava, whereas gas-rich evolved magma types typically erupt explosively. These generalizations are not always correct, however, as some mafic magmas may erupt with great energy and destructive power, while felsic magmas may erupt quietly to form thick lava flows or domes with little impact on the surrounding environment. Magma degassing in shallow chambers and conduits seems to play a significant role in controlling such variations in eruptive behavior. Theoretical modeling of magma degassing consists mainly of thermodynamic calculations, based on mineral composition data combined with experimental determinations of volatiles solubility and phase equilibria, and has mainly been applied to determining degassing kinetics and magma chamber evolution (Carroll and Holloway, 1994 and references therein).

## Eruptive models

Models of volcanic eruptions aim to describe physical aspects of the eruptive process using concepts from thermodynamic and fluid mechanical theory. Ideally, an eruptive model should consider simultaneously the physical processes that occur within the magma chamber, within the volcanic conduit during the ascent of magma, and within the atmosphere after eruption, since the processes that occur in each domain can affect the others. Unfortunately, a successful solution of such a large, coupled problem remains unachievable as yet. A major difficulty is that each region

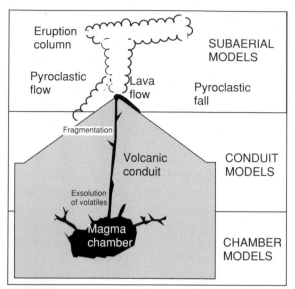

**Fig. 3.14.** Domains of theoretical modeling in physical volcanology. Magma chamber models deal with the dynamics of withdrawal during the course of an eruption. Conduit models focus on the physics of the conduit and aim to describe processes such as exsolution of volatiles, growth of bubbles, or magma fragmentation. Subaerial models focus on the characteristics, dynamics, and emplacement of erupted products and include, among other aspects, models for the eruptive column, pyroclastic and lava flows, or pyroclastic falls.

is considerably different physically and the governing equations for each have different mathematical requirements. Moreover, many of the processes involved are not well understood. To simplify the problem, modeling of eruptive processes has traditionally been considered in the three spatial domains separately: magma chamber, volcanic conduit, and the Earth's surface (Fig. 3.14). *Magma chamber models* aim to describe the dynamics of magma withdrawal over the course of a volcanic eruption. *Conduit models* focus on processes occurring during the ascent of magma in the volcanic conduit, including exsolution of volatiles, nucleation and growth of gas bubbles, and, in the case of explosive volcanism, magma fragmentation. Conduit models also provide information on physical conditions at the vent. Finally, *subaerial models* refer to the characteristics, dynamics, and subsequent emplacement of the erupted materials generally assuming given values for the physical

parameters (volatile content, eruption rate, size of the vent, etc.) of the magma at the vent.

Future eruptive models should address the coupling of these partial models that to date have considered only parts of the general problem. Efforts should focus on the creation of a general model able to describe simultaneously these three spatial domains. Recently, some attempts have been made to partially couple conduit and subaerial models (Dobran *et al.*, 1993; Neri *et al.*, 1998). The idea behind this has been to use as input boundary conditions for the subaerial models the conditions at the vent computed independently from a conduit model. The same idea could be also extended to couple the chamber and the conduit models.

## Magma chamber models

A first group of syn-eruptive magma chamber mathematical models comprises analytical approaches that consider only differences between initial and final stages and do not contemplate temporal variations of the physical parameters during the course of the eruption. Although there are limitations inherent to these models, they provide a first-order approximation to what can be expected from an eruptive event. Thus, for instance, an important question in terms of volcanic hazard concerns the amount of material that can potentially be extruded during a volcanic eruption. Several analytical models aim to answer this question calculating the differences of mass in the chamber between an initial overpressurized stage, associated with the beginning of the eruption, and a final equilibrium stage associated with the end of it (Blake, 1981; Bower and Woods, 1997; Folch *et al.*, 1998) (Fig. 3.15). According to these models, the mass erupted depends critically on whether magma is undersaturated or oversaturated in volatiles prior to the eruption. Undersaturated magma is essentially incompressible, so that the extrusion of small fractions (between 0.1% and 1%) of the chamber contents, is sufficient to relieve the overpressure, bringing the chamber to the lithostatic equilibrium and ending the eruption. The small expansion coefficient of the resident magma and contraction of the chamber walls in response to the pressure decrease therefore restricts the

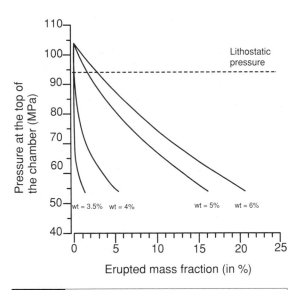

**Fig. 3.15.** Theoretically calculated variations in magma pressure at the chamber roof during a Plinian eruptions as a function of erupted mass fraction for different water contents. The magma chamber is located at 4 km below the Earth surface, equivalent to a lithostatic pressure of 94 Mpa, and the overpressure necessary to trigger the eruption is assumed to be 10 Mpa. The diagram shows the decrease of magma chamber pressure during the eruption. Note that the pressure inside the chamber can fall well below the lithostatic, eventually causing the collapse of the chamber roof if required mechanical conditions are also met. After Martí et al. (2000).

importance in constraining the inflow conditions of conduit models, described in the following section, but have received surprisingly little attention. The first approaches assumed magma to be an incompressible fluid of constant viscosity, and considered only simple chamber geometries (Spera, 1984). Despite these oversimplifications, these models allow the study of the evacuation process and how it depends on the aspect ratio of the magmatic reservoir, by considering evacuation isochrons. An evacuation isochron is a locus of points within the magma chamber marking parcels of magma which will arrive at the conduit entrance at the same time. Successive models have included compositional gradients within the chamber (Spera et al., 1986; Trial et al., 1992), and when combined with geological and geochemical studies of volcanic deposits can lead to the reconstruction of the pattern of pre-eruptive compositional zonation within the chamber.

The assumption of magmatic incompressibility has restricted the applicability of numerical models to eruptions triggered by the injection of fresh magma into the chamber. Recent approaches have tried to overcome this by considering some parts of the chamber filled with vesiculated, i.e., compressible, magma. Such models have allowed simulations of eruptions driven by volatile oversaturation (Folch et al., 1998). Results from these new mathematical simulations allow one to deduce the temporal evolution of the most relevant physical parameters, including pressure, position of the exsolution level, velocity field, eruption rate, and the amount of erupted material, during a Plinian eruption from a shallow, voltile-rich, magma chamber (Fig. 3.16).

amount of material that can be extruded from the chamber under these conditions. However, if magma is initially volatile oversaturated and contains gas bubbles, the magmatic mixture has a much higher compressibility, and the amount of magma that can be extruded from the chamber is much greater (between 1% and 10%). Since such models predict the mass fraction of erupted magma, they can be used a posteriori to constrain chamber size once the volume of material produced by a particular eruptive event is known.

A second group of chamber models looks at temporal variations of the physical parameters and the dynamics of the magma withdrawal process. The introduction of time-dependence necessitates the use of numerical techniques to solve the governing equations. The results and predictions of these models for the dynamics of magma withdrawal from crustal reservoirs are of crucial

## Conduit models

During the course of a volcanic eruption, the ascent of magma up the volcanic conduit and its interaction with the conduit walls exert a strong influence. For this reason, theoretical conduit models are important in anticipating the properties and the potential hazards of the eruption. For example, depending on its composition and content of volatiles, ascending magma can fragment explosively in the conduit. If these parameters

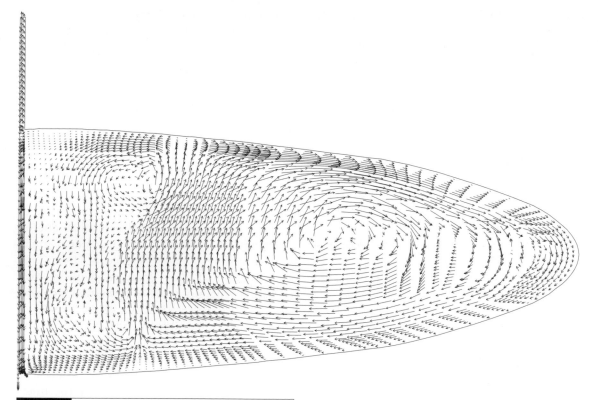

**Fig. 3.16.** Example of a numerical calculation to determine the velocity field for magma inside a chamber of known geometry and volume, after 22 hours of simulated Plinian eruption. The model allows prediction of the movement of a particular parcel of magma inside the chamber at any instant of time. This has implications for the interpretation of eruption products and helps in the understanding of the dynamics of such eruptions. Simplified from Folch *et al.* (1998).

could be known a priori, conduit models could allow prediction of when and where magma will fragment, and what will be the destructive potential of the eruption.

Conduit models involve the numerical solution of the equations describing the mechanics of fluids. These include the continuity, momentum and thermal equations, derived respectively from the general principles of mass, momentum, and energy conservation. In addition, a state law and several constitutive relationships characterizing the physical and rheological properties of the ascending magma require definition. The aim of the models is to find the values of variables such as pressure, density, velocity, temperature,

or gas volume fraction inside the volcanic conduit. Ideally, magma ascent along conduits should be modeled by solving a complex three-dimensional, unsteady (time-dependent), three-phase (gas–liquid mixture plus solid crystals), and non-homogeneous flow problem. Indeed, other effects such as crack formation and propagation, erosion of conduit walls, and the existence of a multicomponent gas phase should also be taken into account. Clearly, however, the complete solution of such problem is difficult while knowledge of the variables involved and their interrelations are only partially known. Therefore, several simplifications including the neglect of certain terms of the governing equations, or the use of a simplified state law and constitutive equations, have typically been made. The great majority of conduit models neglect time-dependent terms of the governing equations and restrict themselves to steady conditions. Steady models do not consider coupling with either chamber dynamics or subaerial processes and are best applied to phases of an eruption during which no significant variation of flow parameters occurs. The different components (i.e., gas bubbles and liquid) of a magmatic mixture have different dynamic behavior and require separate solutions of the governing equations. Homogeneous models simplify the simulation by assuming thermal and mechanical equilibrium between liquid and gas phases, while non-homogeneous models take such mechanical disequilibrium into account. The homogeneous approach is justified if magma viscosity is high enough to prevent relative movement between gas bubbles and liquid, and is a generally accepted approach for chemically evolved magmas in the bubbly flow regime. However, the assumption of homogeneity is not justified above the fragmentation level where the viscosity of the mixture is much lower, allowing for relative motion between liquid and gas phases.

The theoretical basis for conduit models comes from the important contribution of Sparks (1978), who introduced a series of fundamental concepts. These include the concepts of saturation, nucleation, and exsolution levels, the growth of gas bubbles by diffusion and decompression during magma ascent, and the condition of magma fragmentation. These ideas became the basis for a first generation of comprehensive conduit models which, despite their simplicity, gave insights into magma ascent (Wilson et al., 1980; Wilson and Head, 1981; Gilberti and Wilson, 1990). These early models are one-dimensional, steady, and homogeneous, and ignore thermal effects. Their numerical solution gives a first approach to understanding the roles of magma density, velocity, pressure, and conduit radius, and permits effects of volatile content, magma viscosity, and conduit shape on the outflow conditions. Since vent conditions are closely related to eruptive style (Sparks and Wilson, 1976; Wilson 1976), these models were able to explain different explosive regimes in terms of the properties of the ascending magma. For example, a decrease in magma volatile content associated with the extrusion of deeper layers of a stratified magma chamber, or an increase in vent radius due to conduit erosion, produce lower exit velocities, explaining the collapse of Plinian columns and the formation of pyroclastic flows (Fig. 3.17). From a hazard perspective eruptive style can be predicted providing knowledge of the properties

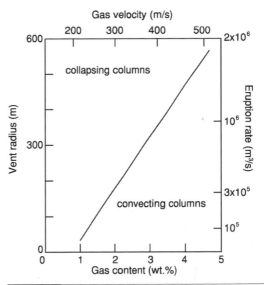

**Fig. 3.17.** General model relating vent radius, magmatic gas content, eruption rate, and gas velocity to convecting and collapsing states of eruption columns. After Wilson et al. (1980).

of the ascending magma in terms of composition, water content, etc.

A more recent and realistic generation of conduit models introduces non-homogeneity (Dobran, 1987, 1992). In this approach separate equations are developed for the liquid and gas phases describing their mechanical relationships. Non-homogeneous models give results that can differ appreciably from homogeneous types, and results depend greatly on the choice of free parameters. A detailed discussion between the differences between these types of conduit models is given by Papale (1996).

Contemporaneous non-homogeneous conduit models consider conduit erosion, through motion of particles against the conduit walls (Macedonio et al., 1994), and incorporate other refinements mainly concerning magma properties (Papale and Dobran, 1993, 1994; Papale et al., 1998). The recent versions (Papale et al., 1998) are the sharpest tools available for predicting flow in volcanic conduits and assessing the roles of magmatic composition on the dynamics of eruptions. These new conduit models represent a significant advance on the development of numerical simulations aimed to predict eruptive behavior.

### Subaerial models

Modeling of subaerial volcanic processes has been one of the most active areas of volcanology over the last two decades, owing to its great importance for the assessment of volcanic hazards. Reviews of the subject are given by Neri and Macedonio, (1996), Sparks *et al.* (1997), Freund and Rosi (1998), and Gilbert and Sparks (1998), who give detailed information on this group of theoretical models. In the context of explosive volcanism, such models solve two-phase flow problems involving gas plus ash particles, for jets, plumes, and gravity currents to find the dynamics of the transported volcanic products. Efforts have focused mainly on modeling the dynamics of eruptive columns including their collapse to produce pyroclastic flows, and the dispersal of pyroclastic material into the atmosphere. Given certain physical conditions at the vent, the models aim to assess pyroclastic transport processes and related volcanic hazards.

The results of this group of theoretical models are related directly to the short- and long-term predictions of the effects of volcanic eruptions. For example, the long-term environmental impact of explosive eruptions is related mainly to climate change, such as temperature decreases, which is of global scale. Climate changes have occurred following the recent eruption of Mt. Pinatubo in 1991; and larger eruptions, such as that of Toba on the island of Sumatra in Indonesia about 74 000 years ago, or more recently that of Tambora Volcano, also in Indonesia, in 1815, are believed to have produced much more dramatic variations (Rampino *et al.*, 1988). Such climate effects occur due to the injection of volcanic aerosols and ash particles into the high atmosphere, carried by the eruption column and associated ash clouds (see Chapter 5). Models of subaerial processes help to predict the transport of volcanic particles in the atmosphere, helping to anticipate potential climate effects of current and future eruptions. The development of theoretical models of pyroclastic flows emplacement and the movement of lava flows and lahars are also important given the short-term

consequences of these types of volcanic processes. Theoretical results can be input into a geographical information system (GIS) package developed specifically to predict the effects of future eruptions, allowing areas likely to be affected by a particular volcanic hazard to be assessed (Gomez-Fernandez, 1998; Gomez-Fernandez and Macedonio, 1998) (Fig. 3.18).

## Volcano monitoring

This chapter has discussed the importance of understanding volcanic phenomena in order to accurately predict the occurrence and effects of future eruptions. How volcanic processes are studied in the field and laboratory, and modeled by experiments and theory have also been covered. However, as mentioned in the introduction, this comprises only part of the information needed to correctly anticipate volcanic eruptions. Information is also required about the present state of a volcano if any changes in its behavior are to be noted. Changes such as deformation of the volcano, seismicity, temperature variation, changes in gas composition, etc. will provide information about hidden movements of magma and related fluids, and a correct interpretation will most likely come from all the precursory phenomena together.

Experience gained from reconstructing historical volcanic eruptions has shown that large volcanoes may remain inactive for hundreds to thousands of years before erupting once more. Reactivation, however, may occur over a very short period of a few months to a few years, showing the importance of a well-equipped volcano observatory to alert attention promptly to signs of renewed activity. Several different types of monitoring equipment are required to detect the various precursory signals of a forthcoming eruption. Traditionally, volcano monitoring has focused on ground-based geophysical and geochemical techniques. However, technological improvements over the last 20 years have brought the addition of sophisticated remote-sensing techniques based on radar or satellite systems to volcano monitoring. Monitoring is one of the main aspects of volcanology and has received continuous attention from the scientific community.

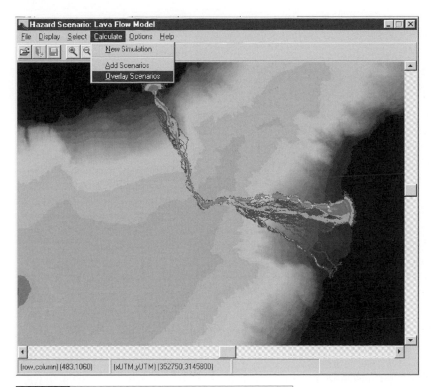

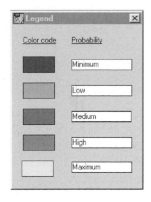

**Fig. 3.18.** Example of volcanic hazard scenario obtained from a computer-based geographical information system (GIS) tool, specifically aimed to volcanic risk assessment. This example shows the simulation of a lava eruption from the Dorsal Ridge volcano in Tenerife, Canary Islands. It allows zones to be defined according to their potential hazard (Gomez-Fernandez, 1998; Gomez-Fernandez and Macedonio, 1998).

Comprehensive reviews containing detailed explanations of the techniques of volcano monitoring have recently been given by McGuire *et al.* (1995), Scarpa and Tilling (1996), Sigurdsson (2000), and Sparks (2003) to which the readers are referred for more information. Early detection and tracking of volcanic pleemes by radar or satellite remote sensing is not discussed here (e.g., see Rose *et al.*, 2000; Lacasse *et al.*, 2004).

## Geophysical and geodetic monitoring

Geophysical and geodetic monitoring techniques include ground-based and remote methods designed to allow changes in the main physical properties of the volcano and its surroundings to be detected. The main physical signals related to mass movement in the interior of the volcano and that can be observed at surface are ground deformation, seismic activity, temperature variations, gravity and stress-field changes, and electromagnetic effects.

Deformation of the ground surface is one of the main precursory signals in active volcanic areas. Ground deformation is commonly attributed to an increase in magma chamber pressure (Mogi, 1958) or the pressurization of a shallow aquifer by a magmatic heat source (Bonafede, 1991). Whichever the case, a pressure increase at depth will cause the surrounding media to deform, which may be measured at the surface in terms of vertical and/or horizontal displacements of previously surveyed reference points. Since pressurization is a major potential eruptive trigger, monitoring and interpreting surface deformation patterns is a key aspect in volcanic eruption forecasting, and has been an active research topic over the last few decades (Van Der Laat, 1996; Dvorak and Dzurisin, 1997 and references therein). Ground deformation has traditionally been measured by precise optical leveling, using electronic distance measurement (EDM) equipment, and tiltmeters, and significant progress is now being made with the use of satellite based systems. Global positioning system (GPS) instruments, using a group of more

**Fig. 3.19.** Seismic monitoring and ground deformation are the main methods used in volcano surveillance. Magma movement inside the volcano may eventually cause volcanic activity, generally accompanied by an increases in seismic activity, surface deformation, changes in gravity and magnetic fields, temperature variations, and changes in the composition of escaping gases. Volcano observatories, using appropriate techniques, record these precursory signals, allowing future eruptions to be anticipated.

than 20 fixed for reference satellites, allow topographic coordinates of latitude, longitude, and elevation to be measured at any point of the Earth's surface to a precision of a few centimeters, irrespectively of topography. Compared to traditional methods GPS allows the monitoring of larger areas and is more efficient and easy to use. Another satellite-based technique being increasingly used in ground deformation monitoring is synthetic aperture radar (SAR) interferometry. This was introduced in 1974 to generate topographic maps, and has since been used to detect changes in the Earth's surface. Differences are detected by superimposing radar images of large areas, taken at different times. The current maximum frequency of sampling is monthly. Observable ground deformation is of the order of a few centimeters.

Seismic activity related to earthquakes or tremors is one of the main indicators of volcanic unrest and seismic monitoring is one of the main tools of geophysical volcano surveillance (McGuire *et al.*, 1995; Scarpa and Tilling, 1996) (Fig. 3.19). Pressure increases and movements of magma and related fluids inside the volcano are normally associated with the opening of fractures, which causes seismic signals usually recorded by short-period seismometers. Data interpretation allows determination of the energy and location of earthquake hypocenters and their evolution with time. Seismic data

combined with measurements from stress meters are occasionally used to determine changes in the accompanying stress-field associated with the seismic events, and provide a method to distinguish volcano-tectonic events from other purely tectonic events unrelated to volcanism.

Gravimetric and magnetic methods of volcano monitoring are also used. The Earth's gravitational and magnetic fields can be disturbed by volcanic activity, owing to the movement of magma beneath the surface. Gravity measurements are usually carried out in combination with observations of ground deformation. This is in order to distinguish small variations in the gravity field due to subsurface mass or density changes from elevation changes caused by magma intrusion or withdrawal (McGuire, 1995). Displacements of fresh magma inside the volcano can also cause small variations of the intensity of the magnetic field, which can be identified as precursory signals of eruptive activity (Zlotnicki, 1995). Finally, geoelectrical methods, including mainly the development of resistivity maps of the volcano's interior, allow any variation attributable to magma intrusion to be detected (Lenat, 1995).

## Geochemical methods

When combined with geophysical methods, the analysis of volcanic gases constitutes a powerful tool for volcano monitoring. The intrusion of fresh magma into the volcanic system, and the ascent of magma toward the Earth's surface, can cause profound changes in the composition of gases escaping as fumaroles or diffusively from the volcano. Continuous monitoring equipment is increasingly being installed at active volcanoes in order to detect any change in state by measuring the concentration of different gas species. The gases ($H_2O$, $CO_2$, Cl, S, $SO_2$, etc.) are released in significant amounts through near-vent, high-temperature fumaroles and through newly opened fractures, when fresh magma reaches shallow depths. These geochemical indicators, occurring with other precursory signals such as thermal anomalies, ground deformation, and seismic activity, would confirm the proximity of a forthcoming eruption. Diffuse degassing, through subsurface fractures, of species such as $CO_2$, helium, and radon, which can be measured in soils or springs, provides information about the overall permeability of the volcanic edifice, the potential for degassing from areas other than the active crater, and the ability of a volcano to release large quantities of volcanic gases diffusively (Stix and Gaonac'H, 2000). Direct measurement of volcanic gas compositions is difficult work, owing to the obvious logistical difficulties and risk involved, and may be complemented by remote sensing of the gas plume. This has the advantage of distance from the volcano, and the greater sampling frequencies that may be obtained than when sampling volcanic gases manually.

Measurements of radionuclides released from volcanoes provide constraints on the rate of deep magma input to shallow depth and the rate of magma degassing. An increase in the first parameter provides evidence of increasing activity. Knowledge of the second parameter helps predict the volume of magma expelled during eruptions, which is of particular interest for studies of explosive activity. Besides providing a geochemical tracer of shallow magma evolution through crystallization and degassing, radionuclide studies offer a unique opportunity to estimate the timescales of magmatic processes, essential for any attempt to predict its future evolution and associated volcanic risk. Analyses of short-lived nuclides released in gases is becoming one of the effective methods of continuous geochemical monitoring at active volcanoes.

## Final remarks

This chapter has briefly summarized the main methods used in contemporaneous volcanology to understand how volcanoes work, and to monitor their current activity. Previous sections have covered the causes and main features of the eruption process, and have described how field and laboratory studies combined with experimental and theoretical modeling constitute the main tools for discovering a particular volcano's characteristics and likely future behavior. We have emphasized mainly the application of mathematical modeling to the study of volcano dynamics, as this is an emerging and powerful way of anticipating volcano behavior. However, all theoretical models need validation by observation, which is provided by field, laboratory, and experimental data. The examples of techniques included here comprise only a brief survey of the wide spectrum available, but cover the main problems of physical volcanology.

A good knowledge of volcanic processes is required to ensure the correct interpretation of eruptive precursors. Geophysical and geochemical monitoring of volcanoes is able to detect precursory signals of eruptive activity. However, volcano forecasting is required not only to detect whether a particular volcano has become active, but also whether renewed activity will culminate in an eruption, and how the eruption will proceed. Mitigation of volcanic hazards will be effective only if monitoring systems are available working currently and if the effects of a forthcoming eruption can be anticipated correctly. Clearly, volcanic activity cannot be stopped but we can avoid its effects and reduce its potential damage if we can predict it well in advance. Physical volcanology and volcano monitoring need, therefore, to work together. Although there is still a long way to go, present knowledge of volcanic processes is adequate for the task of anticipating future activity. Modern GIS have considerably improved the management and mitigation of volcanic risks, particularly when a good knowledge of the volcano and its past activity has been compiled. Integration of data from real-time monitoring and the results of numerical modeling into GIS frameworks is one of modern volcanology's major challenges for the near future.

## References

Alidibirov, M. and Dingwell, D. B. 1996. Magma fragmentation by rapid decompression. *Nature*, **380**, 146–148.

Bagdassarov, N., Dingwell, D., and Wilding, M. 1996. Rhyolite magma degassing: an experimental study of melt vesiculation. *Bulletin of Volcanology*, **57**, 587–601.

Blake, S. 1981. Volcanism and the dynamics of open magma chambers. *Nature*, **289**, 783–785.

1984. Volatile oversaturation during the evolution of silicic magma chambers as an eruption trigger. *Journal of Geophysical Research*, **89**, 8237–8244.

Blake, S. and Fink, J. H. 1987. The dynamics of magma withdrawal from a density stratified dyke. *Earth and Planetary Sciences Letters*, **85**, 516–524.

Bonadonna, C., Ernst, G. G. J., and Sparks, R. S. J. 1998. Thickness variations and volume estimates of tephra fall deposits: the importance of particle Reynolds number. *Journal of Volcanology and Geothermal Research*, **81**, 173–187.

Bonafede, M. 1991. Hot fluid migration: an efficient source of ground deformation: application to the 1982–1984 crisis at Plegrean Fields, Italy. *Journal of Volcanology and Geothermal Research*, **48**, 187–198.

Bower, S. M. and Woods, A. W. 1997. Control of magma volatile content and chamber depth on the mass erupted during explosive volcanic eruptions. *Journal of Geophysical Research*, **102**, 10273–10290.

Bryan, S. E., Cas, R. A. F., and Martí, J. 2000. The 0.57 Ma plinian eruption of the Granadilla Member, Tenerife (Canary Islands): an example of complexity in eruption dynamics and evolution. *Journal of Volcanology and Geothermal Research*, **103**, 209–238.

Campbell, I. H. and Turner, J. S. 1986. The influence of viscosity on fountains in magma chambers. *Journal of Petrology*, **27**, 1–30.

Carey, S. N., Sigurdsson, H., and Sparks, R. S. J. 1988. Experimental studies of particle-laden plumes. *Journal of Geophysical Research*, **93**, 314–328.

Carmichael, I. S. E., Turner, F. J., and Verhoogen, J. 1974. *Igneous Petrology*. New York, McGraw-Hill.

Carroll, M. C. and Holloway, J. R. (eds.) 1994. *Volatiles in Magmas*, Reviews in Mineralogy no. 30. Washington, DC, Mineralogical Society of America.

Cas, R. A. F. and Wright, J. V. 1987. *Volcanic Successions Modern and Ancient*. London, Allen and Unwin.

Condomines, M., Hémond, C. H., and Allègre, C. J. 1988. U–Th–Ra radioactive disequilibria and magmatic processes. *Earth and Planetary Science Letters*, **90**, 243–262.

Condomines, M., Tanguy, J. C., and Michaud, V. 1995. Magma dynamics at Mt. Etna: constraints from U–Th–Ra–Pb radioactive disequilibria and Sr isotopes in historical lavas. *Earth and Planetary Science Letters*, **132**, 25–41.

Dingwell, D. 1998. Magma degassing and fragmentation. In A. Freund and M. Resi (eds.) *Modelling Physical Processes of Exposive Volcanic Eruptions*. Amsterdam, The Netherlands, Elsevier, pp. 1–25.

Dobran, F. 1987. Nonequilibrium modeling of two-phase critical flows in tubes. *Journal of Heat Transfer*, **109**, 731–738.

1992. Nonequilibrium flow in volcanic conduits and application to the eruptions of Mt. St. Helens on May 18, 1980 and Vesuvius in AD 79. *Journal of Volcanology and Geothermal Research*, **49**, 285–311.

Dobran, F., Neri, A., and Macedonio, G. 1993. Numerical simulation of collapsing volcanic columns. *Journal of Geophysical Research*, **98**, 4231–4259.

Donadieu, F. and Merle, O. 1998. Experiments on the indentation process during cryptodome intrusions: new insights into Mount St. Helens deformation. *Geology*, **26**, 79–82.

Dvorak, J. J. and Dzurisin, D. 1997. Volcano geodesy: the search for magma reservoirs and the formation of eruptive vents. *Reviews of Geophysics*, **35**, 343–384.

Ernst, G. G. J., Davis, J. P., and Sparks, R. S. J. 1994. Bifurcation of volcanic plumes in a crosswind. *Bulletin of Volcanology*, **56**, 159–169.

Ernst, G. G. J., Sparks, R. S. S., Carey, S. N., *et al.* 1996. Sedimentation from turbulent jets and plumes. *Journal of Geophysical Research*, **101**, 5575–5590.

Fisher, R. V. and Schmincke, H. U. 1984. *Pyroclastic Rocks*. Berlin, Germany, Springer-Verlag.

Folch, A. and Martí, J. 1998. The generation of overpressure in felsic magma chambers by replenishment. *Earth and Planetary Science Letters*, **163**, 301–314.

Folch, A., Martí, J., Codina, R., et al. 1998. A numerical model for temporal variations during explosive central vent eruptions. *Journal of Geophysical Research*, **103**, 20883–20899.

Freund, A. and Rosi, M. (eds.) 1998. *Developments in Volcanology*, vol. 4, *From Magma to Tephra*. Amsterdam, The Netherlands, Elsevier.

Freund, A. and Tait, S. R. 1986. The entrainment of high-viscosity magma into low-viscosity magma in eruption conduits. *Bulletin of Volcanology*, **48**, 325–339.

Gilbert, G. J. and Sparks, R. S. J. (eds.) 1998. *The Physics of Explosive Volcanic Eruptions*, Geological Society Special Publication no. 145. London, The Geological Society.

Gilberti, G. and Wilson, L. 1990. The influence of geometry on the ascent of magma in open fissures. *Bulletin of Volcanology*, **52**, 515–521.

Gomez-Fernandez, F., 1998. Development of a volcanic risk assessment information system for the prevention and management of volcanic crisis: stating the fundamentals. In C. A. Brebbia and P. Pascolo (eds.) *GIS Technologies and their Environmental Applications*. Computational Mechanics Publications, pp. 111–120.

Gomez-Fernandez, F. and Macedonio, G. 1998. Integration of physical simulation models in the frame of a GIS for the development of a volcanic assessment information system. In C. A. Brebbia, J. L. Rubio, and J. L. Uso (eds.) *Risk Analysis*. Computational Mechanics Publications, pp. 265–274.

Gudmundsson, A. 1988. Effect of tensile stress concentration around magma chambers on intrusion and extrusion frequencies. *Journal of Volcanology and Geothermal Research*, **35**, 179–194.

1998. Magma chambers modelled as cavities explain the formation of rift zone central volcanoes and the eruption and intrusion statistics. *Journal of Geophysical Research*, **103**, 7401–7412

Gudmundsson, A., Martí, J., and Turon, E. 1997. Stress fields generating ring faults in volcanoes. *Geophysical Research Letters*, **24**, 1559–1562.

Gudmundsson, A., Marinoni, L., and Martí, J. 1999. Dyke injection and arrest: implications for volcanic hazards. *Journal of Volcanology and Geothermal Research*, **88**, 1–13.

Hall, A. 1996. *Igneous Petrology*. Chichester, UK, John Wiley.

Huppert, H. E. and Sparks, R. S. J. 1980. The fluid dynamics of basaltic magma chambers replenished by influx of hot, dense ultrabasic magma. *Contributions to Mineralogy and Petrology*, **75**, 279–289.

1988. The fluid dynamics of crustal melting by injection of basaltic sills. *Transactions of the Royal Society of Edinburgh*, **79**, 237–243.

Huppert, H. E., Turner, J. S., and Sparks, R. S. J. 1982a. Replenished magma chambers: effects of compositional zonation and input rates. *Earth and Planetary Science Letters*, **57**, 345–357.

1982b. Effects of volatiles on mixing in calc-alkaline magma systems. *Nature*, **297**, 554–557.

1983. Laboratory investigations of viscous effects in replenished magma chambers. *Earth and Planetary Science Letters*, **65**, 377–381.

Hurwitz, S. and Navon, O. 1994. Bubble nucleation in rhyolitic melts: experiments at high pressure, temperature and water content. *Earth and Planetary Science Letters*, **122**, 267–280.

Komuro, H., Fujita, Y., and Kodama, K. 1984. Numerical and experimental models on the formation mechanism of collapse basins during the Green Tuff orogenesis of Japan. *Bulletin of Volcanology*, **47**, 649–666.

Lacasse, S., Karlsdóttir, S., Larsen, G., et al. 2004. Weather radar observations of the Hekla 2000 eruption cloud, Iceland. *Bulletin of Volcanology*, **66**, 457–473.

Lambert, G., Le Cloarec, M. F., Ardouin, B., et al. 1986. Volcanic emission of radionuclides and magma dynamics. *Earth and Planetary Science Letters*, **76**, 185–192.

Le Cloarec, M. F. and Marty, B. 1991. Volcanic fluxes from volcanoes. *Terra Nova*, **3**, 17–27.

Le Cloarec, M. F., Allard, P., Ardouin, B., et al. 1992. Radioactive isotopes and trace elements in gaseous emissions from White Island, New Zeland. *Earth and Planetary Science Letters*, **108**, 19–28.

Lenat, J. F. 1995. Geoelectrical methods in volcano monitoring. In W. J. McGuire, C. Kilburn and J. Murray (eds.) *Monitoring Active Volcanoes: Strategies, Procedures and Techniques*. London, UCL Press, pp. 248–274.

Linde, A. T. and Sacks, I. S. 1998. Triggering of volcanic eruptions. *Nature*, **395**, 888–890.

Lyakhovsky, V., Hurwitz, S., and Navon, O. 1996. Bubble growth in rhyolitic melts, experimental and numerical investigation. *Bulletin of Volcanology*, **58**, 19–32.

Macedonio, G., Dobran, F., and Neri, A. 1994. Erosion processes in volcanic conduits and application to the AD 79 eruption of Vesuvius. *Earth and Planetary Science Letters*, **121**, 137–152.

Mader, H. M. 1998. Conduit flow and fragmentation. In J. S. Gilbert and R. S. J. Sparks (eds.) *The physics of Explosive Volcanic Eruptions*, Geological Society Special Publication no. 145. London, The Geological Society, pp. 51–71.

Mader, H. M., Phillips, J. C., and Sparks, R. S. J. 1996. Dynamics of explosive degassing of magma: observations of fragmenting two phase flows. *Journal of Geophysical Research*, **101**, 5547–5560.

Mader, H. M., Zhang, Y., Phillips, J., *et al.* 1994. Experimental simulations of explosive degassing of magma. *Nature*, **372**, 85–88.

Marsh, B. D. 1989. Magma chambers. *Annual Reviews of Earth and Planetary Sciences*, **17**, 439–474.

Martí, J. 1993. Paleovolcanismo. In J. Martí and V. Araña (eds.) *La Vulcanología Actual: Nuevas Tendencias*. Madrid, Spain, Consejc Superier de Investigaciones Cientificas, pp. 531–578.

Martí, J. and Araña, V. (eds.), 1993. *La Vulcanología Actual: Nuevas Tendencias*. Madrid, Spain, CSIC.

Martí, J., Ablay, G., Redshaw, L. T., *et al.* 1994. Experimental studies of collapse calderas. *Journal of the Geological Society of London*, **151**, 919–929.

Martí, J., Soriano, C., and Dingwell, D. 1999. Tube pumices: strain markers of a ductile–brittle transition in explosive eruptions. *Nature*, **402**, 650–653.

Martí, J., Folch, A., Neri, A., *et al.* 2000. Pressure evolution during explosive caldera-forming eruptions. *Earth and Planetary Science Letters*, **175**, 275–287.

McBirney, A. R. 1984. *Igneous Petrology*. San Francisco, CA, Freeman, Cooper and Co.

McGuire, W. J., 1995. Monitoring active volcanoes: an introduction. In W. J. McGuire, C. Kilburn, and J. Murray (eds.) *Monitoring Active Volcanoes: Strategies, Procedures and Techniques*. London, UCL Press, pp. 1–31.

McGuire, W. J., Kilburn, C., and Murray, J. (eds.) 1995. *Monitoring Active Volcanoes: Strategies, Procedures and Techniques*. London, UCL Press.

Merle, O. and Borgia, A. 1996. Scaled experiments of volcanic spreading. *Journal of Geophysical Research*, **101**, 13805–13818.

Mogi, K. 1958. Relations of the eruptions of various volcanoes and the deformations of the ground surface around them. *Bulletin of the Earthquake Research Institute*, **36**, 99–134.

Morey, G. W. 1922. The development of pressure in magmas as a result of crystallization. *Journal of the Washington Academy of Science*, **12**, 219–230.

Neri, A. and Macedonio, G. 1996. Physical modelling of collapsing volcanic columns and pyroclastic flows. In R. Scarpa and R. I. Tilling (eds.) *Monitoring and Mitigation of Volcano Hazards*. Berlin, Germany, Springer-Verlag, pp. 389–427.

Neri, A., Papale, P., and Macedonio, G. 1998. The role of magma composition and water content in explosive eruptions. II. Pyroclastic dispersion dynamics. *Journal of Volcanology and Geothermal Research*, **87**, 95–115.

Newhall, C. G. and Punongbayan, R. S. 1996. The narrow margin of successful volcanic-risk mitigation. In R. Scarpa and R. I. Tilling (eds.) *Monitoring and Mitigation of Volcano Hazards*. Berlin, Germany, Springer-Verlag, pp. 807–838.

Pallister, J. S., Hoblitt, R. P., and Reyes, A. G. 1992. A basalt trigger for the 1991 eruptions of Pinatubo volcano. *Nature*, **356**, 426–428.

Papale, P. 1996. Modelling of magma ascent along conduits: a review. In F. Barberi and R. Casale (eds.) *The Mitigation of Volcanic Hazards*. Luxembourg, Office for Official Publications of the European Communities, pp. 3–40.

1998. Strain-induced magma fragmentation in explosive eruptions. *Nature*, **397**, 425–428.

Papale, P. and Dobran, F. 1993. Modeling of the ascent of magma during the plinian eruption of Vesuvius in AD 79. *Journal of Volcanology and Geothermal Research*, **58**, 101–132.

1994. Magma flow along the volcanic conduit during the Plinian and pyroclastic flow phases of the May 18, 1980, Mount St. Helens eruption. *Journal of Geophysical Research*, **99**, 4355–4374.

Papale, P., Neri, A., and Macedonio, G., 1998. The role of magma composition and water content in explosive eruptions. I. Conduit ascent dynamics. *Journal of Volcanology and Geothermal Research*, **87**, 75–93.

Pennisi, M. and Le Cloarec, M. F. 1998. Variations of Cl, F, and S in Mount Etna's plume, Italy, between 1992 and 1995. *Journal of Geophysical Research*, **103**, 5061–5066.

Prousevitch, A. A., Shagian, D. L., and Anderson, A. T. 1993. Dynamics of diffusive bubble growth in magmas: isothermal case. *Journal of Geophysical Research*, **98**, 22283–22307.

Rampino, M. R., Self, S., and Stothers, R. B. 1988. Volcanic winters. *Annual Reviews of Earth and Planetary Sciences*, **16**, 73–99.

Riedel, C., Ernst, G. G. J., and Riley, M. 2003. Controls on the growth and geometry of pyroclastic constructs. *Journal of Volcanology and Geothermal Research*, **127**, 121–152

Roberts, J. L. 1970. The intrusion of magma into brittle rocks. In G. Newall and N. Rast (eds.) *Mechanism of Igneous Intrusion*. Liverpool, UK, Gallery Press, pp. 287–338.

Roche, O., Druitt, T. M., and Merle, O. 2000. Experimental study of caldera formation. *Journal of Geophysical Research*, **105**, 395–416.

Rose, W. I., Bluth, G. J. S., and Ernst, G. G. J. 2000. Integrating retrievals of volcanic cloud characteristics from satellite remote sensors: a summary. *Philosophical Transactions of the Royal Society of London, Series A*, **358**, 1585–1606.

Rosi, M. 1996. Quantitative reconstruction of recent volcanic activity: a contribution to forecasting future eruptions. In R. Scarpa and R. I. Tilling (eds.) *Monitoring and Mitigation of Volcano Hazards*. Berlin, Germany, Springer-Verlag, pp. 631–674.

Rubin, A. 1995. Propagation of magma filled cracks. *Annual Reviews of Earth and Planetary Sciences*, **23**, 287–336.

Scarpa, R. and Tilling, R. I. (eds.) 1996. *Monitoring and Mitigation of Volcano Hazards*. Berlin, Germany, Springer-Verlag.

Sigurdsson, H. (ed.) 2000. *Encyclopedia of Volcanoes*. San Diego, CA, Academic Press.

Smith, R. L. 1979. Ash flow magmatism. *Geological Society of America Special Paper*, **180**, 5–27.

Smith, R. L. and Bailey, R. A. 1968. Resurgent caulderons. *Geological Society of American Memoirs*, **116**, 613–622.

Snyder, D. and Tait, S. 1995. Replenishment of magma chambers: comparison of fluid-mechanic experiments with field relations. *Contributions to Mineralogy and Petrology*, **122**, 230–240.

1996. Magma mixing by convective entrainment. *Nature*, **379**, 529–531.

Sparks, R. S. J. 1978. The dynamics of bubble formation and growth in magmas, a review and analysis. *Journal of Volcanology and Geothermal Research*, **3**, 1–37.

2003. Forecasting volcanic eruptions. *Earth and Planetary Science Letters*, **210**, 1–15.

Sparks, R. S. J. and Marshall, L. A. 1986. Thermal and mechanical constraints on mixing between mafic and silicic magmas. *Journal of Volcanology and Geothermal Research*, **29**, 99–124.

Sparks, R. S. J. and Wilson, L. 1976. A model for the formation of ignimbrite by gravitational column

collapse. *Journal of the Geological Society of London*, **132**, 441–451.

Sparks, R. S. J., Sigurdsson, H., and Wilson, L. 1977. Magma mixing: a mechanism for triggering acid explosive eruptions. *Nature*, **267**, 315–318.

Sparks, R. S. J., Carey, S. N., and Sigurdsson, H., 1991. Sedimentation from gravity currents generated by turbulent plumes. *Sedimentology*, **38**, 839–856.

Sparks, R. S. J., Bursik, M. I., Carey, S. N., *et al.* 1997. *Volcanic Plumes*. New York, John Wiley.

Spera, F. 1984. Some numerical experiments on the withdrawal of magma from crustal reservoirs. *Journal of Geophysical Research*, **89**, 8222–8236.

Spera, F., Yuen, D. A., Greer, J., *et al.* 1986. Dynamics of magma withdrawal from stratified magma chambers. *Geology*, **14**, 723–726.

Stix, J. and Gaonac'H, H. 2000. Gas, plume and thermal monitoring. In H. Sigurdsson (ed.) *Encyclopedia of Volcanoes*. San Diego, CA, Academic Press, pp. 1141–1163.

Sturtevant, B., Kanamori, B., and Brodsky, E. E. 1996. Seismic triggering by rectified diffusion in geothermal systems. *Journal of Geophysical Research*, **101**, 25269–25282.

Tait, S., Jaupart, C., and Vergniolle, S. 1989. Pressure, gas content and eruption periodicity of a shallow crystallizing magma chamber. *Earth and Planetary Science Letters*, **92**, 107–123.

Toramaru, A. 1995. Numerical study of nucleation and growth of bubbles in viscous magmas. *Journal of Geophysical Research*, **100**, 1913–1931.

Trial, A. F., Spera, F., Greer, J., *et al.* 1992. Simulations of magma withdrawal from compositionally zoned bodies. *Journal of Geophysical Research*, **97**, 6713–6733.

Turner, J. S. and Campbell, I. H. 1986. Convection and mixing in magma chambers. *Earth Sciences Reviews*, **23**, 255–352.

Van Der Laat, R. 1996. Ground-deformation methods and results. In R. Scarpa and R. I. Tilling (eds.) *Monitoring and Mitigation of Volcano Hazards*. Berlin, Germany, Springer-Verlag, pp. 147–168.

Walker, G. P. L. 1973. Explosive volcanic eruptions: a new classification scheme. *Geologisctic Rundschall*, **62**, 431–446.

Willams, H. and McBirney, A. R. 1979. *Volcanology*. San Francisco, CA, W. H. Freeman.

Wilson, C. J. N. 1980. The role of fluidization in the emplacement of pyroclastic flows: an experimental approach. *Journal of Volcanology and Geothermal Research*, **8**, 231–249.

1984. The role of fluidization in the emplacements of pyroclasic flows. II. Experimental results and their interpretation. *Journal of Volcanology and Geothermal Research*, **20**, 55–84.

Wilson, L. 1976. Explosive volcanic eruptions. III. Plinian eruption columns. *Geophysical Journal of the Royal Astronomy Society*, **45**, 543–556.

Wilson, L. and Head, J. W. 1981. Ascent and eruption of basaltic magma on the earth and moon. *Journal of Geophysical Research*, **86**, 2971–3001.

Wilson, L., Sparks, R. S. J., and Walker, G. P. L. 1980. Explosive volcanic eruptions. IV. The control of magma properties and conduit geometry on eruption column behaviour. *Geophysical Journal of the Royal Astronomy Society*, **63**, 117–148.

Wilson, M. 1989. *Igneous Petrogenesis: A Global Tectonic Approach*. London, Unwin Hyman.

Wohletz, K. H. 1986. Explosive magma–water interactions: thermodynamics, explosion mechanisms, and field studies. *Bulletin of Volcanology*, **48**, 245–264.

Wohletz, K. H. and Heiken, G., 1992. *Volcanology and Geothermal Energy*. Berkeley, CA, University of California Press.

Wohletz, K. H. and Sheridan, M. F. 1983. Hydrovolcanic explosions II. Evolution of basaltic tuff rings and tuff cones. *American Journal of Science*, **283**, 385–413.

Wood, W. J. and Fraser, D. G. 1976. *Elementary Thermodynamics for Geologists*. Oxford, UK, Oxford University Press.

Woods, A. W. and Bursik, M. I. 1994. A laboratory study of ash flows. *Journal of Geophysical Research*, **99**, 4375–4394.

Woods, A. W., Bursik, M. I., and Kurvatov, A. V. 1998. The interaction of ash flows with ridges. *Bulletin of Volcanology*, **60**, 38–51.

Zimanowski, B. 1998. Phreatomagmatic explosions. In A. Freund and M. Rosi (eds.) *Developments in Volcanology*, vol. 4, *From Magma to Tephra*. Amsterdam, The Netherlands, Elsevier, pp. 25–53.

Zimanowski, B., Büttner, R., Lorenz, V., *et al.* 1997. Fragmentation of basaltic melt in the course of explosive volcanism. *Journal of Geophysical Research*, **102**, 803–814.

Zimanowski, B., Fröhlich, G., and Lorenz, V. 1991. Quantitative experiments on phreatomagmatic explosions. *Journal of Volcanology and Geothermal Research*, **48**: 341–358.

Zlotnicki, J. 1995. Geomagnetic surveying methods. In W. J. McGuire, C. Kilburn, and J. Murray (eds.) *Monitoring Active Volcanoes*. London, UCL Press, pp. 275–300.

# Chapter 4

# Volcanoes and the geological cycle

Ray A. F. Cas

## Introduction

Volcanoes are one of the most exciting landforms on Earth today, and frequent volcanic eruptions around the world every year are a constant reminder of the dynamic nature of the Earth as a planet. With many well-documented major volcanic eruptions in the last two decades (e.g., the eruptions of Mount St. Helens in 1980 in the northwestern United States, El Chichón in 1982 in Mexico, Galungung in 1982 in Indonesia, Hawaii from 1992 to present, Mt. Pinatubo in 1991 in the Philippines, Mt. Unzen from 1991 to 1994 in Japan, and Soufriere Hills, Montserrat from 1996 to present), it may appear that the frequency of eruptions is increasing and that volcanic eruptions are a relatively recent or increasingly significant phenomenon in the Earth's history. However, the remains of ancient volcanic rock successions around the world indicate that volcanic activity has been a fundamental aspect of the evolution of our planet throughout its history. The apparent increase in the frequency of eruptive activity in recent times is probably largely an artefact of improved surveillance opportunities, technology, and media coverage. In this chapter the significance of volcanic activity in the history and evolution of the Earth will be examined. In particular, we will explore how volcanism plays a fundamental role in the geological cycle and the dynamic mechanism that drives the geological cycle, plate tectonics. We will also explore how far back in time modern plate tectonics can be recognized as the driving force for the geological cycle and associated volcanism. We will also consider how volcanism has contributed to the development of the atmosphere, hydrosphere and even life, during the geological cycle.

## The geological cycle

"The geological or rock cycle" is the term that is used for the recycling of rock materials, minerals, molecules, and ions by dynamic natural processes throughout Earth history, involving the exchange of matter between the Earth's interior, its outer shells (lithosphere and crust), the atmosphere, hydrosphere, and even the biosphere (Fig. 4.1). It has been recognized for a long time that the major rock groups are the products of these dynamic recycling processes. The geological or rock cycle works something like this. When magmas are erupted as volcanics on the Earth's surface, they are subjected to chemical and physical weathering and erosional processes, which strip ionic and solid mineral matter off the volcanics and transport them away to be incorporated into sedimentary deposits such as sands and clays, much of which are ultimately deposited on the ocean floor. There, those sediments, and the volcanic rocks making up the ocean crust, are rafted along on lithospheric plates and may eventually be subducted at the oceanic trenches (Carey, Chapter 1, this volume). They then become deeply buried during continued subduction, and are subjected

*Volcanoes and the Environment*, eds. J. Martí and G. G. J. Ernst. Published by Cambridge University Press. © Cambridge University Press 2005.

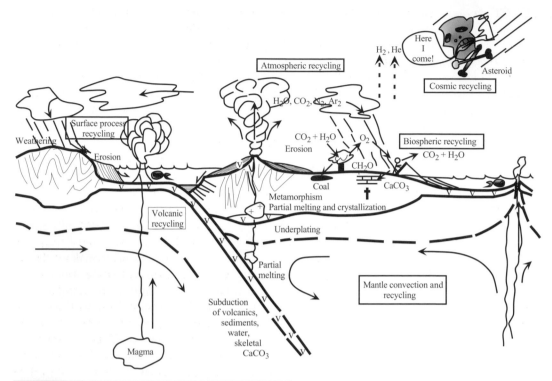

**Fig. 4.1.** A schematic representation of the geological cycle, depicting the interaction between the Earth's four spheres (lithosphere, atmosphere, hydrosphere, and biosphere) as well as the interaction between the Earth and the cosmos.

to metamorphism as pressure and temperature increase. Some of this deeply subducted rock material may then begin to partially melt, and rise towards the Earth's surface as magma. Some of this magma crystallizes in the subsurface as plutonic bodies of rock, often causing the crust to thicken from below by addition of magma to the base of the crust in a process called underplating, but some may be erupted again as volcanic magmas on the Earth's surface.

In addition, sediments and seafloor volcanics may be scraped off onto the leading edge of the overriding lithospheric plate in the trenches, where they added to the highly deformed and metamorphosed accretionary prisms there. At some stage a continental mass may be carried into the subduction zone, leading to major collision between continents or between continents and island arcs. The continental crust on the subducting plate is buoyant because of its low

density compared with mantle rocks, and it therefore resists being subducted. It is underthrust under the overriding continental crust, where it is buckled, folded, faulted, and metamorphosed. The combined crust is thus enormously thickened forming huge, high collision mountain ranges such as the Himalayas. The lower part of the thickened crust, which may be up to 80 or more kilometers thick, is subjected to heat and pressure and may also begin to melt forming granitic magmas. These magmas may also begin to rise because of their buoyancy. Meanwhile, the surface of the uplifted, thickened mountain range is being subjected to very high rates of physical erosion, with huge volumes of sediment being shed into nearby rivers. These rivers will continue to downcut and will begin to erode deeper levels of the crust, including high-grade metamorphic and granitic rocks. The rivers eventually deliver much of the sediment to the deep ocean floor where it again begins its cycle of transport on top of oceanic lithosphere towards a subduction zone.

At the same time atmospheric gases are reacting with the lithosphere in processes such as chemical weathering, rainwater is absorbed into the soils and bedrock, organisms and plants are extracting compounds such as nitrogen, oxygen and carbon dioxide from the atmosphere, hydrosphere, and lithosphere, and fixing them during growth. When these organisms and flora die, their remains become incorporated into the lithosphere (Fig. 4.1), and are thus recycled back through the geological cycle.

Volcanic activity therefore plays a fundamental role in the geological cycle, transferring matter from the mantle and the lower crust to the Earth's surface, atmosphere, and hydrosphere. Subsequently this volcanic material, as well as volcanically derived elements fixed by the biosphere and lithosphere, is recycled when the products of erosion are deposited in the sea and may eventually be subducted back into the Earth's interior. We will now see how volcanic activity has been fundamental in shaping the Earth we live on, as well as other planetary bodies of the solar system.

In this discussion, common time terms will be used, and the reader is referred to Harland *et al.* (1990) for reference. In particular, the Priscoan eon is used for the time interval from the origin of the Earth, 4.57 billion years ago to 4.0 billion years ago, the Archean eon extends from 4.0 to 2.5 billion years ago (some people combine these two eons into one, the Archean), the Proterozoic eon extends from 2.5 billion to 570 million years ago (Ma), and the Phanerozoic eon extends from 570 Ma to the present.

# Volcanic activity in the planetary system: causes and implications for Earth

The Earth and the planetary solar system formed about 4570 Ma through progressive accretion of solid and gaseous matter into planetary bodies in fixed orbits in the solar nebula. Examination of the surfaces of the planets during landings by manned and unmanned spacecraft, from the images from orbiting satellite radar and photographic surveys, and various other remote-sensing techniques indicate that volcanic activity has affected at least all of the rocky planets and

many orbiting moons. Volcanic activity is therefore not just confined to the Earth, and although the plate tectonic model (Carey, Chapter 1, this volume) provides a dynamic framework within which the location of many volcanoes on Earth can be explained, this is clearly not the case for all the planetary bodies, most of which show little or no evidence for plate tectonic activity (Frankel, 1996; Thomas *et al.*, 1997).

In some cases, such as the Moon, some large-volume volcanic events associated with major impact structures, especially of the older lunar highlands, appear to have been related to and probably triggered by the late stages of the major meteorite impact phase of the Moon's early history. Analysis of the history of the Moon (Head, 1976) suggests that the Moon underwent a prolonged phase of major impact activity from its origins to about 3800 Ma, and this is thought to have produced the rugged impact generated terrains of the lunar highlands. This was followed by a declining phase of impacts between 3800 and 3300 Ma. Smaller isolated impact craters which pockmark the lunar surface testify to ongoing isolated events throughout the Moon's history to the present day. The large-volume lunar mare flood basalt fields have been dated to have formed between 3900 and 3100 Ma. Their eruption therefore coincides with the late stages of major impacts, and continues through the waning phase of activity. Originally the eruptions of the mare lavas were thought to have been triggered by impacts, and this could be argued for those whose ages coincide with the period of heavy bombardment. However, it is difficult to argue this for those mare lavas erupted hundreds of millions of years after major bombardments ceased. It also appears that the mare lavas were derived from great depth within the Moon's mantle (Frankel, 1996). Although this does not exclude a major impact origin, it also permits a deep-seated hotspot or plume origin sourced by radioactivity-generated heat (Frankel, 1996).

Mercury appears to have had a similar history to the Moon. Both have a very thick lithosphere, and neither shows any evidence of ever having experienced plate tectonic processes (Thomas *et al.*, 1997). Mercury, like the Moon, shows no evidence of tectonic activity other than that associated with meteorite impacts.

Mars and Venus have experienced more recent volcanic activity than the Moon and Mercury. The volcanism on Venus and Mars is also more diverse in character than on the Moon and Mercury, and includes silicate igneous rocks with compositions unlike any known on Earth. The surface of Venus has relatively few impact craters and many volcanic landforms, indicating that volcanic activity has been active to relatively recent times. In particular, it appears to have experienced a major phase of flood-like lava volcanism around 500 Ma, that smoothed the Venutian landscape and buried earlier impact generated landscapes (Frankel, 1996).

On Mars, impact craters occur and it appears that volcanism stopped perhaps as recently as several tens of millions of years ago as suggested for the giant volcano, Olympus Mons (Frankel, 1996). Mars and Venus show evidence of fault scarps, and rifts, and there has been considerable speculation on the likelihood of plate tectonic systems during their history (e.g., Sleep, 1994). For Mars, plate tectonics could only have occurred in its earliest history when it was still hot, and perhaps subjected to widespread mantle convection. Because Mars is substantially smaller than Earth, its interior would have cooled quickly during its early history, leading to a thick, cold rigid crust and lithosphere (Thomas et al., 1997).

Venus is only slightly smaller than Earth, and there have been numerous suggestions that plate tectonic processes may have operated, based on linear alignments of volcanic edifices (arcs?), extensive fracture zones, and ridge-like features (Frankel, 1996). However, although collision mountain belts exist, subduction zones, mid-ocean spreading ridge-like features, and other plate features do not appear to be present (Thomas et al., 1997). Some of the widespread early volcanism on Mars and Venus may thus be related to the same bombardment events that appear to have caused volcanism on the Moon and Mercury; some may be related to early plate tectonic-like effects, but more recent volcanic and tectonic activity cannot be ascribed to plate tectonics. Relatively recent volcanism can only be attributed to hotspot or plume volcanism, associated with internal convection and thermal instability.

Thomas et al. (1997) suggest that the critical factor determining whether or not plate tectonics operates is the viscosity contrast between the lithosphere and asthenosphere, and not the thickness of the lithosphere. For example, although Venus has a lithosphere about the same thickness as that of the Earth, and has about the same planetary diameter, it does not experience plate tectonics. The Earth's lithosphere is also more rigid than that of Mars, which has a surface temperature of about 450 °C, and yet Mars also does not experience plate tectonics.

On Io, one of the rocky moons of Jupiter, and the most volcanically active planetary body in our solar system, some of the volcanoes are sulfur-erupting volcanoes (Sagan, 1979) and spectacular modern explosive eruption plumes have been imaged from these (Frankel, 1996). The volcanic activity on Io is thought to be triggered by extremely strong tidal forces, generated by the multiple gravitational forces acting on Io from Jupiter and two of the other moons, Ganymede and Europa (Frankel, 1996). These forces are thought to deform and generate frictional melting of the interior of Io, leading to virtually continuous volcanism.

It is thus apparent that volcanism in our solar system cannot be related only to plate tectonics. There are diverse causes. It also follows from this that the volcanism on Earth need not be completely related just to plate tectonic factors, although much of it is. Impacts, especially during Earth's early history, could also be a factor, internal thermal instabilities and plumes could be a factor, and at some times, even tidal forces between Earth and the Moon could be a factor. To illustrate that the Earth is not a completely predictable volcanic entity, although the majority of magmas erupted on Earth have a silicate chemical and mineral composition, eruptions of carbonatite magmas are known, and even iron oxide, magnetite lavas have been found in the Andes, in Chile (e.g., Henriquez and Martin, 1978).

The Earth shows clear evidence of extraterrestrial impact events in the geological past. Impact craters ranging from Precambrian to Pleistocene are known. Some of these also appear to coincide approximately with episodes of regionally extensive and voluminous volcanic activity. For example, both major impact events (Alvarez *et al.*, 1980) and major phases of volcanic activity such as flood basalt eruption phases (Rampino, 1987; Rampino and Stothers, 1988; Courtillot *et al.*, 1990; Rice, 1990) have been proposed as being responsible for mass extinctions, including that of the dinosaurs, at the end of the Cretaceous time period. If major impact events or phases were capable of triggering volcanic activity on other planetary bodies, then they should also have created such volcanic responses during the Earth's history, as proposed by Rampino (1987). This cause–effect relationship may have been especially significant during Earth's early history when the major impact phases of activity affected the solar system (Lowman, 1976). Although the present oxygen-rich nature of the Earth's atmosphere leads to combustion of much incoming extraterrestrial matter due to frictional heating, large meteoritic objects could still impact on the Earth's surface. For example, Meteor Crater in Arizona, USA is only 25 000 years old. In the geological past, when atmospheric oxygen levels were significantly less than at present, the combustion effect would have been minimal, so enhancing the number of impact events and their effects. Such events are random events not easily accommodated by the steady-state geological cycle, but their effects on Earth history could be profound.

In summary, therefore, although the plate tectonic paradigm provides a logical explanation for the occurrence of many of the volcanic patterns on Earth today and in the past, it does not explain them all. Even within the plate tectonic framework volcanism can occur in random localities within plates, and at random times (hotspot activity) (Carey, Chapter 1, this volume). There has also been some debate as to how long plate tectonics has been part of the Earth's structural and tectonic make-up, and its geological cycle (see below). In addition, random volcanic events could also have been associated with extraterrestrial impact events throughout geological time, externally triggered or indigenous deep-level plume events, or even gravitationally or "tidally" induced, especially when the Earth

was young, its surface was hot, internal temperature gradients high, and its outer margins less rigid than today.

## When did volcanic activity begin in Earth history?

The oldest known volcanic rocks on Earth occur in the Isua greenstone belt of Greenland (Appel *et al.*, 1998). These rocks are between 3700 and 3800 Ma old and consist of metamorphosed submarine basalt pillow lavas, massive basalt lavas, subaqueous hyaloclastite (lava-associated breccias), and interbedded fine-grained sedimentary rocks. They indicate that at least as long ago as 3800 Ma, volcanic activity was significant on the Earth's surface. In addition, the oldest dated rocks on Earth, the Acasta gneiss of the Slave Province, Canada, which have an emplacement age of 3962 Ma, consist of metamorphosed granite and tonalite (Bowring *et al.*, 1989), which indicates that subsurface crustal plutonic magmatic activity was well established then, and probably surface volcanism as well. However, the age of these oldest known rocks does not reflect the age of the Earth. The reason why older rocks have not been discovered, even though the age of the Earth is known to be 4570 Ma, lies in the processes of the "geological or rock cycle," as outlined above. Because of the Earth's dynamic nature, different types of rocks are constantly being created, modified, and recycled through the processes of volcanism, subduction, uplift, erosion, sedimentation, burial of sediments, lithification, tectonic deformation, and metamorphism. It thus seems likely that rocks formed during the earliest history of the Earth would have been subject to the processes of the rock cycle and evidence of their existence removed.

Do we have any indication when a solid Earth first formed? Although the oldest dated rock succession occurs in Canada, there are older dated minerals. Detrital crystal fragments of the mineral zircon ($ZrSiO_4$), with a lead isotopic age of up to 4400 Ma, have been collected from a quartzite and conglomerate sedimentary rock sequence from the Archean of Western Australia

(Maas *et al.*, 1992; Wilde *et al.*, 2001). The depositional age of the metasedimentary host rocks is about 3000 Ma. Zircon is a mineral that crystallizes in granitic magmas formed by the melting of continental-type crust. It appears that the dated zircon grains were eroded from 4270 Ma granites exposed on continental crust about 3000 Ma ago. More importantly they indicate that a solid continental crust existed on Earth at least 4270 Ma ago, and that igneous processes, and therefore probably volcanic processes, were operating then. This is also consistent with the ages of the oldest dated samples from the Moon, including lava fragments (4200 Ma), lava dust (4350 Ma), and mantle rock, a peridotite with an age of 4500 Ma. These suggest that the Moon had developed a solid volcanic surface by at least 4350 Ma, and perhaps as early as 4500 Ma. The peridotite represents one of the earliest crystallized rocks in the Moon's (and solar system's) history (Frankel, 1996).

Because the Moon is significantly smaller than the Earth, it would have cooled more quickly than Earth, and it is likely that it formed a crust earlier than Earth. We also know that the Earth's mantle must have been at least 200 °C average hotter than today (Frankel, 1996; Ranalli, 1997), causing more vigorous convection and therefore volcanism than today. It is therefore likely that volcanism was a fundamental, and probably the most dominant, process affecting the Earth's earliest crust and surface.

# The role of volcanic activity in creating the atmosphere, hydrosphere, and life

The origin of the Earth's atmosphere and hydrosphere has been a matter of scientific discussion for a long time. Possible origins include:

- it originated at the time of the accretion of the Earth 4570 Ma ago from gases condensing above the hot surface of the Earth, or
- it and its gaseous component were captured from space during the orbit of the Earth, or it was subjected to periodic "cosmic showers," or

- it originated endogenously from within the Earth during its long term evolution since original accretion.

The arguments against the present atmosphere forming at the time of initial planetary accretion are compelling. Accretion would have produced enormous heat from the friction caused by collision of solids. The Earth was probably a largely molten mass, with its surface temperature extremely high. Evidence lies in the abnormally low concentrations in the Earth's atmosphere of the heavy rare gas xenon, which has an atomic mass of 131. Increasing the temperature of gases causes them to become more energetic and the molecules to circulate with increased velocities. It is envisaged that the temperature of the Earth's surface and atmosphere was so high as to cause atmospheric xenon to escape the Earth's gravity field during and in the period after accretion. It then follows that if the heavy gas xenon escaped, then all lighter gases, including all the common gases making up our present atmosphere, and perhaps even heavier ones as well, would have escaped.

The Earth's atmosphere therefore originated after accretion. Although some theories suggest that "capture" of the atmosphere from space or from "cosmic showers" may have occurred after the Earth cooled sufficiently, and it is likely that there is some contribution from such a source, the most reasonable source for most of the Earth's atmosphere is from the release of volcanic gases from the Earth's interior. As we have already seen, volcanic activity has probably occurred since the Earth formed and so there has been an extremely long time-span of Earth history for accumulation of released gases.

Although modern volcanoes are known mostly for the large volumes of lavas and pyroclastic deposits they erupt (see Cas and Wright (1987) for a discussion of the processes), they also release huge volumes of gases. For example, during the eruption of Mt. Pinatubo in the Philippines on June 15, 1991, 20 Mt of $SO_2$ alone were released, this then converting to 30 Mt of climate-modifying $H_2SO_4/H_2O$ aerosol gas complexes (McCormick *et al.*, 1995). Although the gas types and their abundances vary from volcano to volcano depending on the magma composition, a typical gas composition by volume would include 70–80% $H_2O$, 8–12% $CO_2$, 3–5% $N_2$, 5–8% $SO_2$, and minor proportions of $H_2$, CO, $S_2$, $Cl_2$, and Ar.

Whereas the abundance of most gases in the atmosphere can be explained by progressive long-term accumulation under the influence of nearly continuous volcanic activity through geologic time, the abundance of particularly oxygen, carbon dioxide, and nitrogen cannot be thus explained. Volcanic gases are almost totally devoid of free oxygen, but conversely they contain very large volumes (10% or more) of carbon dioxide. By contrast the atmosphere contains some 21% oxygen and only 0.031% carbon dioxide. The inconsistencies in oxygen and carbon dioxide abundances in volcanic gases and the Earth's atmosphere can only be explained by the influence of photosynthetic and respiratory organisms through geologic time (Cloud, 1988). However, herein lies a conundrum. Modern respiratory organisms require oxygen, yet it is almost certain that the early Earth's atmosphere contained none. Furthermore, the early atmosphere would have been "thin" and easily penetrated by ultraviolet (UV) radiation from the Sun that is harmful to life.

Life could not have evolved until it could be protected from UV radiation. The early atmosphere must have been carbon dioxide- and water-vapor-rich (a super greenhouse), based on their high abundances in volcanic gases, but neither carbon dioxide nor water vapor are as effective in absorbing and screening the Earth's surface from UV as oxygen and ozone. In the absence of the shielding effects of oxygen and ozone, it is likely the initiation of life, by whatever means, had to wait until liquid water bodies had accumulated on the Earth's surface, and that it occurred under anaerobic or reducing conditions. When did this happen?

The oldest supracrustal rocks on Earth, the 3800–3700 Ma volcanic rocks of the Isua greenstone belt of Greenland, provide the answer. These volcanic rocks contain two particular eruptive forms of subaqueous lavas, pillow lava and hyaloclastite (Appel *et al.*, 1998). Pillow lavas form when fluidal magma such as basalt is erupted under water or flows into water at a slow rate. As the surface of the lava comes in contact with the water it causes the water at the interface to boil, forming an insulating layer of steam (Mills, 1984). This allows the lava surface to cool sufficiently slowly to form a plastic skin. As more magma is injected under the skin from the vent it breaks through the skin as a lobe of lava on average about 0.5 to 1 m in diameter. This in turn also forms a skin, preserving a pillow-like form. Many such lobate breakouts lead to a lava whose internal structure is very complex and consists of multiple lobes or pillows stacked on top of each other. Sometimes, however, while the magma is still liquid, the insulating steam layer collapses, in part because of natural instabilities at the steam–water boundary, in part because of the effects of strong convection currents in the surrounding water mass. Water then comes directly in contact with the liquid lava, which chills instantaneously to solid glass, shrinks in volume, and by this shatters to an aggregate of glassy debris called hyaloclastite (Cas and Wright, 1987).

Indirect data even suggest that surface water may have existed as long ago as 4400 Ma (Wilde *et al.*, 2001). Zircon crystals of this age from the Jack Hills of Western Australia have elevated $^{18}O$ values suggesting interaction with surface water.

We therefore know that surface bodies of water, resulting from condensation, atmospheric precipitation, and surface runoff existed by 3800 Ma and possibly as early as 4400 Ma. This in turn tells us that the surface temperature of the Earth and its atmosphere must have fallen below 100 °C by then, otherwise liquid water could not have accumulated.

At just what stage life first evolved is not clear. However, the oldest known organic molecules that represent the first known "life forms" occur in the Pilbara region of Western Australia, in rocks at least 3500 Ma old (Schopf, 1993). The fossil life forms consist of anaerobic cyanobacteria, non-cellular procaryotic organisms some of which were stromatolitic, and some filamentous (Schopf, 1983, 1993). Stromatolitic fossils of algae of 3500 Ma age are known from both the Pilbara region of Western Australia (Walter *et al.*, 1980) and the Barberton Mountain Land, South Africa (Byerly *et al.*, 1986). Nisbet (1995) argues that the earliest organisms may have been chemotrophic bacteria that evolved around subaqueous volcanic hydrothermal vents under

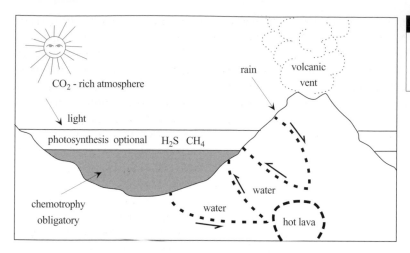

**Fig. 4.2.** Cartoon showing the possible setting for the origin of life around submarine volcanic hydrothermal systems. From Nisbet (1995).

anaerobic conditions, perhaps as long as 3800 Ma ago (Fig. 4.2), consistent with evidence for both aqueous environments and the activity of a carbon cycle at that time (see also Stetter, Chapter 6, this volume). Davies (1998) overcomes the problem of ultraviolet effects on non-cellular organisms by suggesting that life could have evolved underground, within the influence of subterranean, volcanically related, hydrothermal fluid systems. This is based on the discovery that primitive bacteria-like organisms occur even today several kilometers below the Earth's surface. However, the push for life forming earlier and earlier continues, with proposals that evidence for biogenic carbon occurs in the 3800–3700 Ma Isua belt rocks in Greenland (Schidlowski, 1988; Mojzsis *et al.*, 1996; Rosing, 1999).

Some of the earliest organisms may have been photosynthetic organisms (Schopf, 1993), because about this time in Earth history the geological record begins to contain "banded iron formations" (Cloud, 1988). These are fine-grained, laterally very continuous sequences of red siliceous or cherty layers and white cherty layers. The red layers are rich in haematite and magnetite, which are iron oxide minerals rich in $Fe^{3+}$, the oxidized form of iron, whereas the white layers consist of iron-free chert. Banded iron formations indicate that there was abundant soluble $Fe^{2+}$ in the oceans, indicating reducing conditions, but periodically this $Fe^{2+}$ was oxidized to produce the $Fe^{3+}$-rich haematite and magnetite layers, indicating flushes of oxygen in the environment. The $Fe^{2+}$ in the environment acted as a very important oxygen receptor or sink, and was probably

crucial to the survival of the earliest life forms. Being non-cellular, those organisms had no cell walls or membranes to protect them from direct contact with oxygen, which is poisonous to the organic compounds that constituted the first life forms. The source or origin of the iron and silica in banded iron formations is still unclear, but there have been suggestions that they originated from exhalative release of hydrothermal fluids or volcanic volatiles, perhaps from submarine eruptions or volcanic hydrothermal vent systems (Derry and Jacobsen, 1990; Eriksson, 1995).

After this time, oxygen levels appear to have progressively increased, as indicated by the appearance of pervasively oxidized continental "red bed" sedimentary rocks at about 2000 Ma (Eriksson, 1995). Eucaryotic cells, with protective cell membranes or walls and oxygen-mediating enzymes, developed at about 1300 Ma, respiratory metazoans appeared about 700 Ma ago, and organisms with skeletons of $CaCO_3$ appeared about 600 Ma ago. Land plants appeared about 420 Ma ago, indicating high enough oxygen and ozone levels in the atmosphere to shield out sufficient UV to make the land surfaces habitable (Cloud, 1988).

The development of photosynthetic organisms was a very important stage in terms of changing global climate. Up to this time, atmospheric conditions would have been influenced very much by the high carbon dioxide levels created by ongoing volcanic activity, creating a super-greenhouse effect much more severe than the current climatic effects of industrial greenhouse gases. In the process of photosynthesis, plants take in carbon dioxide and water from the environment and convert these into carbohydrate, which becomes fixed as plant tissue. Oxygen is also produced and is released to the atmosphere where it then becomes available to help shield out UV. The removal of carbon dioxide from the atmosphere when photosynthetic organisms developed would thus have made climate more amenable. This effect would have been reversed slightly by respiratory release of carbon dioxide by newly evolved respiratory organisms in the Late Precambrian and through the Phanerozoic.

It is thus clear that volcanism created most of the atmospheric components, and it was thus also responsible for the hydrosphere. It can also be argued that without volcanism there would be no life, because the atmospheric and hydrospheric environment in which life evolved resulted from volcanism, and perhaps the earliest life forms, chemotrophic bacteria, originated in the immediate environment of submarine volcanic hydrothermal vents once surface waters had developed or even within subterranean hydrothermal systems. Even the chemical building-blocks of higher life forms, amino acids and carbohydrates, consist of the chemical components in gases released by volcanoes.

## The modern plate tectonic system as a guide for recognizing the tectonic setting of ancient volcanic successions

As outlined in Chapter 1 (Carey, this volume), the settings in which volcanic activity occurs on the present-day Earth's surface can most readily be related to the modern plate tectonic framework. In this section, the essentials of the modern plate tectonic framework that can be used to identify diagnostic elements in ancient counterparts will be briefly reviewed.

Volcanic activity is semicontinuous along the mid-oceanic ridge (MOR) systems which encircle the globe in a connected network of divergent plate margins. At these ridges, new oceanic crust is being created nearly continuously because MORs are the sites of upwelling limbs of major mantle convection systems. Mantle-derived basaltic magma rises into the fissures along the MORs from the asthenosphere as the lithospheric plates on either side are rafted in opposite directions on the diverging limbs of convection cells. This leads to the nearly continuous eruption of submarine basalt lavas, of both the pillowed and massive sheet types, and also associated hyaloclastite. MOR fissure vents are usually so deep that the hydrostatic pressure due to water depth prevents explosive eruptions from them. MOR volcanism is the most voluminous on Earth, and the basaltic volcanic oceanic crust produced covers almost 70% of the Earth's surface and has a relatively uniform chemical composition. It is called oceanic tholeiite basalt or MORB (Wilson, 1989).

Volcanic activity is also common along the volcanic arcs associated with the oceanic trenches, which are the sites of the subduction or "recycling" of the oceanic lithosphere back into the Earth's interior (Fig. 4.1). As the downgoing oceanic plates plunge back into the mantle, commonly at angles of 30–70°, they are progressively heated as they go deeper into the mantle. Eventually hydrous fluids are released from the downgoing slab and these rise into the overlying wedge of mantle. This causes the melting temperature in the mantle wedge to decrease, causing partial melting, of perhaps as much as 5% of the mantle wedge. The resulting melts and further hydrous fluids rise buoyantly towards the Earth's surface through the crust, and often cause further partial melting of mantle and crust through which they pass, especially if the uprise velocity is low.

Because magma is rising from many localities along the slab in a band parallel to the trench, a distinct line of volcanoes or a volcanic arc forms parallel to the trench. Where the trench and subduction zone are wholly located within a large ocean basin away from continental landmasses, the volcanoes of the arc begin forming as submarine volcanoes, and may rise above sea level to form lines of volcanic islands called volcanic island arcs. Where trenches and subduction zones occur along the margins of major continents, the arcs form on land and are called continental margin volcanic arcs. Island arcs erupt largely tholeiitic island arc basalts (IAB), and lavas predominate over pyroclastic deposits and eruptions. Continental arc volcanoes also erupt some mantle-derived basalts, but mostly derive voluminous magmas from melting of continental crust, resulting in intermediate to high $SiO_2$ magmas such as andesites, dacites, and rhyolites. Arc magmas include a spectrum of geochemical compositional suites such as calc-alkaline, tholeiitic, and alkaline (Wilson, 1989). Although lava-producing volcanic centers such as stratovolcanoes and dome complexes are common, very large explosive volcanic centers, such as calderas, and voluminous, widespread pyroclastic deposits, including huge ignimbrite sheets (the deposits of dynamic pyroclastic flows), are more common than in island arcs (see Cas and Wright (1987) and Sigurdsson (2000) for a discussion of eruption styles and deposit characteristics of stratovolcanoes and calderas).

As lithospheric plates move away from the MORs, they begin to cool. In addition mantle magmas accrete to the bases of the sliding plates causing them to thicken by the addition of dense peridotite from below. The lithospheric slab becomes denser and heavier. As it plunges into the mantle the density and weight of the plate may cause its distal margin in front of the trench to begin to sink, not just slide laterally. As a result, the position of the trench migrates forward, and extensional stresses begin to develop in the leading margin of the overriding plate. Often, the area of the plate behind the volcanic arc begins to rift, or extend in a direction normal to the plate margin (Karig, 1974). This perturbs the configuration of the major mantle convection cell under the overriding plate, and may cause smaller subsidiary convection cells to form beneath the arc, particularly under the influence of uprising magmas feeding the volcanic arc.

As a result, divergent spreading ridges may form behind such volcanic arcs (Karig, 1974) leading to the opening of relatively small back arc basins (e.g., Taylor, 1995). In volcanic island arc systems, such back arc basins develop oceanic crust; in continental margin arc systems such basins are initially continental, but may develop into oceanic basins if extension and rifting of the continental crust proceeds to the spreading stage, producing oceanic crust and lithosphere.

Oceanic crust produced by back arc spreading ridges is dominated by basalts known as back-arc basin basalts (BAB), which may be compositionally distinguishable from MOR basalts, especially in terms of some trace elements and isotopes (Wilson, 1989). Where back arc spreading begins behind continental volcanic arcs, significant volumes of silicic magmas such as andesites, dacites, and rhyolites are produced. Often significant volumes of basalt may also be erupted, producing bimodal (i.e., low $SiO_2$ and high $SiO_2$) associations of magmas.

Such bimodal volcanic associations may also be characteristic of areas of the Earth's continental crust where spreading systems intersect or propogate through continental crust (Wilson, 1989). For example, the Basin and Range Province of the western United States may be the result of the subduction of a segment of the East Pacific MOR under the western United States continental crust (Eaton, 1982). There, magmatism is very diverse in its chemical composition, including alkaline, tholeiitic, and calc-alkaline suites (Christiansen and Lipman, 1972). The East Africa rift zone appears to be a limb of the Indian Ocean system to which it joins in the Afar region of the Red Sea. Although the subsided crust in the East African rift zone is continental, if further extension, or continued subcrustal spreading occurs, then oceanic crust may begin to form. Again there, magma compositions are very diverse, including peralkaline suites of volcanic rocks and early flood basalts (Wilson, 1989).

In addition, volcanism may occur within the interior of lithospheric plates (intraplate) as a result of apparently randomly located mantle "hotspots." These are regions of the mantle which are anomalously hot, and generate large volumes of mantle magmas which rise buoyantly to the Earth's surface as "plumes" of magmas. Such hotspots may remain active for millions or even tens of millions of years. For example, the Hawaiian chain of volcanic islands, and to their north, the Emperor island chain, are thought to have resulted from a reasonably stationary hotspot located under the currently active "big island" of Hawaii, which occurs at the southern end of the chain (Carey, Chapter 1, this volume). The islands become progressively older to the north, a trend that is consistent with the drift of the Pacific lithospheric plate northwestwards over the deep hotspot, which has successively released batches of oceanic island basalt magma (OIB) (Wilson, 1989) through the plate, producing the succession of volcanic islands over tens of millions of years. Other areas of intraplate hotspot activity may also occur within continental masses, and the Tertiary basaltic province of eastern Australia (Johnson, 1989) is a good example. Eruptive products are diverse, ranging from undersaturated basic alkaline to highly silicic peralkaline magmas. Although dominated by mafic lavas, which may be localized to the confines of volcanic edifices or be far-flowing, confined by topography and flowing up to tens of kilometers from the vent, the products also include more evolved lavas and pyroclastic deposits. Pyroclastic deposits may be localized to small monogenetic volcanic centers such as scoria cones, maars, and tuff cones, but may also be well developed in large continental shield volcanoes (see Cas (1989) for a summary of eruption styles, volcano types, and nature of the eruption products).

Large-volume basalt volcanic provinces, called flood basalt or plateau basalt provinces, and large igneous provinces, are also likely to be the products of major mantle hotspots or plumes. These are known from both continental and oceanic crustal settings. They are characterized by multiple, huge-volume basalt lavas that are sheet-like, cover areas of $10^4$–$10^5$ km$^2$, and flow up to 300 km, or more. Huge oceanic plateaus such as the Ontong Java Plateau of the western Pacific Ocean are probably long-lived submarine flood basalt provinces or large igneous provinces. Well-known examples of continental flood basalt provinces include the Deccan Traps, India, the Karoo and Etendecka Basalts, South Africa and Namibia, the Ethiopian Province, and the Parana Basalts of South America. Intriguingly, these flood basalt provinces are associated with the rifted continental margins resulting from the break-up of the Mesozoic supercontinent of Gondwana. The Columbia River Province of the western United States is associated with the northern margin of the extensional Basin and Range rift province.

Subduction of oceanic lithosphere appears to occur readily because of its increasing density due to cooling as it drifts away from MORs. When continental masses are rafted into subduction zones as passengers on subducting lithospheric plates, they resist being subducted into the mantle because of their relatively low density and buoyancy relative to the mantle. Continental masses may thus collide with each other at subduction zones. The downgoing continental mass may be underthrust under the leading edge of the overriding continental mass, and resists being subducted any further, causing an anomalously thickened zone of continental crust. This appears to have happened in the Himalaya mountain range of Asia as a result of the attempted subduction of India under Asia.

Curiously, such continental collisions and attempted subduction do not produce large volumes of volcanic rocks or volcanic arcs. It seems likely, however, that in the lower crust of such collision belts, partial melting and generation of crustal magmas must eventually occur because of the high pressure and temperature conditions at the base of such thickened crust. If such magmas form, they do not appear to migrate up sufficiently to erupt, but must presumably reside or solidify in the crust, forming granite batholiths. Why is volcanic activity not prevalent in continental collision belts?

Examination of settings in which volcanic activity occurs in the modern plate tectonic framework suggests that there is a relationship between settings where voluminous volcanism occurs, and regional tectonic stress fields. Clearly, divergent plate boundaries, where voluminous mid-oceanic spreading ridge volcanism occurs, and in young incipient divergent plate margin settings such as the Red Sea and East Africa rift system, are areas where the regional tectonic stress field is extensional. Under these conditions $\sigma_1$, the maximum principal stress component is (sub)vertical, and $\sigma_3$, the minimum principal stress component is horizontal, leading to normal and lystric faulting and graben formation. These represent the ideal conditions for large volumes of buoyant magma to reach the Earth's surface, utilising the numerous (sub)vertical fault and fracture systems as magma conduits.

It used to be considered that convergent subduction plate settings coincided with widespread compressional stress fields. This is certainly the case in areas of collision of continental masses with each other, or where significant island systems, oceanic plateaus or ridges collide with trench walls or continental masses. In these settings, interestingly, volcanic activity ceases. For example, the Himalayas, the greatest collision mountain belt on Earth, is not marked by a volcanic arc. The simple explanation might be that because subduction of oceanic lithosphere has ceased, partial melting of that lithosphere is no longer taking place. And yet, because of the enormous crustal thickness (>80 km), it is likely that melting of the lower crust is occurring, producing granitic magmas. Why do such magmas not reach the Earth's surface and produce volcanism? Similarly in areas where the Louisiade Ridge is colliding with the Vanuatu Trench, there is an absence of arc volcanoes, whereas away from the collision zone there are numerous arc volcanoes. Similarly along the Peru–Chile subduction margin of the Andean mountain range of South America, where oceanic ridges such as the Carnegie, Nazca, and Juan Fernandez ridges collide with the trench, there is a break in the presence of volcanoes, which elsewhere form a nearly continuous line of arc volcanoes parallel to the trench.

It has now been recognized, however, that many subduction settings, especially where volcanic systems are well developed, show significant evidence of extensional processes within and behind the arc. This is most clearly obvious in oceanic arc systems, such as the Marianas and Tonga–Kermadec arcs, which have experienced back arc basin formation and extension. Such basins have well-defined spreading ridge systems in their center and the arc volcanoes are usually located along the trenchward margin of the extensional back arc basin. In these convergent subduction systems, the volcanic activity is entirely located in the extensional stress field domain of the arc system. This extensional stress field has developed due to passive, gravitationally induced sinking of the subducting plate, allowing the hinge zone to retreat or "roll back" oceanwards, and the overriding plate to extend. A compressional stress field domain occurs only at the leading edge of the overriding plate where

sediments are being scraped off the downgoing plate and where there is some frictional coupling between the two plates.

Even in continental margin subduction settings such as the Andes and the Cascades arc of northwestern America, there appears to be evidence of extension where the volcanic arc volcanoes occur. For example, in the Andes the location of the arc volcanoes coincides closely with major regional grabens such as the Puna and Altiplano, indicating extension. Lamb et al. (1997), however, argue that the grabens are bounded by differentially uplifted "pop-up" fault blocks in a subduction system that is in widespread regional compression for up to 1000 km from the trench. During the Mesozoic, the Andean subduction margin experienced limited rifting and extension, forming small back arc basins which developed oceanic crust (Dalziel et al., 1974). The alternation between compressional tectonics and extensional tectonics that the Andes has experienced can be explained as a function of the relative rates at which the leading edge of the South American plate is advancing westwards, which is controlled by the spreading rate at the Atlantic mid-oceanic spreading ridge, and the rate at which the hinge zone of the subducting Nazca plate in the Pacific is rolling back (Sebrier and Soler, 1991). If the rate of rollback is greater than the spreading rate at the mid-Atlantic ridge, then extension at the Andean margin will occur, accompanied by voluminous volcanism. If the spreading rate exceeds the rollback rate, then the Andean margin will undergo compression, with the likely effect of reducing the level of volcanism. Even under mild compression some volcanism may occur if the magma fluid pressure exceeds the regional compressional stress field or where local transtensional basins form along transverse fault zones.

Interestingly, many of the the relatively young, large-scale continental flood basalt provinces of the world occur at, or very close to, the rifted margins of the major continents.

It thus seems that voluminous volcanic episodes and provinces are commonly related to regions of the Earth's crust experiencing extension or transtension in either divergent oceanic, continental rift, or arc–back arc settings. This guide can then be used to assess the paleotectonic settings for ancient rock successions containing widespread, voluminous volcanic successions.

## Reconstructing the tectonic setting of ancient volcanic successions: the Phanerozoic record

Identifying the original setting of ancient volcanic successions is a challenging task. In the first instance, the task is to identify criteria for recognizing the type of crust the volcanism occurred on. Where volcanism is thought to have occurred on oceanic crust, the subjacent rock succession must preserve the rock succession that makes up oceanic crust. Deep-sea drilling through present-day oceanic crust, as well as onland exposures through ancient oceanic crust which has been uplifted and tectonically emplaced onto land, reveals a consistent stratigraphy and succession of rock units. The uppermost unit, called layer 1, consists of typical open ocean fine-grained sediments, often with abundant pelagic microfossils preserved. This lies above layer 2, consisting of a thick succession of basalt lavas, including both classical submarine pillow lavas, and massive sheet lava (layers 2a, 2b). This passes down into layer 2c, which consists of (sub)vertically oriented sheet-like dykes of basalt and dolerite which may pass down into coarse plutonic rocks such as gabbros of layer 3 (Brown *et al.*, 1989). The base of the oceanic crust, called the oceanic Moho, separates the crust from the coarse olivine rich peridotites of the upper mantle. Many cases of fault emplaced successions of ancient oceanic crust, known as ophiolites are known from continental and/or arc collision belts of Phanerozoic age (i.e. 0–570 Ma) (Coleman, 1977, 1984; Moores, 1982), clearly indicating that oceanic and/or back arc basin oceanic crust existed, that processes of subduction were active during the Phanerozoic, and that periodically fragments of oceanic crust became incorporated into subduction complexes, or were overthrust tectonically onto continental land masses in a process called obduction. Obduction is the process whereby a subduction zone forms involving the attempted subduction of con-

tinental lithosphere under oceanic lithosphere. As argued above, because of the buoyancy of continental crust, the subduction system gets jammed up, the continent buoys up, and a segment of oceanic crust and lithosphere ends up perched on land.

By contrast, the existence of ancient continental crust is characterized by abundant, thick, variably folded and metamorphosed continental and continental shelf and slope sedimentary rock successions, especially those with abundant quartz sand and silt grains. In addition, such metasedimentary rocks will be intruded by regionally extensive granite batholiths, indicating a source of melted silica and aluminum-rich continental crust, and the rock successions may be significantly metamorphosed, producing high temperature, low to moderate pressure metamorphic mineral assemblages indicative of the middle to lower crust.

The influence of a subjacent continental-type crust may also be indicated by the composition of the volcanics. If voluminous andesites, dacites, and rhyolites occur, they also indicate a magma source of partially molten subjacent continental crust. The corollary, however, does not apply: abundant basaltic volcanic rocks do not necessarily indicate a subjacent oceanic crust, unless the previously defined succession of rock units of oceanic crust also occur. Basalts originate from partial melting of mantle, whether it be under oceans, or continents. However of all volcanic rocks, basalts from different settings may have some distinctive geochemical characteristics that may help to narrow the possible settings (Pearce and Cann, 1973; Wilson, 1989).

Beyond establishing the nature of the crust, assessing the tectonic setting in which volcanism occurred is the next step in understanding the role of volcanism in the geological cycle. "Tectonics" refers to the large-scale dynamic lithospheric and crustal processes which affect the Earth's surface and cause crustal deformation.

Of the various settings outlined above, mid-oceanic and back arc basin spreading ridge volcanism are the easiest to recognize. These settings should be easily discernible in the preserved stratigraphy of oceanic crust, including the layer 1 oceanic sediment assemblage, layer 2 massive and pillowed basalt lavas, and sheeted dolerite dyke complex, and layer 3 gabbroic plutonic complex, overlying mantle peridotites. We have clear evidence of such settings occurring in the Phanerozoic, e.g., the Cretaceous Samail ophiolite of Oman and the Jurassic Vourinos ophiolite of Cyprus (Coleman, 1977) and the Early Ordovician ophiolites of Newfoundland, Canada (Dewey and Bird, 1971).

By contrast, subduction settings are more difficult to recognize in the geological record. This is because subduction settings commonly show signs of both compression and extension. Compression is evident at the trench and in the leading edge of the overriding lithospheric plate, where sediment rafted into the trench on the downgoing plate is "scraped off," and added to the leading edge of the overriding plate, forming a highly folded and faulted metasedimentary wedge, which may also include fault slices of dismembered oceanic crust and even seamounts rafted into the subduction zone.

These highly deformed wedges of "off-scraped" sedimentary and volcanic rocks are called accretionary prisms or subduction complexes (Karig and Sharman, 1975). They represent the only physical evidence of the subduction process. Sometimes, the rocks are so deformed that they lose all coherence of the original stratigraphy. They become so dismembered by the very high shear stresses in the subduction zone that they break into highly distorted or strained slivers centimeters to many kilometers in dimension, which become chaotically mixed and highly sheared, and are called "tectonic melange," or "broken formation." Accretionary complexes may also contain domains of rock which have been subducted tens of kilometers below the seafloor in the downgoing plate and have been subjected to very high pressure, relatively low temperature metamorphism, before being jostled back as faulted slices to the surface. The distinctive high pressure, low temperature metamorphism produces distinctive assemblages of metamorphic minerals called blueschists, which include high pressure minerals such as glaucophone, and rock types such as eclogite. The famous Mesozoic Franciscan accretionary prism association of California (Bachman, 1982; Underwood, 1984; Aalto,

1989) is one of the best-documented accretionary prism systems of Phanerozoic age preserved in the geological record.

In addition, ancient subduction settings should be marked by assemblages of volcanic arc rock assemblages, including for young oceanic arcs, island arc tholeiite basalts and basaltic andesites, and subsurface plutonic equivalents, open ocean pelagic and hemipelagic sedimentary rocks, deposited by deep water suspension settling, together with associated volcaniclastic sedimentary rocks which represent aprons of debris derived from the arcs and transported into deep water by mass-flow sedimentary processes. Such a succession should be preserved on a "basement" of old oceanic crust.

For continental margin arcs, although some basaltic rock successions will occur, there would also be a significant association of more "evolved," more felsic volcanic and plutonic rock associations, including andesites and diorites, dacites and granodiorites, and rhyolites and granites which comonly show calc-alkaline geochemical affinities. Sedimentary rock associations will include abundant coarse volcanic conglomerates and sandstones, as well as coarse older continental crust derived bedrock sedimentary debris derived from sources such as older metasedimentary and metaigneous source rocks. Paleoenvironments represented by these sedimentary associations include continental alluvial fans and braided rivers, coarse shoreline deltas, and submarine fan and apron settings.

The geochemical compositions of continental arc volcanic belts vary from being alkaline to tholeiitic to calc-alkaline. Such associations are considered to be diagnostic of subduction arc settings by some people. However, similar associations may occur in major continental rift settings, such as the Cainozoic Basin and Range Province of the western United States, which include examples of almost every chemical association known (e.g., Christiansen and Lipman, 1972).

Behind the arc, there may be a complex of extensional graben basins. In fact, the arc is almost certainly located within intra-arc or back arc extensional basins, in an area of the subduction system subjected to neutral, or extensional regional stresses. For example, volcanoes of the

Cascades arc of the northwestern United States are associated with intra-arc grabens. The volcanoes of the Andes are also closely associated with major grabens. At times extension may even lead to arc rifting, the onset of spreading, and the formation of new oceanic crust and basins, which appears to have occurred along the Andean continental margin during the Jurassic and Cretaceous (Dalziel *et al.*, 1974). The sedimentary associations that may be preserved in these back arc extensional basins could include a complex association of alluvial fans, braided rivers, lakes, deltas, fan deltas, and shallow to deep marine fans and aprons. Such volcanic and sedimentary associations are unfortunately similar to volcanic and sedimentary associations that might be developed in continental rift settings, such as the relatively narrow East Africa rift zone, or the broader Basin and Range Province of the western United States.

This raises a major problem. Unless well-defined regionally extensive (hundreds of kilometers) accretionary complexes are found in association with relatively narrow belts of volcanic and sedimentary rocks that define a linear or curved trace also over hundreds of kilometers long, evidence of subduction settings is poor to say the least. To put this in a true scale context, major arc volcanoes such as stratovolcanoes or calderas are tens of kilometers in diameter, and are separated by kilometers to tens of kilometers. Regrettably, some reconstructions of ancient terrains invoke the existence of volcanic arcs, when the dimensions of belts of volcanic rocks are only on the order of tens of kilometers long, and evidence for accretionary complexes are lacking. Similarly, the existence of calc-alkaline volcanic associations is considered by some to be sufficiently indicative of subduction arc settings, when clearly, calc-alkaline associations also occur in other settings such as rift settings.

As outlined above, continental rift settings may show similar features to continental intra-arc and back arc settings, and in the absence of accretionary prisms, the two may be impossible to distinguish, especially if the rift zone is narrow, or has several narrow branches such as the East Africa rift zone. However, if the zone of volcanism (and associated plutonic activity) is extremely wide (hundreds of kilometers or more) relative to the regional tectonic trend, then it is unlikely that the setting was a subduction-related one, especially where there is no clear accretionary complex identifiable. It has not been demonstrated that very shallow dipping subduction zones will produce very wide volcanic arc belts hundreds or more kilometers wide. Complications occur in reconstructing ancient tectonic settings when late stage strike–slip faulting on a regional scale dismembers an original tectonic system and displaces particular elements along strike. For example, the displacement of a subduction complex away from its arc and back arc system may make such an arc system difficult to distinguish from a rift system. Resolving such a situation cannot be done by guessing. It is better to be objective and list the possibilities and pros and cons rather than inventing, for example, subduction settings, even invoking multiple coexisting subduction zones to explain closely spaced linear volcanic belts, when no real evidence exists.

As an example of the problems of reconstructing ancient configurations, the Paleozoic Lachlan Fold Belt of southeastern Australia evolved from a widespread oceanic paleogeography in the Cambro-Ordovician to a continental paleogeography in the Late Devonian (Cas, 1983). Cambrian and Ordovician rock systems consist of mafic volcanics and intrusions and interbedded deep marine sedimentary rocks indicative of a setting marginal to the paleo-Australian landmass. The volcanic successions are commonly fault bounded, with the faults commonly having experienced the last phases of movement in the Devonian and Carboniferous. There has been prolonged debate on the nature of the crust upon which volcanism and sedimentation occurred, some favoring Cambrian oceanic and associated arc crust (Crawford and Keays, 1978; Gray, 1997), others continental crust (Cas, 1983; Chappell *et al.*, 1988). There have also been various suggestions that volcanism occurred at plate margin arc subduction settings or intraplate rift settings. However, no obvious Cambro-Ordovician accretionary prism successions have been recognized, except for a possible remnant belt of polydeformed metasedimentary and volcanic rocks outcropping locally along the south coast of New

South Wales, probably related to an Ordovician to earliest Silurian arc system.

For the Silurian to Middle Devonian rock systems of the Lachlan Fold Belt, it is clear from the widespread occurrence of granite batholiths of this age throughout the fold belt, that the crust was almost everywhere continental in character, thickened during an end-Ordovician – Early Silurian compressional orogenic event. The system is marked by numerous extensional basins. Most of these appear to have been floored by subsided continental crust, but one may have developed embryonic oceanic or transitional crust. Despite this, a number of people have proposed that multiple subduction zones and associated arcs existed within the Lachlan Fold Belt during the Silurian to Middle Devonian. However, clear accretionary prisms with high pressure metamorphic assemblages have not been identified where the subduction zones are supposed to occur, nor have extensive tracts of remnant oceanic crust, which is a prerequisite for subduction: subduction cannot occur without consuming oceanic lithosphere. Also, the scale of the proposed individual "volcanic arcs" is too small. A more likely scenario is a broad magmatically active rift terrain, perhaps similar to the present day Basin and Range Province, influenced by major mantle activity, perhaps intraplate hotspot or plume activity, and widespread crustal melting. It seems difficult to explain magmatic activity in a belt hundreds of kilometers wide in any other way, especially in the absence of clearly defined subduction system(s), and when geochemical characteristics of ancient volcanic successions may be ambiguous in their significance.

Often attempts are made to evaluate the tectonic setting of ancient volcanic successions by the process of "geochemical fingerprinting." This involves comparing the geochemical characteristics of ancient volcanic successions to those of modern volcanic suites from particular tectonic settings, and by analogy, interpreting the tectonic affinity of the ancient succession based on similarities of geochemistry. This is fraught with dangers because there is nothing sacrosanct about magma geochemistry. The geochemistry of particular magmatic successions is influenced by a large range of variables including the composition and previous history of the source region in the Earth's interior from which the magma is derived, the degree of partial melting the source region undergoes, the pressure and temperature in the source region, the influence of external fluids, the magma ascent rate and the degree of interaction with country rock through which the magma rises, to mention only just a few of the variables.

Reconstructions and interpretations based solely on geochemistry should be based on a probability-based likelihood. For example, plots of various geochemical parameters on tectonic affinity diagrams show a spread of data points, and not uncommonly data from volcanics in one tectonic setting overlap with fields for other tectonic settings. Data for particular tectonic settings should be contoured. Then when data for ancient successions are plotted on such contoured tectonic affinity diagrams, the likelihood of the ancient succession representing one tectonic setting or another can be stated in statistically realistic probability terms, rather than being depicted in a single biased manner. The reason for suggesting such a statistically meaningful approach is that a particular geochemical compositional suite of volcanics does not necessarily uniquely depict a particular setting.

It is clear that tectonic reconstructions of ancient rock systems and their settings cannot be done on the basis of one parameter only, such as structural style, or the geochemical characteristics of igneous suites of rocks. Successful tectonic reconstructions depend on a holistic approach – considering all geological aspects of the entire rock belt in question. So these are some of the problems of reconstructing the setting of ancient volcanic successions and associated rocks. The question then is, how far back in time can we recognize evidence for the existence of plate tectonic settings based on the modern plate tectonic paradigm?

To answer this we must work backwards in time. Clearly plate tectonics can be ascribed with certainty as far back as the age of the oldest oceanic crust on the modern ocean floor, which is Jurassic in age. What evidence is there for plate tectonic processes during the Paleozoic? Cambro-Ordovician ophiolite complexes (slices of ancient oceanic crust), occurring in complex tectonic zones consistent with plate subduction processes is well documented for the Appalachian orogen (Dewey and Bird, 1971) indicating the likelihood of plate subduction processes as long ago as the beginning of the Paleozoic era almost 600 Ma ago.

# Precambrian volcanism and tectonic settings

There can be extreme difficulties in evaluating the tectonic setting of Precambrian volcanic and associated rock successions for several reasons. First, in the absence of zone macrofossils, dating and correlating rocks is difficult and expensive because of the costs of using precision radiometric or isotopic dating methods. Second, even if it is possible to date rocks, the experimental errors, given the ages involved, may be significant and on the order of millions to tens of millions of years. It thus becomes very difficult to establish a complete and exact chronology of events in such a long, distant interval of time. Too often attempts to reconstruct events in the Precambrian lose sight of this and present perspectives couched in modern or recent geological time and rate terms. Third, because of the range of effects of the geological cycle as outlined above, the geological record is often likely to be incomplete, although in some areas of Archean geology it is remarkable how continuous or complete the geological record appears to be.

There is an aura about Precambrian rock systems because of their antiquity, somehow implying that Precambrian processes were abnormal relative to modern Earth processes. However, it is easy to forget that the Priscoan, Archean, and Proterozoic eons represent 87.5% of geological time. On a total Earth timescale, therefore, what happened during the Precambrian was more correctly the norm, and the Phanerozoic rock systems represent very recent evolutionary stages in the history of the Earth.

Were Precambrian processes significantly alien to modern or even Phanerozoic geological processes? While this is almost certainly true in some cases, basic physical and chemical principles were the same. Thurston (1994) and Sylvester et al. (1997) find that many volcanic and associated volcaniclastic sedimentary successions (and therefore the processes) were the same in the Precambrian as in Phanerozoic successions. Common modern-day volcanic products such as viscous felsic lava domes, low viscosity mafic sheet lavas, autobreccias, subaqueous hyaloclastites, and ignimbrites can all be found in the Archean rock record. Exceptions include ultramafic komatiite lavas which are almost exclusively Precambrian in age, and predominantly middle to late Archean, but there are even some Mesozoic ones reported on Gorgona Island in Mexico.

Eriksson (1995) and Eriksson et al. (1997, 1998a) demonstrate that normal sedimentary processes and environments also existed during the Archean and Proterozoic. However, in the absence of vegetation to provide stable floodplains for single channel meandering rivers almost all alluvial plain rivers would have been braided because of the lack of stable banks. As a consequence sediment flux rates would have been high, and in aqueous environments, fan systems would have been common (fan deltas, deltas, and subaqueous fans). In addition, banded iron formations are peculiar to the late Archean and early Proterozoic, not because weird processes occurred, but because particular environmental geochemical conditions existed then that do not today. If those conditions did occur today, banded iron formations would also form today. Perhaps what was different then during much of the Precambrian, were some of the essential conditions that existed both inside and on the Earth's surface.

Many works have been written and compiled on Precambrian geology, including Windley (1976), Kroner (1981), Condie (1992, 1994, 1997), Coward and Ries (1995), de Wit and Ashwal (1997), Eriksson et al. (1998b), Percival and Ludden (1998), and Bleeker (2002) and from these it is clear that there is still a lot of uncertainty and debate about the physical, chemical, and as a result, the geological state of the Earth during the Precambrian. The peculiarities of Precambrian atmospheric evolution and the role of volcanism have already been discussed above. We will now consider the state of the lithosphere at the time, the implications this has for the tectonic regime, and the effects of this on the understanding of Precambrian volcanism.

Since a fundamental requirement for plate tectonics is the existence of, and the subduction of, oceanic lithosphere, it is difficult to advocate plate tectonic settings for Precambrian volcanics unless evidence for oceanic crust associated with accretionary prisms can be found. In this regard, there is considerable debate in the geological literature on whether or not clearly defined ophiolite successions of Precambrian age exist, whether or not plate tectonics as we know it today operated, or if some modified form of global tectonics existed. Bickle et al. (1994) and Hamilton (1998) have proposed that there is no clear evidence for the existence of remnants of oceanic crust (ophiolites) of Archean age. Some authors such as de Wit (1998) and Choukrone et al. (1997) dispute this, arguing that because of the long-term effects of the geological cycle, including subduction, uplift, tectonic dismembering, and erosion of orogenic rock systems of Archean age, the chances of preserving ophiolites are low. De Wit (1998) suggests that there are sufficiently well-preserved mafic volcanic and plutonic complexes with a number of the stratigraphic elements of the ophiolite succession preserved to warrant considering them to be the relicts of true Archean oceanic crust. He cites several examples in the pan-African orogenic belts as good examples of preserved Archean ophiolites.

Archean greenstone–granite belts, which represent the most important architectural component of Archean cratons, commonly consist of associations of basalts, komatiites (which are ultrabasic lavas and intrusions; see below), felsic volcanic rocks, granitic bodies and associated sedimentary successions. The basalts are commonly tholeiitic or hybrids of komatiites, called komatiitic basalts. Although the mafic and ultramafic components of greenstone belts (so called

because metamorphism and alteration have given them a greenish color because of the secondary minerals developed, such as serpentine, chlorite, epidote, and actinolite) have been considered in some instances to represent remnants of oceanic crust, in the absence of ophiolite stratigraphy, this is difficult to argue. Certainly komatiites cannot be considered to represent a component, or product, of normal oceanic crust-forming processes, because they are derived from the deep mantle, not the asthenosphere (Arndt, 1994). Another significant problem in terms of considering the tectonic significance of mafic and ultramafic successions of Archean greenstone belts is that in at least some greenstone belts (e.g., the Norseman–Wiluna belt, Western Australia) the mafic–ultramafic volcanics contain significant zircon xenocrysts, indicating that the magmas passed through granodioritic bodies or continental crust during uprise to the Earth's surface (Squire et al., 1998). It would appear that such greenstone belts were erupted onto continental type crust, and therefore do not represent oceanic crust.

Other uncertainties associated with adopting a modern plate tectonics perspective of Archean tectonics include the absence of any known blueschist facies metamorphic assemblages of Archean age. As previously outlined, blueschists are high pressure metamorphic rocks thought to result from the unique high pressure, deep metamorphic regime of subduction zones. The oldest known blueschist assemblages are 1.8 billion years old from China, and although blueschists are known from Paleozoic successions, they become relatively common from the Mesozoic onwards (Choukrone et al., 1997). Does this mean that subduction processes did not occur during the Archean or that a different subduction regime existed?

It is commonly accepted that the mantle was at least 200 °C hotter in the Archean than at present, resulting mostly from higher levels of heat released from radioactive decay than at present (Choukrone et al., 1997; Ranalli, 1997), but presumably also from residual accretion-generated heat. Higher heat flow would have created more vigorous convection in the mantle than at present and more ductile deformation.

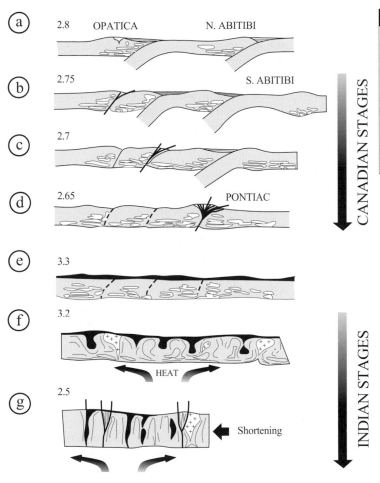

(a) 2.8   OPATICA        N. ABITIBI

(b) 2.75                 S. ABITIBI

(c) 2.7

(d) 2.65                 PONTIAC

CANADIAN STAGES

(e) 3.3

(f) 3.2

HEAT

(g) 2.5

Shortening

INDIAN STAGES

**Fig. 4.3.** Representation of two possible Archean tectonic systems, showing a convergent tectonic model based on the geology of the Superior Province of Canada, and a diapiric tectonic model based on the geology of the Dharwar Craton of India. Successions of cross-sections show possible evolution of tectonic stages. Numbers above each cross-section refer to the time depicted in gigayears. From Choukrone et al. (1997).

Higher rates of convection almost certainly would have led to more rifting and more subduction on smaller scales and under more ductile conditions. Evidence for crustal convergence, compressional deformation on a regional scale, and crustal accretion exists in the nappe and fold and thrust belt terranes of the Archean, such as the Superior Province of Canada (Choukrone et al., 1997) (Fig. 4.3). Although these show clear evidence of large-scale crustal convergence and shortening, and achieve high metamorphic grades indicating burial to perhaps several tens of kilometers, they lack blueschists.

Calculations of the effect of higher heat flow on crustal formation processes suggest that oceanic crust would have been thicker, perhaps greater than 25 km, and lithosphere thinner and more ductile (Ranalli, 1997). The effect of this would have been to make oceanic lithosphere more buoyant and more difficult to subduct than at present. In addition, it is possible that with a thicker crust and under more ductile mantle conditions, during attempted subduction, delamination or flaking of the crust from the upper mantle along the oceanic Moho may have occurred, leading to sinking and subduction of lithospheric mantle but not the crust, which would have been accreted to the overriding crust. Ranalli (1997) calls this type of subduction "flake tectonics." Under a higher temperature, ductile subduction regime, high pressure–low temperature blueschist metamorphic conditions were probably not achieved. What the effect of such a subduction regime may have been on the generation of subduction generated arc magmas is not clear. Hamilton (1998) argues strongly that there is no evidence for plate tectonics in the Archean, and that Archean tectonics was driven by the more ductile and higher thermal regimes inherent in the Earth's early history.

In addition to convergent tectonics (plate or flake), it also appears that in some regions (e.g., Pilbara Craton, Western Australia, or Dharwar Craton, India) the ductile thermal regime in the crust and lithosphere allowed buoyancy driven granite diapiric tectonics (Choukrone et al., 1997). The alternative way of viewing this is as gravitational sinking ("sagduction") of dense, supracrustal flood lava-like greenstones into older, relatively thin and ductile continental type crust which then squeezes up diapirically (Chardon et al., 1998) (Fig. 4.3). Such tectonic domains are marked by multiple, relatively low density, granitoid or gneissic domes which have intruded, and are surrounded by, concentrically deformed, denser, volcanic and sedimentary greenstone belts or structural basins. No linearity or polarity exists, and the dome and basin relationship between granites and greenstones suggests that gravitationally controlled tectonic processes dominated by buoyant granitoid diapirs and their deformational effects on the greenstone belts occurred in these terranes, not plate tectonic processes (see also Bleeker, 2002).

Since modern felsic magmas and basalts can be found in continental crustal settings, and in both subduction and rift settings, and with a wide range of geochemical characteristics, their tectonic significance is also equivocal. Since it appears that mantle convection was more vigorous in the Archean than today, rifting and rift related volcanism would have been common. Felsic volcanics cannot therefore be taken as being indicative of subduction settings; they could equally be rift related, in the dome and basin terranes, they could be related to granite diapiric intrusions, and they could also be related to melting caused by plume activity.

In addition to magmatism associated with convergent, extensional and granite diapiric tectonics, it also seems that ultramafic and mafic plume tectonics was an important process during the Archean, and probably since then to the present day. Modern hotspot activity which forms oceanic islands, basaltic oceanic plateaus, and intraplate basaltic continental volcanic fields, especially flood basalts, can be related to hotspot plume activity (Richards et al., 1989). In the Archean, the clearest representation of this was komatiite and associated basaltic volcanism (Campbell and Hill, 1988; Campbell et al., 1989). As depicted by Campbell and Hill (1988), the final tectonic effects on the crust are similar to those proposed in the diapiric tectonic model (Fig. 4.4).

Komatiites are unusual igneous rocks. They are marked by very low $SiO_2$ (<45 wt.%), and high MgO (>18 wt.%) contents. They are thought to have been erupted as high temperature (1450–1650 °C), and low viscosity (0.1 –? 5 Pa s) lavas capable of high velocity, turbulent flow and of thermally eroding their substrate (Huppert et al., 1984). Komatiite lavas also preserve spectacular crystallization textures, including needle-like and plate-like olivine and clinopyroxene spinifex textures, thought to result from rapid chilling of magma, for example in the surface crust of a flowing lava. They also preserve spectacular cumulate olivine textures, thought to result from the slower cooling, crystallization, and settling of olivine crystals in the interior of a lava, much of it probably after the lava had stopped flowing. Although komatiites are most commonly preserved in Archean rock successions, occurring in two main pulses of activity globally at 3400 and 2700 Ma (Arndt, 1994), some are also known in Proterozoic successions, although these are mostly komatiitic basalts, and minor occurrences are as young as Cretaceous (e.g., Gorgona Island, Mexico). Komatiite lavas are known from all the major Archean cratons of the Earth. Cratons are areas of continental crust which have been tectonically stable for a long period of time, usually one or more eons long. Archean cratons are the nuclei of the large present-day continental land masses and are thought to have resulted from crustal accretion processes during the Archean and have been largely tectonically stable ever since.

Komatiites represent a unique type of volcanism in Earth history. Their source is thought to have been in the deep mantle, and probably involved as much as 20% melting of the source mantle region (Arndt, 1994). They are generally

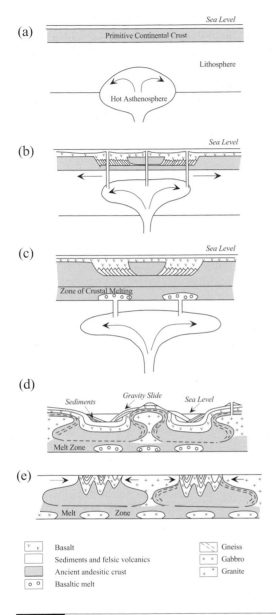

Basalt

Sediments and felsic volcanics

Ancient andesitic crust

Basaltic melt

Gneiss

Gabbro

Granite

**Fig. 4.4.** Sequence of cartoons depicting the uprise of an Archean mantle plume and its effects on the crust and the structure of greenstone belts. (a) Generation and uprise of a mantle plume. (b) Plume spread laterally in upper mantle beneath the crust and generating plume-derived flood volcanism and extensional tectonics. (c) Partial melting of lower crust by regionally high heat flow and underplated plume-derived magmas. (d) Diapiric uprise of lower crustal magmas to high levels in the upper crust and initial supracrustal gravitational deformation effects. (e) Lateral spreading of diapirs in the upper crust and peripheral compressional shortening or deformation of upper crustal strata. From Campbell and Hill (1988).

uncontaminated by upper mantle and crustal rocks, indicating rapid uprise to the Earth's surface. This has given rise to the proposal that komatiite magmas rose as huge mantle diapirs or plumes (Campbell *et al.*, 1989). The reason for their occurrence mostly in the Archean rock system is not clear, especially the occurrence in two distinct pulses at 3400 and 2700 Ma. It presumably relates to the high thermal regime and the enhanced levels of mantle convection during the Archean. Since their age coincides with the last stages of the major meteorite–asteroid bombardment phase in the history of the solar system, their origin is perhaps related to such an event, or several such events. Apart from the outpouring of regionally extensive komatiite and associated basalt lava sheets, it is unclear whether komatiite events and their host plumes caused any significant crustal deformation, either extensional or compressional. It is unclear if komatite volcanism coincided with the formation of extensional basins, which would certainly have enhanced the uprise and eruption of the magmas, as discussed previously. Plume tectonics may thus be represented only by voluminous volcanic outpourings, but it may also be related to crustal deformation, especially crustal extension, or impact events.

It therefore seems unclear whether or not evidence of normal, modern plate tectonic settings is preserved in rock systems of Archean age (see also Bleeker, 2002). The debate about the occurrence of ophiolites, and the absence of high pressure blueschist metamorphic facies casts doubt on whether "normal" subduction occurred, although some form of convergent tectonics certainly did. There is clear evidence (voluminous magmatism) for extensional tectonics, implying extensional stress fields and significant rifting activity. The abundance of granites suggests extensive areas of continental crust, especially in the late Archean. Unusual granite dome and greenstone basin belts also suggest that gravitationally controlled diapiric or sagduction tectonics occurred in some regions of the Earth's crust. Also, pulses of voluminous komatiite and associated basalt volcanism clearly indicate dynamic, deep-seated mantle plume influenced tectonics was also important.

The settings for Proterozoic (2500–600 Ma) volcanism can also be difficult to establish. However, there appears to be greater agreement in the literature that a more clearly defined plate tectonic regime existed during the Proterozoic eon (Condie, 1992). This is supported by the strong evidence for Proterozoic ophiolites from 2000 Ma (Helmstaedt and Scott, 1992), and the existence of accretionary prism-like fold and thrust belts, as well as high pressure blueschist metamorphic assemblages from at least 1800 Ma onwards (Choukrone et al., 1997), indicating perhaps deeper subduction of more rigid lithospheric plates. Volcanic rock successions also appear to be generally similar counterparts, and numerous studies clearly demonstrate that paleoenvironments were similar to today, although some exceptions are indicated by the extensive banded iron formations which extended from the Precambrian to about 1800 Ma.

Such a brief summary gives the impression that the Proterozoic tectonic and volcanic world was similar to today's and was dominated by large rigid plate interaction as today, with all the usual plate tectonic settings for volcanism and deformation.

However, such a simple assessment has to be approached with considerable caution. First, there are extensive Proterozoic orogenic or fold and thrust belts that do not appear to have involved subduction. They originated as intracontinental rifts, never developed to the break-up, spreading and oceanic crust stage, and therefore were not subjected to the tectonic effects of subduction, but were subjected to major regionally extensive orogenic deformation and metamorphism. O'Dea et al. (1997) document the evolution of the Mt. Isa Rift and Fold Belt of northern Australia, an extensive mid-Proterozoic intracontinental rift and fold belt terrain, that experienced multiple episodes of rift basin formation, volcanism, sedimentation, deformation, and uplift, culminating in the compressional Isan Orogeny between 1590 and 1500 Ma. The Mt. Isa Rift and Fold Belt is not an isolated example in the Proterozoic. Etheridge et al. (1987) noted that none of the Proterozoic orogenic belts of Australia show evidence for spreading, oceanic crust, and subduction. Green (1992) notes that globally there are many Proterozoic rift zones not associated with subduction zones. There is therefore a clear case for intracontinental and intraplate orogeny, indicating that magmatism associated with fold and thrust belts does not necessarily have to occur at plate margin subduction zones, an important lesson for other ancient orogenic belts, including younger Phanerozoic ones.

It is also important to stress that this principle applies not just to Proterozoic cases. Even in the modern world there are many examples of intracontinental rift terrains including narrow rift systems such as the East Africa rift zone, the Rhine Graben and North Sea rift systems of Europe, and the Baikal rift of Russia (Olsen, 1995). A very broad extensional and magmatically active rift terrain is of course the Basin and Range Province of the United States. In these settings if regional stress fields changed from extensional to compressional or transpressional, deformation would most easily be taken up by pre-existing normal faults formed during extension, and so causing basin inversion.

The concept of intraplate orogeny is not new (Park and Jaroszewski, 1994; Neil and Houseman, 1999). Neil and Houseman (1999) suggest that if a depression of the base of the lithosphere occurs into the asthenosphere then even a slight lateral compressive stress transmitted through the lithosphere could cause the dense lithosphere to downwell due to Rayleigh–Taylor instability that develops at the boundary. Such downwelling will cause shortening of the crust, leading to thickening and uplift. If detachment of the downwelling mass has occurred, the crust adjusts by extending. Extension and compression can then be achieved.

## Volcanism, the magma cycle, and the growth of the continental crust

It is appropriate to return to the relationship between volcanism and the geological cycle to assess how volcanism has contributed to the growth of the continental crust throughout geological time. Although there has been considerable debate on the nature of the Earth's early

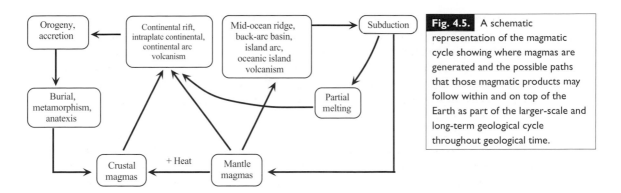

**Fig. 4.5.** A schematic representation of the magmatic cycle showing where magmas are generated and the possible paths that those magmatic products may follow within and on top of the Earth as part of the larger-scale and long-term geological cycle throughout geological time.

crust and how much continental versus oceanic crust there may have been, it seems clear that throughout geological time there has been a net transfer of matter from the mantle to the continental crust through the activity of the magma cycle as part of the geological cycle.

The magma cycle (Fig. 4.5) involves the mantle as the ultimate source for all magmas. Mantle magmas can take several possible paths. First and most obvious is eruption as lavas at mid-oceanic spreading ridges, forming seafloor crust, whereafter that crust is convected to subduction zones. It then undergoes partial melting, which contributes to volcanic arc magmas. The remainder is reassimilated into the mantle and becomes part of the long-term mantle magma cycle again.

Other mantle-derived magmas contribute to oceanic island and oceanic plateau volcanism and edifices, which are also ultimately transported into subduction zones as part of downgoing recycled crust and lithosphere. Some oceanic island/plateau edifices are completely recycled back into the mantle, but some may be decapitated in or collide with the subduction zone and become accreted to the overriding plate, as part of the accretionary prism and crust of the arc system of the overriding plate. Similarly slices of oceanic lithosphere (ophiolites) may become accreted to the arc crust and accretionary prism. In this way, mantle-derived volcanics are directly and physically incorporated into arc crust, which ultimately becomes accreted to continental crust through collison, if in fact the arc system is not already a part of a continental margin.

Mantle magmas may also contribute directly to arc volcanoes if they are derived from the

mantle wedge above subduction zones. If the arc is on a continental margin, then again there is direct transfer of mantle magmas to continental crust. Mantle magmas also contribute directly to the building of continental crust through volcanism in continental rift zones and in intraplate continental hotspot volcanic provinces. Even if rifting is successul to the spreading stage, some mantle-derived volcanics and intrusions will be incorporated into the crust of the newly formed rifted continental margin. If rifting is aborted, then the volcanics become part of the upper crustal succession, emplaced there as part of the rift valley volcanic succession.

In all of these cases, only a small part of the rising mantle magmas reach the Earth's surface. Much of the rising magma is trapped at the base of the crust and is added to it from underneath in a process called underplating, or is emplaced as intrusions within the crust.

Magmas resulting from partial melting or anatexis of continental or evolved arc crust form after input of heat from the mantle either through direct conduction or through heat transfer from the passage of mantle-derived aqueous fluids or magmas through the crust. These magmas rise, and are either emplaced as intrusions higher in the crust or are erupted on the Earth's surface, leading to the surface build-up of the continental crust.

Volcanic systems exposed subaerially on the Earth's surface are weathered and eroded, and the products are ultimately deposited in sedimentary basins, where they are buried, deformed, and incorporated into the crust. Eroded sediment which is deposited at the continental shoreline as part of a growing continental margin shelf–slope complex may simply be buried and become part of the continental margin crust, or may at some later stage become involved in collision in a subduction zone, and then becomes part of the interior collision zone crust of a larger, amalgamated continental mass.

Clearly then volcanic products derived from the mantle have been contributing to the growth of continental crust throughout geological time through the magma cycle and its role in the larger global geological cycle. Various attempts have been made to evaluate the rate of growth of continental crust through geological time. For example, McLennan and Taylor (1982) used isotopic fractionation trends in shales of various ages to evaluate the relative roles of primitive mantle source components relative to recycled supracrustally eroded sources. They found that through most of the Archean, primitive sources predominate, but that from the beginning of the Proterozoic there is an increasing input from recycled "mature" continental crustal sources.

This suggests that areas of continental crust developed during the Archean by a combination of arc terrane accretion at Archean-style subduction or accretion zones, and by rift related and intraplate magmatism, leading to the development of Archean cratons. From the Proterozoic onwards larger continents existed, shedding more sediment into the surrounding oceans than during the Archean. In addition, a more structured plate tectonic regime appears to have developed leading to arc–arc, arc–continent, and continent–continent collisions, and growth of continents by accretion and collision. Rift and intraplate hotspot magmatism would have continued to contribute to growth of continents. Additionally, under a regime of increasing atmospheric oxygen, and therefore increasing rates of weathering and erosion, increasing rates of sediment input to continental margin shelf and slope systems. The process would have become increasingly more significant as the surface area of landmasses exposed above sea level increased, therefore leading to increased surface area subject to weathering and erosion and increased sediment flux rates to the continental margins and oceans. Only when vegetation became widespread on landmasses from the Late Paleozoic onwards would this sediment flux rate have declined or stabilized.

## Magmatic cycles

A fundamental task of science is to recognize patterns, because patterns represent activity that is repeated and suggest there is an underlying scientific cause or mechanism that is systematic in its course and has natural in-built "memory" or a feedback process. There have at times been attempts to identify magmatic cycles in Earth history, or regular patterns, and then seek the underlying tectonic or tectonomagmatic cause. This approach has often been attempted with Archean volcanic successions with proposals that magma cycles begin with voluminous mafic and ultramafic magmatism and conclude with felsic magmatism (e.g., Hutchinson, 1973).

Barley *et al.* (1998) have proposed that there have been specific periods in the Earth's history when there were major increases, or cycles, of magmatism, tectonic activity, and the formation of ore deposits. They cite the interval from about 2.8 to 2.6 billion years ago as one of these cycles and suggest that it was related to the break-up of an Archean supercontinent. They draw analogy with a younger magmatotectonic cycle in the Mesozoic that coincided with the break-up of the Gondwana and Laurasian supercontinents.

On a global scale there is currently constant volcanic activity. Seafloor lavas are constantly being erupted at mid-oceanic ridges, and if we accept that in the past, when the Earth's interior was hotter than now, mantle convection must have been more active than today, then it is likely that there has always been constant seafloor volcanism throughout geological time. However, from time to time, spreading rates and magma production rates may have changed, depending on the total length of spreading ridge axes on the Earth's surface and convection rates.

The most likely settings where cyclicity of magmatism would occur would be in plate margin arc related subduction settings and continental rift settings. At subduction settings magma activity may vary depending on events in the subduction system. For example, during normal passive subduction involving rollback of the hinge zone of the downgoing plate, there should be a constant supply of magmas to the arc volcanic system. However, during periods of collision of oceanic ridges or landmasses with arc massifs, severe compression and cessation of subduction may temporarily interrupt volcanism until the subduction zone re-establishes itself, either adjacent to the old arc and subduction system, or in a totally new location, thus terminating magmatism at the old arc locality completely. Or, during rollback and the onset of arc rifting, back arc spreading, and opening of a new back arc basin, volcanism may be temporarily stopped, or the compositional character of the erupting magmas may change because magmas from both the subduction source and the newly developing spreading ridge are mingling or coming up simultaneously. Arc splitting and back arc spreading may occur several times in the one arc system as has occurred in many of the arcs of the western Pacific (Karig, 1974), and so a definable cyclicity of volcanic activity may occur.

In rift systems, peak volcanic activity usually precedes and accompanies the early phases of rifting. Volcanism may include widespread preliminary basaltic and bimodal volcanism. Once a rift widens to the stage of spreading and formation of oceanic crust, volcanism stops in the initial rift basin and becomes focused in the newly formed spreading mid-oceanic ridge system, unless randomly occurring intraplate hotspot activity occurs after break-up, as occurred in eastern Australia after the break-up of Gondwana in the Cretaceous (Johnson, 1989).

And then, of course, magmatism may just occur randomly in time and in geological setting, caused by hotspot or plume activity, or impact events.

A very nice example of magmatic cyclicity occurs in the Cambrian–Carboniferous Lachlan Fold Belt of southeasten Australia (Cas, 1983). The history of this orogenic belt, already discussed briefly previously in this chapter, is marked by magmatically active time intervals, alternating with time intervals when magmatism waned or stopped. The latter coincides with periods of regional compression and orogenic uplift, the former coincides with periods of extension,

new basin formation, the development of a new regional paleogeography, and presumably a new tectonic regime. Interestingly, volcanism is most prominent in new basins immediately after peak compression, suggesting that magmas rise through the crust in response to a changing regional stress field regime from compressional to extensional. So here magmatic cyclicity appears to be related to changing regional stress fields and tectonic conditions.

So sometimes magmatic cyclicity may occur; in other cases not. Each case has to be evaluated on its merits, and as part of an assessment of the total geological evolution of a rock system. By doing this the causes of the particular occurrence of volcanics can be assessed. Without a comprehensive geological assessment, it cannot.

## Volcanism, tectonics, and global climate change

Although the media and popular scientific literature invariably attribute present-day climate change to industrially derived greenhouse gas emissions, release of ozone damaging chlorofluorocarbons (CFCs), and the mystical powers of El Niño (a universal scapegoat for everything that goes wrong in modern life!), in fact global climate has been constantly changing throughout geological time (Frakes, 1979; Frakes et al., 1992). In part this can be related to the initial high temperature of the Earth immediately after formation and its progressive cooling since. It can also be attributed to the progressively changing composition of the Earth's atmosphere through time, initially being a $CO_2$-rich system as a result of the high $CO_2$ content of volcanic gases, and then changing to an oxygen-rich, $CO_2$-poor system in response to the evolution of life, and the role of photosynthesis and respiration, as discussed previously in this chapter. In addition, we know that volcanic gases and resultant atmospheric aerosols have direct effects on climate change (McCormick et al., 1995). Although volcanic eruptions usually produce short-term climate effects, very large eruptions could have more profound effects, perhaps even triggering glacial stages during ice ages

(Rampino and Self, 1992) and perhaps even changing climate so much as to cause mass extinctions (Rampino and Stothers, 1988; Courtillot *et al.*, 1990; Rice, 1990). This topic of the effects of volcanic eruptions on climate is covered comprehensively by Self (Chapter 5, this volume), and will not be further dealt with here.

In addition to short-term variations to climate, there are also long-term factors which have caused global climate change throughout geological time. These include the variations in the Earth's orbital behavior around the Sun, producing regular fluctuations in insolation to the Earth's atmosphere, called Milankovitch cycles. However, the indirect effects of volcanic activity on global climate change, as part of the Earth's plate tectonic cycle, are rarely considered. These volcano-tectonic processes have probably been more important in controlling long-term global climate change than the relatively short-term effects of major, but short-lived, volcanic eruptions.

The first obvious effect of seafloor spreading and plate tectonics on climate is that most of the continents are constantly being moved over the face of the globe across latitude and climatic zones. By this mechanism alone the climate of most continents is changing, and this has nothing to do with greenhouse gas emissions. Australia, for example, is steadily drifting northwards into hotter latitudes because spreading on the Indian–Australian mid-oceanic ridge in the Southern Ocean is north–south directed. Because Antarctica appears to be stationary, Australia's migration northwards is occurring at the half spreading rate of about 4 to 4.5 cm per year. This means that if the spreading rate remains constant into the future, the wonderful city of Melbourne, which is currently at a latitude of 45° S, will reach the tropics in about 40 Ma.

The second significant volcano-tectonic effect involves changes to spreading rates at mid-oceanic ridges. Increased spreading rates are caused by increased rates of mantle convection beneath the spreading ridges. This causes the ridges to inflate, and because spreading ridges have huge volumes (they are 2000–3000 km wide, several kilometers high, and thousands to tens of thousands of kilometers long), such inflation

causes the displacement of huge volumes of seawater, causing transgression or flooding of all continental shorelines and coastal plains globally. This causes the surface area of the oceans to increase, and the albedo of the Earth (the reflectivity to solar radiation) to increase, causing global cooling. Conversely, when spreading rates decrease, this causes global regression, decreased global albedo (because more land is exposed and land absorbs more solar radiation than seawater), and therefore global warming.

The third significant effect of seafloor spreading and plate tectonics is to move the the continental landmasses over the face of the globe, which has effects on the major oceanic current circulation patterns. Without the presence of landmasses, the oceans would simply circulate in major east–west elongated, latitudinally confined gyres, influenced by the rotation of the globe. Where north–south trending landmasses exist they often deflect warm tropical currents poleward into colder latitudes, as happened to the major Gulf Stream in the western Atlantic Ocean during the Tertiary (Crowell and Frakes, 1970). The ingress of warm currents into polar latitudes increases rates of evaporation off the sea surface, which leads to increased precipitation in polar regions, mostly as snowfall. If this effect is significant over an extended period of time, then ice sheets may form as the snow compacts and thickens, or those ice sheets already present may begin to spread, heralding the beginning of an ice age or a glacial stage of an ice age. As a result global albedo increases, leading to reduced insolation, and global cooling. Sea level drops and global regression of the continental margins occurs.

For ice sheets to develop permanently as major influences on global climate, they require a major landmass(es) around the poles as a seat for the ice sheet to accumulate and grow. This again happened during the late Tertiary when the landmasses of the northern hemisphere became congregated around the north pole as a consequence of drift due to plate tectonic movements. Similarly, during the Permian to Carboniferous ice age of the southern hemisphere, the continental supercontinent of Gondwana had drifted to a polar setting, establishing the ideal conditions for the growth of the huge ice sheet and glacial system that covered much of Gondwana, and caused global cooling, sea level fall, and regression of the world's oceans (Crowell and Frakes, 1970).

Volcanism and associated tectonics can lead to the building of land barriers to major east–west orientated oceanic gyre currents, causing them to be deflected across latitudes in poleward directions. The northward deflection of the Gulf Stream during the late Tertiary referred to above occurred after the seaway between the Atlantic and the Pacific oceans between North and South America was closed as a result of arc volcanism and subduction processes in the Middle America arc and trench system. An arc massif grew above sea level building the continuous Panama isthmus. The Gulf Stream, which used to flow west into the Pacific Ocean, was then blocked and was diverted northwards, causing a fundamental change in global climate patterns and leading to the onset of the Pleistocene ice age.

## Conclusions

The role of volcanic activity during the geological cycle is profound, and involves dramatic effects on a day-to-day basis as well as long-term effects that have shaped the Earth we live on. Volcanism plays a pivotal role in creating a complex of interactions between the lithosphere, atmosphere, hydrosphere, biosphere, and even the planetary sphere.

Volcanic eruptions and their direct hazards (see Tilling, Chapter 2, this volume) are a fundamental influence on the everyday lives of hundreds of millions of people who live within the boundaries of major volcanic provinces. Those of us who do not live in such provinces consider ourselves free of the influences of volcanic activity. However, we have seen in this chapter, as well as other chapters in this volume, because volcanic activity is controlled by global forces, all civilizations are in fact subject to its influences. Major eruptions and their volcanic aerosols can cause significant global climate disturbances. Seafloor spreading rate changes can cause sea levels to rise of fall globally. The climate of almost every conti-

nent is slowly changing year by year as the drift of continents atop moving lithospheric plates shifts the latitudinal position of the continental landmasses, slowly but surely.

Similarly, the blocking effects of migrating landmasses to major oceanic current systems, and the building of land barriers by volcano-tectonic processes at volcanic arc–subduction zone settings, also causing blocking and deflection of major oceanic circulation systems, can lead to major climate changes, even ice ages.

We have also seen that volcanism has had a fundamental effect on the composition of the Earth's atmosphere throughout geological time, and therefore its climate. The early atmosphere was $CO_2$ rich, and this probably caused "super-greenhouse" climatic conditions. The interaction between the atmosphere, hydrosphere, and biosphere, and particularly the effects of photosynthesis and respiration, caused a change in atmospheric composition, and therefore climate.

Although present-day volcanic activity can be related in most cases to logical tectonic settings in today's well-ordered plate tectonic framework, some volcanism occurs in random settings related to mantle hotspot or plume activity. In the geological past, however, it cannot just be assumed that modern plate tectonic processes operated in the same way and at the same scale and rates as today. Ophiolite complexes and blueschist metamorphic assemblages are consistent with the existence of plate-related oceanic spreading and subduction processes at least as far back as the early Proterozoic, about 2000 Ma ago. For the Archean, however, there are no known blueschist metamorphic assemblages, there is debate about whether or not ophiolites and oceanic crust existed, and it is known that the Earth's interior was at least 200 °C hotter than today. This would have caused more vigorous mantle convection than today. Rifting would have been widespread, more ductile styles of subduction and accretion would have occurred, and volcanic activity would have been vigorous in both rift and subduction settings. In addition however, magmatic hotspot, plume, and diapiric tectonic processes would have been important phenomena. The Archean Earth would have been a different one to the present-day one.

Does this mean that the principle of uniformitarianism ("the present is the key to the past") is invalid? It does if the principle is interpreted in a literal way, seeking to match exact conditions and configurations in the Archean as today. The principle is however alive and well if it is interpreted to mean that uniform principles of physics and chemistry applied throughout geological time. If these principles were consistent then the changing conditions and geological configurations throughout time can be explained in logical, scientific ways.

The real challenge for geology is not only to try to ascertain how far back into the past modern plate tectonic configurations can be recognized, but also to acknowledge that for more than half the Earth's history volcanism occurred under different tectonic regimes to the present. How far into the second half of the Earth's history can we recognize Archean-style tectonic influences? Plate tectonics is not the answer to everything in the geological record, and even today it does not explain some volcanic activity.

# References

Aalto, K. 1989. The Franciscan complex of northernmost California: sedimentation and tectonics. *Geological Society of London Special Publication*, **10**, 419–432.

Alvarez, L. W., Alvarez, W., Asaro, F., *et al.* 1980. Extraterrestrial cause for the Cretaceous–Tertiary extinction. *Science*, **208**, 1095–1108.

Appel, P. W. U., Fedo, C. M., Moorbath, S., *et al.* 1998. Well preserved volcanic and sedimentary features from a low-strain domain in the 3.7–3.8 Ga Isua Greenstone Belt, West Greenland. *Terra Nova*, **10**, 57–62.

Arndt, N. T. 1994. Archean komatiites. In K. C. Condie (ed.) *Developments in Precambrian Geology*, vol. 11, *Archean Crustal Evolution*. Amsterdam, Elsevier, pp. 11–44.

Bachman, S. B. 1982. The coastal belt of the Franciscan: youngest phase of northern California subduction. *Geological Society of London Special Publication*, **10**, 401–418.

Barley, M. E., Krapez, B., Groves, D. I., *et al.* 1998. The Late Archaean bonanza: metallogenic and environmental consequences of the interaction between mantle plumes, lithospheric tectonics and global cyclicity. *Precambrian Geology*, **91**, 65–90.

Bickle, M. J., Nisbet, E. G., and Martin, A. 1994. Archaean greenstone belts are not oceanic crust. *Journal of Geology*, **102**, 121–138.

Bleeker, W. 2002. Archean tectonics: a review, with illustrations from the Slave Craton. *Geological Society of London Special Publication*, **199**, 151–181.

Bowring, S. A., Williams, I. S., and Compston, W. 1989. 3.96 Ga gneisses from the Slave Province, Northwest Territories, Canada. *Geology*, **17**, 971–975.

Brown, J., Colling, A., Park, D., *et al.* 1989. *The Ocean Basins: Their Structure and Evolution*. New York, Pergamon Press.

Byerly, G. R., Lowe, D. R., and Walsh, M. M. 1986. Stromatolites from the 3300–3500-Myr Swaziland Supergroup, Barberton Mountain Land, South Africa. *Nature*, **319**, 489–491.

Campbell, R. I. and Hill, R. I. 1988. A two stage model for the formation of the granite-greenstone terrains of the Kalgoorlie–Norseman area, Western Australia. *Earth and Planetary Science Letters*, **90**, 117–130.

Campbell, R. I., Griffiths, R. W., and Hill, R. I. 1989. Melting in an Archaean mantle plume: heads it's basalts, tails it's komatiites. *Nature*, **339**, 697–699.

Cas, R. A. F. 1983. *A Review of the Palaeogeographic and Tectonic Evolution of the Palaeozoic Lachlan Fold Belt, Southeastern Australia*, Special Publication no. 10. Sydney, Australia, Geological Society of Australia.

1989. Physical volcanology in Australian and New Zealand Cainozoic intraplate terrains. In R. W. Johnson (ed.) *Intraplate Volcanism in Australia and New Zealand*. Cambridge, UK, Cambridge University Press, pp. 55–85.

Cas, R. A. F. and Wright, J. V. 1987. *Volcanic Successions: Modern and Ancient*. London, Allen and Unwin.

Chappell, B. W., White, A. J. R., and Hine, R. 1988. Granite provinces and basement terranes in the Lachlan Fold Belt, southeastern Australia. *Australian Journal of Earth Sciences*, **35**, 505–524.

Chardon, D., Choukrone, P., and Jayananda, M. 1998. Sinking of the Dharwar Basin (South India): implications for Archaean tectonics. *Precambrian Research*, **91**, 15–39.

Choukrone, P., Ludden, J. N., Chardon, D., *et al.* 1997. Archaean crustal growth and tectonic processes: a comparison of the Superior Province, Canada and the Dharwar Craton, India. *Geological Society of London Special Publication*, **121**, 63–98.

Christiansen, R. L. and Lipman, P. W. 1972. Cenozoic volcanism and plate tectonic evolution of the western United States. II. Late Cenozoic. *Philosophical Transactions of the Royal Society of London, Series A*, **271**, 249–284.

Cloud, P. E. 1988. *Oasis in Space: Earth History from the Beginning*. New York, W. W. Norton.

Coleman, R. G. 1977. *Ophiolites*. Berlin, Springer-Verlag.

1984. The diversity of ophiolites. *Geologie Mijnbouw*, **63**, 141–150.

Condie, K. C. (ed.) 1992. *Developments in Precambrian Geology*, vol. 10, *Proterozoic Crustal Evolution*. Amsterdam, Elsevier.

(ed.) 1994. *Developments in Precambrian Geology*, vol. 11, *Archean Crustal Evolution*. Amsterdam, Elsevier.

1997. *Plate Tectonics and Crustal Evolution*. Oxford, UK, Butterworth Heinemann.

Courtillot, V., Vandamme, D., Besse, J., *et al.* 1990. Deccan volcanism at the Cretaceous/Tertiary boundary: data and inferences. *Geological Society of America Special Paper*, **247**, 401–410.

Coward, M. P. and Ries, A. C. (eds.) 1995. *Early Precambrian Processes*, Special Publication no. 95. London, Geological Society.

Crawford, A. J. and Keays, R. R. 1978. Cambrian greenstone belts in Victoria: marginal sea-crust slices in the Lachlan Fold Belt of southeastern Australia. *Earth and Planetary Science Letters*, **41**, 197–208.

Crowell, J. C. and Frakes, L. A. 1970. Phanerozoic glaciation and the causes of ice ages. *American Journal of Science*, **268**, 193–224.

Dalziel, I. W. D., de Wit, M. J., and Palmer, K. F. 1974. Fossil marginal basin in the southern Andes. *Nature*, **250**, 291–294.

Davies, P. 1998. *The Fifth Miracle: The Search for the Origin of Life*. Melbourne, Australia, Penguin Books.

Derry, L. A. and Jacobsen, S. B. 1990. The chemical evolution of Precambrian seawater: evidence from REEs in banded iron formations. *Geochimica et Cosmochimica Acta*, **54**, 2965–2977.

Dewey, J. and Bird, J. M. 1971. Origin and emplacement of the ophiolite suite: Appalachian ophiolites in Newfoundland. *Journal of Geophysical Research*, **76**, 3179–3206.

de Wit, M. J. 1998. On Archean granites, greenstones, cratons and tectonics: does the evidence demand a verdict? *Precambrian Geology*, **91**, 181–226.

de Wit, M. J. and Ashwal, L. D. (eds.) 1997. *Greenstone Belts*. Oxford, UK, Clarendon Press.

Eaton, G. P. 1982. The Basin and Range Province: origin and tectonic significance. *Annual Reviews of Earth and Planetary Sciences*, **10**, 409–440.

Eriksson, K. A. 1995. Crustal growth, surface processes, and atmospheric evolution on the early Earth. *Geological Society of London Special Publication*, **95**, 11–25.

Eriksson, K. A., Krapez, B., and Fralic, P. W. 1997. Sedimentological aspects. In M. J. de Wit and L. D. Ashwal (eds.) *Greenstone Belts*. Oxford, UK, Clarendon Press, pp. 33–54.

Eriksson, P. G., Condie, K. C., Tirsgaard, H., *et al.* 1998a. Precambrian clastic sedimentation systems. *Sedimentary Geology*, **120**, 5–54.

Eriksson, P. G., Tirsgaard, H., and Mueller, W. U. (eds.) 1998b. Precambrian clastic sedimentary systems. *Sedimentary Geology (Special Issue)*, **120**, 1–343.

Etheridge, M. A., Rutland, R. W. R., and Wyborn, L. A. 1987. Orogenesis and tectonic processes in the early to middle Proterozoic of northern Australia. *American Geophysical Union, Geodynamics Series*, **17**, 131–147.

Frakes, L. A. 1979. *Climates throughout Geological Time*. Amsterdam, Elsevier.

Frakes, L. A., Francis, J. E., and Syktus, J. I. 1992. *Climate modes of the Phanerozoic*. Cambridge, UK, Cambridge University Press.

Frankel, C. 1996. *Volcanoes of the Solar System*. Cambridge, UK, Cambridge University Press.

Gray, D. R. 1997. Tectonics of southeastern Australian Lachlan Fold belt: structural and thermal aspects. *Geological Society of London Special Publication*, **121**, 149–177.

Green, J. C. 1992. Proterozoic rifts. In K. C. Condie (ed.) *Developments in Precambrian Geology*, vol. 10, *Proterozoic Crustal Evolution*. Amsterdam, Elsevier, pp. 97–136.

Hamilton, W. B. 1998. Archean magmatism and deformation were not products of plate tectonics. *Precambrian Geology*, **91**, 143–180.

Harland, W. B., Armstrong, R. L., Cox, A. V., *et al.* 1990. *A Geologic Time Scale 1989*. Cambridge, UK, Cambridge University Press.

Head, J. W. 1976. Lunar volcanism in time and space. *Reviews of Geophysics and Space Physics*, **14**, 265–300.

Helmstaedt, H. H. and Scott, D. J. 1992. The Proterozoic ophiolite problem. In K. C. Condie (ed.) *Developments in Precambrian Geology*, vol. 10, *Proterozoic Crustal Evolution*. Amsterdam, Elsevier, pp. 55–95.

Henriquez, H. and Martin, R. F. 1978. Crystal growth textures in magnetite flows and feeder dykes, El Laco, Chile. *Canadian Mineralogist*, **16**, 581–589.

Huppert, H. E., Sparks, R. S. J., Turner, J. S., *et al.* 1984. Emplacement and cooling of komatiite lavas. *Nature*, **309**, 19–22.

Hutchinson, R. W. 1973. Volcanogenic sulphide deposits and their metallogenic significance. *Economic Geology*, **68**, 1223–1245.

Johnson, R. W. (ed.) 1989. *Intraplate Volcanism in Australia and New Zealand*. Cambridge, UK, Cambridge University Press.

Karig, D. E. 1974. Evolution of arc systems in the Western Pacific. *Annual Reviews of Earth and Planetary Sciences*, **2**, 51–75.

Karig, D. E. and Sharman, G. F. 1975. Subduction and accretion in trenches. *Geological Society of America Bulletin*, **86**, 377–389.

Kroner, A. (ed.) 1981. *Precambrian Plate Tectonics*. Amsterdam, Elsevier.

Lamb, S., Hoke, L., Lorcan, K., *et al.* 1997. Cenozoic evolution of the Central Andes in Bolivia and northern Chile. *Geological Society of London Special Publication*, **121**, 237–264.

Lowman, P. D. 1976. Crustal evolution in silicate planets: implications for the origin of continents. *Journal of Geology*, **84**, 1–26.

Maas, R., Kinny, P. D., Williams, I. S., *et al.* 1992. The Earth's oldest known crust: a geochronological and geochemical study of 3900–4200 Ma old detrital zircons from Mt. Narryer and Jack Hills, Western Australia. *Geochimica Cosmochimica Acta*, **56**, 1281–1300.

McCormick, M. P., Thomason, L. W., and Trepte, C. R. 1995. Atmospheric effects of the Mt. Pinatubo eruption. *Nature*, **373**, 399–404.

McLennan, S. M. and Taylor, S. R. 1982. Geochemical constraints on the growth of the continental crust. *Journal of Geology*, **90**, 347–361.

Mills, A. A. 1984. Pillow lavas and the Leidenfrost effect. *Journal of the Geological Society of London*, **141**, 183–186.

Mojzsis, S. J., Arrhenius, G., McKeegan, K. D., *et al.* 1996. Evidence for life on Earth before 3,800 million years ago. *Nature*, **384**, 55–59.

Moores, E. M. 1982. Origin and emplacement of ophiolites. *Reviews of Geophysics and Space Physics*, **20**, 735–760.

Neil, E. and Houseman, G. H. 1999. Rayleigh–Taylor instability of the upper mantle and its role in intraplate orogeny. *Geophysical Journal International*, **138**, 89–107.

Nisbet, E. G. 1995. Archaean ecology: a review of evidence for the early development of bacterial biomes, and speculations on the development of a global scale biosphere. *Geological Society of London Special Publication*, **95**, 27–51.

O'Dea, M. G., Lister, G. S., Macready, T., *et al.* 1997. Geodynamic evolution of the Proterozoic Mount Isa terrain. *Geological Society of London Special Publication*, **121**, 19–37.

Olsen, K. H. (ed.) 1995. *Continental Rifts: Evolution, Structure, Tectonics*. Amsterdam, Elsevier.

Park, R. G. and Jaroszewski, W. 1994. Craton tectonics, stress, and seismicity. In P. L. Hancock (ed.) *Continental Deformation*. Tarrytown, NY, Pergamon Press, pp. 200–222.

Pearce, J. A. and Cann, J. R. 1973. Tectonic setting of basic volcanic rocks determined using trace element analyses. *Earth and Planetary Science Letters*, **19**, 290–300.

Percival, J. A. and Ludden, J. N. (eds.) 1998. The Earth's evolution through Precambrian time. *Precambrian Geology (Special Issue)*, **91**, 1–226.

Rampino, M. R. 1987. Impact cratering and flood basalt volcanism. *Nature*, **327**, 468.

Rampino, M. R. and Self, S. 1992. Volcanic winter and accelerated glaciation following the Toba super-eruption. *Nature*, **359**, 50–52.

Rampino, M. R. and Stothers, R. B. 1988. Flood basalt volcanism during the last 250 million years. *Science*, **241**, 663–668.

Ranalli, G. 1997. Rheology of the lithosphere in space and time. *Geological Society of London Special Publication*, **121**, 19–37.

Rice, A. 1990. The role of volcanism in K/T extinctions. *Geological Society of America Special Paper*, **244**, 39–56.

Richards, M., Duncan, R., and Courtillot, V. 1989. Flood basalts and hotspot tracks: plume heads and tails. *Science*, **246**, 103–107.

Rosing, M. T. 1999. $^{13}$C-depleted carbon micro-particles in >3700 Ma sea floor sedimentary rocks from west Greenland. *Science*, **283**, 674–676.

Sagan, C. 1979. Sulphur flows on Io. *Nature*, **280**, 750–753.

Schopf, J. W. (ed.) 1983. *Earth's Earliest Biosphere: Its Origins and Evolution*. Princeton, NJ, Princeton University Press.

Schopf, J. W. 1993. Microfossils of the Early Archean Apex Chert: new evidence of the antiquity of life. *Science*, **260**, 640–646.

Sebrier, M. and Soler, P. 1991. Tectonics and magmatism in the Peruvian Andes from late Oligocene time to the present. *Geological Society of America Special Paper*, **265**, 259–278.

Sigurdsson, H. (ed.), 2000. *Encyclopedia of Volcanoes*. San Diego, Cal, Academic Press.

Sleep, N. H. 1994. Martian plate tectonics. *Journal of Geophysical Research*, **99**, 5639–5655.

Squire, R. J., Cas, R. A. F., Clout, J. F., *et al.* 1998. Volcanology of the Archaean Lunnon Basalt and its relevance to nickel sulfide bearing trough structures at Kambalda, Western Australia. *Australian Journal of Earth Science*, **45**, 695–715.

Schidlowski, M. 1988. A 3,800-million-year isotopic record of life from carbon in sedimentary rocks. *Nature*, **333**, 313–318.

Sylvester, P. J., Harper, G. D., Byerly, G. R., *et al.* 1997. Volcanic aspects. In M. J. de Wit and L. D. Ashwal (eds.). *Greenstone Belts*. Oxford, UK, Clarendon Press, pp. 55–90.

Taylor, B. (ed.) 1995. *Backarc Basins: Tectonics and Magmatism*. New York, Plenum Press.

Thomas, P. G., Allemand, P., and Mangold, N. 1997. Rheology of the planetary lithospheres: a review from impact cratering mechanics. *Geological Society of London Special Publication*, **121**, 39–62.

Thurston, P. C. 1994. Archean volcanic patterns. In K. C. Condie (ed.) *Developments in Precambrian Geology Archean Crustal Evolution*, vol. 11. Amsterdam, Elsevier, pp. 45–84.

Underwood, M. B. 1984. A sedimentologic perspective on stratal disruption within sandstone-rich melange terranes. *Journal of Geology*, **92**, 369–385.

Walter, M. R., Buick, R., and Dunlop, J. S. R. 1980. Stromatolites 3400–3500 Myr old from the North Pole area, Western Australia. *Nature*, **284**, 443–445.

Wilde, S. A., Valley, J. W., Peck, W. H., and Graham, C. M., 2001, Earth's oldest mineral grains suggest an early start for life. *Nature*, **409**, 175–178.

Wilson, M. 1989. *Igneous Petrogenesis*. London, Allen and Unwin.

Windley, B. F. (ed.) 1976. *The Early History of the Earth*. New York, John Wiley.

# Chapter 5

# Effects of volcanic eruptions on the atmosphere and climate

Stephen Self

The bright sun was extinguish'd and
  the stars
Did wander darkling in the eternal
  space,
Rayless, and pathless, and the icy earth
Swung blind and blackening in the
  moonless air;
Morn came and went – and came,
  and brought no day . . .

*A section from "Darkness" by Lord Byron, written in June
1816 on the shores of Lake Geneva in the midst of the
"Year without a Summer," 14 months after the great
eruption of the volcano Tambora in Indonesia.*

## Introduction

Interest in the effects of volcanic activity on atmospheric phenomena, including the importance of volcanism in moderating climate and weather, has a long but patchy history, with the earliest recognition of a connection stretching back to classical times (Forsyth, 1988). The modern era of description began with the flood lava eruption from the Lakagígar (Laki) fissure in Iceland. Benjamin Franklin is usually attributed with being the first to make the connection between reports of an eruption in Iceland and an appalling "dry fog" (a sulfuric acid aerosol cloud) that hung over Europe in the summer of 1783 (Franklin, 1784; see also Thordarson and Self, 2003). From this point on, the role of volcanism in influencing and moderating our climate and weather has been a topic of debate (Self and Rampino, 1988; Robock, 2000) culminating in the past few decades with the need for a detailed understanding of natural influences on, and variability in, our atmosphere. This knowledge will help humankind to deal with serious social issues such as global warming due to anthropogenic causes.

In this chapter we will concentrate on short-term effects and changes to the atmosphere that last from less than one, to a few, years, such as those caused by the Laki eruption. The effects are caused by sulfate aerosol clouds generated from the sulfur (S) gas injected into the atmosphere by explosive volcanic activity. However, the chapter begins with a brief account of volcanic impacts on the atmosphere over much longer time periods and, at the end, we discuss the role that past super-eruptions may have played on climate and the environment.

## In the beginning . . .

After the loss of its primary atmosphere to space, outgassing from the Earth, manifested as volcanic activity, created a significant portion of the secondary atmosphere and continues to modify it. Volcanic gas emissions have probably remained fairly constant in composition throughout time, consisting of water, carbon dioxide, sulfur, halogens, and many other species in small amounts (e.g., nitrogen, argon, trace metals, and perhaps methane). The gaseous composition of the atmosphere has evolved due to several forcings and has been responsible for climate change over long ($10^6$–$10^8$ years) periods of time. Early in Earth history, from 4.5 to 3.5 billion years ago, our atmosphere was dominated by the carbon dioxide ($CO_2$) degassed by volcanism, after which photosynthesis and burial of organic carbon led to the accumulation of oxygen (Kasting, 1993). Volcanic water vapor ($H_2O$) contributed to the oceans. To reach a significant concentration of oxygen took another 2 billion years, and since that time the atmosphere has evolved to its present elemental composition (78 vol.% nitrogen and 21 vol.% oxygen, with all others in trace amounts totaling 1%).

The nitrogen-dominated composition of Earth's atmosphere is due to the inert nature of this gas. Other elements readily take part in biogeochemical cycles which bind them to the solid earth, whereas nitrogen is relatively unreactive and so most has remained in the atmosphere. Despite continued outgassing of carbon dioxide, the $CO_2$ content of the atmosphere (~370 ppm presently (e.g., Caldeira et al., 2003), but 315 ppm before the remarkable rise of the past 50 years due to anthropogenic activities) seems to have been generally decreasing over geologic time. Periods of higher outgassing that corresponded to elevated levels of volcanism, e.g., at mid-ocean ridges during periods of higher lithospheric plate spreading such as during the late Cretaceous period, might have contributed to higher atmospheric $CO_2$ concentrations and periods of warmer climate during Earth history. Both $H_2O$ and $CO_2$ are in such relatively high abundance in the atmosphere that additions of these species by individual eruptions probably do not markedly increase their concentrations, and thus do not influence the short-term greenhouse effect of these gases.

# The importance of volcanic sulfur emissions

Sulfur is the single volcanic gaseous species that has so far been demonstrated to have a profound effect on atmospheric composition (e.g., Rampino and Self, 1984). Injection of other volcanic volatiles such as chlorine (Cl) and $H_2O$ by eruption columns may temporarily alter atmospheric composition. However, evidence for several-year-long residence of Cl compounds after volcanic injections is lacking, and volcanic HCl appears to be rapidly scavenged on ash fallout and by other processes (Rose et al., 1982; Tabazedeh and Turco, 1993).

Sulfur is released into the atmosphere during eruptions mainly as sulfur dioxide ($SO_2$) gas, but also as hydrogen sulfide ($H_2S$) gas from magmas with reducing conditions (e.g., Scaillet et al., 1998). $H_2S$ rapidly oxidizes to $SO_2$ within several days. Both gases are converted in the atmosphere by photochemical oxidation to sulfuric acid ($H_2SO_4$) by combination with hydroxyl radicals via several reactions. $H_2SO_4$ and $H_2O$ are the dominant components ($H_2SO_4$:$H_2O$ ~ 70%:30%) of volcanic sulfate aerosols in the stratosphere. These form by a variety of processes (Fig. 5.1) including nucleation onto tiny ash and ice particles over a period of about a month after a volcanic eruption (Thomason, 1991; Zhao et al., 1996). Tropospheric aerosol-forming processes are faster but the lifetime of the sulfate particles is also considerably shorter (Toon and Pollack, 1980).

Aerosol particles are icy to liquid spheres with size ranges from 0.1 to 1.0 μm, with a modal diameter normally at ~0.5 μm, in midrange wavelength of incoming solar radiation. Stratospheric residence time ranges from weeks to months to several years for regional, hemispheric, and globally distributed aerosol clouds, respectively. Upper tropospheric residence times are poorly known (Graf et al., 1997) but lower tropospheric (<2–3 km, including the boundary layer) residence times for sulfate aerosols are in the order of days and much S gas is deposited by dry deposition before conversion to aerosols (e.g., Stevenson et al., 2003). While a volcanic sulfate aerosol cloud has a potential residence time in the stratosphere of up to several years, most of the fine volcanic ash lofted into the stratosphere by eruption columns settles out of the atmosphere in less than 3 months. After spreading around the Earth, aerosol clouds gradually diminish in concentration by sedimentation of the particles out of the atmosphere. The radiative

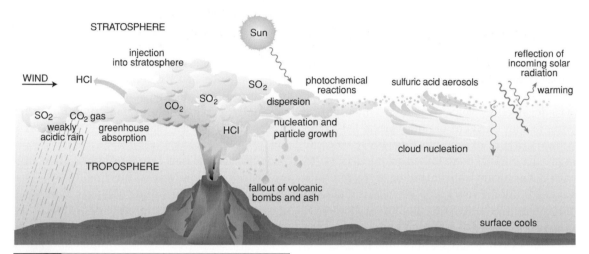

**Fig. 5.1.** Schematic diagram of the various volcanic inputs to the atmosphere and their interaction, fates, and radiative impact, including the formation of sulfuric acid aerosols. After McCormick *et al.* (1995) and Robock (2000).

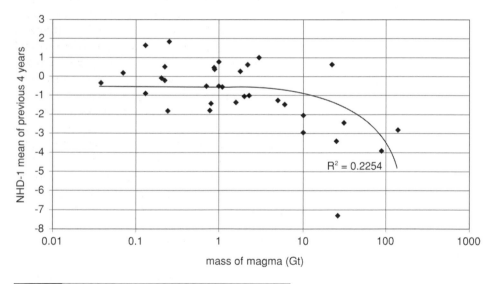

**Fig. 5.2.** The mass of magma erupted for a set of historic eruptions plotted against the deviation from the mean temperature (°C) calculated from the summer mean northern hemisphere temperature deviations of Briffa *et al.* (1998), NHD-1 time series (adapted after Blake, 2003). Note that there are a few warmer-than-average summers compared with cooler-than-average summers associated with eruptions, and that there is a weak relationship between mass of magma erupted and cooling up to 10 Gt (∼4 km³) of magma but a stronger dependency for larger eruptions. Part of the variable signal may be explained by some smaller eruptions releasing comparatively more S gases than others.

effects of the aerosols usually last a maximum of 2–3 years. A considerable amount of the mass of stratospheric aerosol falls onto the polar ice caps, where it increases the normal background acidity of the annual snow and ice fall. These annual layers can be sampled by ice cores and provide an important record of S-rich volcanic events which have been exploited by many studies (e.g., Zielinski, 2000).

Other atmospheric perturbations such as depletion of stratospheric ozone in temperate–polar latitudes are also associated with periods of

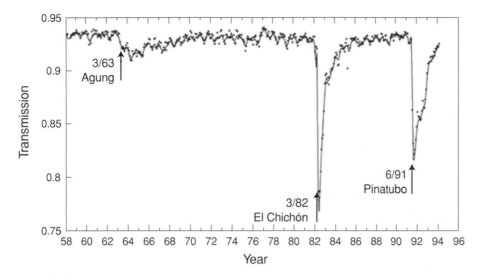

**Fig. 5.3.** The effect of stratospheric aerosols on incoming radiation is shown by this record of the atmospheric transmission of direct solar radiation at Mauna Loa observatory, Hawaii, from 1958 to 1994. The only three significant decreases in radiation received at the instrument are in 1963–4, 1982–4, and 1991–3, corresponding to the stratospheric aerosol clouds formed by the eruptions of Agung (1963), El Chichón (1982), and Pinatubo (1991), respectively. (Data and figure provided by E. Dutton, NOAA-CMDL.)

enhanced volcanic aerosols, including a relationship to increased levels of damaging ultraviolet light (Vogelmann *et al.*, 1992). Processes involving sulfate aerosols as sites of heterogeneous chemical reactions are implicated (Solomon *et al.*, 1996), similar to those thought to cause the ozone holes, but this phenomenon has existed only since man-made emission of chlorofluorocarbons (CFCs) began in the 1950s (Solomon, 1999).

## Climatic effects of volcanic aerosols

Explosive volcanic eruptions and rare large-scale, but still quite explosive, lava-flow-producing events cause short-term but important effects on climate and weather. Sulfuric acid aerosol clouds are the most significant of several intermittent and short-lived natural atmospheric perturbations that influence the radiation budget, surface temperatures, and circulation patterns of the atmosphere. The effects can be zonal, hemispheric, or global depending on the location of the volcano and the atmospheric circulation patterns at the time of the eruption (Robock, 1991; Kelly *et al.*, 1996; Halmer *et al.*, 2002). Tropical to equatorial eruptions will usually produce global aerosol dispersal as long as the eruption columns reach the stratosphere.

Recent studies have established a strong correlation between eruptions that release significant amounts of S gases and temperature change. Some have suggested that there may be a general linear relationship between the S gas yield of a volcanic eruption and the resulting cooling of surface temperature (e.g., Palais and Sigurdsson, 1989; Sigurdsson, 1990). There does seem to be a general relationship between eruption magnitude (mass or volume) and $SO_2$ release, and climate cooling (Fig. 5.2), as would be expected (Blake, 2003). Scattering of incoming short-wave radiation by aerosols resident in the atmosphere after volcanic eruptions leads to periods of lowered incident solar radiation (Fig. 5.3), cooler surface and tropospheric temperatures, and stratospheric warming. Because some of the climatic effects are indirect, the picture is actually quite complex (Robock, 2000).

Volcanic aerosols affect stratospheric stability and influence tropospheric dynamics. Thus one of the major measurable signals of volcanically induced radiative effects, temperature change on land masses, shows a response where eruptions in tropical latitudes produce winter warming in northern hemisphere continents but a cooling in summer (Robock and Mao, 1992). It is, of course, the abnormal, cool summer temperatures that have the greatest impact on humankind, and which feature most strongly in historic documentation of the effects of eruptions. Since the emergence of satellite-borne sensors that can measure $SO_2$, ozone, aerosols, and atmospheric temperature, considerable direct evidence now exists for a connection between individual explosive eruptions, the generation of volcanic aerosols, and climatic cooling. Aerosols, and short-lived ash and ice clouds derived from eruption columns, may provide nuclei for upper tropospheric and stratospheric cirrus clouds that can also lead to cooling of the Earth's surface, while increases in atmospheric water vapor in tropical regions may have a similar effect (Soden et al., 2002).

In reality, however, the magnitude of the signal in the surface temperature record is difficult to detect against background variation, even for the biggest historic eruptions. Compositing monthly to annual temperature records for a number of years spanning the times of significant eruptions has shown that even the largest historic eruptions are associated with coolings of only 0.3–0.8 °C globally, or up to 1 °C hemispherically, for 1–3 years (e.g., Self et al., 1981; Bradley, 1988; Mass and Portman, 1989), even after the removal of opposing influences such as warming induced by the El Niño–Southern Oscillation (ENSO) phenomenon (Angell, 1988, 1997). Some models suggest that eruptions on the scale of huge prehistoric events such as Toba 74 000 years ago (see below) have the potential to change climate to a much greater degree than has been observed following historic events, but simple scaling-up from the effects of smaller eruptions may not be warranted (Oppenheimer, 2002). One suggestion has been that the sulfate aerosol loading in the atmosphere is self-limiting such that the aerosol particles grow bigger with increasing S gas concentrations and sediment out of the atmosphere faster (Pinto et al., 1989). Alternatively, as suggested by Bekki (1995) and applied to the Toba case (Bekki et al., 1996), some models suggest that very large amounts of $SO_2$ will dehydrate

the atmosphere and prolong the lifetime of the gas introduced into the stratosphere thus increasing the duration over which sulfate aerosols will form.

As well as individual eruptions leading to climatic consequences, there is the possibility that decades- to century-long periods of higher-frequency S-emitting eruptions have cooled our climate. Such a period is the Little Ice Age (Grove, 1988) from the 1500s to the early 1800s when the northern hemisphere was 1–2 °C cooler than at present. In this period, for example, British vineyards sadly died out, and the River Thames froze frequently, permitting Frost Fairs to be held on the river almost annually. Atmospheric climate modelers have recently addressed the causes of the Little Ice Age, and have shown that both solar forcing (determined by sunspot cycles) and increased S-rich volcanism (as detected from ice-core records) may have played a part (e.g., Porter, 1986; Crowley and Kim, 1999). Conversely, an almost eruption-free period from 1912 to 1963 may have been associated with the measured average global climatic warming of about 0.5 °C over this period, but the beginnings of anthropogenic global warming make a definitive cause-and-effect for the warm period difficult to determine (Crowley, 2000; Robock, 2000).

# Atmospheric effects of individual volcanic eruptions: from 1783 to 1982

An important question is "How large an influence can volcanic eruptions have on climate by affecting surface temperature, precipitation, and weather patterns?" Although the greater the eruption magnitude the lower the frequency of occurrence (Pyle, 1998), eruptions have a frequency that varies according to type and magnitude of the event. Larger eruptions with much more massive S gas releases than have been witnessed historically are a certainty in the future. However, the data we have with which to assess their potential effects are derived from historic events, and some of the major examples will now be discussed.

In Europe, the year 1783 is often referred to as "Annus Mirabilis" (Year of Awe), because of the coincidence of several large-scale natural disasters and the extraordinary state of the atmosphere that caused great public concern. Benjamin Franklin, US Ambassador in Paris in 1783, is usually credited with first making the association between a volcanic eruption and an atmospheric phenomenon. While he was probably the first to publish in English (Franklin, 1784), German and French scientists had also made the same connection in 1783. Franklin proposed that the awful haze, or dry fog (sulfuric acid aerosols), that hung over the city in the middle of 1783 was caused by an eruption in Iceland. In fact, one of the world's greatest historic lava-producing eruptions had occurred at Lakagígar (usually shortened to Laki), Iceland, between June 1783 and February 1784 (for a review see Thordarson and Self, 2003). In this event, some of the >150 Mt of aerosols that were generated may have been limited to the troposphere. This is one of the only historic examples of a long-duration basaltic fissure eruption maintaining a high concentration of tropospheric aerosols. Under this haze, as with other aerosol veils, the Sun's rays were dimmed, and anomalously warm and cold weather occurred, including bitter cold in August and the coldest winter on record in New England. Crop failures were commonplace the next year, and were also experienced after the next large eruption of Tambora, which occurred in 1815. The effects of this great eruption will be discussed later.

The aftermath of Krakatau's climactic eruption in Indonesia in August 1883 was the first time that scientists first recognized and observed volcanic *dust veils*, because the developed world rapidly learned of the eruption by telegraphic communication. Observers could relate the colorful sunsets and other optical phenomena that spread over the northern hemisphere in the months following to news of the eruption. The Royal Society of London published a volume that documented the eruption and the ensuing dust veil (Symonds, 1888). At that time it was thought that fine-grained particles of silicate dust caused the atmospheric perturbations (as discussed in the classic review by Lamb, 1970). However, after the first recognition of the stratospheric sulfate aerosol layer (Junge *et al.*, 1961) it soon became evident that these particles in the atmosphere were not fine ash but aerosol droplets (Fig. 5.4).

After Krakatau, scientists measured and reported decreases in incoming radiation following volcanic eruptions at Santa Maria (Guatemala) in 1902 and Katmai (Alaska) in 1912. After Katmai very few other eruptions produced a significant stratospheric aerosol veil for a period of 50 years until Gunung Agung erupted on Bali, Indonesia, in 1963 (Self and King, 1996). This small but violent eruption which released about 7 Mt of $SO_2$ occurred at a time of year when the circulation patterns caused most of the aerosol cloud to be dispersed into the southern hemisphere. Immediately following this event, which occurred soon after the stratospheric sulfate aerosol layer had been discovered, the first airborne samples of aerosol droplets were collected between Australia and Bali. Temperature changes in both the stratosphere and troposphere were widely noted and measured for the first time after a volcanic eruption and linked to surface temperature changes. Later it was shown that the tropical troposphere cooled by as much as 0.5 °C, while the stratosphere in the same region warmed by several degrees. This useful data from the period after Agung led to the earliest attempt to reproduce the cooling after an eruption with a climate model (Hansen *et al.*, 1978).

Several other recent eruptions have proven invaluable in improving our understanding of the connection between volcanism, aerosols, and climate change. One of these was a small-volume eruption ($\sim$0.5 km$^3$) of very S-rich trachyandesite magma at El Chichón, Mexico, in 1982. Amongst other firsts, this eruption was the first to be thoroughly documented in terms of the stratospheric aerosols produced, their spread and atmospheric residence, and climatic effects time (Rampino and Self, 1984). El Chichón's large aerosol veil ($\sim$11–13 Mt) was also the first to be tracked by instruments on satellites. How much surface cooling this eruption actually caused has, however, been difficult to pin down due to a simultaneous ENSO event that caused warming in the tropics and which offset the volcanically induced cooling (Angell, 1988).

pre-eruption          post-eruption

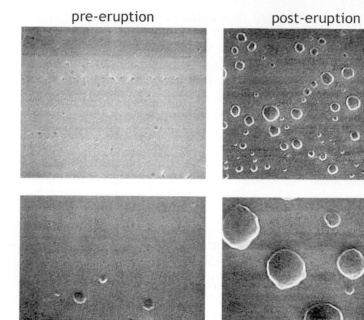

**Fig. 5.4.** An example of aerosol droplets in the 0.5–1 μm size range sampled on filters carried by high-flying aircraft after the El Chichón eruption of 1982. The filters on the left are from pre-eruption flights and those on the right are from post-eruption missions. (Figure adapted from *Eos, Transactions of the American Geophysical Union.*)

## Assessing the atmospheric impact of past eruptions

Satellite-borne instruments have been able to detect the $SO_2$ and aerosol clouds generated by volcanic eruptions only for a little longer than two decades. Several methods have been utilized to determine the output of S from magmas involved in pre-1960s eruptions. One of these, the so-called "petrologic method" (Devine *et al.*, 1984) has the advantage of theoretically being able to be applied to older eruptions but the method, like others, has limitations. The petrologic method involves direct electron microprobe analysis of the S concentration in pristine glass (melt) inclusions within magmatic crystals and in quenched eruption products, which represent un-degassed and degassed melt, respectively. Mass balance equations are used to assess the potential atmospheric S yield from individual eruptions using the mass of erupted melt and the difference in S concentration between un-degassed and degassed melt (Gerlach and Graeber, 1985; Self and King, 1996). The method

gives realistic estimates for divergent margin and hotspot basaltic eruptions (e.g., Thordarson *et al.*, 1996) and can be used to obtain information on the amounts of S released by these types of eruptions as far back in time as well-preserved glassy volcanic material can be found. This is 15–16 million years in the case of Columbia River flood basalt lava eruptions (Thordarson and Self, 1996).

Various studies have shown that the solubility of S in magmas is dependent on the melt composition, temperature, ambient pressure, and the partial pressures of sulfur and oxygen (e.g., Carroll and Webster, 1994). Further, the behavior of S in magmas depends considerably on these parameters. Scaillet *et al.* (1998, 2003) have calibrated and developed the petrologic method by experiments to determine the partitioning of S into an exsolved gas phase for evolved magmas of different compositions and oxidation state. Their results suggest that highly oxidizing dacitic and rhyolitic arc-derived magmas will normally develop an excess (exsolved) fluid phase that contains much of the S from within the magma reservoir (see also Wallace, 2001). This fluid phase may constitute up to 5–6 wt.% of the magma body and may contain up to 7 wt.% S, which is released as $SO_2$ upon eruption. However, melt inclusions trapped in phenocrysts growing in these magmas will not record the volatile concentrations within this sequestered fluid phase, and will thus contain low S concentrations, as found in the 1991 Pinatubo eruption products (Gerlach *et al.*, 1996). Application of the petrologic method to eruptions from this type of magmatic system may give estimates of S releases that are too small, often by one to two orders of magnitude.

Delineating the climatic effects of past eruptions, those that occurred at the beginning of instrumental temperature records or earlier, has proven difficult. These, of course, include eruptions considerably bigger than recent historic examples, and about which we would like to have a more precise picture of the effects. Much work has naturally involved proxy records such as tree-ring studies (e.g., LaMarche and Hirschboeck, 1984; Briffa *et al.*, 1998) and interpretation of historic documents. Ice-core records identify past years in which there was high sulfuric acid fallout for which a volcanic source is the only plausible explanation (Fig. 5.5). Some of these have no known source eruption, such as the event before Tambora in the year AD 1809 or the AD 1258 mystery eruption (Stothers 2000). Estimates of the temperature impact of much larger events include widespread cooling up to 5–10 °C, and some cases will be discussed later. It should be noted that a large acidity "spike" in an ice-core record denotes significant release of S-bearing volcanic gas to the atmosphere, and not necessarily a huge mass or volume of erupted magma. It follows that the "mystery" eruptions may not have been of exceptionally large magnitude, but may have been modest in volume and from S-rich magmatic sources.

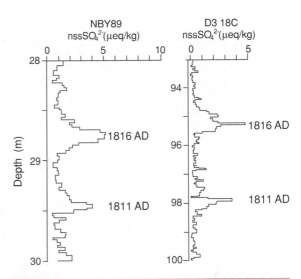

**Fig. 5.5.** Two acidity profiles (expressed as non-sea salt sulfate ions (nss-$SO_4^{2-}$) in microequivalents per kg of ice) from ice cores in the Antarctic (left) and Greenland (right), showing peaks attributed to the fallout of aerosols from the Tambora eruption in 1815 and an "unknown" eruption around 1809. After Langway *et al.* (1995).

## Atmospheric effects of individual volcanic eruptions: Pinatubo 1991 and Tambora 1815

Whilst Tambora is the biggest known eruption in the past few thousand years, and by far the largest recent historic eruption, the June 1991 eruption of Mt. Pinatubo provides us with the most closely monitored, large atmospheric aerosol cloud, the effects of which were carefully monitored. Both cases merit detailed description. Instead of searching for the subtle climatic effects of typical, relatively small historic eruptions, it is instructive to look in detail at the strongest atmospheric perturbations caused by volcanism, so we begin with the one for which we have the most complete picture and then view the Tambora case in hindsight.

### Mount Pinatubo

Mount Pinatubo (15° 07′ N 120° 20′ E), on the island of Luzon in the Philippines, produced the second-largest eruption (~5 km³ of dacite magma) in the twentieth century (after Katmai, Alaska, 1912; 11 km³). It also emitted the largest stratospheric $SO_2$ cloud ever observed by modern instruments (Bluth *et al.*, 1992), which generated an aerosol veil that hung over the Earth for more than 18 months causing spectacular sunsets and sunrises worldwide (reviewed by Self *et al.*, 1996).

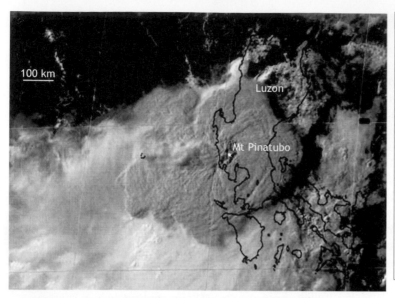

**Fig. 5.6.** Top: Japanese Geostationary Meteorological Satellite (GMS) visible wavelength satellite image of the top of the giant umbrella cloud developing above the eruption column of Mt. Pinatubo near the end of the climactic phase of the eruption on June 15, 1991 (16:40 local time). Scale bar is 100 km long. Bottom: Map of the spreading Pinatubo eruption cloud derived from Japanese GMS satellite images and the SO₂-dominated stratospheric cloud mapped by Total Ozone Mapping Spectrometer Satellite (TOMS). (TOMS data courtesy of Arlin Krueger, NASA Goddard Space Flight Center.)

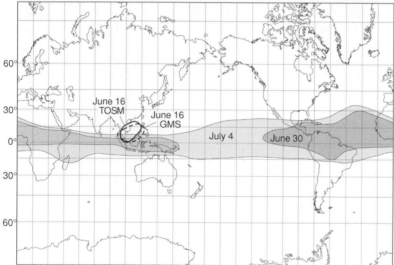

After 10 weeks of precursory activity, Mt. Pinatubo erupted climactically on June 12–15, 1991. Eruption columns of ash and gas reached in excess of 35 km altitude and injected about 17 Mt of $SO_2$ into the stratosphere, about twice the amount produced by the 1982 eruption of El Chichón, Mexico. The $SO_2$ formed a layer of sulfate aerosols containing 28 Mt, the largest perturbation to the stratospheric aerosol layer since the eruption of Krakatau in 1883 (and possibly equal to the post-Krakatau aerosol loading). The early summer date and tropical location of Pinatubo's eruption resulted in rapid spread of the aerosol cloud around the Earth in about 3 weeks (Fig. 5.6). The cloud attained global coverage 6–7 months after the eruption, producing a marked decrease in solar radiation at the Earth's surface, surface and atmospheric temperatures, and changing regional weather patterns.

The widespread dispersal of the aerosol cloud led to many optical effects such as unusual colored sunrises and sunsets, crepuscular rays, and a hazy, whitish appearance of the Sun ("Bishop's Ring"). The author observed these in Hawaii for much of late 1991, and through most of 1992; after a lull in the fall of 1992, they returned in the early months of 1993, finally dying away in about August of that year. In the months following the eruption, optical depth increases of the stratosphere were the highest ever measured by modern techniques, in the order of 0.3 to 0.4. Optical depths remained at higher values above background for a year longer following Pinatubo than following the El Chichón eruption, and the background aerosol concentration was again reached about 4 years after the eruption (McCormick *et al.*, 1995).

Startling decreases in abundance and in rates of ozone destruction were also observed over Antarctica in 1991 and 1992 due in part to the presence of extra aerosol caused by the Mt. Hudson eruption in southern Chile during August 1991 (Prather, 1992). A sharp reduction in ozone at 9–11 km in altitude (approximately at the tropopause) in the austral spring of 1991 was noted at the time of arrival of the Mt. Pinatubo and Mt. Hudson aerosols. The southern hemisphere "ozone hole" increased in 1992 to an unprecedented $27 \times 10^6$ km$^2$ in size, and depletion rates were observed to be faster than ever before recorded. In late 1992, weather patterns caused a shift in the polar vortex, and warm ozone-rich tropical air entered the Antarctic atmosphere to partially halt the ozone depletion.

As observed after several eruptions, including those of Agung in 1963 and El Chichón in 1982, stratospheric warming and lower tropospheric and surface cooling were noticed after the Pinatubo eruption (Fig. 5.7). Warming of up to 2–3 °C in the lower stratosphere at heights of 16–24 km (equivalent to a pressure of 30–100 mbar) occurred within 4–5 months of the eruption between the equator and 20° N latitude, and was also later noticed in middle northern latitudes. The distribution of warming closely mirrored the dispersal pattern of the aerosol cloud, strongly suggesting that temperature increase was due to absorption of incoming solar radiation by the aerosols.

Radiative forcing of the climate system by stratospheric aerosols depends on the geographic distribution, altitude, size distribution, and optical depth of the stratospheric aerosols. The

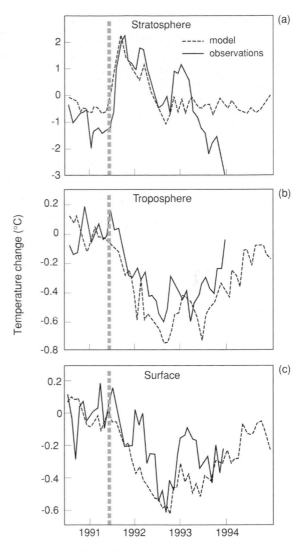

**Fig. 5.7.** Observed and modeled monthly temperature change of (a) stratosphere, (b) troposphere, and (c) surface before and after the Mt. Pinatubo eruption (marked by vertical dashed line). Stratospheric observations are 30-mbar zonal mean temperature at 10 °S latitute; model results are 10–70-mbar layer at 8° to 16 °S latitude. Other results are essentially global, with observed surface temperature derived from a network of meteorological stations. Base period for stratospheric temperatures is 1978–92, while troposphere and surface are referenced to the 12 months preceding the Mt. Pinatubo eruption. Panel (b) shows the model cooling predicted for 1991–2 (Hansen *et al.*, 1993) by observed model air temperature anomalies relative to the 1951–80 mean for northern hemisphere summers of 1991 and 1992. Figures courtesy of James Hansen and Helene Wilson, NASA Goddard Institute for Space Studies, after Hansen *et al.* (1993).

optically dense Pinatubo aerosol cloud caused marked changes in the amount of radiation reaching the Earth's surface. These changes, in turn, affected weather and climate for up to 4 years after the eruption. Several experiments measured the radiative climate forcing of the Pinatubo aerosols. The NASA Earth Radiation Budget Experiment provided the first unambiguous direct measurement of large-scale climate forcing by an eruption, an average radiative cooling of 2.7 W/m$^2$ by August of 1991 (Minnis et al., 1993). Direct solar beam reductions of 25–30% were measured at widely distributed stations, while measurements made by satellite-borne instruments suggested that the globally averaged net radiation at the top of the atmosphere may have decreased by about 2.5 W/m$^2$ in late 1991. These values translate into a global cooling of at least 0.5–0.7 °C, as seen in the global and northern hemisphere temperature records. General circulation models of temperature changes under this aerosol cloud predicted northern hemisphere surface cooling of as much as 0.5 °C (Hansen et al., 1992) and were verified by meteorological data (Hansen et al., 1993) (Fig. 5.7).

Climate models appear to have predicted the cooling after the Pinatubo eruption with a reasonable degree of accuracy. The interest in Pinatubo's atmospheric effects was both as a natural experiment and as a climatic perturbation. It allowed researchers to see what changes in surface temperature and circulation a large volcanic aerosol event can bring about – and it also served as validation for climate models. With a three-dimensional global circulation model, atmospheric scientists were able to predict the global cooling in 1991–3 and then check their results against real surface-temperature trends. Using a forcing in the model equivalent to a global mean optical depth of about 0.15, based on conditions appropriate for the Pinatubo aerosol cloud, yielded a model radiative forcing at the tropopause of −4 W/m$^2$. The maximum average cooling of up to 0.6 °C by late 1992 over high-latitude landmasses (Parker et al., 1996) was in good agreement with the modeled cooling. As well as the notable atmospheric effects of the eruption, Pinatubo's importance to science was to provide a strong natural perturbation for testing climate and atmospheric chemistry models.

Even more than 10 years after the eruption, the Pinatubo data are frequently used to test models, such as for estimating the effects of drying out of the atmosphere and its importance for the stratospheric water vapor budget (Soden et al., 2002).

The superimposition of Pinatubo's aerosol-induced climatic effects on long- and short-term variable atmospheric forcings, such as ENSO and "greenhouse" warming, led to speculation on post-1991 cooling on the current warming trend in the late twentieth and early twenty-first centuries. Global temperature trends show a gradual, unsteady rise from the mid-1970s to present, perhaps due to forcing by greenhouse gases, and this rise led to record high temperatures in the early to mid-1990s. However, cooling after the Pinatubo eruption offset the warming trend considerably, such that cooler than normal conditions dominated the northern hemisphere. The Pinatubo climate forcing was stronger than the opposite warming effects of either the ENSO event or anthropogenic greenhouse gases in the period from 1991 to 1993.

The predicted and observed climatic cooling after Pinatubo resulted in noticeable changes in the local climate and weather. For example, in 1992, the United States had its third coldest and third wettest summer in 77 years. Floods along the Mississippi River in the summer of 1993 and drought in the Sahel area of Africa may be attributable to climatic shifts caused by the Pinatubo aerosols and aerosol-induced temperature changes, and the cooling manifested itself globally as changing regional climate and weather patterns until about mid 1995 (Hansen et al., 1996).

## Tambora

The year 1816, the "Year without a Summer" in northeastern North America and western Europe, is probably the best-known example of a volcanically induced climate cooling event (Stothers, 1984; Harington, 1992). It was largely caused by the great eruption of Tambora, Indonesia, in April 1815, which yielded about 30 km$^3$ of trachyandesite (phonolitic) magma and generated at least 100 Mt of sulfuric acid aerosols in one of the largest recognized eruptions of the past few millennia (Self et al., 1984, 2004; Sigurdsson, and Carey, 1988a, 1988b). Interestingly, many ice-core records from both polar regions contain an annual layer with enhanced acidity content (indicating a period of volcanic aerosol fallout) in 1809–10 (Legrand and Delmas, 1987; Zielinski et al., 1996a) suggesting that another, as yet unknown, eruption prior to Tambora generated a global aerosol cloud (Fig. 5.5). This event may be associated with a small cooling detected in composites of northern hemisphere temperature records (e.g., Rampino and Self, 1982; Angell and Korshover, 1985; Mann et al., 1998) for the period before 1815, but the aerosols should have decayed away before Tambora erupted. It has been also noted that low sunspot numbers (the Maunder Minimum) contributed to reduced solar radiation over this same period (e.g., Lean and Rind, 1999).

After a few violent but minor-volume explosions in early April, on the 11th Tambora erupted for approximately 20 hours during a high-intensity Plinian eruption accompanied by pyroclastic flows and a fine-grained, extremely widespread ash cloud. The eruption rate may not have been much greater than that of Pinatubo 1991, which produced about 5 km$^3$ in 3.5 hours; therefore, as a rough estimate, the eruption cloud may have been of similar height, a conclusion supported by eruption modeling (Sigurdsson and Carey, 1988a). The Tambora phonolitic magma possessed moderately reducing conditions (Scaillet et al., 1998) and thus estimates of the amount of S degassed, which has been estimated to be $1 \times 10^{14}$ g (Palais and Sigurdsson, 1989), agree quite well with independent estimates from ice cores (e.g., Zielinski, 1995). Emission of gas from an exsolved fluid phase coexisting in the magma body prior to 1815 is not required to explain the S degassing from Tambora.

A bipolar ice-core acidity peak indicates that a global aerosol cloud developed in 1815–16. The aerosol veil may have caused cooling of up to 1 °C regionally, and more locally, with effects lasting until the end of 1816 and extending to both hemispheres (Stothers, 1984). Using measurements of the optical extinction of stars from astronomical observatories operating at the time, Stothers estimated that the optical depth of the atmosphere north of about 50° N latitude peaked at about 1.0 in late 1815, compared with a maximum of 0.2 after Pinatubo. This also argues for a large aerosol mass loading – the greatest of the past few

100 years, at least. Not surprisingly, then, the remarkable response of the atmosphere in 1816 caused the classic case of "volcano weather" (Stommel and Stommel, 1983) and associated hardship in society (Post, 1985).

In an area around Hudson Bay, Canada, the reduction in mean daily temperature from the average was 5–6 °C, and the Bay froze over so early that shipping was greatly disrupted through the whole summer of 1816 (Catchpole and Faurer, 1983). Snow and freezing conditions occurred in New England in June 1816 due to northwesterly winds bringing cold polar air southwards, causing the "Year Without a Summer" (recently reviewed by Oppenheimer, 2003). This was due to a greatly strengthened meridional circulation pattern, also noted after Pinatubo to a lesser degree (see Robock, 2000). With such marked temperature drops and freezing level falls for two consecutive summers after Tambora, one wonders how near the situation came to establishing a permanent snow cover and bringing ice/albedo feedbacks into play that could have led to a sharper and longer cold period?

Abnormally cool temperatures in western Europe led to widespread famine and misery, although the connection to volcanic aerosols was not realized at the time despite the widely reported atmospheric optical phenomena. Surprisingly, news of the Tambora eruption took 8 months to reach Europe and by the time it was announced in the press the association with the atmospheric conditions was not made. In other northern hemisphere regions there were average or warmer conditions, a consequence of the summer cooling–winter warming pattern described earlier. It is also not known with certainty whether 1815–16 was an El Niño year, which could perhaps have ameliorated the climate effects in some areas.

## The potential atmospheric effects of super-eruptions: Toba and flood basalt events

The largest historic eruptions have taught us a great deal about the connections between volcanic activity and its atmospheric effects, and have shown that global or hemispheric temperature decreases on the order of 1 °C are the maximum that Earth has experienced in historic times. However, much larger eruptions do infrequently occur, although there have been very few volcanic eruptions with magma outputs in excess of 100 km$^3$ in the past several thousand years (Newhall and Self, 1982). A super-eruption, for instance, of the size of the Toba event some 73 000 years ago (Chesner *et al.*, 1991) happens once approximately every 100 000 years (Pyle, 2000).

Two types of super-eruption have occurred on Earth: huge lava outpourings and great explosive eruptions (Rampino, 2002). Geological processes presently operating within the Earth are capable of producing only the explosive type, but the Laki eruption is considered to be a small-scale example of the type of activity that formed the huge lava flows of the geologic past, known as flood basalt eruptions. We will now examine what we know about the Toba eruption and its effects, and finish with a consideration of what atmospheric changes flood basalt eruptions may have caused in previous geologic epochs.

## The Toba super-eruption

The huge Toba event, named for Lake Toba in Sumatra which fills the caldera caused by the eruption, produced about 2800 km$^3$ of magma, compared with 0.5 km$^3$ for Mount St. Helens in 1980 or 5 km$^3$ for Pinatubo. Studies of a Greenland ice core have identified a marked acidity peak dated at about 71 000 years (Zielinski *et al.*, 1996b). This work suggests that if this peak was caused by the Toba aerosol fallout (the discrepancy is due to the different dating techniques used for estimating the ages of the ash and the ice layers) then aerosols remained in the stratosphere for an unusually long time, about 6 years. Toba erupted during a period of relatively rapid interglacial–glacial cooling and sea-level fall, apparently before the onset of marked cooling into a major glacial period at the end of interstadial 19, and separated from it by a warm event (Zielinski *et al.*, 1996b). The connection between these two events, the great eruption and the abrupt cooling, examined by Rampino and Self (1994), may be coincidental (Oppenheimer, 2002).

However, study of this and other great eruptions led to the concept of volcanic winter (Rampino *et al.*, 1988), a natural parallel to nuclear winter (Turco *et al.*, 1990). It is tempting to imagine that an eruption of the scale of Toba would have had devastating global climatic effects, but the hypothesis is far from proven by the available data. However, an estimated 15 cm layer of ash fell over the entire Indian subcontinent, and similar amounts over much of southeast Asia (Rose and Chesner, 1990), and recently the Toba ash has been found in the South China Sea, implying that several centimeters of ash covered southern China (Song *et al.*, 2000). As little as 1 cm of ash is enough to devastate agricultural activity, and so an eruption of this size would have catastrophic consequences. Billions of lives throughout most of Asia would be threatened if Toba erupted today and there would be consequences for the global economy as well, notwithstanding any serious global climate effects.

Even if we consider a conservative estimate of the amount of stratospheric dust and aerosols (∼1000 Mt), there are arguments that can be made about how Toba aerosols could have accelerated global cooling by causing a volcanic winter. Recent calculations with a general circulation model for a radiative decrease 100 times that of Pinatubo, in agreement with some estimates of the possible magnitude of Toba's impact (Rampino and Self, 1992), yield a northern hemisphere temperature drop of 10 °C (Jones and Stott, 2002). This would severely shorten the growing season at mid to high latitudes for at least a year, with possible longer-term climatic effects. However, the model runs did not produce long-term effects severe enough to precipitate glacial conditions.

Botanical studies suggest that reduction of early growing-season temperatures by ∼10 °C can kill ≥50% of temperate to sub-Arctic evergreen forests, and severely damage the surviving trees. Temperate deciduous trees, and other less cold-hardy vegetation, are expected to fare even worse. Cold-sensitive tropical vegetation might be significantly affected by low-latitude cooling. These results suggest a global ecological crisis, with reductions in standing crops of plants and animals. Although no biotic extinctions are seen at the time of Toba's eruption, human genetic studies indicate that prior to ~50 000 year ago, the human population suffered a severe bottleneck (perhaps reduced to only 4000 to 10 000 individuals), followed by rapid population increase, technological innovations, and spread of humans out of Africa. The predicted environmental and ecological effects of the Toba eruption may lend support to a possible connection between that volcanic event and the human population bottleneck (Rampino and Ambrose, 1999). These important issues are still in the hypothesis stage, and are as yet unproven, because our detailed knowledge of the timing of events around 73 000–70 000 years ago, and of the applicability of current climate models to such scenarios, needs to be significantly improved (Oppenheimer, 2002).

## Flood basalt events and mass extinctions

Extinction events are important factors in the history of life on Earth, and many recent studies suggest catastrophic causes for at least some biotic mass extinctions. The two processes that have been invoked are impacts of asteroids or comets and the occurrence of episodes of flood basalt volcanism. On one hand, the end-Cretaceous (65 Ma) mass extinction has been correlated with the collision of a 10-km diameter comet or asteroid with the Earth (Chapman and Morrison, 1994) which would have had severe environmental consequences (Toon et al., 1997), but convincing evidence of impacts has not as yet been found at times of several other extinction events. On the other hand, the coincidence of the eruption of the Siberian flood basalt lava flow province and the even more severe end-Permian extinctions (250 Ma), and the near-coincidence of the Deccan flood basalt province (India) and the Cretaceous/Tertiary (K/T) extinctions, fostered speculations that flood basalt eruptions have contributed to a number of mass extinctions (Erwin, 1994; Courtillot, 1999; see review by Wignall, 2001). Several workers have compared the dates of extinction events of various magnitudes with dates of flood basalt episodes and found some significant correlations (e.g., Rampino and Stothers, 1988; Stothers, 1993; Courtillot and Renne, 2003), supporting a possible cause-and-effect connection. Thus, it could be that extreme events of

both terrestrial and extraterrestrial origin are responsible for many of the punctuation marks in Earth history. These issues have been widely debated and much of the following is drawn from Rampino and Self (2000).

Continental flood-basalt eruptions are the largest subaerial eruptions of basaltic composition, with known volumes of individual lava flows >2000 km$^3$. A series of these huge eruptions piles up a thick stack of basalt lava flows, creating vast provinces extending to more than 1 million km$^2$ and containing more than 2 million km$^3$ of basaltic lava (Coffin and Eldholm, 1994). The estimated age dates of continental flood basalt events compiled from recent sources (radiometric, paleomagnetic, and stratigraphic data) all indicate that in most cases the largest volume of lava production happened within 1 million years or less. In at least three cases, the Deccan, Newark, and Siberian flood basalts, a direct measure of correlation with major mass extinction boundaries is possible. For example, the Deccan Traps flood basalts of India contain >50 recognized flows in 1200-m sections in the best-exposed region around Mahabaleshwar (Courtillot et al., 1988). Recent age dating of the basalts using $^{40}$Ar/$^{39}$Ar techniques, however, indicates that the vast bulk of the thick stack of flows was erupted over a period of only about 500 000 years, close to the time of the K/T boundary. Furthermore, the K/T boundary asteroid impact fallout layer has been reported to lie in sediments within the Deccan lavas, supporting the radiometric dating and paleomagnetic results that indicate that the eruptions began prior to the K/T boundary.

Several kinds of environmental effects of flood basalt eruptions have been suggested, including climatic cooling from sulfuric acid aerosols, greenhouse warming from $CO_2$ and $SO_2$ gases, and acid rain. Before evaluating these effects, however, the nature of flood basalt volcanism should be taken into account. There are few estimates of the amounts of climatically significant gases these volcanic events can release into the atmosphere. The style of eruption of flood basalt lava flows is an important determinant in the atmospheric and climatic effects of these events. In fissure-fed basaltic lava flow eruptions, the volumetric eruption rate is one critical parameter in determining the height of fire fountains over active vents, the convective rise of the volcanic gas plumes in the atmosphere, and hence the climatic effects of an eruption (Thordarson et al., 1996).

Flood basalt eruptions were originally envisioned as flowing as turbulent sheets 10–100 m thick that covered large areas in a matter of days. By contrast, it is now believed that flood basalts are erupted mainly as fissure-fed pahoehoe flows that inflate to their great thicknesses as and after they reach their great extent (Self et al., 1997). This suggests relatively slow emplacement of the flows over longer periods of time, most likely years to decades. Yet, average eruption rates must still have been higher than in any historic eruption except perhaps the extremes of some Icelandic outpourings. For example, the time required to emplace a large (1500 $km^3$) lava flow at the peak output rate (about $1.4 \times 10^7$ kg/s) of the largest historical lava flow eruption, the 1783 Laki (Iceland) eruption, would be approximately 10 years.

Flood basalt events can potentially influence surface temperatures in two ways, by either cooling or warming the atmosphere by the action of $SO_2$ or $CO_2$, respectively. As discussed above, climatic cooling is primarily a result of the formation and spread of stratospheric $H_2SO_4$ aerosols. From estimates of the S content of basaltic magmas and evidence of degassing of the lava flows, the S release from a large flood basalt eruption of approximately 1000 $km^3$ volume (such as the 14.7 million-year-old Roza Flow of the Columbia River flood basalt province in the USA Pacific Northwest) can be estimated at about 10 000 Mt of $SO_2$ along with significant amounts of HF and HCl (Thordarson and Self, 1996). At maximum eruption rates similar to Laki, the Roza eruption would have produced lava fountains $\geq 1.5$ km in height, and created a convective column rising $\geq 15$ km above the vents. If the Roza eruption lasted for 10 years, the annual aerosol burden in the atmosphere could have been as much as 1800 Mt per year, with a portion injected into the lower stratosphere. Would this be enough to produce a large increase in the aerosol loading of the regional and perhaps hemispheric

atmosphere? We cannot be certain because the balance between the quasi-continuous source of S gas, aerosol formation rates, and deposition rates for the aerosols is something that is only just beginning to be tested by atmospheric chemistry models (Stevenson et al., 2003).

The conversion of such a large amount of $SO_2$ to $H_2SO_4$ might have greatly depleted stratospheric $H_2O$ and OH, which at present consists of about 1000 Mt globally, but a large amount of water could have been injected into the lower stratosphere by the eruption plumes. How this enormous amount of aerosols would have manifested itself on atmospheric chemistry and dynamics cannot be calculated with presently available models. More accurate modeling of dense tropospheric and stratospheric aerosol clouds, and their effects on atmospheric dynamics and chemistry is needed.

Volcanic aerosols should cool the Earth's surface for all altitudes if the aerosol size is in the normal range (Lacis et al., 1992), with maximum cooling for low level aerosols. Therefore, volcanogenic aerosols in the troposphere will cool the surface as well. In most volcanic eruptions, however, tropospheric aerosols have a very short lifetime of about 1 week before they are washed out of the lower atmosphere. A flood basalt eruption, however, could release such large amounts of sulfuric acid aerosols (creating regional optical depths of >10), and perhaps cooling the climate and suppressing atmospheric convection, so that rainout might be less effective at removing the aerosols.

Historic stratospheric aerosol clouds have typical lifetimes of about 1–2 years, so that the ocean/atmosphere climate system does not have a chance to come to equilibrium with the short-lived volcanic perturbation. It follows that historic eruptions which created ~10 Mt of stratospheric aerosols are associated with surface cooling of only a few tenths of a degree Celsius in the 1–2 years following the eruption. Studies using a global climate model (GCM) suggest that if maintained for 50 years, a stratospheric loading of only about 10 Mt of aerosols would lead to a 5 °C global cooling. Therefore, if flood basalt eruptions were continuous at an average activity level for decades, significant cooling of climate is possible but only if a significant portion of the gas enters the stratosphere.

Basaltic eruptions can also have episodes of more explosive activity. Laki, for example, produced ash estimated to have been about 3% of the total erupted volume. Some continental flood basalt eruptions are known to have been accompanied by more explosive volcanism. For example, the North Atlantic Basalts were accompanied by a series of 200 or more basaltic ashes (Knox and Morton, 1988), equivalent to a magma volume of several thousand $km^3$. These explosive basaltic eruptions probably involved hydrovolcanic processes, and the widespread ash in the atmosphere could have led to some cooling (Jolley and Widdowson, 2002).

Greenhouse warming caused by large emissions of $CO_2$ from flood basalt volcanism has also been suggested as a cause of climatic change leading to mass extinctions. For example, based on published estimates of the concentration of $CO_2$ in basaltic magmas, the fraction of $CO_2$ degassed, and the volume and timing of the Deccan eruptions, it has been estimated that from ~6 to 20 × $10^{16}$ moles of $CO_2$ could have been released over a period of several hundred thousand years. Results of models designed to estimate the effects of this increased $CO_2$ on climate and ocean chemistry suggest, however, that the total increase in atmospheric $CO_2$ would have been less than 200 ppm, leading to a predicted global warming of less than 2 °C over the period of eruption of the lavas (Caldeira and Rampino, 1990). It is unlikely that such a small and gradual warming would have been an important factor in mass extinctions of life. In actual fact, each flood basalt eruption would deliver a modest amount of $CO_2$ to the atmosphere (compared to the background mass of the gas) over a period of a few years to, perhaps, a few decades, followed by an hiatus in activity of thousands to tens of thousands of years. Again this would militate against a long-term effect from $CO_2$ because after each eruption the added gas would be recycled before the next eruption occurred.

Another possible source of greenhouse warming from flood basalt eruptions is $SO_2$ gas emissions. The Laki eruption showed that prolonged basaltic fissure eruptions can create large, persistent downwind plumes of $SO_2$ gas and $H_2SO_4$ aerosols in the lower atmosphere. During the summer of 1783 in Europe, while the sulfurous dry fog hung in the air, the weather was stifling. Historical reports of acrid odor, difficulty in breathing, dry deposition of sulfate, and vegetation damage all indicate high $SO_2$ concentrations in the lower atmosphere from mid June to early August, which apparently led to higher-than-average human mortality (Grattan et al., 2003). The July 1783 temperatures in western Europe were up to ~3 °C warmer than the long-term average. The average concentration of $SO_2$ in the atmosphere over western Europe in July 1783 could have been up to several parts per million (background values are ~1 ppb), and calculations suggest that the greenhouse effect of the $SO_2$-rich volcanic haze could have led to significant regional short-term warming. The even greater emissions of $SO_2$ from large-volume flood basalt eruptions could have led to more severe regional warming events that could have affected flora and fauna over wide areas.

Several other less well-constrained extinction mechanisms for flood basalt volcanism have been suggested. In the case of the Siberian flood basalts and the Permian/Triassic extinctions (250 Ma) it has been suggested that the volcanism caused the release of vast quantities of methane from hydrates that had accumulated on adjacent high-latitude shelf areas. In other scenarios, the addition of volcanic $CO_2$ into the atmosphere might contribute to a global warming and overturn of anoxic ocean bottom waters, and perhaps create physiological problems for land animals. As discussed above, however, it is very unlikely that the basaltic volcanism could have produced enough $CO_2$ by itself to lead to these conditions.

In future work it is uncertain whether we will be able to directly measure the amount of S in flood basalt magmas and calculate a degassing budget, as we did for the Roza eruption. Flood basalt events associated with mass extinctions are old enough that finding well-preserved glassy material in which to analyze S concentrations is unlikely (but not impossible). Moreover, flood basalt lavas are often devoid of phenocrysts, so the chance of finding glass inclusions in which to measure pre-eruptive S concentrations is again

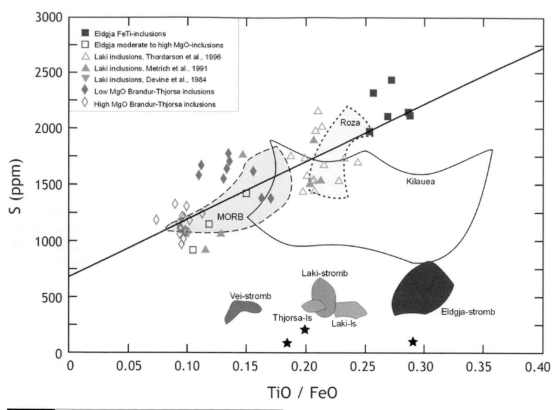

**Fig. 5.8.** Sulfur (S) concentration plotted vs. TiO/FeO ratio for samples from basalt lava eruptions from Iceland and other regions, and of various ages. All data are from electron microprobe analyses. The best-fit line goes through a set of data derived from glass inclusions in crystals from Icelandic eruptions, including Laki 1783, indicating the pre-eruption S concentrations in the magma. The fields of data from glass inclusions from MORB, Kilauea (Hawaii), and the Roza (Columbia River flood basalt province) lava flow samples are also shown. At bottom plot the fields of degassed vent tephra and lava flow samples (stars) for some of the Icelandic eruptions are shown. The difference between these two fields represents the amount (in ppm) of S degassed during basaltic lava-forming eruptions, as a function of Ti/Fe ratio. If this ratio can be measured in ancient flood basalt lavas, a crude proxy indication of the amount of S gas emitted during eruption of those lavas may be obtained. (Data from Thordarson et al., 2003 and papers referenced therein.)

eruptions. Thordarson *et al.* (2003) have shown that most tholeiitic fissure basalt eruptions studied to date have similar degassing characteristics, and we might assume that flood basalt eruptions follow the same pattern; certainly the Roza eruption did. Thus we will be able to make a first-order estimate of the amount of S gas released in flood basalt eruptions (Fig. 5.8), but we also need to know the typical volumes erupted in each lava-forming event, which is also not yet known for provinces older than the Columbia River basalts.

## Conclusions and future studies

Perhaps the most important challenge to the research effort on global change, for the purpose of making policy decisions, will be the early detection of the effects of global warming caused by human activities. It will be essential to recognize and remove effects resulting from natural phenomena amongst which volcanic aerosols are significant. Nevertheless, we are still a long way from understanding fully

reduced. However, most flood basalt provinces consist largely of tholeiitic basalt, which is a common type erupted today. It will hopefully be possible to develop a proxy indication of the initial sulfur concentrations in flood basalt lavas by comparison with more modern tholeiitic the effects of volcanic eruptions on atmospheric composition, climate, and weather. We have an imperfect record of past eruptions and of changes in temperature and weather following these eruptions. Future events will provide natural experiments in which we can study the atmospheric impact of volcanic aerosols, an effort already begun after eruptions such as El Chichón and Mt. Pinatubo. Moreover, knowledge of the impact of volcanic eruptions on climate is highly important for testing and improving climate models (Robock, 2000). New aspects of the impact of volcanic aerosols on the environment are continually being uncovered; for instance, it was recently shown that the post-Pinatubo aerosol cloud improved photosynthesis in trees and enhanced the terrestrial carbon sink for 2–3 years (Gu et al., 2003).

Fortunately, volcanoes tend to erupt less frequently as the magnitude of the eruptions increases. There are only one or two eruptions the size of Mt. Pinatubo (5 km$^3$) or Krakatau (12 km$^3$) every century. Three to five eruptions with a volume of a few tens of cubic kilometers, such as Tambora, occur every thousand years. There is very good evidence from records of tree-rings and meteorological data that eruptions like Tambora in 1815, Krakatau in 1883, and Mt. Pinatubo in 1991 caused measurable cooling of 0.5 to perhaps 1 °C in the northern hemisphere. It is the potential effects of larger eruptions that are still not quantified and should be the topic of future research. The larger the signals the more readily they should be detected. For instance, evidence for abrupt environmental changes in the aftermath of Toba may be detectable in high-resolution records (e.g., lakes and bogs), and in ice cores. Continued climate modeling efforts should also be made using climatologies appropriate for conditions at the time of the eruption, which becomes more challenging for eruptions occurring earlier than the Pleistocene epoch.

A major question regarding any possible relationship between flood basalt lava eruptions and extinction events involves the nature and severity of the environmental effects of the eruptions and their potential impact on life. Although the correlation between some flood basalt episodes and extinctions may implicate volcanism in the extinctions, it is also possible that other factors lead to the apparent association. Flood basalt episodes have been related to the inception of mantle plume activity and thus may represent one facet of a host of geological factors, e.g., changes in sea-floor spreading rates, rifting events, increased tectonism and volcanism, sea-level variations, that tend to be correlated, and may be associated with unusual climatic and environmental fluctuations that could lead to significant faunal changes. It has also been suggested that a coincidence of both a large asteroid impact and a flood basalt episode might be necessary to cause a mass extinction (e.g., Glasby and Kunzendorf, 1996). If flood basalt volcanism did contribute to the end-Cretaceous and other mass extinctions of the geologic past, it did not leave any direct forms of evidence as did asteroid impacts (such as shocked quartz and a layer rich in iridium) and thus the issue is going to be difficult to resolve.

Future explosive eruptions will certainly produce global aerosol clouds in the stratosphere, with much more severe effects than past historic explosive events have had on short-term climatic effects. Thus, it is also worthwhile to study the environmental effects of the largest known (Toba-size) explosive eruptions, despite the fact that eruptions like this occur at a minimum frequency of one every 50 000–60 000 years. Even smaller eruptions such as Laki are devastating but quite rare; however, much can be learnt about their effects that will help future planning and mitigation (e.g., Grattan, 1998). Statistical analysis of the ranked sizes of past eruptions indicate that the next "big" eruption might be about 90 km$^3$ (Pyle, 1998), about twice the magnitude of Tambora. Statistics may be just that, statistics, and while there is no need for panic, it is obvious that the past few thousand years have been relatively devoid of large eruptions and that humankind has little idea of what lies ahead in terms of volcanic impacts on Earth. With evidence of previously unknown eruptions in the 30–300 km$^3$ magnitude range being discovered at the rate of one or two per decade (e.g., Monzier et al., 1994), we may be underestimating the frequency of large explosive events that could severely affect the atmosphere. If this is the case we may be in for a surprise at any time, and we need to have vigorous research programs in place investigating the potential sulfur gas output, aerosol-generating capabilities, and atmospheric effects of the larger classes of volcanic eruption.

# References

Angell, J. K. 1988. Impact of El Niño on the delineation of tropospheric cooling due to volcanic eruptions. *Journal of Geophysical Research*, **93**, 3697–3704.

1997. Stratospheric warming due to Agung, El Chichón and Pinatubo taking into account the quasi-biennial oscillation. *Journal of Geophysical Research*, **102**, 947–948.

Angell, J. K. and Korshover, J. 1985. Surface temperature changes following the six major volcanic episodes between 1780–1980. *Journal of Climate and Applied Meteorology*, **24**, 937–951.

Bekki, S. 1995. Oxidation of volcanic $SO_2$: a sink for stratospheric OH and $H_2O$. *Geophysical Research Letters*, **22**, 913–916.

Bekki, S., Pyle, J. A., Zhong, W., *et al.* 1996. The role of microphysical and chemical processes in prolonging the climate forcing of the Toba eruption. *Geophysical Research Letters*, **23**, 2669–2672.

Blake, S. 2003. Correlations between eruption magnitude, $SO_2$ yield, and surface cooling. *Geological Society, London Special Publications*, **213**, 371–380.

Bluth, G. J. S., Doiron, S. D., Schnetzler, C. C., *et al.* 1992. Global tracking of the $SO_2$ clouds from the June 1991 Mount Pinatubo eruptions. *Geophysical Research Letters*, **19**, 151–154.

Bradley, R. S. 1988. The explosive volcanic eruption signal in Northern Hemisphere continental temperature records. *Climate Change*, **12**, 221–243.

Briffa, K. R., Jones, P. D., Schweingruber, F. H., *et al.* 1998. Influence of volcanic eruptions on Northern Hemisphere summer temperatures over 600 years. *Nature*, **393**, 450–455.

Caldeira, K. and Rampino, M. R. 1990. Carbon dioxide emissions from Deccan volcanism and a K/T boundary greenhouse effect. *Geophysical Research Letters*, **17**, 1299–1302.

Caldeira, K., Jain, A. K., and Hoffert, M. I. 2003. Climate sensitivity uncertainty and the need for energy without $CO_2$ emission. *Science*, **299**, 2052–2054.

Carrol, M. R. and Webster, J. 1994. Solubilities of sulphur, noble gases, nitrogen, fluorine, and chlorine in magmas. *Reviews in Mineralogy*, **30**, 231–279.

Catchpole, A. J. W. and Faurer, M.-A. 1983. Summer sea ice severity in Hudson Strait, 1751–1870. *Climate Change*, **5**, 115–139.

Chapman, C. R. and Morrison, D. 1994. Impacts on the Earth by asteroids and comets: assessing the hazards. *Nature*, **367**, 33–40.

Chesner, C. A., Rose W. I., Deino, A., *et al.* 1991. Eruptive history of Earth's largest Quaternary caldera (Toba, Indonesia) clarified. *Geology*, **19**, 200–203.

Coffin, M. L. and Eldholm, O. 1994. Large igneous provinces: crustal structure, dimensions, and external consequences. *Reviews of Geophysics*, **32**, 1–36.

Courtillot, V. 1999. *Evolutionary Catastrophes*. Cambridge, UK, Cambridge University Press.

Courtillot, V. and Renne, P. 2003. On the ages of flood basalt events. *Comptes Rendus Geoscience*, **335**, 113–140.

Courtillot, V., Feraud, G., Maluski, H., *et al.* 1988. Deccan flood basalts and the Cretaceous/Tertiary boundary. *Nature*, **333**, 843–846.

Crowley, T. J. 2000. Causes of climate change over the past 1000 years. *Science*, **289**, 270–277.

Crowley, T. J. and Kim, K.-Y. 1999. Modelling the temperature response to forced climate change over the last six centuries. *Geophysical Research Letters*, **26**, 1901–1904.

Devine, J. D., Sigurdsson, H., Davis, A. N., *et al.* 1984. Estimates of sulphur and chlorine yield to the atmosphere from volcanic eruptions and potential climatic effects. *Journal of Geophysical Research*, **89**(B7), 6309–6325.

Erwin, D. H. 1994. The Permian–Triassic extinction. *Nature*, **367**, 231–236.

Forsyth, P. Y. 1988. In the wake of Etna, 44 BC. *Classical Antiquity*, **7**, 49–57.

Franklin, B. 1784. Meteorological imaginations and conjectures. *Manchester Literary and Philosophical Society Memoirs and Proceedings*, **2**, 373–377.

Gerlach, T. M. and Graeber, E. J. 1985. Volatile budget of Kilauea volcano. *Nature*, **313**, 273–277.

Gerlach, T. M., Westrich, H. R., and Symonds, R. B. 1996. Pre-eruption vapor in magma of the climactic Mount Pinatubo eruption: source of the giant stratospheric sulfur dioxide cloud. In C. G. Newhall and R. S. Punongbayan (eds.) *Fire and Mud: Eruptions and Lahars of Mount Pinatubo, Philippines*. Quezon City, Philippines, Philippine Institute of Volcanology and Seismology, pp. 415–433.

Glasby, G. and Kunzendorf, H. 1996. Multiple factors in the origin of the Cretaceous/Tertiary boundary: the role of environmental stress and Deccan Trap volcanism. *Geologische Rundschau*, **85**, 191–210.

Graf, H.-F., Feichter J., and Langmann, B. 1997. Volcanic sulfur emissions: Estimates of service strength and its contribution to global sulfate distribution budget. *Journal of Geophysical Research*, **102**, 10727–10738.

Grattan, J. P. 1998. The distal impact of volcanic gases and aerosols in Europe: a review of the 1783 Laki fissure eruption and environmental vulnerability in the late 20th century. *Geological Society of London Special Publications*, **15**, 7–53.

Grattan, J. P., Durand, M., and Taylor, S. 2003. Illness and elevated human mortality coincident with volcanic eruptions. *Geological Society of London Special Publications*, **213**, 401–414.

Grove, J. M. 1988. *The Little Ice Age*. London, Methuen.

Gu, L., Baldocchi, D. D., Wofsy, S. C., *et al.* 2003. Response of a deciduous forest to the Mount Pinatubo eruption: enhanced photosynthesis. *Science*, **299**, 2035–2038.

Halmer, M. M., Schmincke, H.-U., and Graf, H.-U. 2002. The annual volcanic gas input into the stratosphere, in particular into the stratosphere: a global data set for the past 100 years. *Journal of Volcanology and Geothermal Research*, **115**, 511–528.

Hansen, J., Lacis, A., Ruedy, R., *et al.* 1992. Potential climate impact of the Mount Pinatubo eruption. *Geophysical Research Letters*, **19**, 215–218.

1993. How sensitive is the world's climate? *National Geographic Research and Exploration*, **9**, 143–158.

Hansen, J., Sato, M. K. I., Ruedy, R., *et al.* 1996. A Pinatubo climate modelling investigation. In *The Mount Pinatubo Eruption: Effects on the Atmosphere and Climate*, NATO ASI Series no. 142. Heidelberg, Springer-Verlag, pp. 233–272.

Hansen, J. E., Wang, W. C., and Lacis, A. A. 1978. Mount Agung eruption provides test of a global climatic perturbation. *Science*, **199**, 1065–1068.

Harington, C. R. (ed.) 1992. *The Year without a Summer? World Climate in 1816*. Ottawa, Canadian Museum of Nature.

Jolley, D. W. and Widdowson, M. 2002. North Atlantic rifting and Eocene climate cooling. *Abstracts of Volcanic and Magmatic Studies Group, Annual Meeting*, part II, Edinburgh, 16–17 December 2002, p. 10.

Jones, G. S. and Stott, P. A. 2002. Simulation of climate response to a super-eruption. *American Geophysical Union Chapman Conference on Volcanism and the Earth's Atmosphere*, Santorini, Greece, 17–21 July 2002, Abstracts, p. 45.

Junge, C., Chagnon, C. W., and Manson, J. E. 1961. Stratospheric aerosols. *Journal of Meteorology*, **18**, 81–108.

Kasting, J. F. 1993. Earth's early atmosphere. *Science*, **259**, 920–926.

Kelly, P. M., Jones, P. D., and Pengqun, J. 1996. The spatial response of the climate system to explosive volcanic eruptions. *International Journal of Climatology*, **16**, 537–550.

Knox, R. W. O. and Morton, A. C. 1988. The record of early Tertiary N. Atlantic volcanism in sediments of the North Sea basin. *Geological Society of London Special Publication*, **39**, 407–419.

Lacis A., Hansen, J., and Sato, M. K. I. 1992. Climate forcing by stratospheric aerosols. *Geophysical Research Letters*, **19**, 1607–1610.

LaMarche, V. C. and Hirschboeck, K. K. 1984. Frost rings in trees as records of major volcanic eruptions. *Nature*, **307**, 121–126.

Lamb, H. H. 1970. Volcanic dust in the atmosphere: with its chronology and assessment of its meteorological significance. *Philosophical Transactions of the Royal Society London, Series A*, **266**, 425–533.

Langway, C. C. Jr., Osada, K., Clausen, H. B., Hammer, C. U., and Shoji, H. 1995. A 10-century comparison of prominent bipolar volcanic events in ice cores. *Journal of Geophysical Research*, **100**, D8, 16, 211–216, 247.

Lean, J. and Rind, D. 1999. Evaluating Sun–climate relationships since the Little Ice Age. *Journal of Atmospheric and Solar–Terrestrial Physics*, **61**, 25–36.

Legrand, M. and Delmas, R. J. 1987. A 220-year continuous record of volcanic $H_2SO_4$ in the Antarctic Ice Sheet. *Nature*, **327**, 671–676.

Mann, M. E., Bradley, R. S., and Hughes, M. K. 1998. Global-scale temperature patterns and climate forcing over the past six centuries. *Nature*, **392**, 779–787.

Mass, C. F. and Portman, D. A. 1989. Major volcanic eruptions and climate: a critical evaluation. *Journal of Climate*, **2**, 566–593.

McCormick, M. P., Thomason, L. W., and Trepte, C. R. 1995. Atmospheric effects of the Mt. Pinatubo eruption. *Nature*, **272**, 399–404.

Minnis, P., Harrison, E. F., Stowe, L. L., *et al.* 1993. Radiative climate forcing by the Mount Pinatubo eruption. *Science*, **259**, 1411–1415.

Monzier, M., Robin, C., and Eissen, J.-P. 1994. Kuwae (~1425 AD): the forgotten caldera. *Journal of Volcanology and Geothermal Research*, **59**, 207–218.

Newhall, C. G. and Self, S. 1982. The volcanic explosivity index (VEI): an estimate of explosive magnitude for historical volcanism. *Journal of Geophysical Research*, **87**, 1231–1238.

Oppenheimer, C. 2002. Limited global change due to the largest known Quaternary eruption, Toba ~ 74 kyr BP? *Quaternary Science Reviews*, **21**, 1593–1609.

2003. Climatic, environmental and human consequences of the largest known historic eruption: Tambora volcano (Indonesia) 1815. *Progress in Physical Geography*, **27**, 230–259.

Palais, J. M. and Sigurdsson, H. 1989. Petrologic evidence of volatile emissions from major historic and pre-historic volcanic eruptions. In A. Berger, R. E. Dickinson, and J. W. Kidson (eds.) *Understanding Climate Change*. Washington, DC, American Geophysical Union, pp. 31–53.

Parker, D. E., Wilson, H., Jones, P. D., *et al.* 1996. The impact of Mount Pinatubo on world-wide temperatures. *International Journal of Climatology*, **16**, 487–497.

Pinto, J. R., Turco, R. P., and Toon, O. B. 1989. Self-limiting physical and chemical effects in volcanic eruption clouds. *Journal of Geophysical Research*, **94**, 11, 165–171, 174.

Porter, S. C. 1986. Pattern and forcing of northern hemisphere glacier variations during the last millennium. *Quaternary Research*, **26**, 27–48.

Post, J. A. 1985. *The Last Great Subsistence Crisis of the Western World*. Baltimore, MD, Johns Hopkins University Press.

Prather, M. 1992. Catastrophic loss of stratospheric ozone in dense volcanic clouds. *Journal of Geophysical Research*, **97**, 10, 187–191.

Pyle, D. M. 1998. Forecasting sizes and repose times of future extreme volcanic events. *Geology*, **26**, 367–370.

2000. Sizes of volcanic eruptions. In H. Sigurdsson (ed.) *The Encyclopedia of Volcanoes*. London, Academic Press, pp. 263–269.

Rampino, M. R. 2002. Supereruptions as a threat to civilizations on Earth-like planets. *Icarus*, **156**, 562–569.

Rampino, M. R. and Ambrose, S. H. 1999. Volcanic winter in the Garden of Eden: the Toba supereruption and the late Pleistocene human population crash. *Geological Society of America Special Paper*, **345**, 1–12.

Rampino, M. R. and Self, S. 1982. Historic eruptions of Tambora (1815), Krakatoa (1883) and Agung (1963), their stratospheric aerosols and climatic impact. *Quaternary Research*, **18**, 127–163.

1984. The atmospheric impact of El Chichón. *Scientific American*, **250**, 48–57.

1992. Volcanic winter and accelerated glaciation following the Toba supereruption. *Nature*, **359**, 50–52.

1994. Climate-volcanic feedback and the Toba eruption of ~74 000 years ago. *Quaternary Research*, **40**, 69–80.

2000. Volcanism and biotic extinctions. In H. Sigurdsson (ed.) *The Encyclopedia of Volcanoes*. London, Academic Press, pp. 263–269.

Rampino, M. R. and Stothers, R. B. 1988. Flood basalt volcanism during the past 250 million years. *Science*, **241**, 663–668.

Rampino, M. R., Self, S., and Stothers, R. B. 1988. Volcanic winters. *Annual Reviews of Earth and Planetary Sciences*, **16**, 73–99.

Robock, A. 1991. The volcanic contribution to climate change of the past 100 years. In M. E. Schlesinger (ed.) *Greenhouse-Gas-Induced Climate Change: A Critical Appraisal of Simulations and Observations*. Amsterdam, Elsevier, pp. 429–444.

2000. Volcanic eruptions and climate. *Reviews of Geophysics*, **38**, 191–219.

Robock, A. and Mao, J. 1992. Winter warming from large volcanic eruptions. *Geophysical Research Letters*, **19**, 2405–2408.

Rose, W. I. and Chesner, C. A. 1990. Worldwide dispersal of ash and gases from Earth's largest known eruption: Toba, Sumatra, 75 kyr. *Paleogeography, Paleoclimatology, Paleoecology*, **89**, 269–275.

Rose, W. I., Jr., Stoiber, R. E., and Malinconico, L. L. 1982. Eruptive gas compositions and fluxes of explosive volcanoes: budget of S and Cl emitted from Fuego volcano, Guatemala. In R. S. Thorpe (ed.) *Andesites*. Chichester, UK, John Wiley, pp. 669–676.

Scaillet, B., Clemente, B., Evans, B. W., *et al.* 1998. Redox control of sulfur degassing in silicic magmas. *Journal of Geophysical Research*, **103**, 23, 923–937, 949.

Scaillet, B., Luhr, J. F., and Carroll, M. R., 2003. Petrologic and volcanic constraints on volcanic sulfur emissions to the atmosphere. In A. Robock and C. Oppenheimer (eds.) *Volcanism and the Earth's Atmosphere*, Geophysical Memoir no. 180. Washington, DC, American Geophysical Union, pp. 11–40.

Self, S. and King, A. J. 1996. Petrology and sulfur and chlorine emissions of the 1963 eruption of Gunung Agung, Bali, Indonesia. *Bulletin of Volcanology*, **58**, 263–286.

Self, S. and Rampino, M. R. 1988. The relationship between volcanic eruptions and climate change: Still a conundrum? *Eos (Transactions of the American Geophysical Union)* **69**, 74–75, 85–86.

Self, S., Gertisser, R., Thondorson, T., *et al.* 2004. Magma volume, volatile emissions, and stratospheric aerosols from the 1815 eruption of Tambora. *Geophysical Research Letters*, **31**, L20608, doi:1029/2004 GL020925.

Self, S., Rampino, M. R., and Barbera J. J. 1981. The possible effects of large 19th and 20th century volcanic eruptions on zonal and hemispheric surface temperatures. *Journal of Volcanological and Geothermal Research*, **11**, 41–60.

Self, S., Rampino, M. R., Newton, M. R., *et al.* 1984. A volcanological study of the great Tambora eruption of 1815. *Geology*, **12**, 659–673.

Self, S., Thordarson, T., and Keszthelyi L. 1997. Emplacement of continental flood basalt lava flows. In J. J. Mahoney and M. F. Coffin (eds.) *Large Igneous Provinces: Continental, Oceanic, and Planetary Flood Volcanism*, Geophysical Memoir no. 100. Washington, DC, American Geophysical Union, pp. 381–410.

Self, S., Zhao, J.-X., Holasek, R. E., Torres, R. C., *et al.* 1996. The atmospheric impact of the Mount Pinatubo eruption. In C. G. Newhall and R. S. Punongbayan (eds.) *Fire and Mud: Eruptions and Lahars of Mount Pinatubo, Philippines.* Quezon City, Philippines, Philippine Institute of Volcanology and Seismology, pp. 1089–1115.

Sigurdsson, H. 1990. Evidence of volcanic aerosol loading of the atmosphere and climate response. *Paleogeography, Paleoclimatology, Paleoecology*, **89**, 227–289.

Sigurdsson, H. and Carey, S. 1988a. Plinian and co-ignimbrite tephra fall from the 1815 eruption of Tambora volcano. *Bulletin of Volcanology*, **51**, 243–270.

1988b. The far reach of Tambora. *Natural History*, **6**, 66–73.

Soden, B. J., Wetherald, R. T., Stenchikov, G. L., *et al.* 2002. Global cooling following the eruption of Mount Pinatubo: a test of climate feedback by water vapour. *Science*, **296**, 727–730.

Solomon, S. 1999. Stratospheric ozone depletion: a review of concepts and history. *Reviews of Geophysics*, **37**, 275–316.

Solomon, S., Portmann, R. W., Garcia, R. R., *et al.* 1996. The role of aerosol variations in anthropogenic ozone depletion at northern midlatitudes. *Journal of Geophysical Research*, **101**, 6713–6727.

Song, S.-R., Chen, C.-H., Lee, M.-Y., *et al.* 2000. Newly discovered eastern dispersal of the youngest Toba Tuff. *Marine Geology*, **167**, 303–312.

Stevenson, D., Johnson, C., Highwood, E., *et al.* 2003. Atmospheric impact of the 1783–1784 Laki eruption. I. Chemistry modelling. *Atmospheric Chemistry and Physics Discussion*, **3**, 551–596.

Stommel, H. and Stommel, E. 1983. *Volcano Weather*. Newport, RI, Seven Seas Press.

Stothers, R. B. 1984. The great eruption of Tambora and its aftermath. *Science*, **224**, 1191–1198.

1993. Flood basalts and extinction events. *Geophysical Research Letters*, **20**, 1399–1402.

2000. Climatic and demographic consequences of the massive volcanic eruption of 1258. *Climate Change*, **45**, 361–374.

Symonds, G. J. (ed.) 1888. *The Eruption of Krakatoa, and Subsequent Phenomena*. London, Trubner, for the Royal Society.

Tabazadeh, A. and Turco, R. P. 1993. Stratospheric chlorine injection by volcanic eruptions: hydrogen chloride scavenging and implications for ozone. *Science*, **20**, 1082–1086.

Thomason, L. W. 1991. A diagnostic aerosol size distribution inferred from SAGE II measurements. *Journal of Geophysical Research*, **96**, 22, 501–522, 528.

Thordarson, T. and Self, S. 1996. Sulphur, chlorine and fluorine degassing and atmospheric loading by the Roza eruption, Columbia River Basalt group, Washington, USA. *Journal of Volcanological and Geothermal Research*, **74**, 49–73.

2003. Atmospheric and environmental effects of the 1783–84 Laki eruption: a review and re-assessment. *Journal of Geophysical Research*, **108** (D1), 4011. doi:10.1029/2001JD002042.

Thordarson, T., Self, S., Miller, J. D., *et al.* 2003. Sulphur release from flood lava eruptions in the Veidivotn, Grimsvotn, and Katla volcanic systems. *Geological Society of London Special Publications*, **213**, 103–122.

Thordarson, T., Self, S., Skarsson, N., *et al.* 1996. Sulphur, chlorine, and fluorine degassing and atmospheric loading by the 1783–1784 AD Laki (Skaftár Fires) eruption in Iceland. *Bulletin of Volcanology*, **58**, 205–225.

Toon, O. B. and Pollack, J. B. 1980. Atmospheric aerosols and climate. *American Scientist*, **68**, 268–278.

Toon, O. B., Cahnle, K., Morrison, D., Turco, R. P., and Covey, O. 1997. Environmental perturbations caused by the impacts of asteroids and comets. *Reviews of Geophysics*, **35**, 41–78.

Turco, R. P., Toon, O. B., Ackerman, T. P., *et al.* 1990. Nuclear winter: climate and smoke – an appraisal of nuclear winter. *Science*, **247**, 166–176.

Vogelmann, A. M., Ackerman, T. P., and Turco, R. P. 1992. Enhancements in biologically effective ultraviolet radiation following volcanic eruptions. *Nature*, **359**, 47–49.

Wallace, P. J. 2001. Volcanic $SO_2$ emissions and the abundance and distribution of exsolved gas in magma bodies. *Journal of Volcanology and Geothermal Research*, **108**, 85–106.

Wignall, P. B. 2001. Large igneous provinces and mass extinctions. *Earth Science Reviews*, **53**, 1–33.

Zhao, J., Turco, R. P., and Toon, O. B. 1996. A model simulation of Pinatubo volcanic aerosols in the stratosphere. *Journal of Geophysical Research*, **100**, 7315–7328.

Zielinski, G. A. 1995. Stratospheric loading and optical depth estimates of explosive volcanism over the last 2100 years derived from the GISP2 Greenland ice core. *Journal of Geophysical Research*, **100**, 20937–20955.

2000. Use of paleo-records in determining variability within the volcanism–climate system. *Quaternary Science Reviews*, **19**, 417–438.

Zielinski, G. A., Mayewski, P. A., Meeker, L. D. *et al.* 1996a. A 110 000 year record of explosive volcanism from the GISP2 (Greenland) ice core. *Quaternary Research*, **45**, 109–118.

1996b. Potential atmospheric impact of the Toba mega-eruption. *Geophysical Research Letters*, **23**, 837–840.

# Chapter 6

# Volcanoes, hydrothermal venting, and the origin of life

Karl O. Stetter

## Introduction

The first traces of life on Earth date back to the early Archean age. Microfossils of cyanobacteria-like prokaryotes within fossil stromatoliths demonstrate that life already existed 3.5 billion years ago (Awramik *et al.*, 1983; Schopf and Packer, 1987; Schopf, 1993). Life had already originated much earlier, possibly by the end of the major period of meteorite impacts about 3.9 billion years ago (Schopf *et al.*, 1983; Mojzsis *et al.*, 1996). At that time, the Earth is generally assumed to have been much hotter than today (Ernst, 1983). Questions arise about possible physiological properties, modes of energy acquisition, and kinds of carbon sources of the earliest organisms which may have made their living in a world of fire and water.

Today most life forms known are mesophiles adapted to ambient temperatures within a range from 15 to 45 °C. Among bacteria, thermophiles (heat-lovers) have been recognized for some time, which grow optimally (fastest) between 45 and 70 °C. They thrive within Sun-heated soils, self-heated waste dumps, and thermal waters, and are closely related to mesophiles. Since Louis Pasteur's time it has generally been assumed that vegetative (growing) cells of bacteria (including most thermophiles) are quickly killed by temperatures of above 80 °C. In contrast, during recent years, hyperthermophilic bacteria and archaea (formerly the archaebacteria) with unprecedented properties have been isolated mostly from areas of volcanic activity (Stetter, 1986, 1992; Stetter *et al.*, 1990). They grow between 80 and 113 °C and represent the organisms at the upper temperature border of life. Although these hyperthermophiles flourish at temperatures above those used in pasteurization, they are unable to propagate at 50 °C or below. Organisms like *Pyrolobus* are so well adapted to heat that temperatures of 85 °C are still too low to support growth (Blöchl *et al.*, 1997). Hyperthermophiles belong to various phylogenetically distant groups and may represent rather ancient adaptations to their high-temperature environments. In this chapter, I give an insight into the biotopes, modes of life, and phylogeny of hyperthermophiles, and show evidence for their primitiveness and their probable existence since the dawn of life in the early Archean age.

## Biotopes of hyperthermophiles

In nature, hyperthermophiles are found in water-containing volcanically and geothermally heated environments situated mainly along terrestrial and submarine tectonic fracture zones where plates are colliding (subduction) or moving away from each other (spreading). The temperatures in active volcanoes are much too high to support life (e.g., about 1000 °C or more in molten lava), but fumaroles and hot springs associated with volcanic activity provide much lower, more suitable

*Volcanoes and the Environment*, eds. J. Martí and G. G. J. Ernst. Published by Cambridge University Press. © Cambridge University Press 2005.

| | Type of thermal area | |
|---|---|---|
| Characteristics | Terrestrial | Marine |
| Locations | Solfataric fields: steam-heated soils, mud holes, and surface waters; deeply originating hot springs; subterranean oil stratifications | Submarine hot springs, hot sediments and vents ("black smokers"); active seamounts |
| Temperatures | Surface: up to 100 °C[a]; depth: above 100 °C | Up to about 400 °C ("black smokers") |
| Salinity | Usually low (0.1–0.5% salt) | Usually about 3% salt |
| pH | 0 to 10 | 5 to 8.5 (rarely: 3) |
| Major life-supporting gases and nutrients | $H_2O$, $CO_2$, $CO$, $CH_4$, $H_2$, $H_2S$, $S^0$, $S_2O_3^{2-}$, $SO_3^{2-}$, $SO_4^{2-}$),[b] $NH_4^+$, $N_2$, $NO_3^-$, $Fe^{2+}$, $Fe^{3+}$, $O_2$ (surface) | |

**Table 6.1** Biotopes of hyperthermophiles

[a] At sea level, depending on the altitude.
[b] Seawater contains about 30 mmol/l of sulfate.

temperatures. Within saturated or superheated steam life is not possible and liquid water is a fundamental prerequisite. In several hyperthermophiles growth temperatures exceed 100 °C, which is the boiling temperature of water at sea level. However, this is only possible under conditions increasing the boiling point of water (e.g., by elevated atmospheric, hydrostatic, or osmotic pressure) in order to keep water in the liquid phase. For example, by an overpressure of 1 bar (200 kPa), the boiling point of water is raised from 100 to 120 °C. This corresponds to a water depth of merely 10 m.

Due to the presence of reducing gases (Table 6.1) and the low solubility of oxygen at high temperatures, biotopes of hyperthermophiles are essentially anaerobic. A majority of hyperthermophilic organisms depends exclusively on simple inorganic compounds provided by their environment to gain energy and to build up their cell components. Hyperthermophiles have been isolated from terrestrial and marine environments.

## Terrestrial biotopes

Terrestrial biotopes of hyperthermophiles are mainly sulfur-containing solfataric fields, named

after the Solfatara Crater at Pozzuoli (Naples area), Italy. Solfataric fields consist of soils, mud holes, and surface waters heated by volcanic exhalations from magma chambers a few kilometers below which are passing through porous rocks (e.g., liparite) above. Very often, solfataric fields are situated at or in the close neighborhood of active volcanoes and activity is greatly increased during eruption phases. For example, in 1980, 1 week before a fissure eruption, we had taken samples at the Krafla area in Iceland (Fig. 6.1). A huge vigorously gassed mud crater had formed, throwing mud lumps tens of meters high into the air. Solfataric fields exist in many countries all over the world, for example the Campi Flegrei in Pozzuoli, Italy; the Kerlingarfjöll mountains, Iceland; Yellowstone National Park, USA; Caldera Uzon, Kamchatka, Russia; Noboribetsu, Hokkaido, Japan; Tangkuban Prahu, Java, Indonesia; and Rotorua, New Zealand. Depending on the altitude above sea level, the maximum water temperatures are up to 100 °C (Table 6.1). However, there are lakes with hot springs at the bottom the temperatures of which exceed 100 °C (e.g., Lake Yellowstone; Lake Tanganijka, Cape Banza) (Fig. 6.2). The salinity of solfataric fields is usually low. However, there are exceptions if

**Fig. 6.1.** Solfataric field at Krafla, Iceland.

they are situated at the beach (e.g., Faraglione, Vulcano, and Maronti beach, Ischia, both Italy). The chemical composition of solfataric fields is very variable and depends on the site. Steam within the solfataric exhalations is mainly responsible for the heat transfer. $CO_2$ keeps the soils anaerobic and prevents penetration of oxygen into greater depths. In addition, $H_2S$ reduces oxygen to water yielding elemental sulfur. An important gaseous energy source for hyperthermophiles is hydrogen, which may be formed either pyrolytically from water or chemically from FeS and $H_2S$ (Drobner *et al.*, 1990). Many solfataric fields are rich in iron minerals like ferric hydroxides, pyrite, and other ferrous sulfides. In soil profiles (Fig. 6.3), two different zones are visible: an oxidized, strongly acidic (pH 0–3) upper layer of about 15 to 30 cm in thickness with an ochre color caused by ferric hydroxides overlies a reduced lower zone exhibiting a slightly acidic pH of between 4 and 6.5 and a bluish-black color due to ferrous sulfides. Acidity within the surface layer is mainly based on the presence of sulfuric acid which may be formed abiotically by oxidation of $SO_2$ or biotically (see below).

Less-usual compounds may be enriched at some sites like the magnetite or arsenic minerals auripigment and realgar in Geysirnaja Valley and Caldera Uzon, Kamchatka. Sometimes, solfataric fields contain silicate-rich neutral to slightly alkaline (pH 7–10) hot springs originating from the depth (Fig. 6.4). Their content of sulfur compounds is usually low. Under special conditions some of these hot springs may form geysirs like Bunsen's geysir and Strokkur in Iceland (Fig. 6.5), and those of Yellowstone Park, Rotorua, and Geysirnaja valley. Possibly due to periodical exposure to atmospheric oxygen at high temperature, there is no evidence for larger amounts of hyperthermophiles within water of those "jumping springs." Geothermal areas of neutral hot springs with low sulfur contents are situated, for example at Lac Abbé and Lac Assal, Djibouti, and at Pemba and Cape Banza (Fig. 6.2), Lake Tanganijka, Congo, at the African Rift Valley tectonic fracture system (pH 6.5–7.5; 60–103 °C).

Active volcanoes may harbor hot crater lakes which are heated by fumaroles (e.g., Askja, Iceland and Karymsky, Kamchatka) (Fig. 6.6). Usually, those abound in sulfur and are very acidic and represent a further biotope of hyperthermophiles. Nothing is known about possible

**Fig. 6.2.** Hydrothermal system (aragonite hot smoker chimneys) in Lake Tanganyka, Cape Banza, Congo.

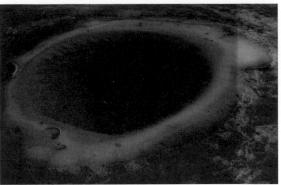

**Fig. 6.4.** Neutral silicate-rich hot spring at Hveravellir, Iceland.

**Fig. 6.3.** Soil profile: solfataric field, Krafla, Iceland.

**Fig. 6.5.** Eruption of Geysir Strokkur, Iceland.

microbial life in the interior of active volcanoes. These mountains are assumed to be "hot sponges" which may contain a lot of aquifers like cracks and holes, possibly providing so far unexplored biotopes for hyperthermophiles. The first evidence for the presence of communities of hyperthermophiles within geothermally heated rocks 3500 m below the surface has recently been demonstrated (Stetter *et al.*, 1993). Soils on the flanks of volcanoes, depending on the interior heat flow, may also harbor hyperthermophiles. For example, at the Tramway Ridge and southern

**Fig. 6.7.** Hot soils at the top of Mt. Melbourne, Antarctica (ice-free areas).

**Fig. 6.6.** Newly formed peninsula with hydrothermal system in Karymsky caldera lake, Kamchatka (background: erupting Karymsky volcano, about 3 km away).

crater on top of Mt. Erebus, and on top of Mt. Melbourne, both in Antarctica, there are wet soils with temperatures between 60 and 65 °C and pH 5–6 at an altitude of 3500 m (Fig. 6.7). They represent "islands" of thermophilic life within a deep-frozen continent.

Very often solfataric fields and hot springs are situated in the neighborhood of, or even within, swamps, grassland, or rainforests. Therefore, in addition to the inorganic nutrients initially present there, leaves, wood, insects, etc. may fall in and may provide complex organic material as possible additional nutrients. Artificial "solfataric fields" are smouldering coal-refuse piles situated in humid areas as for instance at Ronneburg, Thüringen, Germany. Usually, they contain elemental sulfur and exhibit an acidic pH. They harbor some hyperthermophiles simi-

lar to those of natural environments (Fuchs *et al.*, 1996).

## Marine biotopes

Marine biotopes of hyperthermophiles consist of various hydrothermal systems situated at shallow and abyssal depths. Similar to ambient sea water, submarine hydrothermal systems usually contain high concentrations of NaCl and sulfate and exhibit a slightly acidic to alkaline pH (5–8.5). Otherwise, the major gases and life-supporting mineral nutrients may be similar to those in terrestrial thermal areas (Table 6.1).

Shallow submarine hydrothermal systems are found in many parts of the world, mainly on beaches with active volcanism, like at Vulcano Island (Fig. 6.8), Ischia Island and Stufe di Nerone, Naples (all Italy); Ribeira Quente, São Miguel, Azores; Sangeang Island, Indonesia; Obock, Gulf of Tadjura, Dijbouti; and Kunashiri, Japan. The hydrothermal system at Vulcano is situated close to the Porto di Levante harbor between the Fossa and Vulcanello Volcanoes. It consists of $H_2S$-containing submarine fumaroles, hot springs, and hot sandy sediments situated at a depth of 1–10 m with temperatures between 80 and 105 °C (Fig. 6.9). Various novel groups of marine hyperthermophiles have been discovered at this location, some members of which were found later to exist in other shallow and abyssal marine hydrothermal systems also (see below; e.g., Table 6.3). A further recently discovered shallow submarine hydrothermal system is located at a depth of 105 m at the Kolbeinsey Ridge (67° 05′ 29″ N, 18° 42′ 53″ W) which represents the

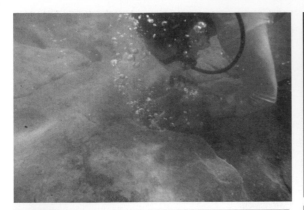

**Fig. 6.8.** Sampling at a shallow submarine hydrothermal vent at Porto di Levante, Vulcano, Italy.

**Fig. 6.10.** Abyssal hot "black smoker" chimneys at the East Pacific Rise, 21° N. Depth: 2500 m, maximal fluid temperature: 365 °C.

**Fig. 6.9.** Outgassing at a shallow submarine hydrothermal system, Porto di Levante, Vulcano, Italy.

**Fig. 6.11.** Emerging US deep-dive submersible Alvin. The working platform contains deep-sea samples.

northern continuation of the mid-Atlantic ridge (Fricke *et al.*, 1989). It had been visited first by the German diving vessel *Geo*. Hot fluids with temperatures between 40 and 131°C and pH 6.5 are venting out from cracks and holes at the seafloor. Shallow hydrothermal plumes can have a complex geometry and dispersal dynamics, as recently documented at the Steinahóll vent site on the Reykjanes Ridge south of Iceland where venting occurs from a depth of 300 meters or so. (Ernst *et al.*, 2000).

Most impressive are the deep-sea "smoker" vents, where mineral-laden hydrothermal fluids with temperatures up to about 400 °C escape into the cold (2.8 °C) surrounding deep-sea water and build up huge rock chimneys (Fig. 6.10). Although these hot fluids are sterile, the sur-

rounding porous smoker rock material appears to contain very steep temperature gradients which provide zones of suitable growth temperatures for hyperthermophiles. Some smoker rocks are teeming with hyperthermophiles (for example $10^8$ cells of *Methanopyrus* per gram of rock inside a mid-Atlantic "Snake Pit" vent chimney). Deep-sea vents are located along submarine tectonic fracture zones and are known from several places which can be visited for sampling by deep-dive submersibles like *Alvin* (USA) (Fig. 6.11), *Nautile* (France), and *Mir* (Russia). Powerful big black smoker systems do exist at the mid-Atlantic ridge deep-sea floor at a depth of between 3000 and 4000 meters at the "TAG," "Snake Pit," "Broken Spur," and "Moose" sites. The temperature of the smoker fluids there is usually between 200 and

**Fig. 6.12.** "Beehive" hot smokers, mid-Atlantic ridge, "Snake Pit" site. Depth: 3600 m.

**Fig. 6.13.** Sampling at the summit of Teahicya Seamount, Polynesia (French deep-dive submersible *Cyana*). Depth: 1500 m.

360 °C and the pH about 9. In addition to regular black smoker vent chimneys, beehive-shaped smoker vents are founde the mid-Atlantic ridge from which fluids with temperatures of up to 244 °C are seeping out (Fig. 6.12). At the base of the mid-Atlantic smokers there are usually hydrothermally heated lava rocks and minor sediments with much lower temperatures between 20 and 100 °C.

Based on the gases and mineral nutrients of the cooled-down venting fluids an enormous chemosynthetic production of microbial biomass by thermophilic and mesophilic vent bacteria and archaea is going on, which represents the starting point for a (aphotic) food web which includes a tight assembly of various higher life forms like deep-sea shrimps (e.g., *Rimicaris excoculata*), crabs, mussels, and fishes. In contrast, the surrounding deep-sea floor is low in nutrients and, therefore very poor in animals.

Further deep-sea vents are situated in the Pacific, for example at 21° N on the East Pacific Rise at a depth of about 2500 m (Fig. 6.10). Many hot smokers with temperatures up to 400 °C are found there, at a lava-covered deep-sea floor without major sediments. Again, based on microbial biomass there is an "island" of very rich deep-sea life including giant clams (*Calyptogena magnifica*), giant tube worms (*Riftia pachyptila*) up to 4 m long and many crabs and fishes.

Another extensive abyssal vent system is within the Guaymas Basin. It is located at the southern continuation of the San Andreas fault at a depth of 1500 m between the Baja California peninsula and the mainland of Mexico. The bottom of this ocean arm is covered with mighty organic sediments (up to 400 m in thickness) which are heated hydrothermally. The temperature gradients within the sediments are rather steep. For example, we have measured there at several sites temperatures of 85, 105, 132, and 150 °C in sediment depths of 5, 10, 15, and 35 cm, respectively. In addition to the heated sediments, assemblies of active black and white smoker chimneys and smoker flanches are present exhibiting fluid temperatures of between 308 and 326 °C (pH 7). The heated flanch rocks, similar to walls of smoker vents, contain various species of hyperthermophiles.

As has recently been recognized, a further type of submarine high-temperature environment is provided by the active seamounts. Close to Tahiti, there is a group of active seamounts which represents hotspot volcanoes situated upon the Pacific tectonic plate. At a depth of 1500 m there is the summit crater of Teahicya Seamount which abounds in ferric iron hydroxide minerals and harbors extensive hydrothermal venting systems (Fig. 6.13). In the same area, there is another huge abyssal volcano, Macdonald Seamount (28° 58.7′ S, 140° 15.5′ W) the summit of which is situated approximately 40 m below the sea surface. In 1989, during a French–German expedition, just as we approached the Macdonald area, the seamount began to erupt. We were therefore able to explore for the first time an erupting submarine seamount and its

| Table 6.2 | Hyperthermophilic members of archaeal genera present in samples during 1989 Macdonald Seamount eruption |

| Sample | Members of genus (viable cells/ml$^{-1}$)$^a$ | | | |
| --- | --- | --- | --- | --- |
| | Pyrodictium | Pyrococcus | Thermococcus | Archaeoglobus |
| Seawater, 2 km distance from active zone | 0 | 0 | 0 | 0 |
| Metallic gray slick floating in patches | 10 | 10$^3$ | 10$^3$ | 10 |
| Seawater, green discoloration, strong odour of H$_2$S, 1 km distance from active zone | + | + | + | + |
| Seawater, green discoloration, strong odour of H$_2$S, 0.3 km distance from active site | 0 | 0 | + | 0 |
| Lava rocks with orange–red precipitates from the active crater wall | 0 | + | + | + |

$^a$ +, quantitatively determined ($\geq 1$ cell/ml); 0, not present.

surroundings for the presence of hyperthermophilic archaea (Huber *et al.*, 1990a). On the first day of eruption (January 24, 1989) we took hydrocast samples about 2–4 km away from the active site. The pH was between 8.30 and 8.02 and methane concentration varied between 41.5 and 62.2 nl/l, which are normal values for open surface seawater. After one quiescent day eruptions began again, and large patches of grayish slick of metallic appearance floated at the surface. Some of this material was sampled. Next day there were further strong eruptions, interrupted by periods of silence. Hydrocast samples were taken from water showing green discoloration at the surface. They had abnormally low pH values of around 6 and extremely high methane concentrations of up to 7800 nl/l, indicating that they came from the submarine plume of the eruption. On the same day during a more quiet period, sampling within the active crater was attempted by the submersible *Cyana*. Some lava rocks with orange–red hydrothermal precipitates from the crater wall were collected, but sampling ended when the volcano erupted again. No hyperthermophiles could be enriched from seawater samples taken on the first day at a distance of 2 km from the active zone. In contrast, samples taken from the floating metallic gray-looking slick, the same as the seawater samples taken from the submarine eruption plume and from rocks from the active crater, contained high concentrations of viable hyperthermophiles (Table 6.2). As expected, none of the

hyperthermophiles enriched was able to grow at 60 °C or below, and therefore within ambient seawater. This demonstrates the presence of these organisms in active seamounts and their release during submarine volcanic eruptions.

## Phylogeny of hyperthermophiles

During the last few years, powerful molecular techniques have been developed in order to investigate phylogenetic relationships of living organisms (Zuckerkandl and Pauling, 1965; Woese *et al.*, 1976, 1983). In order to synthesize proteins, living beings contain ribosomes. Ribosomes of different organisms exhibit the same overall structure which is made up of a large and a small subunit, both consisting of homologous proteins and ribonucleic acids (rRNAs). Phylogenetic distances can be determined from sequence differences of the corresponding macromolecule. Based on the pioneering work of Carl Woese, 16S rRNA (harbored by the small ribosomal subunit) is widely used in phylogenetic studies of prokaryotes (Woese *et al.*, 1976). It consists of about 1500 bases and is homologous to the eukaryotic 18S rRNA. Due to 16S/18S rRNA sequence comparisons, a universal phylogenetic tree is now available (Woese *et al.*, 1990). It exhibits a tripartite division of the living world into the bacterial (formerly: "eubacterial"), archaeal (formerly: "archaebacterial"),

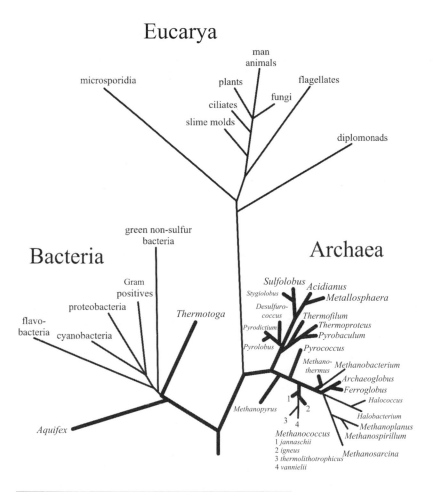

Eucarya

man
animals

microsporidia

plants            flagellates

ciliates      fungi

slime molds

diplomonads

green non-sulfur
bacteria

Bacteria                                    Archaea

Gram            Sulfolobus  Acidianus
positives      Stygiolobus    Metallosphaera
proteobacteria            Desulfuro-   Thermofilum
flavo-                      coccus    Thermoproteus
bacteria                  Pyrodictium   Pyrobaculum
        cyanobacteria                Pyrococcus
                     Pyrolobus   Methano-  Methanobacterium
                              thermus   Archaeoglobus
                                        Ferroglobus
                          1    2         Halococcus
                Methanopyrus            Halobacterium
                     3  4                Methanoplanus
Aquifex            Methanococcus  Methanospirillum
                  1 jannaschii
                  2 igneus
                  3 thermolithotrophicus  Methanosarcina
                  4 vannielii

*Thermotoga*

**Fig. 6.14.** Hyperthermophiles within the 16S rRNA-based phylogenetic tree. Bold lines represent hyperthermophiles. Schematically redrawn and modified from Woese *et al.* (1990) and Stetter (1992).

and eucaryal (formerly: "eukaryotic") domains (Woese and Fox, 1977; Woese *et al.*, 1990) (Fig. 6.14). The root was inferred from phylogenetic trees of duplicated genes of ATPase subunits and elongation factors Tu and G (Gogarten *et al.*, 1989; Iwabe *et al.*, 1989). Accumulating biochemical and molecular evidence strongly supports Woese's 16S rRNA-based classification and adds considerable detail about the archaea. For example, the total genome sequence of the archaeon *Methanococcus jannaschii* revealed homologies of genes of only 15% with bacteria and 30% with eucaryotes (Bult *et al.*, 1996). More than 50% appear to be characteristic of the archaea, therefore yielding evidence

for the correctness of their classification as a second independent domain of prokaryotes. Within the tree (Fig. 6.14), deep branches are evidence for early separation. For example, the separation of the bacteria from the stem common to archaea and eucarya represents the deepest and earliest branching point. Short phylogenetic branches indicate a rather slow rate of evolution. In contrast to the eucarya, the bacterial and archaeal domains within the universal phylogenetic tree exhibit some extremely deep and short branches. Based on the unique phylogenetic position of *Thermotoga,* a thermophilic ancestry of the bacteria had been taken into consideration (Achenbach-Richter *et al.*, 1987). Surprisingly, hyperthermophiles are represented by all short and deep phylogenetic branches which form a cluster around the phylogenetic root (Fig. 6.14, bold lines). The deepest and shortest phylogenetic branches are represented by *Aquifex* and

*Thermotoga* within the bacteria and *Methanopyrus, Pyrodictium,* and *Pyrolobus* within the archaea. A further, even more deeply branching off lineage of very tiny archaea has been discovered in a hot submarine vent, recently. It has been tentatively named *Nanoarchaeota* (Huber *et al.,* 2002). In the laboratory, *Nanoarchaeota* can be cultivated as a co-culture with *Ignicoccus.* They exhibit the smallest genome of a living organism (500 kb). On the other hand, mesophilic and moderately thermophilic bacteria and archaea, as a rule represent long lineages within the phylogenetic tree and had a fast rate of evolution (e.g., Gram-positives; *Proteobacteria; Halobacterium; Methanosarcina*) (Fig. 6.14). Some deep phylogenetic branches like those representing the Thermotogales, Methanobacteriales, and Methanococcales in addition to hyperthermophiles contain extremely thermophilic (e.g., *Thermotoga elfii, Methanobacterium thermoautotrophicum, Methanococcus thermolithotrophicus*) and even mesophilic (e.g., *Methanobacterium uliginosum, Methanococcus vanniellii*) species. However, those exhibit a faster rate of evolution as indicated by longer lineages (Fig. 6.14). Based on that rate of evolution, they have already been able to adapt to their colder environments. Hyperthermophilic eucaryotes are unknown. If existing at all, similar to hyperthermophilic prokaryotes they should represent very deep and short phylogenetic lineages which are so far lacking within eucarya.

## Taxonomy of hyperthermophiles

So far, 66 species of hyperthermophilic bacteria and archaea are known. These have been isolated from different terrestrial and marine thermal areas of the world (Table 6.3). The first hyperthermophile described was the archaeon *Sulfolobus acidocaldarius* discovered by Thomas Brock in Yellowstone National Park hot springs, USA. Hyperthermophiles are very divergent, both in terms of their phylogeny and physiological properties, and are grouped into 29 genera in 10 orders. At present 16S rRNA sequence-based classification of prokaryotes appear to be imperative for the recognition and characterization of novel taxonomic groups. In addition, more

traditional taxonomic features such as GC contents of DNA, DNA–DNA homology, morphology, and physiological features may be used as separating characters in order to obtain a high resolution of taxonomy within members of a phylogenetic lineage (e.g., description of different species and strains). Within the bacteria, *Aquifex pyrophilus, Aquifex aeolicus* and *Thermotoga maritima, Thermotoga neapolitana,* and *Thermotoga hypogea* exhibit the highest growth temperatures (between 95 and 90 °C, respectively) (Table 6.3). Within the archaea, the organisms with the highest growth temperatures (between 102 and 113 °C) are found within both kingdoms, the Crenarchaeota (the former "sulfur-metabolizers") and the Euryarchaeota (the former "methanogens–halophiles") (Burggraf *et al.,* 1997) (Fig. 6.15). They are members of the crenarchaeal genera *Pyrolobus, Pyrodictium, Hyperthermus, Pyrobaculum, Ignicoccus,* and *Stetteria* and the euryarchaeal genera *Methanopyrus* and *Pyrococcus.* At the species level, if separating features are lacking, taxonomic classification may not always reflect phylogenetic distances. For example, within the Sulfolobales, the species *Sulfolobus acidocaldarius, Sulfolobus solfataricus,* and *Sulfolobus metallicus* are members of distant phylogenetic lineages each one representing different (still undescribed) genera, while *Sulfolobus shibatae* is a very close relative of *Sulfolobus solfataricus* (Fuchs *et al.,* 1996). The Methanobacteriales and Methanococcales, were first described based on mesophilic members (Balch *et al.,* 1979). Hyperthermophiles and extreme thermophiles are *Methanothermus fervidus* and *Methanothermus sociabilis* among the Methanobacteriales and *Methanococcus thermolithotrophicus, Methanococcus jannaschii,* and *Methanococcus igneus* among the genus *Methanococcus* (Table 6.3).

In the future, more taxonomic features will be available, from a deeper understanding of the genotype after total genome sequencing. At present, sequencing of total genomes of the hyperthermophiles *Thermotoga maritima, Aquifex aeolicus, Sulfolobus solfataricus, Pyrobaculum aerophilum, Pyrolobus fumarii, Pyrococcus furiosus, Archaeoglobus fulgidus,* and *Methanococcus jannaschii* is in progress or has already been completed (Bult *et al.,* 1996; Klenk *et al.,* 1997).

**Table 6.3** | Taxonomy and upper growth temperatures of hyperthermophiles

| Order | Place of first discovery of genus[a] | Genus | Species | $T$ max (°C) | Reference |
|---|---|---|---|---|---|
| **Domain: BACTERIA** | | | | | |
| Thermotogales | Vulcano[P] | Thermotoga | T. maritima[T] | 90 | Huber et al. (1986) |
| | | | T. neapolitana | 90 | Jannasch et al. (1988) |
| | | | T. thermarum | 84 | Windberger et al. (1989) |
| | | | T. subterranea | 75* | Jeanthon et al. (1995) |
| | | | T. elfii | 72* | Ravot et al. (1995) |
| | | | T. hypogea | 90 | Fardeau et al. (1997) |
| | Djibouti | Thermosipho | T. africanus[T] | 77* | Huber et al. (1989b) |
| | Rotorua | Fervidobacterium | F. nodosum[T] | 80* | Patel et al. (1985) |
| | | | F. islandicum | 80* | Huber et al. (1990b) |
| Aquificales | Kolbeinsey | Aquifex | A. pyrophilus[T] | 95 | Huber et al. (1992) |
| | | | A. aeolicus | 93 | Huber and Stetter (1999) |
| | Yellowstone | Thermocrinis | T. ruber[T] | 89 | Huber et al. (1998) |
| **Domain: ARCHAEA** | | | | | |
| **I. Kingdom: Crenarchaeota** | | | | | |
| Sulfolobales | Yellowstone | Sulfolobus | S. acidocaldarius[T] | 85 | Brock et al. (1972) Zillig et al. (1980) |
| | | | S. solfataricus | 87 | |
| | | | S. shibatae | 86 | Grogan et al. (1990) |
| | | | S. metallicus | 75[b] | Huber and Stetter (1991) |
| | Naples[P] | Metallosphaera | M. sedula[T] | 80[b] | Huber et al. (1989a) |
| | | | M. prunae | 80[b] | Fuchs et al. (1995) |
| | Naples[S] | Acidianus | A. infernus[T] | 95 | Segerer et al. (1986) |
| | | | A. brierleyi | 75[b] | Brierley and Brierley (1973); Segerer et al. (1986) |
| | | | A. ambivalens | 95 | Zillig et al. (1987b); Fuchs et al. (1996) |
| | Furnas | Stygiolobus | S. azoricus[T] | 89 | Segerer et al. (1991) |
| Thermoproteales | Krafla[S] | Thermoproteus | T. tenax[T] | 97 | Zillig et al. (1981) |
| | | | T. neutrophilus | 97 | Stetter (1986) |
| | | | T. uzoniensis | 97 | Bonch-Osmolovskaya et al. (1990) |
| | Krafla[G] | Pyrobaculum | P. islandicum[T] | 103 | Huber et al. (1987b) |
| | | | P. organotrophum | 103 | Huber et al. (1987a) |
| | | | P. aerophilum | 104 | Völkl et al.(1993) |
| | Iceland | Thermofilum | T. pendens[T] | 95 | Zillig et al. (1983a) |
| | | | T. librum | 95 | Stetter (1986) |
| Desulfurococcales | Iceland | Desulfurococcus | D. mucosus[T] | 97 | Zillig et al. (1982) |
| | | | D. mobilis | 95 | Zillig et al. (1982) |
| | | | D. saccharovorans | 97 | Stetter (1986) |
| | | | D. amylolyticus | 97 | Bonch-Osmolovskaya et al. (1985) |
| | Vulcano[F] | Staphylothermus | S. marinus[T] | 98 | Fiala et al. (1986) |
| | Hverageröi | Sulfophobococcus | S. zilligii[T] | 95 | Hensel et al. (1997) |
| | Milos | Stetteria | S. hydrogenophila[T] | 102 | Jochimsen et al. (1997) |

(cont.)

**Table 6.3** *(cont.)*

| Order | Place of first discovery of genus[a] | Genus | Species | T max (°C) | Reference |
|---|---|---|---|---|---|
| | Kodakara | *Aeropyrum* | *A. pernix*.[T] | 100 | Sako *et al.* (1996) |
| | Kolbeinsey | *Ignicoccus* | *I. islandicus*[T] | 103 | Burggraf *et al.* (1997) |
| | Yellowstone | *Thermosphaera* | *T. aggregans*[T] | 90 | Huber *et al.* (1997) |
| | Vulcano[P] | *Thermodiscus* | *T. maritimus*[T] | 98 | Stetter (1986) |
| | Vulcano[P] | *Pyrodictium* | *P. occultum*[T] | 110 | Stetter (1982); Stetter *et al.* (1983) |
| | | | *P. brockii* | 110 | Stetter *et al.* (1983) |
| | | | *P. abyssi* | 110 | Pley *et al.* (1991) |
| | São Miguel | *Hyperthermus* | *H. butylicus*[T] | 108 | Zillig *et al.* (1990) |
| | Mid-Atlantic | *Pyrolobus* | *P. fumarii*[T] | 113 | Blöchl *et al.* (1997) |
| *II. Kingdom: Euryarchaeota* | | | | | |
| Thermococcales | Vulcano[P] | *Thermococcus* | *T. celer*[T] | 93 | Zillig *et al.* (1983b) |
| | | | *T. litoralis* | 98 | Neuner *et al.* (1990) |
| | | | *T. stetteri* | 98 | Miroshnichenko *et al.* (1989) |
| | | | *T. profundus* | 90 | Kobayashi *et al.* (1994) |
| | | | *T. alcaliphilus* | 90 | Keller *et al.* (1995) |
| | | | *T. chitonophagus* | 93 | Huber *et al.* (1995b) |
| | | | *T. fumicolans* | 103 | Godfroy *et al.* (1996) |
| | | | *T. peptonophilus* | 100 | Gonzáles *et al.* (1995) |
| | | | *T. acidaminovorans* | 93 | Dirmeier *et al.* (1998) |
| | Vulcano[P] | *Pyrococcus* | *P. furiosus*[T] | 103 | Fiala and Stetter (1986) |
| | | | *P. woesei* | 103 | Zillig *et al.* (1987a) |
| | | | *P. abyssi* | 102 | Erauso *et al.* (1993) |
| Archaeoglobales | Vulcano[P] | *Archaeoglobus* | *A. fulgidus*[T] | 92 | Stetter *et al.* (1987); Stetter (1988) |
| | | | *A. profundus* | 92 | Burggraf *et al.* (1990b) |
| | | | *A. veneficus* | 88 | Huber *et al.* (1997) |
| | Vulcano[P] | *Ferroglobus* | *F. placidus*[T] | 95 | Hafenbradl *et al.* (1996) |
| Methanobacteriales | Kerlingarfjöll | *Methanothermus* | *M. fervidus*[T] | 97 | Stetter *et al.* (1981) |
| | | | *M. sociabilis* | 97 | Laurer *et al.* (1986) |
| Methanococcales | Naples[N] | *Methanococcus* | *M. thermolitho-trophicus*[c]) | 70 | Huber *et al.* (1982) |
| | | | *M. jannaschii* | 86 | Jones *et al.* (1983) |
| | | | *M. igneus* | 91 | Burggraf *et al.* (1990a) |
| Methanopyrales | Guaymas | *Methanopyrus* | *M. kandleri*[T] | 110 | Kurr *et al.* (1991) |

[a] Djibouti, marine hot vents, Obock, Djibouti, Africa; Furnas, Furnas Caldeiras, São Miguel, Azores; Guaymas, hot deep sea sediments, Guaymas Basin, Mexico; Hveragerði, hot alkaline spring, Hveragerði, Iceland; Iceland, solfataric hot spring, Iceland; Kerlingarfjöll, Kerlingarfjöll Solfataras, Iceland; Kodakara, coastal solfataric vent, Kodakara-Jima Island, Japan; Kolbeinsey, submarine hot vents, Kolbeinsey, Iceland; Krafla[G], Krafla Geothermal Power Plant, Iceland; Krafla[S], Krafla Solfataras, Iceland; Mid-Atlantic, TAG site hot deep sea vents, mid-Atlantic ridge; Milos, shallow submarine hot brine seeps, Paleohori Bay, Milos, Greece; Naples[N], Stufe di Nerone marine hot vents, Naples, Italy; Naples[P], Pisciarelli Solfatara, Naples, Italy; Naples[S], Solfatara Crater, Pozzuoli, Naples, Italy; Rotorua, hot spring, Rotorua, New Zealand; São Miguel, submarine hot vents, São Miguel, Azores; Vulcano[F], Faraglione marine hot vents, Vulcano, Italy; Vulcano[P], Porto di Levante marine hot vents, Vulcano, Italy; Yellowstone, Yellowstone National Park hot springs, USA.

[b] Extreme thermophiles related to hyperthermophiles.

[c] First thermophilic member of the genus.

[T] Type species.

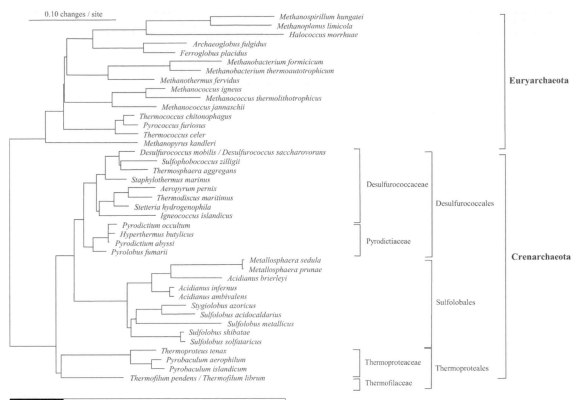

**Fig. 6.15.** Phylogenetic relationships among the archaeal kingdoms, based on 16S rRNA, using maximum parsimony method (ARB program) (Ludwig and Strunk, 1997). Positions of *Archaeoglobus fulgidus* and *Ferroglobus placidus* have been determined by transversion analysis (Woese *et al.*, 1991).

## Sampling and isolation of hyperthermophiles

In order to isolate hyperthermophiles, anaerobic and aerobic samples of hot waters, soils, rocks, muds, and sediments may be taken. Water usually contains only low concentrations of hyperthermophiles. In this case, microbes can be concentrated by ultrafiltration. During collection, the samples are as a rule exposed to oxygen, which at high temperatures is hazardous to anaerobic hyperthermophiles. Therefore, after filling the samples into tightly stoppered glass bottles (e.g., 100-ml storage bottles) (Fig. 6.16), either shock cooling or immediate oxygen reduction (e.g., by injection of $H_2S$ or sodium dithionite) of the samples is recommended in order to avoid cell inactivation (Fig. 6.17). In laboratory experiments, cultures of *Pyrobaculum islandicum* ($10^8$ viable cells/ml) died when exposed to air for 40 min. (Huber *et al.*, 1987a). The samples should be carried to the laboratory at ambient temperature after cooling down. Cultivation experiments with parallel samples transported at the high *in situ* temperature and at ambient temperature yielded only cultures from samples carried at ambient temperature. Most likely, organisms within heated samples had died due to the lack of nutrients, while those in the samples carried at low temperature survived in a kind of dormant stage. In the laboratory, anaerobic samples may be processed under protective gas within an anaerobic chamber (Fig. 6.18). Aerobic and anaerobic enrichments can be attempted on various possible substrates, considering the composition and temperatures of the biotope (Table 6.1). In addition, hyperthermophiles may be enriched already *in situ* by injecting possible substrates into the hot environment. In view of the presence of complex high temperature communities, all kinds of energy-yielding couples of electron donors and acceptors may be possible (e.g., Thauer *et al.*, 1977). In addition, deep-sea and

**Fig. 6.16.** Anaerobic samples enclosed in tightly stoppered 100-ml storage bottles and 25-ml roll tubes.

**Fig. 6.17.** Sampling of anaerobes at Lac Abbé, Djibouti, Africa.

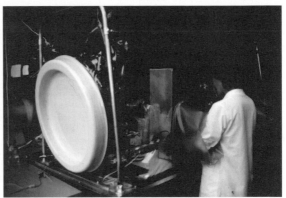

**Fig. 6.18.** Anaerobic chamber.

deep subterranean organisms may be able to use energy-yielding reactions only favorable under the high *in situ* pressure. In the laboratory, as a rule, after inoculation with original samples, positive enrichment cultures of hyperthermophiles can be detected after an incubation time of 1 day to 1 week, depending on the initial cell concentration and the doubling time of the organisms. For a deeper understanding of the organisms, pure cultures are an imperative prerequisite. They can be obtained by cloning of single species from enrichment cultures. The traditional way to establish pure cultures is by obtaining single colonies on agar-solidified media. As a rule, however, for hyperthermophiles agar is not suitable as a solidifying agent owing to its melting point of only about 95 °C and its instability on hydrolysis at high temperatures. Instead of agar, gellan gum (Gelrite, Kelco, USA) or polysilicate may be used for solidification (e.g., Völkl

et al., 1993). However, due to the high incubation temperatures, surfaces of solid media dry out rapidly and many hyperthermophiles are unable to grow there and form colonies. A novel direct and rapid cloning procedure suitable for hyperthermophiles has been developed by us (Ashikin et al., 1987; Huber et al., 1995a). It consists of a computer-controlled inverse microscope equipped with a strongly focused infrared laser ("optical tweezers") (Fig. 6.19). Viable cells were separated directly under visual control from the enrichment mixture within a square capillary connected to a sterile syringe (Huber et al., 1995a). Viable cells are selected employing membrane-potential-active fluorescent dyes (Beck and Huber, 1997).

| Table 6.4 | Energy-yielding reactions in chemolithoautotrophic hyperthermophiles |
|---|---|
| Energy-yielding reaction | Genera |
| $2S^0 + 3O_2 + 2H_2O \rightarrow 2H_2SO_4$ | *Sulfolobus,*[a] *Acidianus,*[a] *Metallosphaera*[a] |
| $2FeS_2 + 7O_2 + 2H_2O$ | *Aquifex* |
| $\quad \rightarrow 2FeSO_4 + 2H_2SO_4 =$ "metal leaching" | |
| $H_2 + O_2 \rightarrow H_2O$ | *Aquifex, Acidianus,*[a] *Metallosphaera,*[a] |
| | *Pyrobaculum,*[a] *Sulfolobus,*[a] *Thermocrinis*[a] |
| $H_2 + HNO_3 \rightarrow HNO_2 + H_2O$ | *Aquifex, Pyrobaculum*[a] |
| $4H_2 + HNO_3 \rightarrow NH_4OH + 2H_2O$ | *Pyrolobus* |
| $2FeCO_3 + HNO_3 + 5H_2O$ | *Ferroglobus* |
| $\quad \rightarrow 2Fe(OH)_3 + HNO_2 + 2H_2CO_3$ | |
| $4H_2 + H_2SO_4 \rightarrow H_2S + 4H_2O$ | *Archaeoglobus*[a] |
| $H_2 + S^0 \rightarrow H_2S$ | *Acidianus, Stygiolobus, Pyrobaculum,*[a] |
| | *Thermoproteus,*[a] *Pyrodictium,*[a] *Ignicoccus* |
| $H_2 + 6FeO(OH) \rightarrow 2Fe_3O_4 + 4H_2O$ | *Pyrobaculum* |
| $4H_2 + CO_2 \rightarrow CH_4 + 2H_2O$ | *Methanopyrus, Methanothermus,* |
| | *Methanococcus* |

[a] Facultatively heterotrophic.

**Fig. 6.19.** "Optical tweezers" cloning facility. Left side: Neodym-YAG infrared laser; middle: inverse microscope with videocamera; right side: joystick and computer for manual and automatic manipulation of the microscope stage.

## Strategies of life and environmental adaptations of hyperthermophiles

Hyperthermophiles are adapted to distinct environmental factors including composition of minerals and gases, pH, redox potential, salinity, and temperature.

## General metabolic potentialities

Most hyperthermophiles exhibit a chemolithoautotrophic mode of nutrition: inorganic redox reactions serve as energy sources (chemolithotrophic) and $CO_2$ is the only carbon source in order to build up organic cell material (autotrophic). Therefore, these organisms are fixing $CO_2$ by chemosynthesis and are designated as chemolithoautotrophs. The energy-yielding reactions in chemolithoautotrophic hyperthermophiles are anaerobic and aerobic types of respiration (Table 6.4). Molecular hydrogen serves as an important electron donor. Within the natural environment, it may be a component of volcanic gases or may originate via anaerobic pyrite formation (Drobner *et al.*, 1990). Other electron donors are sulfide, sulfur, and ferrous iron (Table 6.4). As in mesophilic respiratory organisms, in some hyperthermophiles oxygen may serve as an electron acceptor. In contrast, however, oxygen-respiring hyperthermophiles are microaerophilic growing only at reduced oxygen concentrations (even as low as 10 ppm) (Blöchl *et al.*, 1997). Anaerobic respiration types are the nitrate, sulfate, sulfur, and carbon dioxide respirations (Table 6.4). While chemolithoautotrophic hyperthermophiles produce organic

| Table 6.5 | Energy-yielding reactions in chemoorganoheterotrophic hyperthermophiles | | |
|---|---|---|---|
| Type of metabolism | External electron acceptor | Energy-yielding reaction | Genera |
| Respiration | $S^0$ | $2[H] + S^0 \rightarrow H_2S$ | *Pyrodictium*[a] *Thermoproteus*[a] *Pyrobaculum*[a] *Thermofilum* *Desulfurococcus* *Thermodiscus* *Thermosphaera* |
| | $SO_4^{2-}$ $(S_2O_3^{2-}; SO_3^{2-})$ $NO_3^-$ $O_2$ | $8[H] + H_2SO_4 \rightarrow H_2S + 4H_2O$ $2[H] + HNO_3 \rightarrow HNO_2 + H_2O$ $2[H] + O_2 \rightarrow H_2O$ | *Archaeoglobus*[a] *Pyrobaculum* *Sulfolobus*[a] *Metallosphaera*[a] *Acidianus*[a] *Aeropyrum* |
| Fermentation | — | Glucose $\rightarrow$ L(+) – lactate + acetate + $H_2$ + $CO_2$ | *Thermotoga* *Thermosipho* *Fervidobacterium* *Pyrodictium*[a] |
| | — | Peptides $\rightarrow$ isovalerate, isobutyrate, butanol, $CO_2$, $H_2$, etc. | *Hyperthermus* *Thermoproteus*[a] *Staphylothermus* |
| | | Yeast extract | *Desulfurococcus* *Thermococcus* *Sulfophobococcus* |
| | | Pyruvate $\rightarrow$ acetate + $H_2$ + $CO_2$ | *Pyrococcus* |

[a] Facultative chemolithoautotrophs.

[H], carrier-bound hydrogen derived from organic material.

matter, there are hyperthermophiles which depend on organic material as energy and carbon sources (Table 6.5). They are designated as chemoorganoheterotrophs (or, in short, heterotrophs). Several chemolithoautotrophic hyperthermophiles are facultative heterotrophs (Table 6.4), which are able to use organic material as an alternative to inorganic nutrients whenever it is provided by the environment (e.g., by decaying cells). Heterotrophic hyperthermophiles gain energy either by aerobic or different types of anaerobic respiration, using organic material as electron donors or by fermentation (Table 6.5).

## Physiological properties of the different groups of hyperthermophiles
### Terrestrial hyperthermophiles
The acidic hot oxygen-rich surface layer of terrestrial solfataric fields almost exclusively harbors extremely acidophilic hyperthermophiles. They consist of coccoid-shaped (Fig. 6.20) aerobes, facultative aerobes, and strict anaerobes growing between pH 1 and 5 with an optimum around pH 3 (Table 6.6). Phylogenetically, they belong to the archaeal genera *Sulfolobus*, *Metallosphaera*, *Acidianus*, and *Stygiolobus*. Together, these genera form the Sulfolobales order (Fig. 6.15, Table 6.3). Members of *Sulfolobus* are strict aerobes

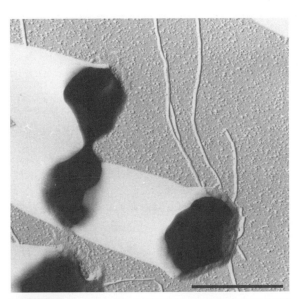

**Fig. 6.20.** Cells of *Metallosphaera prunae*, dividing and flagellate. Electron micrograph with platinum shadowing. Scale bar = 1 μm.

growing autotrophically, heterotrophically, or facultatively heterotrophically (Brock *et al.*, 1972). During autotrophic growth, $S^0$, $S^{2-}$, and $H_2$ (Huber *et al.*, 1992) may be oxidized, yielding sulfuric acid or water as end products (Table 6.4). Under microaerophilic conditions, some *Sulfolobus* isolates are able to reduce ferric iron and molybdate (Brierley and Brierley, 1982). During heterotrophic growth, sugars, yeast extract, and peptone may serve as energy sources (Brock, 1978). Growth of *Sulfolobus* requires low ionic strength. In accordance with this, *Sulfolobus* has never been found so far in marine environments. Several members of the Sulfolabales, like *Sulfolobus metallicus*, *Metallosphaera sedula*, *Acidianus infernus*, and *Acidianus brierleyi*, are powerful "metal-leachers" able to gain energy by oxidation of sulfidic ores like pyrite, chalcopyrite, and sphalerite, forming sulfuric acid and solubilizing the heavy metal ions (Brierley and Brierley, 1973; Huber and Stetter, 1991) (Table 6.4). Members of *Acidianus* are able to grow by anaerobic and aerobic oxidation of $H_2$, using $S^0$ and $O_2$ as electron acceptors, respectively (Segerer *et al.*, 1986). Alternatively, *Acidianus* is able to grow by $S^0$-oxidation (Table 6.4). Members of *Acidianus* are capable of growing in the presence of up

to 4% salt. They have been isolated from low-salt terrestrial environments as well as from a submarine solfataric field (Segerer *et al.*, 1986). *Stygiolobus* represents a genus of strictly anaerobic, thermoacidophilic sulfidogens, gaining energy by reduction of $S^0$ by $H_2$ (Segerer *et al.*, 1991).

Terrestrial hot springs with low salinity and the depth of solfataric fields harbor slightly acidophilic and neutrophilic hyperthermophiles which are usually strict anaerobes. They are members of the genera *Pyrobaculum*, *Thermoproteus*, *Thermofilum*, *Desulfurococcus*, *Sulfophobococcus*, and *Thermosphaera* (Table 6.3). Cells of *Pyrobaculum*, *Thermoproteus*, and *Thermofilum* are stiff, regular rods with almost rectangular edges (Fig. 6.21). During the exponential growth phase, under the light microscope, spheres become visible at the ends ("golf clubs") (Fig. 6.22) and at the middle of the cells, the function of which is unknown (Zillig *et al.*, 1981; Huber *et al.*, 1987a). Cells of *Pyrobaculum* and *Thermoproteus* are about 0.5 μm in diameter whereas those of *Thermofilum* ("the hot thread") are only about 0.17–0.35 μm. *Thermoproteus tenax*, *Thermoproteus neutrophilus*, and *Pyrobaculum islandicum* are able to grow chemolithoautotrophically by anaerobic formation of $H_2S$ from $H_2$ and $S^0$ (Fischer *et al.*, 1983) (Table 6.4). *Thermoproteus tenax* and *Pyrobaculum islandicum* are facultative heterotrophs, while *Pyrobaculum organotrophum*, *Thermoproteus uzoniensis*, *Thermofilum librum*, and *Thermofilum pendens* are obligate heterotrophs growing by sulfur respiration on organic substrates like prokaryotic cell extracts or yeast extract (Table 6.5). Members of *Desulfurococcus* and *Thermosphaera* are coccoid-shaped, strictly heterotrophic sulfur-respirers while *Sulfophobococcus* is a purely fermentive coccoid organism. *Thermosphaera aggregans* grows in aggregates shaped like a bunch of grapes (Fig. 6.23). Exclusively from the depth of solfataric fields in the southwest of Iceland (Kerlingarfjöll mountains), members of the genus *Methanothermus* have been isolated which appear to be endemic species of this area. So far, the species *Methanothermus fervidus* and *Methanothermus sociabilis* are known. They are highly oxygen-sensitive strict chemolithoautotrophic methanogens gaining energy by reduction of $CO_2$ by $H_2$ (Table 6.4) and growing at

| Table 6.6 | Growth conditions and morphology of hyperthermophiles | | | | | |
|---|---|---|---|---|---|---|
| | Growth conditions | | | | | |
| Species | Minimum temperature (°C) | Optimum temperature (°C) | Maximum temperature (°C) | pH | Aerobic (ae) versus anaerobic (an) | Morphology |
| *Sulfolobus acidocaldarius* | 60 | 75 | 85 | 1–5 | ae | Lobed cocci |
| *Metallosphaera sedula* | 50 | 75 | 80 | 1–4.5 | ae | Cocci |
| *Acidianus infernus* | 60 | 88 | 95 | 1.5–5 | ae/an | Lobed cocci |
| *Stygiolobus azoricus* | 57 | 80 | 89 | 1–5.5 | an | Lobed cocci |
| *Thermoproteus tenax* | 70 | 88 | 97 | 2.5–6 | an | Regular rods |
| *Pyrobaculum islandicum* | 74 | 100 | 103 | 5–7 | an | Regular rods |
| *Pyrobaculum aerophilum* | 75 | 100 | 104 | 5.8–9 | ae/an | Regular rods |
| *Thermofilum pendens* | 70 | 88 | 95 | 4–6.5 | an | Slender regular rods |
| *Desulfurococcus mobilis* | 70 | 85 | 95 | 4.5–7 | an | Cocci |
| *Thermosphaera aggregans* | 67 | 85 | 90 | 5–7 | an | Cocci in aggregates |
| *Sulfophobococcus zilligii* | 70 | 85 | 95 | 6.5–8.5 | an | Cocci |
| *Staphylothermus marinus* | 65 | 92 | 98 | 4.5–8.5 | an | Cocci in aggregates |
| *Thermodiscus maritimus* | 75 | 88 | 98 | 5–7 | an | Disks |
| *Aeropyrum pernix* | 70 | 90 | 100 | 5–9 | ae | Irregular cocci |
| *Stetteria hydrogenophila* | 70 | 95 | 102 | 4.5–7 | an | Irregular disks |
| *Ignicoccus islandicus* | 65 | 90 | 100 | 3.9–6.3 | an | Irregular cocci |
| *Pyrodictium occultum* | 82 | 105 | 110 | 5–7 | an | Disks with cannulae |
| *Hyperthermus butylicus* | 80 | 101 | 108 | 7 | an | Lobed cocci |
| *Pyrolobus fumarii* | 90 | 106 | 113 | 4–6.5 | ae/an | Lobed cocci |
| *Thermococcus celer* | 75 | 87 | 93 | 4–7 | an | Cocci |
| *Pyrococcus furiosus* | 70 | 100 | 105 | 5–9 | an | Cocci |
| *Archaeoglobus fulgidus* | 60 | 83 | 95 | 5.5–7.5 | an | Irregular cocci |
| *Ferroglobus placidus* | 65 | 85 | 95 | 6–8.5 | an | Irregular cocci |
| *Methanothermus sociabilis* | 65 | 88 | 97 | 5.5–7.5 | an | Rods in clusters |
| *Methanopyrus kandleri* | 84 | 98 | 110 | 5.5–7 | an | Rods in chains |
| *Methanococcus igneus* | 45 | 88 | 91 | 5–7.5 | an | Irregular cocci |
| *Thermotoga maritima* | 55 | 80 | 90 | 5.5–9 | an | Rods with sheath |
| *Aquifex pyrophilus* | 67 | 85 | 95 | 5.4–7.5 | ae | Rods |

temperatures between 65 and 97 °C (Table 6.6). From terrestrial neutral low-salinity hot springs situated at the base of an evaporite mound at Lac Abbé, Djibouti, Africa, *Thermotoga thermarum* has been isolated. Members of *Thermotoga* are rod-shaped bacterial hyperthermophiles. Usually they thrive in marine hydrothermal systems where they gain energy by fermentation of carbohydrates (see next section). As an exception, *Thermotoga thermarum* grows only at low ionic strength (up to 0.55% NaCl) and is therefore adapted to terrestrial low-salinity hot springs.

### Marine hyperthermophiles

A variety of hyperthermophiles are adapted to the high salinity of seawater (≈3% salt). They are represented by members of the archaeal genera *Pyrolobus*, *Pyrodictium*, *Hyperthermus*,

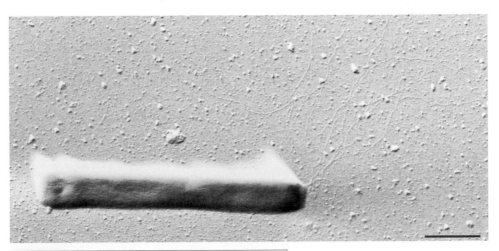

**Fig. 6.21.** Cells of *Pyrobaculum aerophilum*.
Electron micrograph with platinum shadowing.
Scale bar = 1 μm.

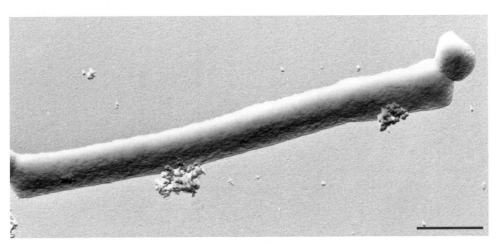

**Fig. 6.22.** Cell of *Thermoproteus tenax*, showing terminal "golf-club."
Electron micrograph with platinum shadowing. Scale bar = 1 μm.

*Stetteria, Thermodiscus, Ignicoccus, Staphylothermus, Aeropyrum, Pyrobaculum* (only one halophilic species), *Methanopyrus, Pyrococcus, Thermococcus, Archaeoglobus,* and *Ferroglobus,* and of the bacterial genera *Aquifex* and *Thermotoga* (Table 6.3). The organism with the highest growth temperature is *Pyrolobus fumarii* exhibiting an upper temperature border of growth above 113 °C (Blöchl *et al.,* 1997) (Table 6.6). *Pyrolobus* has been isolated from the walls of an active deep sea smoker chimney at the mid-Atlantic ridge. It is so dependent on high temperatures that it is unable to grow below 90 °C. Cells of *Pyrolobus* are lobed cocci about 0.7–2.5 μm in diameter (Fig. 6.24). *Pyrolobus fumarii* gains energy by chemolithoautotrophic nitrate reduction, forming ammonia as end product. Alternatively, under microaerophilic conditions, it shows weak but significant growth on $H_2$ and traces of oxygen (0.05%). Close relatives to *Pyrolobus* are members of *Pyrodictium,* cells of which are disks (0.2 μm thick and up to 3 μm in diameter) which are usually connected by networks of hollow cannulae, about 25 nm in diameter (König *et al.,* 1988; Rieger *et al.,* 1995) (Fig. 6.25). Cannulae have been observed so far only in *Pyrodictium.* Representatives of

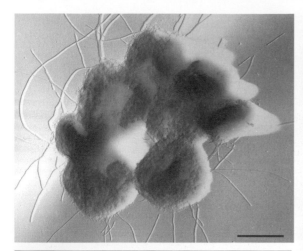

**Fig. 6.23.** *Thermosphaera aggregans*: aggregate of cells shaped like a bunch of grapes, with flagellae. Electron micrograph with platinum shadowing. Scale bar = 1 μm.

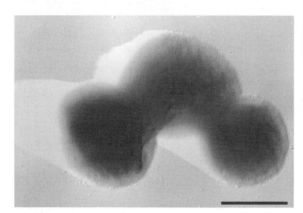

**Fig. 6.24.** *Pyrolobus fumarii*: lobed cells. Electron micrograph with platinum shadowing. Scale bar = 1 μm.

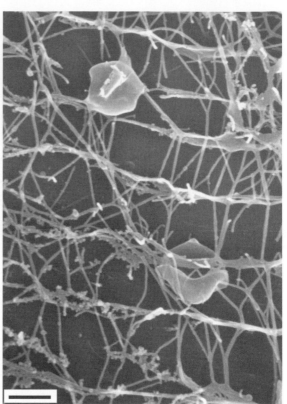

**Fig. 6.25.** *Pyrodictium abyssi*: cells and network of cannulae. Scanning electron micrograph. Sclae bar = 1 μm.

*Pyrodictium* have been isolated from shallow submarine hydrothermal systems as well as from deep-sea hot vents (Stetter *et al.*, 1983; Pley *et al.*, 1991). As a rule, members of *Pyrodictium* are chemolithoautotrophs gaining energy by reduction of $S^0$ by $H_2$. In addition, some isolates are able to grow by reduction of sulfite and thiosulfate (Huber *et al.*, 1987a). As an exception, *Pyrodictium abyssi* is able to grow heterotrophically by peptide fermentation (Pley *et al.*, 1991). The upper temperature border of growth of *Pyrodictium* is 110 °C. By its 16S rRNA, a very close relative to *Pyrodictium* is *Hyperthermus* (Fig. 6.15).

However, in contrast to *Pyrodictium*, *Hyperthermus butylicus* is a purely fermentative hyperthermophile which does not form cannulae and does not grow at 110 °C (Zillig *et al.*, 1990). *Stetteria hydrogenophila* represents a group of disk-shaped hyperthermophiles which grow mixotrophically on a combination of $H_2$ and peptides (Jochimsen *et al.*, 1997). Elemental sulfur or thiosulfate may serve as electron acceptors. *Ignicoccus islandicus* is a coccoid-shaped strictly chemolithoautotrophic member of the Desulfurococcaceae (Fig. 6.15) which gains energy by sulfur respiration (Huber *et al.*, 2000). So far, *Pyrobaculum aerophilum* is the only marine member of the Thermoproteales. It is a rod-shaped strictly chemolithoautotrophic hyperthermophile. *Pyrobaculum aerophilum* gains energy either anaerobically by nitrate reduction or microaerobically by reduction of $O_2$ (traces). Molecular hydrogen serves as electron donor. *Methanopyrus* represents the deepest phylogenetic branch within the Euryarchaeota. It is a

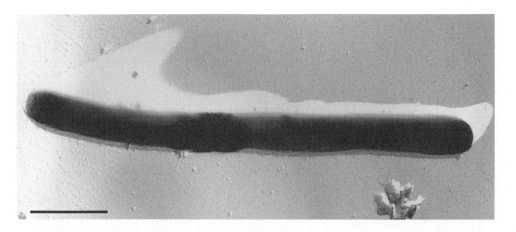

**Fig. 6.26.** Cell of *Methanopyrus kandleri*.
Electron micrograph with platinum shadowing. Scale bar = 1 μm.

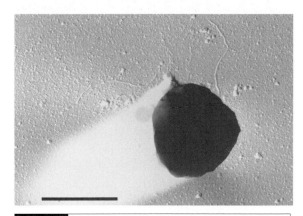

**Fig. 6.27.** Cell of *Methanococcus igneus*.
Electron micrograph with platinum shadowing. Scale bar = 1 μm.

rod-shaped chemolithoautotrophic methanogen (Fig. 6.26). *Methanopyrus kandleri* occurs both within shallow and abyssal submarine hot vents and exhibits the highest growth temperatures in methanogens (up to 110 °C). Further marine hyperthermophilic methanogens are the coccoid shaped *Methanococcus jannaschii* and *Methanococcus igneus* (Fig. 6.27) within the Methanococcales (Fig. 6.14) where they represent the shortest phylogenetic branch-offs. *Methanococcus thermolithotrophicus* is a less extreme thermophile, growing at temperatures up to 70 °C (Huber *et al.*, 1982). Archaeal sulfate reducers are represented by members of *Archaeoglobus* (Table 6.3) (Stetter *et al.*, 1987; Stetter, 1988). *Archaeoglobus fulgidus*

and *Archaeoglobus lithotrophicus* are chemolithoautotrophs able to grow by reduction of $SO_4^{2-}$ and $S_2O_3^{2-}$ by $H_2$ (Table 6.4), while *Archaeoglobus veneficus* is only able to reduce $S_2O_3^{2-}$. *Archaeoglobus fulgidus* is a facultative heterotroph growing on a variety of organic substrates like formate, sugars, starch, proteins, and cell extracts (Fig. 6.5) (Stetter, 1988). Sequencing of the total genome of this organism revealed the presence of genes for the fatty acid β-oxidation pathway, indicating that its physiological properties may be even broader than initially described (Klenk *et al.*, 1997). *Archaeoglobus profundus* is an obligate heterotroph. Cells of members of *Archaeoglobus* are coccoid to triangular-shaped (Fig. 6.28). Like methanogens, they show a blue–green fluorescence at 420 nm under the UV microscope. In agreement, *Archaeoglobus* possesses several coenzymes that had been thought to be unique for methanogens (e.g., $F_{420}$, methanopterin, tetrahydromethanopterin, methanofurane). Usually, members of *Archaeoglobus* are found in shallow and abyssal submarine hydrothermal systems. In addition, the same species have been detected in deep geothermally heated subterranean oil reservoirs and may be responsible for some of the $H_2S$ formation there ("reservoir souring") (Stetter *et al.*, 1987, 1993). *Ferroglobus* represents a further genus within the Archaeoglobales. Under the UV microscope, *Ferroglobus placidus* has a similar appearance to that of *Archaeoglobus* (Fig. 6.29). However, it is a strict chemolithoautotroph, gaining energy anaerobically by reduction of $NO_3^-$ with $Fe^{2+}$ (Table 6.4), $H_2$, and $H_2S$ as electron donors. Alternatively, *Ferroglobus placidus* is able to reduce $S_2O_3^{2-}$ to $H_2S$ with $H_2$ as

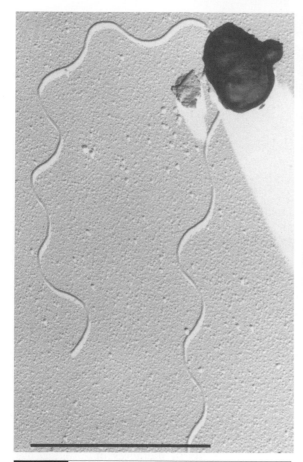

**Fig. 6.28.** Cell of *Archaeoglobus veneficus*, with flagellae. Electron micrograph with platinum shadowing. Scale bar = 1 μm.

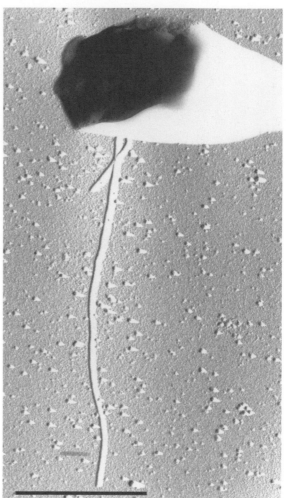

**Fig. 6.29.** Cell of *Ferroglobus placidus*, flagellate. Electron micrograph with platinum shadowing. Scale bar = 1 μm.

electron donor (Hafenbradl *et al.*, 1996). The unique metabolism of *Ferroglobus placidus* may offer a biological mechanism for the anaerobic formation of $Fe^{3+}$ which may have operated during banded iron formations (BIFs) under the anoxic high-temperature conditions on the early Earth up to 3.8 billion years ago (Appel, 1980). Within the Bacteria domain, the deepest phylogenetic branch is represented by *Aquifex*. *Aquifex pyrophilus* is a motile rod-shaped strict chemolithoautotroph (Fig. 6.30). It is a facultative (micro) aerobe. Under anaerobic conditions, *Aquifex pyrophilus* grows by nitrate reduction with $H_2$ and $S^0$ as electron donors. Alternatively, at very low oxygen concentrations (up to 0.5%, after adaptation) it is able to gain energy by oxidation of $H_2$ and $S^0$. Members of *Aquifex* are found in

shallow submarine vents. *Aquifex pyrophilus* grows at temperatures up to 95 °C, the highest growth temperatures found so far within the bacterial domain (Table 6.6).

Groups of strictly heterotrophic hyperthermophiles are also thriving in submarine vents. *Thermodiscus maritimus* is a disk-shaped archaeal heterotroph growing by sulfur respiration on yeast extract and prokaryotic cell homogenates (Tables 6.3, 6.5, 6.6). In the absence of $S^0$, it is able to gain energy by fermentation (Stetter, 1986). Cells of *Staphylothermus marinus* are coccoid and arranged in aggregates like bunches of grapes.

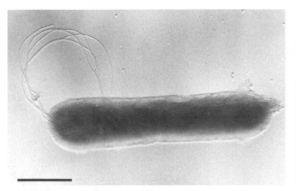

**Fig. 6.30.** Cell of *Aquifex pyrophilus* with bundle of flagellae. Electron micrograph with platinum shadowing. Scale bar = 1 μm.

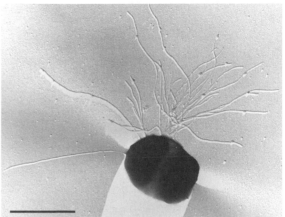

**Fig. 6.31.** Dividing cell of *Pyrococcus furiosus* with bundle of flagellae.
Electron micrograph with platinum shadowing. Scale bar = 1 μm.

They are highly variable in diameter between 0.5 and 15 μm. *Staphylothermus marinus* is able to ferment peptides, forming fatty acids, alcohols, $H_2$, and $CO_2$ (Table 6.5). In rich culture medium, its optimal and maximal temperatures of growth are raised by 6–7 °C compared to minimal medium (Fiala *et al.*, 1986). *Aeropyrum pernix* is a strictly aerobic spherical-shaped marine hyperthermophile growing optimally at neutral pH. It gains energy in metabolizing complex compounds like yeast extract and peptone (Sako *et al.*, 1996). Members of *Pyrococcus* and *Thermococcus* are submarine coccoid species, cells of which occur frequently in pairs (Fig. 6.31). They gain energy by fermentation of peptides, amino acids, and sugars, forming fatty acids, $CO_2$ and $H_2$ (Table 6.5). Hydrogen may inhibit growth and can be removed by gassing with $N_2$ (Fiala and Stetter, 1986). Alternatively, inhibition by $H_2$ can be prevented by addition of $S^0$, whereupon $H_2S$ is formed instead of $H_2$. *Pyrococcus furiosus* grows optimally at 100 °C (Table 6.6). It is able to ferment pyruvate, forming acetate, $H_2$, and $CO_2$ (Schäfer and Schönheit, 1992). In addition to their marine environment, species of *Pyrococcus* and *Thermococcus* are found in geothermally heated oil reservoirs of high salinity (together with *Archaeoglobus*; see above) and are able to grow in the presence of crude oil (Stetter *et al.*, 1993). Many submarine hydrothermal systems contain hydrothermophilic members of *Thermotoga*, which represent the second deepest phylogenetic branch within the bacterial domain (Fig. 6.14). Cells of *Thermotoga* are rod-shaped,

thriving together with archaeal hyperthermophiles within the same environment. They show a characteristic "toga," a sheath-like structure surrounding cells and overballooning at the ends (Fig. 6.32). *Thermotoga* ferments various carbohydrates and proteins. As end products, $H_2$, $CO_2$, acetate and L(+)-lactate are formed (Table 6.5). Hydrogen is inhibitory to growth. During cultivation on a large scale (e.g., 300 I), $H_2$ has to be continuously removed. In the presence of $S^0$, $H_2S$ is formed instead of $H_2$ which does not inhibit growth. Other genera within the Thermotogales are members of *Thermosipho* and *Fervidobacterium* which are strict heterotrophs, too. Similar to *Thermotoga*, cells possess a toga. However, they are less extremely thermophilic (Table 6.3).

# Distribution of species and complexity in hyperthermophilic ecosystems

The global distribution of species of hyperthermophiles and their modes of spreading are so far unknown. Hyperthermophiles do not form spores and are unable to grow at ambient temperatures. In addition, most of them are strict anaerobes. Usually, there are long cold

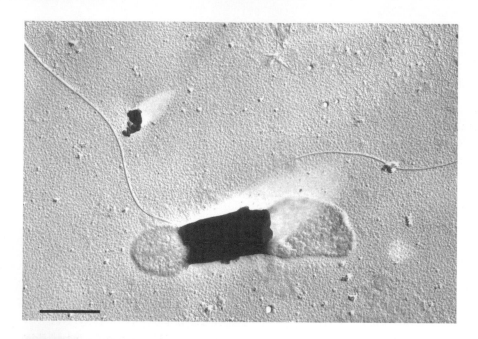

**Fig. 6.32.** Cell of *Thermotoga maritima* with toga and one flagellum.
Electron micrograph with platinum shadowing. Scale bar = 1 μm.

oxygen-rich areas in between different high-temperature biotopes which may act as barriers against dissemination and, therefore may favor the formation of endemic species. This appears to be the case with members of the genus *Methanothermus* which are hypersensitive to oxygen and which have been found so far only in the southwest of Iceland (Stetter *et al.*, 1981; Lauerer *et al.*, 1986). On the other hand, it was surprising to find the same species of hyperthermophiles in hydrothermal systems thousands of kilometers away from each other. For example, we detected the strictly anaerobic *Pyrodictium occultum*, *Pyrococcus furiosus*, *Thermococcus celer*, and *Archaeoglobus fulgidus* at active Macdonald Seamount, Polynesia, as well as at Vulcano Island, Italy. Furthermore, we were able to isolate *Methanopyrus kandleri* at Kolbeinsey, Iceland and also at Guaymas, Mexico. Therefore, even strictly anaerobic hyperthermophiles are able to spread over long distances. In agreement, for several strictly anaerobic hyperthermophiles, oxygen appears to be toxic only at growth temperatures (Huber *et al.*, 1987a). In the cold (e.g., 4 °C) they survive exposure to oxygen for years. Based on this capability, within aerosols and volcanic dust (formed for example within solfataric fields and during volcanic eruptions) hyperthermophiles may be carried around the globe by atmospheric currents. Although direct experimental evidence is lacking so far, spreading by air could explain observations of rapid colonization of newly created hot environments: from a smouldering uranium mine at Ronneburg, Germany, just less than 10 years after its ignition, we were able to isolate the hyperthermophile *Sulfolobus acidocaldarius* which had been isolated originally from Yellowstone Park (Huber and Stetter, unpublished data) (Table 6.3). Moreover, during the 1997 eruption of Karymsky Volcano (Kamchatka, Eastern Siberia), a new peninsula bearing a hydrothermal system had formed within a cold lake within Karymsky caldera, about 3 km away from the active volcano. Samples taken from this hydrothermal system only 9 months after its formation were teeming with rod- and coccoid-shaped hyperthermophiles (Eis, Forster, and Stetter, unpublished data) indicating quick inoculation.

It has been generally assumed that microbial ecosystems become simpler with increasing environmental temperatures (Castenholz,

1979). In contrast, isolation and survey of hyperthermophiles demonstrates an unanticipated physiological and phylogenetic diversity within high-temperature ecosystems. Direct microscopic inspection of the biotopes indicates that there are many other morphotypes yet to be cultivated, suggesting that we have studied only the "tip of a hot iceberg" so far. Moreover, DNA extraction directly from a hot spring in Yellowstone Park and amplification, cloning, and sequencing of the corresponding 16S rRNA genes revealed an extremely rich phylogenetic diversity, containing members of the Archaea with many unknown branches (Barns *et al.*, 1994). A procedure to obtain pure cultures of unknown organisms harboring novel 16S rRNA sequences identified within the environment has been described (Huber *et al.*, 1995a). It combines visual recognition of single cells by "phylogenetic staining" (DeLong *et al.*, 1989) and cloning by "optical tweezers" (Ashikin *et al.*, 1987; Huber *et al.*, 1995a).

# Basis of heat stability and the upper temperature limit for life

Due to the small size of cells of hyperthermophiles, any protection by insulation against the hot environment appears impossible. All cell components, therefore, have to be heat resistant. The molecular basis is so far unknown and still under investigation. Hyperthermophiles belong to two phylogenetically very different domains of life, the Bacteria and Archaea. Therefore, the strategies of molecular mechanisms including heat adaptation may be rather different depending on the phylogenetic position of the corresponding organism.

Cell components such as lipids, nucleic acids, and proteins are usually rather heat-sensitive. Membrane lipids of the bacterial hyperthermophile *Thermotoga maritima* contain a novel glycerol ether lipid, 15,16-dimethyl-30-glyceryloxy-triacontanedioic acid. In contrast to the ester lipids known in mesophiles, it may significantly increase stability of membranes against hydrolysis at high temperatures (De Rosa *et al.*, 1989). In contrast, membranes of all Archaea (including even mesophiles) contain ether lipids derived from diphytanyl-glycerol or its dimer di(biphytanyl)-diglycerol, which exhibit a remarkable resistance against hydrolysis at high temperatures and acidic pH (Kates, 1992). Thermal resistance of the DNA double helix appears to be improved in hyperthermophiles by reverse gyrase, a unique type I DNA topoisomerase that causes positive supertwists for stabilization (Kikuchi and Asai, 1984; Bouthier de la Tour *et al.*, 1990). In addition, archaeal hyperthermophiles possess histones phylogenetically related to the eucaryotic core histones (e.g., H2A, H2B, H3, and H4). *In vitro* addition of histones to purified DNA increased its melting temperature drastically (Thomm *et al.*, 1982; Reddy and Suryanarayana, 1988; Neuner, 1990; Sandman *et al.*, 1990). The secondary structure of ribonucleic acids appears to be stabilized against thermal destruction by an increased content of GC base pairs within the stem areas and by post-transcriptional modification (Edmonds *et al.*, 1991; Woese *et al.*, 1991; Kawai *et al.*, 1992). Purified enzymes of hyperthermophiles usually show an extraordinary heat stability *in vitro*. For example, an amylase from *Pyrococcus woesei* is still active at 130 °C (Koch *et al.*, 1990). Deeper insights of hyperthermophilic proteins will most likely be obtained after comparison of three-dimensional structures with homologous mesophilic enzymes. At the upper temperature border of growth of hyperthermophiles, the function of heat-shock proteins appears to become essential. At 108 °C, about 80% of the soluble protein of a crude extract of *Pyrodictium occultum* consists of a heat-inducible molecular chaperone designated *Thermosome* (Phipps *et al.*, 1993). With the *Thermosome* fully induced, cultures of *Pyrodictium occultum* were able to survive 1 hour autoclaving at 121 °C (Hoffmann, 1993).

The upper temperature border of life is still unknown and depends on the stability of the biomolecules. At temperatures in the order of 100 °C already some low-molecular-weight compounds such as ATP and NAD hydrolyze quite rapidly (half life below 30 min *in vitro*) and thermolabile amino acids like cystein and glutamic acid are decomposing (Bernhardt *et al.*, 1984). The survival of organisms growing at these temperatures may be ensured by rapid resynthesis of thermosensitive compounds. In addition, at higher temperatures, proton permeability of membranes may become too high to maintain electrochemical proton gradients in order to gain energy (van de Vossenberg et al., 1998; W. Konings, personal communication). When carefully sampled to avoid contamination by loose material from the smoker walls, the hot deep-sea smoker fluids (200–350 °C), in contrast to an earlier report, have proved to be sterile. This finding is in line with the dramatic destruction of the building-blocks of life at those temperatures and pressures (Bernhardt et al., 1984; Trent et al., 1984; White, 1984). The maximal growth temperature at which microbial life can exist may possibly be found between 113 and 150 °C. Within this temperature range, heat-sensitive biomolecules could possibly still be resynthesized at biologically feasible rates.

## Conclusions: hyperthermophiles in the history of life

Although still nothing is known about the prerequisites and mechanisms that led to the first living cell, investigations on recent hyperthermophiles yield surprising results enabling us to draw conclusions about possible features of the last common ancestor of life. Within the (16S rRNA-based) universal phylogenetic tree of life, hyperthermophiles form a cluster around the root, occupying all the short and deep phylogenetic branches (Fig. 6.3, bold lines). This is true for both the archaeal and bacterial domains. As a rule, members of the deepest and shortest lineages exhibit the highest growth temperatures ("the shorter and deeper, the hotter"). From their 16S rRNA, these slowly evolving organisms appear to be still closest to the root (common ancestor) and therefore are the most primitive ones still existing. In general, the shortest and deepest lineages exhibit a chemolithoautotrophic mode of nutrition (e.g., Pyrolobus, Pyrodictium, Methanopyrus, Aquifex). In an ecological sense, they are primary producers of organic matter (e.g., their own cell material) which gives rise to complex hyperthermophilic microcosms. These chemolithoautotrophs are able to use various

mixtures of oxidized and reduced minerals and gases as energy sources and to assimilate carbon from $CO_2$. In addition, for growth they just need liquid water, trace minerals, and heat. Hydrogen and sulfurous compounds are widely used inorganic energy sources which mainly occur within volcanic environments. Most hyperthermophiles do not require or even tolerate oxygen for growth. They are completely independent of sunlight. In view of a possible close similarity of these recent hyperthermophiles to ancestral life, a chemolithoautotrophic hyperthermophilic common ancestor appears most probable. In addition, growth conditions of recent hyperthermophiles fit well with our view of the primitive Earth about 3900 Ma ago, when life could have originated: the atmosphere was overall reducing and there was a much stronger volcanism (Ernst, 1983). In addition, Earth's oceans were continuously heated by heavy impacts of meteorites (Davies, 1996). Therefore, within that scenario, early life had to be heat resistant to survive at all and ancestral hyperthermophiles should even have dominated in the Early Archean age. Heavy impacts, in addition, caused significant material exchange between planets of our solar system. Based on their ability to survive for long times in the cold, even at −140 °C, hyperthermophiles could have successfully disseminated between the planets: when ejected into space by heavy impacts, inside pieces of rock and in a dormant state, they should have been able to pass long distances to other planets. After landing, they could have inoculated them. Since early planets like Mars are assumed to have harbored volcanic and hydrothermal activity (Carr, 1996), colonization by primitive hyperthermophiles appears rather probable. Concerning Mars, at present its surface is too cold and neither liquid water nor active volcanism have been detected. Therefore, hyperthermophiles could not grow there any more. The discovery of deep geothermally heated subterranean communities several kilometers below Earth's surface (Stetter et al., 1993) makes it imaginable that active microcosms of hyperthermophiles could still exist on Mars deep below its surface, assuming that water, heat, and simple nutrients are present. In view of a possible early impact material exchange scenario between the planets it is uncertain whether life originated on Earth at all. Therefore, the search for life on Mars could give insights for a deeper understanding of its origin, and hyperthermophiles could play a key role in such an investigation.

## Acknowledgments

I wish to thank Harald Huber for establishing phylogenetic 16S rRNA trees and Reinhard Rachel for electron micrographs. The work presented from my laboratory was supported by grants of the Deutsche Forschungsgemeinschaft, the Bundesministerium für Bildung, Wissenschaft, Forschung und Technologie, the European Commission and the Fonds der Chemischen Industrie.

## References

Achenbach-Richter, L., Gupta, R., Stetter, K. O., et al. 1987. Were the original eubacteria thermophiles? Systematic and Applied Microbiology, 9, 34–39.

Appel, P. W. U. 1980. On the early archaean isua iron-formation, West Greenland. Precambrian Research, 11, 73–78.

Ashikin, A., Dziedzic, J. M., and Yamane, T. 1987. Optical trapping and manipulation of single cells using infrared laser beams. Nature, 300, 769–771.

Awramik, S. M., Schopf, J. W., and Walter, M. R. 1983. Filamentous fossil bacteria from the Archaean of Western Australia. Precambrian Research, 20, 357–374.

Balch, W. E., Fox, G. E., Magrum, L. J., et al. 1979. Methanogens: re-evalution of a unique biological group. Microbiology Reviews, 250–296.

Barns, S. M., Funyga, R. E., Jeffries, M. W., et al. 1994. Remarkable archaeal diversity detected in a Yellowstone National Park hot spring environment. Proceedings of the National Academy of Sciences of the USA, 91, 1609–1613.

Beck, P. and Huber, R. 1997. Detection of cell viability in cultures of hyperthermophiles. FEMS Microbiology Letters, 147, 11–14.

Bernhardt, G., Lüdemann, H.-D., Jaenicke, R., et al. 1984. Biomolecules are unstable under "Black Smoker" conditions. Naturwissenschaften, 71, 583–585.

Blöchl, E., Rachel, R., Burggraf, S., et al. 1997. Pyrolobus fumarii, gen. and sp. nov., represents a novel group of archaea, extending the upper temperature limit for life to 113 °C. Extremophiles, 1, 14–21.

Bonch-Osmolovskaya, E. A., Miroshnichenko, M. L., Kostrikina, N. A., et al. 1990. Thermoproteus uzoniensis sp. nov., a new extemely thermophilic archaebacterium from Kamchatka continental hot springs. Archives of Microbiology, 154, 556–559.

Bonch-Osmolovskaya, E. A., Slesarev, A. I., Miroshnichenko, M. L., et al. 1985. Characteristics of Desulfurococcus amylolyticus n. sp., a new extreme thermophilic archaebacterium from hot volcanic vents of Kamchatka and Kunashir. Microbiologyia, 57, 78–85. (in Russian)

Bouthier de la Tour, C., Portemer, C., Nadal, M., et al. 1990. Reverse gyrase, a hallmark of the hyperthermophilic archaebacteria. Journal of Bacteriology, 172, 6803–6808.

Brierley, C. L. and Brierley, J. A. 1973. A chemolithoautotrophic and thermophilic microorganism isolated from an acidic hot spring. Canadian Journal of Microbiology, 19, 183–188.

1982. Anaerobic reduction of Sulfolobus species. Zentralblatt für Bakteriologie, Mikrobiologie and Hygiene, I, Abteilung Originale, C3, 289–294.

Brock, T. D. 1978. Thermophilic Microorganisms and Life at High Temperatures. New York, Springer-Verlag.

Brock, T. D., Brock, K. M., Belly, R. T., et al. 1972. Sulfolobus: a new genus of sulfur-oxidizing bacteria living at low pH and high temperature. Archives of Microbiology, 84, 54–68.

Bult, C. J., White, W., Olsen, G. J., et al. 1996. Complete genome sequence of the methanogenic archaeon, Methanococcus jannaschii. Science, 273, 1058–1073.

Burggraf, S., Fricke, H., Neuner, A., et al. 1990a. Methanococcus igneus sp. nov., a novel hyperthermophilic methanogen from a shallow submarine hydrothermal system. Systematic and Applied Microbiology, 13, 263–269.

Burggraf, S., Huber, H., and Stetter, K. O. 1997. Reclassification of the cranarchaeal orders and families in accordance with 16S rRNA sequence data. International Journal of Systematic Bacteriology, 47, 657–660.

Burggraf, S., Jannasch, H. W., Nicolaus, B., et al. 1990b. Archaeoglobus profundus sp. nov. represents a new species within the sulfate-reducing archaebacteria. Systematic and Applied Microbiology, 13, 24–28.

Carr, M. H. 1996. Water on early Mars. In G. R. Bock and J. A. Goode (eds.) Evolution of Pydrothermal Ecosystems on Earth (and Mars?), Ciba Foundation Symposium, no. 202. Chichester, UK, John Wiley, pp. 249–267.

Castenholz, R. W. 1979. Evolution and ecology of thermophilic microorganisms. In M. Shilo (ed.)

*Strategies of Microbial Life in Extreme Environments.* Berlin, Dahlem Konferenzen, pp. 373–392.

Davies, P. C. W. 1996. The transfer of viable microorganisms between planets. In G. R. Bock and J. A. Goode (eds.) *Evolution and Pydrothermal Ecosystems on Earth (and Mars?)*, Ciba Foundation Symposium no. 202. Chichester, UK, John Wiley, pp. 304–317.

DeLong, E. F., Wickham, G. S., and Pace, N. R. 1989. Phylogenetic stains: ribosomal RNA-based probes for the identification of single cells. *Science*, **243**, 1360–1363.

De Rosa, M., Gambacorta, A., Huber, R., *et al.* 1989. Lipid structures in *Thermotoga maritima*. In M. S. da Costa, J. C. Duarte, and R. A. D. Williams (eds.) *Microbiology of Extreme Environments and its Potential for Biotechnology.* London, Elsevier, pp. 167–173.

Dirmeier, R., Keller, M., Hafenbradl, D., *et al.* 1998. *Thermococcus acidaminovorans* sp. nov., a new hyperthermophilic elkalophilic archaeon growing on amino acids. *Extremophiles*, **2**, 109–114.

Drobner, E., Huber, H., Wächtershäuser, G., *et al.* 1990. Pyrite formation linked with hydrogen evolution under anaerobic conditions. *Nature*, **346**, 742–744.

Edmonds, C. G., Crain, P. F., Gupta, R., *et al.* 1991. Posttranscriptional modification of tRNA in thermophilic archaea (archaebacteria). *Journal of Bacteriology*, **173**, 3138–3148.

Erauso, G., Reysenbach, A.-L., Godfroy, A., *et al.* 1993. *Pyrococcus abyssi* sp. nov., a new hyperthermophilic archaeon isolated from a deep-sea hydrothermal vent. *Archives of Microbiology*, **160**, 338–349.

Ernst, W. G. 1983. The early earth and the archaean rock record. In J. W. Schopf (ed.) *Earth's Earliest Biosphere: Its Origin and Evolution*. Princeton, NJ, Princeton University Press, pp. 41–52.

Ernst, G. G. J., Cave, R. R., German, C. R., *et al.* 2000. Vertical and lateral splitting of a hydrothermal plume at Steinahóll, Reykjanes Ridge, Iceland. *Earth and Planetary Science Letters*, **179**, 529–537.

Fardeau, M.-L., Ollivier, B., Patel, B. K. C., *et al.* 1997. *Thermotoga hypogea* sp. nov., a xylanolytic, thermophilic bacterium from an oil-producing well. *International Journal of Systematic Bacteriology*, **47**, 1013–1019.

Fiala, G. and Stetter, K. O. 1986. *Pyrococcus furiosus* sp. nov. represents a novel group of marine heterotrophic archaebacteria growing optimally at 100 °C. *Archives of Microbiology*, **145**, 56–61.

Fiala, G., Stetter, K. O., Jannasch, H. W., *et al.* 1986. *Staphylothermus marinus* sp. nov. represents a novel genus of extremely thermophilic submarine heterotropic archaebacteria growing up to 98 °C. *Systematic and Applied Microbiology*, **8**, 106–113.

Fischer, F., Zillig, W., Stetter, K. O., *et al.* 1983. Chemolithoautotrophic metabolism of anaerobic extremely thermophilic archaebacteria. *Nature*, **301**, 511–513.

Fricke, H., Giere, O., Stetter, K. O., *et al.* 1989. Hydrothermal vent communities at the shallow subpolar Mid-Atlantic ridge. *Marine Biology*, **102**, 425–429.

Fuchs, T., Huber, H., Burggraf, S., *et al.* 1996. 16SrDNA-based phylogeny of the archaeal order Sulfolobales and reclassification of *Desulfurolobus ambivalens* as *Acidianus ambivalens* comb. nov. *Systematic and Applied Microbiology*, **19**, 56–60.

Fuchs, T., Huber, H., Teiner, K., *et al.* 1995. *Metallosphaera prunae*, sp. nov., a novel metal-mobilizing, thermoacidophilic archaeum, isolated from a uranium mine in Germany. *Systematic and Applied Microbiology*, **18**, 560–566.

Godfroy, A., Meunier, J.-R., Guezennec, J., *et al.* 1996. *Thermococcus fumicolans* sp. nov., a new hyperthermophilic archaeon isolated from a deep-sea hydrothermal vent in the North Fiji Basin. *International Journal of Systematic Bacteriology*, **46**, 1113–1119.

Gogarten, J. P., Kibak, H., Dittrich, P., *et al.* 1989. Evolution of the vacuolar H$^+$-ATPase: implications for the origin of eukaryotes. *Proceedings of the National Academy of Sciences of the USA* **86**, 6661–6665.

Gonzáles, J. M., Kato, C., and Horikoshi, K. 1995. *Thermococcus peptonophilus* sp. nov., a fast-growing extremely thermophilic archaebacterium isolated from deep-sea hydrothermal vents. *Archives of Microbiology*, **164**, 159–164.

Grogan, D., Palm, P., and Zillig, W. 1990. Isolate B 12, which harbours a virus-like element, represents a new species of the archaebacterial genus *Sulfolobus, Sulfolobus shibatae*, sp. nov. *Archives of Microbiology*, **154**, 594–599.

Hafenbradl, D., Keller, M., Dirmeier, R., *et al.* 1996. *Ferroglobus placidus* gen. nov., sp. nov., a novel hyperthermophilic archaeum that oxidizes Fe$^{2+}$ at neutral pH under anoxic conditions. *Archives of Microbiology*, **199**, 308–314.

Hensel, R., Matussek, K., Michalke, K., *et al.* 1997. *Sulfophobococcus zilligii* gen. nov., spec. nov. a novel hyperthermophilic Archaeum isolated from hot alkaline springs of Iceland. *Systematic and Applied Microbiology*, **20**, 102–110.

Hoffmann, A. 1993. Reinigung, Charakterisierung und partielle Sequenzierung einer ATPase aus dem hyperthermophilen Archaeon *Pyrodictium occultum*. Ph.D. thesis, University of Regensburg, Germany.

Huber, G. and Stetter, K. O. 1991. *Sulfolobus metallicus* sp. nov., a novel strictly chemolithoautotrophic thermophilic archaeal species of metal-mobilzers. *Systematic and Applied Microbiology*, **14**, 372–378.

1999. Aquiticales. In *Encyclopedia of Life Sciences*, London, Nature Publishing Group.

Huber, G., Spinnler, C., Gambacorta, A., *et al.* 1989. *Metallosphaera sedula*, gen. and sp. nov. represents a new genus of aerobic, metal-mobilizing, thermoacidophilic archaebacteria. *Systematic and Applied Microbiology*, **12**, 38–47.

Huber, G., Drobner, E., Huber, H., *et al.* 1992. Growth by aerobic oxidation of molecular hydrogen in *Archaea*: a metabolic property so far unknown for this domain. *Systematic and Applied Microbiology*, **15**, 502–504.

Huber, H., Burggraf, S., Mayer, T., *et al.* 2000. *Ignicoccus* gen. nov., a novel genus of hyperthermophilic, chemolithoautotrophic *Archaea*, represented by two new species, *Ignicoccus islandicus* sp. nov. and *Ignicoccus pacificus* sp. nov. *International Journal of Systematic and Evolutionary Microbiology*, **50**, 2093–2100.

Huber, H., Jannasch, H., Rachel, R., *et al.* 1997. *Archaeoglobus veneficus* sp. nov., a novel facultative chemolithoautotrophic hyperthermophilic sulfite reducer, isolated from abyssal black smokers. *Systematic and Applied Microbiology*, **20**, 374–380.

Huber, H., Thomm, M., König, H., *et al.* 1982. *Methanococcus thermolithotrophicus*, a novel thermophilic lithotrophic methanogen. *Archives of Microbiology*, **132**, 47–50.

Huber, R., Burggraf, S., Mayer, T., *et al.* 1995a. Isolation of a hyperthermophilic archaeum predicted by *in situ* RNA analysis. *Nature*, **376**, 57–58.

Huber, R., Dyba, D., Huber, H., *et al.* 1998. Sulfur-inhibited *Thermosphaera aggregans* sp. nov., a new genus of hyperthermophilic archaea isolated after its prediction from environmentally derived 16S rRNA sequences. *International Journal of Systematic Bacteriology*, **48**, 31–38.

Huber, R., Hohn, M., Rachel, R., *et al.* 2002. A new phylum of Archaea represented by a nanosized symbiont, *Nature*, **417**, 63–67,

Huber, R., Huber, G., Segerer, A. *et al.* 1987a. Aerobic and anaerobic extremely thermophilic autotrophs. In H. W. van Verseveld and J. A. Duine (eds.) *Microbial Growth on $C_1$ Compounds*, Proceedings on the 5th International Symposium. Dordrecht, The Netherlands, Martinus Nijhoff, pp. 44–51.

Huber, R., Kristjansson, J. K., and Stetter, K. O. 1987b. *Pyrobaculum* gen. nov., a new genus of neutrophilic, rod-shaped archaebacteria from continental solfataras growing optimally at 100 °C. *Archives of Microbiology*, **149**, 95–101.

Huber, R., Langworthy, T. A., König, H., *et al.* 1986. *Thermotoga maritima* sp. nov. represents a new genus of unique extremely thermophilic eubacteria growing up to 90 °C. *Archives of Microbiology*, **144**, 324–333.

Huber, R., Stöhr, J., Hohenhaus, S., *et al.* 1995b. *Thermococcus chitonophagus* sp. nov., a novel, chitin-degrading, hyperthermophilic archaeum from a deep-sea hydrothermal vent environment. *Archives of Microbiology*, **164**, 255–264.

Huber, R., Stoffers, P., Cheminee, J. L., *et al.* 1990a. Hyperthermophilic archaebacteria within the crater and open-sea plume of erupting Macdonald Seamount. *Nature*, **345**, 179–181.

Huber, R., Wilharm, T., Huber, D., *et al.* 1992. *Aquifex pyrophilus* gen. nov. sp. nov., represents a novel group of marine hyperthermophilic hydrogen-oxidizing bacteria. *Systematic and Applied Microbiology*, **15**, 340–351.

Huber, R., Woese, C. R., Langworthy, T. A., *et al.* 1989. *Thermosipho africanus* gen. nov., represents a new genus of thermophilic eubacteria within the "Thermogales". *Systematic and Applied Microbiology*, **12**, 32–37.

1990b. *Fervidobacterium islandicum* sp. nov., a new extremely thermophilic eubacterium belonging to the "Thermotogales". *Archives of Microbiology*, **154**, 105–111.

Iwabe, N., Kuma, K., Hasegawa, M., *et al.* 1989. Evolutionary relationship of Archaebacteria, Eubacteria, and eukaryotes inferred from phylogenetic trees of duplicated genes. *Proceedings of the National Academy of Sciences of the USA*, **86**, 9355–9359.

Jannasch, H. W., Huber, R., Belkin, S., *et al.* 1988. *Thermotoga neapolitana*, sp. nov. of the extremely thermophilic, eubacterial genus *Thermotoga*. *Archives of Microbiology*, **150**, 103–104.

Jeanthon, C., Reysenbach, A.-L., Haridon, S. L., *et al.* 1995. *Thermotoga subterranea* sp. nov., a new thermophilic bacterium isolated from a continental oil reservoir. *Archives of Microbiology*, **164**, 91–97.

Jochimsen, B., Peinemann-Simon, S., Völker, H., *et al.* 1997. *Stetteria hydrogenophila*, gen. nov. and sp. nov., a novel mixotrophic sulfur-dependent *crenarchaeote* isolated from Milos, Greece. *Extremophiles*, **1**, 67–73.

Jones, W. J., Leigh, J. A., Mayer, F., *et al.* 1983. *Methanococcus jannaschii* sp. nov., an extremely thermophilic methanogen from a submarine hydrothermal vent. *Archives of Microbiology*, **136**, 254–261.

Kates, M. 1992. Archaebacterial lipids: structure, biosynthesis and function. In M. J. Danson, D. W. Hough, and G. G. Lunt (eds.) *The Archaebacteria: Biochemistry and Biotechnology*. London, Portland Press, pp. 51–72.

Kawai, G., Hushizume, T., Yasuda, M., *et al.* 1992. Conformational rigidity of $N^4$-acetyl-2'-*O*-methylcytidine found in tRNA of extremely thermophilic archaebacteria (Archaea). *Nucleosides and Nucleotides*, **11**, 759–771.

Keller, M., Braun, F.-J., Dirmeier, R., *et al.* 1995. *Thermococcus alcaliphilus* sp. nov., a new hyperthermophilic archaeum growing on polysulfide at alkaline pH. *Archives of Microbiology*, **164**, 390–395.

Kikuchi, A. and Asai, K. 1984. Reverse gyrase: a topoisomerase which introduces positive superhelical turns into DNA. *Nature*, **309**, 677–681.

Klenk, H.-P., Clayton, R. A., Tomb, J. F., *et al.* 1997. The complete genome sequence of the hyperthermophilic, sulphate-reducing archaeon *Archaeoglobus fulgidus*. *Nature*, **390**, 364–370.

Kobayashi, T., Kwak, Y. S., Akiba, T., *et al.* 1994. *Thermococcus profundus* sp. nov., a new hyperthermophilic archaeon isolated from deep-sea hydrothermal vent. *Systematic and Applied Microbiology*, **17**, 232–236.

Koch, R., Zabowski, P., Spreinat, A., *et al.* 1990. Extremely thermostable aylolytic enzyme from the archaebacterium *Pyrococcus furiosus*. *FEMS Microbiology Letters*, **71**, 21–26.

König, H., Messner, P., and Stetter, K. O. 1998. The fine structure of the fibers of *Pyrodictium occultum*. *FEMS Microbiology Letters*, **49**, 207–212.

Kurr, M., Huber, R., König, H., *et al.* 1991. *Methanopyrus kandleri*, gen. and sp. nov. represents a novel group of hyperthermophilic methanogens, growing at 110 °C. *Archives of Microbiology*, **156**, 239–247.

Lauerer, G., Kristjansson, J. K., Langworthy, T. A., *et al.* 1986. *Methanothermus sociabilis* sp. nov., a second species within the *Methanothermaceae* growing at 97 °C. *Systematic and Applied Microbiology*, **8**, 100–105.

Ludwig, L. and Strunk, O. 1997. ARB: a software environment for sequence data. Available online at: http://www.arb-home.de

Miroshnichenko, M. L., Bonch-Osmolovskaya, E. A., Neuner, A., *et al.* 1989. *Thermococcus stetteri* sp. nov. a new extremely thermophilic marine sulfur-metabolizing archaebacterium. *Systematic and Applied Microbiology*, **12**, 257–262.

Mojzsis, S. J., Arrhenius, G., McKeegan, K. D., *et al.* 1996. Evidence for life on Earth before 3,800 million years ago. *Nature*, **384**, 55–59.

Neuner, A. 1990. Isolierung, Charakterisierung und taxonomische Einordnung coccoider mariner hyperthermophiler Archaebakterien. Ph.D. thesis, University of Regensburg, Germany.

Neuner, A., Jannasch, H. W., Belkin, S., *et al.* 1990. *Thermococcus litoralis* sp. nov. a new species of extremely thermophilic marine archaebacteria. *Archives of Microbiology*, **153**, 205–207.

Patel, B. K. C., Morgan, H. W., and Daniel, R. M. 1985. *Fervidobacterium nodosum* gen. nov. and spec. nov., a new chemoorganotrophic, caldoactive, anaerobic bacterium. *Archives of Microbiology*, **141**, 63–69.

Phipps, B. M., Typke, D., Hegerl, R., *et al.* 1993. Structure of a molecular chaperone from a thermophilic archaebacterium. *Nature*, **361**, 475–477.

Pley, U., Schipka, J., Gambacorta, A., *et al.* 1991. *Pyrodictium abyssi* sp. nov. represents a novel heterotrophic marine archaeal hyperthermophile growing at 110 °C. *Systematic and Applied Microbiology*, **14**, 245–253.

Ravot, G., Magot, M., Fardeau, M. L., *et al.* 1995. *Thermotoga elfii* sp. nov., a novel thermophile bacterium from an African oil-producing well. *International Journal of Systematic Bacteriology*, **45**, 308–314.

Reddy, T. and Suryanarayana, T. 1988. Novel histone-like DNA-binding proteins in the nucleoid from the acidothermophilic archaebacterium *Sulfolobus acidocaldarius* that protect DNA against thermal denaturation. *Biochimica Biophysica Acta*, **949**, 87–96.

Rieger, G., Rachel, R., Herrmann, R., *et al.* 1995. Ultrastructure of the hyperthermophilic archaeon *Pyrodictium abyssi*. *Journal of Structural Biology*, **115**, 78–87.

Sako, Y., Nomura, N., Uchida, A., *et al.* 1996. *Aeropyrum pernix* gen. nov. sp. nov., a novel aerobic hyperthermophilic Archaeon growing at temperatures up to 100 °C. *International Journal of Systematic Bacteriology*, **46**, 1070–1077.

Sandman, K., Krzycki, J. A., Dobrinski, B., *et al.* 1990. DNA binding protein HMf isolated from the hyperthermophilic archaeon *Methanothermus fervidus*, is most closely related to histones. *Proceedings of the National Academy of Sciences of the USA*, **87**, 5788–5791.

Schäfer, T. and Schönheit, P. 1992. Maltose fermentation to acetate, $CO_2$ and $H_2$ in the anaerobic hyperthermophilic archaeon *Pyrococcus furiosus*: evidence for the operation of a novel sugar fermentation pathway. *Archives of Microbiology*, **158**, 188–202.

Schopf, J. W. 1993. Microfossils of the Early Archaean Apex Chert: new evidence of the antiquity of life. *Science*, **260**, 640–646.

Schopf, J. W. and Packer, B. M. 1987. Early Archaean (3.3.-billion to 3.5-billion-year-old) microfossils from Warrawoona Group, Australia. *Science*, **237**, 70–73.

Schopf, J. W., Hayes, J. M., and Walter, M. R. 1983. Evolution of Earth's earliest ecosystems: recent progress and unsolved problems. In J. W. Schopf (ed.) *Earth's Earliest Biosphere: Its Origin and Evolution*. Princeton NJ, Princeton University Press, pp. 361–384.

Segerer, A. H., Neuner A., Kristjansson, J. K., *et al.* 1986. *Acidianus infernus* gen. nov., sp. nov., and *Acidianus brierleyi* comb. nov.: Facultatively aerobic, extremely acidophilic thermophilic sulfur-metabolizing archaebacteria. *International Journal of Systematic Bacteriology*, **36**, 559–564.

Segerer, A. H., Trincone, A., Gahrtz, M., *et al.* 1991. *Stygiolobus azoricus* gen. nov., sp. nov. represents a novel genus of anaerobic, extremely thermoacidophilic archaebacteria of the order *Sulfolobales*. *International Journal of Systematic Bacteriology*, **41**, 495–501.

Stetter, K. O. 1982. Ultrathin mycelia-forming organisms from submarine volcanic areas having an optimum growth temperature of 105 °C. *Nature*, **300**, 258–260.

1986. Diversity of extremely thermophilic archaebacteria. In T. D. Brock (ed.) *Thermophiles: General, Molecular and Applied Microbiology*, New York, John Wiley, pp. 39–74.

1988. *Archaeoglobus fulgidus* gen. nov., sp. nov.: a new taxon of extremely thermophilic archaebacteria. *Systematic and Applied Microbiology*, **10**, 172–173.

1992. Life at the upper temperature border. In J. and K. Trân Thanh Vân, J. C. Mounolou, *et al.* (eds.) *Colloque Interdisciplinaire du Comité National de la Recherche Scientifique*. Gif-sur-Yvette, France, Editions Frontières, pp. 195–219.

Stetter, K. O., Fiala, G., Huber, G., *et al.* 1990. Hyperthermophilic microorganisms. *FEMS Microbiology Reviews*, **75**, 117–124.

Stetter, K. O., Huber, R., Blöchl, E., *et al.* 1993. Hyperthermophilic archaea are thriving in deep North Sea and Alaskan oil reservoirs. *Nature*, **365**, 743–745.

Stetter, K. O., König, H., and Stackebrandt, E. 1983. *Pyrodictium* gen. nov., a new genus of submarine disc-shaped sulphur reducing archaebacteria growing optimally at 105 °C. *Systematic and Applied Microbiology*, **4**, 535–551.

Stetter, K. O., Lauerer, G., Thomm, M., *et al.* 1987. Isolation of extremely thermophilic sulfate reducers: evidence for a novel branch of archaebacteria. *Science*, **236**, 822–824.

Stetter, K. O., Thomm, M., Winter, J., *et al.* 1981. *Methanothermus fervidus*, sp. nov., a novel extremely thermophilic methanogen isolated from an Icelandic hot spring. *Zbl. Bakt. Hyg., I. Abt. Orig. C2*, 166–178.

Thauer, R. K., Jungermann, K., and Decker, K. 1977. Energy conservation in chemotrophic anaerobic bacteria. *Bacteriology Reviews*, **41**, 100–180.

Thomm, M., Stetter, K. O., and Zillig, W. 1982. Histone-like proteins in eu- and archaebacteria. *Zbl. Bakt. Hyg. I. Abt. Orig. C3*, 128–139.

Trent, J. D., Chastain, R. A., and Yayanos, A. A. 1984. Possible artfictural basis for apparent bacterial growth at 250 °C. *Nature*, **207**, 737–740.

van de Vossenberg, J. L. C. M., Driessen, A. J. M., and Konings, W. N. 1998. The essence of being extremophilic: the role of the unique archaeal membrane lipids. *Extremophiles*, **2**, 163–170.

Völkl, P., Huber, R., Drobner, E., *et al.* 1993. *Pyrobaculum aerophilum* sp. nov., a novel nitrate-reducing hyperthermophilic archaeum. *Applied and Environmental Microbiology*, **59**, 2918–2926.

White, R. H. 1984. Hydrolytic stability of biomolecules at high temperatures and its implication for life at 250 °C. *Nature*, **310**, 430–431.

Windberger, E., Huber, R., Trincone, A., *et al.* 1989. *Thermotoga thermarum* sp. nov. and *Thermotoga neapolitana* occuring in African continental solfataric springs. *Archives of Microbiology*, **151**, 506–512.

Woese, C. R. and Fox, G. E. 1977. Phylogenetic structure of the prokaryotic domain: the primary kingdoms. *Proceedings of the National Academy of Sciences of the USA*, **74**, 5088–5090.

Woese, C. R., Achenbach, L., Rouviere, P., *et al.* 1991. Archaeal phylogeny: re-examination of the phylogenetic position of *Archaeoglobus fulgidus* in light of certain composition-induced artifacts. *Systematic and Applied Microbiology*, **14**, 364–371.

Woese, C. R., Gutell, R., Gupta, R., *et al.* 1983. Detailed analysis of the higher-order structure of 16S-like ribosomal ribonucleic acids. *Microbiological Reviews*, **47**, 621–669.

Woese, C. R., Kandler, O., and Wheelis, M. L. 1990. Towards a natural system of organisms: proposal for the domain Archaea, Bacteria, and Eucarya. *Proceedings of the National Academy of Sciences of the USA*, **87**, 4576–4579.

Woese, C. R., Sogin, M., Stahl, D. A., *et al.* 1976. A comparison of the 16S ribosomal RNAs from mesophilic and thermophilic bacilli. *Journal of Molecular Evolution*, **7**, 197–213.

Zillig, W., Gierl, A., Schreiber, G., *et al.* 1983a. The archaebacterium *Thermofilum pendens* represents a novel genus of the thermophilic, anaerobic sulfur respiring *Thermoproteales*. *Systematic and Applied Microbiology*, **4**, 79–87.

Zillig, W., Holz, I., Janecovic, D., *et al.* 1983b. The archaebacterium *Thermococcus celer* represents a novel genus within the thermophilic branch of the Archaebacteria. *Systematic and Applied Microbiology*, **4**, 88–94.

1990. *Hyperthermus butylicus*, a hyperthermophilic sulfur-reducing archaebacterium that ferments peptides. *Journal of Bacteriology*, **172**, 3959–3965.

Zillig, W., Holz, I., Klenk, H. P., *et al.* 1987a. *Pyrococcus woesei* sp. nov., an ultra-thermophilic marine archaebacterium, represents a novel order, *Thermococcales*. *Systematic and Applied Microbiology*, **9**, 62–70.

Zillig, W., Stetter, K. O., Prangishvilli, D., *et al.* 1982. *Desulfurococcaceae*, the second family of the extremely thermophilic, anaerobic, sulfur-respiring *Thermoproteales*. *Zbl. Bakt. Hyg., I. Abt. Orig. C3*, 304–317.

Zillig, W., Stetter, K. O., Schäfer, W., *et al.* 1981. *Thermoproteales*: a novel type of extremely thermoacidophilic anaerobic archaebacteria isolated from Icelandic solfataras. *Zbl. Bakt. Hyg., I. Abt. Orig. C2*, 205–227.

Zillig, W., Stetter, K. O., Wunderl, S., *et al.* 1980. The *Sulfolobus–Caldariella* group: taxonomy on the basis of the structure of DNA-dependent RNA polymerases. *Archives of Microbiology*, **125**, 259–269.

Zillig, W., Yeates, S., Holz, I., *et al.* 1987b. *Desulfurolobus ambivalens*, gen. nov. sp. nov., an autotrophic archaebacterium facultatively oxidizing or reducing sulfur. *Systematic and Applied Microbiology*, **8**, 197–203.

Zuckerkandl, E. and Pauling, L. 1965. Molecules as documents of evolutionary history. *Journal of Theoretical Biology*, **8**, 357–366.

# Chapter 7

# Volcanism and mass extinctions

## Paul B. Wignall

## Mass extinctions

Mass extinction events are brief intervals of geological time marked by the loss of numerous species from diverse environments around the globe. Around a dozen such events have been identified and the biggest, known as the "big five," mark some of the main boundaries of the geological column. The most famous mass extinction is that which brought the 160 Ma reign of the dinosaurs to a close at the Cretaceous–Tertiary boundary: an event that is universally abbreviated to the K–T event (not C–T because this is an abbrevation for Cenomanian–Turonian, a stage boundary in the Upper Cretaceous which, incidentally, is also an extinction horizon). The other four major extinction events happened at the end of the Ordovician, within the Late Devonian, at the end of the Permian (this was the greatest of them all), and at the end of the Triassic (Hallam and Wignall, 1997).

Mass extinctions are of fundamental importance in the history of life because they often eliminate dominant groups from an environment thereby clearing the way for the radiation of previously minor groups. The best-known such changeover concerns the replacement of dinosaurs by mammals as the dominant terrestrial vertebrates in the aftermath of the K–T event. Both mammals and dinosaurs enjoyed a long evolutionary history prior to the end-Cretaceous calamity, but only the dinosaurs occupied the large terrestrial vertebrate niches. The complete extinction of dinosaurs allowed the mammals to evolve to the larger body sizes we see today.

Despite their importance, mass extinctions have only been the subject of major investigation since 1980. This year was marked by the publication of a paper by Luis Alvarez, a nobel laureate in physics, and co-workers which purported to have discovered the cause of the K–T mass extinction (Alvarez et al., 1980). Having analyzed a K–T boundary section in Italy, and discovered anomalously high concentrations of the trace metal iridium in a thin boundary clay, Alvarez et al. (1980) made two distinct, but related claims; first, that the Ir was derived from a meteorite that had impacted the Earth and second, that such an impact was responsible for the mass extinction.

That a large meteorite impact (sometimes referred to as a bolide impact) could cause mass extinction had previously been suggested by McLaren (1970), who proposed that a tsunami, many kilometers high, could have caused the extinctions. Contentious evidence for a large tsunami has been detected in some K–T boundary sections (Alvarez et al., 1992), but the principal kill mechanism has been ascribed to an interval of global darkness caused by the injection of vast amounts of dust into the higher atmosphere (Alvarez et al., 1980). It is postulated that such dust, the pulverized remnants of the bolide and the target rock, would remain in the atmosphere for several years causing global

*Volcanoes and the Environment*, eds. J. Martí and G. G. J. Ernst. Published by Cambridge University Press. © Cambridge University Press 2005.

darkness, cooling, and a total shutdown of all photosynthetic activity with obvious dire consequences for most ecosystems. High concentrations of soot in a K–T boundary clay from Denmark suggest that widespread forest fires may have followed impact, thereby exacerbating the intensity and duration of darkness (Wolbach et al., 1990).

The original catastrophic kill mechanism of Alvarez and colleagues envisaged mass extinction occurring within days to years of bolide impact, i.e., in a geological instant. The recent discovery of the probable K–T crater at Chicxulub in the Yucatan Peninsula of Mexico has lead to suggestions that the damaging effects of the impact may have lasted for thousands of years. The Chicxulub crater is developed in limestones and evaporites (calcium sulfate). Vaporization of these target rocks during impact would produce huge volumes of carbon dioxide and sulfur dioxide (Pope et al., 1994). The latter would react with atmospheric water vapor producing sulfuric acid aerosols thereby contributing to the blanket of darkness and causing acid rain for several years after impact. Carbon dioxide has a longer residence time in the atmosphere and this greenhouse gas may have caused the short period of global refrigeration to be followed by global warming. Average global temperatures are suggested to have risen by 4 °C after the impact with further dire consequences for life on Earth (Pope et al., 1994).

The ensuing years since 1980 have seen the study of mass extinctions and their causes grow into one of the most interdisciplinary fields of geological research. Debate has been intense but the bolide impact–mass extinction link has gained widespread support for the K–T event (Glen (1994) contains interesting reviews and perspectives on the history of this debate), but not for other mass extinctions. Not all geologists are convinced of the lethality of bolide impact and alternative, Earth-bound causes of extinction have been sought. In particular, a small, but vociferous, group has highlighted volcanic activity as a potential source of global environmental damage and extinction (e.g., Keith, 1982; Officer and Drake, 1983; Rampino and Stothers, 1988; Courtillot, 1999).

## Lethal volcanism?

That massive volcanic eruptions could cause global extinctions was first mooted by Budyko and Pivivariva (1967). In a paper that was remarkably prescient of future developments, they proposed that ten closely spaced, Krakatau-scale eruptions could produce several years of global darkening with catastrophic consequences for photosynthesizing plants and the food chain. The damage done by volcanic eruptions of recent times at least partly corroborates this scenario (see also Chapter 5). The largest historical eruptions have been observed to cause global cooling due to their ability to inject dust and sulfate aerosols into the stratosphere (Devine et al., 1984). Thus, the eruption of Mt. Pinatubo in 1991 is blamed for the unusually cool winter of that year. Rather surprisingly, the most harmful environmental effect of this eruption appears to have been felt in the Red Sea, 6000 km from the eruption site (Genin et al., 1995). The cold, stormy winter of 1994/95 in the region caused vigorous mixing of the water column and upwelling of deep-water nutrients with the resulting growth of filamentous algae. These smothered the corals of the Red Sea and lead to the demise of many reefs. However, rather than demonstrating the harmful effects of explosive volcanism, this phenomenon is perhaps most significant because of its demonstration of the extremely delicate balance of the Red Sea ecosystem.

Neither the eruption of Mt. Pinatubo nor the larger eruptions of Tambora in 1815 and Krakatau in 1883 are known to have caused the extinction of any species, despite producing notable global cooling. Slightly further back in geological time, the Toba eruption, approximately 75 000 years ago, is thought to have blasted 400 times more ash and dust into the atmosphere than the Krakatau eruption and yet this too caused no extinctions (Kent, 1981). Somewhat surprisingly, the most lethal volcanic eruptions appear not to be related to the explosive activity of acidic and calc–alkaline volcanoes, but rather to the much more quiescent eruptions of basalts from fissures.

The largest and most harmful eruption of historical times occurred in Iceland during 1783–4

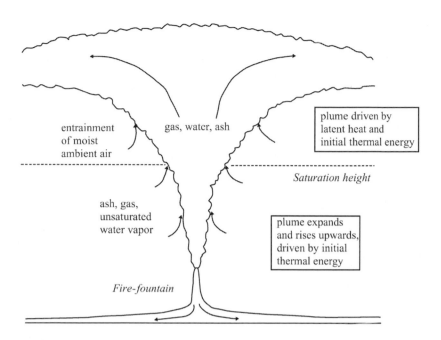

entrainment
of moist
ambient air

gas, water, ash

plume driven by
latent heat and
initial thermal energy

*Saturation height*

ash, gas,
unsaturated
water vapor

plume expands
and rises upwards,
driven by initial
thermal energy

*Fire-fountain*

**Fig. 7.1.** Mechanism for the injection of gas and fine ash into high levels of the atmosphere from fire-fountains associated with fissure eruptions. This mechanism is thought to be responsible for the global cooling experienced after the Laki eruptions in Iceland in 1783–4. After Woods (1993).

at Laki. This was a fissure eruption which produced 15 km³ of basaltic lava, much of it within the first 40 days of eruption (Sigurdsson, 1982; Thordarson *et al.*, 1996). Unlike acidic volcanism, such eruptions are associated with the release of huge volumes of sulfur dioxide (Devine *et al.*, 1984) and it is this gas, which reacts with water vapor to form sulfate aerosols, that appears to be the most environmentally harmful volcanic product. The Laki eruption is calculated to have released 122 Mt of sulfur dioxide (Thordarson *et al.*, 1996), which substantially contributed to the dense shroud of blue haze that hung over Iceland for the next year. As a result the harvest in Iceland completely failed in 1783 and lead to the death by starvation of three-quarters of the island's livestock and nearly a quarter of the human population (Sigurdsson, 1982).

The effects of the Laki eruption were experienced throughout the northern hemisphere in the severe winter of 1783–4. One of the best records of this event was kept by Benjamin Franklin who was then the young American

ambassador to the court of Louis XVI in France. Franklin brilliantly deduced that the severe cold was probably the result of a volcanic eruption that had blown dust into the atmosphere, thus partially obscuring sunlight. This was probably the first occasion that the link between volcanism and climate had been made (Sigurdsson, 1982). The widespread cooling caused by the Laki eruption suggests that ejecta from the eruption reached the stratosphere, allowing it to be circulated around the globe by stratospheric winds.

Clearly, although fissure eruptions are relatively non-violent compared with volcanic eruptions like those of Krakatau and Tambora, they must be capable of propelling gas and fine ash to a considerable height. The answer to this dilemma may come from an understanding of the physics of the lava-fountains which punctuate the length of fissure eruptions (Thordarson *et al.*, 1996).

Woods (1993) has modeled the nature of lava-fountain eruptions (Fig. 7.1). Lava within lava-fountains degasses and releases large volumes of sulfur dioxide, carbon dioxide, and lesser amounts of chlorine and fluorine, along with fine, particulate ash. This gas–ash cloud rises as a plume above the lava-fountain entraining and heating up water vapor from the atmosphere. As the plume rises, atmospheric pressure decreases until a point is reached at which the water vapor

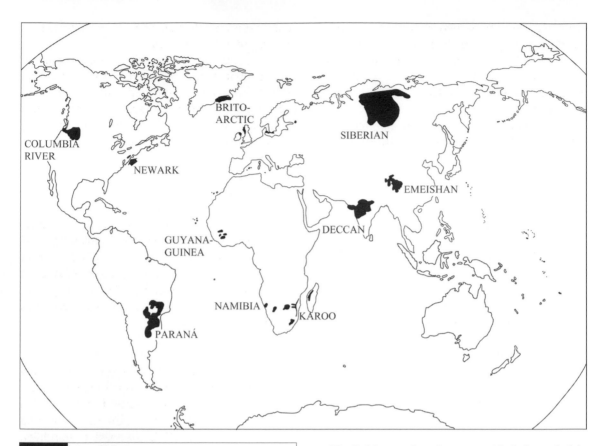

**Fig. 7.2.** Global distribution of flood basalt provinces. The apparent area of provinces today is not a good guide to the original area covered by flood basalts. Thus, the Siberian Traps are well exposed and extensively eroded whereas most of the lavas of the Brito-Arctic Province are buried beneath younger sediments or hidden beneath the waters of the North Atlantic.

reaches its saturation point. This causes condensation, the release of latent heat, and a consequent decrease of plume buoyancy which is therefore able to rise to higher levels in the atmosphere. By this mechanism fissure eruptions may be capable of injecting gas and ash into the lower stratosphere (Fig. 7.1). Factors such as the humidity of the lower atmosphere and the proportion of fine ash in the plume are therefore deemed important controls of the environmental damage wrought during fissure eruptions. The Laki eruption is thought to have been particularly ash-rich (Woods, 1993); such material is more efficient at transferring its heat energy to water vapor than volcanic gases.

The Laki eruptions have provided the only historical example of one of the most dramatic and voluminous manifestations of volcanic activity: flood basalts. Eight major, rapidly erupted, continental flood basalt provinces are known from around the world (Fig. 7.2). These range in age from the Late Permian to the Miocene, although the presence of several, giant, radiating dyke swarms from older Paleozoic and Proterozoic regions may be the eroded remnants of older continental flood basalt provinces (Ernst and Buchan, 1997). Dyke swarms and other intrusions characterize many provinces and volumetrically they are often far more significant than the lava flows themselves. However, individual flows within flood basalt provinces dwarf the scale of the Laki flow. Typically they cover a few hundred square kilometers, range from a few tens to a few hundreds of meters thick and can commonly comprise more than 1000 km$^3$ of basalt (e.g., Self et al., 1997). Given the environmentally devastating (albeit localized) effects of the Laki eruption, it is reasonable to speculate that flows involving more than two orders of magnitude more lava may have had dire global consequences. The link between mass extinctions and volcanism has therefore focused on continental flood basalts. Initial emphasis has been on the Deccan Traps, a large province in northwest India (Fig. 7.3), known to have been erupted sometime in the latest Cretaceous – earliest Tertiary interval.

# The Deccan Traps and the death of the dinosaurs

Investigations into the link between flood basalts and extinctions fall into two categories. First, ever more sophisticated attempts at absolute dating have attempted to demonstrate the synchronicity of eruptions and extinctions. Second, speculation on the causal link between volcanism and extinction (i.e., the kill mechanisms) has attemped to explain the selectivity and rates of the extinctions with predicted, volcanically induced environmental changes.

The breakthrough in the Deccan Traps – K–T mass extinction link came with a geomagnetic study. This revealed that the lavas were erupted during two reversals of the magnetic field in an interval of less than 2 million years (Courtillot et al., 1988). Most of the basalt, perhaps as much as 80%, was emplaced in less than 1 million years during chron C29R (Courtillot, 1990). This date has a double significance because it indicates that around 1 million $km^3$ of lava was erupted at precisely the same time as the mass extinction. Argon–argon dating has confirmed these results. Duncan and Pyle (1988), sampling lavas from the base and top of a 2-km-thick volcanic pile, produced ages of 68.5 Ma and 66.6 Ma respectively. Within error these ages are virtually identical and imply very rapid eruption rates indeed. Thus, it has been estimated that lava flows involving up to 10 000 $km^3$ of basalt (Courtillot, 1990) occurred on average every 5000 years (Cox, 1988) in the Deccan Traps. As commonly found in many flood basalt provinces, the later stages of the Deccan Traps eruptions were characterized by explosive acidic eruptions that may have provided the final *coup de grâce* (White, 1989).

In a memorable sentence on the amount of lava erupted, Archibald (1996, p. 143) noted that "the Deccan Traps contain almost enough basalt to cover both Alaska and Texas to a depth of 2,000 ft [600 m]." With such a massive "smoking gun" now available, the Deccan Traps rapidly became established as the principal alternative culprit to bolide impact (Mclean, 1985; Stothers et al., 1986; Courtillot et al., 1988; Courtillot, 1990, 1999). Officer et al. (1987) were the first to propose, in detail, the full sequence of events that are now widely regarded as the volcanic kill mechanism (Fig. 7.4). Much of the blame is placed on the emission of volcanic gases, particularly the large volumes of sulfur dioxide emitted during fissure eruptions. The extinctions are considered to have occurred in two distinct phases.

## Phase One

Sulfate aerosols injected into the lower stratosphere cause global darkening, freezing, and a virtual shutdown of photosynthetic activity, with the consequent collapse of numerous marine and terrestrial ecosystems. This situation has been termed "volcanic winter" and is similar to the "nuclear winter" scenario – the predicted consequences of a full-scale nuclear war. From knowledge of the Laki eruption and other major eruptions of historical times, it is likely that sulfate aerosols are only likely to remain in the stratosphere for 1–2 years before being rained-out (Sigurdsson, 1982; Devine et al., 1984). The subsequent acid rain may have been a further potentially damaging effect, particularly in freshwater ecosystems (Officer et al., 1987). As will be apparent, this short-lived phase of the extinction scenario (1–2 years) is essentially identical in its cause and predicted effects to the bolide impact model and it is commonly considered the main cause of extinction. The second phase is also similar to the proposed later stage effects of bolide impact at Chicxulub.

## Phase Two

Global warming occurs as the greenhouse effects of the increased atmospheric $CO_2$ concentrations are felt (Fig. 7.4). Whereas Phase One may only

**Fig. 7.3.** Deccan Traps scenery in Mahabaleshwar, India.
Photograph courtesy of M. Widdowson.

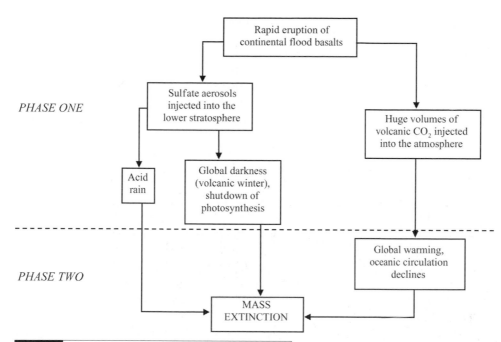

**Fig. 7.4.** Flow diagram showing the two distinct phases attributed to a volcanically induced mass extinction event. Phase One is very brief and may only last a few years whilst Phase Two spans tens to hundreds of thousands of years.

last a few years the effects of increased $CO_2$ are likely to have spanned 10 000 to 100 000 years (Courtillot, 1990). As a final addition to this list, it is also possible that the emission of chlorine gas during the eruptions may have triggered depletion of the ozone layer (Sigurdsson, 1990; Officer, 1993).

Officer *et al.* (1987) sought to increase the potential lethality of their kill mechanism by inferring much shorter recurrence intervals between eruptions, perhaps as little as an eruption every few hundred years or even less. If this was the case, such close spacing would have allowed little time for recovery between events, with the result that environmental conditions may have progressively deteriorated over a few thousand years. However, as noted above, the majority of authors favor much longer recurrence intervals.

The volcanic kill mechanism has strong similarities with that proposed for the bolide impact, making it difficult to distinguish between the two alternatives: some have attempted to unite the two causes. Rampino (1987), and more recently Jones *et al.* (2002), suggested that large bolide impacts may be able to generate flood basalt provinces by decompressive melting at the site of the crater. However, this is theoretically impossible: even excavation of a 20-km-deep crater (which would require a stupendously large impact) would be insufficient to cause melting (White, 1989). Meteorite impact and volcanism must be considered as two separate extinction mechanisms. One potential means of distinguishing between them lies in the proposed duration of the extinction event. Bolide impact proponents envisage extinction occurring in a geological instant, whereas the volcanic extinction mechanism is spread over several thousand years. Unfortunately, the resolution of the fossil record is such that it is barely possible to distinguish between these two durations.

The duration of the K–T extinction crisis is a subject of ongoing debate. Due to their rarity, dinosaurs are rather poor fossils for assessing the rapidity of extinction. However, detailed study of dinosaur ranges from the Hell Creek Formation of Montana and North Dakota, one of the more fossiliferous dinosaur-yielding stratigraphic units, suggests that they disappeared abruptly at the end of the Cretaceous (Sheehan *et al.*, 1991). Unfortunately, no comparable details are available from dinosaur beds elsewhere in the world.

The rather patchy fossil record from the Deccan provides intriguing insights into the faunal crisis. The flood basalts rest on the Lameta Group, a terrestrial unit that contains a diverse Upper Cretaceous fauna, including dinosaurs (Vianey-Liaud et al., 1987). Vertebrate remains are also common in the sediments interbedded with the lava flows and it is notable that the fauna is essentially unchanged at any level. In particular, the fish and amphibian faunas from lacustrine intertrappean sediments are virtually identical to those from the Lameta Group (Jaeger et al., 1989), although dinosaur diversity undergoes some decline (Prasad and Khajuria, 1995). These observations are perhaps surprising considering the widely acknowledged vertebrate extinctions of the K–T event, but it serves to emphasize the fact that most of the evidence for the end-Cretaceous demise of the dinosaurs (and other terrestrial vertebrates) comes from North America; little is known of the details of the timing and intensity of extinctions from elsewhere in the world (Archibald, 1996). Freshwater ecosystems are particularly susceptible to the effects of acid rain, and the observation that the local lacustrine fauna is little affected by the Deccan eruptions suggests that the environmental consequences of volcanic $SO_2$ eruptions were unimportant.

Planktonic foraminifera provide a much more valuable record of diversity changes during the K–T crisis because they are abundant in most marine rocks, particularly deep-sea sediments, where a complete sedimentary record can accumulate. Planktonic foraminifera suffered a catastrophic extinction event at the end of the Cretaceous with perhaps only one species surviving (Smit, 1982). Unfortunately, but perhaps not unexpectedly, foraminiferal experts cannot agree on whether this extinction was instantaneous or spread over a million years or so. Even following a blind test on samples taken from a K–T boundary section in Tunisia, no consensus could be reached as to the rapidity of the extinction. The majority of workers favor an instantaneous extinction event (e.g., Olsson and Liu, 1993), but a few favor a more protracted crisis spread over

a few hundred thousand years (e.g., Keller, 1996). Only this minority view accords with a Deccan Traps-induced extinction.

More direct evidence of the relationship between the foraminifera extinction and Deccan Traps eruption comes from offshore wells from the east coast of India. These record a major foraminiferal extinction event at the K–T boundary (Jaiprakash et al., 1993), but this does not appear to be intimately related to the eruption of basalts. In this region basalt flows occur in the Upper Cretaceous, prior to the extinctions, and in the Early Tertiary, after the extinctions and during the subsequent foraminiferan radiation event. The mass extinction occurs during a 6-Ma quiescent interval that punctuates the basalt eruptions in this region (Jaiprakash et al., 1993).

Recent investigations of Deccan Traps geology have further complicated the story relating to extinction. Pyroclastic deposits have been discovered that indicate that at least some of the eruptions were considerably more explosive than modern fissure outpourings (Widdowson et al., 1997) – thus Phase One of the extinction scenario may have been more intense. However, numerous, intensely weathered soil horizons (boles) have also been found within the Deccan Traps. These indicate that the eruptions were separated by 10 000–100 000-year quiescent intervals (Widdowson et al., 1997), rather than the much shorter durations suggested by Officer et al. (1987). Furthermore, palynological evidence from the boles suggests that the time between eruptions was more than adequate for the complete recovery and revegetation of the landscape (M. Widdowson, 1997, personal communication). If there was sufficient recovery time, the question therefore arises, were individual flows in the Deccan Traps able to cause extinctions across the globe (Cox, 1988)?

The link between the K–T mass extinction and the Deccan Traps is far from compelling, but the presence of the lavas may offer an explanation for the distinctive sea-level variations at this time. Approximately 1 million years before the end of the Cretaceous sea level fell dramatically and was followed by a rapid transgression immediately prior to the K–T boundary (Hallam, 1987). Modeling of the mantle plumes, which are thought to be responsible for the eruption of flood basalt provinces, predicts the occurrence of such rapid regression–transgression couplets. Immediately prior to the onset of eruptions, widespread doming and sea-level retreat marks the arrival of the plume-head at the base of the lithosphere. The subsequent burst of volcanism causes the plume-head to collapse and triggers widespread subsidence and thus transgression (Campbell and Griffiths, 1990; Courtillot, 1990). Some extinctions have been related to the rapid environmental changes associated with such sea-level changes (Hallam, 1987; Archibald, 1996), thus contributing another indirect, kill mechanism to the list of volcanically induced hazards.

## Flood basalts and other mass extinctions

By the late 1980s several geologists had realized that, if there was a link (albeit an enigmatic one) between the Deccan Traps and the K–T mass extinction, then other such biotic crises may also be related to the eruption of continental flood basalt provinces (Loper *et al.*, 1988; Rampino and Stothers, 1988). In a detailed examination of post-Paleozoic mass extinctions, Rampino and Stothers (1988) made two bold claims: first, all mass extinctions of the last 250 Ma coincide with flood basalt provinces and second, that both show a distinct periodicity. The possibility that all post-Paleozoic mass extinctions showed a 26 Ma periodicity was first proposed by Raup and Sepkoski (1984), although more recent compilations of extinctions and a more up-to-date timescale suggest that the periodicity signal was probably spurious (Benton, 1995; Hallam and Wignall, 1997). In their study, Rampino and Stothers (1988) were only able to achieve a periodical signal in continental flood basalt eruption times by removing two provinces from their analysis. Ironically, one of these was the Siberian Traps, perhaps one of the more significant extinction-related volcanic provinces (see below).

Rampino and Stothers' first conclusion, that all flood basalts coincide with mass extinctions, also failed the test of detailed scrutiny because,

at that time, the dating of many provinces indicated that they were not coincidental with extinction events. Nonetheless, in a later analysis, Stothers (1993) found a statistically significant correlation between extinctions and flood basalts. This was only achieved by recognizing Bajocian, Tithonian (end Jurassic), Aptian, Middle Miocene, and Pliocene mass extinctions, none of which is genuine (cf. Hallam and Wignall, 1997). In an appraisal of his results, Stothers concluded that, whilst some flood basalt provinces undoubtedly coincide with mass extinction events, they may be "only accidental flukes of timing" (Stothers, 1993, p. 1399). In fact, recent work suggests that Stothers may have been unduly pessimistic. Since the publication of Rampino and Stothers' (1988) paper, all the major flood basalts have been the subject of improved dating efforts, principally using the argon–argon technique. This has revealed a much closer correspondence between the ages of most flood basalt provinces and extinction events than even Rampino and Stothers originally appreciated (Wignall, 2001) (Fig. 7.5).

So, does this mean that the eruption of major flood basalt provinces causes mass extinction? Each individual case is examined below.

### The Emeishan Basalts and the end-Maokouan mass extinction

Only recognized in the past few years, the end-Maokouan marked a major mass extinction for marine life (Jin *et al.*, 1994; Stanley and Yang, 1994). This level occurs at the end of the Middle Permian and, in the complicated stage nomenclature of the Permian, it is also known as the end of the Guadalupian. Shallow marine faunas in the tropics were particularly badly affected, notably brachiopods and giant foraminifera known as fusulinids, but the event appears not to have been important at higher latitudes (Hallam and Wignall, 1997). The extinction event approximately coincides with the eruption of the Emeishan basalts, a flood basalt province now folded in the mountains of southwest China (Yin *et al.*, 1992) (Fig. 7.2). The absolute, radiometric age of this province is only imprecisely known, but biostratigraphic data suggests that the onset of eruption began in the latest Maokouan (Jin and

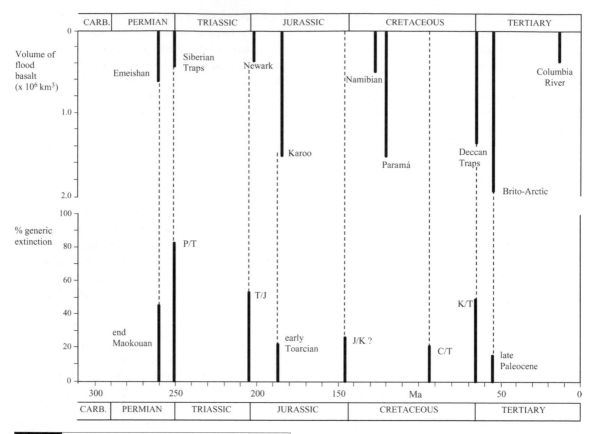

**Fig. 7.5.** Comparison of the magnitudes and timing of post-Paleozoic mass extinction events and flood basalt provinces. There is a good correlation in time for six out of nine flood basalt provinces although there is no correlation between the magnitude of extinction and volume of lavas erupted.

Shang, 2000), precisely contemporaneous with the marine mass extinction. However, magnetostratigraphic evidence suggests the main phase of basaltic volcanism predated the extinction with only the later stage, volumetrically less significant pyroclastic volcanism occurring during the marine crisis (Ali *et al.*, 2002).

## The Siberian Traps and the Permian–Triassic (P–T) mass extinction

The P–T mass extinction was the most severe in the history of life: in excess of 90% of marine and terrestrial species disappeared (see review in Hallam and Wignall, 1997). Species at all latitudes succumbed and, perhaps significantly, the forests of high southern latitudes, with their characteristic glossopterid flora, were entirely eliminated (Retallack, 1995). The few survivors of this holocaust were especially widespread. On land the medium-sized, mammal-like reptile *Lystrosaurus* occurs at all latitudes, whilst in the sea a few genera of bivalves have a similar global distribution (Hallam and Wignall, 1997). In the oceans the extinction appears closely related to the development of oxygen-poor to anoxic conditions. Initially these conditions were only developed in deep-ocean settings (Isozaki, 1994), but towards the close of the Permian anoxic conditions became rapidly established even in very shallow-water settings (Wignall and Twitchett, 1996). This oceanographic event has been attributed to a phase of extreme global warming, associated with the development of a much reduced latitudinal temperature gradient. Terrestrial evidence from fossil soils also indicates warming in high latitude sites (Retallack, 1997).

Until the last decade volcanism did not figure in many P–T mass extinction scenarios, but redating of the Siberian Traps, a major flood basalt province in western Siberia (Fig. 7.2), has revealed that their eruption closely coincides with this event. This discovery has given considerable impetus to the study of volcanism-related mass extinction.

Determining the age of the Siberian Traps has proved rather elusive. Baksi and Farrar (1991), in their review of published ages, noted that they ranged from 290 to 160 Ma, although their own argon–argon dates suggested a 5–10 Ma eruption interval beginning in the Middle Triassic (238.4 ± 1.4 Ma). These ages suggested that the Siberian Traps were of little relevance to the study of the mass extinction at the end of the Permian. This situation abruptly changed with the publication of new argon–argon dates that indicated the rapid eruption rate (in as little as 800 000 years), of the entire province at 248 Ma – the assumed age of the P–T boundary (Renne and Basu, 1991). One of the best available ages for a boundary section is 251 Ma derived from a tuff that precisely marks the main level of extinction in South China (Claoué-Long et al., 1991; Bowring et al., 1998). This is slightly too old to correspond with Renne and Basu's (1991) age for the Siberian Traps. However, Renne et al. (1995) recalculated the argon–argon ages, after recalibrating their standard, and produced an age of 250.0 ± 1.6 Ma thereby avoiding the potential embarrassment that the Siberian Traps may have been slightly too young to be implicated in the P–T mass extinction. This by no means marks the end of the debate on the age of the Siberian Traps.

The Siberian Traps actually consist of four distinct regions which have distinct and different eruption histories. Recent argon–argon dates from the Noril'sk region, in the northwest of the province, suggest eruption either spanned the interval from 248 to 247 Ma (Venkatesan et al., 1997), too recent to be implicated in the mass extinction, or it began at 251.2 ± 0.3 Ma (Kamo et al., 1996), precisely the time of the extinction. Reports of Lystrosaurus, a post-extinction survivor, from intertrappean sediments at the base of the lava pile (Venkatesan et al., 1997), is strong evidence that the eruptions postdated the extinction in the Noril'sk region at least. In contrast, dating of the volcanism of the Maimecha–Kotui region, in the northeast of the province, indicates eruption at 253.3 ± 2.6 Ma (Basu et al., 1995), significantly before the extinction event.

Reports on the geology of the Siberian Traps have only gradually become available to western scientists over the past few years. These reveal that the Traps are a rather unusual flood basalt province. There are few significant fossil soil horizons developed within the lava pile suggesting, in accord with the radiometric dates, that the eruptions were closely spaced (Sharma, 1997). This is a typical feature of flood basalt provinces, but the individual flows appear to have been unusually small, rarely extending more than 100 km in extent and never exceeding 150 m in thickness (Zolotukhin and Al'mukhamevdov, 1988). Volumetrically the flows were an order of magnitude smaller than those of the Deccan Traps. However, the presence of basaltic tuffs, which locally reach 700 m thick, suggests that the volcanism was extraordinarily explosive by flood basalt standards (Sharma, 1997). The total volume of volcanic rocks within the province is difficult to assess. Widely stated in the literature to be the largest volcanic province in the world, the present-day volume of lavas and tuffs is actually only a modest $3.4 \times 10^5$ km$^3$ (Sharma, 1997), although some have estimated the original volume of volcanics to be more than twice that amount (e.g., White, 1989). It is of course possible that the original area of lava was originally much greater – Triassic lavas underlie the Jurassic cover of the West Siberian Basin for example (Westphal et al., 1998) and appear to be contemporaneous with the main outcrops (Reichow et al., 2002). The Siberian Traps are often said to contain $2 \times 10^6$ km$^3$ of lava, but this figure also includes the substantial volume of intrusive rocks, mainly dykes, that constitute much of the extent of the province. In conclusion, estimating the original size of the Siberian Traps is a rather intractable problem due to the extensive amount of post-Triassic erosion of the province and current uncertainty as to the volume of lavas beneath the West Siberian Basin.

Despite the above reservations concerning both the age and magnitude of the Siberian Traps, the province has featured as one of the main causes of the P–T event in many recent extinction scenarios. The proposed kill mechanisms are identical to those previously attributed to the Deccan Traps (Fig. 7.4); some authors suggest that Phase One was the most lethal (Yin *et al.*, 1992; Renne *et al.*, 1995), others suggest Phase Two (Veevers *et al.*, 1994; Wignall and Twitchett, 1996) and others prefer the "double whammy" of Phase One and Two as the cause of the extinctions (Campbell *et al.*, 1992; Conaghan *et al.*, 1994). The proponents of Phase One point to the unusually high proportion of pyroclastic deposits in the Siberian Traps as evidence that huge amounts of dust and sulphate aerosols were probably injected into the stratosphere during the eruption of the province (Renne *et al.*, 1995). The Siberian Traps were ideally placed to cause maximum environmental damage because, at their eruption site, close to the paleo-North Pole, the atmosphere is at its thinnest, thus making the injection of volcanic material into the stratosphere a relatively easy proposition. It is proposed that the P–T volcanic winter may have lasted in excess of 500 000 years and caused a total collapse of the food chain, a severe glaciation, and a major marine regression (Renne *et al.*, 1995). The second phase of the extinction mechanism need hardly be invoked if such a calamity occurred during the first phase.

It is difficult to comprehend how the Siberian Traps could have caused such continuous darkness given that the dust and aerosols of modern eruptions only remain in the stratosphere for a year or so (Devine *et al.*, 1984). Furthermore, field data in support of the Phase One scenario is limited. No sea-level fall occurs at the P–T boundary; in fact this was a time of rapid sea-level rise (Hallam and Wignall, 1999). The presence of "stone rolls" in paleosols from Australia are given as evidence for cooling – these are a feature formed during intense freezing (Conaghan *et al.*, 1994). However, Australia was at paleolatitudes in excess of 70° S in the Late Permian and it is not therefore surprising to find evidence of freezing in this region. More significant is the loss of stone rolls in the earliest Triassic which suggests significant high-latitude warming. Thus, despite the predictions noted above, the sum total of evidence for Phase One of the volcanic extinction mechanism is meager.

As noted above, field evidence overwhelmingly indicates that the P–T event was associated with rapid warming, implying that Phase Two may have been of greater significance. The evidence includes the loss of cold-adapted *Glossopteris* forests in high southerly paleolatitudes and the development of red beds formed in warm, arid conditions at such sites (Veevers *et al.*, 1994; Retallack, 1997). Contemporaneous oxygen isotope fluctuations from carbonates record a major temperature rise in tropical regions (Holser *et al.*, 1991). The most significant effects of this global warming may have been a decline in oceanic circulation, as the equator-to-pole thermal gradient declined, and a decrease in dissolved oxygen levels (Wignall and Twitchett, 1996). This proximate cause of extinction may be an ultimate consequence of the eruption of the Siberian Traps, although other causes of atmospheric $CO_2$ increase, such as oxidation of coals, could be equally important (e.g., Faure *et al.*, 1995).

## The Newark flood basalts and the Triassic–Jurassic (T–J) mass extinction

The T–J mass extinction is one of the big five extinction events of the Phanerozoic. It principally affected marine life, with most reef faunas going extinct along with a broad range of other invertebrates, notably ammonites and bivalves (Hallam and Wignall, 1997). Major turnovers of terrestrial faunas also occurred around this time, with the result that dinosaurs rose to dominance in the early Jurassic. However, the precise synchrony of the terrestrial extinctions has been questioned, and it appears that the crisis on land may have been substantially earlier than that in the seas (Benton, 1994).

Major paleoenvironmental changes occurred around the T–J boundary including a major warming pulse at the base of the Jurassic, indicated by evidence from fossil leaves (McElwain *et al.*, 1999). Major sea-level oscillations also characterized the boundary interval with a brief regression at the close of the Triassic followed by a spectacular transgression, associated with the spread of anoxic waters, in the earliest Jurassic (Hallam, 1997). Because of the large end-Triassic regression, complete T–J marine records are rare and this makes it difficult to precisely pinpoint the level of the extinction. However, the available evidence indicates that many extinctions probably occurred at the peak of regression in the latest Triassic (or even somewhat earlier), and prior to the earliest Jurassic anoxic event and global warming (Hallam, 1990, 1997; Hallam and Wignall, 1999).

Until recently there was little need to implicate volcanism in this extinction crisis, but new radiometric dating of several flood basalt provinces associated with the initial break-up along the central and North Atlantic margins has changed this perspective. The Newark Supergroup of eastern North America contains within it a substantial flood basalt province (Fig. 7.2). New U–Pb dates indicate a brief eruption at 201 ± 1 Ma, an interval within the Hettangian, the basalmost stage of the Jurassic (Weems and Olsen, 1997). This age is confirmed by biostratigraphic evidence from the interbedded sediments and indicates that the eruptions may have postdated the extinctions, but only by as little as 60 000 years (Olsen et al., 1990). Recent work has demonstrated that the Newark flood basalts formed the northwestern edge of a vast province of flood basalts termed the Central Atlantic Magmatic Province (Marzoli et al., 1999). Scattered outcrops of flood basalts cover substantial areas of Guyana, Guinea, Surinam, and northern Brazil and more substantial amounts may be buried beneath the sediments of offshore central Africa (Deckart et al., 1997). Dating of these lavas is rather imprecise, but it appears that a short period of intense eruptions occurred sometime between 204 and 195 Ma, an interval that brackets the T–J boundary. Like the Siberian Traps, the principal manifestations of igneous activity in the Guyana–Guinea Province are swarms of dykes (Deckart et al., 1997).

It is difficult to link this volcanism directly to the T–J extinction, but, like the P–T anoxic event, the widespread early Jurassic anoxia may be related to global warming triggered by volcanic $CO_2$ release. The distinctive T–J regression-transgression couplet could reflect the regional doming of the northern hemisphere continents, followed by dome collapse and the widespread establishment of a tensional regime that permitted the sea to flood the interior of the continents (Hallam and Wignall, 1999). Thus, if the marine extinctions were caused by the severe sea-level fall at the end of the Triassic, then the volcanism may be held indirectly responsible.

## The Karoo Igneous Event and the early Toarcian extinction

The early Toarcian witnessed a modest marine extinction event that primarily affected mollusk groups (ammonites, belemnites, and bivalves). Initially detected in northwest European seaways (Hallam, 1961), it has since been identified in South America (Aberhan and Fürsich, 1996) indicating that it is probably a global crisis. The close association between the extinctions and the widespread development of black shales strongly suggests that the proximate cause of extinction was the spread of anoxic bottom waters (Hallam, 1996). Evidence that volcanism may also be implicated has come, once again, from improved dating of a flood basalt province, this time the Karoo volcanics of South Africa (Pálfy and Smith, 2000) (Fig. 7.2). This is an especially large province (Fig. 7.5), and initial dating suggested eruption was around the beginning of the Toarcian (Fitch and Miller, 1984). More recent dating suggest a brief eruption interval around 183 Ma (Duncan et al., 1997; Pálfy and Smith, 2000), a date within the middle of the Toarcian. The correspondence is clearly close to the early Toarcian extinction event, although it implies that it was the onset of flood basalt eruptions, rather than their acme, which triggered the environmental catastrophe. A potential feedback, suggested by Hesselbo et al. (2000), suggests oceanic warming, triggered by volcanic $CO_2$ release, may have caused the release of methane trapped in gas hydrates on the continental slope, a mechanism suggested for other extinction intervals (e.g., Erwin, 1994; Dickens et al., 1995). The consequent exacerbation of the warming trend would have contributed to the global stagnation and anoxia.

## Flood basalts but no extinctions in the Early Cretaceous

A glance at Fig. 7.5 reveals that, whereas there is generally a good correlation between flood basalts and major extinction events, the relationship breaks down in the Early Cretaceous.

There may be several reasons for this. The end-Jurassic (J–K) mass extinction, often known as the Tithonian extinction, has appeared in many compilations of extinctions (Hallam and Wignall, 1997). However, it is probably an artefact of such compilations and not a real mass extinction. The Tithonian Stage is an unusually long stage and consequently it may contain more background extinctions than other, shorter stages. The Tithonian fossil record is also especially good compared to the succeeding Cretaceous stages, a factor which also creates elevated extinction rates (Benton, 1995).

The Lower Cretaceous is marked by the presence of two major flood basalt provinces in the southern hemisphere, the Namibian and Paraná Provinces; two halves of a single province now separated by the South Atlantic (Fig. 7.2). Their eruption does not correspond with elevated extinction rates, despite the fact that the Paraná basalts constitute one of the most voluminous of all volcanic provinces. Peate (1997) has speculated that individual eruptions may have been separated by long quiescent intervals thereby not allowing any cumulative effects to develop. Paraná eruptions occurred during the Valanginian Stage of the Early Cretaceous, an interval of marked global temperature rise (Price et al., 2000). In the well-studied sections of northwest Europe the Valanginian is also marked by the humid depositional style of the Wealden facies, which are in stark contrast to the more arid depositional styles seen both before and after this interval (Ruffell and Rawson, 1994). As with the eruption of other large igneous provinces, it is tempting to see the hallmarks of volcanic $CO_2$-induced warming in these climatic events, although the impact on the world's biota appears negligible.

## Cenomanian–Turonian extinctions and oceanic plateaus

The Cenomanian–Turonian (C–T) is a modest extinction event comparable in magnitude and cause to the early Toarcian event. Marine mollusks were the principal victims (along with a few species of planktonic foraminifera), and their demise is related to the spread of anoxic waters and black shale development once again (Hallam and Wignall, 1997). Unlike the early Toarcian event, there is no associated continental flood basalt province but intraoceanic volcanism, responsible for the Caribbean–Columbian oceanic plateaus and probably part of the Ontong–Java oceanic plateau, dates approximately from this interval (Kerr, 1998). Due to their inaccessibility it is difficult to quantify the volume of extrusives associated with these plateaus, but it may be as high as 20 million $km^3$. Global warming has been postulated as the main consequence of this voluminous volcanism, with the consequent changes in oceanic circulation leading to the marine extinction (Kerr, 1998).

## Brito-Arctic flood basalts and the end-Paleocene extinction

Although minor, perhaps the single, best causal link between volcanism and extinction can be made for the late Paleocene crisis (Coffin and Eldholm, 1993; Sutherland, 1994). Deep-sea benthic foraminifera were the principal victims of this event (Kaiho, 1991), and their demise has been attributed to global warming and the replacement of cold, oxygenated deep-water masses with warmer, more oxygen-poor water (Dickens et al., 1995).

The onset of global warming coincides closely with the rapid eruption of the vast volumes of flood basalt during the break-up of the North Atlantic (Coffin and Eldholm, 1993; Eldholm and Thomas, 1993; Sutherland, 1994). These now constitute the Brito-Arctic Province (also known as the North Atlantic Igneous Province) which covers large areas of northern Britain and southeastern Greenland (Fig. 7.2). Even greater volumes of flood basalt are buried beneath younger sediments of the northern European continental margin and it is estimated that, when these obscured lavas are included, there may be as much as $2 \times 10^6$ $km^3$ of flood basalts in the province (Saunders et al., 1997). Not all this lava was necessarily erupted in a single, late Paleocene pulse – Saunders et al. (1997) suggest that there was a minor, precursor phase of volcanism in the earliest Palaeocene – but it is clear that the Brito-Arctic Province is one of the greatest volcanic episodes in the Earth's history. Recent work in the Inner Hebrides of Scotland has revealed that many so-called red soil horizons identified in previous studies are actually tuffs (Emeleus et al., 1996). Thus, the eruption of the province may have been both more geologically rapid and explosive than previously appreciated, with little time available between eruptions for weathering and soil formation. As noted above, the eruptions coincide with a well-documented, global warming event suggesting that the release of volcanic $CO_2$ may have been responsible for global environmental change. The effects of this warming may have included the dissolution of gas hydrates in the sediments of the world's oceans (Dickens et al., 1995). This icy substance is only stable at low temperatures close to freezing and, on warming, it releases water and methane – a highly effective greenhouse gas. Thus, once triggered by the release of volcanic $CO_2$, a positive-feedback mechanism may act to accelerate the rate of global warming. Such a mechanism has also been proposed to act during other warming events such as that associated with the P–T mass extinction (Erwin, 1994).

The Brito-Arctic eruptions were not the only major volcanic activity occurring in the late Paleocene. Bralower et al. (1997) have documented a rapid increase of explosive volcanism in the Caribbean region at this time, which they suggest may also have contributed to the increase in atmospheric $CO_2$ concentrations.

## Summary

From the foregoing discussion, a number of interesting and intriguing observations can be made about the link between flood basalt eruptions and mass extinctions.

First, there is no direct correlation between the magnitudes of the eruptions and the scale of the extinction crisis (Fig. 7.5). Three of the greatest flood basalt provinces, the Karoo, Paraná, and Brito-Arctic, correlate with either minor extinctions or no extinction at all. There may be several explanations for this. Most obviously, the estimates of basalt volume depicted in Fig. 7.5 may be grossly inaccurate. Thus, the apparently minor volume of the Siberian Traps may reflect substantial erosion of a previously much

more extensive province, whereas estimates of the size of the huge Brito-Arctic Province may inadvertently include younger lavas associated with North Atlantic rifting. Alternatively, other factors, rather than the simple volume of lava, may be more important in generating an extinction crisis. The recurrence times between flows may be important. Thus, it has been suggested that the relatively prolonged eruption history of the Paraná Province could have prevented any cumulative effects from developing. The size of individual flows may also be important; they appear to be especially large in the Deccan Traps for example. However, other large flows, such as those in the Columbia River Province were probably much larger than those in the Siberian Traps and yet the eruption of this North American province is not associated with extinctions (Self et al., 1997). Much more work is needed on the detailed emplacement history of continental flood basalt provinces.

Rather than the flow size and volume, the location of the eruptions and their violence may be important factors. The Siberian Traps were erupted at high northern paleolatitudes, thus facilitating injection of material into the stratosphere, and they contain unusually high proportions of pyroclastic deposits by the standards of flood basalt provinces. This implies that the main environmental damage of flood basalt eruptions may relate to global cooling due to the injection of sulfate aerosols into the stratosphere. However, geological evidence suggests that this implication may be wrong.

A second important observation is that the eruption of flood basalt provinces often coincides with global warming and/or the development of oxygen-poor marine conditions (e.g., Siberian Traps, Central Atlantic Magmatic Province, Karoo Province, Paraná Province, Brito-Arctic Province). This implies (but does not prove) that the release of volcanic $CO_2$ may be the most harmful effect of such eruptions, with the resultant change in oceanic circulation responsible for the development of oxygen-poor oceans. Thus, although the study of modern eruptions has focused attention on the short-term climatic cooling caused by stratospheric sulphate aerosols and dust, the most significant and long-term damage may be caused by global warming (Phase Two of the extinction scenario of Fig. 7.4).

Clearly there is much scope for further work on the link between volcanism and extinction, but the fact that every extinction crisis of the past 300 Ma coincides with the formation of a large igneous province indicates that a cause-and-effect relationship should be sought. Intriguingly, the best correlation comes from the mid-Phanerozoic examples. After the Jurassic the only clear-cut correspondence is between the end-Cretaceous extinction and the Deccan Traps, but in this instance bolide impact provides a more compelling culprit.

## Postscript: plume eruptions and extinctions

This chapter has focused on the link between mass extinctions and flood basalts because only they are considered to be of sufficient magnitude to cause global environmental change. Flood basalt provinces are generally thought to be the record of the initial burst of volcanism associated with plume eruptions. Not all plume eruptions are associated with the formation of flood basalt provinces, but it may be that the volcanic activity associated with other plume eruptions may be sufficiently violent to trigger an environmental crisis (Larson, 1991). Such a connection has been made with the Frasnian–Famennian mass extinction of the late Devonian – one of the big five mass extinction events (Larson, 1991; Garzanti, 1993; Becker and House, 1994). Until recently the location of this plume was a matter of conjecture, and Becker and House (1994) speculated that it may have been in an intraoceanic site. However, Wilson and Lyashkevich (1996) have identified a major phase of Frasnian–Famennian volcanism and rifting in the Pripyat–Dnieper–Donets rift system of the Ukraine which they attribute to the impact of multiple plumes on the base of the lithosphere around this time. The volume of magma extruded was modest, by flood basalt standards (5400 km$^3$), but the bulk of it (70–90%) was the product of explosive volcanism (Wilson and Lyashkevich, 1996). Racki (1998) has postu-lated that this volcanic activity in the Ukraine may be the local expression of a global phase of plate reorganization and volcanism that is somehow linked to this late Devonian faunal crisis. Once again a close, but not perfect, temporal link between volcanicity and extinction has been established, but the influence of such activity on the ocean–climate system is far from established.

## References

Aberhan, M. and Fürsich, F. T. 1996. Diversity analysis of Lower Jurassic bivalves of the Andean Basin and the Pliensbachian-Toarcian mass extinction. *Lethaia*, **29**, 181–195.

Ali, J. R., Thompson, G. M., Song X.-Y., *et al.* 2002. Emeishan Basalts (SW China) and the "end-Guadalupian" crisis: magnetobiostratigraphic constraints. *Journal of the Geological Society of London*, **159**, 21–29.

Alvarez, L. W., Alvarez, W., Asaro, F., *et al.* 1980. Extraterrestrial cause for the Cretaceous–Tertiary extinction: experimental results and theoretical interpretation. *Science*, **208**, 1095–1108.

Alvarez, W., Smit, J., Lowrie, W., *et al.* 1992. Proximal impact deposits at the Cretaceous–Tertiary boundary in the Gulf of Mexico: a restudy of DSDP Leg 77 Sites 536 and 540. *Geology*, **20**, 697–700.

Archibald, J. D. 1996. *Dinosaur Extinction and the End of an Era: What the Fossils Say*. New York, Columbia University Press.

Baksi, A. and Farrar, E. 1991. $^{40}$Ar/$^{39}$Ar dating of the Siberian Traps, USSR: evaluation of the ages of the two major extinction events relative to episodes of flood-basalt volcanism in the USSR and the Deccan Traps, India. *Geology*, **19**, 461–464.

Basu, A. R., Poreda, R. J., Renne, P. R., *et al.* 1995. High He-3 plume origin and temporal-spatial evolution of the Siberian flood basalts. *Science*, **269**, 822–825.

Becker, R. T. and House, M. R. 1994. Kellwasser events and goniatite successions in the Devonian of the Montagne Noire with comments on possible causations. *Courier Forschungsinstitut Senckenberg*, **16**, 45–77.

Benton, M. J. 1994. Late Triassic to Middle Jurassic extinctions among continental tetrapods: testing the patterns. In N. C. Fraser and H.-D. Sues (eds.) *In the Shadow of the Dinosaurs*. Cambridge, UK, Cambridge University Press, pp. 366–397.

1995. Diversification and extinction in the history of life. *Science*, **268**, 52–58.

Bowring, S. A., Erwin, D. H., Jin, Y., *et al.* 1998. U/Pb zircon geochronology and tempo of the end-Permian mass extinction. *Science*, **280**, 1039–1045.

Bralower, T. J., Thomas, D. J., Zachos, J. C., *et al.* 1997. High resolution records of the late Paleocene thermal maximum and circum-Caribbean volcanism: is there a causal link? *Geology*, **25**, 963–966.

Budyko, M. I. and Pivivariva, Z. I. 1967. The influence of volcanic eruptions on solar radiation incoming to the Earth's surface. *Meteorologiya i Gidrologiya*, **10**, 3–7.

Campbell, I. H. and Griffiths, R. W. 1990. Implications of mantle plume structure for the evolution of flood basalts. *Earth and Planetary Science Letters*, **99**, 79–93.

Campbell, I. H., Czamanske, G. K., Fedorenko, V. A., *et al.* 1992. Synchronism of the Siberian Traps and the Permian–Triassic boundary. *Science*, **258**, 1760–1763.

Claoué-Long, J. C., Zhang, Z., Ma, G., *et al.* 1991. The age of the Permian–Triassic boundary. *Earth and Planetary Science Letters*, **105**, 182–190.

Coffin, M. F. and Eldholm, O. 1993. Scratching the surface: estimating dimensions of large igneous provinces. *Geology*, **21**, 515–518.

Conaghan, P. J., Shaw, S. E., and Veevers, J. J. 1994. Sedimentary evidence of the Permian/Triassic global crisis induced by the Siberian hotspot. *Memoir of the Canadian Society of Petroleum Geology*, **17**, 785–795.

Courtillot, V. E. 1990. What caused the mass extinction? A volcanic eruption. *Scientific American*, **2634**, 53–60.

1999. *Evolutionary Catastrophes: The Science of Mass Extinction*. Cambridge, UK, Cambridge University Press.

Courtillot, V. E., Besse, J., Vandamme, D., *et al.* 1988. Deccan flood basalts and the Cretaceous/Tertiary boundary. *Nature*, **333**, 843–846.

Cox, K. G. 1988. Gradual volcanic catastrophes? *Nature*, **333**, 802.

Deckart, K., Féraud, G., and Bertrand, H. 1997. Age of Jurassic continental tholeiites of French Guyana, Surinam and Guinea: implications for the initial opening of the Central Atlantic Ocean. *Earth and Planetary Science Letters*, **150**, 205–220.

Devine, J. D., Sigurdsson, H., and Davis, A. N. 1984. Major eruptions cause short-term global cooling because of sulphate aerosols. *Journal of Geophysical Research*, **89**, 6309–6325.

Dickens, G. R., O'Neill, J. R., Rea, D. K., *et al.* 1995. Dissolution of oceanic methane hydrate as a cause of the carbon isotope excursion at the end of the Paleocene. *Paleoceanography*, **10**, 965.

Duncan, R. A., Hooper, P. R., Rehacek, J., *et al.* 1997. The timing and duration of the Karoo igneous event, southern Gondwana. *Journal of Geophysical Research*, **102**, 18127–18138.

Duncan, R. A. and Pyle, D. G. 1988. Rapid eruption of the Deccan flood basalts at the Cretaceous/Tertiary boundary. *Nature*, **333**, 841–843.

Eldholm, O. and Thomas, E. 1993. Environmental impact of volcanic margin formation. *Earth and Planetary Science Letters*, **117**, 319–329.

Emeleus, C. H., Allwright, E. A., Kerr, A. C., *et al.* 1996. Red tuffs in the Palaeocene lava succession of the Inner Hebrides. *Scottish Journal of Geology*, **32**, 83–89.

Ernst, R. E. and Buchan, H. L. 1997. Giant radiating dike swarms: their use in identifying pre-Mesozoic large igneous provinces and mantle plumes. In J. J. Mahoney and M. F. Coffin (eds.) *Large Igneous Provinces*, Monograph no. 100. Washington, DC, American Geophysical Union, pp. 297–334.

Erwin, D. H. 1994. The Permo-Triassic extinction. *Nature*, **367**, 231–236.

Faure, K., de Wit, M. J., and Willis, J. P. 1995. Late Permian global coal hiatus linked to $^{13}$C-depleted $CO_2$ flux into the atmosphere during the final consolidation of Pangea. *Geology*, **23**, 507–510.

Fitch, F. J. and Miller, J. A. 1984. Dating Karoo igneous rocks by the conventional K–Ar and $^{40}$Ar/$^{39}$Ar age spectrum methods. *Geological Society of South Africa Special Publication*, **13**, 247–266.

Garzanti, E. 1993. Himalayan ironstones, "superplumes," and the breakup of Gondwana. *Geology*, **21**, 105–108.

Genin, A., Lazar, B., and Brenner, S. 1995. Vertical mixing and coral death in the Red Sea following the eruption of Mount Pinatubo. *Nature*, **377**, 507–510.

Glen, W. (ed.) 1994. *The Mass-Extinction Debates: How Science Works in a Crisis*. Stanford, CA, Stanford University Press.

Hallam, A. 1961. Cyclothers, transgressions and faunal change in the Lias of north west Europe. *Transactions of the Edinburgh Geological Society*, **18**, 132–174.

1987. End-Cretaceous mass extinction event: argument for terrestrial conditions. *Science*, **238**, 1237–1242.

1990. The end-Triassic mass extinction event. *Geological Society of America Special Paper*, **247**, 577–583.

1996. Major bio-events in the Triassic and Jurassic. In O. H. Walliser (ed.) *Global Events and Event Stratigraphy*. Berlin, Springer-Verlag, pp. 265–284.

1997. Estimates of the amount and rate of sea-level change across the Rhaetian–Hettangian and Pliensbachian–Toarcian boundaries (latest Triassic to early Jurassic). *Journal of the Geological Society of London*, **154**, 773–779.

Hallam, A. and Wignall, P. B. 1997. *Mass Extinctions and their Aftermath*. Oxford, UK, Oxford University Press.

1999. Mass extinctions and sea-level changes. *Earth-Science Reviews*, **48**, 217–250.

Hesselbo, S. P., Gröcke, D. R., Jenkyns, H. C., *et al.* 2000. Massive dissociation of gas hydrate during a Jurassic oceanic anoxic event. *Nature*, **406**, 392–395.

Holser, W. T., Schönlaub, H. P., Boeckelmann, K., *et al.* 1991. The Permian–Triassic of the Gartnerkofel-1 Core (Carnic Alps, Austria): synthesis and conclusions. *Abhandlungen der Geologischen Bundesanstalt*, **45**, 213–232.

Isozaki, Y. 1994. Superanoxia across the Permo-Triassic boundary: recorded in accreted deep-sea pelagic chert in Japan. *Memoir of the Canadian Society of Petroleum Geologists*, **17**, 805–812.

Jaeger, J.-J., Courtillot, V., and Tapponier, P. 1989. Paleontological view of the ages of the Deccan Traps, the Cretaceous/Tertiary boundary, and the India–Asia collision. *Geology*, **17**, 316–319.

Jaiprakash, B. C., Singh, J., and Raju, D. S. N. 1993. Foraminiferal events across the K/T boundary and age of Deccan volcanism in Palakollu area, Krishna–Godavari basin, India. *Journal of the Geological Society of India*, **41**, 105–117.

Jin, Y. and Shang, J. 2000. The Permian of China and its interregional correlation. *Developments in Palaeontology and Stratigraphy*, **18**, 71–98.

Jin, Y., Zhang, J., and Shang, Q. 1994. Two phases of the end-Permian mass extinction, *Canadian Society of Petroleum Geologists Memoir*, **17**, 813–822.

Jones, A. P., Price, G. D., Price, N. J., *et al.* 2002. Impact induced melting and the development of large igneous provinces. *Earth and Planetary Science Letters*, **202**, 551–561.

Kaiho, K. 1991. Global changes of Paleogene aerobic/anaerobic benthic foraminifera and deep-sea circulation. *Paleogeography, Paleoclimatology, Paleoecology*, **83**, 65–85.

Kamo, S. L., Czamanske, G. K., and Krogh, T. E. 1996. A minimum U–Pb age for Siberian flood basalt volcanism. *Geochimica Cosmochimica Acta*, **60**, 3505–3511.

Keith, M. L. 1982. Violent volcanism, stagnant oceans and some inferences regarding petroleum, strata-bound ores and mass extinctions. *Geochimica Cosmochimica Acta*, **46**, 2621–2637.

Keller, G. 1996. The Cretaceous–Tertiary mass extinction in planktonic foraminifera: biotic constraints for catastrophe theories. In N. Macleaod and G. Keller (eds.) *Cretaceous–Tertiary Mass Extinctions*. New York, W. W. Norton, pp. 49–84.

Kent, D. V. 1981. Asteroid extinction hypothesis. *Science*, **211**, 648–650.

Kerr, A. C. 1998. Oceanic plateau formation: a cause of mass extinction and black shale deposition around the Cenomanian–Turonian boundary. *Journal of the Geological Society of London*, **155**, 619–626.

Larson, R. L. 1991. Geological consequences of superplumes. *Geology*, **19**, 963–966.

Loper, D. E., McCartney, K., and Buznya, G. 1988. A model of correlated episodicity in magnetic-field reversals, climate, and mass extinctions. *Journal of Geology*, **96**, 1–15.

Marzoli, A., Renne, P. R., Piccirillo, E. M., *et al.* 1999. Extensive 200-million-year-old continental flood basalts of the central Atlantic magmatic province. *Science*, **284**, 616–618.

McElwain, J. C., Beerling, D. J., and Woodward, F. I. 1999. Fossil plants and global warming at the Triassic–Jurassic boundary. *Science*, **285**, 1386–1390.

McLaren, D. J. 1970. Time, life, and boundaries. *Journal of Paleontology*, **44**, 801–815.

Mclean, D. M. 1985. Deccan Traps mantle degassing in the terminal Cretaceous marine extinctions. *Cretaceous Research*, **6**, 235–259.

Officer, C. B. 1993. Victims of volcanoes. *New Scientist*, February 20, 34–38.

Officer, C. B. and Drake, C. L. 1983. The Cretaceous–Tertiary transition. *Science*, **219**, 1383–1390.

Officer, C. B., Hallam, A., Drake, C. L., *et al.* 1987. Late Cretaceous and paroxysmal Cretaceous/Tertiary extinctions. *Nature*, **326**, 143–149.

Olsen, P. E., Fowell, S. J., and Corent, B. 1990. The Triassic/Jurassic boundary in continental rocks of eastern North America: a progress report. *Geological Society of America Special Paper*, **247**, 585–594.

Olsson, R. K. and Liu, C. 1993. Controversies on the placement of the Cretaceous–Palaeogene boundary and the K/P mass extinction of planktonic foraminifera. *Palaios*, **8**, 127–139.

Pálfy, J. and Smith, P. L. 2000. Synchrony between Early Jurassic extinction, oceanic anoxic event, and the Karoo–Ferrar flood basalt volcanism. *Geology*, **28**, 747–750.

Peate, D. W. 1997. The Paraná–Etendeka Province. In J. J. Mahoney and M. F. Coffin (eds.) *Large Igneous Provinces*, Monograph no. 100. Washington, DC, American Geophysical Union, pp. 217–245.

Pope, K. O., Baines, K. H., Ocampo, A. C., *et al.* 1994. Impact winter and the Cretaceous/Tertiary extinctions: results of a Chicxulub asteroid impact model. *Earth and Planetary Science Letters*, **128**, 719–725.

Prasad, G. V. R. and Khajuria, C. K. 1995. Implications of the infra- and inter-trappean biota from the Deccan, India, for the role of volcanism in Cretaceous–Tertiary boundary extinctions. *Journal of the Geological Society of London*, **152**, 289–296.

Price, G. D., Ruffell, A. H., Jones, C. E., *et al.* 2000. Isotopic evidence for temperature variation during the early Cretaceous (Late Ryazanian – mid-Hauterivian). *Journal of the Geological Society of London*, **157**, 335–343.

Racki, G. 1998. Frasnian–Famennian biotic crisis: undervalued tectonic control? *Paleogeography, Paleoclimatology, Paleoecology*, **141**, 171–198.

Rampino, M. R. 1987. Impact cratering and flood basalt volcanism. *Nature*, **327**, 468.

Rampino, M. R. and Stothers, R. B. 1988. Flood basalt volcanism during the past 250 million years. *Science*, **241**, 663–668.

Raup, D. M. and Sepkoski, J. J., Jr. 1984. Periodicity of extinctions in the geological past. *Proceedings of the National Academy of Sciences of the USA*, **81**, 801–805.

Reichow, M. K., Saunders, A. D., White, R. V., *et al.* 2002. $^{40}Ar/^{39}Ar$ dates from the West Siberian Basin: Siberian flood basalt province doubled. *Science*, **296**, 1846–1849.

Renne, P. R. and Basu, A. R. 1991. Rapid eruption of the Siberian Traps flood basalts at the Permo-Triassic boundary. *Science*, **253**, 176–179.

Renne, P. R., Zhang, Z., Richards, M. A., *et al.* 1995. Synchrony and causal relations between Permian–Triassic boundary crises and Siberian flood volcanism. *Science*, **269**, 1413–1416.

Retallack, G. J. 1995. Permian–Triassic life crisis on land. *Science*, **267**, 77–80.

1997. Palaeosols in the Upper Narrabeen Group of New South Wales as evidence of Early Triassic palaeoenvironments without exact modern analogues. *Australian Journal of Earth Sciences*, **44**, 185–201.

Ruffell, A. H. and Rawson, P. F. 1994. Palaeoclimate control on sequence stratigraphic patterns in the late Jurassic to mid-Cretaceous, with a case study from Eastern England. *Paleogeography, Paleoclimatology, Paleoecology*, **110**, 43–54.

Saunders, A. D., Fitton, J. G., Kerr, A. C., *et al.* 1997. The North Atlantic Igneous Province. In J. J. Mahoney and M. F. Coffin (eds.) *Large Igneous Provinces*, Monograph no. 100. Washington, DC, American Geophysical Union, pp. 45–93.

Self, S., Thordarson, T., and Keszthelyi, L. 1997. Emplacement of continental flood basalt lava flows. In J. J. Mahoney and M. F. Coffin (eds.) *Large Igneous Provinces*, Monograph no. 100. Washington, DC, American Geophysical Union, pp. 381–410.

Sharma, M. 1997. Siberian Traps. In J. J. Mahoney and M. F. Coffin (eds.) *Large Igneous Provinces*, Monograph no. 100. Washington, DC, American Geophysical Union, pp. 273–295.

Sheehan, P. M., Fastovsky, D. E., Hoffman, R. G., *et al.* 1991. Sudden extinction of the dinosaurs: latest Cretaceous, upper Great Plains, USA. *Science*, **254**, 835–839.

Sigurdsson, H. 1982. Volcanic pollution and climate: the 1783 Laki eruption. *Eos*, **63**, 601–602.

1990. Evidence for volcanic loading of the atmosphere and climatic response. *Paleogeography, Paleoclimatology, Paleoecology*, **89**, 277–289.

Smit, J. 1982. Extinction and evolution of planktonic foraminifera after a major impact at the Cretaceous/Tertiary boundary. *Geological Society of America Special Paper*, **190**, 329–352.

Stanley, S. M. and Yang, X. 1994. A double mass extinction at the end of the Paleozoic era. *Science*, **266**, 1340–1344.

Stothers, R. B. 1993. Flood basalts and extinction events. *Geophysics Research Letters*, **20**, 1399–1402.

Stothers, R. B., Wolff, J. A., Self, S., *et al.* 1986. Basaltic fissure eruptions, plume heights, and atmospheric aerosols. *Geophysics Research Letters*, **13**, 725–728.

Sutherland, F. L. 1994. Volcanism around K/T boundary time: its role in an impact scenario for the K/T extinction events. *Earth Science Reviews*, **36**, 1–26.

Thordarson, T., Self, S., Oskarsson, N., *et al.* 1996. Sulfur, chlorine, and fluorine degassing and atmospheric loading by the 1783–1784 AD Laki (Skaftár Fires) eruption in Iceland. *Bulletin of Volcanology*, **58**, 205–225.

Veevers, J. J., Conaghan, P. J., and Shaw, S. E. 1994. Turning point in Pangean environmental history at the Permian/Triassic (P/Tr) boundary. *Geological Society of America Special Paper*, **288**, 187–196.

Venkatesan, T. R., Kumar, A., Gopalan, K., *et al.* 1997. $^{40}$Ar–$^{39}$Ar age of Siberian basaltic volcanism. *Chemical Geology*, **138**, 303–310.

Vianey-Liaud, M., Jain, S. L., and Sahni, A. 1987. Dinosaur egg shells (Saurischia) from the Late Cretaceous Intertrappean and Lameta Formations (Deccan, India). *Journal of Vertebrate Paleontology*, **7**, 408–424.

Weems, R. E. and Olsen, P. E. 1997. Synthesis and revision of groups within the Newark Supergroup, eastern North America. *Geological Society of America Bulletin*, **109**, 195–209.

Westphal, M., Gurevitch, E. L., Sansanov, B. V., *et al.* 1998. Magnetostratigraphy of the lower Triassic volcanics from deep drill SG6 in western Siberia: evidence for long-lasting Permo-Triassic volcanic activity. *Geophysical Journal International*, **134**, 254–266.

White, R. S. 1989. Igneous outbursts and mass extinctions. *Eos*, 1480–1482.

Widdowson, M., Walsh, J. N., and Subbarao, K. V. 1997. The geochemistry of Indian bole horizons: palaeoenvrionmental implications of Deccan intravolcanic surfaces. *Geological Society of London Special Publication*, **120**, 269–281.

Wignall, P. B. 2001 Large igneous provinces and mass extinctions. *Earth-Science Reviews*, **53**, 1–33.

Wignall, P. B. and Twitchett, R. J. 1996. Oceanic anoxia and the End Permian mass extinction. *Science*, **272**, 1155–1158.

Wilson, M. and Lyashkevich, Z. M. 1996. Magmatism and the geodynamics of rifting of the Pripyat–Dnieper–Donets rift, East European Platform. *Tectonophysics*, **268**, 65–81.

Wolbach, W. S., Gilmour, I., and Anders, E. 1990. Major wildfires at the Cretaceous/Tertiary boundary. *Geological Society of America Special Paper*, **247**, 391–400.

Woods, A. W. 1993. A model of plumes above basaltic fissure eruptions. *Geophysical Research Letters*, **20**, 1115–1118.

Yin, H., Huang, S., Zhang, K., *et al.* 1992. The effects of volcanism on the Permo-Triassic mass extinction in South China. In W. C. Sweet *et al.* (eds.) *Permo-Triassic Events in the Eastern Tethys.* Cambridge, UK, Cambridge University Press, pp. 146–157.

Zolotukhin, V. V. and Al'mukhamevdov, A. I. 1988. Traps of the Siberian platform. In J. D. Macdougall (ed.) *Continental Flood Basalts.* Amsterdam, Kluwer, pp. 273–310.

# Chapter 8

# Effects of modern volcanic eruptions on vegetation

Virginia H. Dale, Johanna Delgado-Acevedo, and James MacMahon

## Introduction

In any one year, approximately 60 volcanoes erupt on the Earth. Even though about 80% of these eruptions occur under the oceans, the terrestrial volcanic events are common enough to have major impacts on nearby vegetation, often over large areas (e.g., Bilderback, 1987). Volcanic activity both destroys or modifies existing vegetation and creates new geological substrates upon which vegetation can re-establish. The types of plants surviving and recovering after volcanic activity largely depend upon the type of activity that takes place, the nutrient content of material ejected or moved by the volcano, the distance from the volcanic activity, and the types of vegetation propagules that survive in place or are transported from adjacent areas. The resulting changes in the vegetation abundance and patterning can have dramatic effects on the social and economic conditions of the humans in the areas surrounding volcanoes.

## Impacts of volcanoes on existing flora

### Physical impacts

#### Primary impacts

The primary impacts of volcanic activity on vegetation correlate to the specific type of volcanic activity (Table 8.2). In associating impacts with types of volcanic activity, we refer to the many studies on vegetation survival and reestablishment that have been conducted on volcanoes (Table 8.1). We divide the volcanic activities into six categories: lava formation, pyroclastic flows, debris avalanches, mudflows, tephra and ash depositions, and blowdowns. Over the time course of an eruptive cycle, most volcanoes produce more than one type of volcanic activity and can have complex effects upon the vegetation (e.g., Lawrence and Ripple, 2000; Dale et al., 2005).

*Lava formations* occur when molten rock comes to the surface, moves varying distances, and cools, forming solid rock. The behavior of a flow depends largely on the viscosity of the lava, which is affected by its silica content, temperature, and the amounts of dissolved gases and solids it contains. The extrusion of new lava mechanically destroys all vegetation in its path. Lava flows typically follow topography, often forming isolated patches of surviving vegetation. The flows can extend many kilometers from the main volcano. The heat and gases associated with lava can also have deleterious effects upon the surrounding vegetation. A very abrupt transition is often observed between healthy, surviving vegetation and the newly formed black rock. Frequently, fires burn vegetation in strips along the perimeters of lava flows (Smathers and Mueller-Dombois,

The submitted manuscript has been authored by a contractor of the U.S. Government under contract No. DE-AC05-96OR22464. Accordingly the U.S. Government retains a nonexclusive, royalty-free license to publish or reproduce the published form of this contribution, or allow others to do so, for U.S. Government purposes.

*Volcanoes and the Environment*, eds. J. Martí and G. G. J. Ernst. Published by Cambridge University Press. © Cambridge University Press 2005.

**Table 8.1** | Impact of volcanic events from a vegetation floristic perspective

| Type of volcanic disturbance | Characteristics of disturbance | | | | Impact on vegetation | | |
|---|---|---|---|---|---|---|---|
| | Areal extent affected[a] | Intensity[b] (removal or burial of vegetation and propagules) | Duration of effect on vegetation | Direct effect on vegetation | Trajectory of vegetation recovery compared to that typical for the area | Ultimate vegetation as compared to the climax vegetation expected for region in the absence of disturbance |
| Lava | Small – medium | High | Centuries | Buries or burns | May alter because no residuals occur | May be different because primary succession occurs |
| Pyroclastic flows | Small | High | Decades to centuries | Buries | May alter because no residuals occur | May be different because primary succession occurs |
| Debris avalanches | Medium | Moderate – high | Decades to centuries | Buries | May alter if few residuals | May have alternative stable states |
| Mudflows | Medium | Low – moderate | Years | Flows over and may bury herbs and shrubs | Does not alter | Does not alter vegetation (because trajectory does not change) |
| Tephra and ash deposition | Large | Low – high | Years to decades | Buries herbs and shrubs | Does not alter | Does not alter vegetation (because trajectory does not change) |
| Blowdowns | Medium | Medium | Years to decades | Blows down tall vegetation, residuals may be abundant | Does not alter | Does not alter vegetation (because trajectory does not change) |

[a] A small area affected refers to the site immediately in the vicinity of the volcano. Medium effects impact sites adjacent to the volcano, and large effects encompass quite a large area (e.g., some tephra is transported around the world).

[b] High-intensity impacts kill most or all of the vegetation in the vicinity. Moderate-intensity events kill or damage many of the plants, but there is some recovery. Low-intensity events only damage the vegetation, and most plants recover.

**Table 8.2** | Summary of studies of effects of volcanic activity on vegetation

| Types of physical impacts | Volcano | Location | Dates of eruption | Reference |
|---|---|---|---|---|
| Lava | Mt. Wellington | New Zealand | 9000 yrs ago | Newnham and Lowe (1991) |
| | Mt. Fuji | Japan | 1000 | Hirose and Tateno (1984); Ohsawa (1984); Masuzawa (1985); Nakamura (1985) |
| | Rangitoto | New Zealand | 1300, 1500, 1800 | Clarkson (1990) |
| | Mt. Ngauruhoe and Mt. Tongariro | New Zealand | 1550+ | Clarkson (1990) |
| | Snake River Plains | Idaho, USA | ~1720 | Eggler (1971) |
| | Jorullo | Mexico | 1759 | Eggler (1959) |
| | Ksudah | Kamchatka, Russia | 1907 | Grishin et al. (1996) |
| | Waiowa | New Guinea | 1943 | Taylor (1957) |
| | Kilauea Iki and Mauna Loa | Hawaii, USA | 1959 | Fosberg (1959); Smathers and Mueller-Dombois (1974); Kitayama et al. (1995); Aplet et al. (1998); Baruch and Goldstein (1999); Huebert et al. (1999) |
| | Surtsey | Iceland | 1963 | Fridriksson (1987); Fridriksson and Magnusson (1992) |
| | Isla Fernandina | Galapagos, Equador | 1968 | Hendrix (1981) |
| | Hudson | Argentina | 1991 | Inbar et al. (1995) |
| | Krakatau | Indonesia | 1992, 1993 | Whittaker et al. (1989, 1992); Partomihardjo et al. (1992); Thorton (1996); Whittaker et al. (1998) |
| Pyroclastic flows | Vesuvius | Italy | 79 | Mazzoleni and Ricciardi (1993) |
| | Kilauea Iki | Hawaii, USA | 1750, 1840, 1955 | Atkinson (1970) |
| | El Paracutin | Mexico | 1943 | Eggler (1948, 1959, 1963); Rejmanek et al. (1982) |
| | Mount St. Helens | Washington, USA | 1980 | Del Moral and Wood (1988a, 1988b, 1993b); Wood and del Moral (1988); Wood and Morris (1990); Halvorson et al. (1991, 1992); Chapin (1995); del Moral et al. (1995); Halvorson and Smith (1995); Tsuyuzaki and del Moral (1995); Tsuyuzaki and Titus (1996); Tsuyuzaki et al. (1997); Titus and del Moral (1998a, 1998b); Tu et al. (1998) |
| Avalanches | Mt. Taranaki | New Zealand | 1550 | Clarkson (1990) |
| | Ksudach | Kamchatka, Russia | 1907 | Grishin (1994); Grishin et al. (1996) |
| | Mt. Katmai | Alaska, USA | 1912 | Griggs (1918a, 1918b, 1919, 1933) |
| | Mount St. Helens | Washington, USA | 1980 | Adams et al. (1987); Dale (1989, 1991) |
| | Ontake | Japan | 1984 | Nakashizuka et al. (1993) |

| Disturbance | Site | Location | Year | References |
| --- | --- | --- | --- | --- |
| Mudflows | Krakatau | Indonesia | 1883 | Tagawa et al. (1985) |
| | Mt. Lassen | California, USA | 1914–15 | Heath (1967) |
| | Mt. Rainier | Washington, USA | 1947 | Frenzen et al. (1988) |
| | Mt. Lamington | New Guinea | 1951 | Taylor (1957) |
| | Mount St. Helens | Washington, USA | 1980 | Halpern and Harmon (1983) |
| | Mt. Pinatubo | Philippines | 1991 | Mizuno and Kimura (1996) |
| Tephra and ash deposition | Auckland Isthmus | New Zealand | ~9500 yrs ago | Newnham and Lowe (1991) |
| | Laacher Volcano | Germany | ~12 900 yrs ago | Schmincke et al. (1999) |
| | M. Mazama | Oregon, USA | ~6000 yrs ago | Horn (1968) |
| | Laguna Miranda | Chile | ~4800 yrs ago | Haberle et al. (2000) |
| | Craters of the Moon | Idaho, USA | ~2200 yrs ago | Eggler (1941); Day and Wright (1989) |
| | Vesuvius | Italy | 79 | Dobran et al. (1994) |
| | Mt. Taranaki | New Zealand | 1655 | Clarkson (1990) |
| | Jorullo | Mexico | 1759 | Eggler (1959) |
| | Mt. Victory | New Guinea | 1870 | Taylor (1957) |
| | Krakatau | Indonesia | ~1880 | Whittaker et al. (1998) |
| | Krakatau | Indonesia | 1883 | Bush et al. (1992); Thorton (1996) |
| | Mt. Tarawera | New Zealand | 1886 | Clarkson and Clarkson (1983); Clarkson (1990) |
| | Soufrière | St. Vincent | 1902 | Beard (1976) |
| | Mt. Katmai | Alaska, USA | 1912 | Griggs (1918a) |
| | Popocatapetl | Mexico | 1920 | Beamon (1962) |
| | Mt. Lamington | New Guinea | 1951 | Taylor (1957) |
| | Kilauea Iki | Hawaii, USA | 1959 | Smathers and Mueller-Dombois (1974) |
| | Isla Fumandina | Galapagos, Ecuador | 1968 | Hendrix (1981) |
| | Mt. Usu | Japan | 1977–1978 | Riviere (1982); Tsuysaki (1987, 1989, 1991, 1996); Tsuyuzaki and del Moral (1994) Haruki and Tsuyuzaki (2001) |
| | Mount St. Helens | Washington, USA | 1980 | Mack (1981); Antos and Zobel (1982, 1984, 1985a, b, c, 1986); del Moral (1983, 1993); Seymour et al. (1983); Hinckley et al. (1984); del Moral and Clampitt (1985); Zobel and Antos (1986, 1991, 1992; Adams et al. (1987); Harris et al. (1987); Wood and del Moral (1987); Chapin and Bliss (1988, 1989); Pfitsch and Bliss (1988); del Moral and Bliss (1993); Foster et al. (1998) |
| Blowdowns | El Chichón | Mexico | 1982 | Burnham (1994) |
| | Hudson | Argentina | 1991 | Inbar et al. (1994) |
| | Lascar Volcano | Chile | 1993 | Risacher and Alonso (2001) |
| | Mt. Lamington | New Guinea | 1951 | Taylor (1957) |
| | Mount St. Helens | Washington, USA | 1980 | Franklin et al. (1985); Halpern et al. (1990) |

1974). These fires ignite the vegetation or built structures in the path of the lava flow, and such fires can spread over large regions. The mechanical breakage of trees by moving lava is also quite destructive, as was observed at Parícutin in Mexico (Inbar *et al.*, 1994). The recent eruptions at Kilauea Iki in Hawaii provide the best current examples of the diverse effects of lava on vegetation (Smathers and Mueller-Dombois, 1974).

Pyroclastic flows occur when extremely hot (often over 700 °C), incandescent material escapes from the volcanic crater. The denser part of the ejected material hugs the ground and follows topography silently, moving with great force and speed (up to 200 km/hr). The flows produce an unsorted and heterogeneous mass that typically destroys all vegetation in its path by either the toxic gases that are released, the heat of the release, or burial. Pyroclastic flows are associated with Plinian eruptions and have been observed, for example, at volcanoes in Mexico, Mount St. Helens in Washington state in the USA, and Vesuvius in Italy. Although pyroclastic eruptions are not as common as lava flows, their impacts on the vegetation are just as immediate and destructive. Nevertheless, there are some differences. The edges of pyroclastic flows are not as abrupt as with lava. Although pyroclastic flows generally follow topography, they have enough force to move uphill and "jump" over ridges. These flows occur close to the volcano and move outward.

Debris avalanches occur with the partial collapse of the volcano. As the volcanic structure fails, the rocks, soil, prior vegetation, glaciers, and any other material on the volcano are moved with great force down the mountain. Debris avalanches are usually not associated with any heat but can bury vegetation with as much as 200 m of material. Recent analyses have shown that as many as 18 debris avalanches have occurred at volcanoes on the Hawaiian islands but had not been detected previously because most of the material flowed into the ocean. Following the debris avalanche at Mt. Katmai in Alaska, Griggs (1918a) found that the depth of the material greatly influenced the survival of vegetation. The largest recorded debris avalanche occurred at Mount St. Helens and resulted in a diverse topography of steep slopes in some places and smooth, flat areas in other locations where mudflows subsequently swept over the debris avalanche. The few surviving plants grew from fragments that were swept down the mountain in the flow (Dale, 1989). Vegetation on the slopes of volcanoes on the Kamchatka Peninsula in Russia have also been buried by debris avalanches (Grishin *et al.*, 2000). After these avalanches, the undulating and steep macrotopography can last centuries into the future. Debris avalanches tend to follow topography, have abrupt edges, and occur close to the mountain.

Mudflows differ from debris avalanches by being composed of more than 50% water. Thus, mudflows usually move as a rather viscous fluid, can flow around existing vegetation, and will not necessarily knock over trees unless the mud is moving rapidly. One hazard, however, is that mudflows gather material in their path, and this material may be destructive. For instance, in the 1980 eruption of Mount St. Helens, the mudflow that moved down the North Fork of the Toutle River encountered a logging camp and picked up the previously cut logs. When this material struck the downstream vegetation and structures, it was extremely destructive and eliminated 17 bridges and most of the forest in its pathway. At some distance from the mountain, many large trees did survive, although all vegetation smaller than 1 m was demolished. Studies of mudflow effects have also been carried out at Mt. Rainier in Washington state, Krakatau in Indonesia, Mt. Lassen in California, and Mt. Lamington in New Guinea. In each case, the area covered by the mudflow depended on the slope and the amount of material that was moved. In many ways, these mudflows are similar to alluvial fan deposits. Mudflows typically follow existing stream channels although they may also create new channels (Fig. 8.1). The edges are not as abrupt as the disturbances discussed above, and there is much survival of vegetation where plants are buried by only a few centimeters of mud.

Tephra refers to the solid, fragmented matter that is ejected by a volcano during eruption. It is typically composed of ash that is less than 2 mm in diameter; lapilli, which are stones 2 to 64 mm across; bombs or rounded stones; and blocks, which are angular stones more than

**Fig. 8.1.** The mudflow down the North Fork of the Toutle River instigated by the 1980 eruption of Mount St. Helens left intact the adjoining forest but inundated the stream channel and adjacent forests.

64 mm across. Tephra is usually ejected during violent eruptions but may be ejected during minor eruptions, as well. The tephra from prehistoric eruptions that occurred in Yellowstone National Park covered most of the USA. The eruption of Mt. Mazama in Oregon nearly 6000 years ago deposited ash layers that can still be seen in roadcuts in Canada more than 600 km from the crater. The 1883 eruption of Krakatau distributed fine ash around the world. Typically, the ash deposits are heaviest closest to the volcano and follow the prevailing wind patterns outward from the mountain. Eruptions in Alaska and Mexico have shown that the effect of tephra on vegetation is related to the depth of the material (Griggs, 1918c; Eggler, 1963). Close to a volcano, the ash deposits may bury the vegetation so deeply that it cannot grow through the ash to the surface. Further away, tephra deposits on leaves may alter photosynthesis and plant growth (Cochran *et al.*, 1983; Bilderback and Carlson, 1987) or influence the morphology of underground plant structures (Zobel and Antos, 1987), and wind-borne tephra may abrade leaves causing abscissions (Black and Mack, 1984).

Blowdowns of existing vegetation by the shock wave and winds that come with an explosive eruption are uncommon. The blowdown that occurred in a perimeter around the north portion of the 1980 eruption of Mount St. Helens is currently the best illustration of a blowdown. The 1980 eruption of Mount St. Helens was a rare lateral blast due to the earthquake-generated landslide off the north flank of the volcano relieving pressure in the mountain and initiating the eruption. The force of the blast from the new vent was greatly influenced by topography, such that trees on the leeward side of the mountain were more protected. Based on blowdown patterns at Mount St. Helens, patterns of prehistoric blowdowns have been related to eruptions at Yellowstone. At Mount St. Helens, the blowdown area was subsequently covered by ash deposits, and sites protected by blown down trees were more likely to have the herbaceous vegetation survive the heavy ash deposits (Franklin *et al.*, 1985) (Fig. 8.2).

### Secondary impacts

Volcanic eruptions produce a number of secondary effects on vegetation including climate changes, soil changes, and increases in carbon dioxide. The best-known climate change caused by volcanic activity followed the 1815 eruption of Tambora. Indeed, 1816 is commonly known as the "Year without a Summer" because the ash that encircled the globe reduced the sunlight penetrating to the Earth. It even snowed in the United States during July of that year. This effect was a short-term climatic change that caused a decline in agricultural production. The eruption of Parícutin in Mexico increased the precipitation in the region (Eggler, 1948). Another volcanic impact on climate resulted from the eruptions of Kilauea Iki in Hawaii, changing typical humid conditions to summer drought. Precipitation in the area decreased from about 2400 to 1300 mm per year (Smathers and Mueller-Dombois, 1974).

Soil changes are probably the most dramatic and long-term impact of volcanic eruptions. The eruptions cause new material to be put in place in the form of lava, debris, mud, or tephra. The material can act as a mulch that reduces competition among other plants. However, the hardness of some material can be a detriment to invasion and establishment of new plants. Lava flows present a most inhospitable habitat for plant colonization, as was noted at Isla Fernandina in Ecuador (Hendrix, 1981). Another example occurred at Wizard Island in central Oregon, USA, where the severe conditions on the exposed lava crag have excluded all plants except crustose lichens (Jackson and Faller, 1973). At the same volcano, however, the narrow interstices within the angular blocks of lava on the small flank flows provide increased opportunities for soil development and plant growth (Fig. 8.3). As a result, forest development has occurred on the latter sites (Jackson and Faller, 1973). The nutrient status and physiological processes of the plants that invade and grow on the volcanic soils influence changes in soil properties (Matson, 1990). The situation experienced at Parícutin in Mexico may be typical of many volcanic eruptions (Eggler, 1948). The lava flows buried much of the old soil. Following the eruption, the new surface varied in texture from dust to boulders, but most of the new material was the size of sand and gravel. In some areas, the ash was so deep that all influence of the old soil was lost. Furthermore, erosion by water can be considerable in steep areas and can cause much loss of material. Also, following extensive eruptions most of the soils are deficient in nutrients, particularly phosphorus and nitrogen, and have low water-holding capacity. This condition was documented at Mt. Ngauruhoe and Mt. Tongariro in New Zealand (Clarkson, 1990). Finally, volcanic disturbances have been known to eliminate mycorrhizae, soil fungi responsible for assimilation of nutrients from soil into plant roots of many species (Allen, 1987; Allen and MacMahon, 1988).

Increases in the concentration of atmospheric gases such as carbon dioxide can result from volcanic eruptions and have an impact on plants in a local area or even worldwide. Because carbon dioxide availability is generally thought to limit plant growth, a local increase in carbon dioxide may actually enhance growth and reproduction of surviving or colonizing plants. Global impacts of increased carbon dioxide can occur if the changes are sufficient to alter global conditions. During earth-forming processes, such activity occurred, but modern eruptions like that of Mt. Pinatubo in the Philippines have also affected global carbon dioxide conditions. On a local level, volcanic activity can increase atmospheric concentration of sulfur dioxide which injures plants (Winner and Mooney, 1980). Worldwide changes in the chemical composition of the atmosphere and of the amount of particulate matter, both consequences of volcanic activity, may have altered past weather, climate, and the welfare of organisms (Webb and Bartlein, 1992).

### Factors affecting survival

Both the event and the vegetation have characteristics that affect the ability of plants

**Fig. 8.2.** Trees blown down by the 1980 eruption of Mount St. Helens harbored some surviving young conifers.

**Fig. 8.3.** A pioneering pine tree on volcanic material at Newberry Caldera in Oregon, USA.

to survive volcanic eruptions. The important characteristics of the event include the temperature of the material, chemical characteristics of the eruption, depth of burial, hardness of the new surface, and water and wind forces. When the eruption is extremely hot, all plants and propagules may be destroyed, or fires may be started. Chemical features of an eruption that affect vegetation most often are associated with gases of pyroclastic flows; however, volcanoes can alter nutrient conditions via the soils formed by eruptive activities. The depth of the material has a clear effect on survival, for plants buried by more than 300 cm cannot grow through to the surface (Griggs, 1918a; Smathers and Mueller-Dombois, 1974), even though the plants may live at least eight seasons under the tephra (Zobel and Antos, 1992, 1997). The depth of burial is influenced not only by the amount of material ejected from the volcano but also by the topographic features, snow conditions, and existing vegetation. For instance, during the eruption of Mount St. Helens, sites that were under snow were less affected by the tephra deposition, because with the melting of the snow, the tephra was not distributed in a uniform layer (Franklin *et al.*, 1985). Furthermore, herbaceous vegetation located under tree canopies received less tephra deposition than herbaceous vegetation in the open air. The hardness of the new surface clearly affects plant survival. In extreme disturbances like lava formations, virtually no buried plants survive (Smathers and Mueller-Dombois, 1974).

The force of the water and wind can have an extreme effect upon the vegetation in the path of a volcanic eruption. The blast of wind coming from the 1980 eruption of Mount St. Helens was strong enough to break off trees more than 40 m tall so that they looked like a pile of "pickup sticks." A distinct edge exists where this effect ceased to occur as the winds dissipated. Water force can also exert influence as it does in mudflows, which can knock over small trees and shrubs, as occurred at Mt. Rainier (Frehner, 1957). Volcanic eruptions can also change the ways in which water or wind relate to the vegetation. In at least one case on the Auckland isthmus in New Zealand, the accumulation of ash prevented drainage of a swamp and thus maintained

habitat for wetland species (Clarkson, 1990). Thus, these physical events can cause much disturbance to existing vegetation or even create the conditions for the establishment of new vegetation associations.

Often, a number of plants survive volcanic eruptions. This survival is highly correlated with distance from the volcanic crater. Because the impact of an eruption is typically most severe close to the crater, more vegetation is killed nearer the origin. Exceptions occur when linear features, such as lava flows or mudflows, move quite a distance from the volcano and destroy or severely damage vegetation in their paths. These flows do not always originate at the crater.

Plants also have characteristics that are important in their survival. Smathers and Mueller-Dombois (1974) found that each type of volcanic disturbance on Kilauea Iki in Hawaii resulted in establishment of a unique sequence of algae, moss, lichens, native woody seed plants, non-native woody seed plants, grasses and sedges, and forbs. Most clearly, the position of the *perennating* organs can be a predictor of survival in the face of different types of disturbances. The perennating organ refers to that part of a plant that survives from year to year to begin the new growth of a plant. For a tree, this organ is the leaf buds, which are normally quite high off the ground. For plants called geophytes, year-to-year survival occurs in the form of a bulb under the ground. Raunkiaer (1934) had associated the position of the perennating organ with climate changes that occur with latitude on the Earth and with elevation on a mountain. Adams, *et al.* (1987) found that the position of the bud also related to the type and intensity of the volcanic impact. For example, trees with their buds in the air usually survive mudflows, and plants with their buds below the ground were the sole survivors of heavy ash deposits and debris avalanches at Mount St. Helens (Fig. 8.1).

Deciduousness can have a positive influence on survival, as well. For example, after the leaves were burned from all trees in the singe perimeter surrounding Mount St. Helens, the deciduous trees survived, for they are adapted to experience leaf loss (Adams *et al.*, 1987). The timing of the 1980 eruption also had an impact, for in May tent caterpillars often defoliate deciduous trees in the Pacific Northwestern USA after which the trees typically releaf.

Plants also have a great deal of flexibility in their size and number of parts. This so-called *plasticity* is a major feature that allows plants to exist in stressful environments, such as those created by volcanic eruptions. For example, following the eruption of Mount St. Helens *Senecio sylvaticus* plants which normally produce many flowers and are typically 50 cm tall only had one flower and were about 10 cm tall (Dale and Adams, 2003). In addition, some perennials (e.g., *Montia parviflora*) completed their life cycle within a year under the stressful conditions on the new substrate and thus behaved like annual plants.

The size of the organism can influence survival. For instance, plants that can project above mudflows and heavy ash deposits are more likely to survive those disturbances. Trees that survived in the splatter area of the lava from the 1968 eruption of Kilauea Iki volcano in Hawaii had relatively large basal diameters of 20 cm or more (Smathers and Muller-Dombois, 1974). An important survival factor associated with plant size was the layer of epiphytic moss surrounding the stem base that protected the larger trees. Also, with heavy pumice deposition, only the larger trees survived the deeper deposit. Smaller-diameter *Metrosideros* spp. trees survived only under the shallower pumice blanket (Smathers and Muller-Dombois, 1974). A survival strategy is exhibited by large plants that exist in the face of extreme stress: they can defer growth until conditions become more favorable. For example, at Mt. Katmai in Alaska, Griggs (1919) found that all plants except trees and bushes had been destroyed. Similarly on the mudflows at Mount St. Helens, trees that rose above the new material tended to survive.

# The recovery process

Plant recovery in the aftermath of any disturbance occurs via two pathways. Plants that survive the event grow and reproduce, or new plant colonists are transported into an area by seeds, which germinate, grow, and reproduce.

Over time, one species will gradually be replaced by another by the process of *succession*, which refers to the non-seasonal, directional, and continuous pattern of colonization and extinction on a site by species populations (Clements, 1916; Drury and Nisbet, 1973; Connell and Slatyer, 1977; Grime, 1977). The early invaders are termed *pioneer* species, and the long-term expected vegetation is called *climax*, although in many cases the climax is never reached because of intervening disturbances. *Primary succession* is initiated on new substrates that lack propagules or surviving plants. The rate of early, primary succession is generally slow, but as soils develop and recruitment shifts from long-distance dispersal to locally produced seeds, recovery accelerates. Primary succession in stressful environments, such as lava beds, may take many centuries to unfold, but significant and complex events can occur within a decade on more hospitable substrates, such as the mudflows of Mt. Rainier and Mount St. Helens. Contrary to ecological theory (Connell and Slatyer, 1977), on volcanic substrates, pioneer and climax species can grow together with only shifts in species dominance marking the successional years (Dale and Adams, 2003). In addition, random plant survival or re-establishment can profoundly affect succession for many years. Once fortuitously established plants set seed, they may overwhelm any effects of dispersal from outside the habitat (del Moral and Wood, 1993a, 1993b). Plant succession after a volcanic eruption may be much slower than recovery after other large disturbances because of the intensity of the impact (Turner *et al.*, 1997, 1998; Turner and Dale, 1998). The recovery of plants following a volcanic eruption is influenced by four major factors: physical and biological legacies, plant recruitment, survival, and time to maturity of the vegetation.

## Influence of legacies
The term *legacy* refers to both the physical and biological elements that are left in place after an eruption (Foster *et al.*, 1998). Physical legacies includes the macro- and microtopography in the aftermath of the eruption and changes in the soil conditions, such as moisture, nutrients, and particle size. Soil water can be limiting, as is the case for the Kula Volcano in

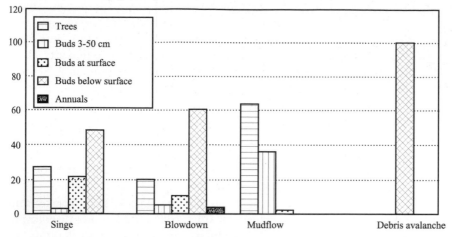

Percent of surviving species by life form

Type of disturbance

**Fig. 8.4.** Percentage of surviving life forms for various disturbances that occurred in the 1980 eruption of Mount St. Helens. Based on data in Adams and Dale (1987).

Turkey, where insufficient soil water slows the growth of shrub species (Oner and Oflas, 1977). At Mt. Ngauruhoe and Mt. Tongariro in New Zealand, the floristic richness is associated with altitude and the proportions of boulders, stones, gravel, and ash present (Clarkson, 1990). Lava flow texture and age as well as precipitation and temperature were found to affect vegetation biomass and species composition on Mauna Loa, Hawaii (Aplet *et al.*, 1998). Furthermore, many of the soils available following the eruptions are deficient in nutrients or have very low water-holding capacity. Levels of soil nitrogen, organic matter, and cation-exchange capacity appear to be correlated with the degree of plant-community development in the four mudflow vegetation types on Mt. Rainier, Washington state, USA (Frenzen *et al.*, 1988). Similar trends in soil organic matter concentrations, nitrogen concentrations, and cation-exchange capacities were noted among the four mudflow communities with values consistently lower than those in the adjacent non-disturbed areas. Because of the low nitrogen-holding capacity of many volcanic soils, plants associated with nitrogen-fixing bacteria are frequent. They were, for example, observed at Volcano Usu in

Japan (Tsuyasaki, 1994), Mt. Katmai in Alaska, USA (Griggs, 1918c), and Mount St. Helens in Washington state, USA (Wood and del Moral, 1988; Dale, 1991). However, in other cases (such as at Jorullo, Mexico) the lava or ash itself contains fixed nitrogen, notably in the form of ammonium chloride (Eggler, 1959). It is not known whether this nutrient could have been part of the rock as it cooled from magma or was concentrated from gases that escaped the fumaroles. Thus, physical legacies influence the vegetation in many different ways just as they influence animals (Edwards, Chapter 9, this volume).

Biological legacies include surviving plants, animals, seeds, spores of microorganisms, and plant or animal organic matter such as needles or pieces of roots that contribute to soil properties. Survival of seeds is largely determined by the depth of the material, since the newly emplaced volcanic material is typically barren. Most studies find few propagules in the new material (e.g., Tsuyuzaki, 1994). Thus, seeds growing on newly created substrates are typically transported from the surrounding areas. At Mount St. Helens, the surviving plants were largely correlated to their life form (Fig. 8.4). In some cases, only a fragment of the plant survived, and a whole plant grew from that fragment (Fig. 8.5). Animals that survive eruptions can have a great influence on the vegetation recovery. For example, pocket gophers that survived in the blowdown area at Mount

**Fig. 8.5.** Only a root fragment of this carex plant survived the mudflow instigated by the 1980 eruption of Mount St. Helens, but shortly after the mudflow these two plants grew from that single root fragment.

St. Helens dug down to original soil and brought viable seeds up to the new surface (Anderson and MacMahon, 1985a, 1985b). However, subsequent to eruption, little food is available, so herbivores can be destructive consumers of the few surviving plants. Nevertheless, the animals add nutrients to the soils through their feces. Survival of mycorrhyzal spores can enhance plant survival and growth.

## Factors affecting plant recruitment

Numerous factors affect plant recruitment. Distance to the seed source is of primary importance because most seeds come from outside the devastated area. Physical processes, such as the movement of wind or water, transport many new recruits and are greatly affected by distance and prevailing wind or water patterns. The dispersal agent transferring seeds is also critical. Even on newly created islands, such as at Krakatau, over time wind has become more important than water in dispersing seeds (Whittaker *et al.*, 1992) (Fig. 8.6). Soil deposition from upland sediments onto volcanic material may provide a source of seeds (Tu *et al.*, 1998). As animals move to a newly colonized site, not only are they important to herbivory and nutrient cycling, but they also serve as agents of dispersal. For example at Surtsey, plants have grown out of bird droppings are well as bird carcasses (Fridriksson, 1975). Animals may be attracted to a volcanic site by water, food, nesting sites, cover, and escape from predation or hunting. Humans can also be important agents for moving seeds into an area either indirectly or through deliberate seeding or transplanting efforts. At both Mount St. Helens and Mt. Usu non-native seeds were deliberately introduced with the intent to control erosion (Tsuyusaki, 1987; Dale, 1989). However, at Mount St. Helens these introduced plants

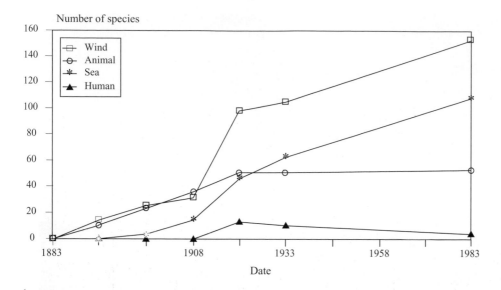

**Fig. 8.6.** Spectra of modes of dispersal of vascular plants persisting on Krakatau. Based on the data in Whittaker *et al.* (1992).

did not effectively reduce erosion (Franklin *et al.*, 1988) and appear to be competing with native vegetation, for there was a decline in native plant diversity and an increase in conifer mortality where non-native plants became established (Dale, 1989).

The physical characteristics of the seeds have an important influence on plant recruitment because dispersal is so critical. Plants with plumed seeds were more likely to have been early colonizers at Surtsey (Fridriksson, 1975). Dale (1989) showed that wind-tunnel studies of the relationship between achene diameter and plume diameter can successfully predict the types of seeds that were recruited onto the Mount St. Helens debris avalanche. Subsequent studies at Mount St. Helens found that seed size increases over time; smaller seeds and seeds with appendages for dispersal were more common in the early years. Plants with heavy seeds tended to become more common over time on volcanic substrates. For newly created volcanic islands (e.g., Surtsey and Krakatau), mechanisms for water dispersal of seeds are important (Fridriksson, 1975; Thorton, 1996).

Finally, the existence of *safe sites* for successful seedling establishment is crucial. These safe sites occur where physical conditions enhance seedlings survival. Many examples exist of successful plant establishment on volcanic sites occurring in the shade or in crevices where moisture and temperature conditions are more favorable (Dale, 1989; Tsuyuzaki *et al.*, 1997; Titus and del Moral, 1998a) (Fig. 8.7). Dale (1989) showed that spider webs are important, not only for trapping seeds but also for trapping moisture that is critical for early seedling survival. At Mt. Katmai in Alaska, the surviving vegetation served as a windbreak in the shelter of which new seedlings established (Griggs, 1919). Plants at Surtsey established in a mixture of tephra and nitrogen-rich decaying seaweed (Fridriksson, 1975). In 1886, a botanist observed a thin gelatinous film of blue–green algae on ash substrate at Krakatau and surmised that this hygroscopic layer provided an appropriate medium for germination and growth of the first ferns to colonize the site (Thorton, 1996). At Plateau Dome in New Zealand, boulders, cracks, and crevices trap windblown materials and plant litter which then retain water for long periods, providing a more favorable microclimate for seeds to germinate and establish (Clarkson and Clarkson, 1983). Gullies on Volcano Usu in Japan created a unique habitat where vegetation survival and recovery differed from areas outside the gullies (Tsuyasaki, 1994). Severe

**Fig. 8.7.** A carex on the debris avalanche created by the 1980 eruption of Mount St. Helens with the mountain and crater seen in the background. This young carex plant was able to colonize and survive in a safe site within the rocks left after the debris avalanche.

erosion in the gullies meant that plants with stoloniferous roots were relatively more common there.

## Factors affecting plant survival

Once the plants have been recruited into an area, they must survive long enough to reproduce in order to leave progeny. But plants that do not survive to maturity still contribute to the ecosystem by providing cover and humus for the few that do. Factors that affect survival of the newly recruited vegetation include herbivory, drought, subsequent eruptions, and human activities. Herbivory can have a major influence on survival. For example, a few sphinx moth caterpillars were observed voraciously eating fireweed, *Epilobium angustifolium*, within a small patch on

a debris avalanche at Mount St. Helens 2 years after the eruption. These few caterpillars quickly consumed the fireweed within that site and eliminated that population. Ungulates can be important herbivores in volcanic areas. Again at Mount St. Helens, the elk population has become abundant, partially because the Mount St. Helens National Monument now serves as a refuge from hunting. These elk selectively influence the presence of trees. For example, the only cherry that has been seen on the debris avalanche had no leaves, and most alders have been nibbled. Plants that are preferred by elks tend not to survive as well (Russell, 1986). Introduced feral goats affected species diversity, composition, and communitiy structure at different altitudes of the Volcan Acedo of the Galápagos Islands (Desender et al., 1999).

Drought can be a major hazard to the survival of plants, particularly because of the sandy soils that are typically found in volcanic areas. In the first years of life, young trees are especially susceptible to droughty conditions. Therefore, pioneer plants are often limited to sites where soil moisture is amenable to growth. At Jorullo, Mexico, where plant survival was high, moisture tended to collect on the rim of the crater (Eggler, 1959).

Soil conditions that affect plant survival include soil texture, pH, and the presence of mycorrhizae and soil nutrients. However, for volcanic sites with limited nutrient availability, mycorrhizae may not significantly affect plant biomass (Titus and del Moral, 1998b, 1998c). Often leguminose species are prevalent on newly formed volcanic substrates because of the paucity of nitrogen and the association of these plants with nitrogen-fixing bacteria. Active lava flows can be a source of nitrogen via cloud interception of volcanic NO (Huebert et al., 1999).

Subsequent eruptions and other disturbances can have a major effect upon survival; they may bury or remove competitors, but they can also remove individual plants that are more susceptible to disturbance at their younger and smaller stages. On the Auckland isthmus, New Zealand, a series of dates on plant remains associated with volcanic material indicate that, at various times, the vegetation was overwhelmed by basaltic lavas or showered with basaltic ash or lapilli (Clarkson, 1990). In the Hekla area of Iceland, most of the trees have a large thickening at the base, indicating both a long life and frequent damage. The few recorded seedlings of birch (Betula pubescens) on the three oldest lava fields all occur in small ruptures. It was particularly interesting to note that they were growing at the edges of the tephra patches but not in deep depressions, fissures, or on the thick moss carpet where willow (Salix spp.) tends to germinate and grow (Bjarnason, 1991). Subsequent eruptions of a satellite volcano at Surtsey resulted in tephra deposits that buried new seedlings (Fridriksson, 1975). Similarly, eruptions of volcanoes near Krakatau have influenced forest stand dynamics on that volcanic island, but storm damage and, in particular, lightning strikes have also caused tree mortality there (Whittaker et al., 1998). On Mount St. Helens, minor subsequent eruptions caused mudflows and tephra depositions that destroyed vegetation.

Human activities, of course, have a major influence on plant survival. Often, humans try to protect themselves from future volcanic impacts and, in the process, may not consider the vegetation recovery. For example, the mass of dredging that occurred following the Mount St. Helens eruption, in order to avoid downstream inundation, buried recovering plants under dredged material. Human activities not only directly affect the vegetation but also can indirectly affect the people who rely on the plants for their food or livelihood. At Parícutin Volcano, agriculture had been the chief means of livelihood for the people, and turpentine was the second largest source of income (Eggler, 1948). After the initial eruption, pine trees were tapped at the lower altitudes around the site of the volcano. This industry was then brought to a standstill by the volcano. With loss of their fields and of the turpentine industry, many people turned to the cutting of dead and dying timber for railroad ties and sawlogs. However, they also cut too many living trees along with the dead ones and thus compromised the future of the forest and their own livelihood of tapping for turpentine (Eggler, 1948). Another common human impact to sites undergoing vegetation recovery after a volcanic eruption is the introduction of non-native species. Studies on the Mauna Loa volcano in Hawaii found that invasive non-native plant species can use resources more efficiently than native species and thus have higher growth rates (Baruch and Goldstein, 1999).

## Factors affecting the time to maturity of the vegetation

When one considers time to recovery to mature vegetation, one must carefully evaluate the definition of mature vegetation. That definition can be either 100% cover of vegetation or the presence of plants that are considered to be climax for the region. The time to 100% recovery of cover is typically much earlier than the time to reach a climax community. In some cases, the climax is seldom attained because of the influence of frequent disturbances in the area. The time to reach 100% cover greatly depends on the soil substrate conditions, presence of legacies, distance to surviving vegetation, and other factors affecting survival and recovery mentioned previously. The variable lifetime of successional species also influences the recovery process. For instance, in the Pacific Northwest of the USA, the early successional tree Douglas fir can live as long as 1000 years, and thus the climax stage may not be reached for more than a millennium (but given the typical fire interval of 400 to 600 years, this would be rare). Volcanic areas are subject to repeated volcanic activity (especially along the Pacific Rim) and thus attainment of the climax vegetation is infrequent.

## Social and economic consequences of vegetation changes resulting from volcanic activity

Vegetation destruction and recovery have a variety of social and economic consequences following a volcanic eruption. These impacts can affect agriculture, forest products, and tourism. In many cases, agriculture is enhanced by tephra deposition because the tephra can serve as a mulch and thus deter competing species. However, for sites close to the mountain, many agricultural areas are buried by lava, pyroclastic flows, mudflows, or tephra deposits, such as occurred at Hudson Volcano in Argentina where the 1991 eruption impacted an area of 1000 km in radius (Inbar et al., 1995), at Mt. Pinatubo in the Philippines, and at many other sites. Thus managers of sites on or near volcanoes should include potential volcanic impacts in their resource management plans (Dale et al., 1998).

The diversity of social and economic impacts related to vegetation after a volcanic eruption can be exemplified by the 1980 eruption of Mount St. Helens. Forest devastation from the eruption resulted in a dramatic loss of forest products. Following the 1980 Mount St. Helens eruption, even though many trees were not buried, they were blown over by the eruptive force. The forest industry was concerned that these trees should be harvested quickly before beetle outbreaks occurred, but later analysis showed that the trees were too dry for beetles to have been attracted to them. Nevertheless, massive salvaging operations were quickly begun to lessen the economic loss resulting from the eruption's devastation of forested

lands. Crops losses following the eruption of Mount St. Helens were much less than expected but still amounted to about US$100 million in 1980 (Cook et al., 1981). Favorable weather conditions subsequent to the eruption may have benefited some crops. Although the ash reduced photosynthesis by as much as 90%, it eventually washed off most of the plants (Cook et al., 1981). However, crop production costs were increased because the ash necessitated machinery repairs and increased tillage.

Tourism can be both increased and decreased by volcanic activity. Some people are attracted to ongoing volcanoes and want to be as close to the mountain as possible. Others avoid tourist attractions until the mountain is declared safe. When plans were being put in place to create the Mount St. Helens National Monument with the idea that this would be a long-term monument to the forces of nature, one Washington Congressman noted that as the vegetation recovers the impact of the volcano may not be as evident and thus the tourist draw may diminish. The Visitors' Center built at Mount St. Helens has become a major attraction in the area and is an economic boon to the region that was severely affected by the 1980 eruption. Tourist sites throughout the world exist where people come to visit the hot springs associated with some volcanic activity (as in New Zealand and Iceland) or to see the impacts on the mountain or to visit "drive-in volcanoes" (as advertised in the Caribbean).

## Summary

Many factors influence survival and recovery of plants after volcanic eruptions. The most clearcut effects upon the vegetation directly correlated to the type and intensity of the eruption and the characteristics of the eruptive material. Vegetation recovery patterns are influenced by plant survival, distance from the mountain, specific substrate types, and the time since the eruption. The characteristics of the vegetation itself and local climate conditions affect rates and patterns of recovery. Thus, based on the type of eruption, the prevailing climate, and the general vegetation in the region, the numerous studies of vegetation survival and re-establishment can be used to make reasonable predictions of the general patterns of recovery likely to occur at any one volcano. Site-specific predictions are more difficult because each location may have undergone a variety of different impacts during the volcanic sequence and have unique initial vegetation, soil conditions, and legacies.

## Acknowledgments

This research was partially sponsored by Minority Biomedical Research Support program, University of Puerto Rico, Cayey Campus and by Oak Ridge Institute of Science and Education and Oak Ridge National Laboratory, Oak Ridge, Tennessee. The National Geographic Society supported some of the field work by Dale at Mount St. Helens. Wendy Adams, John Edwards, Robert O'Neill, and Patricia Parr provided useful reviews of the paper. Roger del Moral suggested many useful references. We appreciate the careful editing of Frederick M. O'Hara and Linda O'Hara. Oak Ridge National Laboratory is managed by UT-Battelle, LLC, for the US Department of Energy under contract DE-AC05-00OR22725. This paper is Environmental Sciences Publication number 4822.

## References

Adams, A. B. and Dale, V. H. 1987. Comparisons of vegetative succession following glacial and volcanic disturbances. In D. E. Bilderback (ed.) *Mount St. Helens 1980: Botanical Consequences of the Explosive Eruptions*. Berkeley, CA, University of California Press, pp. 70–147.

Adams, A. B., Dale, V. H., Kruckeberg, A. R., *et al.* 1987. Plant survival, growth form and regeneration following the May 18, 1980, eruption of Mount St. Helens, Washington. *Northwest Science*, **61**, 160–170.

Allen, M. F. 1987. Re-establishment of mycorrhizae on Mount St. Helens: migration vectors. *Transactions of the British Mycological Society*, **88**, 413–417.

Allen, M. F. and MacMahon, J. A. 1988. Direct VA mycorrhizal inoculation of colonizing plants by pocket gophers (*Thomomys talpoides*) on Mount St. Helens. *Mycologia*, **80**, 754–756.

Anderson, D. C. and MacMahon, J. A. 1985a. Plant succession following the Mount St. Helens volcanic eruption: facilitation by a burrowing rodent, *Thomomys talpoides. American Midland Naturalist*, **114**, 62–69.

1985b. The effects of catastrophic ecosystem disturbance: the residual mammals at Mount St. Helens. *Journal of Mammalogy*, **66**, 581–589.

Antos, J. A. and Zobel, D. B. 1982. Snowpack modification of volcanic tephra effects on forest understory plants near Mount St. Helens. *Ecology*, **63**, 1969–1972.

1984. Ecological implications of belowground morphology on nine coniferous forest herbs. *Botanical Gazette*, **145**, 508–517.

1985a. Plant form, developmental plasticity and survival following burial by volcanic tephra. *Canadian Journal of Botany*, **63**, 2083–2090.

1985b. Upward movement of underground plant parts into deposits of tephra from Mount St. Helens. *Canadian Journal of Botany*, **63**, 2091–2096.

1985c. Recovery of forest understories buried by tephra from Mount St. Helens. *Vegetatio*, **64**, 105–114.

1986. Seedling establishment in forests affected by tephra from Mount St. Helens. *American Journal of Botany*, **73**, 495–499.

Aplet, G. H., Hughes, R. F., and Vitousek, P. M. 1998. Ecosystem development on Hawaiian lava flows: biomass and species composition. *Journal of Vegetation Science*, **9**, 17–26.

Atkinson, I. A. E. 1970. Successional trends in the coastal and lowland forest of Mauna Loa and Kilauea volcanoes, Hawaii. *Pacific Science*, **24**, 387–400.

Baruch, Z. and Goldstein, G. 1999. Leaf construction cost, nutrient concentration, and net $CO_2$ assimilation of native and invasive species in Hawaii. *Oecologia*, **121**, 183–192.

Beard, J. S. 1976. The progress of plant succession on the Soufrière of St. Vincent: observations in 1972. *Vegetatio*, **31**, 69–77.

Bilderback, D. E. (ed.) 1987. *Mount St. Helens 1980: Botanical Consequences of the Explosive Eruptions*. Berkeley, CA, University of California Press.

Bilderback, D. E. and Carlson, C. E. 1987. Effects of persistent volcanic ash on Douglas-fir in Northern Idaho. *US Department of Agriculture Forest Service Intermountain Research Station Research Paper*, **380**, 1–3.

Bjarnason, A. H. 1991. Vegetation on lava fields in the Hekla area, Iceland. *Acta Phytogeographica Suecica*, **77**, 97–104.

Black, R. A. and Mack, R. N. 1984. Aseasonal leaf abscission in *Populus* induced by volcanic ash. *Oecologia*, **46**, 295–299.

Burnham, R. 1994. Plant deposition in modern volcanic environments. *Transactions of the Royal Society Edinburgh, Earth Sciences*, **84**, 275–281.

Bush, M. B., Whittaker, R. J., and Partomihardjo, T. 1992. Forest development on Rakata, Panjang and Sertung: contemporary dynamics (1979–1989). *GeoJournal*, **28**, 185–199.

Chapin, D. M. 1995. Physiological and morphological attributes of two colonizing plant species on Mount St. Helens. *American Midland Naturalist*, **133**, 76–87.

Chapin, D. M. and Bliss, L. C. 1988. Soil–plant water relations of two subalpine herbs from Mount St. Helens. *Canadian Journal of Botany*, **66**, 809–818.

1989. Seedling growth, physiology, and survivorship in a subalpine, volcanic environment. *Ecology*, **70**, 1325–1334.

Clarkson, B. D. 1990. A review of vegetation development following recent (<450 years) volcanic disturbance in North Island, New Zealand. *New Zealand Journal of Ecology*, **14**, 59–71.

Clarkson, B. R. and Clarkson, B. D. 1983. Mt. Tarawera. II. Rates of change in the vegetation and flora of the high domes. *New Zealand Journal of Ecology*, **6**, 107–119.

Clements, F. E. 1916. *Plant Succession: An Analysis of the Development of Vegetation*. Washington, DC, Carnegie Institute.

Cochran, V. L., Bezdicek, D. F., Elliott, L. F., *et al.* 1983. The effect of Mount St. Helens' volcanic ash on plant growth and mineral uptake. *Journal of Environmental Quality*, **12**, 415–418.

Connell, J. H. and Slatyer, R. O. 1977. Mechanisms of succession in natural communities and their role in community stability and organizations. *American Naturalist*, **111**, 1119–1144.

Cook, R. J., Barron, J. C., Papendick, R. I., *et al.* 1981. Impact on agriculture of the Mount St. Helens eruptions. *Science*, **211**, 16–18.

Dale, V. H. 1989. Wind dispersed seeds and plant recovery on Mount St. Helens debris avalanche. *Canadian Journal of Botany*, **67**, 1434–1441.

1991. Revegetation of Mount St. Helens debris avalanche 10 years posteruptive. *National Geographic Research and Exploration*, **7**, 328–341.

Dale, V. H. and Adams, W. M. 2003. Plant reestablishment 15 years after the debris avalanche at Mount St. Helens, Washington. *Science of the Total Environment*, **313**, 101–113.

Dale, V. H., Lugo, A., MacMahon, J., *et al.* 1998. Ecosystem management in the context of large, infrequent disturbances. *Ecosystems*, **1**, 546–557.

Dale, V. H., Swanson, F. J., and Crisafulli, C. M. (eds.) 2005. *Ecology Responses to the 1980 Eruption of Mount St. Helens*. New York, Springer-Verlag.

Day, T. A. and Wright, R. G. 1989. Positive plant spatial association with *Eriogonum ovalifolium* in primary succession on cinder cones: seed trapping nurse plants. *Vegetatio*, **80**, 37–45.

del Moral, R. 1983. Initial recovery of subalpine vegetation on Mount St. Helens. *American Midland Naturalist*, **109**, 72–80.

1993. Mechanisms of primary succession on volcanoes: a view from Mount St. Helens. In J. Miles and D. H. Walton (eds.) *Primary Succession on Land*. London, Blackwell Scientific Publications, pp. 79–100.

1998. Early succession on lahars spawned by Mount St. Helens. *American Journal of Botany*, **85**, 820–828.

1999. Plant succession on pumice at Mount St. Helens, Washington. *American Midland Naturalist*, **141**, 101–114.

del Moral, R. and Bliss, L. C. 1993. Mechanisms of primary succession: insights resulting from the eruption of Mount St. Helens. *Advances in Ecological Research*, **24**, 1–66.

del Moral, R. and Clampitt, C. A. 1985. Growth of native plant species on recent volcanic substrates from Mount St. Helens. *American Midland Naturalist*, **114**, 374–383.

del Moral, R. and Wood, D. M. 1988a. Dynamics of herbaceous vegetation recovery on Mount St. Helens, Washington, USA, after a volcanic eruption. *Vegetatio*, **74**, 11–27.

1988b. The high elevation flora of Mount St. Helens. *Madrona*, **35**, 309–319.

1993a. Early primary succession on the volcano Mount St. Helens. *Journal of Vegetation Science*, **4**, 223–234.

1993b. Early primary succession on a barren volcanic plain at Mount St. Helens, Washington. *American Journal of Botany*, **80**, 981–992.

del Moral, R., Titus, J. H., and Cook, A. M. 1995. Early primary succession on Mount St. Helens, Washington, USA. *Journal of Vegetation Science*, **6**, 107–120.

Desender, K., Baert, L., Maelfait, J. P., *et al.* 1999. Conservation on Volcan Alcedo (Galapagos): terrestrial invertebrates and the impact of introduced feral goats. *Biological Conservation*, **87**, 303–310.

Dobran, F., Neri, A., and Todesco, M. 1994. Assessing the pyroclastic flow hazard of Vesuvius. *Nature*, **367**, 551–554.

Drury, W. H. and Nisbet, I. C. T. 1973. Succession. *Journal of the Arnold Arboretum, Harvard University*, **54**, 331–368.

Eggler, W. A. 1941. Primary succession on volcanic deposits in southern Idaho. *Ecological Monographs*, **11**, 277–298.

1948. Plant communities in the vicinity of the volcano El Parícutin, Mexico, after two and a half years of eruption. *Ecology*, **29**, 415–436.

1959. Manner of invasion of volcanic deposits by plants with further evidence from Parícutin and Jurullo. *Ecological Monographs*, **29**, 267–284.

1963. Plant life of Parícutin volcano, Mexico, eight years after activity ceased. *American Midland Naturalist*, **69**, 38–68.

1971. Quantitative studies of vegetation on sixteen young lava flows on the island of Hawaii. *Tropical Ecology*, **12**, 66–100.

Fosberg, R. F. 1959. Upper limits of vegetation on Mauna Loa, Hawaii. *Ecology*, **40**, 144–146.

Foster, D. R., Knight, D. H., and Franklin, J. F. 1998. Landscape patterns and legacies resulting from large, infrequent forest disturbances. *Ecosystems*, **1**, 497–510.

Franklin, J. F., Frenzen, P. M., and Swanson, F. J. 1988. Re-creation of ecosystems at Mount St. Helens: contrasts in artificial and natural approaches. In J. Cairnes (ed.) *Rehabilitating Damaged Ecosystems*, vol. 2. Philadelphia, PA, CRC Press, pp. 1–37.

Franklin, J. F., MacMahon, J. A., Swanson, F. J., *et al.* 1985. Ecosystem responses to the eruption of Mount St. Helens. *National Geographic Research*, **1**, 198–216.

Frehner, H. F. 1957. Development of soil and vegetation on Kautz Creek flood deposit in Mount Rainier National Park. M.S. thesis, University of Washington, Seattle.

Frenzen, P. M., Krasney, M. E., and Rigney, L. P. 1988. Thirty-three years of plant succession on the Kautz Creek mudflow, Mount Rainier National Park, Washington. *Canadian Journal of Botany*, **66**, 130–137.

Fridriksson, S. 1975. *Surtsey: Evolution of Life on a Volcanic Island*. London, Butterworth.

1987. Plant colonization of a volcanic island, Surtsey, Iceland. *Arctic and Alpine Research*, **19**, 425–431.

Fridriksson, S. and Magnusson, B. 1992. Development of the ecosystem on Surtsey with reference to Anak Krakatau. *GeoJournal*, **28**, 287–291.

Griggs, R. F. 1918a. The recovery of vegetation at Kodiak. *Ohio Journal of Science*, **19**, 1–57.

1918b. The great hot mudflow of the Valley of 10,000 Smokes. *Ohio Journal of Science*, **19**, 117–142.

1918c. The Valley of Ten Thousand Smokes: an account of the discovery and exploration of the most wonderful volcanic region in the world. *National Geographic Magazine*, **33**, 10–68.

1919. The beginnings of revegetation in Katmai Valley. *Ohio Journal of Science*, **19**, 318–342.

1933. The colonization of the Katmai ash, a new and inorganic "soil." *American Journal of Botany*, **20**, 92–111.

Grime, J. P. 1977. Evidence for the existence of three primary strategies in plants and its relevance to ecological and evolutionary theory. *American Naturalist*, **111**, 1169–1194.

Grishin, S. Y. 1994. Role of *Pinus pumila* in primary succession on the lava flows of volcanoes of Kamchatka. In W. C. Schmidt and F.-K. Holtmeier (eds.) *Proceedings of International Workshop on Subalpine Stone Pines and their Environment: The Status of our Knowledge*, Forest Service Gen. Tech. Rep. no. INT-GTR-309. Washington, DC, US Department of Agriculture, pp. 240–250.

Grishin, S. Y., del Moral, R., Krestov, P., *et al.* 1996. Succession following the catastrophic eruption of Ksudach volcano (Kamchatka, 1907). *Vegetatio*, **127**, 129–153.

Grishin, S. Y., Krestov, P., and Verkholat, P. 2000. Influence of the 1996 eruptions in the Karymsky Volcano Group, Kamchatka, on vegetation. *National History Research*, **7**, 39–40.

Haberle, S. G., Szeicz, J. M., and Bennett, K. D. 2000. Late Holocene vegetation dynamics and lake geochemistry at Laguna Miranda Region, Chile. *Revista Chilena de Historia Natural*, **73**, 655–669.

Halpern, C. B. and Harmon, M. E. 1983. Early plant succession on the Muddy River mudflow, Mount St. Helens. *American Midland Naturalist*, **110**, 97–106.

Halpern, C. B., Frenzen, P. M., Means, J. E., *et al.* 1990. Plant succession in areas of scorched and blown-down forest after the 1980 eruption of Mount St. Helens, Washington. *Journal of Vegetation Science*, **1**, 181–194.

Halvorson, J. J. and Smith, E. H. 1995. Decomposition of lupine biomass by soil microorganisms in developing Mount St. Helens pyroclastic soils. *Soil Biology and Biochemistry*, **27**, 983–992.

Halvorson, J. J., Franz, E. H., Smith, J. L., *et al.* 1992. Nitrogenase activity, nitrogen fixation and nitrogen inputs by lupines at Mount St. Helens. *Ecology*, **73**, 87–98.

Halvorson, J. J., Smith, J. L., and Franz, E. H. 1991. Lupine influence on soil carbon, nitrogen and microbial activity in developing ecosystems at Mount St. Helens. *Oecologia*, **87**, 162–170.

Harris, E., Mack, R. N., and Ku, M. S. B. 1987. Death of steppe cryptogams under the ash from Mount St. Helens. *American Journal of Botany*, **74**, 1249–1253.

Haruki, M. and Tsuyuzaki, S. 2001. Woody plant establishment during the early stages of volcanic succession on Mount Usu, northern Japan. *Ecological Research*, **16**, 451–457.

Heath, J. P. 1967. Primary conifer succession, Lassen Volcanic National Park. *Ecology*, **48**, 270–275.

Hendrix, L. B. 1981. Post-eruption succession on Isla Fernandina, Galapagos. *Madrono*, **28**, 242–254.

Hinckley, T. M., Imoto, H., Lee, K., *et al.* 1984. Impact of tephra deposition on growth in conifers: the year of the eruption. *Canadian Journal of Forest Research*, **14**, 731–739.

Hirose, T. and Tateno, M. 1984. Soil nitrogen patterns induced by colonization of *Polygonum cuspidatum* on Mt. Fuji. *Oecologia*, **61**, 218–223.

Horn, E. M. 1968. Ecology of the pumice desert, Crater Lake National Park. *Northwest Science*, **42**, 141–149.

Huebert, B., Vitousek, P., Sutton, J., *et al.* 1999. Volcano fixes nitrogen into plant-available forms. *Biogeochemistry*, **47**, 111–118.

Inbar, M., Hubp, J. L., and Ruiz, L. V. 1994. The geomorphological evolution of the Parícutin cone and lava flows, Mexico, 1943–1990. *Geomorphology*, **9**, 57–76.

Inbar, M., Ostera, H. A., Parica, C. A., *et al.* 1995. Environmental assessment of 1991 Hudson volcano eruption ashfall effects on southern Patagonia region, Argentina. *Environmental Geology*, **25**, 119–125.

Jackson, M. T. and Faller, A. 1973. Structural analysis and dynamics of the plant communities of Wizard Island, Crater Lake National Park. *Ecological Monographs*, **43**, 441–461.

Kitayama, K., Mueller-Dombois, D., and Vitousek, P. M. 1995. Primary succession of Hawaiian montane rain forest on a chronosequence of

eight lava flows. *Journal of Vegetation Science*, **6**, 211–222.

Lawrence, R. L. and W. J. Ripple. 2000. Fifteen years of revegetation of Mount St. Helens: a landscape-scale analysis. *Ecology*, **18**, 2742–2752.

Mack, R. N. 1981. Initial effects of ashfall from Mount St. Helens on vegetation in eastern Washingon and adjacent Idaho. *Science*, **213**, 537–539.

Masuzawa, T. 1985. Ecological studies on the timberline of Mount Fuji. I. Structure of plant community and soil development on the timberline. *Botanical Magazine, Tokyo*, **98**, 15–28.

Matson, P. 1990. Plant–soil interactions in primary succession at Hawaii Volcanoes National Park. *Oecologia*, **85**, 241–246.

Mazzoleni, S. and Ricciardi, M. 1993. Primary succession on the cone of Vesuvius. In J. Miles and D. W. H. Walton (eds.) *Primary Succession on Land*. London, Blackwell Scientific Publications, pp. 101–112.

Mizuno, N. and Kimura, K. 1996. Vegetational recovery in the mud flow (lahar) area. In M. Nanjo (ed.) *Restoration of Agriculture in Pinatubo Lahar Areas*, Research Report (Project 07 044174). Tohoku, Japan, International Research of the Faculty of Agriculture, Tohoku University.

Nakamura, T. 1985. Forest succession in the subalpine region of Mt. Fuji, Japan. *Vegetatio*, **64**, 15–27.

Nakashizuka, T., Iida, S., Suzuki, W., *et al.* 1993. Seed dispersal and vegetation development on a debris avalanche on the Ontake volcano, Central Japan. *Journal of Vegetation Science*, **4**, 537–542.

Newnham, R. M. and Lowe, D. J. 1991. Holocene vegetation and volcanic activity, Auckland Isthmus, New Zealand. *Journal of Quaternary Science*, **6**, 177–193.

Ohsawa, M. 1984. Differentiation of vegetation zones and species strategies in the subalpine region of Mt. Fuji. *Vegetatio*, **57**, 15–52.

Oner, M. and Oflas, S. 1977. Plant succession on the Kula volcano in Turkey. *Vegetatio*, **34**, 55–62.

Partomihardjo, T., Mirmanto, E., and Whittaker, R. J. 1992. Anak Krakatau's vegetation and flora circa 1991, with observations on a decade of development and change. *GeoJournal*, **28**, 233–248.

Pfitsch, W. A. and Bliss, L. C. 1988. Recovery of net primary production in subalpine meadows of Mount St. Helens following the 1980 eruption. *Canadian Journal of Botany*, **66**, 989–997.

Raunkiaer, C. 1934. *The Life Forms of Plants and Statistical Plant Geography*. Oxford, UK, Clarendon Press.

Rejmanek, M., Haagerova, R., and Haager, J., 1992. Progress of plant succession on the Parícutin Volcano: 25 years after activity ceased. *American Midland Naturalist*, **108**, 194–198.

Risacher, F. and Alonso, H. 2001. Geochemistry of ash leachates from the 1993 Lascar eruption, northern Chile: implication for recycling of ancient evaporites. *Journal of Volcanology and Geothermal Research*, **109**, 319–337.

Riviere, A. 1982. Plant recovery and seed invasion on a volcanic desert, the crater basin of USU-san, Hokkaido. *Ecological Congress, Sapporo, Seed Ecology*, **13**, 11–18.

Russell, K. 1986. Revegetation trends in a Mount St. Helens eruption debris flow. In S. A. C. Keller (ed.) *Mount St. Helens: Five Years Later*. Cheney, WA, Eastern Washington University Press, pp. 231–248.

Schmincke, H. U., Park, C., and Harms, E. 1999. Evolution and environmental impacts of the eruption of Laacher See Volcano (Germany) 12 900 a BP. *Quaternary International*, **61**, 61–72.

Seymour, V. A., Hinckley, T. M., Morikawa, Y., *et al.* 1983. Foliage damage in coniferous trees following volcanic ashfall from Mt. St. Helens. *Oecolgia*, **59**, 339–343.

Smathers, G. A. and Mueller-Dombois, D. 1974. *Invasion and Recovery of Vegetation after a Volcanic Eruption in Hawaii*, National Park Service Science Monograph Series no. 5. Washington, DC, Government Printing Office.

Tagawa, H., Suzuki, E., Partomihardjo, T., *et al.* 1985. Vegetation and succession on the Krakatau Islands, Indonesia. *Vegetatio*, **60**, 131–145.

Taylor, B. W. 1957. Plant succession on recent volcanoes in Papua. *Journal of Ecology*, **45**, 233–243.

Thorton, I. 1996. *Krakatau: The Destruction and Reassembly of an Island Ecosystem*. Cambridge, UK, Cambridge University Press.

Titus, J. H. and del Moral, R. 1998a. Seedling establishment in different microsites on Mount St. Helens, Washington, USA. *Plant Ecology*, **134**, 13–26.

1998b. The role of mycorrhizal fungi and microsites in primary succession on Mount St. Helens. *American Journal of Botany*, **85**, 370–375.

1998c. Vesicular–arbuscular mycorrhizae influence Mount St. Helens pioneer species in greenhouse experiments. *Oikos*, **81**, 495–510.

Tsuyuzaki, S. 1987. Origin of plants recovering on the volcano Usu, Northern Japan, since the eruptions of 1977 and 1978. *Vegetatio*, **73**, 53–58.

1989. Analysis of revegetation dynamics on the volcano Usu, northern Japan, deforested by 1977–1978 eruptions. *American Journal of Botany*, **68**, 1468–1477.

1991. Species turnover and diversity during early stages of vegetation recovery on the volcano Usu, northern Japan. *Journal of Vegetation Science*, **2**, 301–306.

1994. Fate of plants from buried seeds on Volcano Usu, Japan, after the 1977–1978 eruptions. *American Journal of Botany*, **81**, 395–399.

1996. Species diversity analyzed by density and cover in an early volcanic succession. *Vegetatio*, **122**, 151–156.

Tsuyuzaki, S. and del Moral, R. 1994. Canonical correspondence analysis of early volcanic succession on Mt. Usu, Japan. *Ecological Research*, **9**, 143–150.

1995. Species attributes in early primary succession. *Journal of Vegetation Science*, **6**, 517–522.

Tsuyuzaki, S. and Titus, J. T. 1996. Vegetation development patterns in erosive areas on the Pumice Plains of Mount St. Helens. *American Midland Naturalist*, **135**, 172–177.

Tsuyuzaki, S., Titus, J. H., and del Moral, R. 1997. Seedling establishment patterns on the Pumice Plain, Mount St. Helens, Washington. *Journal of Vegetation Science*, **8**, 727–734.

Tu, M., Titus, J. H., Tsuyuzaki, S., *et al.* 1998. Composition and dynamics of wetland seed banks on Mount St. Helens, Washington, USA. *Folia Geobotanica*, **33**, 3–16.

Turner, M. G. and V. H. Dale. 1998. What have we learned from large, infrequent disturbances? *Ecosystems*, **1**, 493–496.

Turner, M. G., Baker, W. I., Peterson, C. J., *et al.* 1998. Factors influencing succession: lessons from large, infrequent natural disturbances. *Ecosystems*, **1**, 511–523.

Turner, M. G., Dale, V. H., and Everham, E. H. 1997. Crown fires, hurricanes and volcanoes: a comparison among large-scale disturbances. *BioScience*, **47**, 758–768.

Webb, T., III, and Bartlein, P. J. 1992. Global changes during the last three million years: climatic controls and biotic responses. *Annual Reviews of Ecology and Systematics*, **23**, 141–173.

Whittaker, R. J., Bush, M. B., Partomihardjo, T., *et al.* 1992. Ecological aspects of plant colonization of the Krakatau Islands. *GeoJournal*, **28**, 201–210.

Whittaker, R. J., Bush, M. B., and Richards, K. 1989. Plant recolonization and vegetation succession on the Krakatau Islands, Indonesia. *Ecological Monographs*, **59**, 59–123.

Whittaker, R. J., Schmitt, S. F., Jones, S. H., *et al.* 1998. Stand biomass and tree mortality from permanent forest plots on Krakatau, Indonesia, 1989–1995. *Biotropica*, **30**, 519–529.

Winner, W. E. and Mooney, H. A. 1980. Responses of Hawaiian plants to volcanic sulfur dioxide: stomatal behavior and foliar injury. *Science*, **210**, 789–791.

Wood, D. M. and del Moral, R. 1987. Mechanisms of early primary succession in subalpine habitats on Mount St. Helens. *Ecology*, **68**, 780–790.

1988. Colonizing plants on the Pumice Plains, Mount St. Helens, Washington. *American Journal of Botany*, **75**, 1228–1237.

Wood, D. M. and Morris, W. F. 1990. Ecological constraints to seedling establishment on the Pumice Plains, Mount St. Helens, Washington. *American Journal of Botany*, **77**, 1411–1418.

Zobel, D. B. and Antos, J. A. 1986. Survival of prolonged burial by subalpine forest understory plants. *American Midland Naturalist*, **115**, 282–287.

1987. Composition of rhizomes of forest herbaceous plants in relation to morphology, ecology, and burial by tephra. *Botanical Gazette*, **148**, 490–500.

1991. 1980 tephra from Mount St. Helens: spatial and temporal variation beneath forest canopies. *Biology and Fertility of Soils*, **12**, 60–66.

1992. Survival of plants buried for eight growing seasons by volcanic tephra. *Ecology*, **73**, 698–701.

1997. A decade of recovery of understory vegetation buried by volcanic tephra from Mount St. Helens. *Ecological Monographs*, **67**, 317–344.

# Chapter 9

# Animals and volcanoes: survival and revival

## John S. Edwards

*Let's take a positive view. The mountain is an emblem of all the forms of wholesale death: the deluge, the great conflagration (sterminator Vesevo, as the great poet was to say), but also of survival, of human persistence. In this instance, nature run amok also makes culture, makes artifacts by murdering, petrifying history. In such disasters there is much to appreciate.*

*Susan Sontag, The Volcano Lover*

## Introduction

The vivid memory of a photograph in my grammar-school Latin textbook that showed the body of a dog, cast in a convulsed pose since the Plinian eruption of Vesuvius in antiquity, serves always to remind me that volcanic eruptions can be harmful to the health of animals, ourselves included. It is stating the obvious to allude to the lethal hazards of catastrophic vulcanicity, be they lava, mud or pyroclastic flows, or deep tephra, for living systems. The broader biological interest of vulcanicity is not so much in its lethality as in the statistics of survival and the modes of recolonization in devastated areas. Thus my focus in what follows is on survival and revival of animal communities; survival as a facet of the perennial debate concerning the role of refugia, and revival

as part of the little-understood process of primary colonization by animals. My emphasis will be on observations on arthropods following the 1980 eruption of Mount St. Helens since this has been my principal research experience, and because a sweep of the literature makes it very clear that our knowledge of the animal ecology of volcanoes generally lags far behind vegetation studies, except for the intensive studies following the Mount St. Helens eruption, made possible because of its proximity to major research centers.

Mount St. Helens, a young, active, snow- and ice-capped stratovolcano located in the Cascade Mountains of southwest Washington State, last erupted on May 18, 1980, when an earthquake caused the collapse of the north-facing slopes of its cone. The ensuing landslide released the explosive force of the eruption in a lateral, northward direction. Although the energy released was a mere 5% of that of Krakatau, the laterally directed blast of incandescent ejecta had profound ecological impact. An area of 600 km$^2$ was devastated, within which 350 km$^2$ of conifer forest was removed or flattened. Powdery tephra was spread over $2 \times 10^5$ km$^2$ to the north-northeast of the volcano, of which $10^5$ km$^2$ were covered to a depth greater than 10 mm. The eruption was lethal to large animals caught in its vicinity. Fifty-seven humans died, and Mohlenbrock (1990) quotes Washington State Department of Game estimates of other vertebrate deaths at 11 million fish, 27 000 grouse, 11 000 hares, 6000 black-tailed deer, 1400 coyotes, 300 bobcats, 200 black bear, and 15 mountain lions. An estimated 1600

*Volcanoes and the Environment*, eds. J. Martí and G. G. J. Ernst. Published by Cambridge University Press. © Cambridge University Press 2005.

Roosevelt elk (*Cervus elaphus roosevelti*) were killed by the explosion, but survivors re-entered the devastated area within days and the combination of rapid vegetation regrowth at the periphery of the impacted area, subsequent mild winter conditions, and restricted hunting harvest allowed a rapid recolonization of the northwest sector of the blast zone within the first 5 years after the eruption (Merrill *et al.*, 1986). Total mortality in populations of small birds, reptiles, amphibians, and invertebrates can only be imagined but, as will be discussed below, surprising numbers of smaller animals survived thanks to local topography, snowpack, and their seasonality.

A significant consequence of the horizontal, north-directed blast of the 1980 eruption at Mount St. Helens is the contrast in landscape between the area to the south of the cone where the landscape was impacted by tephra fallout and mudflows, while the northern aspect comprises massive pyroclastic flows (Fig. 9.1) and mudflows, ringed by blowndown forest. The impacts on animal populations contrast accordingly. Thus in the following pages I address the various consequences of tephra fallout for animals, then consider issues of animal survival in the vicinity of catastrophic eruption, the recolonization by animals of devastated landscapes, and prospects for future studies of the effects of volcanoes on animals.

## Effects of tephra deposits on animals

Tephra deposits clearly affect the biota in proportion to their depth and temperature of emplacement but also in relation to the seasonal cycle. It was our experience on the southern slopes of Mount St. Helens, away from the north-directed blast, that subalpine meadows received the springtime tephra fallout of 1980 essentially as a harmless mulch at depths up to about 20 cm. Vegetation, though not necessarily all of the pre-existing species, penetrated this sandy tephra layer later during summer of the same year and large vertebrates, birds, and mobile insects recolonized the area within months as plant life reappeared so that the ecosystem had, to all intents and purposes, returned to normal within months

of the eruption. That is not to say that there were no long-term effects on vegetation or on animal activity in the habitat but without pre-eruption baseline data one can only infer normality from general experience in the habitat. The profound importance of altitude and seasonality cannot be overemphasized in this context: a dormant community under snow, as at Mount St. Helens, will be impacted far less by tephra fallout than an active unprotected summer one.

Surprisingly little quantitative data has been published on the impact of tephra on small mammals. Deaths of squirrels, marmots, and mice due to 2–10 cm of tephra near Lake Iliamna Alaska (Jagger, 1945), and loss of vertebrates in tephra-impacted areas around Volcan Parícutin in Mexico (Burt, 1961) exemplify lethality. On the other hand the survival of northern pocket gophers (*Thomomys talpoides*) on and around Mount St. Helens in areas of heavy tephra fallout illustrates the survival values of occupying a subterranean habitat, feeding on subterranean vegetation, and in the case of the May 1980 eruption of Mount St. Helens, benefiting from the initial protection of snow cover (Anderson, 1982; Anderson and MacMahon, 1985). At least 14 species of small mammal survived the initial eruption and its immediate consequences. Many of them survived the blast because their fossorial habit, the protective snow cover, and the time of day (0832 PDT) of the eruption together had the effect of shielding them from the direct effects of the blast. By 1987, out of a hypothetical list of 32 species of resident small mammal, 6 had returned to the pyroclastic/debris flow zone, 15 species occupied the tree blowdown areas, and 22 species were present in the tephra fall zone (MacMahon *et al.*, 1989).

Burrowing animals can hasten the return of soil fertility through their earth-turning excavations; pocket gophers played a significant role in the restoration of soil fertility as a result of their burrowing activity, bringing old fertile soil above the sterile tephra (MacMahon and Warner, 1984) and by acting as major vectors of mycorrhizal fungi as they inoculated plants that established anew on volcanic debris (Allen *et al.*, 1992). Comparable soil movement and the creation of disturbed microtopography caused by

**Fig. 9.1.** Volcanic landscapes in the North American Pacific Northwest. (a) 1984 view of the Pumice Plain of Mount St Helens formed after the 1980 eruption, looking south toward the breached crater. Three years after the eruption this area was devoid of vegetation but supported an arthropod fauna dependent on wind-borne organic fallout. (b) Mount Rainier, southeast slopes adjacent to Muir snowfield. Foreground is at 2500 m. Sparse localized vegetation is present but resident arthropod populations are widespread up to at least 3500 m. Photographs by Dr. P. R. Sugg, from Sugg et al. (1994).

land iguanas in the Galápagos were observed in tephra deposited by a 1968 eruption on the western caldera rim of Isla Fernandina (Hendrix, 1981). Collapsed iguana tunnels also provided settling sites for the wind-borne seeds of pioneer plants and iguana feces containing viable seed were also seen to be foci for seed germination on the most barren tephra surfaces. At Mount St. Helens willows acquired mycorrhizae only after the arrival of several species of rodent such as *Peromyscus maniculatus orcas* and *Microtus* sp.; it was also shown that fungal spores remained viable after passage through the gut of North American elk (*Cervus elaphus* Merriam.) and could thus be transported long distances across the post-eruption landscape (Allen *et al.*, 1992). Elk, perhaps disoriented by changes to the landscape, returned to the blast zone only weeks after the eruption so the potential was there, almost immediately, for the spread of microorganisms into the sterilized surface.

Quantitative studies of bird diversity and abundance in the subalpine zone of Mount St. Helens (Manuwal *et al.*, 1987) showed that while birds must have been killed or driven from the areas of ashfall during and immediately after the eruption there appeared to be no long-term impact on the quickly returning populations except where trees had been seared and killed by hot gases or buried by the debris avalanche.

The invertebrate fauna of heavily impacted streams recovered rapidly during the first 5 years after the eruption, and continued at a steady rate so that by 1990, 10 years after the eruption, aquatic invertebrate faunal diversity had reached 80% of that to be expected for a comparable undisturbed stream (Anderson, 1992). Aquatic insects such as stonefly, caddis fly, and mayfly larvae subjected to tephra in an experimental simulation of ashfall into a stream appeared to suffer no short-term effects from accumulation of ash on their exoskeleton, and most were able to shed the deposits within 24 hr when placed in clear water (Gersich and Brusven, 1982). Again, fish and amphibia were extirpated from most stream systems around Mount St. Helens but within 5 years sculpins and tailed frogs had returned, even to heavily disrupted streams (Hawkins *et al.*, 1988). Refugia for these aquatic animals were afforded by ice-covered lakes (Crawford, 1986). They would undoubtedly have succumbed in greater numbers had the eruption occurred just 2 or 3 months later, when snow and ice no longer afforded protection. A surprising conclusion reached by aquatic ecologists who followed the regeneration of stream communities is that the state of the surrounding landscape was not necessarily a reliable indicator of the condition of the stream habitat for even where terrestrial surfaces remained barren, streams had recovered to the extent that they supported diverse organisms. This was especially surprising in the case of amphibians which are considered to be very sensitive to temperature, to insolation, to humidity in the case of adults, and to the availability of breeding sites, and which were thus expected to have been entirely eliminated by the eruption. Contrary to

expectations, amphibia reappeared even in the pyroclastic zone, and populations in some cases rebounded rapidly (MacMahon, 1982; Zalisko and Sites, 1989). The situation is different for anadromous fish as evaluated by Martin *et al.* (1986). The removal of riparian vegetation and woody debris from streams impacted by the eruption had adverse effects on coho salmon fingerlings as a result of the loss of sheltering debris and raised water temperatures. The recovery of riparian vegetation reduces stream temperature and adds woody debris to the water. Tree growth data suggest that trees would be tall enough to shade smaller streams after 5–20 years, but that 50–75 years of growth would be necessary to generate the large woody debris that provides refuge during winter storms.

The short-term impacts of tephra fallout further afield in the agricultural areas of Washington state were diverse. As reported by Cook *et al.* (1981), crop losses were estimated to be about US$100 million, that is about 7% of the normal crop value, but the figure was less than initially predicted due to a very favorable growing season in 1980 for wheat, potatoes, and apples. Alfalfa hay was heavily contaminated with ash but it proved to be non-toxic to livestock even at levels up to 10% of ash by weight in the fodder. Toxic elements such as arsenic, cadmium, copper, and molybdenum were present only in trace amounts and had no adverse effects, but tests on the effect of the tephra on the growth of day-old chicks showed a reduction by 6% for each 10% increase in the ash content of their diet, while sand, as a control, caused a 4% incremental reduction under comparable treatments. Mortality was reported to be unaffected by the addition of up to 30% weight of tephra to the chick diet. Milk production by cattle and growth rate of dairy calves was said to be unaffected by the presence of tephra in the diet. The effects of the tephra fallout on insects were however very different and are discussed below.

## Effects of tephra on insects

The lethal effects of deep blanketing and heat aside, tephra can have variable effects on

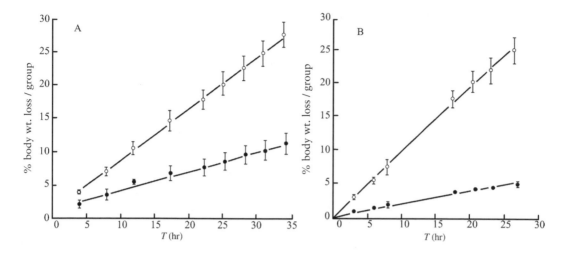

**Fig. 9.2.** The lethal effect of fine textured Mount St. Helens tephra composed of 0.5–100 μm angular granules on rate of water loss from insects. Water loss from groups of (A) house cricket *Acheta domesticus* and (B) cockroach *Periplaneta americana* exposed to Mount St. Helens tephra (open circles) compared with groups treated with worn river sand (solid circles) at 28 °C, at 62–65% relative humidity. Means ± standard errors for six crickets, and three cockroaches per treatment, with two replicates. Animals were not fed or given access to water during or after the period of exposure. Weights were taken of each individual until 50% mortality was reached in the experimental groups. None of the control animals on sand died. Differences are significant in both species to $p < 0.0001$. From Edwards and Schwartz (1981).

arthropods, depending on its physical texture. One of the physiological features that enable insects, with their relatively large surface-to-volume ratio, to occupy dry habitats (and thus a key to their evolutionary success) is their capacity to conserve water by means of a superficial layer of lipid, generally waxes, on their cuticular exoskeleton (Hadley, 1994). So long as this layer is intact and the ambient temperature is not above the melting point of their wax, insects and to a lesser extent spiders, can maintain a viable water balance under extremely adverse desiccating conditions. However that lipid layer is susceptible to damage by scarifying agents such as dusts with angular or acicular particles, for example the silica gels widely used in insect control. Thus it might be expected that fine-textured tephra in the form of powders or grit might be harmful to

arthropods as a result of damage to the cuticular lipid layer.

The powdery tephra from Mount St. Helens, composed of pulverized dacite and andesite together with volcanic glass, proved to be highly abrasive to insect cuticles, and thus potentially lethal where surface contact was extensive. This factor was evaluated by comparing the effect of exposure to newly fallen tephra from Mount St. Helens with that of smooth-grained river sand. Mount St. Helens tephra greatly increased the rate of water loss of crickets and cockroaches under laboratory conditions (Edwards and Schwartz, 1980), leading to death within 30 hr (Fig. 9.2). Similar results were reported for other insects such as houseflies (*Musca domestica*), mason bees (*Osmia lignaria*), and yellowjackets (*Vespula* sp.) (Cook *et al.* 1981). It is likely that most insects that were already active in early May were impacted by desiccation throughout the extensive area of tephra fallout to the northeast of the volcano, and this presumably was a major factor in the effects noted by Cook *et al.* (1981) in agricultural land to the lee of the eruption (Fig. 9.3). Among the agricultural pest insects in the area the Colorado potato beetle (*Leptinotarsa decemlineata*) did not recover from 24 hr of exposure to the tephra, but they did survive contact with the tephra where they were washed in plots watered by an overhead irrigation system. Similarly, the pear psylla (*Psylla pyricola*), a serious orchard pest, was little affected by light (6 mm) falls of ash, especially where the ash was

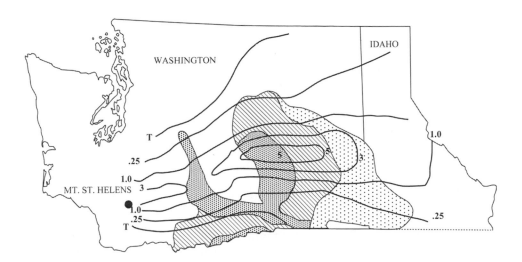

**Fig. 9.3.** Pattern of tephra fallout over Washington and Idaho states following the eruption of Mount St. Helens on May 18, 1980, showing depth of uncompacted tephra in cm. Lethal impact of the tephra on insects within the tephra fallout area is discussed in the text. From Cook *et al.* (1981).

washed off the foliage by irrigation systems, but 2.5–4.5 cm depths of ashfall depressed both the pear psylla population and that of their predators (Fye, 1983). The effect on grasshopper populations was dramatic. Economically significant populations of grasshoppers (*Melanoplus* sp.) of up to 10–12/m$^2$ were present in eastern Washington in early May 1980, as estimated by the Department of Agriculture Animal and Plant Health Inspection Service (APHIS), and plans had been made to aerial-spray 120 000 acres of rangeland as a control measure, but the eruption intervened, the grasshoppers were eliminated, and plans for aerial spraying were abandoned. For comparison, 130 000 acres (525cc ha) had been sprayed in Washington in 1979, while in Oregon, where the tephra did not fall, 150 000 acres (606cc ha) were sprayed in 1980. Cook *et al.* (1981) also document the damage to economically beneficial insects as a result of tephra fallout from Mount St. Helens. Honeybees (*Apis mellifera*) and other pollinators were annihilated from the ashfall area. Orchard mason bees returned to their colonies as the sky darkened due to the approaching tephra cloud, and were thus sheltered from exposure to the airborne material, but the following morning they flew out to forage, encountered the

tephra-coated vegetation and never returned. Johansen *et al.* (1981) estimated that four-fifths of the 15 000 honeybee colonies in the irrigated parts of the Columbia Basin were destroyed or severely impacted. The effects would have been far more severe had the eruption occurred 2–3 weeks earlier, while pollination was in progress in the extensive apple orchard areas near the towns of Yakima and Wenatchee in eastern Washington, both of which received heavy tephra fallout.

New *et al.* (1988) note that volcanic ash on foliage has proven to be lethal to lepidopteran larvae in the vicinity of Indonesian volcanoes, but in striking contrast to the events described above, grasshoppers, carpenter bees, dragonflies, and other arthropods were observed to be active on Sertung Island in the Krakatau group during September 1996 when persistent daily fallout of basaltic tephra from the nearby active volcano on Anak Krakatau covered the vegetation (JSE, unpublished data). It seems likely that this contrast in lethality is attributable to differences in the physical texture of the tephras between Mount St. Helens and Sertung Island.

In evaluating the impact of recent vulcanism on butterflies in the Pacific Northwest Pyle (1984) draws attention to lethal impacts of tephra for larvae and draws a distinction between those which feed on plants with hairy leaves, such as *Viola* spp., that hold tephra, in contrast to grasses which shed it. While local extinctions were detected after the eruption of Mount

St. Helens, the resilience of Lepidoptera was evident in the rapid return of many species.

# The question of animal survival after eruption

Intuition and its anthropocentric view of the world tells us that life surely cannot survive the impact of a catastrophic eruption. Billows of molten lava, layers of tephra at 600–800 °C, pressure blasts, and toxic gases are obviously lethal to all forms of life, especially large mammals such as ourselves. And yet, in the complex topography of volcanic landscapes, and given seasonal effects such as deep snowpack to the lee of ridges, there remains the possibility of the survival of microorganisms and, in principle, plant propagules and small animals. The issue of local *survival* versus total *extirpation* of life is a critical one for an understanding of patterns and processes of recolonization after volcanic eruptions, and both cases have been vigorously argued, especially since the eruption of Krakatau. Indeed the debate has come to be known as the "Krakatau Problem." The classic debate that persisted through several decades is discussed in detail by Thornton (1996) in his excellent monograph on Krakatau and it is sufficient here to briefly summarize the arguments as they apply to animals.

Melchior Treub (1888), the first biologist to visit the remains of Krakatau after the cataclysmic eruption of 1883, was certain that no living vestige of plant or animal life could have survived in the immediate vicinity of the eruption. He viewed the subsequent rapid process of plant recolonization as proceeding by immigration upon a *tabula rasa*. The majority of early visitors to the remainder of old Krakatau (now called Rakata), for example Verbeek, the geologist who first surveyed the post-eruption site, and Dammerman who first visited Rakata in 1919 and whose 1948 monograph on the animal life of post-eruption Rakata is a classic, all agreed with Treub that no living thing could have survived there. But Backer, a botanist who arrived in Indonesia in 1901, and who visited Rakata and the neighboring islands on several occasions

later, came to champion opposite views, following those of some other botanists (e.g., Lotsy, 1908; Lloyd Praeger, 1915). He issued a voluminous and in places acerbically *ad hominem* monograph (Backer, 1929) advancing arguments for the survival of seeds first buried by tsunami flood then bared by erosion. Zoologists, too, came forward with cases for the survival of earthworms (Michaelsen, 1924) and other animals (Scharff, 1925) in sheltered sites where floods could conceivably have cooled the tephra sufficiently to allow the persistence of pockets of survivors. Thus debate lines were drawn. Dammerman (1929) marshaled a convincing set of arguments against the survival of any animal species on Rakata, based on the depth of pyroclastic deposits, the evidence from carbonized tree trunks of extremely high temperatures during tephra emplacement, and the known dispersive capacities of early animal colonists, both vertebrate and invertebrate. Dammerman further raised the issue as to what animal survivors might have found to feed on in the immediately post-eruption landscape. While the majority of recent biologists agree that the complete extirpation of life on Krakatau is most probable (e.g., Thornton and New, 1988) nematologists consider the survival of soil nematodes a possibility (Winoto Suatmaji *et al.*, 1988) and the proven long-term survival of seeds buried during the early phase of the building of Anak Krakatau (Whittaker *et al.*, 1995) leaves the possibility still open of limited survival.

History repeated itself after the 1980 eruption of Mount St. Helens when the question of survival once again arose, but this time under the very key element. Despite the estimates of pyroclastic temperatures of 600–800 °C (Banks and Hoblitt, 1981), the early appearance of angiosperms such as lupines on the cooled pyroclastic deposits led to suggestions that localized survival of seed or viable plant fragments may have been possible in sheltered sites. In the case of Mount St. Helens, however, the role of long-distance transport by wind of seeds, especially while still attached to their capsules, over wind-packed snow surfaces cannot be discounted. The evidence for survival of insects can only be based on the distribution of less mobile, wingless species, for most winged species include an obligate dispersive phase, sometimes extending hundreds of kilometers, in their life histories. It is a seemingly wasteful process since vast numbers fail to find a viable habitat (see below) but it does ensure that new opportunities in the spatial and temporal mosaic of the landscape will be found and colonized. Further, local flight, on a scale of kilometers, by winged insects is a universal phenomenon. Rescue helicopter pilots who entered the blast area of Mount St. Helens in search of human survivors a day after the eruption reported seeing yellowjackets (vespid wasps) and "blowflies" (Diptera) flying over the barren mineral surfaces.

The capture in pitfall traps of worker ants (*Formica rufa*) in the pyroclastic flow zone later in 1980 might be taken to signify the improbable survival of ant colonies, but is explicable as the consequence of wind transport since high winds frequently rake the landscape. A more curious case is the single capture in 1983 at a pyroclastic site of a small carabid beetle, *Bembidion oblongulum*, that is subterranean in habitat and accordingly has reduced eyes and wings (Sugg and Edwards, 1998). The possibility that this species survived in an area where it was protected by local topography and snowpack and quickly re-exposed by erosion of tephra cannot be excluded but it seems more likely on balance that it was again the product of wind transport. A final example of observed insect survival in the blast zone is instructive since it is relevant to Dammerman's (1929) question concerning the possibility of animal survival in the absence of adequate food resources on the posteruption landscape. Large numbers of carpenter ants (*Camponotus* sp.) were observed foraging on rain-caked tephra surfaces at the border of the blast zone 1 month after the eruption of Mount St. Helens (JSE, unpublished data). These ants must have surfaced from a subterranean colony, perhaps beneath a log buried deep under snowpack on the lee side of a protective ridge. Windborne insect bodies and localized fungal blooms at seepages on the tephra surface appeared to be their sole food source, for no nectar-bearing plants were to be found within several kilometers. The absence of ants from this site in subsequent years suggests that these survivors of the cataclysm found insufficient forage, especially carbohydrate, in the absence of flowering plants during the summer following the eruption and that the colonies thus succumbed. Kuwayama (1929) found a comparable pattern after the June 1929 eruption of Mt. Komagatake. The ants *Formica fusca* and *F. truncorum* were active 9 days after tephra fall, but colonies were dead 11 days later.

## The process of recolonization by animals

Six case histories that exemplify the initial stages of colonization of new surfaces form the substance of this section: Rakata, Hawaii, the Canary Islands, Motmot, Surtsey, and Mt. Etna. (Mount St. Helens will be discussed in more detail in the next section.) They are geographically diverse but despite their regional and climatic differences there are some generalities to be made. It is to Dammerman (1948) that we can turn for the first clear formulation of the basic pattern of events of primary colonization on such new volcanic surfaces. Assuming survivors to be absent, or ecologically insignificant after a major eruption, what are the initial events of primary succession on the new mineral surface following a volcanic eruption? Orthodoxy, as reflected by the majority of textbooks, has it that plants are the primary colonists, after or together with microorganisms such as cyanobacteria, and perhaps fungi, lichens, and bryophytes. Primary producers in this sequence of events provide for the invasion of animals, led by herbivores, and also for the later explosive diversification of the parasite, predator, and scavenger faunas. But the observations of Dammerman and our own studies at Mount St. Helens tell a different story. Apart from microorganisms the textbook sequence is reversed: predators and scavengers arrive first, pioneer plants only later. In the absence of primary production the pioneer animals must depend on imports that may come by water or by air. The pioneer animals of mini islands on the Australian Great Barrier Reef (Heatwole, 1971) and of Surtsey near Iceland (Lindroth *et al.*, 1973) utilized marine debris. Thus for islands generally flotsam is a critical commodity in the establishment of supralittoral colonists. The evidence collected by Dammerman (1929, 1948) and subsequently by Thornton and colleagues (Thornton, 1996) at Rakata in the Krakatau group and by Ball at Motmot (Ball and Glucksman, 1975) reflect the importance of rafting as a source of colonists too. Earthworms, mollusks, and diverse arthropods are disproportionately represented among the first colonists, reflecting the fauna of logs in mainland forest, and are thus indicative of naturally rafted vegetation as a source of colonists. The termites of Rakata provide a strong case for this interpretation: of the nine species on Rakata, not one is a soil-dweller, that is all of them nest in trees rather than in or on soil and were in a position to be transported on floating vegetation (Thornton, 1996).

The air too can provide a major source of input. Wind-borne materials provide the resource base in a wide variety of terrestrial environments collectively termed, together with caves and abyssal marine habitats, the allobiosphere by G. Evelyn Hutchinson (1965), where animal assemblages thrive in the absence of local primary production. Examples of habitats where imported resources support animal populations are sand dunes, river bars, bare volcanic surfaces of lava or tephra (Edwards, 1988), and the high alpine zone, termed the aeolian zone by Swan (1963) to denote the dependence of high alpine communities on wind-borne organic resources. Ever since Alexander von Humboldt first recorded his observations of insects on the high snows of the Ecuadorian volcano Chimborazo (Humboldt, 1808), reports have accumulated of alpine animal communities, including arthropods, reptiles, birds, and mammals, that depend largely or entirely on aeolian arthropod fallout (review: Edwards, 1987). Whymper (1892), in his account of travels in the High Andes, distinguished between the resident insects on Andean volcanoes, and the "stragglers" observed by Humboldt: "At the greatest heights [insects] were found less upon the surface than *in* the soil, sometimes living amongst stones imbedded in ice, in such situations and numbers as to preclude the idea that they were stragglers." Whymper collected earwigs high on the bare surfaces

of the volcanoes Chimborazo and Cayambe, and spiders and beetles on the summits of Corazón (4785 m) and Pichincha (4794 m).

Newly formed, seemingly barren and forbidding lava fields too have their predator/scavenger guild of arthropods, termed by Howarth (1987) the neogeoaeolian biota. Some examples of the characteristics of the arthropod fallout will suffice to convey the quantity and diversity of this little-known phenomenon before presenting a more detailed account of the events of recolonization at Mount St. Helens.

## Example 1: Rakata

Dammerman (1929) estimated that between 1908 and 1933, 92% of the total immigrant fauna of Rakata, the remnant of Krakatau, had arrived by air. The first living organism found on Rakata, just months after the eruption, was a solitary spider, reflecting the propensity of many spiders to disperse by means of ballooning on their own silken thread (Darwin, 1839; Crawford *et al.*, 1995), which will carry them passively on the wind sometimes for prodigious distances and sometimes in enormous numbers. On Anak Krakatau, the new volcano that arose in 1930 from the caldera formed by the eruption of Krakatau. Thornton *et al.* (1988), during 10 days in August 1985, collected more than 70 species of arthropod, representing ten orders and including spiders, flies (Diptera – the most numerous component of the total catch), mirid and lygaeid bugs, aphids, cicadellids, psocids, moths, and representatives of ten families of beetle in traps set 1.5m above the surface of lava flows. The majority of these arthropods were thought to have been derived from the "aerial plankton," and to have reached the island of Anak Krakatau from neighboring islands. With a fallout rate in the aerial traps of 20 individuals per square meter per day Thornton made a conservative estimate that between 0.5 million and 5 million insects arrived on the 2.34 km$^2$ island from sources at least 4 km distant during the trapping period. Traps deployed on the barren lava surface yielded numbers of an omnivorous cricket *Speonemobius* sp. (New and Thornton, 1988), lycosid spiders, ants, earwigs, and a species of mantid, all of them predators and/or scavengers, and evidently entirely dependent on the wind-borne arthropod fallout for their sustenance. As on Surtsey and Motmot Islands, the shoreline of Anak Krakatau also has a pioneer detritivore community composed of collembolans, chloropid flies, and anthicid and tenebrionid beetles whose resource is seaborne organic debris and which had established a foothold within 6 months of the emergence of the island from the sea.

## Example 2: Hawaii

The highest points in Hawaii – Mauna Kea (4205 m) and Mauna Loa (4169 m) on the island of Hawaii, and Haleakala (3055 m) on Maui – are alpine aeolian sites comprising stone deserts with low annual precipitation, often persisting much of the year as snow. This forbidding habitat nonetheless harbors mosses, lichens, and resident arthropods (Howarth, 1987). The resource base for the resident arthropod fauna is fallout derived from insects and spiders that initiated flight and were carried up from fertile lowlands on orographic winds. Some members of the resident arthropod population amounting to 18 species are biologically noteworthy. Howarth (1987) discusses an extraordinary lygaeid bug, *Nysius wekiuicola*, that subsists on moribund insects around the summit of Mauna Kea. True bugs of the diverse genus *Nysius* are represented by a group of species on Hawaii that are typical of the almost universal seed feeding habit but *N. wekiuicola*, a flightless, long-legged form, is unique in its predator/scavenger habit. It has presumably evolved on Hawaii from a more typical seed-eating progenitor. An aberrant flightless moth on Haleakala is also part of an aeolian community. The larvae of *Thyrocopa apatela* build silken webs under rocks about the summit of Haleakala where they feed on wind-borne debris, principally fragments of dried leaves of the endemic alpine shrub *Dubautia menziesii*. Lycosid spiders (wolf spiders) also occur in the aeolian zone on both islands.

Lowland lava flows are also home to scavenging arthropods. As Howarth (1987) points out, these lava fields are among the most inhospitable habitats on Earth, with daily temperature fluctuations on the black mineral surfaces approaching 50 °C and with rapid drainage assuring a xeric environment. Bare lava flows at about 1000 m on Kilauea nonetheless have their own fauna, termed the neogeoaeolian ecosystem.

A remarkable faunistic parallel to Anak Krakatau is described by Howarth (1979) from these lava fields on the island of Hawaii, where a different species of nemobiine cricket, *Caconemobius fori*, shares the arthropod fallout resource with a lycosid spider, an earwig, and a mantid. Here too, the predator/scavenger population of the bare lava depends on wind-borne organic fallout. The genus *Caconemobius* is endemic to the Hawaiian islands, where most of the species frequent moist base rock along the coast on lava flows and cliffs and in lava caves. *Caconemobius fori* is thought to be most closely related to *C. varius* which frequents the numerous lava caves. The lycosid spiders, too, have exploited both the lava cave habitat and the open lava flows. Indeed the subterranean arthropod fauna constitutes a distinct ecosystem associated with the extensive lava caves on five of the Hawaiian islands: Kilauea, Oahu, Molokai, Maui, and Hawaii. Their diverse fauna includes amphipod and isopod crustaceans, at least six species of spider, a pseudoscorpion, centipede, and millipede. The hexapods include Collembola, two gryllids, an earwig, a cixiid bug, a carabid beetle, and at least two families of Lepidoptera, all of which are dependent on the import of organic material to the subterranean voids of the lava fields.

## Example 3: the Canary Islands

Historically recent lava flows on Tenerife and Lanzarote in the Canary Islands, some of them at high altitude, with little or no vegetation and extreme daily temperature ranges as on Hawaii, have a resident predatory and scavenging arthropod fauna that includes at least seven families of spider, a phalangid, mites, an isopod, a centipede, Collembola, and five orders of insect, including crickets (though no *Caconemobius* species), an earwig, a predatory bug, and a melyrid beetle, all of which are predators and/or scavengers (Ashmole and Ashmole, 1987; Ashmole *et al*, 1990). These primary colonists of bare volcanic surfaces depend on the resource of airborne organic fallout. Their small size allows them to make use of shade and shelter in the uneven texture of both pahoehoe and aa lava flows. Ashmole and Ashmole (1987) carried out quantitative studies on the flux of arthropods on the snow surfaces of Pico Tiede (3718 m), the central stratovolcano of Tenerife. They found arthropods on the snow surfaces at densities up to $18/m^2$, with a diversity comprising at least 37 arthropod families among which the Diptera contributed 17 families. Aphids and psyllids dominated the samples numerically with 53% of the total catch. Comparable fallout is doubtless also occurring on the bare lava flows, and it is this fallout that constitutes the resource base for the predator/scavenger guild of the new lava.

## Example 4: Motmot Island

Motmot Island first appeared in 1968 during an eruption in Lake Wisdom, a large caldera lake on Long Island, Papua New Guinea. Ball and Glucksman (1975), who have followed the history of Motmot, found that this small volcanic island had acquired, within 4 years, a fauna composed of scavenging and predatory beetles, ants, earwigs, and lycosid spiders. Most of the colonists were associated with the lake shore line but lycosid spiders and earwigs were the pioneers of the central areas of Motmot. A further eruption in 1973 destroyed the pioneer plant life of Motmot but evidently had little effect on the invertebrates, whose diet must have been based on flotsam along the strand and on air-borne materials derived from the surrounding Long Island vegetation across Lake Wisdom (Fig. 9.4).

A brief plant survey of Motmot Island in 1988 (Osborne and Murphey, 1989) revealed a minor increase in plant diversity, with a total of 19 species of fern and spermatophyte comprising some losses and some gains, mainly ferns and figs, compared with the earlier report of Ball and Glucksman (1975). A recent resurvey of the Motmot Island fauna in July 1999 (Edwards and Thornton, 2001; Thornton, 2001) revealed no significant increase in arthropod diversity from that reported by Ball and Glucksman (1975). In 1999 two species of spider, a lycosid and a linyphiid, present in great numbers, and one ant species dominated the entire island, exploiting the landfall of midges, caddis flies, damselflies, and dragonflies derived from the surrounding Lake Wisdom. Earwigs and beetles, noted as prominent scavengers in the 1970s, are now insignificant and localized members of the simple, Arctic-like ecosystem. Despite a relatively diverse input of wind-borne arthropods from the encircling Long Island forest, demonstrated by Malaise-trap captures, the sparse vegetation of the lava fields supports few herbivores. The large predator/scavenger population is clearly dependent on the import of allochthonous material, and thus resembles the early stages of succession observed on Mount St. Helens.

## Example 5: Surtsey Island

Surtsey, at 63° N, near the south coast of Iceland, arose from the sea in November 1963. Within 6 months a strand microcosm dependent on the flotsam composed mainly of algae, logs, and carcasses of birds and fish, had established, and by 1970, 112 species of insect from 14 orders (excluding ectoparasites of vertebrates), and 24 species of arachnid had been collected along the perimeter of Surtsey (Lindroth et al., 1973). The majority of these immigrants must have arrived from Iceland, 35–40 km distant, and from Heimaey, 20–25 km away. Further inland on the bare volcanic landscape arthropods were slow to establish, but potential colonists were there, for linyphiid spiders were observed alighting on the lava (Fridricksson, 1975). Flies (Diptera) were the most diverse element of the early fauna on Surtsey, while one of the usual pioneering groups of bare ground, the Coleoptera (beetles) were represented by only three species up to 1970. Clearly the flux of air-borne arthropods contributing to the organic fallout on Surtsey must be less than at lower latitudes, especially at sites where sources are closer. But as plants establish on the surface around the coastal areas there will doubtless be an increase in the fallout from local sources, leading to the establishment at higher altitudes of predatory and scavenging pioneers on bare mineral surfaces. The pattern is much the same as at Krakatau but the tempo is slower.

## Example 6: Etna

The famous Sicilian volcano Etna has a predator/scavenger guild of arthropods associated with

**Fig. 9.4.** Motmot Island and its recolonization, showing an example of colonization by rafting. (a) A floating tree trunk (*Albizia falcataria* or *Prosopis insularum*) which had drifted from the far shore of Lake Wisdom in the caldera of Long Island. From the presence of sprouts on the trunk and the appearance of upper branches it appeared that the tree had lain on its side partially in the water for some time before floating into the lake. It was carrying numerous invertebrates (see text). (b) Motmot erupting on 2 May 1973. The eruption obliterated vegetation but had little effect on invertebrate diversity. From Ball and Glucksman (1975).

bare mineral surfaces, that depends on upwelling insects, especially aphids and beetles, from the fertile agricultural lowlands surrounding the volcano. On aa lava flows the pioneer lichen *Stereocaulon vesuvianum* supports larvae of the butterfly *Luffia*. Otherwise the only inhabitants are predators such as pholcid and salticid spiders. High on the cone, where snowfields persist through June, the beetles characteristic of snowfield margins around the world, such as the carabid beetles *Nebria, Bembidion*, and *Trechus*, depend on allochthonous materials. The question of their persistence through late summer when the slopes are dry and hot awaits an answer. Elsewhere on the cone, time and opportunity has led to the development of vegetation which supports a diverse arthropod fauna (Wurmli, 1971, 1974).

## Recolonization in the blast zone at Mount St. Helens

Situated close to major research centers, Mount St. Helens was the first catastrophic eruption to be studied intensively by scientists representing many different disciplines from the moment of eruption. The pattern of pioneer primary recolonization found in the blast zone summarized below supported our hypothesis, based on our experience with arthropod life on other Cascades volcanoes, that the pioneers of the subalpine pyroclastic flow surfaces would be predators and scavengers, dependent on wind-borne materials (Fig. 9.1). The resource base for potential arthropod colonists was surveyed intensively from 1981 to 1985 by means of pitfall traps and fallout collectors (Edwards, 1986) (Fig. 9.5). Representatives of at least 150 different families of insect from 18 orders arrived on the wind at the survey sites in the blast zone at altitudes between 1000 and 1200 m. These arthropods had traveled at least several kilometers on the wind from peripheral vegetation and aquatic habitats. Diptera dominated the fallout diversity, with 42 families, Coleoptera followed with 30, Hemiptera–Homoptera with 18, Lepidoptera with 15, and Hymenoptera with 10. Other insect orders were represented by 1 to 6 families, while spiders contributed 9 families.

Among these immigrants only 19 species showed evidence of establishing breeding populations within the first 3 years following the eruption. Of these 11 were carabid beetles, accompanied by agyrtid, tenebrionid, and trachypachid beetles, all of them scavengers and small enough (2–3 mm) to be able to find shelter among the pumice pebbles of the pyroclastic flow surface. Later immigrant insects included hemipterans (Lygaeidae and Saldidae), Orthoptera (Gryllacrididae), and grylloblattids (Sugg and Edwards, 1998). Spiders were well represented in the early fallout:

from 1981 to 1986 ballooning spiders made up 23% of the air-borne arthropods that reached the sampling sites. About half of them were wolf spiders (Lycosidae), while linyphiids comprised 34% of the catch. By 1986 six species of spider (two lycosids and four linyphiids) had established breeding populations, but in contrast to the beetles most spiders were taken at sites already colonized by pioneer plants, mainly lupines. The earliest case of an immigrant that had arrived on foot at any of the pyroclastic flow sampling sites occurred in 1984 when flightless gryllacridids were taken in pitfall traps. Then, a year later an amaurobiid spider, a non-ballooning species, and harvestmen (phalangids) also returned on foot.

The pattern of spider arrival and survival as monitored at Mount St. Helens has broad implications for spider population dynamics for it demonstrated unequivocally that large numbers of immigrants had arrived from sites at least 50 km distant. The significance of ballooning as a means of long-distance immigration has been widely underestimated (e.g., Decae, 1987; Wise, 1993); the results from Mount St. Helens leave no doubt about its importance.

The principal roles as true pioneers in Act 1, Scene 1 of the succession play were taken by small carabid beetles of the genus *Bembidion*. *Bembidion* is a diverse genus commonly associated with disturbed or barren habitats such as the margins of braided rivers, avalanche chutes, landslides, periglacial retreat surfaces and alpine snowfield margins. Three species, *Bembidion planatum* (Fig. 9.6), *B. improvidens,* and *B. obscurellum* were taken in 1981, during the spring following the eruption, and they were the first to show population growth attributable to local breeding. They were, other than microorganisms, the first settlers.

All of the pioneer arthropods faced the challenge of a seemingly exceedingly rigorous habitat that was snow covered for 6–7 months of the year, and subject to dusty wind, high temperatures, and desiccation during the summer. Few animals could survive such conditions but the small size (2–3 mm in length) of the carabids enables them to exploit the protection afforded by the pebbly texture of the surface where high temperatures

**Fig. 9.5.** (a) Fallout collectors as used to estimate organic fallout flux on Mount St. Helens. The array of close-packed balls is held in a 0.1 m² frame underlain by fine nylon mesh that allows drainage but retains material that falls into the dead space around the lower half of the balls. Windblown tephra accumulates in the collectors as well as organic material. The latter was separated from the tephra by flotation then dried and weighed before identification of fragments was made as far as possible. In the example shown, situated on a recent larva flow on Anak Krakatau, a stone was placed on the collector to prevent loss to high winds. (b) An example of fallout material taken in fallout collector deployed on the pyroclastic flow of Mount St. Helens. Numerous insect, plant, and lichen fragments are evident in the sample derived from a 0.1 m² collector exposed for 10 days.

can be avoided by retreating under stones or to their shadows during the heat of the day.

The influx of spiders proved to be a surprisingly large component of the arthropod fallout. From 1981 to 1986 ballooning spiders made up over 23% of the fallout fauna, and contributed about 100 individuals per square meter per year. As noted above most of the spiders were lycosids (wolf spiders), but linyphiids were also a significant component, accounting for 50% of the 125 species diversity. By 1986 six spider species, both lycosids and linyphiids, had established reproducing populations on the pyroclastic flow.

Estimates of fallout biomass were made with fallout collectors designed to simulate the pyroclastic flow surface and to give an index of the true flux of organic material at the surface. The fallout collectors were emptied at approximately 10 day intervals during the active period of the year from late May to early October. After the first year known predator species were removed from samples before biomass determinations were made. Thus all data reflect the arrival of imported material but underestimate it to the extent that some immigrant predators and scavengers were rejected. Further, large insects

**Fig. 9.6.** An example of a pioneer insect, a carabid beetle (*Bembidion planatum*), that is among the first pioneers to colonize the tephra surfaces of the newly erupted Mount St. Helens. Scale bar = 1 mm. Photograph by Dr. P. R. Sugg.

such as grasshoppers and butterflies that did not fall into the lower spaces of the collectors were subject to capture by birds such as crows. This factor further biases the data toward underestimates of the true arthropod flux. We recorded inputs as high as 18 mg/m$^2$ per day for dried arthropod bodies and 26 mg/m$^2$ per day for non-arthropod material, mainly consisting of plant and lichen fragments. On average the combined organic fallout of arthropod and plant material amounted to about 5–15 mg(dry wt.)/m$^2$ per day for about 100 days encompassing the summer months (Fig. 9.7). Of this material 2–10 mg comprised arthropod bodies of non-resident species such as flies, aphids, and a broad spectrum of other arthropods. Broadly similar biomass figures were obtained by Ashmole and Ashmole (1987) on Tenerife, and Heiniger (1989) in the Bernese Oberland.

The principal source for the wind-borne organic fallout on Mount St. Helens is the fertile lowlands to the west and southwest comprising agricultural land and a mosaic of forested and logged areas. Nearer the pyroclastic flows are great areas of blown down or standing dead trees and mudflows where decaying logs, recolonizing herbs, pools, and streams also provide habitat for a diverse arthropod fauna. Dispersal flights take the arthropods into the air column from which they may then be deposited by local winds

on the mountain. Local orographic effects may have been significant in determining local fallout fluxes, for significant differences were found between sampling sites on different parts of the pyroclastic flow areas.

As the pioneer resident predator/scavenger populations built up, increasing quantities of the immigrant arthropods were taken as live prey or were scavenged (Sugg, 1989). For example, carabid beetles bearing aphids or midges in the mandibles were frequently encountered, particularly during night hours when they were most active. Similarly, small scavenging tenebrionid beetles would make entry to the bodies and remove all but the most heavily sclerotized parts of dead grasshoppers.

Our measurements of the quantity of arthropod fallout on Mount St. Helens were not entirely unexpected on the basis of data from other sites (Edwards, 1987) but given the low nutrient content of the mineral deposits left by the eruption, the possibility arose that the arthropod fallout might not only be a resource base for the pioneer arthropods but also a contributor to the net nutrient import to the area. Accordingly experiments were carried out to determine the rate of release of nutrients from insect bodies. Mesh bags containing dried adult *Drosophila* were placed on the pyroclastic flow and lightly covered with tephra. The bags were subsequently removed

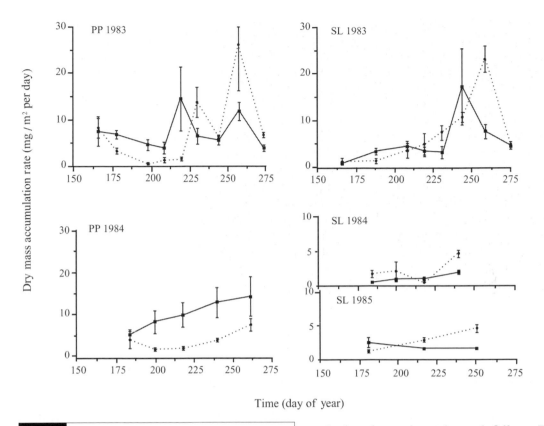

Time (day of year)

**Fig. 9.7.** Biomass accumulation of arthropods (solid line) and vascular plant fragments (broken line) in fallout collectors set in pyroclastic flow surfaces on Mount St. Helens from spring through summer, 1983–5 at two sites, PP (Pumice Plain) and SL (Spirit Lake). Time is denoted as julian day. Data are means +/− 1 SE. From Edwards and Sugg (1993).

after intervals of 17 days to 2 years and their nitrogen and phosphorus content compared with the original samples (Edwards and Sugg, 1993). Even though the exoskeletons of many of the flies retrieved after 2 years looked more or less intact, Kjeldahl-extractable nitrogen decreased by 25% within the first 17 days, and thereafter remained more or less constant, presumably due the resistance of chitin to decay, and perhaps to the lack of chitin-digesting microorganisms in the habitat. The phosphorus content of the fly bodies, on the other hand, decreased by 73% in the first 17 days and thereafter a further 10%, giving a total phosphorus yield over 2 years of 83% of the original body content. On the basis of these data, and others for the mean content of nitrogen

and phosphorus in arthropod fallout (Edwards and Sugg, 1993), we estimate that the Mount St. Helens arthropod fallout cumulatively contributes about 80 mg of fixed nitrogen and 5.5 mg phosphorus as net import per year to the mineral surface. As noted above, we know that these figures are underestimates, and possibly gross underestimates since larger arthropod bodies such as bumblebees and grasshoppers that were not concealed by the fallout collectors are known, from our direct observation, to have been taken by birds and are not taken into account. The original pyroclastic flow materials and tephra at Mount St. Helens contained very little carbon or nitrogen and it is clear that the arthropod fallout was a significant source of nutrient. The total organic carbon and total Kjeldahl nitrogen content of 1980 samples of pyroclastic flow material in the vicinity of our arthropod sampling sites were reported to be zero (Engle, 1983). By 1985 the levels were still relatively low (Nuhn, 1987) with 0.5–1 g/kg organic material, 10–90 mg/kg Kjeldahl-extractable nitrogen, and 300–400 mg/kg phosphorus. This increase can only have been derived from imported material, of which allochthonous arthropods was a significant fraction, as shown by our fallout collectors. The estimated accumulation rates of total organic carbon (97.7 mg/kg per year) and nitrogen (2.8 mg/kg per year) in plant-free pyroclastic material from Mount St. Helens (Halvorsen et al., 1991) are comparable to estimates for other volcanoes: Mt. Katmai (Griggs, 1933) and Mt. Shasta (Dickson and Crocker, 1953). The potential sources of nutrient in these two latter cases were not discussed, but the source of nitrogen which is a limiting nutrient in the tephra of Parícutin was considered to be ammonium chloride from fumaroles. It seems likely that in all cases arthropod fallout could have contributed to the nutrient content of the new surfaces. And since plant seeds tend to lodge in small declivities that also trap dead arthropod bodies, and on spider webs (Dale, 1989), it could be said that many seeds germinate in a bed of arthropod compost. We conclude from this study that arthropod fallout, which is a widespread feature of raised topography, is a hitherto underestimated medium for nutrient transport at the landscape scale. It might be argued that these figures reflect only the pattern of events on a temperate volcano surrounded by fertile lowlands, but the presence of arthropod fallout from Alaskan tundra snow patches to tropical islands suggests that it is a general phenomenon and that the aerial transport of nutrients should be taken into account in modeling the patterns of recolonization of volcanoes.

Transport of nutrients within the blast zone was reflected in the occurrence of mosquitoes in our fallout traps several kilometers from the nearest bodies of water that could support aquatic larvae. Lakes such as Spirit Lake and ponds of the post-eruption landscape contained rich organic brews of cooked vegetation that provided the food for bacterial blooms (Baross et al., 1982). The bacterial content of Spirit Lake water reached nearly 0.5 billion cells per milliliter, a concentration described by Larson (1993) as "possibly unprecendented in the annals of environmental microbiology." This bloom in turn supported large populations of mosquito larvae which were able to thrive in this turbid anoxic environment because of their air-breathing "snorkel" mode of respiration. The transient but prolific production of adult mosquitoes was reflected for several years in our arthropod fallout collections made several kilometers from the nearest aquatic source.

Birds returned to the open landscape of the blast zone as quickly as the arthropods (Manuwal et al., 1987). Water pipits, gray-crowned rosy finches, and corvids commonly exploit the arthropod fallout on snow surfaces in the subalpine and alpine zones from California to Alaska and were quick to return to feed in the open landscape of the pyroclastic flows and upper slopes of Mount St. Helens where they scavenged the arthropod fallout. Indeed the crows on occasions took advantage of the resources provided by our arthropod fallout collectors. For birds of open land such as killdeer plover (Charadrius vociferus) the post-eruption landscape provided extensive new habitat.

With the establishment of plant cover, especially where springs and streams provide a water source, the arthropod faunal diversity increases explosively and begins to lose its post-eruption character. Many of the pioneer arthropod species then disappear for unknown reasons perhaps related to competition with later immigrants. But they can still be found in the more limited habitats afforded by bare disturbed ground such as braided river beds, alpine screes and landslides, waiting for the bonanza of the next eruption. They fit Hutchinson's (1951) concept of the fugitive species: "They are forever on the move, always becoming extinct in one locality as they succumb to competition and always surviving by re-establishing themselves in some other locality as a new niche opens."

## Later events in animal recolonization

So far I have emphasized the pioneering events of recolonization, the role of the "parachute troops" (Lindroth, 1970) and the water-borne "rafters" in establishing a bridgehead after an eruption. As soon as plants establish on the surface and primary production gets under way new linkages and interactions develop, new niches arise and complexity increases. The urge to understand these patterns and the processes of recoloniza-tion has challenged ecologists for decades, and a large body of theory has accumulated that is pri-marily directed to the dynamics of revegetation, especially on bare ground such as the forelands of retreating glaciers (Matthews, 1992). But animals have also attracted attention and are the focus of the equilibrium theory of island biogeography (MacArthur and Wilson, 1967) that has provided the conceptual basis for ecological, biogeograph-ical, and evolutionary studies.

Volcanoes create islands, whether set in the water of seas and lakes, or on land, where their elevation offers habitats that may be sepa-rated by great distances from the nearest similar upland environments. They are in effect ecolog-ical islands, and are thus ideal subjects for the test of island biogeographical theory, especially where repeated eruptions wipe the slate and a replay of the successional saga begins again.

Krakatau is the classic volcano for such stud-ies. Even though early expeditions were infre-quent, they were thorough and now, together with more recent intensive studies led by Thorn-ton (1996), a vivid picture spanning more than a hundred years has emerged.

Island biogeography theory proposes that as organisms colonize a new island the *rate* of addi-tion of new species will eventually decline until extinctions equal arrivals and an equilibrium is achieved that depends primarily on the area of the island and the locale. Species good at dispers-ing get there easily, while more sedentary forms will take longer. Then there is the filter of sur-vival, for arrival does not guarantee a place in the developing food web: newly arrived mice, for example, will not establish without seeds to eat, and rafting frogs will leave no heirs unless they find a pond.

Dammerman (1948), who first addressed the animal biogeography of Krakatau, anticipated modern equilibrium theory in comparing the fauna of Krakatau with that of other islands in the region. He recognized factors such as mobil-ity and dispersive capacity (high for birds and flying insects but low for earthworms), the dis-tance from nearest sources of colonists, and the dependence of animal diversity on plant diver-sity. From a scene of total devastation in August

1883, Rakata Island (the remnant of Krakatau) had acquired grasses a year later, 49 species of plant 14 years later, nearly 300 species by 1928, and now to the untrained eye appears to be entirely clothed with tropical rainforest like that of the adjacent mainland of Java and Suma-tra, but which botanical reconnaisance shows still lacks many forest elements (Thornton, 1996). It has taken a century to acquire the present 30 species of resident birds, in a sequence that had losses as well as gains on its way to what may now be close to the equilibrium diversity. The vertebrate fauna also now includes snakes, lizards, bats, and rats, and there are now diverse invertebrates well in excess of 600 species. The non-flying animals have all presumably arrived by swimming between islands as has been observed in the case of the large monitor lizards, or by rafting on the sometimes massive jumbles of vegetation released by landslides into main-land rivers and so into the adjoining seas. The islands have long been used as bases for fish-ing and as a rich source of edible land crabs; the human element has doubtless contributed more than its share too. The process of diversifica-tion continues today, and under continuing close scrutiny provides the data for an increasingly robust test of the equilibrium theory of island biogeography.

## Consequences of human intervention in post-eruptive ecosystem restoration

Economic considerations prompted human inter-ventions on the post-eruption Mount St. Helens landscape, many of which had impacts on ani-mal populations. For example, large areas were seeded with grass and legumes in an attempt to retard erosion. Establishment was very uneven, due in part to the impoverished nutrient status of the tephra, and to the instability of steeper ter-rain. But where grasses did establish, peak popu-lations of mice (*Peromyscus miniculatus*), built on the available seed source, turned to the bark of conifers during the winter, resulting in high tree mortality until increased concentration of raptors and coyotes reduced the mouse population (Dale, 1991). Salvage logging operations which removed much large stable woody debris from watercourses are thought to have had a negative impact on the quality and quantity of winter habitat for anadromous fish such as coho salmon (Franklin *et al.*, 1995). Extensive restocking of several fish species has been successful where fine sediment has been naturally cleared from the gravels of spawning areas (Franklin *et al.*, 1995).

## Final remarks

Such is the frequency of major eruptions in relation to the human lifespan that we are inclined to see them as unique events, but it is important to bear in mind that, taken on a broader evolutionary timescale, the ecosystems of volcanic areas have seen it all before many times over. On the Pacific "Rim of Fire," for example, Mount St. Helens has erupted at least 20 times during the last 4500 years (Crandall and Mullineaux, 1978) and was still recovering from a previous eruption 180 years earlier when the eruption of 1980 provided the latest great natural experiment for volcano ecologists. With every eruptive cycle habitat is destroyed or disrupted, forests burn but then regenerate, and rivers silt up, rendering them turbid and sterile, but within decades salmon, and the invertebrates their young will feed on, return yet again. Here, as on volcanoes around the Earth, the cycle repeats.

### The role of arthropods as primary colonizers

I have attempted in the foregoing to build a picture of events and processes in recolonization that seem to be widespread, if not universal, in the recolonization of post-eruption volcanic terrain. In particular I have emphasized the pioneering role of arthropods, both as primary colonists, and as a medium of nutrient transport. Little noticed but ubiquitous, the "aerial plankton," composed of prodigious numbers of insects and spiders undergoing their dispersive behavior patterns, provides both the primary colonizers,

the fugitive species that specialize as pioneers on barren mineral surfaces, and the arthropod fallout that is their resource base. Cumulatively the arthropod fallout contributes nutrients to the new surfaces and in doing so facilitates the establishment of pioneer plants. Other animals have their place in the post-eruption succession, but most of them must await the maturation of the landscape and its vegetation before they are able to return.

### The refugia question

The role of refugia in the process of recolonization after volcanic events has no general solution; each case must be examined on its merits. So many factors are involved that each case must be assessed according to the energy released by the eruption and the history of vulcanicity at the site, the topography of surfaces generated, their temperature at the time of deposition, their chemical composition, and the climate, especially seasonal effects such as snowpack and the incidence of dormancy or diapause in organisms at the site. Krakatau and Mount St. Helens stand in contrast here: there is general concensus that all life was annihilated at Krakatau, but it is clear that microrefugia outside the immediate blast zone at Mount St. Helens played a significant role in the provision of colonizers of the post-eruption landscape.

### Tempo and pattern

Another pervasive theme is the relatively rapid tempo of recolonization. Krakatau, Surtsey, and Mount St. Helens tell the same story of pioneers gaining an early entry, transients establishing an early bridgehead, and permanent colonization proceeding from these sites. Recovery of animal populations, especially the majority of them that are dependent on revegetation, is slower than that following other major disturbances such as fire or hurricane (Turner *et al.*, 1997) especially where great areas of mineral surface are the starting point. Their tempi may differ on the scales of annual or decadal rates, but on a geological or evolutionary timescale they are all rapid, as they replay cycles that must have been evolving since the earliest appearance of the biosphere.

## Prospects

### The need for biodiversity studies of volcanic sites

Catastrophic eruptions are sufficiently infrequent in relation to the timescale and accessibility to the individual volcano ecologist that very few individuals will have the opportunity to follow the early responses of animals to more than one volcanic event. It is then necessarily a case of building on the work of others. One lesson we have learned from our work on Mount St. Helens is the need for baseline data. We now have detailed knowledge of the biodiversity on the post-eruption landscape at Mount St. Helens, but other than on higher plants, vertebrates, and butterflies we have almost no pre-eruption data. Systematic biodiversity surveys, even without full taxonomic analysis, of areas surrounding volcanoes with potential for eruption would provide baseline data so necessary for making a distinction between putative survivors and immigrants in order to build a picture of the recolonizing process. The original congressional mandate of the US National Parks Service includes the preparation of inventories of flora and fauna but in the 100 years of their existence this charge has not been realized for any of the country's national parks. This absence is particularly relevant to volcanoes such as Mt. Rainier (Fig. 9.1), which has the potential for volcanic and mudslide activity but for which there are as yet only very partial biological inventories.

### Ecophysiological studies of pioneers in post-eruptive sites

What are the special features of pioneer insect species that enable them to endure the rigors of life on open mineral surfaces where fine, abrasive dusts, high winds, extreme insolation, and desiccation limit the survival of immigrants to a few species? Further, why do these fugitive pioneer species disappear from the scene with the advent of vegetation? Do they succumb to competition with later arrivals, or does the changed physical environment impact aspects of their developmental physiology? These questions are amenable to experimental examination.

### Interactions between animals, plants, and microorganisms during early recolonization of devastated post-eruption landscapes

Examples of the importance of animals in the redistribution of mycorrhizal fungi discussed above point to a complex web of changing interactions during early succession, for which many more cases are doubtless waiting to be discovered. Further, the great unknown of microbiological diversity, particularly among the extremophile bacteria of volcanic areas (see Stetter, Chapter 6, this volume) will doubtless prove to have significance for the biology of animal colonists. For how long after an eruption are plants and animals simply "doing their own thing"? How soon do the interactions between species generate what can be recognized as an ecosystem?

## References

Allen, M. F., Crisafulli, C., Friese, C. F., *et al.* 1992. Re-formation of mycorrhizal symbioses on Mount St. Helens 1980–1990: interactions of rodents and mycorrhizal fungi. *Mycological Research*, **96**, 447–453.

Anderson, D. C. 1982. Observations on *Thomomys talpoides* in the region affected by the eruption of Mount St. Helens. *Journal of Mammalogy*, **63**, 652–655.

Anderson, D. C. and MacMahon, J. A. 1985. Plant succession following the Mount St. Helens volcanic eruption: facilitation by a burrowing rodent *Thomomys talpoides*. *American Midland Naturalist*, **114**, 62–69.

Anderson, N. H. 1992. Influence of disturbance on insect communities in Pacific Northwest streams. *Hydrobiologia*, **248**, 79–92.

Ashmole, N. P. and Ashmole, M. J. 1987 Arthropod communities supported by biological fallout on recent lava flows in the Canary Islands. *Entomologica Scandinarica (Suppl.)*, **32**, 67–88.

Ashmole, N. P., Ashmole, M. J., and Oromi, P. 1990. Arthropods of recent flows on Lanzarote. *Vieraea*, **18**, 171–187.

Backer, C. A. 1929. *The Problem of Krakatao as Seen by a Botanist*. Surabaya, Weltvreden.

Ball, E. and Glucksman, J. 1975. Biological colonization of Motmot, a recently created tropical island. *Proceedings of the Royal Society of London*, Series B, **190**, 421–442.

Banks, N. G. and Hoblitt, R. P. 1981. In P. W. Lipman and D. R. Mullineaux (eds.) *The 1980 Eruption of Mount St. Helens, Washington*, US Geological Survey Professional Paper no. 1250. Washington, DC, US Government Printing Office.

Baross, J. A., Dahm, C. N., Ward, A. K., *et al.* 1982. Initial microbiological response in lakes to the Mt. St. Helens eruption. *Nature*, **296**, 49–52.

Burt, W. H. 1961. Some effects of Volcan Paricutin on vertebrates. *Occasional Papers of the Museum of Zoology, University of Michigan*, **620**, 1–24.

Cook, R. J., Barron J. C., Papendick R. I., *et al.* 1981. Impact on agriculture of the Mount St. Helens eruption. *Science*, **211**, 16–22.

Crandall, D. R. and Mullineaux, D. R. 1978. *Potential Hazards from Future Eruptions of Mount St. Helens Volcano*. US Geological Survey Bulletin no. 1383-C. Washington, DC, US Government Printing Office.

Crawford, B. A. 1986. The recovery of surviving fish populations within the Mount St. Helens National Volcanic Monument and adjacent area. In S. A. C. Keller (ed.) *Mount St. Helens: Five Years Later*. Chiney, WA, Eastern Washington University Press, pp. 293–296.

Crawford, R. L., Sugg, P. R., and Edwards, J. S. 1995. Spider arrival and primary establishment on terrain depopulated by volcanic eruption at Mount St. Helens, Washington. *American Midland Naturalist*, **133**, 60–75.

Dale, V. H. 1989. Wind dispersed seeds and plant recovery on the Mount St. Helens debris avalanche. *Canadian Journal of Botany*, **67**, 1434–1441.

1991. Revegetation of Mount St. Helens debris avalanche ten years post-eruptive. *National Geographic Research and Exploration*, **7**, 328–341.

Dammerman, K. W. 1929. Krakatau's new fauna. *Proceedings of the 4th Pan-Pacific Scientific Congress, Java*, **37**, 83–118.

1948. The fauna of Krakatau 1883–1933. *Verhandelingen Koniklijke Nederlansche Akademie van Wetenschappen, Afdeling Natuurkunde II*, **44**, 1–594.

Darwin, C. 1839. *Journal of Researches into the Geology and Natural History of the Various Countries Visited by H.M.S. Beagle*. London, Colburn.

Decae, A. E. 1987. Dispersal: ballooning and other mechanisms. In W. Nentwig (ed.) *Ecophysiology of Spiders*. Berlin, Springer-Verlag, pp. 348–356.

Dickson, B. A. and Crocker, R. L. 1953. A chronosequence of soils and vegetation near Mt Shasta, California. II. The development of the forest floor and carbon and nitrogen profiles of the soils. *Journal of Soil Science*, **4**, 142–156.

Edwards, J. S. 1986. Derelicts of dispersal: arthropod fallout on Pacific Northwest volcanoes. In W. Danthanarayana (ed.) *Insect Flight: Dispersal and Migration*. New York, Springer-Verlag, pp. 196–203.

1987. Arthropods of aeolian ecosystems. *Annual Review of Entomology*, **32**, 163–179.

1988. Life in the allobiosphere. *Trends in Ecology and Evolution*, **3**, 111–114.

Edwards, J. S. and Schwartz, L. M. 1981. Mount St. Helens ash: a natural insecticide. *Canadian Journal of Zoology*, **59**, 714–715.

Edwards, J. S. and Sugg, P. R. 1993. Arthropod fallout as a resource in the recolonization of Mount St. Helens. *Ecology*, **74**, 954–958.

Edwards, J. S. and Thornton, I. W. B. 2001. Colonization of an island volcano, Long Island, Papua New Guinea, and an emergent island, Motmot, in its caldera lake. VI. The pioneer arthropod community of Motmot. *Journal of Biogeography*, **28**, 1379–1388.

Engle, M. S. 1983. Carbon, nitrogen, and microbial colonization of volcanic debris on Mount St. Helens. Thesis, Washington State University, Pullman Washington, USA.

Franklin, J. F., Frenzen, P. M., and Swanson, F. J. 1995. Re-creation of ecosystems at Mount St. Helens: contrasts in artificial and natural approaches. In J. Cairns (ed.) *Rehabilitating Damaged Ecosystems*. Boca Raton, FL, Lewis Publishers, pp. 287–333.

Fridricksson, S. 1975. *Surtsey: Evolution of Life on a Volcanic Island*. London, Butterworths.

Fye, R. E. 1983. Impact of volcanic ash on pear psylla (*Homoptera: Psyllidae*) and associated predators. *Environmental Entomology*, **12**, 222–226.

Gersich, F. M. and Brusven, A. M. 1982. Volcanic ash accumulation and ash-voiding mechanisms of aquatic insects. *Journal of the Kansas Entomological Society*, **55**, 290–296.

Griggs, R. F. 1933. The colonization of Katmai ash, a new and "inorganic" soil. *American Journal of Botany*, **20**, 92–113.

Hadley, N. F. 1994. *Water Relations of Terrestrial Invertebrates*. San Diego, CA, Academic Press.

Halvorsen, J. H., Smith, J. L., and Franz, E. H. 1991. Lupine influence on soil carbon, nitrogen, and bacterial activity in developing ecosystems at Mount St. Helens. *Oecologia*, **87**, 162–170.

Hawkins, C. P., Gottschalk, L. J., and Brown, S. S. 1988. Densities and habitat of tailed frog tadpoles in small streams near Mount St. Helens following the 1980 eruption. *Journal of the North American Benthological Society*, **7**, 246–252.

Heatwole, H. 1971. Marine-dependent terrestrial biotic communities on some cays in the Coral Sea. *Ecology*, **52**, 363–366.

Heiniger, P. H. 1989. Arthropoden auf Schneefelden und in schneefreien Habitaten im Jungfraugebiet (Berner Oberland, Schweiz). *Mitteilungen Schweizes Entomologische Gesellschaft*, **62**, 375–386.

Hendrix, L. B. 1981. Post-eruption succession on Isla Fernadina, Galapagos. *Madrono*, **28**, 242–254.

Howarth, F. G. 1979. Neogeoaeolian habitats on new lava flows on Hawaii Island: an ecosystem supported by windborne debris. *Pacific Insect*, **20**, 133–144.

  1987. Evolutionary ecology of aeolian and subterranean habitats in Hawaii. *Trends in Ecology and Evolution*, **2**, 220–223.

Humboldt, A. von 1808. *Ansichten der Natur mit Wissenschaftlichen Elauterungen*. Stuttgart, Germany, Cotta Verlag.

Hutchinson, G. E. 1951. Copepodology for the ornithologist. *Ecology*, **32**, 571–577.

  1965. *The Ecological Theater and the Evolutionary Play*. New Haven, CT, Yale University Press.

Jagger, T. A. 1945. *Volcanoes Declare war: Logistics and Strategy of Pacific Volcano Science*. Honolulu, HI, Paradise Pacific Ltd.

Johansen, C. A., Eves, J. D., Mayer, D. F., *et al.* 1981. Effects of ash from Mt St. Helens on bees. *Melanderia*, **37**, 20–29.

Kuwayama, S. 1929. Eruption of Mt. Komagatake and insects. *Kontyu*, **3**, 271–273.

Larson, D. 1993. The recovery of Spirit Lake. *American Scientist*, **81**, 166–177.

Lindroth, C. H. 1970. Survival of animals and plants in ice-free refugia during the Pleistocene glaciation. *Endeavour*, **29**, 129–134.

Lindroth, C. H., Andersson, H., Bodvarsson, H., *et al.* 1973. Surtsey, Iceland: the development of a new fauna 1963–1970. Terrestrial invertebrates. *Entomologia Scandinavica (Suppl.)*, **5**, 1–280.

Lloyd Praeger, R. 1915. Clare Island survey. *Part X. Proceedings of the Royal Irish Academy*, **31**, 92–94.

Lotsy, J. P. 1908. *Vorlesungen ueber Deszendenztheorien* vol. 2. Jena.

MacArthur, R. H. and Wilson, E. O. 1967. *The Theory of Island Biogeography*. New York, John Wiley.

MacMahon, J. A. 1982. Mount St. Helens revisited. *Natural History*, **91**, 19–23.

MacMahon, J. A. and Warner, N. A. 1984. Dispersal of mycorrhizal fungi: process and agents. In S. Williams and M. Allen (eds.) *VA Mycorrhizae and Reclamation of Arid and Semiarid Lands*. Laramie, WY, University of Wyoming Press, pp. 28–41.

MacMahon, J. A., Parmentier, R. R., Johnson, K. A., *et al.* 1989. Small mammal recolonization on the Mount St. Helens Volcano: 1980–1987. *American Midland Naturalist*, **122**, 365–387.

Manuwal, D. A., Huff, M. H., Bauer, M. R., *et al.* 1987. Summer birds of the upper subalpine zone of Mount Adams, Mount Rainier and Mount St. Helens, Washington. *Northwest Science*, **61**, 82–92.

Martin, D. J., Wasserman, L. J., Dale, L. J., *et al.* 1986. Influence of riparian vegetation on posteruption survival of coho salman fingerlings in the westside streams of Mount St. Helens, Washington. *North American Journal of Fisheries Management*, **6**, 1–8.

Matthews, J. A. 1992. *The Ecology of Recently Deglaciated Terrain*. Cambridge, UK, Cambridge University Press.

Merrill, E. M., Raedeke, K. J., Knutson, K. L., *et al.* 1986. Elk recolonization and population dynamics in the northwest portion of the Mount St. Helens blast zone. In S. A. C. Keller (ed.) *Mount St. Helens: Five Years Later*. Cheney, WA, Eastern Washington University Press, pp. 359–368.

Michaelsen, W. 1924. Oligochaeten von Niederlandisch-Indien. *Treubia*, **5**, 379–401.

Mohlenbrock, R. H. 1990. Mount St. Helens. *Natural History*, **99**, 27–29.

New, T. R. and Thornton, I. W. B. 1988. A prevegetation population of crickets subsisting on allochthonous aeolian debris on Anak Krakatau. *Philosophical Transactions of the Royal Society of London*, Series B, **322**, 481–485.

New, T. R., Bush, M. B., Thornton, I. W. B., *et al.* 1988. The butterfly fauna of the Krakatau Islands after a century of colonization. *Philosophical Transactions of the Royal Society of London*, Series B, **322**, 445–457.

Nuhn, W. W. 1987. Soil genesis on the 1980 pyroclastic flow of Mount St. Helens. Thesis, University of Washington, Seattle, WA, USA.

Osborne, P. L. and Murphey, R. 1989. Botanical colonization of Motmot island, Lake Wisdom, Madang Province. *Science in New Guinea*, **15**, 57–63.

Pyle, R. M. 1984. The impact of recent vulcanism on Lepidoptera. In R. I. Vane-Wright and P. R. Ackery (eds.) *The Biology of Butterflies*. London, Academic Press, pp. 323–336.

Scharff, R. E. 1925. Sur la problème de l'île de Krakatau. *Comptes Rendus du Congrès de l'Association française pour l'Avancement des Sciences*, **49**, 746–750.

Sontag, S. 1992. *The Volcano Lover*. New York, Farrar Straus Giroux.

Sugg, P. M. 1989. Arthropod populations at Mount St. Helens: survival and revival. Ph.D. thesis, University of Washington, Seattle, WA, USA.

Sugg, P. M. and Edwards, J. S. 1998. Pioneer aeolian community development on pyroclastic flows after the eruption of Mount St. Helens. *Arctic and Alpine Research*, **30**, 400–407.

Sugg, P. M., Greve, L., and Edwards, J. S. 1994. Neuropteroidea from Mount St. Helens and Mount Rainier: dispersal and immigration in volcanic landscapes. *PanPacific Entomologist*, **70**, 212–221.

Swan, L. W. 1967. Aeolian zone. *Science*, **140**, 77–78.

Thornton, I. W. B. 1996. *Krakatau: The Destruction and Reassembly of an Island Ecosystem*. Cambridge, MA, Harvard University Press.

2001. Colonization of an island volcano, Long Island, Papua New Guinea, and an emergent island, Motmot, in its caldera lake. I. General introduction. *Journal of Biogeography*, **28**, 1299–1310.

Thornton, I. W. B. and New, T. R. 1988. Krakatau invertebrates: the 1980s fauna in the context of a century of recolonization. *Philosophical Transactions of the Royal Society of London*, Series B, **322**, 493–522.

Thornton, I. W. B., New, T. R., McLaren, D. A., *et al.* 1988. Air-borne arthropod fallout on Anak Krakatau and a possible pre-vegetation pioneer community. *Philosophical Transactions of the Royal Society of London*, Series B, **322**, 481–485.

Treub, M. 1888. Notice sur la nouvelle flore de Krakatau. *Annales du Jardin Botanique de Buitenzorg*, **7**, 213–223.

Turner, M. G., Dale, V. H., and Everham, E. H. 1997. Fires, hurricanes, and volcanoes: comparing large disturbances. *BioScience*, **47**, 758–768.

Whittaker, R. J., Partomihardjo, T., and Riswan, S. 1995. Surface and buried seed banks from Krakatau, Indonesia: implications for the sterilization hypothesis. *Biotropica*, **27**, 345–354.

Whymper, E. 1892. *Travels amongst the Great Andes of the Equator*. 1987 reprint: Salt Lake City, UT, Gibbs M. Smith Inc.

Winoto Suatmaji, R. A., Coomans, A., Rashid, F., *et al.* 1988. Nematodes of the Krakatau archipelago, Indonesia: a preliminary overview. *Philosophical Transactions of the Royal Society of London*, Series B, **322**, 369–378.

Wise, D. H. 1993. *Spiders in Ecological Webs*. Cambridge, UK, Cambridge University Press.

Wurmli, M. 1971. Zur pflanzlichen und tierischen Besaidlung der rezenten Laven und Tephrata des Aetna, unter besonderer Berucksichtigung der Makrofauna und Struktureller Aspekte. Thesis, University of Vienna.

1974. Biocenoses and their successions on the lava and ash of Mount Etna. *Image Roche*, **59**, 32–40; **60**, 2–7.

Zalisko, E. J. and Sites, R. W. 1989. Salamander occurrences within Mount St. Helens blast zone. *Herpetological Review*, **20**, 84–85.

# Chapter 10

# Human impacts of volcanoes

## Peter J. Baxter

## Introduction

Volcanoes can hold a deep fascination. Images of erupting volcanoes grab our attention as we marvel at the sight of the Earth in violent movement, and tourists flock to view steaming craters to sense the enormous energy lying dormant beneath their feet. Volcanoes are often striking features in landscapes of great beauty, and people have been drawn over the centuries to live on their flanks with the promise of verdant agricultural land. But many communities have learned that years of peace can be brutally interrupted by the return of volcanic activity, and in some parts of the world such as Hawaii and Indonesia, volcanoes have even been granted the status of gods.

In most active volcanic areas, however, burgeoning populations have no memory of past eruptions when they recur with intervals of hundreds or thousands of years, and no feeling for the disaster that can lie ahead when the sleeping giant awakes. The destruction of Pompeii and Herculaneum in the eruption of Vesuvius in AD 79 has the same hold on the popular imagination as the sinking of the Titanic by an iceberg, with the spectacle of normal living being abruptly halted by catastrophe and the evidence of extinguished life locked deep beneath the ground or sea. The collapse of the Minoan culture after the eruption of Santorini some 3600 years ago is perhaps the stuff of legend, but a huge eruption did occur which buried or swept away

the settlements on the island and had impacts on other islands of the Aegean, such as Crete. Excavations at one of the buried towns, Akrotiri, have revealed a remarkable culture, a kind of Pompeii of the Aegean. Unlike at Pompeii and Herculaneum, no human remains have been found, suggesting that the islanders had managed to flee the island in time.

Amongst the range of natural hazards of sudden onset which can cause major destruction and loss of life, such as earthquakes, hurricanes, and floods, major volcanic eruptions are relatively uncommon but have the potential to be the most destructive. In terms of scale of devastation on Earth, the largest eruptions are only surpassed by the exceedingly rare impact of large meteorites. Fortunately for us, such events probably occur only once every several tens of thousands of years, the last one being the caldera collapse at Lake Toba about 75 000 years ago which devastated the center of the island of Sumatra. Such an eruption would have immense global consequences on climate and life. The largest eruption in historic times occurred at Tambora in Indonesia in 1815, the resulting perturbation of climate leading to famine in some countries in the northern hemisphere; the occurrence of eruptions with this impact has a probability of once every 2000 years, and with the planet having to support a much larger population today a repetition of such an event could have catastrophic effects on the world economy. In contrast, eruptions of the size of Mount St. Helens in 1980 can be expected once in 20 years on average

*Volcanoes and the Environment*, eds. J. Martí and G. G. J. Ernst. Published by Cambridge University Press. © Cambridge University Press 2005.

| Table 10.1 | Scale of major explosive eruptions | | | |
|---|---|---|---|---|
| | Ejecta volume (km$^3$) | Casualities | Global occurrence rate (years) |
| Toba, Indonesia, 75 000 years ago | 2000 | – | 80 000 |
| Tambora, Indonesia, 1815 | 100 | 60 000 | 2000 |
| Krakatau, Indonesia, 1883 | 15–20 | 36 000 | 300–400 |
| Mount St. Helens, USA, 1980 | 1 | 58 | 15–20 |

amongst the over 700 subaerial volcanoes around the world (Table 10.1).

Yet scientific interest in how volcanoes behave has really only grown in the latter half of the twentieth century. Studies of the impacts of eruptions on humans and the environment are even more recent, whilst popular awareness of the hazards of living in areas of active volcanism remains very limited up to the present time. As we unlock the secrets of how the inhabitants of Pompeii and Herculaneum were killed, we are learning how to use our knowledge to reduce volcanic risk around Vesuvius itself as well as in other densely populated areas of volcanism. We shall now summarize some of these lessons for human health and survival from recent eruptions around the world, starting with some less-well-known impacts of volcanoes in their repose periods.

## Volcanoes at rest

Active volcanoes can lie dormant for hundreds or even thousands of years, and major eruptions like the one at Mt. Lamington, Papua New Guinea in 1951, have even occurred at mountains that were not previously recognized to be volcanoes. But signs of life can often be detected by volcanologists, such as the occasional fumarole or hot spring, or evidence of magmatic gases emanating from the soil at fault lines. By collecting and analyzing samples of the gases and waters monitoring of the volcano can begin, and changes in their temperature and composition may yield important clues to the onset of renewed activity. In some volcanoes, evidence of strong degassing from the crater or the soils may be obvious, and without it necessarily presaging the ascent of magma and a violent eruption. Nevertheless, these seemingly quiet periods of degassing can disrupt and endanger the lives of the communities that live nearby by causing air pollution of the outdoor and indoor air, when the gases carbon dioxide and radon may find their way inside dwellings. Invisible contamination of springs used for drinking water may also occur.

Other ways in which living in volcanic areas can affect health arises from the use of volcanic rocks for building or other industrial purposes. Tuffs, for example, have been exploited in the building of large cities such as Rome, Naples, and Mexico City because of the availability of the material and ease of use. Tuffs are erupted deposits bonded by natural cements or are naturally welded, as in ignimbrite rocks which are composed of pumice and glass shards sintered by the intense heat. In some countries houses are dug into the walls of cliffs formed by tuffs, the most famous example being in Anatolia, Turkey. Zeolites are minerals found in tuffs and have many specific uses in the chemical industry because of their fibrous nature and associated physical properties. However, the dust formed by the small zeolite mineral fibres, erionite, may resemble asbestos in size and shape and pose a risk of mesothelioma, an incurable form of cancer, in the inhabitants of the region who are exposed through inhaling the dust as the tuff rock weathers to form the soil on which the agricultural communities depend (Baris et al., 1981). Tuffs used for building homes in parts of Italy (e.g., Orvieto and Naples) have been found to have an elevated uranium content and to release radon, a radioactive gas whose decay products or radioactive daughters can cause lung cancer, into the indoor air (IARC, 1988). All rocks can contain uranium and all building materials contain some radium-226, itself a decay product of uranium and the main source of radon in the global environment. Pumice is widely used as an abrasive cleaner and has a variety of other industrial uses such as in the production of "stone-washed" jeans: but miners and workers handling pumice-stone from the island of Lipari, near Sicily, were reported to develop a serious form of pneumoconiosis as a result of high exposure to the dust in the past (Pancheri and Zanetti, 1960).

Some examples of the ways in which the latent activity of volcanoes in repose can affect the health of those living on their flanks now follow.

## Masaya volcano, Nicaragua: the environmental effects of sulfur dioxide and hydrogen chloride

The main gas emitted by volcanoes is water vapor, followed by carbon dioxide, sulfur dioxide, hydrogen chloride, hydrogen fluoride, and hydrogen sulfide, with lesser amounts of radon, helium, hydrogen, and carbon monoxide. Volcanologists monitoring volcanoes study the output and composition of gaseous emissions from craters, fumaroles, and ground fractures, but the results of this work are rarely applied to studying the effects of the emissions on people or the environment. Masaya volcano provides an important opportunity for interdisciplinary studies of its emissions during its periodic degassing phases.

Traveling northwards along the Pan-American Highway on the way to Nicaragua's capital Managua, one sees that the lush vegetation and coffee fields are suddenly replaced by an area of what appears like scrubland, with stunted grass and bushes and a few stunted trees, which lasts for about 8 km along the road. There is no clue as to what has caused this striking appearance but as the road passes along a raised ridge (the Llano Pacaya) the traveler would be right in thinking that the effect was related in some way to the strong prevailing wind which sweeps the area. In fact, the wind draws the plume from Masaya Volcano, which lies 20 km away, over the ridge and down the other side for another 30 km to the Pacific Ocean. Years of fumigation by volcanic gases and associated acid rain have left an agricultural wasteland of 500 km$^2$ in the plume's

path. About 50 000 people live in, or close to, this area and in its poorest parts the houses stand out against a bleak landscape (Fig. 10.1).

Periods of strong degassing and repeated filling of lava lakes have occurred frequently over the centuries at Masaya's craters; the early Spanish explorers thought the volcano was a gateway to hell, and erected a large cross over the then-active crater. Today, a cross still stands guard over Santiago crater, the source of several intense degassing crises this century. Masaya is the most active of all the volcanoes in Central America. Modern interest in its degassing began after a crisis started in 1924, when local coffee-growers watched helplessly as their crops were devastated by the gases. A scheme was proposed to seal over the vent and divert the gases to make sulfuric acid in a plant on the crater rim, but the plan was abandoned when after an attempt to seal off the crater with dynamite in 1927, the emissions ceased. Many important plantations were abandoned after this, but the coffee-growers' resistance to the volcano was renewed when degassing recommenced in 1950–9. One plan that was seriously considered was to cap the crater with a 300-m-high stack to carry the gases away over the Sierras. Then in 1953 the Nicaraguan air force was deployed to drop two bombs into the crater but without any effect. Even using an atomic bomb was considered (McBirney, 1956). These attempts to control the unpredictable activity of volcanoes now seem quixotic, but in the face of the destruction of the local agricultural economy they were desperate measures. In the next degassing event in 1979–86 the emissions and their effects began to be better studied but work was curtailed when civil war broke out in 1983.

Volcanologists routinely monitor active volcanoes using a special mobile instrument, the correlation spectrometer (COSPEC), to estimate the daily emissions of sulfur dioxide. It can be driven under a plume or mounted on fixed-wing aircraft to fly below the plume. A rapid increase in output of the gas can signify to volcanologists a rise in magma as a precursor to an eruption. This instrument has been in use since the 1960s to monitor the output of sulfur dioxide from chimneys and power-station stacks where the source of the gas is the combustion of coal and oil fuels.

**Fig. 10.1.** The Llano Pacaya area downwind of the Masaya Volcano, Nicaragua. The lush tropical vegetation has been replaced with non-productive scrubland as a result of fumigation by the plume and acid rainfall during successive degassing crises.

Sulfur dioxide is an important air pollutant in industrialized countries and it is highly appropriate to apply our knowledge from these settings to volcanoes. At its height of degassing in the early 1980s the average flux of sulfur dioxide from Masaya was 1275 t/day, hydrogen chloride 830 t/day, hydrogen fluoride 16 t/day, with hydrogen sulfide and sulfates as only minor species. The key to the environmental impacts of this volcano seemed to lie in the very high emission of sulfur dioxide and hydrogen chloride (Baxter *et al.*, 1982).

Sulfur dioxide and hydrogen chloride in the plume will have a direct destructive effect on contact (dry deposition) with vegetation if their concentrations during fumigation episodes are high enough. Flowers of the coffee and other crops are vulnerable, as are the young coffee beans and the leaves of other sensitive species. Effects can be

readily seen on the sides of trees facing the volcano – the leaves appear burnt and discolored around the edges, with the branches often bare of leaves altogether.

Hydrogen chloride is much more soluble than sulfur dioxide and readily combines with moisture to form hydrochloric acid. Volcanic plumes are rich in water vapor which readily condenses to make the plume visible from afar; hydrochloric acid will dissolve in the droplets to form acid aerosols, but these will rapidly fall out of the plume and deposit on vegetation downwind of the crater. Just how far these aerosols are transported is not known at Masaya, but the distance is going to be substantially shorter than the range of damaging concentrations of sulfur dioxide in the air. Sulfur dioxide in volcanic plumes, as in emissions from fossil-fuel combustion sources such as power stations, usually takes hours to oxidize and to form sulfuric acid aerosols (which, incidentally, are the main contributors to the formation of anthropogenic acid rain). These evaporate and become smaller in diameter; instead of falling out of the plume they can be transported for long distances. Whether sulfate aerosols are formed in sufficient quantities to damage vegetation or trigger respiratory problems before the plume reaches the ocean is not yet known, but the role of acid aerosols will be considered again below in the studies at Poás and the Montserrat volcano.

Thus, in addition to the dry deposition of hydrochloric acid, the key to the environmental and health impacts of the plume at Masaya, as it was studied by scientists in the 1980s, rests with sulfur dioxide concentrations in the plume and the acid rain formed when rain falls through the plume and takes up the hydrogen chloride gas and the hydrochloric acid aerosols, with hydrogen fluoride (also present almost entirely as an aerosol) adding a minor contribution. The pH of rainwater in the downwind sector is very variable but typical values were in the range of 2.6–3.5. Aside from direct damage to vegetation, the chronic acid input appears to have exceeded the buffering capacity of the soil and only a few resistant shrub species are able to flourish.

In the 1980s the concentrations of sulfur dioxide in the fumigated areas were intermittently above the then World Health Organization (WHO) air-quality standard as the plume deviated with the wind conditions. Degassing recommenced in 1996 and measurements of sulfur dioxide using diffusion tubes on two occasions has shown that levels throughout the area of vegetation damage are again elevated to levels likely to be intermittently exceeding air quality standards (the UK air quality standard for sulfur dioxide is 100 ppb over a 15-min averaging period (Department of Health, 1997); in one community this value was *the average* over *2 weeks* of measurement with diffusion tubes). The effect of sulfur dioxide is to worsen the condition of people suffering from asthma and other chronic lung diseases. Asthma attacks can be triggered and deaths brought forward in people with severe forms of pneumonia and bronchitis, and heart disorders (Department of Health, 1997).

Anthropogenic acid rain is not in itself a health problem, but acidification could mobilize toxic elements such as aluminum and lead in surface waters such as lakes used as reservoirs for drinking water. Volcanic acid rain, however, may be collected directly for use as drinking water so

its composition merits careful interest. Around Masaya, as in many other tropical areas, rainwater is collected on the roofs of houses and stored in large concrete cisterns. Contamination may arise from toxic elements such as fluoride in the plume, metals dissolved from galvanized steel roofs, and lead in old pipes. Studies of acid rain distribution and drinking water quality are therefore required in addition to measuring sulfur dioxide concentrations in the ambient air.

Masaya has played an important part in alerting us to the environmental hazard of air pollution from degassing volcanoes. It also shows the importance of combining studies of vegetation damage and health, as vegetation cover plays an important part in acting as windscreens and in absorbing pollutants such as sulfur dioxide. Allowing the continuing destruction of vegetation in the area will inevitably increase exposure to gases, not only to humans but also to the remaining trees and shrubs. The volcano cannot be capped, but further efforts to find ways of reducing the impact of the gases are warranted.

## Poás volcano, Costa Rica: degassing volcanoes with crater lakes

Of about 700 subaerial volcanoes in the world about 80 have crater lakes. Craters of the right size and shape can act as excellent collectors of rainwater and can become like a huge test tube or reactor vessel on top of an active volcano. Heat and gases may emanate from crater wall fumaroles filling from the magma system which is buffered by a hydrothermal system between it and the lake. Thus crater lakes act as condensers, traps, and calorimeters for magmatic volatiles and heat. Measuring the temperature and chemical composition of the water can give important clues as to what is happening to the magmatic intrusion beneath. Close monitoring of crater lakes may therefore contribute to forecasting volcanic events (Kusakabe, 1996). It is not surprising that volcanologists are eager to study crater lakes wherever possible and to identify the key chemical parameter for eruptive forecasting. Unfortunately, active crater lakes can become serious hazards in their own right when an eruption at the bottom of the lake can cause the ejection of the water to form disastrous lahars (volcanic mudflows) as has happened at Mt. Kelet and Mt. Kawah-ijen, Indonesia, and at Mt. Ruapehu, New Zealand. But Poás volcano in Costa Rica has shown in its recent activity how a periodic drying-up of the lake during the dry season may greatly enhance the environmental impact of its degassing.

Poás is one of several volcanoes that overlook the Central Valley of Costa Rica where most of the country's population resides. The summit rises 1300 m above its base at 1400 m and reaching it is a comfortable hour's drive from San José, the country's capital. The eastern flanks of the volcano are mostly tropical rainforest whilst on the west side the coffee plantations are amongst the most important in Costa Rica. The active crater is very large with a diameter of about 1000 m and the crater bottom is about 150 m below the rim. For some years the visible degassing took place from fumaroles in the lava dome and on the sides of the crater. However, on clear days a plume from the volcano was usually not visible from afar and the volcano's activity during the 1980s seemed to be quite innocuous.

But in 1986 the temperature of the lake water increased to 65 °C and the lake level began to fall. It was in 1988 during the dry season that the lake dried out for the first time and health complaints surfaced in the small communities located on the bottom of the west side of the volcano. The nearest farm to the crater was 8 km away and there extensive vegetation damage began at the same time as coffee plants were being impacted in fields further down. The strong prevailing wind blew the invisible plume along three main ridges into the Central Valley area and it was along valley ridges that most of the vegetation damage and corrosion of metalwork took place. Soon after the rainy season began in 1989 the lake started to refill and the health complaints subsided. The cycle of drying and reforming went on until 1993, but in 1994 the advent of the wet season failed to cause any refilling of the lake. The gas emissions appeared to increase, as did the aerosols, and vegetation damage began to get worse again. After some minor explosive activity within the crater, and the occasional slight ashfall, the volcano suddenly ceased its activity in August 1994, whereafter the lake began to refill again.

Volcanologists inferred that the lake cycle of drying up and reforming again in the wet season was due to an intrusion of magma, the heat from which had caused a greater rate of evaporation of the lake (Brantley et al., 1987; Brown et al., 1989). Crater lakes are invariably acid from the gases passing through them and Poás was no exception: with a pH of zero, or less (negative values) it was probably the most acidic natural lake in the world at that time. During the weeks when the lake dried out gas emissions would fail to be scrubbed by the lake water, but in addition intense aerosol generation would take place from the acid pools on the crater floor (Fig. 10.2). The gas emissions were mainly carbon dioxide, with sulfur dioxide, hydrogen sulfide, and hydrogen chloride in similar amounts. As at Masaya, sulfur dioxide was the main air pollutant in the affected villages whilst the acid rain which fell during the wet season over most of the western flanks of the volcano was produced by raindrops dissolving the hydrogen chloride. Studies using sulfur dioxide diffusion tubes in three affected villages showed that the levels were indeed higher in the dry season when the lake was empty with mean concentrations over 2–3 weeks of 50–270 ppb compared with 20–110 ppb in the wet season. Unfortunately it was not possible to study the aerosol concentrations in the ambient air.

Epidemiological studies in the local communities suggested that the respiratory health of children had not been affected by the emissions whilst there was some corroboration of an effect on adults for asthma-type symptoms compared with a control agricultural area further south of the country.

There is a strong suspicion that the acid aerosols contributed in a major way to the symptoms of eye and respiratory tract irritation and the damage to the vegetation, including the flowers and leaves of the coffee plant. An experiment was conducted between scientists of the University of Heredia and the British Geological Survey in which a collecting device was suspended over the lake and in the path of the plume. At the same time the lake waters were also sampled and compared with the collected condensate. The lake water did show a hyperconcentrating effect in the dry season, with a big increase in the sulfate and

**Fig. 10.2.** Poás Volcano, Costa Rica, 1992. Intense aerosol degeneration from the dried-up crater lake (a) forms a plume of gas and acid aerosols which mixes with rain clouds (b) as it is blown downwind of the crater.

a smaller increase in hydrogen chloride with only a slightly discernible increase in fluoride. The condensate reflected the lake water composition, except that it contained enhanced amounts of chloride and fluoride. This result showed that the aerosols were largely produced by the mechanical formation and bursting of bubbles rather than evaporation, though there was further enhancement of acidity with the soluble gases hydrogen chloride and hydrogen fluoride dissolving into the aerosols above the lake water. The combination of the acid aerosols (with an even lower pH than the lake) and sulfur dioxide will have contributed to respiratory irritation and asthma-like symptoms in the impacted areas downwind. As the flux of sulfur dioxide was relatively low (perhaps only 200–300 t/day) the effects of the plume were most likely attributable to the highly acidic aerosols. About 10 000 people lived in the affected villages with many more thousands potentially at risk further down in the Central Valley area.

Analogous to the escape of gases into the environment is the leakage of lake water from the crater, a common feature of crater lakes. At Poás the acid waters of the Agrio (pH 2.3) and Desague rivers have high concentrations of toxic contaminants (e.g., manganese, aluminum, and fluoride) which may pollute groundwater as well as surface waters. Aluminum and manganese are poisons to the brain and nervous system, whilst fluoride excess can lead to fluorosis, a crippling bone disease, but there is no evidence of these effects arising in the local population. The water of the Rio Agrio tastes like the effluent of a chemical plant, and it is not likely to be directly used for drinking purposes – though it is locally believed to have medicinal properties.

Other famous examples of crater lakes are Kawah-ijen and Kelut in Indonesia. The lake at Kawah-ijen is the largest accumulation of acid water in the world, being 1 km wide and 180 m deep, but is mostly known for its sulfur mine. About 250 workers carry blocks of elemental sulfur from a fumarolic part of the crater up its 400 km wall and then walk 15 km before shedding their 50–80-kg load. About 10 t/day are removed in this way which routinely exposes the unprotected men to the sulfurous gases. An eruption at Kelut in 1919 drained the lake and formed a vast lahar which spread over 150 km$^2$ killing at least 5000 people and destroyed 100 villages. Afterwards tunnels were hand-dug to keep down the water level, and these have had to be re-excavated after each subsequent eruption. This was one of the world's first engineering feats to control a volcanic hazard: the original work took 7 years and cost the lives of numerous workers who died from accidental flooding along the tunnels or from exposure to the high temperatures inside them (A. Bernard, personal communication).

## Furnas volcano, San Miguel Island, Azores: ground gas emissions

The present village of Furnas is probably no more than 200 years old though the same area was already populated by early Portuguese colonizers when the Furnas volcano had a major eruption in 1630. The present population of about 1000 has no traditional memory of such an eruption and is unaware of the precariousness of its situation inside a volcanic crater. The village is surrounded by the towering walls of the caldera which form a beautiful but threatening barrier to the rest of the island. There are numerous thermal and cold-water springs around and above the village and the fumarolic hydrothermal activity has given the village its reputation as a spa. The fumaroles are situated in the village and their waters are led by pipes downhill to the spa for use in warm baths, though in the past the waters also used to be drunk for their medicinal properties. The fumarole waters and some of the springs contain elevated amounts of mobilized toxic elements such as aluminum, manganese, and arsenic. Until recently the spring water feeding the drinking supply for the local fishing village, Ribeira Quente, contained mildly elevated levels of fluoride (5 ppm) sufficient to produce dental fluorosis which is visible in all the adult population (Fig. 10.3). The volcanic nature of the place is also revealed when the visitor becomes aware that the floors of some houses are hot to touch, whilst in a number of the old houses small fumaroles are dug in the gardens and used to cook in: the gentle steam and carbon

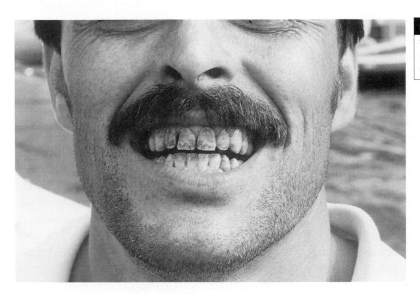

**Fig. 10.3.** Adult with dental fluorosis from drinking spring water containing elevated fluoride levels when a child.

dioxide discharges are excellent for cooking stews! But the main evidence of volcanic activity comes from the copious ground emissions of carbon dioxide which acts as the carrier gas for radon, both of which are derived from the magma below (Baxter *et al.*, 1999).

In the early 1990s Furnas was made one of the EC Laboratory Volcanoes and a multidisciplinary team worked on the volcanic and passive gas emission hazards (Baxter *et al.*, 1999). Soil sampling and mapping showed that about one-third of the houses of the town were built on a carbon dioxide anomaly. Very high outpourings of carbon dioxide occurred in these areas from drains in the road and vents below the wooden floors of houses. Inside, elevated concentrations of carbon dioxide could be found in cupboards and corners of rooms at floor level or entry points of pipes in kitchens and bathrooms. Surprisingly enough, the local people were not at all cognizant of the dangers, even though a few people knew that pits or trenches dug into the ground could carry risks of unconsciousness or even death if entered. Radon concentrations were also markedly elevated in some of the houses sampled. Radon acts in a synergistic way with tobacco smoking to cause lung cancer (in the United States around 10 000 cases of lung cancer annually are estimated to be associated with indoor radon exposure) and so many houses are

likely to need remedying to prevent the ingress of both radon and carbon dioxide.

The charming people of Furnas seem to be stoically resigned to the potential risks they face. Evidence was found of the possibility of undetectable surges of carbon dioxide into houses which could be sufficient to cause asphyxiation without warning. Geochemical studies also indicated that there could be a substantial quantity of carbon dioxide stored in the hydrothermal system below the village, leaving the people at risk of one day experiencing a massive release of carbon dioxide through cracks opening up in the ground in response to an earthquake or as a precursor event to a major eruption. Although this hazard cannot be readily confirmed by any feasible study it should nevertheless be a prominent concern for emergency planning in the area, in addition to the more usual eruptive hazards (Baxter *et al.*, 1999a).

Undoubtedly there are many other populated volcanic areas around the world where soil gas emissions can present similar threats to health and safety. Other well-known examples are Vulcano Island, Italy (Baxter *et al.*, 1990), and Mammoth Mountain (Farrar *et al.*, 1995), in the Long Valley Caldera area of California. At Vesuvius caves and depressed land areas are known to fill up with carbon dioxide before and after eruptions of the volcano. More systematic studies are also needed of radon emissions in volcanic and hydrothermal areas.

## The eruptive phase

Eruptions can be classified as two main types: effusive, involving mainly lava flows, and explosive, which are characterized by sudden and explosive releases of energy and tephra in the form of pyroclastic flows and ash and pumice falls. Gas and acid aerosol emissions occur in both. The largest effusive eruption in modern times was at the 25-km-long Laki fissure in Iceland in 1873 which produced a basaltic lava flow of about 12 km$^3$ over 8 months before the eruption stopped, but during this time the enormous gas and aerosol emissions, which included hydrogen fluoride, together with small amounts of ash, led to the decimation of vegetation and the loss of grazing animals. The ensuing famine cost the lives of a fifth of the Icelandic population (Thordarson and Self, 1993). In contrast, the largest explosive eruption in modern history was at Tambora Volcano, Indonesia, in April 1815, with massive pyroclastic flows and ash fallout as far as 1300 km away. As many as 90 000 people may have been killed in the general vicinity of the volcano, but the eruption became notable for the year 1816 being known as the "Year without a Summer" in the northern hemisphere, with crop failures and delayed harvests in many parts of the world, leading to localized famines (Francis, 1993). An especially cold winter hit the northern hemisphere in 1873–4 and the Laki eruption has been held responsible for this. Much interest is now being shown in the climatic effects of eruptions, which is more than of academic interest – a repeat of a Tambora-scale eruption in today's much more populated and interdependent world might have a far greater global impact on food supplies.

The nature of most explosive volcanoes in the world is explained by their location in subduction zones, such as the Pacific "Ring of Fire." The magma is formed with the fluxing action of water and calcium carbonate in the oceanic sediments dragged down below the converging plates. As well as lowering the melting temperature of rock, the water and calcium carbonate increase the amount of gas available in the magma as it rises to the surface in an eruption, and these factors lead to more explosive eruptions. At rift zones, the volcanoes occur along the seams of diverging plates, and their magma is more fluid as it also has a lower silica content. Thus the viscous magma of Mount St. Helens, a subduction volcano, was estimated to contain about 6% by weight of gases, mainly carbon dioxide and water; at Kilauea, Hawaii, the gas content amounts to about 1%. The viscous magma is also much more likely to confine the movement of gases inside the volcano, which can lead to pressurization and explosions. The contained gases may be "volatiles" from the ascending magma, or groundwater vaporized by the magma as it rises inside the volcano, like a steam boiler (1 liter of water occupies 1400 liters when converted to steam at normal pressure). Lava flows are only rarely a human hazard, as they move slowly and can be readily avoided, but explosive eruptions produce a range of injury agents which we shall consider below. It is the explosivity of an eruption, driven by the energy stored in the 900–1200 °C temperature of the magma, along with factors determining the style, intensity, and magnitude of the event, such as magma composition, diameter of the conduit, etc., that are the critical factors in determining the health impacts. Advances in understanding and predicting volcanic hazards in the future are likely to emerge as a consequence of scientists being able to successfully model these processes using sophisticated computerized simulations. This section begins, however, with a rare type of event, the gas burst from Lake Nyos, a crater lake of an inactive volcano, and then goes on to consider more typical eruptive phenomena.

### Lake Nyos, Cameroon: gas bursts from crater lakes

The dramatic story of Lake Nyos is about how a beautiful lake can transform without warning into a bowl of deadly gas; it also highlights how the natural world can continue to surprise us with previously unrecognized catastrophic hazards. On August 21, 1986 about a quarter of a million tonnes of carbon dioxide

**Fig. 10.4.** Lake Nyos, Cameroon, after the gas burst on August 21, 1986. The promontory on the side of the lake was overcapped by lake water during the overturning of the lake waters and release of carbon dioxide gas.

were suddenly discharged from Lake Nyos forming a dense cloud which flowed down into the adjacent valleys killing about 1700 people who lived in this remote part of Cameroon (Fig. 10.4). The gas cloud was mysteriously released between 21:00 and 22:00 h on a calm night with a light southerly wind when people were still around talking and completing the day's business or preparing to go to bed. The gas had been stored in the depths of the lake under hydrostatic pressure and was probably released from the lake with a fine water aerosol, the cooling effect of the expansion of the gas combining with the mass of the aerosol to make the carbon dioxide cloud dense enough to flow along the valley floors for as far as 20 km. Photographic evidence and the accounts of eye witnesses who survived indicated that the gas advanced with very little warning and people lost consciousness almost immediately

(Baxter *et al.*, 1989). The gas lay in the valleys at least until 06:30 the next morning when some people began to recover consciousness to find others lying dead alongside them. Animals, birds, and insects were also wiped out. It took at least 24 hr before the outside world appreciated that a most unusual disaster had occurred, and a rescue team did not enter the lifeless zone until 36 hr after the release. The anxiety of members of this team could only be imagined as they had no idea what had happened, but as they advanced along the road towards Nyos they would have observed the animal and human corpses all around whilst remaining fearful that pockets of toxic gas (whatever it was) could still be around. It was about a week before international experts arrived at the scene and for a while there was considerable confusion as to what had happened. However, tests of the lake water soon showed that carbon dioxide was abundantly present in the lake and that a huge disturbance of the waters had occurred which could either have been due to an overturning of the water and release of stored carbon dioxide or, as some volcanologists insisted, a volcanic eruption which had forced gases upward through the lake. The crater in which the lake stood no longer belonged to an active volcano and indeed subsequent seismic monitoring revealed the absence of any significant earthquake activity. The controversy over the mechanisms involved blurred scientific opinion but was soon cleared up by further studies of the density-stratified lake waters which showed that the 200-m-deep lake was recharging from a soda water spring at its bottom, the gas having dissolved and been stored in groundwater, whilst the stability of its waters permitted the accumulation of carbon dioxide at depth, with the strata at the surface supporting fish and other normal biota (Kusakabe, 1996).

Remarkably, a similar event had occurred in 1984 at Lake Monoun, only 100 km south of Lake Nyos (Sigurdsson, 1988). At 90 m Lake Monoun is only half the depth of Lake Nyos and so cannot store as much carbon dioxide. However, a landslide into the lake may have caused its overturning on the night of August 16 killing 35 people who were enveloped by the gas cloud in a marshy area at the bottom of a gentle incline from the lake less than 1 km away. They were walking along a path to a nearby town soon after 03:00 h in order to go to the local market and to start work early that day. Two others died who had come upon the scene unawares in their van at about 05:00 h. When rescuers got to the scene a white, acid smelling cloud hung over the area 0–3 m above ground, and this did not disperse until about 10:30 am when the rescuers could approach. The gas had also killed numerous wild and domestic animals in the area. Studies begun 8 months later revealed that carbon dioxide was the predominant gas which effervesced from depressurized deep water samples, as shown later at Lake Nyos. These lakes contrast with the hot, active crater lakes such as Poás, as their deep-water temperatures were normal, and their composition at the surface unremarkable.

The disaster at Lake Nyos was shocking and became widely reported around the world. This unusual natural phenomenon quite rightly attracted considerable interest, not least for the unprecedented amount of gas released. It was fortunate that the area was sparsely populated, otherwise the death toll could have been in many tens of thousands. Apart from Lake Kivu in the Democratic Republic of Congo, where the origin of the gas is from both organic and volcanic processes, no other lakes storing massive amounts of carbon dioxide have been yet identified. These natural disasters involving carbon dioxide also focused interest on the toxic properties of this gas. Death from carbon dioxide poisoning is a well-known hazard in breweries and wineries and the trick of monitoring for the gas using a candle flame (the flame is extinguished at 8–10% $CO_2$ in air) was well known to the Romans. However, the natural setting at Nyos would have led to mixing of the carbon dioxide with air with the formation of pockets of high and low concentrations of the gas and diffusion occurring only very slowly as the gas cloud remained suspended over low-lying areas on a calm night. Carbon dioxide has a narcotizing or anesthetic effect in concentrations of 10–30% and so people could have lain on the ground in a deep sleep or coma which could either have gone on to death or recovery, depending upon the dispersion of the gas cloud. Prolonged coma is also possible with carbon monoxide and *burn* marks and blistering of the skin have been recognized to arise in patients poisoned by this gas (Baxter *et al.*, 1989). *Burn* marks were also mentioned as being visible in the casualties at Lake Monoun, but a striking finding at Lake Nyos was the presence of large blisters on some of the dead bodies and even on the skin of survivors. These skin manifestations may also have been produced in an analogous way to those in carbon monoxide poisoning; although the mechanism is not understood, it may be related to prolonged hypoxia and stasis of blood in the skin. The lesions could not actually have been caused by chemical burns, such as acid gases found in eruptive discharges, as they would have been too dilute by the time they reached the lower valleys and in any case there was no evidence of acid damage to the vegetation.

## Dieng Plateau, Java: gas bursts from hydrothermal systems

The first report of a major incident involving volcanic gases in this century was the gas burst at Dieng Plateau in central Java in 1979 when 134 people were killed whilst fleeing from a phreatic eruption (Le Guern *et al.*, 1982). Dieng Plateau (2000 m asl) is a remote and fascinating agricultural area where the ruins of the oldest Hindu temples in Java are found. Asphyxiating accumulations of carbon dioxide in depressions and hollows are a recognized hazard there. The disaster began with seismic swarms appearing beneath the area on February 16 and 19 followed by further seismic activity and felt earthquakes in advance of eruptive explosions from the Sinila crater on February 20, which produced a 3.5-km mudflow that passed the nearby Kepucukan village. Frightened by the noise, and realizing that they were trapped by the mudflow to the east, at about 03:00 h the villagers started to flee along the path to the west where they were engulfed by a cloud of carbon dioxide at a place 1.5 km downslope from the Timbang crater. Other villagers escaped to the south across the fields. In this event there was probably very little opportunity for the mixing of carbon dioxide with air and the victims were immediately asphyxiated in the dense gas cloud.

This disaster is of great importance for emergency planners as it provides a rare and important precedent for the hazard of carbon dioxide stored in hydrothermal systems under pressure in active hydrothermal or volcanic areas anywhere in the world. A hydrothermal system is the circulation of heated water through fractures within a volcano. We have already discussed this possible hazard arising at Furnas Volcano and the need to incorporate the potential threat in volcano emergency planning on the island.

## Kilauea volcano, Hawaii: gas releases from lava flows

Air-polluting plumes of gas from the activity of Kilauea volcano have been almost continuous, but variable, since 1983 to the present (Sutton and Elias, 1993). Magma in the reservoir beneath the summit caldera emits a carbon-dioxide-rich gas plume, but during effusive eruptions magma depleted of most of its carbon dioxide travels to the eruptive site at the Pu'u O'o cone where sulfur-dioxide-rich gas is released. From mid 1986 and through 1991 the volcano released as much as 1800 t/day of sulfur dioxide, since when the levels have fallen to around 350 t/day and which rise to 1850 t/day when the volcano is actively erupting. The two emissions combine to form a plume that is visible as a haze under certain weather conditions and is known as vog (volcanic fog or smog). Another type of gas plume is generated when the lava, passing through the lava tube system, is discharged to the coast. Molten lava violently boils seawater to dryness and the steam hydrolysis of magnesium chloride salts releases hydrogen chloride which forms a voluminous cloud containing a mixture of hydrochloric acid and concentrated seawater. This localized lava haze is known as laze and contains as much as 10–15 ppm of hydrochloric acid. This concentration of the acid and the gas itself are likely to be of little health significance other than to those standing close to its source, but it will act as a major contributor to acid rain formation.

The gaseous emissions do not attract the same attention on the Big Island as the swathes of lava emitted by the Pu'u O'o cone on Kilauea's East Rift Zone, which present more immediate hazards to property and livelihoods, and undoubtedly a great attraction to tourists.

## Mount St. Helens eruption 1980

The eruption of Mount St. Helens on May 18, 1980, was only of average size for this volcano which erupts approximately every 150 years, and was very small by Tambora's standards – about 1 km$^3$ of tephra fallout at Mount St. Helens compared with 150 km$^3$ ejected at the 1815 eruption of Tambora – but the eruption in Washington state, USA, attracted popular and scientific attention to volcanoes like no other event in recent times (Lipman and Mullineaux, 1981). Affecting the mainland United States, it also enabled new and detailed studies to be undertaken on explosive volcanic phenomena and their effects, including the impacts of pyroclastic flows and surges on humans, the health effects of ash falls, and the behavior of lahars.

### Pyroclastic flows and surges

The volcano was under intense surveillance by the US Geological Survey and exclusion zones had been set in the wilderness area at risk, but even so the size of the main eruption when it came was unexpected. The weeks before the big event had been marked by degassing and ash emissions from a small vent at the summit, and weak volcanogenic (occasionally magnitude 4) earthquakes around the volcano. At 08:32 h on May 18 the northern flank was destabilized by a magnitude 5 earthquake originating from magma movement, and began to collapse to form a massive avalanche. As the pressure in the volcanic edifice was suddenly released, the internal superheated groundwater flashed into steam and at the same time the magma was decompressed creating a lateral blast surge which, together with the avalanche, devastated about 600 km$^2$ in a wedge-shaped sector north of the volcano extending for as far as 30 km. Fifty-eight people were killed, or were missing and presumed dead. Seventeen people who had been in the edge of the tree blow-down area, or in the area of singed trees only, had survived, though 12 received second-degree burns or suffered respiratory effects from inhalation of hot ash. The mortality rate in the zone of devastation and damage from the surge was estimated to have been 74%. Altogether, 110 people had been in or on the edge of the surge cloud, with a survival rate of 47%. This finding, showing that it was possible to survive in the open in such an eruption, was surprising, and led to further investigation of the experiences of survivors. Post mortem examinations were conducted on 25 victims, the first time that autopsies had been performed in a volcanic eruption to establish the cause of death (Eisele *et al.*, 1981; Bernstein *et al.*, 1986).

Most deaths in volcanic eruptions are caused by pyroclastic flows and surges, and related phenomena. They are Nature's answer to the nuclear weapon, being capable of causing complete destruction in minutes over very large areas around a volcano. But comparisons with atomic bombs should not be taken too literally, as the forms in which the two types of explosion release their enormous energy are quite different. Scientific interest in pyroclastic flows dates back to the beginning of the twentieth century, to two eruptions in the West Indies in 1902 which occurred within a day of one another. On May 8 28 000 people died in an eruption at Mt. Pelée which destroyed the city of St. Pierre on the island of Martinique, and only a day earlier 1500 people

were killed on the island of St Vincent from the eruption of the Soufrière Volcano. Pyroclastic flows had been recorded by Pliny at the eruption of Vesuvius in AD 79 and in contemporary accounts of the major eruption of Vesuvius in AD 1631, but later commentators, unaware of pyroclastic flow phenomena, had misinterpreted them as lava flows. In the eruption of the island of Krakatau in 1883, hot pyroclastic surges traveled 80 km on top of the sea in the Sunda Straights to kill at least 1000 people on southeast Sumatra (Francis, 1993). Studies of the eruption of Katmai in remote Alaska in 1912 showed that massive sheets of silicic or igneous material could be laid down by gas mobilized flows as at St. Pierre, and the name ignimbrites was later given to these. Pre-historic ignimbrite flow deposits have reached the area where Auckland now stands from catastrophic eruptions as far as 250 km away in the center of the North Island of New Zealand. Pyroclastic flows are hot, denser-than-air currents consisting of gas and rock particles that move rapidly down a volcano's slopes and can behave like a fluid or a dense gas. A pyroclastic surge can be regarded as a dilute and turbulent form of flow and which can be an extension of a flow or it can move independently of it. Thus the lateral blast at Mount St. Helens is now regarded as a type of surge. Flows and surges commonly travel 5–15 km from their vent, but in the rarer, large events may exceed a runout of over 100 km. Not all the energy is derived from that stored inside the heat of the magma and its explosive power: potential energy is converted into kinetic energy when pyroclastic flows and surges gather momentum through the force of gravity which is often the main determinant of their velocity. If the temperature of a flow is governed by the extent to which it is diluted by air entrapped in its path, so the velocity mainly depends upon the height of the crater (or other starting point of the flow such as a Plinian eruption column). The temperature of most pyroclastic flows is in the range 600–900 °C, and surges can be as low as 100–200 °C, whilst velocities of both flows and surges are typically 50–150 km/hr (Sparks *et al.*, 1997).

Using the results of the studies at Mount St. Helens and the sparser findings at other

**Fig. 10.5.** Volcanologists investigate the impact of a major surge on Montserrat which occurred in a lava dome collapse on December 26, 1997. The area had been evacuated, but many houses were left with their furniture. In this part of the surge, the hot ash cloud forced its way through buildings, removing windows and part of the roof. All the house contents and roof were burnt away by fires caused by the intensely hot ash (>400 °C) deposited inside the buildings. Survival would have been impossible at this location. (Note the fence posts bent down by the force of the surge moving from left to right.)

volcanoes, we can obtain a picture of what happens inside pyroclastic flows and surges and how people are killed and injured. Close to the source, obliteration, burial, and dismemberment will be inevitable, whilst further out the effects of heat and the trauma from the dynamic pressure, causing bodies to be thrown along hitting the ground or obstacles such as trees and walls, are the main causes of death. Further out, asphyxia from ash obstructing or lining the airways is most likely, whilst indoors people are killed by the collapse of buildings and the inpouring of ash through damaged windows and other parts of the structure (Fig. 10.5). As the flow becomes more dilute, deaths and injuries from impacts of missiles, such as flying corrugated roof sheets, tiles, and building elements, as well as collapsing trees and structures, becomes more evident, but severe burns to the skin or airways are also found in survivors. At the margins of the surge a sense of suffocation with or without singeing of the hair may be all that is experienced. Being inside sturdy, non-wooden buildings with resistant window protection will enable some people to survive who would otherwise die if they were outside and exposed to the dynamic pressure (the dense, fast wind effect) and rapid heat transfer of a surge (Baxter, 1990; Baxter et al., 1998).

Mount St. Helens gave us a picture of the impact of a surge in a wilderness area, but the last time a surge hit a densely populated area was in Martinique in 1902, when there were only two survivors in the town of St. Pierre, the remaining 28 000 occupants being killed either directly or in the fires that consumed the buildings (Baxter, 1990). We have no examples of impacts in modern cities, and so the impact in the Naples Bay area of a moderate-sized eruption of a surge from

Vesuvius cannot be easily envisaged. In future eruptions evacuation of populations is the only sensible option.

## Ashfall

A massive vertical eruption also took place at Mount St. Helens after the magma was uncapped in the landslide, and this continued for most of the day (Lipman and Mullineaux, 1981). The direction of the plume was governed by high-altitude winds, and these blew to the northeast all day. Ash began to fall in Yakima, a city located 135 km northeast of the volcano, and small communities in central Washington state by mid-morning, and total darkness enveloped the area for 16 hr. People waking from a Saturday night out thought it was still night-time, some even sleeping on in the gloom until Monday (the birth rate in the area peaked 9 months later!). The extent of the ashfall had not been anticipated by the authorities, nor the way in which the normal arid conditions in this agricultural area would affect the levels of ash in the air. For 5 days all road, rail, and air traffic was paralyzed by low visibility as the ash became resuspended by winds and traffic; motor vehicle accidents were common amongst those who ventured out. Some people began to panic over the possible health effects of the dust clouds which enveloped them. Power outages occurred as moist ash shorted outside insulators. Water supplies were halted as pumping stopped. Water-treatment plants clogged up with the immense amount of ash, and untreated sewage was discharged into rivers. Fortunately, the air was cleared when an unseasonal fall of rain occurred on May 26, enabling life to resume again, and clean-up operations began.

A great deal of interest was shown in the composition of the ash, with chemical laboratories in the ashfall areas eager to scoop up and test samples of the material which lay as deep as 10 cm in places such as Ritzville. The ensuing results and various "expert" interpretations of their meaning were avidly and indiscriminately reported by the media. Residents were alarmed by the irritant effects of the ash on the eyes and respiratory tract and worried about the possible adverse effects on water quality, agricultural products, farm animals, and machines. Health officials were worried because there was virtually no information that they could find in the medical literature on volcanoes or volcanic ash, and so did not know how best to enlighten the population, though some provided the requisite bland reassurance for the TV cameras with all the impact of someone trying to placate a terrified public on hearing the news of an alien invasion. To fill the knowledge vacuum as soon as possible, a co-ordinated study of the composition of the ash and an epidemiological surveillance system based on attendances at hospital emergency rooms in the entire area of visible ashfall was rapidly established (Baxter et al., 1981).

What were the most important health aspects of the ash itself which needed study and what were the lessons for future ash eruptions at other volcanoes (Buist and Bernstein, 1986)? First, the freshly fallen ash *smelt* of having been in a fire – the same smell of ash after a wood or coal fire – and this emanated from the combustion of vegetation on the volcano in the heat of the eruption. A sulfurous smell is due to outgassing of sulfur dioxide from the particles. Polycyclic aromatic hydrocarbons are produced when organic matter burns and these chemicals get swept up into the eruption column. The ash can be made *acid* by acid aerosols derived from the gases in the plume adhering to the particles as they fall through the plume. To test for this, in any eruption, one part of ash by weight is mixed with ten parts of water and the dissolved substances (leachates) analyzed and the pH measured. Large amounts of ash can temporarily acidify rivers and lakes and kill fish, and the acid readily damages leaves. Fluoride contamination (from hydrogen fluoride emitted in the plume) is detected by leachate studies, and a high level of adsorbed fluoride would be dangerous to grazing animals; in Iceland sheep can collapse and die after eating grass with only a fine covering of ash from Hekla volcano, which erupts frequently enough for farmers and local veterinarians to deal almost routinely with the hazard.

## Particle size

Particle size is the next health consideration. Explosive eruptions generate fine ash, but it is the material of respirable size that is of greatest concern. Conventionally, this is expressed by weight rather than by the number of particles in a given volume of air. At Mount St. Helens, about 10 wt.% of particles were $PM_{10}$, or less than 10 $\mu m$ in diameter, although 90% of the particles by count were of this size. Nowadays, the cut-off for respiratory size is about 3.5 $\mu m$, but $PM_{10}$ is still used for regulatory standards of air quality. Respirable ash can be breathed into the small airways and air sacs (alveoli) of the lungs, and can give rise to immediate adverse effects in people who suffer from asthma and chronic lung disease. The composition of the ash needs to be studied for its crystalline silica content, which can be present in the form of quartz, cristobalite, and trydimite, as this gives the potential for the respirable material to induce silicosis, a chronic lung disease which can lead to disabling respiratory symptoms, usually after many years of occupational exposure to toxic dusts in quarries and mines, etc., where the dust is generated by machine tools. At Mount St. Helens the dust was inconveniently delivered to people's homes! The respirable ash was also screened for forms of fibers, just in case the fine shards had been produced in the shape of asbestos particles, and so would be potentially capable of causing asbestosis and lung cancer if inhaled in sufficient quantities.

All these and many other considerations went through the minds of the investigators at Mount St. Helens and the findings of the extensive investigations were eventually compiled in a medical journal which should be referred to for further details (Buist and Bernstein, 1986). About 6 wt.% of the respirable ash fraction was crystalline silica, roughly half as quartz and half as cristobalite. The respiratory toxicity of the ash on the lung was evaluated in the laboratory in a variety of studies, including the effects on two types of cells derived from the alveolar region of the human lung. The responses of the lungs of laboratory animals were observed by administering the ash by inhalation and by intratracheal injection. The results of the studies at Mount St. Helens were consistent and pointed to the ash being less toxic than quartz but more active than inert dust in its potential to cause silicosis. This implied that there was very little risk of lung damage unless exposure was very heavy

and prolonged over years. Measurements of exposure to ash in a wide selection of outdoor occupations and levels inside homes and other buildings showed that levels of fine ash in the air were elevated but mostly within safe limits, and the ash deposits became incorporated into topsoil or were removed by the wind and rain after a few months, even in Ritzville and Moses Lake where the deepest deposits formed (10 cm). One occupational group was studied in detail as they were believed to be at the highest risk of suffering health effects from ash inhalation – loggers. The US National Institute for Occupational Safety and Health (NIOSH) set up an epidemiological study of a large group of workers employed in the local logging industry. In the early stages huge clouds of ash would be produced when the cut trees fell into vegetation infiltrated with ash deposits, and much resuspension of ash (as well as road dust) occurred when driving the logging crews along the forest tracks. After 4 years of follow-up, it was concluded that the risk of silicosis in loggers was low and transient changes in lung function were suggestive of very mild obstruction to airflow, probably due to increased mucus or inflammation, and not restriction of the lung due to fibrosis (silicosis). The principal finding in the community studies was the increase in the hospital emergency room attendances for asthma and bronchitis in the immediate aftermath of the ashfall, and patients suffering from chronic lung disease had worsening of their symptoms for several months after the eruption if they lived in the worst ashfall areas. There was no obvious effect on mortality in the population of Washington state.

Leachate studies on the ash were reassuring, with only mild acidity being demonstrated and the levels of fluoride were low. Drinking water was not affected. The volcano has remained active with lava dome growth for some years, but no further major ashfalls occurred after June 12.

### Lahars

The third major hazard of the Mount St. Helens eruption on May 18, 1980 was wet debris flows or lahars (volcanic mudflows). The eruption triggered the melting of the glaciers around the shattered summit, the water mixing with the ash and other debris lying loose on the volcano to form mudflows which began very shortly after the eruption. These "non-cohesive" flows are also gravity-driven and descend along valleys bulking or entraining loose debris as they go, starting as meltwater flood surges and growing to the consistency of wet concrete, so that their sheer size and "plug flow" characteristics (they behave like a bulldozer when they strike a resistant object) makes them highly hazardous and destructive. The lahars did not cause any loss of life, but destroyed roads, bridges, and other structures in their paths, ending up in the Columbia River where they obstructed shipping channels. The eruption marked a transition in our understanding of lahar processes, but the true horror of their destructiveness was not brought home to scientists until the disaster at Nevado del Ruiz 5 years later (see below).

## Nevado del Ruiz, 1985

On November 13, 1985, at 21:00–21:15 h, a moderate-sized eruption produced small pyroclastic flows and surges near the summit of the Colombian volcano of Nevado del Ruiz (5400 m), melting about 10% of the glacier and generating four major lahars which headed for the river valleys towards the towns of Armero and Chinchina. The population had no warning and over 23 000 people were killed. The tragedy was the biggest volcanic disaster of the twentieth century after the destruction of St. Pierre, Martinique by Mt. Pelée in 1902 and gave an important impetus to the UN declaration of the 1990s as the International Decade for Natural Disaster Reduction (IDNDR). As it took 2 hr between the eruption starting and the lahar reaching Armero, a warning system could have saved most of the lives. The lack of emergency planning and the contributory factors leading up to the disaster have been the subject of much debate (Voight, 1996). Volcanic activity had commenced a year before and emergency planning for a major eruption was proceeding very slowly until a strong phreatic eruption occurred at the summit on September 11 when a 25-km lahar flowed down the Rio Azufrado. The risk of major lahar formation was obvious and a concerted effort was begun by the authorities, aided by experts from outside countries and international agencies. Local scientists prepared volcanic hazard maps which accurately predicted the actual paths taken by the deadly lahars, and for Armero showed that most of the population would have to travel over 1 km to reach higher ground on the edge of the zone of inundation (subsequently commentators viewing the devastation put the distance to safety on higher ground at only 100–200 m). What went wrong? Too little was done too late: the authorities were ultimately unwilling to bear the political and economic costs of early evacuation and were only prepared to take action at the last possible minute, and so the necessary technical and community preparedness measures were not completed in time (Voight, 1996). Despite the evidence many local and national officials considered a disaster "wouldn't happen."

The evening was quiet and it rained heavily. Just before 23:00 h ash started falling again. A previous ashfall had occurred earlier in the afternoon, but the people had been informed on the radio that it was nothing serious and they should remain calm. The radio suddenly went off the air and then the electrical power failed. A loud noise could be heard in the air as the lahar approached. The main lahar struck Armero at around 23:30 h and was preceded by a river of water that flowed along the streets and was deep and fast enough in places to overturn cars and sweep away people. The terrified population thought they were being flooded when the wall of mud slammed into the town. About 1900 deaths occurred near Chinchina, mainly occupants of poor housing built along the banks of the Chinchina river. The depth of the lahar as it left the Langunillas River valley and headed towards Armero was about 30 m above the river bed and it traveled at an estimated velocity of 12 m/s and lasted for 10–20 min. A second pulse came at half the velocity of the first, and lasted for about 30 min. Several smaller pulses were followed by a third major pulse, the lahar lasting about 2 hr altogether. Buildings collapsed and broke up, with survivors clinging to the debris or being miraculously swept along on top of the mud.

As much as 85% of the town of Armero was left covered in 3–4 m of mud. Official figures put the death toll in the town at 21 015 out of a total population of 29 170. Most people would have been killed immediately by the severe trauma caused by the collapse of buildings, flying debris, and burial by the slurry. The head of the lahar would have been turbulent, about three times higher than the final depth of the lahar deposit, and contained cobbles and boulders. As well as the risk of being engulfed, bodies would have been driven against stationary objects or contorted and crushed by entrained debris such as trees and collapsed parts of buildings, resulting in mutilation or fractures of the limbs. In the shallower and slower-moving parts of the flow stones and other sharp objects cut into the skin forming deep lacerations and mud forced its way into the eyes, mouth, ears, and open wounds. The pressure of the mud against the chest would make it difficult if not impossible to breath if buried up to the neck, and death would follow from traumatic asphyxia unless the pressure was relieved by the movement in the flow within one or two minutes.

Thus the severely injured who had survived the initial impact of the lahar would be at further risk of dying within the first few hours from such traumatic conditions as skull and brain injuries, internal organ damage from chest and abdominal trauma, blood loss and slow suffocation from the mud and water which had entered the lungs. There was also the risk of trapped survivors dying within hours from the effects of hypothermia unless the temperature of the surrounding slurry was at least 20 °C. In fact, there were numerous reports that the mud was warm and even hot in many places (the heat had been injected into the water from the summit eruption), which explained the survival of people who had remained trapped for up to 3–4 days before being rescued. The mud was at first very soft, which made rescue difficult, but hardened by the third day, when it became more arduous to dig out the remaining survivors. As the first survivors were not rescued until 12 hr after the disaster struck, most of those with serious injuries would have already died by then for lack of surgical treatment. Figures showed that 1244 patients were treated in hospital of which 150 died during their admission. As would be expected, the majority of these deaths were due to wound infections and associated complications. Most of these would have been preventable if antibiotic treatment could have been given as soon as possible after injury occurred (unlikely given that the soft mud hampered rescue attempts), and if lacerations had been adequately cleaned of mud and left open, with suturing delayed for 5 days (an important surgical rule which always needs reinforcing in disasters).

The rescue and treatment of the survivors was badly hampered by the lack of search-and-rescue equipment. Before this catastrophe little if any thought had been given to a worst foreseeable scenario and how the population should move in response to warnings and the requirements of the emergency services in the event of deaths and injuries in a lahar. The experience at Armero was one of unmitigated horror, and stands as the ultimate example of how volcanic disaster preparedness and mitigation must involve the whole community from the outset if volcanic catastrophes are not to be repeated in other parts of the world in the future.

## Mount Pinatubo, Philippines, 1991

Mount Pinatubo had not erupted for over 600 years and was not even considered as warranting special monitoring. Then, in April 1991, intense steaming began from several vents which formed and was accompanied by earthquake activity. A rapid stratigraphic assessment showed that the volcano was indeed capable of very large eruptions, and in early June the activity began to rapidly escalate. Evacuation of the population in villages closest to the volcano and non-essential personnel and dependants at the Clark Air Base (a US military establishment) was begun on June 9, and the evacuation area was extended on June 12 as major explosive activity began. On the afternoon of June 15 the climactic eruption began, one of the largest of the century. An area of 40 km radius had been placed under evacuation, so that between 60 000 and 65 000 people had relocated successfully. As many as 20 000 people had been saved from death in the pyroclastic flows which filled the deep valleys on the volcano (Newhall and Punongbayan, 1996a). Most of the 200–300 deaths and approximately 300 serious injuries occurred outside the 30 km radius

**Fig. 10.6.** The roof of this hospital in Olangapo collapsed in the ashfall of Mt. Pinatubo, Philippines, 1991, killing eight patients inside. About 300 people died from roof collapse in the ashfall of June 15–16 in one of the largest eruptions of the century.

as a result of roofs collapsing under the weight of the ash, which by a most unfortunate coincidence was made even denser by the addition of rain from a typhoon (Typhoon Yunya) which passed on the afternoon of June 15. This hazard to those who took shelter from the rain and ashfall had not been specifically foreseen (Fig. 10.6). In Olangapo City, about 35 km from the summit of Pinatubo, damage to roofs was widespread and its 300 000 population spent a harrowing time on the night of June 15–16 listening to roofs breaking and the accumulated ash sliding inside. The official depth of ash here was 15 cm, and the loading on roofs would probably have exceeded 2.0 kN/m$^2$ (approximately 200 kg/m$^2$); typical designed loads for pitched roofs would rarely approach this level even for engineered structures. Longer-span roofing fared worse than domestic roofs, an important point when large

structures such as halls and churches may be used as mass shelters: one church roof collapsed killing 35 people. This eruption was the first to be studied for this type of hazard, but unfortunately it was not possible to collate the causes of death or the types of injury incurred. It can be surmised, however, that the failure of roofs resulting in fatalities would most likely be sudden and with roof members falling on to the occupants. These could cause fatal head, chest, and abdominal injuries, or even in the case of a mild impact causing simple concussion, a trapped person would suffocate by being buried in the ash. The roofs in the Philippines are made principally with wind forces in mind, and are not able to bear heavy loads as might be the case in countries which routinely experience heavy snowfall, for example. The eruption highlighted the need to study the roof characteristics in densely populated localities as part of any volcanic risk assessment (Spence *et al.*, 1996).

Around 10 000 Aetas, the indigenous mountain people, lived on the slopes of Mt. Pinatubo, and most of these were successfully evacuated. They avoided contact with lowlanders and believed the volcano was the home to their god, Apo Mallari (Newhall and Punongbayan, 1996b). Their isolation made them vulnerable to measles, and several hundred died in the aftermath of the eruption from this condition, as well as diarrhea and respiratory infections, contracted in the less than satisfactory conditions in the evacuation camps. Despite this, the overall management of the disaster was viewed as a notable success and an important example of how the timely monitoring and forecasting of natural hazards by scientists could save many thousands of lives. Unfortunately, this achievement was followed by the intractable problems presented by the lahars which formed in the rainy seasons from the extensive ash and pyroclastic flow deposits on the volcano's slopes. Thus in the 3 months after the eruption over 100 deaths were attributed to the lahars or related floods, and the hazard and the disruption it has caused to the lives of many thousands of people has persisted for years after.

## Soufrière Hills Volcano, Montserrat

The last eruption of the Montserrat volcano was over 300 years ago and preceded colonization of the island. Only 17 km long and 9 km wide, this island's big neighbors are Antigua and Guadeloupe, with the latter having major hospital facilities to support its quarter of a million population. Eruptions on small islands present special problems, as the options for relocation of a threatened population are small, and the infrastructure for dealing with emergencies limited. Most of the 12 000 inhabitants of the island had no idea that the mountain which overlooked the capital town of Plymouth was a volcano, until it began erupting in July 1995 with steam emissions from cracks in the crater area. As in most eruptions it was difficult to assess what the outcome was going to be, but any doubts that magma was on the move were finally dispelled by November, when a lava dome appeared inside the crater. The subsequent eruptive activity was mainly linked to the growth and collapse of the andesite lava dome, with associated explosive events.

As at Mt. Unzen in Japan, which had started its eruption in 1990, collapse of a growing lava dome endangered the nearby population by the formation of pyroclastic flows from the hot debris. When pyroclastic flows started at Montserrat in April 1996, they triggered the main evacuation of Plymouth and the beginning of a gradual withdrawal of the population to the north of the island as the growth of the lava dome continued. Explosive activity began with an eruption on September 17, 1996, and activity gradually escalated until culminating in a large pyroclastic flow on June 25, 1997 which killed 19 people and injured eight others who, along with at least 50 others, were working or still living in the officially designated Exclusion Zone around the volcano against all advice. Most of the victims were buried under the two main flows, but five bodies recovered from the surge area had been subjected to temperatures of hundreds of degrees at the time of death and were badly burned. The event showed the importance of having an emergency plan for burns cases, as five of the injured required emergency medical evacuation by helicopter to nearby Guadeloupe according to an arrangement which had been set up for injured survivors of hot surges. The most frequent injury was deep burns to the feet from trying to escape across the hot deposits, and one survivor needed surgical amputation of both feet. All burns in volcanic eruptions must be regarded as serious however seemingly trivial because of the extremely high temperatures that may be involved with only brief contact with hot ash.

Frequent pyroclastic flow activity in early 1997 raised concerns about the health effects of the volcanic ash which would be blown by the prevailing wind from each flow towards the still-occupied areas to the west of the volcano. Investigations showed that the ash was even finer than at Mount St. Helens, with 13–20 wt.% being in the $PM_{10}$ range and the majority of the particles being under 3–4 $\mu m$ in diameter, and that it contained a surprisingly large amount of crystalline silica in the form of cristobalite (10–24 wt.%), a mineral which should be regarded as twice as toxic to the lungs as the more commonly found quartz. Exposure measurements were performed on outdoor workers, and inside houses and schools to assess the risk of silicosis. By mid August 1997 it was clear that the risk of developing silicosis in some outdoor workers and even in the general population from the growing quantities of ash in the environment was too high in the southern and central areas which had received most of the ash, and it was timely when these areas were finally evacuated to protect the people against death and injury from the continuing eruptions.

However, after another major event on September 21, the volcano began a series of small and regular explosions and during some of these ash would be blown by high altitude winds to the north of the island for the first time. This could have meant that the north would then also be at a growing threat from the ash, but tests showed that the ash from explosions, although it contained 13–14 wt.% as $PM_{10}$, contained much lower levels of cristobalite (3–6 wt.%), with the composition of the material much more closely resembling that of the magma. Thus the ash from the lava dome was being enriched with cristobalite and the motion of the pyroclastic flows was fragmenting the groundmass containing the cristobalite so that it became even further concentrated in the finer fractions. There were two main conclusions which followed this discovery, namely that any risk assessment on the Exclusion Zone needed to include the risk of silicosis from ash deposits as well as the risk of death and injury, and that this health hazard needs to be looked for at future lava dome eruptions near populated areas.

## Galeras Volcano, Colombia, lava dome eruption in 1993: hazards to volcanologists

An international meeting of scientists was held in January 1993 at Pasto, a city of 300 000 inhabitants situated only 7 km below the crater of Galeras, an impressive volcano located high in the Colombian Andes. Ironically, the meeting was intended to promote research on the volcano to protect the population, as it had been designated a Decade Volcano under the IDNDR and warranting special study. The reactivation of the volcano in 1989 had caused a social and economic crisis in the region. The crater contained an explosive lava dome, which had formed in late 1991, and which had last erupted explosively in July 1992, sending ballistic projectiles several kilometers and generating a shock wave that was felt in most of the

towns surrounding the volcano. As part of the meeting's field activities a team of 12 scientists climbed down into the crater to get to the lava dome to obtain gas samples and make microgravity measurements. Without warning the dome exploded, instantly killing six of the scientists and three tourists who had gone with them (Baxter and Gresham, 1997). The most probable explanation is that the quiet appearance of the dome concealed a pressurization of gases (magmatic or from heating and partial vaporization of groundwater) that was occurring undetectable to the scientists, and the rock capping the conduit failed at the moment they were completing their work and starting to leave. The cause of death was the explosive fusillade of rocks of all sizes that were flung out from the dome. Explosions releasing ballistic ejecta can occur in various stages of eruptions, the maximum range being 5 km from the source. Overpressure waves from eruptions have not been recorded to injure people with the exception of rare reports of pains in the ears, etc. Despite being Colombia's most active volcano, and the seemingly precarious location of Pasto, the only deaths during its 500 years of known history were those of the volcanologists.

Between 1979 and 1993, 15 volcanologists were killed whilst working on volcanoes, including those at the Galeras eruption, a figure which makes volcanology one of the most hazardous scientific occupations. Only 2 months after the Galeras eruption the battered bodies of two more volcanologists killed by a phreatic explosion were found at the crater rim of Guagua Pichincha Volcano, Ecuador. In 1991, the husband and wife team of Maurice and Katia Krafft, who were famous for their thrilling films of eruptive phenomena taken at close range, were killed along with Harry Glicken, a US volcanologist, by a pyroclastic surge at Mt. Unzen. A colleague of Glicken's, David Johnston, a 30-year-old volcanologist with the US Geological Survey, was caught at an observation post which had been previously considered safe by the pyroclastic surge in the May 18, 1980 eruption at Mount St. Helens. As well as at Galeras, tourists have come to grief from crater explosions at Mt. Etna, Italy (nine deaths, 1979); Yasur, Vanuatu (two deaths,

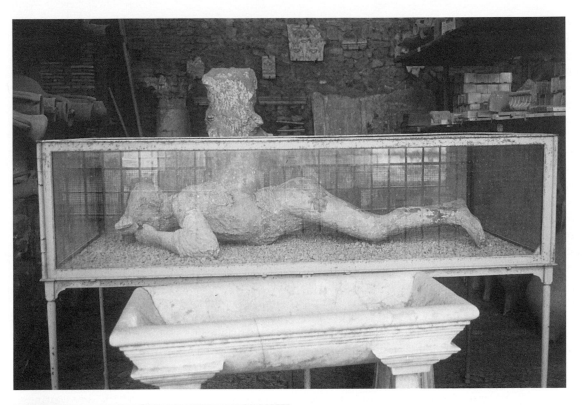

**Fig. 10.7.** One of the plaster casts on display at Pompeii. Most contain bones, the only human remains inside cavities formed by the original body shape.

1994); Popocatepetl, Mexico (five deaths, 1996), and Semeru, eastern Java (two deaths, 1997).

## Vesuvius, AD 79 and 1631

Have studies of eruptions around the world advanced our understanding of eruptive processes, and can we now reconstruct the events at the eruption of Vesuvius in AD 79 which buried Pompeii and Herculaneum? The answer is an emphatic yes, and volcanologists can now forecast the style and size of eruptions at particular volcanoes, even if they are usually unable to say when and whether a major eruption will occur. This information is essential for emergency planning, as we shall discuss below. Vesuvius is one of the world's most dangerous volcanoes, but it is also one of the best studied. The knowledge gained by scientists over recent years may make a radical difference to the lives of those hundreds of thousands of people in the Naples

Bay area who are most at risk when Vesuvius reawakes.

How Pompeii came to be buried perplexed many great minds. When Herculaneum started to be excavated in 1709, and Pompeii in 1754, the startling findings at the two towns began to grip the imagination of eighteenth-century Europe. The area was an obligatory touring ground for many artists, writers, and gentlemen. Goethe wrote in 1787: "the volcanic debris which buried the city cannot have been driven here by the explosive force of the eruption or by a strong wind: my own conjecture is that the stone and ashes must have remained suspended in the air for some time, like clouds, before they descended on the unfortunate city . . . (like a) mountain village buried in the snow." An intriguing feature of the excavations was the invention in 1863 of the technique of plaster-casting that preserved the positions of the Pompeiians in the moment of death. This involves passing liquid plaster of Paris through a small hole of the cavity which had once been filled by the victim's body and contained only the remains of the skeleton (Fig. 10.7). For many decades it was believed that the victims had succumbed to rapid burial from an incessant deluge of airfall pumice and ash. But scientific views on the behavior of pyroclastic flows and surges moved rapidly forward after the eruption of Mount St. Helens in 1980, putting the interpretation of the eruptive deposits in a new light (Sigurdsson *et al.*, 1985).

The eruption of AD 79 is believed to have been in two main phases: a Plinian airfall phase and a pyroclastic surge phase, with most of the deaths and destruction of property occurring in the second phase. The eruption began with the onset of earthquakes in the region on August 20 which grew in frequency over the next 4 days as magma forced its way to the surface. At about 13:00 h on August 24 the eruption began with an ash and pumice vertical column which went as high as 20 km and continued for the next 18 hr. Most of the pumice was made up of small fragments, but pieces of limestone and old volcanic rock 9–11 cm diameter were present and these could have caused extensive damage to clay-tiled roofs and caused serious injuries, including severe impacts to the skull. Many roofs must have caved in under the accumulated ash, causing further fatalities and injuries. The persistent fallout would have reduced visibility with daytime transformed to impenetrable gloom, making escape even more difficult. The eruptive material grew to a depth of 2.4 m, leaving only the walls of two- and three-storey buildings standing above the bleak wasteland. At 07:30 h a surge which left a deposit of only 10–20 cm swept over Pompeii killing all those still present. This was followed immediately by a dense pyroclastic flow which buried the victims in a layer 50–200 cm deep. About half an hour later, an even larger surge flowed the 17 km towards Stabia across the Bay of Naples, cooling and slowing as it went, to the place where Pliny the Elder had taken refuge, enveloping him and his companions in a dense cloud of suffocating ash which killed him without causing burns. In the words of Pliny the Younger writing of his uncle's death:

Then the flames and smell of sulphur which gave warning of the approaching fire drove the others to take flight and roused him to stand up. He stood leaning on two slaves and then suddenly collapsed, I imagine because the dense fumes choked his breathing by blocking his windpipe which was constitutionally weak and narrow and often inflamed. When daylight returned on the 26th – two days after the last day he had seen – his body was found intact and uninjured, still fully clothed and looking more like sleep than death.

Pliny the Younger's full account was the first description of a volcanic eruption.

Herculaneum had been destroyed in the first surge and pyroclastic flow from the volcano when its style of activity changed at around 01:00 h. During the preceding 12-hr period of activity the city had only received a light sprinkling of ashfall, less than 1 cm deep. The surge would have taken no more than 4 min to reach Herculaneum and would have ripped off tiles and carried away building rubble. It left walls intact but was hot enough to carbonize dry, but not green, wood. The townsfolk had already fled, but around 100 skeletons were found in 1982 during excavations of chambers at the waterfront. Many more victims may be found in future excavations, but it seems that those in the chambers were waiting at the beach to be rescued when they were struck by the surge which swirled into the chambers. The victims appear to have been huddled together, many remains being those of young people and children. The city and the chambers with their bodies were covered over with 20 m of pyroclastic flow deposits (Sigurdsson *et al.*, 1985).

At the excavations in Pompeii, 1100 skeletons have been identified so far, with an estimated 1500 when the excavations will one day be complete (de Carolis *et al.*, 1998). The casts method could only be used in 40–50 skeletons. An estimated 400 were presumed to be killed by roof collapse or in the pumice fallout phase, and 600 in the pyroclastic surge. Some houses were protected from the fallout and death in 25% was caused by the ingress of the surge into cellars where people sheltered. Another 25% were found at the first level of the houses where the roof was still intact. The remaining 50% were killed by the surge outside in the streets. What were these people still doing in Pompeii? We have no idea, but perhaps they were the last to attempt to flee but were held back by the dense fallout and resuspension of ash causing an impenetrable blanket in which to move about.

The AD 79 eruption is the most famous Plinian-type eruption (named after Pliny because of his famous description), but a major sub-Plinian eruption of Vesuvius occurred in 1631 in which at least 4000 people died (Rosi *et al.*, 1993). Many people were killed in what recently have been shown to be pyroclastic flows, but other principal injury agents were tephra fallout causing collapse of roofs, and lahars and floods. The reconstruction of these catastrophic eruptions of Vesuvius have only been made possible by recent advances in our understanding of eruptive phenomena, with the 1631 eruption being used as the reference event for the next eruption of this volcano.

## Deaths in eruptions: a summary

A recent review of victims of volcanic eruptions since AD 1783 has shown that more than 220 000 people have been killed (Tanguy *et al.*, 1998). Most of the fatalities resulted from post-eruption famine and epidemic disease (30%), pyroclastic flows and surges (27%), mudflows or lahars (17%), and volcanogenic tsunamis (17%). As in all natural disasters, recording of deaths and injuries can be highly variable as precise studies are not usually undertaken, especially before recent developments in supplying international relief to disaster areas. Using such data to predict future trends is also likely to be unreliable, as a few major events in densely populated areas can wildly skew the data. Above all the world has experienced rapid population growth, environmental degradation and urbanization over recent decades, which may alter the pattern of human vulnerability in the future. In addition, the intervention of volcanologists may greatly help in protecting populations from lahars and other volcanic hazards, whilst international relief efforts can help to forestall epidemics and alleviate food shortages.

## Health aspects of volcanic crisis management

Over 500 million people live in areas of active volcanism and at least 25 of the world's largest cities are located within the destructive range of volcanoes. A densely populated area has not been struck by a pyroclastic flow in recent times, but the possibility of an eruption devastating a region supporting tens or even hundreds of thousands of people has become a reality in a world of burgeoning populations and rapid urbanization. Today, Popocatépetl has resumed its activity, a volcano which has a Plinian eruption every 1000–3000 years with 5–10 km$^3$ of ejecta and which today could cause hundreds of thousands of casualties in a few hours of erupting. A lava dome appeared in the crater in 1995 and has exploded and reformed about twice a year, with moderate ash emissions falling in the area. The population of Mexico has become increasingly centralized around Mexico City and its environs, with the volcano casting its huge shadow over the area. In Italy, 2 million people live in areas of volcanic hazard, with one of the most densely populated areas in the world lying close to Vesuvius in the Bay of Naples area where at least 600 000 people currently are at risk from a resumption of its violent activity which is now increasingly overdue. But even the eruptive activity of the Soufrière Hills volcano on Montserrat, small in comparison with the scenarios of Popocatépetl or Vesuvius, can completely disrupt life for years on a small island. One striking feature of volcanic eruptions is that, in contrast with other, more common natural disasters they can totally and irrevocably change the lives of those under threat, as a result of their extreme destructiveness and time evolution (it is always difficult to know when an eruption is once and for all over).

The re-awakening of a volcano should be regarded as a public health and economic emergency. Nevertheless, officials and scientists should appreciate from the outset that the guiding principle must be to save lives. Any appearance of cost–benefit analysis, or trading the risk of casualties in order to maintain economic order, will lose the confidence of the population who will soon cease to view the pronouncements of officials as being free from self-interest. The health and safety of the population must, ultimately, be the prime concern. The only exception to this approach is when the future economic viability of the community is at stake, when extraordinary measures are needed and putting lives at risk may be permissible, e.g., attempting to divert the lava flow at Heimay in 1973 by cooling it with pumped seawater as it threatened to overwhelm the harbor. The earliest recorded example of such heroism was in the 1669 eruption of Mt. Etna (Sicily) when the lava flow overflowed the walls of Catania and reached the sea. In order to save the city several dozens of men covered with wet cowhides (excellent protective clothing against heat) dug a hole in the side of the lava flow with iron bars, allowing a stream of lava to escape and to relieve the pressure on the front of the lava flowing towards Catania. Even so, the majority of these heroic stances against a volcano are likely to prove futile.

Virtually all eruptions will be preceded by a period of time, such as weeks, months, or even years, in which there are premonitory signs from the volcano and these will require monitoring by a team of volcanologists based in a local observatory. Indeed, the problem may be that there are too many ambiguous signs, which can be confusing. Volcanologists are unable to predict the exact timing and size of an eruption, and the capacity to provide warning of a large eruption may be only within 24–48 hr or even less. The type of eruption that is likely to occur can only be surmised from stratigraphic studies to determine the history of the volcano over the last 10 000–20 000 years, and relying on the past as a guide to future events, a by no means infallible approach.

When a volcano stirs, the first question is whether magma is involved and if so whether it is rising or intruding to the surface. The magma is capable of fueling a major eruption, but in the early stages it may not be possible to determine if events signal a magmatic intrusion or are produced by hydrothermal or tectonic processes which can give rise to small phreatic events. No one has directly visualized a magma chamber inside a volcano, but the flow of magma from a store 6–10 km below the surface results in an overpressure which produces certain physical and geochemical phenomena. The edifice of the volcano starts to inflate, and the deformation allows gases and water to escape through rock fissures to the surface; fumaroles will form in fissures for the first time or existing ones will increase their discharge and alter in their gas composition. Earthquakes will be detectable on seismographs as the magma moves and the rock deforms. Gases and waters can be sampled for changes indicative of magmatic involvement. Uranium is present in all rocks to varying degrees, and carbon dioxide serves to carry radon gas towards the surface, as we have seen at Furnas and Mammoth Mountain, for example. Probes can be located on the volcano to monitor the various phenomena and connected to a central observatory. Some or all these developments may result in changes in air and water quality at the surface and present health hazards to communities. Paradoxically, perhaps, most magma intrusions are *not* followed by eruptions, so there can be long periods of uncertainty that will add to the social and economic problems the threat of activity already poses.

The first steps in a crisis are to construct a hazard map and to define alert levels, as timely evacuation of the population at risk is the only feasible protective measure. Uncertainty surrounding when an eruption will occur and how large it will be are added to by numerous social factors which determine whether a population will heed a warning of an impending eruption and then go on to evacuate completely. Most people who have never experienced a volcanic crisis cannot readily understand why this should be so, but anyone who relies on a "technological fix" approach of prediction by scientists and rapid response by a population is in for a rude awakening, especially if the crisis becomes protracted. Populations rarely have the capacity to rapidly and totally evacuate an area unless under the most imminent threat. With volcanic eruptions, by the time the gravity of the situation is understood it may already be too late to take avoiding action. Thus volcanologists can only forecast or know in advance what will happen in broad terms in space and time, rather than predict in any scientific use of the term, i.e., describe correctly what will happen and the result on the impacted area.

During the premonitory period, emergency plans have to be laid and the various government sectors, including health, need to draw up their own plans that can be brought together as one response plan. It is essential that all the different sectors work together. The size of event planned for should be a representative reasonably foreseeable case, but with inherent flexibility to deal with less or more serious events, as in planning for technological disasters; it is counterproductive to plan for maximum expected events if the probability of these is very small. Health planning should focus on casualty management and public-health measures in the broadest, preventative sense. Medical management of casualties will depend upon the anticipated eruptive phenomena and the types of injuries these may cause, in the knowledge that evacuation measures or drawing up exclusion zones may not be completely effective in preventing people from being impacted. In the Montserrat crisis plans were devised with the authorities on Guadeloupe for a French SAMU team to arrive by helicopter to treat and evacuate burns casualties in the event of pyroclastic flows and surges, as described earlier.

The specific pre-eruption measures should include:

- Designating areas for evacuation, including plans and facilities for emergency relocation. The health and welfare of people relocated either temporarily or for long periods is a public-health matter and usually requires input by workers experienced in health planning for evacuees.
- Developing search and rescue plans including designating sites for emergency field casualty stations.
- Staging mock drills to test the readiness of responders, including the local hospital and casualty teams.
- Providing emergency air monitoring equipment for detecting toxic gases in the outdoor air from crater emissions and indoor air, where applicable, arising from soil degassing.
- Establishing in advance an epidemiological surveillance system.

Ashfalls may impinge upon much wider areas than other eruptive phenomena, so specific measures may be needed to assess the health hazards and to protect the population:

- Prepare laboratory facilities for collecting and analyzing ash for leachable toxic elements, and arrange for regular water analyses, especially from sources of drinking water.
- Arrange for ash samples to be collected and analyzed for particle size and crystalline silica content.
- Provide equipment for monitoring exposures to air-borne ash in the community and amongst people who work outdoors.
- Stockpile lightweight, high-efficiency masks for distribution to the population if ashfalls occur; emergency workers may need visors as well to protect the eyes from ash causing corneal abrasions.
- Prepare for possible temporary breakdowns of water supplies and sewage treatment plants as a result of heavy ash fallout.
- Maintain emergency health services and hospitals.
- Provide emergency shelter and food relief.
- Advise residents of affected communities about protective measures to combat overloading and collapse of roofs in heavy ashfalls.

Community preparedness is a crucial element in all this. The population should become knowledgeable about volcanoes and their eruptive effects and steps need to be taken to improve first aid facilities, for example. Education programs in schools are thought to be particularly effective.

Engineering solutions to problems are quite limited in volcanic eruptions, but the classical examples are diversionary barriers to protect areas against lahars, or explosive diversions of lava flows, e.g., on Mt. Etna in 1993. In the long term land-use planning to restrict the numbers of people and limit key infrastructure in areas at risk is essential. In acute crises, duplication of facilities such as hospitals and police stations may be necessary.

## Risk assessment

Hazard assessment can be done in two ways: a deterministic approach based on stratigraphic deposits and using these to define a maximum impacted area for an eruptive phenomenon such as a pyroclastic flow or ashfall, or the

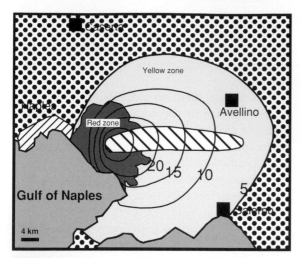

**Fig. 10.8.** Overview of the hazard zones at Vesuvius. The "Red zone" includes the area expected to be affected by pyroclastic flows and surges from collapse of the eruption column: about 600 000 people live in this area. The "Yellow zone" is anticipated to be at risk of significant ashfall and contains about 1 000 000 people. Depth of ash is given in centimeters. Modified from Santacroce (1996).

probabilistic approach based on the probability that a given area will be affected. The latter scheme usually requires using a modeling method which can be computerized and handled with a geographical information system (GIS) to analyze for topographically protected and vulnerable areas. On top of this is risk assessment, or taking probabilities and vulnerability factors to estimate the chances of death and injury.

At Vesuvius, planners have taken the last major eruption of the volcano in 1631 as the reference eruption for the next event. In addition, evidence from studies of 14 previous eruptions has been evaluated to determine likely ashfall distributions, for example. As a result a "red" area of potential complete destruction from pyroclastic flows has been described, which contains 600 000 people who will all need to be evacuated and accommodation found for them in other parts of Italy. A "yellow" area containing about 1 000 000 people is at high risk from damaging ashfalls. A "blue" area indicates where there is a high risk of floods from heavy rainfall induced by the eruption (Fig. 10.8). The problems presented by even

this simple overview can be seen to be daunting (Santacroce, 1996).

Towards the end of 1997 the UK Government became concerned that the escalating activity of the Soufrière Hills volcano would sooner or later require the evacuation of the island population of Montserrat for safety. A full hazard and risk analysis was undertaken by a group of scientists, incorporating the risk of deaths and injury for each possible eruptive scenario. An expert elicitation exercise was performed using subjective probabilities of each event given by the scientists, together with the likely proportions of severe casualties, depending upon the distribution of the population at the time. The data were put through a Monte Carlo simulation. The summary results are shown in Fig. 10.9 in terms of individual risk. Societal risk was also looked at, i.e., the probability of five or more severe casualties being the most important end point. This was chosen as the maximum acceptable number of casualties on an island with minimal hospital or other treatment facilities, and the chance of such a casualty event was much reduced if the population was kept out of Areas 3–5. This approach underpinned the judgement of the scientists that the north of the island (Areas 1 and 2) was likely to be at minimal risk and the political decision was taken not to evacuate the island (Montserrat Volcano Observatory, 1998). This formulation of the risk was only possible as a result of the volcanologists learning a great deal about the size of the volcanic system and its behavior over 2 years of eruptive events – it would not have been possible to construct such a risk assessment at the outset, even though a hazard analysis had been performed in the late 1980s (which had been ignored by the authorities).

## Final remarks

This chapter has drawn from a select group of volcanoes to illustrate the ways in which volcanic phenomena can impact on human health. Undoubtedly there are many other environmental aspects which will be revealed to us as we learn more about the behavior of volcanoes in the future. The prospect of hundreds of

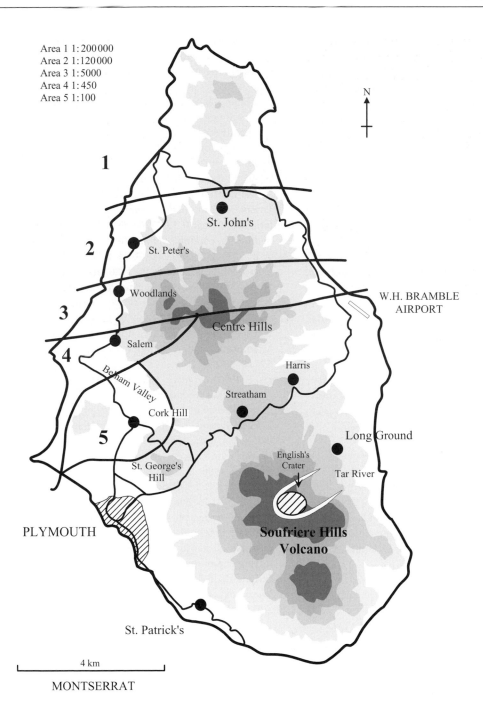

Area 1 1:200 000
Area 2 1:120 000
Area 3 1:5000
Area 4 1:450
Area 5 1:100

N

1

St. John's

2

St. Peter's

Woodlands

3

Centre Hills

W.H. BRAMBLE
AIRPORT

Salem

4

Belham Valley

Harris

Streatham

Cork Hill

Long Ground

5

English's
Crater

Tar River

St. George's
Hill

Soufriere Hills
Volcano

PLYMOUTH

St. Patrick's

4 km

MONTSERRAT

**Fig. 10.9.** Montserrat, West Indies. This map shows the main population areas in December 1997, though Areas 4 and 5 had already been evacuated and were part of the Exclusion Zone around the volcano. The activity was by then impacting on two-thirds of the island. The key shows the estimated individual risks of death (numerical odds) from activity in the next 6 months (annualized) for each of the five areas.

thousands of people being killed in a violent eruption in our time is no longer fanciful, and even volcanic activity on a small scale will have increasingly dramatic political and social consequences in the global village. The global impact of large eruptions raises ecological and climatic issues which can also impinge on human health.

New interdisciplinary approaches and research methods are going to be needed to analyze the complex systems involved and to develop the multidimensional risk assessments needed for improving environmental and public health interventions in natural disasters and advancing mitigation measures.

## References

Baris, Y. I., Saracci, R., Simonato, L., *et al.* 1981. Malignant mesothelioma and radiological chest abnormalities in two villages in Central Turkey. *Lancet*, **1**, 984–987.

Baxter, P. J. 1990. Medical effects of volcanic eruptions. I. Main causes of death and injury. *Bulletin of Volcanology*, **52**, 532–544.

Baxter, P. J. and Gresham, A. 1997. Deaths and injuries in the eruption of Galeras Volcano, Colombia, 14 January 1993. *Journal of Volcanology and Geothermal Research*, **77**, 325–338.

Baxter, P. J., Baubron, J.-C., and Coutinho, R. 1999. Health hazards and disaster potential of ground gas emissions at Furnas Volcano, São Miguel, The Azores. *Journal of Volcanology and Geothermal Research*, **92**, 95–106.

Baxter, P. J., Ing, R., Falk, H., *et al.* 1981. Mount St. Helens eruptions May 18 – June 12, 1980: an overview of the acute health impact. *Journal of the American Medical Association*, **246**, 2585–2589.

Baxter, P. J., Kapila, M., and Mfonfu, D. 1989. Lake Nyos disaster, Cameroon, 1986: the medical effects of large scale emission of carbon dioxide? *British Medical Journal*, **298**, 1437–1441.

Baxter, P. J., Neri, A., and Todesco, M. 1998. Physical modelling and human survival in pyroclastic flows. *Natural Hazards*, **17**, 163–176.

Baxter, P. J., Stoiber, R. E., and Williams, S. N. 1982. Volcanic gases and health: Masaya Volcano, Nicaragua. *Lancet*, **2**, 150–151.

Baxter, P. J., Tedesco, D., Miele, G., *et al.* 1990. Health hazards of volcanic gases. *Lancet*, **2**, 176.

Bernstein, R. S., Baxter, P. J., Falk, H., *et al.* 1986. Immediate public health concerns and actions in volcanic eruptions: lessons from the Mount St. Helens eruptions, May 18 – October 18 1980. *American Journal of Public Health,* **76**(Suppl.), 25–38.

Brantley, S. L., Borgia, A., Rowe, G., *et al.* 1987. Poas Volcano crater lake acts as a condenser for acid-rich brine. *Nature*, **330**, 470–472.

Brown, G., Rymer, H., Dowden, J., *et al.* 1989. Energy budget analysis for Poas crater lake: implications for predictive volcanic activity. *Nature*, **339**, 370–373.

Buist, A. S. and Bernstein, R. S. (eds.) 1986. Health effects of volcanoes: an approach to evaluating the health effects of an environmental hazard. *American Journal of Public Health*, **76**(Suppl.), 1–90.

de Carolis, E., Patricelli, G., and Ciarallo, A. 1998. Rinveimenti di corpi umani nell'area urbana di Pompei. *Rivista di Studi Pompeiani*, **8**.

Department of Health 1997. *Handbook on Air Pollution and Health*. London, HMSO.

Eisele, J. W., O'Halloran, R. L., Reay, D. T., *et al.* 1981. Deaths during the May 18, 1980, eruption of Mount St. Helens. *New England Journal of Medicine*, **305**, 931–936.

Farrar, C. D., Sorey, M. L., Evans, W. C., *et al.* 1995. Forest-killing diffuse $CO_2$ emissions at Mammoth Mountain as a sign of magmatic unrest. *Nature*, **376**, 675–677.

Francis, P. 1993. *Volcanoes: A Planetary Perspective.* Oxford, UK, Clarendon Press.

IARC 1988. *Monographs on the Evaluation of Carcinogenic Risks to Humans*, vol. 43, *Man Made Mineral Fibres and Radon*. Lyon, France, International Agency for Research on Cancer.

Kusakabe, M. 1996. Hazardous crater lakes. In R. Scarpa and R. I. Tilling (eds.) *Monitoring and Mitigation of Volcano Hazards*. Berlin, Springer-Verlag, pp. 573–598.

Le Guern, F., Tafieff, H., and Faivre Pierret, R. 1982. An example of health hazard: people killed by gas during the phreatic eruption: Dieng Plateau (Java, Indonesia), February 20, 1979. *Bulletin of Volcanology*, **45**, 153–156.

Lipman, P. W. and Mullineaux, D. R. 1981. *The 1980 Eruptions of Mount St. Helens, Washington*, Geological Survey Professional Paper no. 1250. Washington, DC, US Government Printing Office.

McBirney, A. R. 1956. The Nicaraguan volcano Masaya and its caldera. *Translations of the American Geophysical Union*, **37**, 83–96.

Montserrat Volcano Observatory 1998. *Preliminary Assessment of Volcanic Risk on Montserrat.* Chesney, Montserrat, Government of Montserrat.

Newhall, C. G. and Punongbayan, R. S. (eds.) 1996a. *Fire and Mud: Eruptions and Lahars of Mount Pinatubo, Philippines.* Quezon City, Philippines, Philippine Institute of Volcanology and Seismology.

1996b. The narrow margin of successful volcanic-risk mitigation. In R. Scarpa and R. I. Tilling (eds.) *Monitoring and Mitigation of Volcano Hazards*. Berlin, Springer-Verlag, pp. 807–838.

Pancheri, G. and Zanetti, E. 1960. Liparosis-silicosis from pumice dust. *Proceedings of the 13th International Congress on Occupational Health*, July 25–29, New York, pp. 767–768.

Rosi, M., Principe, C., and Vecci, R. 1993. The 1631 Vesuvius eruption: a reconstruction based on historical and stratigraphical data. *Journal of Volcanology and Geothermal Research*, **58**, 151–182.

Santacroce, R. 1996. Preparing Naples for Vesuvius. *IAVCEI News*, 1996 1/2, 5–7.

Sigurdsson, H. 1988. Gas bursts from Cameroon crater lakes: a new natural hazard. *Disasters*, **12**, 131–146.

Sigurdsson, H., Carey, S., Cornell, W., *et al.* 1985. The eruption of Vesuvius in AD 79. *National Geographic Research*, **1**, 332–387.

Sparks, R. S. J., Bursik, M. I., Carey, S. N., *et al.* 1997. *Volcanic Plumes*. Chichester, UK, John Wiley.

Spence, R. S. J., Pomonis, A., Baxter, P. J., *et al.* 1996. Building damage caused by the Mount Pinatubo eruption of June 15, 1991. In C. G. Newhall and R. S. Punongbayan (eds.) *Fire and Mud: Eruptions and Lahars of Mount Pinatubo, Philippines*. Quezon City, Philippines, Philippine Institute of Volcanology and Seismology, pp. 1053–1061.

Sutton, J. and Elias, T. 1993. Volcanic gases create air pollution on the island of Hawaii. *Earthquakes and Volcanoes*, **24**, 178–196.

Tanguy, J.-C., Ribiere, C., Scarth, A., *et al.* 1998. Victims from volcanic eruptions: a revised data base. *Bulletin of Volcanology*, **60**, 137–144.

Thordarson, T. L. and Self, S. 1993. The Laki (Skaftár Fires) and Grímsvötn eruptions in 1783–1785. *Bulletin of Volcanology*, **55**, 233–236.

Voight, B. 1996. The management of volcanic emergencies: Nevado del Ruiz. In R. Scarpa and R. I. Tilling (eds.) *Monitoring and Mitigation of Volcano Hazards*. Berlin, Springer-Verlag, pp. 719–769.

# Chapter 11

# Volcanoes, geothermal energy, and the environment

Wendell A. Duffield

## Introduction

Volcanoes and volcanism in all forms produce a variety of impacts on the environment. Some impacts are desirable, for example the eye-pleasing shape of a majestic volcanic edifice outlined across a horizon and the sight of bountiful crops growing in rich soil developed on volcanic ash. Others, however, are detrimental; no one is pleased when a volcanic eruption damages or destroys humans or works of humans. Depending on volume, chemical composition, and explosivity, the detrimental impacts of an eruption can range in aerial extent from local to global.

An indirect but substantial impact on the environment occurs through a link between volcanism (and its associated magmatism) and a resultant partial redistribution of Earth's internal thermal energy into concentrations that are resources for potential human consumption (Duffield and Sass, 2003). Over broad regions of the Earth, temperature increases almost monotonically downward in the crust beneath a few-meters-thick zone that responds to annual temperature fluctuations. Outside of volcanically active regions, this global background thermal condition of the crust is characterized by a conduction-dominated geotherm in the range of 10–50 °C/km. However, when magma enters the crust, for example, as a shallow intrusion beneath a volcano, this background condition is perturbed locally as temperature rises around the intrusion. The amplitude, volume, and longevity of such a thermal anomaly depend mostly on the temperature and volume of the intruded melt.

A shallow intrusion of magma is a potent source of thermal energy for heating both surrounding rocks and groundwater as the melt solidifies and cools. In fact, with sufficient country-rock permeability and available groundwater, magma bodies can initiate and sustain large hydrothermal-convection systems, which are the fundamental stuff of geothermal resources. Water and associated steam from the hottest parts of such systems can be harnessed to generate electricity. Non-electrical uses of geothermal fluids, too cool to drive a turbine, also are common (Lindal, 1973). Both types of geothermal developments carry some environmental consequences (US Department of Energy, 1994).

The non-electrical application of geothermal energy to human endeavors has ancient origins. Long before recorded history, some peoples must have made chance discoveries of hissing steam vents, erupting geysers, boiling mud pots, and bubbling hot springs. One can only speculate on their reactions to such impressive natural phenomena, but some combination of fear, awe, and appreciation seems likely. Later, historical records indicate that hot springs and other geothermal features were used by early peoples for food preparation and for bathing. The geothermally heated spas of the ancient Greeks, Romans, and Japanese have been imitated throughout history, and today their modern counterparts

*Volcanoes and the Environment*, eds. J. Martí and G. G. J. Ernst. Published by Cambridge University Press. © Cambridge University Press 2005.

attract many visitors for recreational and medical reasons.

Prehistoric and early historical use of geothermal energy was effectively limited to non-electrical applications of natural manifestations found at the Earth's surface. With rare exception, surface manifestations produce water or steam with temperatures of less than 100 °C (the boiling point of water at sea level); their relatively low temperatures restrict the variety of possible applications. Lack of knowledge and technical limitations prohibited any attempt to develop deeper, hotter geothermal energy. Still, many early civilizations benefited from the geothermal resources with which they were provided by nature.

Geothermal steam was first harnessed to generate electricity in 1904, at Larderello, Italy. This project was small, short-lived and drew steam from very shallow depth. With subsequent advances in technology, drills became available to penetrate thousands of meters into the Earth in search of hotter and higher-pressured geothermal resources. Such drilling has resulted in the discovery of geothermal fluids as hot as 400 °C, which can provide a resource of high-pressure steam to drive turbine generators at the Earth's surface. The traditional, ancient uses of geothermal fluids continue to have considerable scenic and recreational value, and the present-day capability to produce high-temperature fluid through drilled wells opens the door to diverse utilization of geothermal energy over a broad range of temperatures (Muffler and White, 1972).

Whether developed for electricity or for lower-temperature direct uses, exploited geothermal resources are broadly compatible with the environment. Some change to a pre-existing environmental condition is inevitable for any human enterprise, but compared with the environmental effects of harnessing most other types of energy resources, geothermal carries little negative impact. The types of environmental impacts and means to mitigate them are discussed in detail in a later section. To appreciate the spectrum of possible environmental impacts, one must understand the nature of the resource and the many different technologies that can be applied to the exploitation of geothermal energy.

# A primer on geothermal energy

Geothermal energy is simply the naturally occurring heat within the Earth, and our planet is a bountiful source of such heat. The formation of Earth through planetary accretion about 4.6 billion years ago, followed shortly thereafter by periods of intense bombardment with interstellar debris, produced an initially hot body. In addition, Earth continuously produces heat internally by the decay of naturally radioactive chemical elements (now principally uranium, thorium, and potassium) that occur in small amounts in all rocks. Being a basically hot body adrift in the cold space of the surrounding universe, the Earth is constantly losing, as well as generating, some of its heat. In moving toward thermal equilibrium with its surroundings, Earth heat rises to the surface and is dispersed into outer space.

Scientists can accurately measure the rate at which heat moves through the Earth's crust. Heat flow is simply the vertical temperature gradient multiplied by the thermal conductivity of rocks that record this gradient, and for conduction-dominated regimes, the average regional background value for Earth is about 60 mW/m$^2$. Heat flow is generally somewhat higher than this average in tectonically active areas and somewhat lower in stable cratonic areas (Fig. 11.1). The annual heat loss from the Earth is enormous – equivalent to ten times the annual energy consumption of the United States and more than that needed to power all nations of the world, if it could be fully harnessed. Even if only 1% of the thermal energy contained within the uppermost 10 km of our planet could be tapped, this amount would be 500 times that contained in all oil and gas resources of the world. However, analogous to common fuel and mineral resources, geothermal energy is most attractive for use by humans only where it is concentrated into anomalously rich deposits. For the highest-temperature geothermal resources, magma is the heat engine, and volcanoes are signposts to the possible presence of magma in the upper crust. Learning to interpret what these signposts say about a potential energy resource is the stuff that helps form the science (perhaps

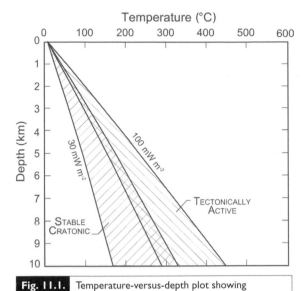

**Fig. 11.1.** Temperature-versus-depth plot showing approximate ranges of conductive heat flow in milliwatts per square meter (mW/m²) for stable cratonic and tectonically active environments.

interspersed with a bit of art) of geothermal exploration.

## The plate tectonic connection

Given the link to volcanoes, it's an easy step to deduce that most high-temperature geothermal resources are along or near the boundaries of the dozen or so lithospheric plates that make up the brittle outermost rind of the Earth (Muffler, 1979). Even though these plate-boundary zones comprise less than 10% of the Earth's surface, their potential to contribute to world energy supply is substantial and widespread (Fig. 11.2).

The "Ring of Fire" that surrounds the Pacific Ocean basin, mostly as subduction-related volcanic arcs, is replete with high-temperature geothermal resources; these make significant, and growing, contributions to the energy demands of many circum-Pacific countries. Beginning in the 1950s, New Zealand and Japan were the first in the region to harness geothermal energy for generating electricity. Today, geothermal-electric projects in the Philippines and Indonesia are the largest and most active worldwide, while smaller-scale but similar projects are in various stages of completion in Central America. In Iceland, along the sub-

aerial part of the volcanically active mid-Atlantic plate-tectonic spreading zone, virtually all space heating is accomplished with geothermal energy, and a huge reserve of electrical-grade geothermal energy is developed piecemeal as needed to supplement abundant hydroelectric resources. To complete the linked triad of plate boundaries, volcanoes, and geothermal resources, the world's largest developed geothermal field, The Geysers in California, is associated with an extensive area of Quaternary volcanism adjacent to the San Andreas transform fault zone.

As all well-trained earth scientists know, some volcanoes and their associated geothermal energy occur within tectonic plates, rather than along their boundaries. Kilauea, near the center of the Pacific Plate on the island of Hawaii, is perhaps the best-known example of a volcano in this tectonic setting. One geothermal power plant exists on Kilauea, and interest in further exploration continues. The huge Quaternary volcanic field at Yellowstone, the site of (arguably) the world's largest geothermal resource, lies deep within the interior of the North American tectonic plate. However, this resource will not be developed, because it is protected by National Park status. The key notion to remember is that high-temperature geothermal resources are found in areas of active, or geologically very young, magmatism wherever such melt becomes lodged in the crust.

## Reservoirs and uses of geothermal energy

Looking at a level of detail beyond the first-order link to volcanoes (magma), deposits of geothermal energy within the crust occur in a variety of geologic reservoirs, or settings. These commonly are classified by temperature and amount of fluid – water and (or) steam – available for carrying thermal energy to the Earth's surface (Fig. 11.3). Because geothermal heat is always on the move as the Earth cools toward thermal equilibrium with outer space, no reservoir of geothermal energy is completely closed to its surroundings. Nonetheless, for purposes of delineating fundamentally

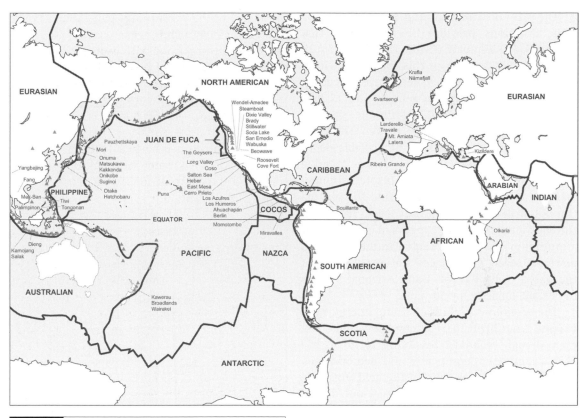

**Fig. 11.2.** Map of Earth's tectonic plates, with schematic distribution of active subaerial volcanoes (triangles) and developed high-temperature geothermal-resources areas.

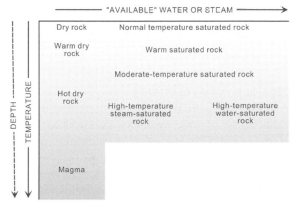

**Fig. 11.3.** Schematic representation of geothermal environments in the upper crust. Generally temperature increases with depth, but the depth line is dashed to indicate that the rate of temperature increase varies within the crust. "Available" refers to water and steam that can be tapped by a well and produced at the Earth's surface.

differing geologic homes for deposits of geothermal energy, the reservoir concept is useful.

## Magma reservoirs

Magma is the highest-temperature and a relatively water-poor part of the geothermal classification by reservoir type. Depending on its chemical composition, magma ranges in temperature from about 650 to 1300 °C. For comparison, common steel melts at about 1500 °C. The most volatile-rich magmas contain no more than a few weight percent of $H_2O$, an amount insufficient and effectively unavailable for geothermal use. Even if the magmatic gases of high-pressure volcano-related fumaroles could be routed to a turbine, caustic and non-condensible constituents would tend to greatly decrease turbine efficiency and cause continuously recurring problems due to chemical corrosion. Thus, the magma reservoir of geothermal energy remains a topic for research into ways to harness this high-grade resource, while serving as a heat source for various other geothermal reservoirs.

The size, distribution, and frequency of volcanic eruptions provide direct evidence that magma is widespread within the crust (Smith and Shaw, 1979). Much of this magma is within about 5 km of the surface – depths easily reached using current drilling technology. The 15 km³ of magma erupted in 1912 to produce the volcanic deposits in The Valley of Ten Thousand Smokes, Alaska, was probably at drillable depth before eruption. Had it been possible to mine this magma's thermal energy before eruption, using heat exchangers lowered into the magma through drill holes, the output would have been the equivalent of about 7000 MW of electricity for a minimum of 30 years – equal to the combined power of seven hydropower stations the size of the one at Glen Canyon Dam, Arizona, producing at top capacity for 30 years.

The economic mining of useful energy from magma is not likely to be achieved any time soon, even though technical feasibility to do so has been examined theoretically and demonstrated both in the laboratory and in the field (Fig. 11.4). The field experiments were carried out during the 1980s within the still-molten core of a lava lake that formed in Kilauea Iki Crater, Hawaii, in 1959. The long-term economics of mining heat from magma are unknown (Armstead and Tester, 1987), at least in part because of uncertainties in the expectable lifespan of materials in contact with magma during the mining process. Finding materials that can withstand extremely high temperatures and the corrosive nature of magma is one of the main obstacles to overcome before magma-energy technology can be widely tested.

Another impediment to tapping the thermal energy contained in magma is in unambiguously locating the magma body before drilling begins. Although experts agree that many bodies of magma are within the drillable part of the Earth's crust (Shaw, 1985), their exact locations can be verified only by drilling, and drilling is expensive. The possibility of the drill missing the target adds to the uncertainties related to longevity of materials in a hot corrosive environment. For the present, these uncertainties definitely make mining of magma heat a high-risk endeavor in using this highest grade of geothermal ore.

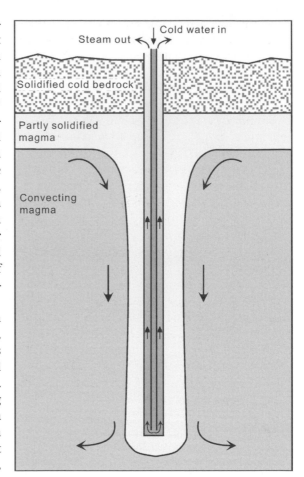

**Fig. 11.4.** Schematic drawing of how thermal energy can be mined from magma. Concentric pipes are emplaced into magma. Water pumped down the inner pipe returns as steam in the outer pipe.

## Hot-dry-rock reservoirs

A hot-dry-rock (HDR) geothermal reservoir contains little, if any, available geothermal fluid. HDR commonly is regarded as being hot enough for potential electrical applications but too dry and (or) impermeable to produce useful amounts of energy-transporting geothermal fluids for such applications; lower temperature HDR reservoirs also exist. The "dry" part of HDR signifies a paucity of available geothermal fluids, but it does not necessarily imply negligible porosity and permeability. Whether porous and permeable or not, HDR is of no current commercial value because of the high cost of drilling wells and circulating water through a reservoir created around these wells. Thus, similar to geothermal energy contained in magma, that in HDR is mainly of academic interest.

Research into HDR is under way at several locations worldwide, and the technical feasibility of mining heat from high-temperature, water-poor rocks has already been demonstrated, but the costs of doing so are still outside the profitable market place. Necessary permeability can be, and has been, created with various methods (called hydraulic fracturing or hydrofracturing) used in the petroleum industry to enhance the recovery of oil and gas. Hydrofracturing involves the high-pressure injection of a fluid (usually water) into a reservoir to crack rock and (or) enlarge pre-existing openings. Possible sources for the water used in hydrofracturing range from tapping nearby shallow, cool groundwater or surface water to using treated sewage effluent.

Scientists of the Los Alamos Scientific Laboratory, New Mexico, initiated pioneering research and development in HDR technology in 1971. To date, the Los Alamos field experiment has involved drilling a pair of wells (injection and production); the creation of a permeable reservoir, through hydraulic fracturing, in rocks surrounding the bottom parts of the wells; and the completion of several flow tests at moderate (as much as 5 MW thermal) power output to demonstrate the feasibility of the technology (Fig. 11.5). This experiment was conducted at Fenton Hill, near the edge of the volcanically young Valles Caldera, near Los Alamos, New Mexico.

In common with other HDR experiments, the Fenton Hill project was bedeviled by engineering difficulties. Perhaps the most troublesome was the very high pumping pressure needed to move water through the reservoir. With this and other nagging problems in mind, emphasis and research priorities evolved toward a concept called "enhanced geothermal systems" in which some of the techniques developed in HDR experiments are applied to hydrothermal systems of low productivity. Productivity is enhanced by some combination of hydrofracturing, targeted injection, and drilling directional (non-vertical) wells toward specific permeable targets.

A field experiment of this type was conducted at Dixie Valley, Nevada, during the period of 1996–9 by a consortium of industry, government, and university researchers (Duffield and Sass, 2003). At Dixie Valley, the well field for a geothermal power plant includes hot-but-dry wells within an area of very productive wells. Researchers investigated the roles of regional and local stress fields in creating and maintaining fracture permeability. Results demonstrate a robust correlation between the orientation of highly permeable fractures and the direction of principal stress within the crustal rocks. The ultimate aim is to improve the rationale for siting wells, both for production and injection purposes, and to develop hydrofracturing techniques and practices that optimize creation of productive zones within otherwise non-productive parts of hydrothermal systems. This approach may result in bringing additional geothermal resources "on line" more rapidly than experiments in areas totally devoid of hydrothermal activity.

## H₂O-rich reservoirs

In the business world of geothermal-energy deposits that are amenable to exploitation, the presence of abundant geothermal fluids and sufficient permeability for their mobility through host rocks are absolutely essential (Norton and Knight, 1977). For even though most geothermal energy is stored within rocks and magma, water and steam are the only natural carriers capable of transporting this energy to the surface where current technologies permit conversion to forms of work that satisfy human needs. This fluid-carrier limitation to the exploitation of geothermal energy will obtain until, or unless, advances in technology permit the use of thermal energy contained in rock and magma. As noted above, research is under way to accomplish this objective. Meanwhile, a magma bed may sound simply like a warmer version of a waterbed, but mattress technology is far from producing a safe container for such a bed-warming fluid.

With increasing amounts of available water comes a broad category of H₂O-saturated reservoirs that encompasses a range of porosity, permeability, and temperature. Within this category

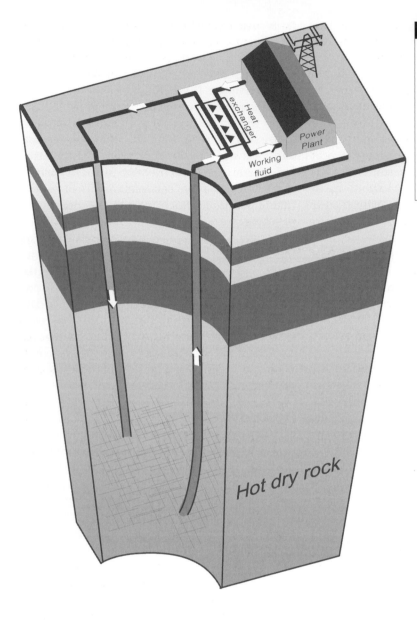

**Fig. 11.5.** Schematic drawing of how thermal energy can be mined from hot dry rock (HDR). Two wells are drilled, and a permeable zone between and around them is created by hydraulic fracturing. Cold water is pumped down one well and returns as hot water through the other well. At the surface, thermal energy in extracted through a heat exchanger, and the cooled water begins another circulation cycle.

are relatively high-temperature rocks, saturated either with steam or liquid water or a mixture of both, and rocks saturated only with water at increasingly lower temperatures. Geothermal reservoirs through which available water circulates freely are called hydrothermal-convection, or simply, hydrothermal systems. Under current technologies, these are the only geothermal reservoirs of commercial interest. The style of exploitation varies, mostly as a function of reservoir temperature and fluid chemistry (Muffler, 1981). Three principal styles are practiced for generating

electricity, and each impacts the environment in somewhat different ways.

### Vapor-dominated reservoirs

As this name implies, steam is the geothermal fluid that saturates the reservoir for this type of hydrothermal system. Some liquid water is present in discrete separated volumes, but steam is the continuous fluid, and fluid pressure is in a vapostatic condition. Vapor-dominated hydrothermal systems are rare, relative to water-saturated systems. From a commercial perspective, vapor-dominated is the most desirable type of electrical-grade hydrothermal system, because steam can be routed directly from a well into a turbine (Fig. 11.6a). Downstream from the outlet of a turbine, very little condensate accumulates for disposal.

The origin of a vapor-dominated reservoir is poorly understood, but requires a potent heat source and a restricted degree of water recharge. A key point to remember in trying to understand the origin and functioning of any hydrothermal system is that it is dynamic on a human-lifetime scale and is open, to varying degrees, to its surroundings. Thus, some appropriate but restricted combination of heat source, porosity/permeability, and water recharge results in a vapor-dominated, rather than a water-saturated, reservoir.

Vapor-dominated "caps" have been created at the tops of water-saturated hydrothermal reservoirs under production, where none originally existed. These are situations where the natural condition was perturbed by fluid production through wells in a way that moved the uppermost part of a water-saturated hydrothermal system into a combination of the above variables that is appropriate for vapor domination. Presumably, with cessation of production and sufficient time for recovery, the natural condition would be re-established.

### Hot-water reservoirs

As the name implies, water saturates this type of hydrothermal system. The profile of fluid pressure reflects a hydrostatic situation. Ambient conditions of heat source, water recharge, and porosity/permeability maintain water saturation within the reservoir. Hot-water dominated is by far the most common type of hydrothermal system known. This is the bread and butter of the geothermal industry, worldwide.

Depending on temperature, hot-water reservoirs are exploited for electricity in two different ways. For those at temperatures above about 200 °C, water is allowed to boil to steam at the inlet pressure of a turbine, after which the separated steam enters the turbine and the residual "warm" water is diverted for disposal. A typical

exploitation scheme is illustrated in Fig. 11.6b. Relative to a vapor-dominated electrical facility, additional hardware (separator) is needed and voluminous warm water must be disposed of as a waste product.

For somewhat cooler systems, the hydrothermal water is routed through a heat exchanger to transfer thermal energy to a fluid whose boiling temperature is considerably below that of water. The cooled geothermal water is then pumped back into the subsurface while the vapor of the "working" fluid drives a turbine. A typical installation for this method of energy conversion is illustrated in Fig. 11.6c. Because the exchange of thermal energy between fluids of two chemical compositions is involved, this method of electrical generation is known as binary.

### Warm-water reservoirs

Reservoirs of available geothermal water too cool to generate electricity can be used directly in a host of applications whose number is limited only by one's imagination of how to use the water's thermal energy. For example, thermal water too cool to produce electricity can still furnish energy for swimming pools and spas, heating soil for enhanced crop production at cool-climate latitudes, and heating buildings (Table 11.1). However, low-temperature geothermal water is a relatively low-grade "fuel" that generally cannot be transported far without considerable thermal-energy loss, unless piping is extremely well insulated and the volumetric rate of flow through the piping is high. Still, about one-third of the geothermal energy used today is consumed in direct-use applications. The thermal energy in reservoirs of warm water – the most widely distributed of the hydrothermal systems – can locally complement or supplant that available from conventional energy sources.

Extensive development of the warm-water reservoirs, commonly found both in volcanic and non-volcanic areas, can significantly improve the energy balance of a region or even an entire nation. For example, the use of geothermal water for space heating and other direct-use applications in Iceland substantially benefits the economy of that nation. Somewhat more than 45% of the country's primary energy use comes from

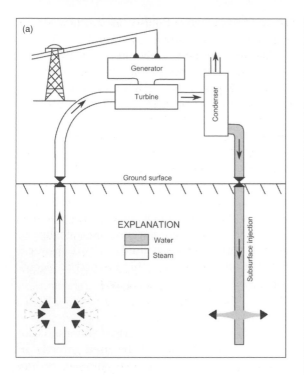

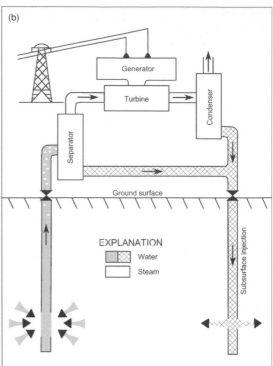

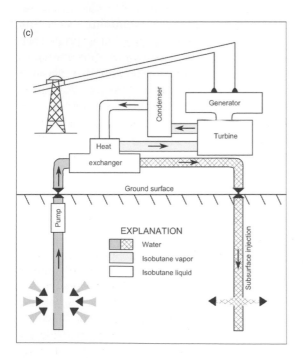

**Fig. 11.6.** Diagrams showing how electricity is generated from hydrothermal-convection systems. In (a), the hydrothermal system produces only steam, which is routed directly into a turbine generator. In (b), the hydrothermal system produces a mixture of hot water and steam. Once separated, the steam is routed into a turbine generator. In (c), only hot water is produced. Its thermal energy is used to boil isobutane, whose vapor drives a turbine generator. In all three cases, geothermal effluent, the water downstream of the heat extraction step, is injected back into the subsurface.

Table 11.1 | Common non-electric uses of geothermal water as a function of temperature

| Temperature (°C) | Use |
|---|---|
| 90 | Drying fish, de-icing |
| 80 | Space heating, milk pasteurization |
| 70 | Refrigeration, distillation of ethanol |
| 60 | Combined space and soil heating for greenhouses |
| 50 | Mushroom-growing |
| 40 | Enhanced oil recovery |
| 30 | Water for winter mining in cold climates |
| 20 | Spas, fish hatching and raising |

geothermal sources. Hydropower provides nearly 17%, whereas oil and coal account for the balance.

Situated within one of the globe-girdling belts of active volcanoes, Iceland has abundant geothermal resources of both electrical and non-electrical grade. The country also has abundant surface-water resources, and most of the nation's electricity is generated from hydropower. Nearly all buildings, however, are heated with geothermal water. The geothermal water (or fresh surface water heated by geothermal water through a heat exchanger) for space heating in Reykjavík, the national capital, is piped as far as 25 km from well fields before being routed to radiators in buildings. Very little thermal energy is lost in transit because a high rate of flow is maintained through well-insulated, large-diameter pipes.

Typically, the water delivered to homes and other buildings in Iceland is of a quality sufficient for other direct uses (for example, bathing and food preparation) in addition to space heating, and therefore is also piped to taps. The size of space-heating developments varies from large enough to serve the entire national capital (about 50% of the country's population) to appropriately small enough for a single-family residence in a rural setting. Geothermal water also is used in

Iceland to heat greenhouses, so that flowers and vegetables can be grown year round.

Geothermal heat can be the dominant source of energy for direct-use applications, such as space heating and industrial processing as in Iceland, or it can play a smaller scale yet important role. In the region around Paris, France, some 10 000 to 15 000 apartment-housing units are heated by 60–80 °C water obtained from geothermal wells about 1.5–2 km deep. Once heat has been extracted at heat exchangers, the geothermal water is pumped down a well to an aquifer where it can be reheated by contact with warm rocks as it circulates back toward the production well. Such a production/injection pair of wells is called a doublet, and each pair is carefully engineered to try to maximize the amount of thermal energy that can be harvested without quick local cooling ("short circuiting") of the productive geothermal aquifer.

The city of Klamath Falls, Oregon, is in an area of abundant near-surface geothermal water of temperatures appropriate for direct-use applications. For more than a century, citizens of this community have used this resource in innovative ways. Hundreds of residents have drilled shallow wells on their property to tap geothermal energy for home heating. Commonly, no hot water is pumped to the surface, but heat is instead extracted by circulating cold city water through a loop of pipe lowered into the well. This technique greatly extends the life of the geothermal resource, because no water is removed from the hydrothermal system. During winter, many residents circulate warm water through pipes embedded in the concrete of driveways and sidewalks to melt the abundant snow that Klamath Falls usually receives. A large-scale direct-use system provides heat for several government and commercial buildings in the downtown area of Klamath Falls, including the Oregon Institute of Technology (OIT), an internationally known center for the development of creative ways to use low-temperature geothermal resources.

A comprehensive list of direct uses internationally would go on and on. As of the early 1990s, such uses were the thermal-energy equivalent of about 21 million barrels of petroleum annually.

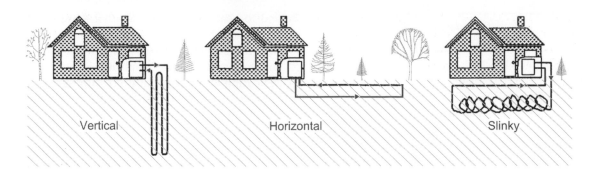

Vertical          Horizontal          Slinky

**Fig. 11.7.** Schematic drawings of three common designs for ground-coupled geothermal heat pumps. Alternative designs for geothermal heat pumps are coupled to shallow groundwater or a surface pond of water.

## Geopressured reservoirs

As a by-product of drilling for petroleum, reservoirs of warm to hot geothermal water have been discovered within sequences of sedimentary rocks that are hydrologically isolated, or nearly so, from their surroundings (Muffler, 1979). Typically, such reservoirs are bounded by an enveloping configuration of very-low-permeability faults and strata of fine-grained sediments, which are parts of marine basins where sediment has rapidly accumulated.

The presence of hydrologically isolated volumes of water-saturated sediment within an environment of rapid sedimentation results in fluid pressures that approach lithostatic, rather than hydrostatic, thus prompting the "geopressured" identifier. Fundamentally, the increasingly deeper-buried sediments can not dewater as rapidly as the new overlying sedimentary load accumulates. As a result, a geopressured reservoir transmits some mechanical energy to a well head, stemming from the fluid overpressure, in addition to the thermal energy of the geothermal water. Moreover, the reservoir also contains potential for combustion energy, because considerable methane is commonly dissolved in the geothermal water, presumably a result of the petroleum-field association. The bulk of the thermal energy of geopressured reservoirs is accounted for by roughly equal contributions from the temperature of the water and the dissolved methane. The most intensely studied area containing geopressured-geothermal energy is the northern part of the Gulf of Mexico, which has been drilled extensively for petroleum.

## Normal-temperature reservoirs

Though not directly associated with volcanism and related magma for their thermal energy, shallow geothermal reservoirs of normal-geotherm temperature are included for completeness. This type of reservoir consists of ordinary near-surface rock and soil, ranging from dry to water saturated (Fig. 11.7). Some argue that this shallow environment should not carry a geothermal label. Yet, it is the condition established there by the flow of heat from the Earth to outer space that has created and sustains a thermal environment that can be put to use in a rapidly growing industry that designs and installs geothermal heat pumps (sometimes referred to as ground-source heat pumps). The term "GeoExchange" is commonly used to describe the process of tapping this source of thermal energy.

A heat pump is simply a machine that causes thermal energy to flow up temperature, that is, opposite the direction it would flow naturally without some intervention. Thus, a heat pump is commonly used for space heating and cooling, when outside ambient air temperature is uncomfortably cold or hot, respectively. The cooling and heating functions require the input of "extra" work (usually electrical energy) in order to force heat to flow upstream. And the greater the "lift," or difference in temperature between the interior of a building and the outside, the more work needed to accomplish the function. A geothermal heat pump increases the efficiency of the heating and cooling functions, by substantially decreasing the thermal lift.

Because rocks and soils are good insulators, they respond little to wide daily temperature fluctuations and instead maintain a nearly constant temperature that reflects the mean integrated over many years. Thus, at latitudes and elevations where most people live, the temperature of rocks and soil only a few meters beneath the surface typically stays within the range of 5–10 °C.

For purposes of discussion, consider the functioning of a conventional air-source heat pump in a single-family residence, a system that exchanges thermal energy between air indoors and outdoors. Whereas such a heat pump must remove heat from cold outside air in the winter and deliver heat to hot outside air in the summer, a geothermal heat pump exchanges heat with a medium that remains at about 8 °C throughout the year. As a result, the geothermal-based unit is almost always pumping heat over a temperature lift much smaller than that for an air-source unit, leading to higher efficiency through less "extra" energy needed to accomplish the lift.

Some consumer resistance to geothermal heat pumps exists because initial purchase-and-installation cost is greater than that for an air-source system. The additional cost comes mostly from the need to bury piping through which fluid (water or antifreeze) is circulated to exchange heat with the ground, or by drilling a shallow well to use groundwater as the heat source/sink. Additional cost varies with the capacity and subsurface design of a given system. Experience to date indicates that the extra expense can be amortized in as little as 3 or 4 years for some systems. Other systems carry a longer pay-off period, but eventually all geothermal heat pumps provide savings that accrue as lower than "normal" utility bills.

## Geothermal resources

Geothermal-energy reservoirs of different types commonly occur in close proximity, perhaps as crudely nested volumes of outward and upward decreasing temperature. For example, a body of magma in the Earth's crust may be enveloped by HDR, which may in turn be overlain by a hydrothermal-convection system. With or without an associated body of magma, HDR and fluid-saturated, high-temperature hydrothermal reservoirs often coexist, the distinction between them being in the amount of *available* fluid. In fact, one of the major challenges for geothermal developers is in siting wells to encounter the high-temperature hydrothermal, rather than the HDR, part of a thermal anomaly in the shallow crust. Low- to moderate-temperature hydrothermal systems may occur in volcanic regions, or away from volcanic areas, if their water is heated solely by deep circulation through crustal rocks that exhibit a normal background geotherm.

Each type of geothermal reservoir can occur over a range of depths, depending upon the geologic characteristics of a given geographic site. Many volcano-related magma bodies in the Earth's crust are estimated to be about 5 km deep, or somewhat deeper, yet some bodies of molten lava pond in craters at the Earth's surface during volcanic eruptions and cool and solidify there under their own thin but growing crust. Similarly, hydrothermal systems as hot as 250 °C may be discovered at depths from somewhat less than one to several kilometers, depending upon the local geothermal gradient and the vigor of upward flow of convecting hot fluids.

Determining whether or not heat can be extracted from a particular hydrothermal reservoir is critically dependent on its depth. The pertinent question here is whether the hydrothermal target is at a drillable depth, no more than roughly 5–6 km with current drilling technology. Determining whether or not heat can be extracted *profitably* from a particular hydrothermal reservoir adds yet another condition to satisfy. An understanding of how much geothermal energy is being, or may soon be, *profitably* extracted is important when assessing the level of environmental impacts to be expected from the use of this thermal-energy source. To avoid confusion, a few terms are defined.

*Resource base* all the thermal energy in the Earth's crust beneath a specified area, referenced to local mean annual temperature

*Accessible resource base* the thermal energy at depths shallow enough to be tapped by drilling using technology available at present or within the foreseeable future

**Resource** that part of the accessible resource base that is producible given reasonable assumptions about future economics and technology

**Reserve** that part of the resource that is identified and is producible with existing technology and under present economic conditions.

A graphical representation of this terminology and some of its underlying concepts is shown in Fig. 11.8. The vertical axis represents the degree of economic feasibility of exploitation, whereas the horizontal axis represents the degree of geologic certainty about whether or not a given geothermal reservoir exists. The vertical axis is subdivided into four categories, with increasingly favorable economic viability toward the top, whereas the horizontal axis recognizes only known and undiscovered reservoirs.

This scheme for categorizing geothermal energy was the basis for a national assessment of resources by the US Geological Survey (Muffler, 1979). Different schemes are used by some others in the United States and abroad (Muffler and Cataldi, 1978). Understanding the precise meaning of resource terminology, though perhaps not too intellectually stimulating in development and application, is absolutely critical in resource assessment and resultant resource-use planning. Without such understanding, one organization's resource may well be another's resource base, which can lead to gross distortions and errors when judgements are made on how much geothermal energy is available for use in a given time frame.

Although students of earth science agree that the Earth's internal heat is immense, the part that is a resource as defined above is rather small. The US Geological Survey (Muffler, 1979) has estimated the geothermal resources of the United States at temperatures greater than 150 °C are sufficient to generate 23 000 MW of electricity for 30 years. This translates to 30 years of electricity at average urban rates of use for about 23 million people in the United States. Obviously, this electrical energy could serve more people in a less-industrialized country. In addition, identified geothermal resources at temperatures less than 150 °C are estimated to be 280 000 MW of thermal energy for 30 years, the equivalent of almost 20 years of oil imports by the United States at 1992 levels. These <150 °C resources are appropriate for non-electrical uses such as space heating, industrial processing, and other direct-use applications, although some with temperatures above about 120 °C could also be used to generate electricity using binary technology.

The above resource estimate is a minimum value of what geothermal can contribute to the USA's energy consumption. If undiscovered hydrothermal resources are considered, and certainly undiscovered resources exist, the total increases several-fold. Moreover, a much larger amount of thermal energy resides in HDR, which currently lies in a category that spans the subeconomic and residual parts of the accessible resource base (Fig. 11.8). If, and some would say when, production techniques move HDR into the resource category, geothermal will become more than a bit contributor to the world's energy-resource inventory. Muffler (1981) has estimated that geothermal energy associated with igneous systems in the USA is about 50% in magma, 43% in HDR, and only 7% in hydrothermal systems.

Even the most optimistic advocates doubt that geothermal energy will replace fossil fuels as a dominant energy source, new technologies notwithstanding. Still, this type of energy is already the principal source of energy locally to regionally, and it can make a significant contribution to the energy mix of many nations, thereby decreasing the chance of collapse of energy supply that attends reliance on a single source. To the extent that future environmental costs are factored into the use of hydrocarbon-based energy, hydrothermal and the emerging use of HDR will become increasingly competitive over large regions.

## Exploring for high-temperature geothermal resources

The following discussion is based on the premise that exploration is aimed at identifying a resource capable of being developed to generate electricity. Within this framework, the described exploration strategy is the "maximum" case.

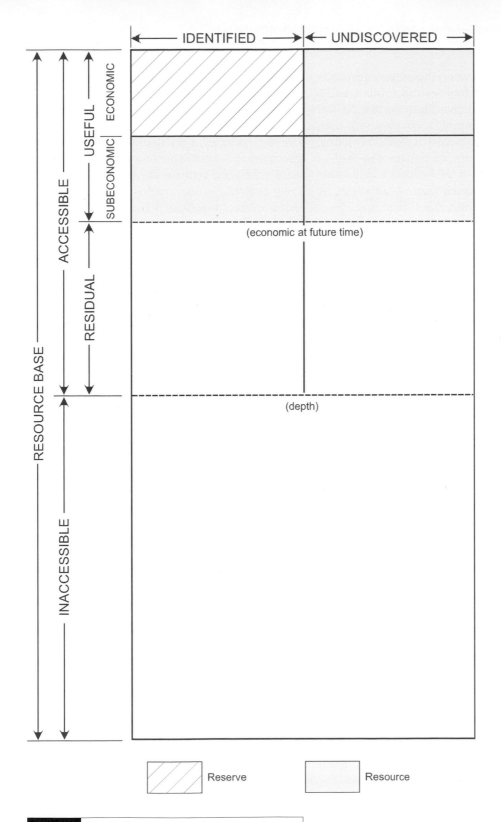

**Fig. 11.8.** Graphical representation of concepts and terminology commonly used in geothermal-resource assessment. The vertical axis represents the degree of economic feasibility of exploitation. The horizontal axis represents the degree of geologic certainty.

Exploration for a lower-temperature resource, perhaps warm water for heating a small building, would entail, and could only justify, far less effort. The reader should keep in mind that, while exploration for an electrical-grade resource may cost many millions of dollars, the cost of developing warm water for heating may include only installation of piping from a hot spring or very shallow well to the heated structure.

If the first and primary link in the exploration chain is the presence of active or geologically young volcanoes (Shaw, 1985), following links build from the volcano presence to all recognizable indicators of underlying crustal magma bodies or recently crystallized, and therefore hot, plutons. Historical volcanism in a volcanic field is a clear indicator of anomalous heat in the underlying crust, and a dormant volcano whose surface and surroundings are pierced by fumaroles and hot springs is an almost equally definitive and obvious indicator. One of the greatest challenges to explorationists is to be able to identify resource potential in geologically young yet extinct volcanoes, especially those in hydrologic settings that preclude fumaroles and thermal springs.

## The "when," "how much," "where," and "what" of volcanism

Any geothermal assessment of a volcanic area should include fundamental information on the time–volume–space–composition features of the volcanic rocks (Duffield, 1992). Placing constraints on these parameters forms the basis for producing a three-dimensional geologic model for the magmatic system. The age, or range in age, of volcanism is an obvious indicator of the emplacement age and therefore probable current temperature of any underlying co-volcanic magma in the crust. The volume of the magma body is of nearly equal importance in assessing the present condition of the crustal heat source. Many workers subscribe to the inference that for every erupted volume of magma, about ten volumes remain in the crustal reservoir. The importance of space is apparent, since widely distributed volcanic vents suggest multiple and equally widespread magma reservoirs, perhaps too diffuse to serve as a substantial and coherent geothermal heat source, whereas a single vent or many closely spaced penecontemporaneous vents suggest a single magma reservoir.

The "what" part of the heading for this section, refers to the chemical composition of the volcanic rocks in a particular study area. This information is important in a geothermal assessment primarily because the presence of chemically evolved, "silicic" rocks is strong evidence that a magma reservoir is indeed present (or at least was, at the time of eruption) in the upper crust beneath a volcano, or volcanic field. The geologic justification that underlies this conclusion is simple, rather compelling, and can be drawn from the publications of many earth scientists.

Abundant data argue that all lithospheric magmatism is fundamentally basaltic, in the sense that mantle-derived basaltic melt supplies heat and mass to the crust, where partial melting, assimilation, fractionation, and mixing produce the spectrum of intermediate to silicic magmas that define the chemical diversity present in many volcanic fields. The relatively thicker and compositionally more diverse continental lithosphere favors more extensive production of silicic magmas than does the thinner basaltic oceanic lithosphere. At a level of detail inappropriate for discussion here, this petrogenetic model encompasses and accommodates many if not all of the first-order geologic variables in magma production and evolution. Distilled down to a high-impact and practical guide for geothermal prospecting, the key concept is that silicic volcanic rocks imply formation in and eruption from a magma reservoir in the crust. Thus, the age of the youngest silicic rocks of a volcanic field implies underlying crustal magma of the same age. And computer-assisted time-series modeling of such magma can help guide conclusions about current associated geothermal resources.

While the presence of silicic volcanic rocks can be interpreted, with some confidence, to imply a crustal magma reservoir, lack of silicic rocks need not imply lack of a reservoir. Some obvious examples of the latter include active volcanoes such as Kilauea and Piton de la Fournaise, where lithologically monotonous eruptions of nothing but basalt are fed from magma reservoirs whose crustal positions are well defined by

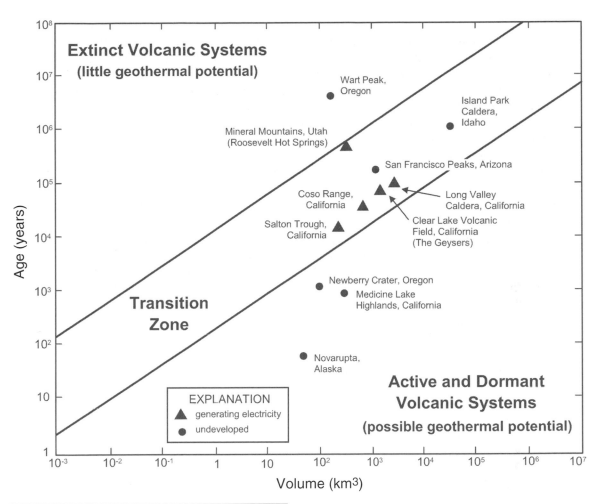

**Fig. 11.9.** Ages of eruptions versus volumes of associated "residual" magma bodies for some volcanic areas of the United States. This plot provides a means of evaluating the viability of a magmatic heat source for a geothermal resource. As of 1997, information gained through exploration drilling at Newberry Crater, Oregon, and at Medicine Lake Highlands, California, indicates high probability of commercial geothermal resources there. Novarupta is protected from commercial exploration by National Park status.

measurements of magma-caused ground deformation, and the distribution of magma-triggered earthquakes.

Once the inferred age, volume, and location of a crustal intrusion are established, translation into the present temperature regime in adjacent country rocks depends on the particular cooling model one adopts. Many are available in the literature. In assessing igneous-related geothermal resources, some researchers have analyzed the problem of pluton cooling through conduction of heat into country rocks, both with and without convection of magma within the intrusion itself. This type of analysis recognizes that hydrothermal convection within country rock would reduce conduction-only cooling times, and deals with this situation qualitatively by noting the offsetting effect of serial magma intrusion, which is suggested by the serial eruptions so typical of volcanic fields. Within this framework, igneous-related geothermal systems can be ranked into three broad categories (Fig. 11.9).

Other researchers have dealt quantitatively with the cooling effect of hydrothermal circulation around and within a solidifying pluton. (Cathles, 1977; Norton and Knight, 1977; Hayba and Ingebritsen, 1997). The recent availability of inexpensive and powerful computers has permitted the high-resolution and accurate calculation of temperature versus depth through time over intrusions of various sizes and initial temperatures. Undoubtedly, increasingly powerful computers of the future will permit an even more detailed analysis of the problem.

Perhaps not surprisingly, calculations show that host-rock permeability is the single most important factor in determining the temperature-versus-time history for a hydrothermal system heated by a magma body. The hottest and most powerful hydrothermal systems develop in reservoirs whose permeability is about $10^{-15}$ m$^2$. The maximum temperature and longevity of a hydrothermal system decrease if permeability is greater than this benchmark value. Thus, whereas permeability is absolutely necessary for the success of a commercial geothermal-electrical development, too much permeability can bring reduced power output, or failure.

Though far less rigorous in evaluating the effects as hydrothermal circulation and changes in rock properties with time, the approach illustrated by Fig. 11.9 is especially useful for geothermal purposes simply because it uses geologic observations to constrain estimates of magma age, volume, depth, and initial temperature for specific volcanic fields.

Erupting magma commonly contains bits and pieces of pre-existing rock that were caught up in the magma as it flowed through the crust. These rocks fragments, called xenoliths, often contain information useful to the geothermal explorationist (Wohletz and Heiken, 1992). For example, xenoliths may include samples of rocks from a hydrothermal reservoir, thus providing fundamental information that might otherwise be available only from rather expensive drilling. Furthermore, xenoliths may also include samples of minerals deposited from water circulating within a hydrothermal reservoir. The type and chemical composition of such minerals plus the nature of fluid inclusions trapped therein can provide information about the temperature in the hydrothermal system. This, too, is important information that might otherwise be available only from more-expensive exploration techniques.

In summary, field studies that define the age(s), vent location(s), volume(s), chemical composition(s), and nature of xenoliths provide a first-order filter in evaluating geothermal resources that may be associated with a volcanic field. Other types of study complement this approach.

## Calculation of subsurface temperature

Other factors being equal, the single most important piece of information in deciding whether to carry a geothermal development project into a drilling phase, is the maximum temperature that may be within reach of the drill bit. Justification should be strong for spending a million dollars or so on a production-size well. Fortunately, many geothermal areas come complete with proxy recorders of the desired information, one of which was mentioned just above (Fournier, 1981).

Hot-spring water and fumarolic gases are nothing more than fluids that leak to the surface from hydrothermal-convection systems. If these geothermal fluids attained chemical equilibrium with their rock container in the subsurface and then quickly rose to their surface vents, with little if any chemical re-equilibration along the way, their compositions should record the subsurface temperature. An extensive data base, compiled from many temperatures measured deep in drill holes, has shown that this path of fluid migration is common (Fig. 11.10).

In practice, a roughly linear relationship between subsurface temperature and composition of geothermal fluids collected at the surface is empirical. However, this empirical geothermometry has a sound thermodynamic basis. Hot water circulating in the Earth's crust dissolves some of the rock through which it flows. The amounts and proportions of dissolved constituents in the water are a direct function of temperature. Quartz is a common mineral in the Earth's crust and has almost ideal characteristics for chemical geothermometry. The solubility of quartz is strongly temperature dependent, dissolved quartz does not readily react with other common dissolved constituents, and dissolved quartz precipitates very slowly as temperature is lowered. The quartz geothermometer is especially useful in the range of about 150–240 °C.

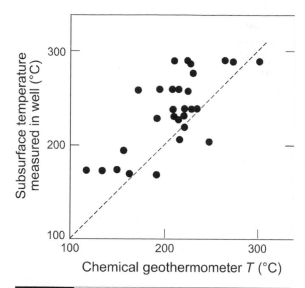

**Fig. 11.10.** Graph of temperatures measured in wells drilled into hydrothermal systems versus temperatures calculated from chemical compositions of hot-spring water or fumarolic vapor, before drilling. Dashed line shows where points should plot if measured and calculated temperatures are equal. The fact that most dots plot above this line indicates that calculated temperatures tend to underestimate measured temperatures, perhaps due to partial re-equilibration of geothermal fluids to changing environment as they flow to the surface.

If upflowing geothermal water mixes with cool shallow groundwater, or partly boils in response to decreasing pressure, appropriate adjustments must be made to the calculated subsurface temperature.

Feldspars – the most common minerals in the Earth's crust – contribute the elements sodium, potassium, and calcium to geothermal waters, in ratios also dependent on temperature. An advantage of dealing with ratios of elements is that dilution by mixing with cool shallow groundwater and boiling changes the ratios little, if any. The ratio between the elements magnesium and lithium, which are also introduced into geothermal water from dissolution of rock, has also proved to be an accurate geothermometer.

Isotopes of individual elements also distribute themselves according to temperature. For example, virtually all hydrothermal waters contain dissolved sulfate, and isotopes of oxygen distribute themselves between the sulfate and its solvent, water, according to a temperature-dependent relationship that can also indicate the subsurface temperature of a hydrothermal system.

In practice, scientists generally use a combination of geothermometers, rather than relying on any single one, to calculate subsurface temperature from the chemical composition of surface geothermal fluids. In so doing, uncertainties associated with results obtained using only one method may be reduced or resolved with reference to the results from other methods. One commonly practiced technique is to compare silica-concentration temperature with that based on a ratio of dissolved constituents; a lower silica temperature very likely reflects dilution by shallow groundwater during upflow.

## Other indirect probes into the subsurface

Several geophysical techniques are applied to geothermal studies aimed at helping to identify magma bodies, hydrothermal-convection systems and geologic structures that may be drilling targets (Lumb, 1981). Geophysical methods have long been used to explore for minerals and fossil fuels; however, their application to geothermal exploration is a relatively recent endeavor that has required some modifications to "conventional" techniques.

The ability of the Earth's crust to conduct electricity depends mostly on the constituent minerals of rocks, pore spaces in rocks, and the chemical compositions of fluids within those pores. Because the saline waters of hydrothermal-convection systems and many hydrothermal minerals are some of the best electrical conductors in the crust, geoelectrical techniques have been a standard geothermal exploration tool from the beginning. By examining a broad spectrum of electrical frequencies and taking advantage of natural electrical currents (for example, those induced by lightning strikes) as well as human-introduced ones, geophysicists can map the electrical structure of the crust to several kilometers depth. Hydrothermal systems within the upper crust typically show as anomalies of 10 $\Omega$ m or less, far more conductive than background crustal values. With an appropriate experimental design, an electrical field survey can define both the aerial extent of and depth to a hydrothermal system. The temperature significance of a conductive volume in the crust should be evaluated in concert with other geologic and geophysical data.

Seismology has been adapted to search for parts of the crust that exhibit anomalously slow seismic velocity, because these anomalies may result from the presence of magma or rock near melting temperature, or both. Early experiments in volcanic/geothermal areas were described as "P-wave delay" studies, because aerial variations in travel times for compressional waves were mapped. Seismic tomography is the more general technique that produces a three-dimensional image of the seismic structure of the Earth's subsurface. Tomography is widely used in medicine where, for example, sound waves are used to image internal parts of the human body. Similarly, to obtain an image of the Earth's interior, seismic waves – from natural and artificial sources – are recorded by seismometers distributed throughout an area of interest. Seismic tomographic experiments, which generally last for weeks to months, have been conducted at several volcanic and (or) geothermal areas (for example, Medicine Lake, Long Valley, and Coso, California), where zones of slow seismic velocity provide evidence for crustal magma bodies or hot plutons or a mixture of the two. Knowledge of an extant magmatic heat source is reassuring to the geothermal explorationist, but one must remember that the target of drilling is a hydrothermal system, not its heat source.

Geomagnetic and gravity surveys can be useful in geothermal exploration, particularly in defining deep and relatively large-scale features. For example, common magnetic minerals lose their magnetization at about 580 °C, the Curie temperature, and the depth to crust with this temperature can be calculated from geomagnetic data. Regions characterized by shallow Curie-temperature depths suggest relatively high geothermal potential, and these are logical targets of more focused and detailed exploration. In a similar vein, magma is generally less dense than its host rock and thus a gravity low coincident with a volcanic field or a large volcanic structure such as a caldera may indicate underlying magma. However, as noted above, while a high

level of confidence that a magmatic heat source underlies an area of exploration adds a degree of comfort, the essential ingredient to success is the presence of a hydrothermal-convection system at drillable depth.

A number of publications (see references) provide a comprehensive summary of geophysical techniques and their limitations, as applied to geothermal exploration, including many case studies. In summary, my judgement is that, by far, the single most important piece of information for the explorationist is subsurface temperature calculated from geothermal-fluid geothermometry. If such geothermometry indicates that a low-temperature hydrothermal system is all to be expected at drillable depth, only exceptionally compelling data of other types could justify drilling expensive holes in hopes of discovering an electrical-grade resource. Drilling is the ultimate test of a working model of the subsurface developed from the various types of surface studies, but it is also a rather expensive test, compared with surface techniques.

## Drilling and drilling

Because drilling is relatively expensive, it commonly is applied in phases staged to accomplish different ends and to probe increasingly deeper, and therefore costly, levels. Early on, an exploration program may include the drilling of small-diameter holes to 100 m or somewhat greater depth for determining temperature gradients and heat flow. In each hole a temperature profile is measured after drilling-related thermal disturbance has dissipated. Areal distribution of thermal gradients can help define the shape, position, and limits of a thermal anomaly, and is a first-order guide for selecting locations for deeper exploration holes. The gradients can be combined with thermal conductivities measured on core or cuttings to provide values of heat flow, which allow a more quantitative estimate of the possible resource than gradients alone.

Heat flow measured in other ways can also help indicate whether a geothermal prospect is worthy of continued exploration. For example, total thermal energy discharged by hot springs and fumaroles can be calculated from temperature and mass flow rates measured at individual vents and integrated over a given area. One researcher developed an alternative and innovative technique to define heat flow by timing the melting of newly fallen snow of known thickness over a geothermal area. But whether by rate of snowmelt, hot-spring output, or data from drill holes, a measure of thermal flux relative to the regional background value helps in prospect assessment.

Once all the surface studies and the less-costly drilling have been completed, the decision is taken to drill (or not) wells intended for the commercial production of geothermal fluids. Most such wells are between 20 and 30 cm in diameter at depths intended for production, and in the range of 1–3 km total depth. Though usually described as exploration holes, the intention and hope is that each will be a production hole. Nonetheless, all such holes are exploration in the sense that at least some new and unanticipated subsurface condition or material will be encountered in each hole; moreover, there is no assurance that a given hole will produce geothermal fluids, even when drilled within a field of pre-existing productive wells.

During drilling, rock cuttings are collected and studied for basic geologic information and for temperature information that may be gleaned from contained suites of hydrothermal minerals. In addition, downhole rock temperature is estimated from that of drilling mud circulated back to the surface, and from measurements made at the bottom of the hole when pauses in drilling permit collection of such data. Depths to highly permeable zones are charted as intervals where drilling fluid is lost to the rock formation.

In the post-drilling phase, wells that hold promise of being productive are tested by injecting water into them, or by flow-testing them, or a combination of the two. Such tests provide records of reservoir pressure versus time, and thus give pretty accurate indications of permeability and size of the encountered hydrothermal reservoir. A suite of well-logging methods may be applied, as decided by drilling and reservoir engineers. Finally, the well is put into production to generate electricity.

Of course, many stumbling blocks and temporary setbacks can be and often are encountered along the way to the successful completion of a typical geothermal electrical project. But for purposes of this chapter, such details can be swept under a rug or flushed down a well. The story now arrives at the point where many successful geothermal-energy projects, electrical and non-electrical alike, are on line and helping to satisfy human energy demands. What are the impacts on the environment from using geothermal energy? How do these compare with impacts associated with other sources of energy? And how can this information help guide planners along wisely charted long-term paths of sustainable human existence?

## The environmental factor

Geothermal resources are similar to many mineral and energy resources. A mineral (including coal and petroleum) deposit is generally evaluated in terms of the quality or purity (grade) and amount (volume or tonnage) of material that can be mined profitably. Such grade and size criteria also can be applied to deposits of geothermal energy. Grade would be roughly analogous to temperature, and size would correspond to the volume of heat-containing material that can be tapped.

However, geothermal resources differ in important ways from many other natural resources. For example, the exploitation of metallic minerals generally involves digging, crushing, and processing huge amounts of rock to recover a relatively small amount of a particular element. The steps in mining coal are similar, though the bulk crushed rock is put to use, rather than a minor concentrate. In contrast, geothermal energy is tapped by means of a liquid carrier – generally the water in the pores and fractures of rocks – that either naturally reaches the surface at hot springs, or can readily be brought to the surface through drilled wells. Thus, the extraction of geothermal energy avoids the large-scale, landscape-scarring movement of rock involved in mining operations that create mine shafts, tunnels, open pits, and waste heaps.

Geothermal energy encompasses another important advantage. It is usable over a very wide spectrum of volume and temperature, whereas the benefits of most other energy and metallic resources can be reaped only if a deposit exceeds some minimum size and (or) grade for profitable exploitation or efficiency of operation. Perhaps the most striking difference, as explained in the earlier discussion about heat pumps, is the fact that geothermal energy can be profitably harnessed at normal background concentrations virtually anywhere, at just a few meters depth. Some metallic ores, for example those of copper and gold, can be mined profitably at a remarkably low grade, though not as low as background crustal concentrations. However, when the environmental impacts caused by open-pit mining of low-grade ore are compared with those associated with any extraction of geothermal heat, the contrast is visually stunning, to the credit of geothermal.

Nonetheless, the extraction and use of geothermal energy are not without environmental consequences. The extraction of hot water and steam, the hardware needed to transform geothermal fluid to electricity and other useful forms of work, and the disposal of "spent" geothermal fluids all impact the environment. Whether one judges these to be negative, neutral, or positive is often the point of departure for sometimes lively debate. One approach at helping to resolve such contentious issues is to compare environmental consequences that stem from alternative ways to accomplish a common goal. For example, one can compare how much $CO_2$ is released to the atmosphere from generating a unit of electrical power from coal versus petroleum versus natural gas versus geothermal steam. With this strategy in mind, the reader can better evaluate the merits of various energy sources with regard to relative environmental consequences.

## Air quality

Gaseous geothermal effluent ranges widely in both composition and concentration. Such variability from hydrothermal system to hydrothermal system originates from variability in the contribution from a magmatic heat source,

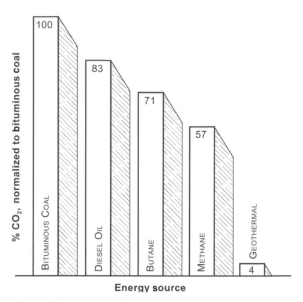

**Fig. 11.11.** The relative amounts of carbon dioxide released to the atmosphere per unit of electricity generated, by various types of power plants. Binary-system geothermal plants (see Fig. 11.6c) emit no carbon dioxide.

variability in the degree of interaction between hydrothermal fluids and their rock container, and variability in the composition of the rock container itself. Production records show that even the gases of an individual system may vary through time, presumably in response to production-caused perturbations to a natural system.

However, large variability notwithstanding, the amount of gaseous effluent released at electrical plants powered by geothermal energy pales to insignificance when compared to gases spewed into the atmosphere at plants powered by conventional fossil fuels (Fig. 11.11). Benign water vapor is by far the most abundant geothermal gas released. In addition, a plant powered by geothermal steam emits, on average, only about 1% of the sulfur dioxide, less than 1% of the nitrous oxides, and 4% of the carbon dioxide emitted by a coal-fired plant of equal size. With binary geothermal technology, described earlier, no gases are released to the atmosphere, because the geothermal fluid remains sealed within piping, from production casing, through a heat exchanger, to injection back into a subsurface aquifer.

The most problematic of released gases has been H$_2$S, which can range from deadly to simply an obnoxious odor, depending upon concentration. Threat to animal life is naturally mitigated by almost instant reduction to safe concentrations through mixing with the ambient atmosphere and oxidation to SO$_2$ somewhat further downwind. However, the foul odor of low concentrations of sulfurous gases experienced by people near and downwind of some geothermal power plants has led to the development of techniques to remove the offending gas at the plant, to a degree that downwind mixing and oxidation are beside the point.

Though not a gas itself, fine particulate matter that is a product of combustion at plants fueled by hydrocarbons becomes suspended in the atmosphere and thus reduces air quality. Since no fuel is burned at a geothermal plant, this type of environmental impact is avoided.

## Water

Because geothermal developments are fundamentally about the use of water (and steam), many environmental impacts are directly or indirectly related to water. And since most geothermal fluids are simply groundwater, albeit anomalously hot and briny, many of the environmental consequences associated with pumping groundwater also accompany a geothermal development. Some environmental impacts, though, are unique to the high temperature of geothermal fluids.

### Brines

Depending mostly on temperature, geothermal waters range from potable to extremely concentrated ore-forming fluids. As a result, problems associated with their use vary widely. Generally speaking, the hotter the brine the more severe and recalcitrant the associated problems, simply because the solubility of most constituents leached from reservoir rocks increases with temperature. Some high-temperature geothermal waters are several times more saline than seawater.

From the developer's perspective, constituents dissolved in geothermal waters are an environmental problem both during and after the energy-extraction cycle. During energy extraction, temperature and pressure are reduced, sometimes substantially, relative to original subsurface values; this often prompts deposition or scaling in piping and perhaps even in a turbine. Moreover, in some high-temperature geothermal fields, boiling of thermal water can begin within the hydrothermal reservoir, if pressure is reduced too much in the production well. Because scaling and the resultant loss in porosity and permeability are extremely difficult to reverse within the reservoir, whenever possible production pressure is adjusted to suppress this problem.

Scaling within pipes is dealt with in two contrasting ways. An increasingly popular method is to avoid scaling through chemical treatment of the geothermal fluid as it is produced. For example, inhibitors may be introduced within the well bore at or below the level where boiling begins, to either suppress deposition of scale, or to promote the formation of such small and oddly shaped crystals that attachment to piping walls is effectively suppressed. Alternatively, or in conjunction with chemical inhibitors, scale is periodically removed by abrasion, a mechanical scrubbing and routing of the scale-clad surface by a device appropriately called a pig.

Beyond the powerhouse, outside the heated building and once the spa's pool water is drained, "waste" geothermal fluid becomes an environmental problem for developer and the average concerned citizen alike. In fact, "acceptable" disposal of geothermal effluent has arguably been the single most contentious issue of the geothermal industry. Effluent of low-temperature, direct-use applications may carry little, if any, negative environmental impact. If the effluent is broadly of normal surface-water temperature and chemical quality, disposal at the surface is generally both legal and benign. However, effluent of higher temperature and briny beyond legally defined limits presents an altogether different set of disposal problems. Even lacking pertinent laws, plants and animals may vote against surface disposal of geothermal effluent through poor health, if not death.

Historically, before environmental awareness rose to current levels and before developers and engineers understood the basics of hydrothermal-reservoir performance under production, hot and briny geothermal effluent was simply stored in an evaporation pond or dumped into a river. This out-of-sight–out-of-mind approach became untenable as adverse effects on aquatic plant and animal life appeared and as reservoir power output began to drop at well heads. With motivations coming from both the environmental and profit-margin fronts, the geothermal industry began to modify standard operating procedures rather rapidly in a way that is environmentally friendly and more profitable, all at once.

Fundamentally, the single solution for the two problems is to inject geothermal effluent back underground into the hydrothermal reservoir that is being exploited, or into adjacent ground that is in hydrologic communication with the reservoir. By this procedure, the problem of thermal and chemical contamination of surface waters is instantly solved, while the problem of decaying reservoir performance has become the subject of experimentation in ways to recharge the fluid of and/or maintain the pressure within a reservoir more effectively than occurs naturally. Injection (commonly and inaccurately referred to as reinjection), though in common practice today, is prudent but is by no means a guarantee that reservoir power output will be maintained over time.

A completely successful program of injection of geothermal effluent requires knowledge of the hydrothermal reservoir characteristics – the scale and distribution of permeability and the chemistry of hydrothermal fluids and reservoir rocks – in order properly to locate and design the injection wells. Obviously, geothermal effluent injected into a highly permeable zone connected to a production well represents the hydraulic version of an electrical short circuit, which will reduce power production through a rapid temperature decrease. Similarly, geothermal effluent whose thermodynamic state promotes precipitation of calcite, silica, or other solids when injected into a reservoir can reduce power output through a decrease in permeability. Some precipitation is generally to be expected, because the effluent is cool and perhaps degassed relative to the pre-existing subsurface fluid.

The current status of injection practices is probably best described as experimental. Both "returned" chemical tracers and temperature effects have demonstrated that injection can indeed feed into and through a reservoir under production. However, far more time and experience are needed properly to evaluate the efficacy of injection in terms of extending the productive life of a reservoir.

One innovative experiment under way at The Geysers, in California, is the use of human effluent (treated sewage) to recharge the reservoir for the world's largest geothermal electrical development. Injection of geothermal effluent from this vapor-dominated reservoir is insufficient to maintain pressure, and natural recharge does not begin to keep pace with the rate of production, to the point that power output dropped about 50% during the late 1980s. Still, the amount of heat stored in the reservoir rocks is enormous, and the introduction of "outside" water may make this energy available, by creating the equivalent of a continuous-motion, energy-gathering broom that sweeps through the reservoir. This use of human effluent has prompted the concept of "flush to flash," a procedure that holds the potential to address dual problems in win–win fashion. Results to date are positive and promising. Treated sewage is functioning as turbine-driving steam.

### Hot springs and fumaroles

By their very nature, most high-temperature and many lower-temperature geothermal areas include hot springs or fumaroles or both. Boiling mud pots and geysers may also be present. All of these thermal manifestations are simply leaks from hydrothermal-convection systems. They are often the focus of traditional customs and ceremonies that grow from generations of observance, and can carry a special and sacred significance that may lead to recognition as a national park or its equivalent.

The intensity (temperature and mass rate of outflow) of a thermal manifestation changes naturally, at both fast and slow rates, depending mostly on changes in hydrologic and tectonic conditions. In an extreme case, the creation or destruction of a manifestation may be essentially instantaneous, especially where earthquake shaking and related ground fracturing perturb a hydrothermal reservoir. Most natural change, though, seems to occur at a rate imperceptible, or nearly so, over a human generation or two, so that people tend to view a thermal manifestation as permanent.

Unfortunately, human exploitation of a hydrothermal reservoir that results in subsurface loss of fluid in excess of natural steam or water outflow may impact the vigor and character of manifestations within just a few years or even less. Typically, hot springs affected by exploitation of their source hydrothermal reservoir decrease in volumetric rate of outflow, including decrease to zero in the extreme limiting case. Counter to most intuition, fumaroles commonly react by becoming more vigorous, especially those fed by boiling at the top of a hot-water hydrothermal-convection system. Apparently, drawdown at the top of a hydrothermal reservoir exposes additional "boiling temperature" rock and simultaneously reduces subsurface pressure such that an increase in boiling is triggered. By contrast, fumaroles fed by an underlying vapor-dominated hydrothermal-convection system typically decrease in vigor with increase in degree of commercial exploitation of the system. But regardless of whether a manifestation becomes more or less vigorous, any human-induced change may be deemed unacceptable, depending on societal values.

Steps can be taken to mitigate or minimize human-caused change. One obvious technique is to position production wells as far from manifestations as possible. At the very least, this will extend the lag time between fluid production and change. At best, distant wells may in fact be hydraulically isolated from manifestations. Alternatively, or in addition and as already alluded to above, subsurface injection of geothermal effluent or "make-up" water from another source, may sustain and maintain pre-development subsurface fluid pressure. Given the typical unequal distribution of permeability and extreme difficulty in exactly replacing production fluids, injection may be a partly successful mitigation technique at best. If a society or other governing body is not willing to risk inducing change to manifesta-

tions, the only sure technique is to avoid human intrusion.

## Ground subsidence

Ground subsidence and impacts on thermal manifestations tend to go hand in hand as two reactions to a single action. Both are reactions to the removal of fluid from a hydrothermal reservoir and the attendant decrease in fluid pressure within the reservoir. The hydrothermal fluid and its solid rock container subside simultaneously, though at different rates, as the net removal of fluid from the reservoir increases.

In developed geothermal areas where repeated precise leveling surveys have been carried out, some ground subsidence has been documented. The same situation likely obtains for non-surveyed areas. In extreme situations, such as Wairakei in New Zealand, subsidence has accumulated to several meters since production began in the 1950s. However, elsewhere, for example at The Geysers in California, the rate of subsidence is an order of magnitude less. The rate of subsidence is controlled principally by interplay between the rate of drawdown of reservoir pressure and the strength of the reservoir rocks. Reservoir rocks at The Geysers are thoroughly consolidated graywacke, whereas those at Wairakei include substantial amounts of poorly consolidated fragmental volcaniclastic deposits.

The principal and obvious negative environmental impact of ground subsidence is the effect of tilting pre-existing structures whose integrity depends on a stable foundation. Even geometrically simple and very subtle changes to land slope may be detrimental, especially in areas of gravitationally driven irrigation systems. Subsidence of coastal areas can lead to obvious problems associated with newly flooded land.

The physics of subsidence that results from withdrawl of geothermal fluid is fundamentally the same as that associated with withdrawl of other fluids from the upper crust, such as normal-temperature groundwater and petroleum. Examples abound. In the American southwest, groundwater pumping has led to more than 8 m of ground subsidence in parts of the San Joaquin Valley of California, and to both subsidence and the formation of gaping fissures near Tucson, Arizona. Petroleum pumping at Long Beach, California, resulted in substantial subsidence and initiated coastal flooding during the 1920s. Excessive groundwater pumping is thought to be the culprit for flooding of parts of the near-sea-level city of Venice, Italy, poignantly illustrated by the increased frequency of waves of the Adriatic Sea lapping at the steps into St. Mark's Cathedral.

For geothermal developments, the common practice of injecting geothermal effluent back underground helps mitigate the problem of subsidence. Analogous injection is not regular practice in the oil patch, and withdrawn normal-temperature groundwater is typically "consumed" and thus not available for injection.

## Induced seismicity

During the 1960s at Rocky Mountain Arsenal, Colorado, happenstance, followed by nearby controlled experiments, showed that injecting fluid into the subsurface can trigger earthquakes (Evans, 1966). In this particular instance, injection was being used to dispose of water contaminated during production of chemical weapons of warfare. The combination of such vile waste and earthquakes led to a quick change in disposal practices.

Since the injection of fluid adds to subsurface fluid pressure and thus reduces the strength of the fluid's rock container, it is not surprising that seismicity is induced. Earlier theoretical analysis and experiments by M. King Hubbert (1951) and his colleagues had in fact shown that, in an extreme situation, with sufficient increase in subsurface fluid pressure, internal friction of coherent rock, or friction across a fault plane, can be reduced to the point that decollement or other low-angle fault offset is to be expected.

Not surprisingly then, the practice of injecting geothermal effluent causes earthquakes, but mostly quakes of such small magnitude that concern for damage to surface structures and features in not warranted. In fact, such induced seismicity may be beneficial in the sense that when accurately charted, the distribution of earthquakes may be a proxy for a map of the principal subsurface channelways of permeability. Geothermal production may be equally, or even more, beneficial in the sense that the distribution of seismicity induced by reduced subsurface fluid pressure may be a proxy for a map of the tapped hydrothermal reservoir.

Some detractors suggest that seismicity induced by fluid injection and production poses a substantial earthquake hazard, locally if not regionally. This is general nonsense; no earthquake triggered by geothermal production or injection has caused damage to property or people. Few have been of sufficient magnitude to be felt by anything other than a sensitive nearby seismometer. The real earthquake hazard associated with geothermal developments is that posed by the tectonic setting typical of active volcanic areas. Geothermal areas within island arcs above subduction zones perhaps carry the maximum potential for earthquake hazards. However, geothermally induced Earth vibrations, of either the production or injection mode, are simply minor and beneficial background music within the global tectonic symphony.

Other potential natural hazards within many high-temperature geothermal areas include volcanic eruptions and landslides, but these, too, are products of the tectonic setting (commonly boundaries between lithospheric plates), not problems that are caused by exploiting geothermal energy.

## Sight and sound

As with most human endeavors that employ things, a geothermal development changes the look of the landscape and occupies ground that might be put to alternative uses (or left as is); it also creates some noise above natural ambient levels. Because the generation of electricity is the most environmentally disruptive of geothermal projects, I discuss this as the worst-case scenario. The reader's imagination can slide along the slope of intermediate situations. At the low-impact end, consider the minimal sight-and-sound environmental impact associated with an individual therapeutic soak in a natural thermal pool.

An electrical development consists fundamentally of a well field (source of fuel), a system of fuel collection (piping for hot water/steam), and a plant where thermal energy is converted

| Table 11.2 | Comparative land use per unit of electrical power generated | |
| --- | --- |

| Technology | Land use relative to coal (%) |
| --- | --- |
| Coal[a] | 100 |
| Solar thermal | 100 |
| Photovoltaics | 89 |
| Wind | 37 |
| Geothermal | 11 |

[a] Includes area of mining.

to electrical energy (turbine generator). Because hot water and steam can not be transported far without considerable loss of thermal energy, an entire development is quite compact, in terms of land area per unit of capacity to generate electricity. A typical 100-MW development occupies in the range of 200–2000 ha, depending mostly on the distribution of production wells. Comparable data for coal, solar thermal, wind, and photovoltaics indicate that geothermal on average occupies an order of magnitude less ground per unit of electrical output (Table 11.2). A similar accounting in favor of geothermal probably also obtains for petroleum, natural gas, and nuclear energy, especially when the areas of "mining" for the pertinent fuels are incorporated into the comparisons. The land area flooded by a large hydroelectric development is in the range of about 3–30 times the land area occupied by a geothermal development, per unit of capacity to generate electricity. For example, when full, the Lake Powell reservoir behind Glen Canyon Dam on the Colorado River in Arizona covers about 66 000 ha; the electrical capacity of turbine generators in Glen Canyon Dam is about 1000 MW, i.e, 3–33 times more consumptive of land than a comparable geothermal development.

Any system that generates electricity by using high-pressure vapor to drive a turbine creates noise from turbine spin and the downstream system of vapor condensation and capture. Thus, generating electricity from geothermal steam is essentially as noisy as generating from steam produced by nuclear and conventional fossil fuels. Spinning turbines of a hydroelectric project also produce noise, though for many developments

this noise is muffled from the public by being contained within a dam. Wind energy produces noise from rotation of the wind-driven generator, but mostly from interactions between the wind and the wind-driven blade(s). Photovoltaic cells do their business silently.

## The environmental balance sheet

In discussions of human uses of energy, geothermal is commonly grouped with so-called alternative or non-conventional sources of energy, if it garners mention at all. Lists that omit geothermal may reflect uncertainty about just how this type should be categorized. Indeed, geothermal is an alternative to such conventional fuels as petroleum, coal, natural gas, nuclear, and hydroelectric, but its contribution to human energy needs is not adequately or accurately captured when described as non-conventional. The technology and hardware for developing and harnessing geothermal energy are "off the shelf," and have been so for several decades. Moreover, uses of geothermal energy, whether in direct or electrical applications, are competitive in today's marketplace.

What is perhaps not adequately acknowledged in most public discussion is that geothermal energy is even more competitive than popularly conceived, if the costs of environmental impacts are factored into comparisons with other energy sources. Such environmental costs are sometimes referred to as externalities. Whatever name is used to describe them, they are real costs, which can be hidden only temporarily from human consciousness and a need for remedial action, if remedies are even possible.

Consider the environmental impact of simply transporting, to say nothing of burning, petroleum. No matter how carefully the means of transportation are designed and used, accidents happen. Thus, petroleum spills at sea and on land recur, and their environmental costs and consequences are significant. These costs tend to be shared by private and public funds, but are not fully reflected by the price of petroleum products in the marketplace.

Potentially much more costly, if not insidious to the long-term continuation of human existence, is the production of atmosphere-altering amounts of carbon dioxide from the combustion of petroleum products. There is no serious debate over the validity of data that indicate a substantial increase of atmospheric carbon dioxide since humans began large-scale use of carbon-based fuels. And there is little disagreement that combustion of such fuels is directly responsible for this increase. What is very contentious, though, is whether or not increased concentrations of carbon dioxide and other greenhouse gases will be detrimental to human existence through triggering climate change or some other as yet unforeseen global impact. An attempt to answer this question is far beyond the scope of this essay, but the fact remains that use of geothermal energy adds virtually no carbon dioxide, or other greenhouse gas, to the atmosphere.

The environmental balance tilts similarly in geothermal's favor when one considers the impacts associated with the use of coal, natural gas, and fuel for nuclear reactors. The disposal and safe storage of "spent" nuclear fuel pose serious and very-long-term environmental problems, which many nations are struggling to cope with in a variety of ways of questionable effectiveness. Many abandoned underground coal-mine workings worldwide, some of which undergo spontaneous combustion and/or collapse, pose their own array of environmental problems. With combustion, coal, and even natural gas, add far more carbon dioxide to the atmosphere than geothermal energy. Moreover, during transportation to sites of combustion, coal and gas suffer their share of accidental spills. Other anecdotal and generic instances of environmental impacts associated with various energy sources abound. The bottom line is that, for most categories of potential impacts, geothermal is far less intrusive on the environment; for some it is equally intrusive; and overall, geothermal is downright benign compared with the current principal sources of energy for the industrialized world.

## Closing thoughts

Although geothermal energy is sometimes referred to as a renewable energy resource, this term is somewhat misleading because the available hot water, steam, and heat in any given hydrothermal system can be withdrawn faster than they are replenished naturally. It is more accurate to consider geothermal energy as a sustainable resource, one whose usefulness can be prolonged or sustained by optimum production strategy and methods, as described in earlier sections. Indeed, the concept of sustainable – versus renewable – production of geothermal resources is the current focus of studies by many scientists and engineers. Major questions being addressed are: how many hundreds or thousands of years are required to replenish naturally a hydrothermal system? What is the most effective method of human intervention for replenishing and sustaining a system to increase its longevity?

In practice, choices must be made between maximizing the rate of fluid withdrawal (energy production) for a short period of time versus sustaining a lower rate for a longer period of time. For example, the decline of steam pressure in wells at the world's largest geothermal development, The Geysers in California, is a direct result of overly rapid development of this field during the 1980s. Nonetheless, with the injection of treated human effluent now under way, it is anticipated that production of steam at The Geysers will be able to power 1000 MW electric for decades to come. Incremental development of a hydrothermal system, coupled with monitoring for possible production-induced hydrologic and chemical changes, is the best way to determine the optimum rate of production for maximizing the longevity of a hydrothermal system. As noted earlier, the common practice of injecting geothermal effluent and/or other "make-up" water back underground can extend the productive life of a hydrothermal system.

Humans worldwide face many uncertainties about the nature and adequacy of a future energy supply. With some obvious regional exceptions, most of the energy consumed during the twentieth century has come from the fossil fuels coal, petroleum, and natural gas. Nuclear fuel gives an erratically growing addition to this list. Without question, this history of energy consumption has resulted in considerable environmental degradation, while making possible the present level of industrialization. One certainty is that dramatically increasing world population and the continued spread of industrialization will result in increased energy demands worldwide, attempts to conserve energy resources notwithstanding. Another certainty is that the traditional fossil-fuel resources are finite and cannot power the world for another century.

How can these increasing energy demands of the future be met in ways that are both environmentally sound and economically beneficial? A reconfigured energy mix is needed. And geothermal energy can contribute significantly to such a mix. It is already known that, using current technology, geothermal energy can contribute as much as 10% to the USAs' energy supply, if all known resources were fully developed. It probably is reasonable to project this amount to a global basis. Moreover, reasonably expectable advances in both exploration and resource-development technologies would allow exploitation of the huge storehouse of geothermal heat that is known to exist in the Earth's crust, but is not in the form of natural hydrothermal systems reachable by the drill bit.

Enthusiasm for geothermal, and any other source of energy, should be tempered by the fact that each of these sources is finite and unevenly distributed around the globe. Thus, natural and political realities must be included if a long-term comprehensive energy-use plan is to be effective. Other factors being equal, it would seem wise for any single nation, group of co-operating nations, or groups that share a common geography to maintain a balanced mix of energy sources, rather than relying on one source, whose "domestic" supply is apt to be less than equal to "domestic" demand. The inordinate impact of policies and politics of the petroleum-rich Middle East on the United States and a few other industrial nations should convince planners of the value of a balanced energy mix. However, having championed the notion of maintaining a balanced energy mix, I close with an example of how a single source of energy may contribute substantially to both local and world economic and social stability.

Consider the geographic distribution of electrical-grade geothermal resources. Their association with active volcanoes and plate-tectonic boundaries coincides with many developing nations. Many of these nations now devote substantial financial resources to importing petroleum and/or coal for primary energy sources. Increasingly, however, many are developing their geothermal resources, which frees a significant part of a budget for much-needed domestic programs. The beauty of geothermal energy is that once the wells are drilled and the turbine is in place, the cost of turbine-driving fuel is zero. Indonesia, the Philippines, Costa Rica, Nicaragua, El Salvador, Guatemala, and Mexico are some of the Pacific Rim countries taking increasing advantage of this free fuel. The fuel cost, or lack thereof, in series with environmental compatibility are strong incentives for developing indigenous geothermal resources. Industrialized nations should encourage and support foreign geothermal programs. In the short term, parties with direct geothermal-development interests benefit. In the longer term, the entire world benefits.

## Acknowledgments

Robert Fournier, Arthur Lachenbruch, Patrick Muffler, Herbert Shaw, Robert Smith, and Donald White, all US Geological Survey scientists, helped me attain whatever knowledge I have of geothermal energy and its relations to magmatism/volcanism/tectonism. I thank them for being patient and helpful during the past 30 years. Parts of this paper were essentially ghost written by my colleagues, John Sass and Michael Sorey. I thank Grant Heiken, John Lund, Sue Priest, Kevin Rafferty, John Sass, and Karl Stetter for critically reviewing an earlier version of the paper.

## References

Armstead, H. C. H. and Tester, J. W. 1987. *Heat Mining: A New Source of Energy*. London, E. and F. M. Spon.

Cathles, L. M. 1977. An analysis of the cooling of intrusives by groundwater convection which includes boiling. *Economic Geology*, **72**, 804–826.

Duffield, W. A. 1992. A tale of three prospects. *Geothermal Resources Council Transactions*, **16**, 145–152.

Duffield, W. A. and Sass, J. H. 2003. *Geothermal Energy: Clean Power from the Earth's Heat*, US Geological

Survey Circular no. 1125. Washington, DC, US Government Printing Office.

Evans, D. M. 1966. The Denver area earthquakes and the Rocky Mountain arsenal disposal well. *Mountain Geologist*, **3**, 23–36.

Fournier, R. O. 1981. Application of water geochemistry to geothermal exploration and reservoir engineering. In L. Rybach and L. J. P. Muffler (eds.) *Geothermal Systems: Principles and Case Histories*. New York, John Wiley, pp. 109–143.

Hayba, D. O. and Ingebritsen, S. E. 1997. Multiphase groundwater flow near cooling plutons. *Journal of Geophysical Research*, **102**, pp. 12235–12252.

Hubbert, M. K. 1951. Mechanical basis for certain geologic structures. *Geological Society of America Bulletin*, **62**, 355–374.

Lindal, B. 1973. Industrial and other applications of geothermal energy. In H. C. H. Armstead (ed.) *Geothermal Energy*. Paris, Unesco, pp. 135–148.

Lumb, J. T. 1981. Prospecting for geothermal resources. In L. Rybach and L. J. P. Muffler (eds.) *Geothermal Systems: Principles and Case Histories*. New York, John Wiley, pp. 77–108.

Muffler, L. J. P. (ed.) 1979. *Assessment of Geothermal Resources of the United States: 1978*, US Geological Survey Circular no. 790. Washington, DC, US Government Printing Office.

1981. Geothermal resource assessment. In L. Rybach and L. J. P. Muffler (eds.) *Geothermal Systems: Principles and Case Histories*. New York, John Wiley, pp. 181–198.

Muffler, L. J. P. and Cataldi, R. 1978. Methods of regional geothermal resource assessment. *Geothermics*, **7**, 53–89.

Muffler, L. J. P. and White, D. E. 1972. Geothermal energy. *The Science Teacher*, **39**(2), pp. 40–43.

Norton, D. and Knight, J. E. 1977. Transport phenomena in hydrothermal systems: cooling plutons. *American Journal of Science*, **277**, 937–981.

Shaw, H. R. 1985. Links between magma-tectonic rate balances, plutonism and volcanism. *Journal of Geophysical Research*, **90**, 11275–11288.

Smith, R. L. and Shaw, H. R. 1979. Igneous-related geothermal systems. In L. J. P. Muffler (ed.) *Assessment of Geothermal Resources of the United States: 1978*, US Geological Survey Circular no. 790. Washington, DC, US Government Printing Office, pp. 12–17.

US Department of Energy 1994. *Geothermal Fact Sheet: Environmental Aspects*. Washington, DC, US Government Printing Office.

Wohletz, K. and Heiken, G. 1992. *Volcanology and Geothermal Energy*. Berkeley, CA, University of California Press.

# Chapter 12

# Volcano-hosted ore deposits

Harold L. Gibson

## Introduction, definitions, and classification

Volcanic rocks host significant base and precious metal ore deposits. But what is an ore deposit? The term ore deposit is an economic, not a geological term, and refers to a naturally occurring material which can be extracted, processed, and delivered to the marketplace or technology at a reasonable profit. The term mineral deposit bears no profitability implications. Ore not only refers to metals, or metal-bearing minerals (metallic ores), but also to many non-metallic minerals valued for their own specific physical or chemical properties, such as fluorite and asbestos that are classified as industrial minerals. In this definition water can also be classified as an ore. Considering water as an ore may not be as outlandish as it first appears. With the world's supply of clean, fresh water ever decreasing, countries with an abundance of fresh water are in a position to "mine" this commodity and offer it to countries that require this resource for either agriculture or consumption. The question, in this case, is not whether it can be done but, because of environmental concerns and the potential for natural habitat destruction, should it be done?

The size of an ore deposit is measured or defined by its reserves. In the case of water, reserves are measured in liters and, when dealing with geothermal energy, temperature is also an important unit of measure. With respect to metallic and non-metallic ores, a deposit's size or reserve is measured by the number of tonnes it contains and its average metal content referred to as its grade (metric tonnes = volume × specific gravity; average grade measured in wt.%, ounces (oz/t) or grams (g/t) per tonne of economic minerals or constituents). Ore reserves are classified on the basis of certainty, which is directly related to the amount of information available on their size and grade, as proven (measured), probable (indicated), and possible (inferred).

There are numerous factors that must be assessed before a mineral deposit becomes ore and is mined. These factors are grouped into four categories: technical, economic, political, and environmental. Technical factors include considerations of the value of the deposit (size and metal content) versus costs inherent in mining (open pit versus underground; transportation) and the recovery of metals (metallurgical complexity and processing). Economic factors include metal prices and price stability (recycling), energy availability and cost, availability of a skilled workforce and labor costs. Political factors include taxation, political stability, and nationalization. Environmental factors include the cost of meeting environmental regulations and assessing potential costs associated with uncertainties regarding changing legislation and land withdrawals. Technical factors are the easiest to evaluate and teams of geologists and engineers normally assess the potential, or technical viability of a mineral deposit. Costs associated with economic, political, and environmental factors are the most difficult to

*Volcanoes and the Environment*, eds. J. Martí and G. G. J. Ernst. Published by Cambridge University Press. © Cambridge University Press 2005.

assess, and significant, unforeseen changes in any of these factors may result in a mineral deposit becoming ore (increase in metal prices) or an ore deposit becoming uneconomic (decrease in metal price, or an increase in the cost of energy or labor).

Volcanic ore deposits include metallic ores, non-metallic ores, or industrial minerals (clays, zeolites) and for that matter, water, in its natural form (drinking, agricultural) or as geothermal energy. This chapter will be limited to metallic ore deposits that are subdivided into two types, magmatic and hydrothermal. These deposits formed throughout geological time and in a wide variety of volcanic and plate tectonic settings. Volcanic ores discussed in this chapter include Ni–Cu PGE (platinum group elements) deposits associated with subaqueous ultramafic flows, and Au–Ag and Zn–Cu–Pb (Au, Ag) deposits associated with subaerial (epithermal deposits) and subaqueous (volcanic-associated massive sulfide) volcanic centers, respectively.

There is no doubt that metallic ores of Cu, Ni, Zn, Au, Ag, and PGE hosted by volcanic rocks are, and will continue to be, required for our technological development. This is an undeniable fact, and for many countries such as Canada, Australia, and Chile to name but a few, the export of metals represents a significant proportion of their gross domestic product and is essential to their economic and social well-being. However, it is perhaps not as well recognized that the quest for these ores, essentially exploration, has had a dramatic impact on humankind's social development. For example, the need for copper (industrial) and desire for gold, in part, fueled the expansion of the Roman Empire as revealed by ancient slags (>2500 BC) in Cyprus and Turkey where ancient miners extracted these metals from volcanic ores. The fevered quest for the elusive riches of the "far east," which included gold derived from volcanic ores (epithermal deposits), allowed daring discovers such as Columbus to convince investors to support a voyage that challenged conventional wisdom and to sail west, in a more direct route to the east! This historic voyage in 1492 led to the discovery and the eventual settlement of North and South America by Europeans. Even in our most recent history the settlement of the American west and Canadian north was, in part, driven by the lure of quick fortunes derived from the discovery and mining of gold during the Californian and Yukon Gold "Rushes."

This chapter describes the volcanic environment and processes that are fundamental to the formation and location of volcano-hosted magmatic and hydrothermal ore deposits. For it is only through an understanding and recognition of the different volcanic environments that host these ores and of the processes responsible for their formation, that we can discover the new resources that are essential for our future.

## Magmatic Ni–Cu (PGE) deposits

Magmatic or magmatic segregation deposits form during the crystallization of magmas. In magmatic deposits the ore minerals may be products of early or late fractional crystallization and gravitative settling, or the separation of an immiscible sulfide and/or oxide liquid from a silicate melt. In magmatic ores the ore minerals are hosted by, and are part of the igneous rocks themselves. Magmatic deposits considered in this chapter are Ni–Cu sulfide ores that occur within komatiitic flows erupted within ancient, Precambrian oceans. Examples of this deposit type include those at Kambalda, Western Australia described herein, as well as the Alexo, Langmuir, Marbridge, Raglan, and Thompson deposits, Canada (Lesher, 1989; Naldrett, 1989).

Komatiites are ultramafic and mafic volcanic rocks that constitute a distinct magnesium-rich magma series (Lesher, 1989; Naldrett, 1989). Except for one example of Ordovician basaltic komatiites, they are normally restricted to Precambrian volcanic successions. Komatiites range from dunite (>40% MgO) through peridotite (20–40% MgO), pyroxene–peridotite (20–30% MgO), pyroxenite (12–20% MgO), and basalt (10–18% MgO) and occur as flows, tuffs, hyaloclastites, autobreccias, and intrusions. Komatiites are interpreted to have crystallized from parental magmas containing up to 35% MgO. Spinifex, a quench texture, occurs in the upper part of some flows and near surface sills (Figs. 12.1 and

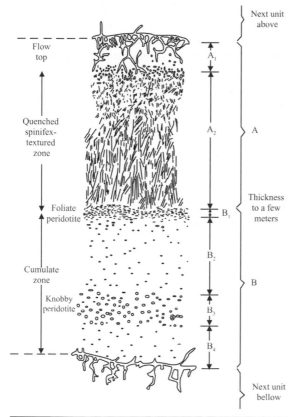

**Next unit above**

A$_1$

A$_2$  A

Flow top

Quenched spinifex-textured zone

Foliate peridotite

Cumulate zone

Knobby peridotite

B$_1$  Thickness to a few meters

B$_2$

B  B$_3$

B$_4$

**Next unit bellow**

**Fig. 12.1.** Section through a Munro-type spinifex-textured flow. A$_1$ refers to the polygonal, fractured flow top, A$_2$ to the spinifex zone which coarsens downward, B$_1$ to the foliated spinifex zone, B$_2$ and B$_4$ to medium- and fine-grained peridotite, and B$_3$ to the "knobby peridotite" or olivine–pyroxene cumulate zone. (Reprinted with permission from the *Geological Society of America Bulletin*, **84**, Fig. 8, p. 963, Pyke, Naldrett, and Eckstrand, 1973.)

12.2). Spinifex texture describes skeletal lath-like olivine and/or pyroxene crystals that resemble an Australian grass bearing that name. Spinifex texture is a product of rapid cooling of the komatiitic flow, cooling that is enhanced by its subaqueous emplacement.

Ultramafic komatiitic flows are interpreted to have been erupted at temperatures in the 1400–1700 °C range, and to have had very low viscosities (Huppert and Sparks, 1985; Williams *et al.*, 1998). This combination of high eruption temperature and low viscosity results in several unique aspects to their emplacement. First, the high eruption temperature of ultramafic komatiites is permissive of thermal erosion, a process whereby

the lava is interpreted to melt and "thermally erode" underlying rocks (Huppert and Sparks, 1985; Williams *et al.*, 1998). Second, calculated Reynolds numbers much greater than the critical value of 2000, and the massive, structureless nature of adcumulate, mesocumulate, and orthocumulate flows, although not unequivocal, suggest that many ultramafic flows were turbulent in their channelized, proximal facies. Turbulent flow inhibits the development of a thermal gradient, which increases the thermal erosion capabilities of the flow and, therefore, the degree of channelization (Huppert and Sparks, 1985; Lesher, 1989). Thin, organized, spinifex-textured flows such as depicted in Figs. 12.1 and 12.2 were clearly not turbulent, but a product of laminar flow as most recently suggested by Cas and Beresford (2001). The low viscosity of komatiitic flows also accounts for their extensive, sheet-like nature such as those within the Archean, Kidd–Munro Assemblage of the Abitibi Subprovince of the Canadian Shield where ultramafic lavas within this assemblage have a lateral extent of several hundred kilometers (Jackson and Fyon, 1992).

Hill *et al.* (1995) integrated the textural range exhibited by komatiites into a model for the geometry of a large komatiitic flow formed by rapid extrusion (high eruption rate and volume) as illustrated in Fig. 12.3. Although there are many uncertainties, this idealized model, which is essentially a modified model for tube-fed, pahoehoe basalt flows, illustrates, in a general way, that proximal to the eruption site (fissure) flows are channelized and form extensive adcumulate dunite flows (up to 35 × 150 km and several hundreds of meters thick). Hill *et al.* (1995) interpreted the adcumulate texture to the direct growth of olivine from overlying, hot, turbulently flowing magma that had only undergone a low degree of supercooling. Still proximal, but further from the vent channelized mesocumulate flows hundreds of meters wide and up to 100 m thick define the main channels and are separated by orthocumulate sheet flows with well-developed spinifex texture (Kambalda-type facies). Sheet flows with orthocumulate rather than adcumulate dunite develop in more distal or lateral environments, where the flow velocity

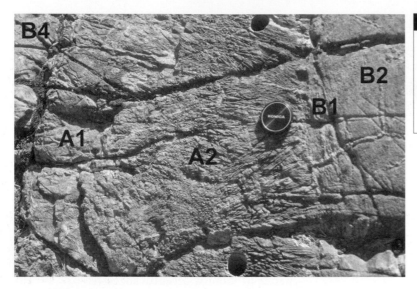

**Fig. 12.2.** Spinifex-textured Munro-type flow showing a polygonal fractured flow top (A1), underlain by spinifex (A2), foliated spinifex (B1), and massive peridotite (B2). The flow is overlain by massive periodite (B4) of the next flow (refer to Fig. 12.1 for subdivisions).

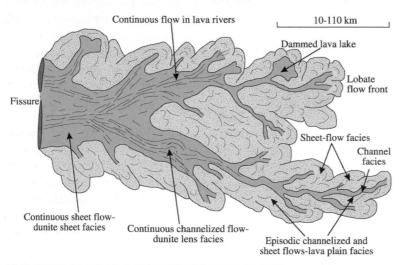

**Fig. 12.3.** Schematic illustration of flow facies within a regional komatiitic flow formed by rapid extrusion. Dunitic lava channels or rivers dominated by adcumulate and, with increasing distance from the fissure, mesocumulus dunite that are separated by levees of orthocumulate and/or Munro-type spinifex-textured flows that also constitute the most distal facies. (Reprinted with modification from *Lithos*, **34**, Hill, R. E. T., Barnes, S. J., Gole, M. J., and Dowling, S. E., The volcanology of komatiites as deduced from field relationships in the Norseman-Wiluna greenstone belt, Western Australia, pp. 159–188, © 1995, with permission from Elsevier Science.)

is less rapid, laminar, channelized and episodic. Furthest from the vent, or as a levee facies, a greater degree of cooling and supercooling combined with decreasing flow velocity results in the development of thin (10 cm to 10 m thick) spinifex-textured flows analogous to small-scale lava tubes formed at low flow rates (Munro-type facies) (Figs. 12.1 and 12.2). As stressed by Hill *et al.* (1995), as an eruption wanes the eruption rate (volume of lava per unit time) decreases resulting in the superposition of more distal facies flows on more proximal facies flows.

## Komatiitic Ore Types

Two distinct types of Ni–Cu (PGE) ore deposits occur within komatiitic flow sequences (Naldrett, 1989). The deposit types differ in their morphology, percent sulfide, grade, and location within different flow facies. Komatiite hosted Ni–Cu sulfide ores contain approximately 25% of the world's known Ni resource (Lesher, 1989).

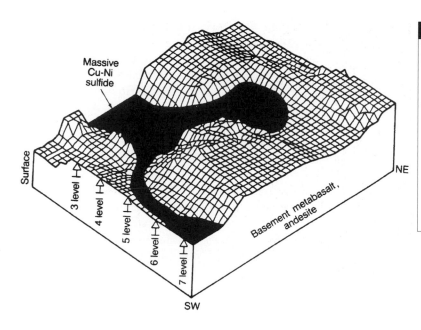

Massive Cu-Ni sulfide

Surface

3 level
4 level
5 level
6 level
7 level

Basement metabasalt, andesite

NE

SW

**Fig. 12.4.** Three-dimensional view of the Langmuir Cu–Ni sulfide deposits rotated to an original horizontal flow surface. Note the pronounced paleotopographic control on the location of massive sulfide. The original basement topography may have been accentuated by thermal erosion during komatiite emplacement. (Reprinted with permission of *Economic Geology 75 Anniversary Volume* 1905–1980, p. 646, Fig. 18, Naldrett, A. J., 1981.)

Alexo Mine, Ontario

Lunnon Orebody, Kambalda, Western Australia

Peridotite
Low grade ore

Peridotite Weak mineralization

Sharp cut-off

Net-textured ore

Sharp cut-off

Massive ore

Pyroxene Peridotite

1 m

Disseminated (net-textured) zone — 2.13 m

Pyrite zone — 5 cm

Banded ore — 1.07 m

Stringer zone Metabasalt

Average thickness and grade (Lunnon)

| | Ni% | Cu% | Co% |
|---|---|---|---|
| | 4.7 | 0.5 | 0.1 |
| | 3.6 | 1.7 | 0.5 |
| | 8.6 | 0.4 | 0.2 |
| | 0.7 | 0.5 | 0.04 |

**Fig. 12.5.** Typical sections through komatiitic Ni–Cu PGE deposits showing the vertical zonation from massive sulfide upward into net-textured sulfide and finally to blebby disseminated sulfide along with average thicknesses and grades (from Naldrett, 1973). (Reprinted from *Ore Geology and Industrial Minerals: An Introduction*, Fig. 11.3, p. 146, Evans, A. M., © 1993, with permission of Blackwell Publishing.)

## Massive ore

Massive Ni–Cu (PGE) sulfide deposits are those deposits that contain greater than 50% total sulfides and commonly greater than 90%. The principal sulfide minerals in order of decreasing abundance are pyrrhotite, pentlandite, pyrite, chalcopyrite, millerite, and violarite. The typical oxide minerals are magnetite and chromite. The grade or chemical composition (primarily expressed as wt.% Ni, Cu, and as ppm for PGE) reflects mineralogical variations. In general the deposits are typically small ($<5 \times 10^6$ tonnes), high grade (1.5–3.5 wt.% Ni), occur at the base of mesocumulus dunitic flows, and they have a lens shape (often referred to as shoots) with a pronounced long axis that reflects their accumulation within lava channelways (Lesher, 1989). Figure 12.4 illustrates morphology typical of a massive sulfide lens, and Fig. 12.5 illustrates the vertical variation

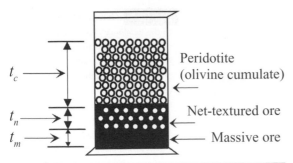

**Fig. 12.6.** The billiard-ball model illustrating the development of sulfide textures in Fig. 12.5 that results from density differences between massive sulfide liquid, silicate liquid, and olivine crystals (from Naldrett, 1989). If porosity of net ore is 50% and of olivine cumulate is 40% then $t_n \times 0.5 \times (\rho_s - \rho_o) = t_c \times 0.6 \times (\rho_o - \rho_l)$ but $\rho_s - \rho_o \approx 0.7$ and $\rho_o - \rho_l \approx 0.35$ thus $t_c \sim 1.7 \times t_n$ where $\rho_s$ = density of sulfide liquid; $\rho_o$ = density of olivine; $\rho_l$ = density of silicate liquid.

in sulfide content and textures through two massive sulfide lenses. Massive ore containing >90% sulfides occurs at the base of the sulfide lens and in most cases represents the economic, or mined portion of the sulfide lens. Massive ore is homogeneous in appearance and/or banded with the banding generally interpreted to be a product of recrystallization and segregation of the sulfides during deformation. Massive ore typically has a sharp contact with overlying net-textured ore where sulfide minerals enclose olivine grains to produce a "matrix or net-textured ore" that is often recovered (Fig. 12.5). Net-textured ore grades upward through decreasing sulfide content into blebby or disseminated sulfides where "blebs" are isolated in a silicate, primarily olivine matrix (Fig. 12.5).

The upward change from massive, to net-textured, to blebby sulfides suggests that the sulfides formed by gravitative settling of an immiscible sulfide liquid. Naldrett (1989) used the analogy of billiard balls in a cylinder of mercury and water to illustrate gravitative settling and to explain the vertical variation in sulfide textures of Fig. 12.5. In the billiard-ball model, illustrated in Fig. 12.6, the balls represent olivine crystals that are denser than the water, which represents the silicate magma, but float on the mercury, which represents the sulfide liquid. In this model

sulfide (mercury) settles to the bottom to form the massive sulfide, the olivine (balls) suspended in sulfide magma (mercury) represent net-textured sulfides, and isolated droplets of sulfide (mercury) within the silicate liquid (water) represent disseminated, blebby sulfides. Although this static model is too simplistic to represent settling and accumulation of sulfides within a dynamic environment, such as a lava channelway, it does illustrate settling processes and resulting sulfide textures. Massive sulfide that lies in direct contact with the footwall rocks is interpreted to have totally "displaced" the silicate liquid and crystals, whereas massive sulfide found at the base, but within the flow, did not (Fig. 12.5). Massive sulfide stringers that extend from the base of the massive sulfide lens into underlying strata may be primary sulfide injections emplaced into the irregular channel floor during surges of lava or are a product of deformation.

### Disseminated ore

Disseminated Ni–Cu ore deposits have sulfide and oxide mineralogy identical to their massive counterparts. They occur as large (100–250 × $10^6$ tonnes), low-grade deposits (<1.0 wt.% Ni) of finely disseminated sulfide in channel-like lenses of adcumulate dunite and dunitic intrusions (Naldrett, 1989).

### Examples

The Kambalda mining camp, Western Australia, provides a well-documented example of komatiite hosted Ni–Cu deposits. The Kambalda mining camp lies within the Eastern Goldfields of the Archean Yilgarn block of Western Australia. At Kambalda the mafic–ultramafic komatiite succession consists of a lower unit of pillowed basalt (>200 m thick), the Lunnon Basalts, overlain by the ultramafic Kambalda Formation (1000 m thick) consisting of the lower Silver Lake member and upper Tripod Hill member. Komatiitic basalts of the Devon Consuls Formation overlie the Kambalda Formation. The volcanic succession has been dated at 2700 Ma (Chauvel et al., 1985).

The volcanic succession and Ni–Cu deposits are exposed around a structural dome produced by refolding, about an east–west axis, of north-northwest trending folds (Figs. 12.7 and 12.8).

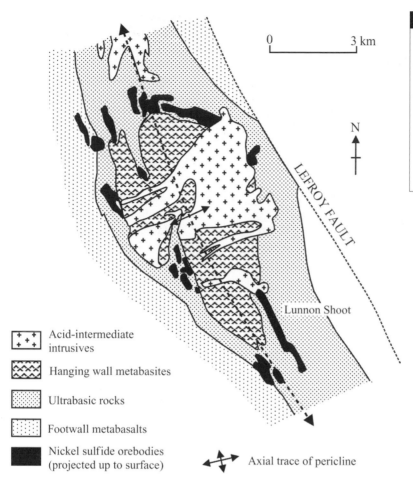

0 ———— 3 km

N

**Fig. 12.7.** Generalized geology of the Kambalda area and location of Ni–Cu sulfide deposits at the contact between footwall metabasalts and ultramafic rocks as exposed within a structural dome (modified from Ross and Hopkins, 1975). (Reprinted from *Ore Geology and Industrial Minerals: An Introduction*, Fig. 11.5, p. 148, Evans, A. M., © 1993, with permission of Blackwell Publishing.)

LEFROY FAULT

Lunnon Shoot

| | |
|---|---|
| ⊞ | Acid-intermediate intrusives |
| ▨ | Hanging wall metabasites |
| ▦ | Ultrabasic rocks |
| ▦ | Footwall metabasalts |
| ■ | Nickel sulfide orebodies (projected up to surface) |

✣  Axial trace of pericline

Archean granites, felsic to mafic dykes, and sills intrude the volcanic succession.

Most of the ore deposits (>70%) are localized within depressions in the underlying basalt flows or by north-northwest trending structural grabens (Lesher, 1989) (Fig. 12.8). These depressions have length-to-width ratios of about 10:1 and extend as much as 100 m into the underlying basalts. This ore environment is one of mesocumulus channel flow facies (MSF) (Fig. 12.3) where the elongated channels forming this facies are separated by lava plains where orthocumulate sheet flows (OSF) (Fig. 12.3) and Munrotype flows (Figs. 12.1, 12.2, 12.3) predominate. Lesher and Arndt (1995) have shown that channelized mesocumulus flows containing massive Ni–Cu orebodies crystallized from the most magnesian magma that shows little evidence of crustal contamination. In contrast, ultramafic lava composing the flanking sheet flow facies includes the most contaminated magma and remnants of melted footwall sediments referred to as xenomelts (Lesher and Burnham, 2002). The lower degree of contamination in the channel facies as compared to the more crustally contaminated sheet flow facies is attributed to the continuous flushing or flow of magma through the channel or tube. This resulted in the replacement of initial contaminated magma by aliquots of new, less contaminated lava during subsequent eruptions. Evidence for contamination is recorded in the composition of the sheet flow facies, which is interpreted as a product of channel overflow during the development of lava levees and buried lava streams during channel development. The channels themselves reflect the concentration of lava within paleotopographic irregularities of a block-faulted pre-komatiite topography

West                                        East

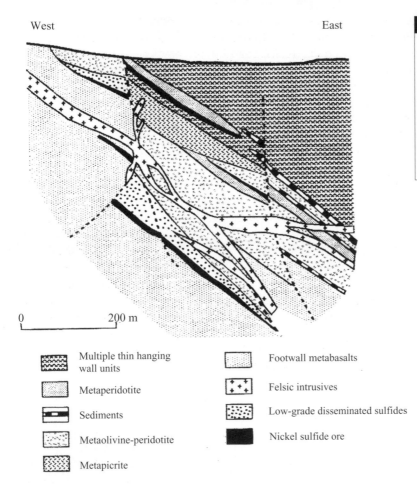

0          200 m

Multiple thin hanging
wall units

Metaperidotite

Sediments

Metaolivine-peridotite

Metapicrite

Footwall metabasalts

Felsic intrusives

Low-grade disseminated sulfides

Nickel sulfide ore

**Fig. 12.8.** Section through Lunnon ore body (shoot) showing the preferential location of massive sulfide within a fault controlled basement depression, enhanced by thermal erosion (modified from Ross and Hopkins, 1975). (Reprinted from *Ore Geology and Industrial Minerals: An Introduction*, Fig. 11.6, p. 149, Evans, A. M., © 1993, with permission of Blackwell Publishing.)

enhanced by thermal erosion of the footwall rocks (Fig. 12.3). It is significant to note that the lowermost flows of the lava plain environment contain interflow volcanic and exhalative (chemical) sediments that are noticeably absent in the channelized ore environment presumably due to thermal erosion (Gresham and Loftus-Hills, 1981). Thermal erosion and assimilation of sulfide-bearing interflow sediments have been suggested as a possible source of sulfur for the magmatic sulfides found within channelized komatiitic flows (Lesher, 1989; Lahaye *et al.*, 2001; Lesher *et al.*, 2001; Lesher and Burnham, 2002).

The progressive increase in crustal contamination upwards in the Kambalda Formation cannot be explained by the thermal erosion of pre-existing komatiitic flows. This systematic vertical contamination of the komatiitic flow stratigraphy

is attributed to contamination within magma conduits due to their progressive "warming" with time (Lesher and Arndt, 1995).

As is typical of deposits of this type, most of the Ni–Cu orebodies at Kambalda occur at or near the base of the komatiitic sequence within thermally eroded, paleotopographic flow channels in basalts that underlie the Kambalda Formation (Lesher, 1989). Massive sulfide orebodies also occur within similar environments within the lower three flows of the Kambalda Formation. These "hanging wall ores," characterized by lower grades, but higher nickel tenor, have been observed to grade laterally into interflow sediments that, in some cases, contain nickeliferous sulfides (Gresham, 1986).

The Six-Mile deposit within the Agnew–Wiluna greenstone belt of the Yakabindie area, Western Australia as described by Hill *et al.* (1989)

is an example of low-grade, disseminated Ni–Cu mineralization. The Ni–Cu sulfides occur within olivine–sulfide adcumulate that is interlayered with olivine adcumulate, the channelized facies, which in turn is overlain by olivine and olivine sulfide orthocumulate and by cyclical layers of olivine mesocumulate to orthocumulate. Hill *et al.* (1989) interpreted the olivine adcumulate to have developed in a channelized flow where turbulently flowing komatiitic lava swept away fractionated magma from the top of the growing pile of olivine cumulates allowing fresh, hot, and unfractionated komatiitic lava to be in contact with the cumulates promoting adcumulate growth (Fig. 12.3). Sulfides, representing gravitative settling of immiscible sulfide liquid droplets, accumulated with the dense olivine crystals and comprise cumulates within the lava channel or tube. Hill *et al.* (1989) attributed the lens-like shape of adcumulate-filled channels overlying felsic volcanics, as compared to the more sheet-like morphology of adcumulate bodies overlying mafic volcanics, to greater thermal erosion of the underlying footwall in the former.

## Genesis

Komatiitic Ni–Cu deposits, like their intrusion-hosted counterparts, owe their formation to: (1) the Ni-, Cu-, and PGE-rich character of their parental magma (source of metals); (2) early sulfur saturation of a primary sulfur undersaturated komatiitic magma and separation of an immiscible sulfide liquid; (3) the high affinity of metals for sulfide liquid versus silicate magma (high partition coefficients); and (4) the high density of a sulfide liquid relative to silicate magma (and crystals) which allows its concentration by gravitative settling. In addition to these factors, komatiitic Ni–Cu deposits also owe their formation to physical–volcanological aspects of their emplacement. Most notable are their high eruption temperature, low viscosity, and turbulent flow when channelized which results in contamination of the magma due to thermal erosion of underlying rocks, and accumulation of sulfides within topographic depressions that are in part primary, and in part a product of thermal erosion within lava channels.

## Composition

Certain ores are characteristic of specific igneous rocks. For example ores found with siliceous or felsic igneous rocks include U, Zr, and Sn whereas ores of Ni, Cu, PGE, and Cr are typical of mafic and ultramafic compositions. Thus the Ni, Cu, and PGEs found within komatiitic Ni–Cu sulfide deposits are interpreted to have been derived from parental mantle-derived magma (Naldrett, 1989).

The source of the sulfur required to form the sulfide orebodies is not as simple. Komatiitic magmas must achieve sulfur saturation and separation of an immiscible sulfide liquid in order for sulfide deposits to form. Like all magmatic Ni–Cu PGE deposits sulfur saturation must occur early, before significant silicate crystallization in order to allow: (1) early partitioning of metals into the sulfide liquid before crystallization of silicate phases such as olivine which readily accommodates Ni into its structure, thus depleting the metal content of the magma; and (2) gravitative settling of the sulfide liquid before crystallization and increasing viscosity inhibit such settling and accumulation.

This raises the question as to what factors control the solubility of sulfur in magmas. There is no doubt that magmas of mafic and ultramafic composition contain more sulfur (up to $\times 3$) than felsic magmas (Naldrett, 1989). The sulfur content of magma depends on several factors, the most important being temperature and composition, and the least important being pressure. Buchanan *et al.* (1983) have shown that for sulfur undersaturated basalt magma the solubility of sulfur increases with increasing temperature. Their experimental work indicates that the dissolved sulfur content increases by a factor of $\times 8.5$ per 100 °C at 1000 °C. At temperatures above 1400 °C this factor decreases to $\times 3$ per 100 °C. Although data pertaining to the effect of increasing temperature on the sulfur content at sulfide saturation (SCSS) of magmas is limited, experimental work by Haughton *et al.* (1974), Shima and Naldrett (1975), and Wendlandt (1982) suggest an increase in SCSS of $\times 3$ to $\times 5$ from 1200 °C to 1450 °C. Thus, the high temperature of ultramafic komatiites favors greater sulfur solubility and a higher SCSS.

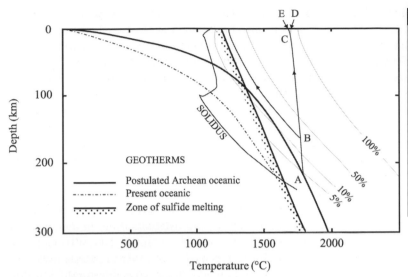

**Fig. 12.9.** A temperature-versus-depth (pressure) diagram showing the relationship between estimates of the modern and Archean oceanic geotherms and melting relationships of possible mantle material drawn to illustrate the generation of komatiitic magma at point A (after Naldrett and Cabri, 1976). Reprinted from *Ore Geology and Industrial Minerals: an Introduction*, Fig. 11, p. 144, Evans, A. M., © 1993, with permission of Blackwell Publishing.)

The FeO, S, and O content are the main "compositional" factors controlling the solubility of sulfur in magmas. With respect to the Fe–S–O–$SiO_2$ system MacLean (1969) has shown that the sulfur content of a silicate melt in equilibrium with a sulfide-rich liquid (SCSS) decreases with increasing oxygen content. MacLean attributed this to the fact that sulfur dissolves by displacing oxygen bonded to $Fe^{2+}$ and that increasing $fO_2$ results in an increase in $Fe^{3+}$ at the expense of $Fe^{2+}$. This relationship between FeO content, $fS$, and $fO_2$ is illustrated by the following reaction where $a$ is the activity or effective concentration, $f$ the fugacity of oxygen and sulfur, and $K$ is the equilibrium constant:

$$FeO(\text{in melt}) + 0.5S_2 = FeS(\text{in melt}) + 0.5O_2$$
(12.1)

$$K = \frac{aFeS \times fO_2^{0.5}}{aFeO \times fS_2^{0.5}}$$
(12.2)

$$aFeS = \frac{K \times aFeO \times fS_2^{0.5}}{fO_2^{0.5}}$$
(12.3)

From Eq. (12.3) it follows that the concentration of FeS (expressed as activity) in a silicate magma at SCSS and at constant temperature and pressure, will increase with increasing FeO content and low $fO_2$.

Experimental work by Wendlandt (1982) and Huang and Williams (1980) indicate that the solubility of sulfur and the SCSS decrease with increasing pressure. Thus, during ascent, a magma's ability to dissolve sulfur increases and it may not achieve sulfur saturation; however, a decrease in temperature during ascent may counterbalance the effect of pressure (Naldrett, 1989).

Notwithstanding the uncertainties in the experimental data (derived mainly for basaltic compositions), ultramafic komatiitic magmas because of their high temperatures ($>1450$ °C), FeO content (relative to magmas of felsic and intermediate composition), and low $fO_2$ content as indicated by their low water content (virtually anhydrous magmas) have the potential to contain dissolved sulfur. The question is: did these magmas contain enough sulfur to become sulfur saturated during their ascent? There were two schools of thought regarding the processes responsible for sulfur saturation in komatiitic magmas. The first, proposed by Naldrett and Cabri (1976), envisaged that ultramafic komatiitic magmas acquired high sulfur content during partial melting of a sulfide-enriched portion of the Archean mantle. In their model (Fig. 12.9) the zone of sulfide melting intersects the Archean geothermal gradient at or about 100 km depth. Below this depth sulfides are interpreted to be molten and, as a result of their higher specific gravity, to have percolated downwards leaving a mantle depleted in sulfides (and sulfur) and produced a deeper zone within the mantle enriched in sulfides (and sulfur). Ultramafic komatiitic magmas were interpreted to have been derived from partial melting of this sulfur-enriched mantle (point A) and to have achieved sulfur saturation and separation of an immiscible sulfide-rich liquid during their ascent due to cooling.

The second and currently favored model proposed that ultramafic komatiitic magmas formed during sulfur-undersaturated conditions and that sulfur saturation and separation of an immiscible sulfide-rich liquid occurred at or near the surface due to assimilation of sulfur-rich rocks. The sulfur-undersaturated, assimilation model first proposed by Keays *et al.* (1982) to explain the low Pd/Ir ratios of komatiites is consistent with the "missing sediments," the light rare earth enrichment (LREE) and high field strength elements (HFSE) within levee facies flows indicative of crustal contamination, the typically chalcophile element depleted character of these flows, the similar sulfur isotopic composition of the ores and sediments, and the Re–Os concentrations and Os isotopic heterogeneity of the sulfides that, collectively, are best explained by assimilation of sulfur-bearing sediments (Lesher *et al.*, 1989, 2001; Naldrett, 1989; Lesher and Arndt, 1995; Williams *et al.*, 1998; Lahaye *et al.*, 2001; Lesher and Burnham, 2002).

### Partitioning of metals

The partitioning of Cu, Ni, and PGE between a sulfide liquid and a silicate magma is expressed in the same manner as trace element partitioning between a silicate magma and a phenocryst. In both cases, the equilibrium is defined by the Nernst partition coefficient $D_i$, for a metal $i$:

$$D_i^{(\text{sulfide melt/silicate magma})}$$
$$= \frac{\text{wt.\% of metal } i \text{ in sulfide liquid}}{\text{wt.\% of metal } i \text{ in silicate magma}} \quad (12.4)$$

When the $D_i$ is $\gg 1$ the metal in question is preferentially partitioned into the sulfide liquid and for $D_i < 1$ the metal is preferentially partitioned into the silicate magma. Although there are always experimental uncertainties in establishing partition coefficients for metals there is no doubt that with $D_i$ for Cu, Ni, Co, and PGE in the range of 100–600, 100–450, 50–100, and $10^5$ respectively, that these metals are preferentially concentrated in the sulfide liquid (Brugmann *et al.*, 1989; Lesher and Campbell, 1993; Peach and Mathez, 1993). Thus once a sulfide liquid separates, Ni, Cu, Co, and PGE are concentrated in the sulfide liquid and the degree of

concentration, expressed as the grade or tenor of the ore, is controlled by the partition coefficients, the original concentration of metals in the coexisting silicate liquid, and the R factor.

The R factor, or the ratio of silicate magma to sulfide liquid, is an important factor governing the final concentration of metals and grade of the sulfide mineralization (Campbell and Naldrett, 1979). The original Ni, Cu, and PGE content of a mafic–ultramafic magma may be in the range of 350 ppm, 250 ppm, and 5 ppb respectively. For very large R factors, the sulfide liquid will not significantly deplete the silicate magma in Ni, Cu, or for that matter PGE, which have a much lower initial concentration, and the composition of the sulfide liquid, and therefore the sulfide ores, can be approximated by the $D_i$ for the metal in question (Eq. 12.4). However, for lower R factors the preferential partitioning of Ni, Cu, and especially PGE into the sulfide liquid will result in a significant drop in the concentration of these metals, particularly the PGE, in the silicate magma under equilibrium conditions. In this case "$Y_i$" expresses the metal content of the sulfide liquid and therefore, sulfide ores:

$$Y_i = \frac{D_i \times C_{o_i} \times (R + 1)}{(R + D_i)} \quad (12.5)$$

where R is the ratio of the mass of silicate magma to the mass of sulfide liquid and $C_{o_i}$ is the original concentration of metal $i$ in the silicate magma (Campbell and Naldrett, 1979). With similar orders of magnitude for the $D_i$ of Cu, Ni, and PGE in both mafic and ultramafic magmas the low PGE content of komatiitic Cu–Ni deposits is interpreted to be, in part, related to low R factors in the order of 100 to 500 (Lesher and Campbell, 1993). In contrast, the highly elevated Ni, Cu, Co, and especially PGE content of sulfides segregating from a basaltic parental magma for the Bushveld Complex requires very high R factors, on the order of 10 000 to 100 000 (Naldrett, 1989).

### Komatiitic flow emplacement and massive sulfide deposits

Chemical and physical characteristics of komatiitic flows such as high eruption temperature, low viscosity, turbulent flow, and thermal erosion are also critical to the formation of komatiitic Ni–Cu deposits. As illustrated in Figs. 12.3 and 12.10, the

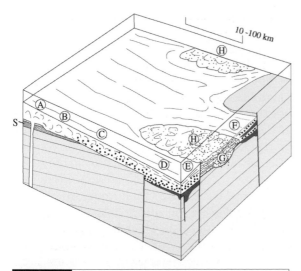

**Fig. 12.10.** Diagrammatic reconstruction of a fissure fed (A), channelized komatiitic flow illustrating turbulent flow (B), thermal erosion (C), sulfur saturation (C) and sulfide accumulation within thermal erosion enhanced paleotopographic channels during laminar flow (D, E, F). Munro-type spinifex-textured flows and orthocumulate sheet flows form a flanking levee facies (G, H).

eruption of komatiitic magma is interpreted to be fissure controlled (A in Fig. 12.10) and the lava moves rapidly away from the fissure to form topographically controlled adcumulate sheet flows (B in Fig. 12.10). In this model the initial komatiitic magma is sulfur undersaturated and sulfur saturation is attributed to the assimilation of sulfur from underlying metalliferous sediments during thermal erosion of the footwall rocks (S in Fig. 12.10). The high eruption temperature and low $SiO_2$ content of the flows results in their extremely low viscosity and turbulent flow, which enhances the thermal erosive capabilities of the komatiitic flow. During assimilation both the silicate and sulfide components have melting temperatures lower than the solidus temperature of the komatiitic magmas (1200 °C) (Arndt, 1976; Lesher and Campbell, 1993) and both components initially dissolve in the komatiitic lavas. However, because sulfur is no longer able to dissolve once the komatiitic magma has attained sulfur saturation the continuing erosion of the sulfide-rich sediment results in continuing assimilation of the silicate component and melting of the sulfide and its incorporation as immiscible sulfide liquid "droplets" directly into the

turbulent moving flow (Lesher and Campbell, 1993). Sulfide liquid droplets derived through separation of an immiscible sulfide liquid, sulfide liquid incorporated directly from melted sedimentary sulfides, and silicate xenomelts are turbulently suspended and carried within, but presumably concentrated near the base of, the rapidly moving komatiitic flow (C in Fig. 12.10). The partitioning of Ni and Cu (PGE) into the sulfide liquid during flow is favored by turbulent mixing which allows a relatively small volume of sulfide liquid to be exposed to a larger volume of silicate magma from which to extract Cu and Ni. Finite-element modeling of turbulent komatiitic lava over channel floor irregularities such as the troughs that contain the sulfide ores suggests that vigorous eddies develop within the turbulent lava on the upstream edges of the troughs (Rice and Moore, 2001). These eddies are capable of suspending sulfide droplets and mixing them with larger volumes of lava thus locally increasing the $R$ factor and presumably the metal content of the sulfides.

As the flow continues its rapid advance away from its feeding fissure it becomes cooler and the flow regime changes from turbulent to laminar. This change in flow regime may also be a product of decreasing channelization, perhaps due to a smoother paleotopography, resulting in a decrease in the flow front velocity and laminar flow. Mesocumulate to orthocumulate channelized sheet flows and levees composed of Munro-type spinifex-textured flows (H in Fig. 12.10) are interpreted to be products of transitional to laminar flow regimes (Figs. 12.3 and 12.10). The change to laminar flow may result in the rapid settling of dense, suspended immiscible sulfide droplets that are "reworked" by local turbulent eddies (location D in Fig. 12.10). The long, linear shape of the massive ore lenses reflects settling and deposition of the immiscible sulfide liquid within troughs in the flow channels or tubes that are, in part, primary and in part a product of thermal erosion within lava channels (E and F in Fig. 12.10). The distance from the feeding fissure (A) to the site of sulfide accumulation (E) is unknown and may be in the order of kilometers to hundreds of kilometers. Although the distances appear long, a komatiitic flow moving at an average velocity of 10 km/hr would cover

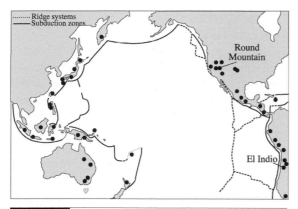

**Fig. 12.11.** Location of epithermal precious metal deposits within different subduction environments and subaerial volcanoes of the Pacific Rim. Round Mountain and El Indio are examples of low sulfidation (LS) and high sulfidation (HS) Au deposits described in the text, and in Figs. 12.15 to 12.18. (Reprinted with permission of The Society of Resource Geology, Special Publication Number 1, Fig. 2.1, Hedenquist, J. W., Izawa, E., Arribas, A., and White, N. C., 1996.)

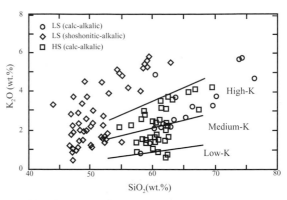

**Fig. 12.12.** Composition of volcanic rocks hosting low sulfidation (LS) and high sulfidation (HS) Au deposits. (Reprinted with permission of The Society of Resource Geology, Special Publication Number 1, Fig. 2.4, Hedenquist, J. W., Izawa, E., Arribas, A., and White, N. C., 1996.)

distances of up to 100 km in as little as 10 hours. Thus, during eruption, the sulfur-undersaturated komatiitic flow in Figs. 12.3 and 12.10 must achieve sulfur saturation through thermal erosion of underlying sedimentary or sulfur bearing rocks, separate out an immiscible sulfide liquid, and mix it turbulently with the komatiitic lava in a brief time interval. This may, in part, account for the lower $R$ factor and therefore lower PGE content of many komatiitic Ni–Cu (PGE) deposits.

# Epithermal precious metal deposits

## Introduction and definition

Epithermal precious metal deposits are products of volcanism-related hydrothermal activity at shallow crustal depths (<2 km) and low temperatures (50–300 °C). Ores include those of Au, Ag, and to a lesser extent, Zn, Pb, and Cu. This chapter focuses on deposits where Au is the dominant economic metal.

Epithermal precious metal deposits occur inboard of island and continental arcs associated with subduction zones, such as volcano-plutonic arcs of the Pacific Rim (Fig. 12.11). Epithermal deposits are typically hosted by subaerial volcanic rocks and show a strong spatial association with volcanic centers and synvolcanic structures

that often localize mineralization. Calc-alkaline, and andesitic–dacitic–rhyolitic rocks are the most common host rocks in continental arcs, whereas epithermal deposits in mature intraoceanic arcs are often associated with alkaline volcanic rocks (Hedenquist *et al.*, 1996) (Fig. 12.12).

As epithermal deposits form in the near-surface environment they are typically restricted to younger volcanic successions that have not undergone significant erosion, and there is a distinct association between the age of host rocks, rate of erosion, and tectonic environment (Hedenquist *et al.*, 1996). For example, epithermal deposits within oceanic arcs of the Western Pacific are typically young, Miocene to Recent, whereas those found in continental arcs are typically older, Cretaceous to Miocene in age. This age dependence with tectonic environment is interpreted to be a product of differing erosion rates with oceanic arcs (0.1–1.0 mm/yr) having a higher rate of erosion than continental arcs (0.01 mm/yr) (Hedenquist *et al.*, 1996).

The size and grade characteristics of epithermal deposits, illustrated in Fig. 12.13, are variable. Deposits range from <10 Mt to greater than 1000 Mt. Grades range from 1 g/t to 80 g/t Au. As these deposits form near surface and are commonly exposed through erosion, they are mined by low-cost methods such as open pits or ramps, which contributes to the generally higher profitability associated with these deposits.

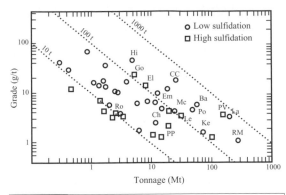

**Fig. 12.13.** Average tonnes and grade of low sulfidation (LS) and high sulfidation (HS) Au deposits. LS: Ba, Baguio; CC, Cripple Creek; Mc, McLaughlin; RM, Round Mountain; Em, Emperor; Hi, Hishikari; Ke, Kelian; La, Ladolam; Po, Polgera. HS: Ch, Chinkuasgih; El, El Indio; Go, Goldfield; PP, Paradise Peak; Le, Lepanto; PV, Pueblo Viejo; Ro, Rodalquilar. (Reprinted with permission of The Society of Resource Geology, Special Publication Number 1, Fig. 2.2 , Hedenquist, J. W., Izawa, E., Arribas, A., and White, N. C., 1996.)

## Classification

Epithermal precious metal deposits are subdivided into high sulfidation (HS) and low sulfidation (LS) types. HS and LS deposit types form in different subaerial volcanic environments from fluids of different compositions that result in different deposit morphology as well as alteration and sulfide assemblages (Head *et al.*, 1987). Calderas are important structural settings for both types of deposits, although the connection in both space and time with volcanism is not always evident in LS deposits.

LS deposits also referred to as adularia–sericite or low-sulfur deposits, contain low-sulfidation-state sulfide assemblages (low sulfur/metal ratios) in association with sericite or adularia alteration and intermediate argillic alteration. HS deposits are also referred to as acid sulfate, alunite–kaolinite, enargite–gold and high-sulfur deposits. They contain high-sulfidation-state sulfide assemblages (high sulfur/metal ratio) surrounded by advanced argillic alteration assemblages that are dominated by alunite and, at deeper levels, pyrophyllite. Characteristics of LS and HS deposits are summarized in Table 12.1.

## Characteristics

### Mineralization and geometry

The extreme variability in the form and geometry of epithermal deposits is illustrated in Fig. 12.14. This variability reflects the low-pressure, hydrostatic conditions under which these deposits formed coupled with differences in the primary permeability of the host rocks, differences that are often modified structurally or hydrothermally (Sillitoe, 1995).

Like all hydrothermal ore deposits epithermal precious metal mineralization is structurally controlled and is confined to extensional or transtensional environments where the mineralization often occurs as discrete veins or vein stockworks. This structural control reflects the necessity of structures, commonly steeply inclined normal or oblique slip or strike slip faults, which act as conduits that allow fluid ascent into the near-surface and surface environments (Sillitoe, 1993). These structures are often reactivated synvolcanic structures that also controlled volcanism and often subsidence within, or proximal to, volcanic centers.

Vein morphology is dominant where the permeability of the surrounding host rock is low, either reflecting a low primary permeability such as a flow or dome, or low permeability achieved through a decrease in permeability due to alteration and self-sealing (Fig. 12.14). Fluid overpressure (Sibson, 1987) creates permeability in two main ways. First, it may trigger synmineralization faulting thus providing conduits for focused fluid flow. Second, where fluid pressure greatly and instantaneously exceeds the lithostatic pressure by an amount greater than the tensile strength of the rock, the host may explosively fragment producing a hydrothermal breccia and or stockwork. In the surface environment this may result in the formation of hydrothermal explosion craters that have a distinctive champagne-glass morphology and attain diameters in excess of 1 km (Muffler *et al.*, 1971; Hedenquist and Henley, 1985). In the subsurface environment silicification of the host rocks and precipitation of silica within the fluid channelways contribute to fluid overpressure resulting in fragmentation and the generation of complex, multigenerational veins and breccias.

| Table 12.1 | Characteristics of low and high sulfidation epithermal Au deposits | |
|---|---|---|
| | Low sulfidation | High sulfidation |
| Regional structural setting | Calderas and other volcanic and sedimentary environments | Calderas, felsic domes |
| Local structural setting | Complex system of faults and or fractures; multigeneration structures | Complex system of faults and or fractures; multigeneration structures |
| Dimensions | 12–190 km$^2$, vertical extent 100–1000 m | Smaller in area, vertical extent often <500 m |
| Timing relationships | Usually younger than volcanic host rocks (>1 Ma) | Mineralization and host rock of similar age (<0.5 Ma) |
| Associated volcanic rocks | Predominantly andesite–dacite–rhyolite; associated intrusion and sediments | Predominately andesite–rhyodacite; commonly domes and ashflow tuffs |
| Alteration | Commonly restricted and visually subtle. Sericite or illite with or without adularia; roscoelite with alkalic rocks; chorite occasionally; outer zone of propylitic alteration | Areally extensive and visually prominent; crystalline alunite, kaolinite–advanced argillic alteration; pyrophylite at deeper levels; outer zone of propylitic alteration |
| Quartz gangue | Crustiform or colloform banded quartz and or chalcedony; cockade and carbonate replacement textures; open space filling | Fine-grained, replacement quartz; residual, slaggy or vuggy quartz host to mineralization |
| Carbonate gangue | Ubiquitous, commonly mangoan | Absent |
| Other gangue | Barite and / or fluorite locally; barite typically above mineralization | Barite occurs with mineralization; native sulfur as open-space filling |
| Sulfide abundance | Typically <5%, but up to 20% by volume locally; pyrite is the predominant sulfide | 10–90% by volume; fine-grained to laminated sulfide |
| Key sulfides | Sphalerite, galena, and tetrahedrite common; Cu occurs as chalcopyrite; arsenopyrite common | Cu sulfosalts such as enargite and luzonite; Cu + Cu–Fe sulfides: chalcocite, covellite and bornite; Cu minerals generally later than pyrite; arsenopyrite rare |
| Common metal | Au and/or Ag (Zn, Pb, Cu) | Cu, Au, As (Ag, Pb) |
| Other metals | Mo, Sb, As (Te, Se, Hg) | Bi, Sb, Mo, Sn, Zn, Te (Hg) |

*Source:* Modified after Sillitoe (1993) and Evans (1993).

Extensive leaching of the wall rocks under acid conditions typical of HS environments results in the nearly complete base leaching of the rock, a significant volume reduction and an increase in permeability (Figs. 12.14 and 12.15). The common product of this process is a vuggy, porous, residual silica-rich deposit, referred to as vuggy quartz, that is often a host for Au (Fig. 12.15). In this case the hydrothermal leaching of the wall rock acted as ground preparation enhancing subsequent fluid flow, trapment, and Au deposition.

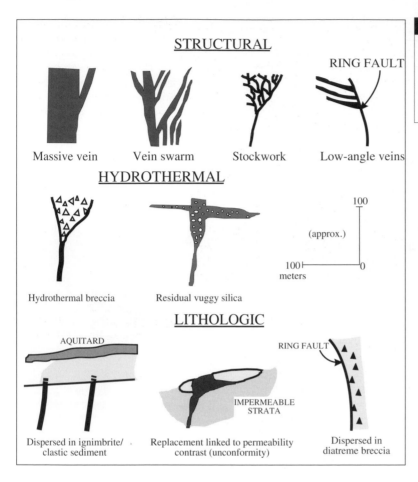

**Fig. 12.14.** The different morphologies of epithermal Au deposits illustrating structural, lithologic, and hydrothermal controls on mineralization and alteration. Modified after Sillitoe (1993).

Primary permeability of the host rock can result in the formation of strata-bound orebodies particularly in the shallow parts of epithermal systems (Fig. 12.14). Poorly lithified, porous sediments and pyroclastic rocks make ideal host rocks resulting in pronounced lateral fluid flow, which may favor the formation of larger deposits (Figs. 12.14 and 12.19 (below)). Lateral fluid flow can be enhanced by overlying less permeable units, such as a welded tuff or flow, which act as an aquitard that confines fluid flow and restricts Au deposition to the more permeable aquifer (Figs. 12.14 and 12.19 (below)). In some instances Au deposition is controlled by regional unconformities (or contacts) separating less permeable footwall strata from more permeable hanging wall units. Less commonly the composition of the host rocks may control Au deposition such as the sulfidation of iron-bearing oxide minerals. This processes is particularly well-documented in mesothermal, Archean iron-formation-hosted Au deposits such as the Lupin deposit, North West Territories, Canada. At Lupin, the replacement of primary magnetite by pyrite (arsenopyrite) consumes sulfur thus destabilizing the Au bisulfide complex, which results in the precipitation of Au (Lhotka and Nesbitt, 1989).

Although there is considerable overlap in the morphological characteristics of LS and HS deposits there are some clear differences in the ore mineralogy and associated gangue minerals (Table 12.1). High-sulfidation state minerals like enargite and luzonite, and tennantite typify high sulfidation deposits whereas arsenopyrite and sphalerite are common in low sulfidation deposits. Although many gangue minerals such as quartz are common to both deposit types, adularia and calcite are restricted to low-sulfidation deposits and alteration minerals such as kaolinite, dickite, and alunite typify high-sulfidation deposits.

**Fig. 12.15.** Silicified heterolithic volcaniclastic deposits that overly an advanced argillic alteration zone at the HS Porcato Au prospect, Peru. Note the vuggy porous nature of the altered volcaniclastics that are the principal host to Au mineralization.

**Fig. 12.16.** Silica sinter deposit typical of LS Au deposits, Steamboat Springs, Nevada.

Ore and gangue mineral textures also differ between the two deposit types. For instance cavities filled by crustiform quartz, colloform banded and botyroidal quartz (carbonate) veins and multiple generations of brecciation and veining typify LS deposits, whereas leached, massive bodies of residual, vuggy quartz typify HS deposits (Fig. 12.15). Sinter aprons or terraces form where hydrothermal fluids discharge at surface (Fig. 12.16). Sinter deposits are only associated with LS deposits as silica precipitation is inhibited by low-pH fluids typical of HS deposits (Fournier, 1985). The sinter is composed of delicately laminated opal, tridymite and/or chalcedony depending on temperature and degree of diagenesis. They are typically devoid of any precious metal values but may contain anomalously high mercury as at the McLaughlin mine (California) where cinnabar was mined from the silica sinter long before the underlying epithermal Au deposit was discovered (Lehrman, 1986).

## Alteration

In HS deposits minerals stable under acidic conditions such as alunite, kaolinite, dickite, pyrophyllite, and diaspore comprise the advanced argillic alteration zone, which forms beneath the

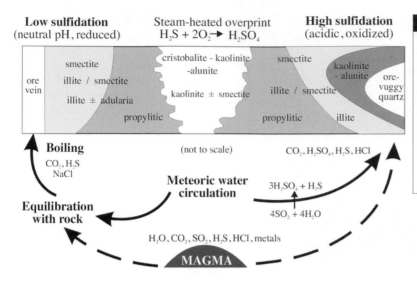

**Fig. 12.17.** Alteration assemblages and their geometry in LS and HS Au deposits. The inferred role of meteoric and magmatic waters in the generation of LS and HS deposits and alteration assemblages is also illustrated. (Reprinted with permission of The Society of Resource Geology, Special Publication No. 1, Fig. 3.2, Hedenquist, J. W., Izawa, E., Arribas, A., and White, N. C., 1996.)

water table by extensive leaching of the rocks by acidic fluids that condensed from acid volatiles. The advanced argillic envelope grades outward to a halo of illite, illite/smectite or smectite (Fig. 12.17).

In contrast, near-neutral pH waters interpreted to be responsible for LS deposits result in alteration assemblages characterized by smectite, illite–smectite, and illite–adularia that zone outward depending on the paleotemperature (Fig. 12.17). However, argillic alteration characterized by lower temperature acid-stable minerals such as cristobalite, kaolinite, and alunite can also form above the water table by groundwater absorption of $H_2S$ containing steam that originates from the boiling of deep, neutral-pH fluids. The steam-heated waters are generated in the vadose zone where $H_2S$ is oxidized by atmospheric oxygen to sulfuric acid ($H_2S + 2 O_2 = H_2SO_4$), which produces acid-leached zones above both LS and HS systems. Downward movement of such acid fluids can result in the superposition of argillic alteration on LS mineralization sometimes leading to confusion regarding the type of mineralization present (Sillitoe, 1995) (Fig. 12.17).

## Examples of LS and HS deposits

The Round Mountain deposit, located in southwest Nevada, contains some 277 Mt grading 1.2 g/t Au (Hedenquist *et al.*, 1996). The high-grade Au occurs within quartz veins, fissures, and cooling joints within a 26.7 Ma densely welded ash-flow tuff (Figs. 12.18 and 12.19). The higher-grade "vein" mineralization is rooted within large, thick (150 m) zones of lower-grade, stratabound mineralization in a poorly welded ashflow tuff (Fig. 12.19). The deposit is located near a caldera ring fracture which, along with ancillary parallel and contemporaneous structures coupled with the primary permeability of the ashflow tuff, likely controlled the location and development of the orebody.

Hydrothermal alteration is characterized by a district-wide pervasive propylitic alteration overprinted by quartz–sericite–adularia alteration in proximity to the Au mineralization (Fig. 12.18b). Silicic and argillic alteration associated with breccia and high-grade, cockscomb quartz–chalcedony–pyrite veins developed late within the upper part of the system (Hedenquist *et al.*, 1996).

The El Indio Au–Ag–Cu vein deposits located within the Chilean Andes are an excellent example of a vein-type HS deposit. From 1980 to 1993, El Indio (and Tambo) deposits have produced some 126 t of Au, 630 t of Ag, and 0.32 Mt of Cu with reserves as of 1993 standing at 121 t of Au, 1075 t of Ag, and 0.61 Mt of Cu (including Tambo) (Hedenquist *et al.*, 1996) (Fig. 12.20a).

Dacitic to rhyolitic ash flow and ashfall tuffs (27–25 Ma), intruded by dacitic stocks are the primary host rocks to the El Indio vein system (Fig. 12.20a). The mineralization is contained within a pervasively altered area some 2 km$^2$

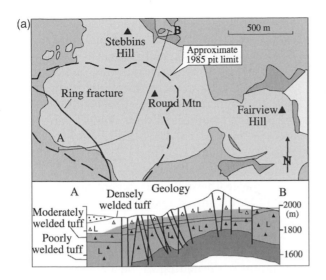

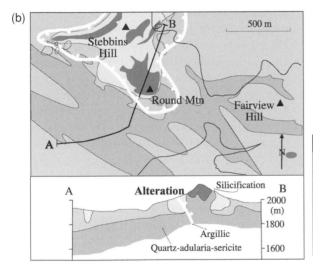

**Fig. 12.18.** (a) Geologic plan map and cross-section through the Round Mountain LS Au deposit with the open pit outline. (b) Alteration plan map and cross-section through the Round Mountain LS Au deposit with the white line showing the limit of angillic alteration. (Reprinted with permission of The Society of Resource Geology, Special Publication Number 1, (a) Fig. 7.1, (b) Fig. 7.2, Hedenquist, J. W., Izawa, E., Arribas, A., and White, N. C., 1996.)

(Fig. 12.20b). The paragenetic sequence is one of early Cu mineralization, dominated by massive enargite veins enveloped by advanced argillic and quartz–sericite alteration, followed by quartz–Au (tennantite) veins mantled by quartz–illite alteration (Hedenquist *et al.*, 1996) (Fig. 12.20b).

Mineralization at Tambo is hosted by hydrothermal and tectonic breccias within a 5 km² area of advanced argillic alteration (barite–alunite–quartz) (Fig. 12.21). Fine-grained, friable alunite–kaolinite–cristobalite–native sulfur, presumably formed by steam-heated acid–sulfate waters, caps the higher parts of the Tambo system (Hedenquist *et al.*, 1996).

## Genesis

Epithermal ore deposits form over the temperature range of <150 °C to 300 °C at shallow crustal levels (<2 km). Low-sulfidation deposits are interpreted to have formed from neutral-pH, reduced geothermal waters that, based on the $\partial^{18}O$ and $\partial D$ composition, appear to have equilibrated with their host rocks as indicated by a positive $\partial^{18}O$ shift, at constant $\partial D$, of geothermal waters from the meteoric water line in Fig. 12.22. The waters are low salinity (<1 wt.% equivalent NaCl), relatively gas-rich (1–2 wt.% $CO_2$, minor $H_2S$), and dominantly meteoric. The Au, Ag, and base metals are interpreted to have been derived

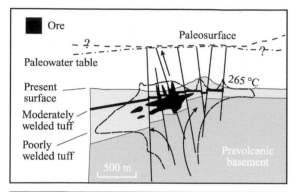

**Fig. 12.19.** Distribution of Au ore within the Round Mountain LS Au deposit with higher-grade structurally controlled vein-ore developed within moderately welded tuff and lower grade, disseminated ore within poorly welded tuff. (Reprinted with permission of The Society of Resource Geology, Special Publication Number 1, Fig. 7.5 , Hedenquist, J. W., Izawa, E., Arribas, A., and White, N. C., 1996.)

from magmatic fluids that were introduced into an active geothermal system (Buchanan, 1981; Berger and Eimon, 1983; Head *et al.*, 1987; Hedenquist, 1987, 2000; Berger and Henley, 1989; Henley, 1991; Sillitoe, 1995; Hedenquist *et al.*, 1996).

High-sulfidation deposits, on the other hand, are interpreted to have formed from acid hydrothermal fluids within an active volcanic–hydrothermal system. The hydrothermal fluids are interpreted to have had a significant magmatic component throughout their history. For example, the isotopic composition of waters that formed alunite in HS deposits indicates a dominant magmatic water signature; however, there are mixing trends towards local meteoric waters as illustrated in Fig. 12.22 (Berger and Henley, 1989; Henley, 1991; Giggenbach, 1992a, 1992b; Sillitoe, 1995; Hedenquist *et al.*, 1996). In HS deposits, unlike LS deposits, the meteoric waters, based on their $\partial^{18}O$ and $\partial D$ composition, have not equilibrated with their host rocks (no positive $\partial^{18}O$ shift in Fig. 12.22).

The solubility of Au in the hydrothermal environment is dominated by its ability to complex with either $HS^-$ (bisulfide) or $Cl^-$ (chloride) ligands. Au bisulfide complexes ($AuHS_2$) are favored by lower temperature (<300 °C), low salinity (<2 wt.% equivalent NaCl), and reduced hydrothermal fluids, whereas Au chloride complexes ($AuCl_2$)

are favored by higher temperature (>300 °C), saline, and weakly oxidized hydrothermal fluids (Seward, 1989). Thus, $AuCl_2$ complexes dominate in the magmatic hydrothermal environment and Au, like Cu and Mo, is preferentially partitioned into the high temperature (500–700 °C), saline magmatic hydrothermal fluid phase where, like the base metals, it is carried as a chloride complex with oxidized ($SO_2$) and reduced ($H_2S$) sulfur (Candella, 1989; Ballhaus *et al.*, 1994).

In the geothermal environment illustrated in Fig. 12.24, the introduced, high-temperature, Au- and base-metal-bearing, magmatic hydrothermal fluid is overwhelmed by the large volume of convecting meteoric water. Mixing of these two fluids results in the neutralization and rapid cooling of the magmatic fluid, the disproportionation of $SO_2$, destabilization of the chloride complex, and precipitation of base metals. Although some of the Au may be deposited with base metals much of it is interpreted to have been transferred to convecting meteoric water as a stable bisulfide complex thus "decoupling" Au from base metals, which may account for the lower overall base metal content of LS Au deposits. The meteoric and magmatic fluid mixture ascends to surface where it discharges and forms shallow LS Au deposits. Other cations, in particular silica, which is ubiquitous to LS deposits where it occurs as veins and sinters, are interpreted to have been leached from the aquifer rocks by convecting meteoric water (Giggenbach, 1992b; Hedenquist and Lowenstern, 1994).

The processes of the emplacement and deposition in the volcanic-hydrothermal or HS environment are not as well understood, but they are interpreted to represent, at minimum, a two-stage process (Fig. 12.24). The first stage is one of ground preparation whereby the permeability and porosity of near-surface rocks are increased by the development of vuggy quartz and advanced argillic–alunite alteration zones. In this first stage, hot, saline, magmatic hydrothermal fluid exsolved from a high-level intrusion (1–3 km depth) separates into a dense, hypersaline and presumably base-metal- and Au-rich residual fluid, and a low density, acid, magmatic vapor containing HCl, $SO_2$ and minor HF (Fig. 12.24). The buoyant, low-density, acidic magmatic vapor rises and separates from the hypersaline

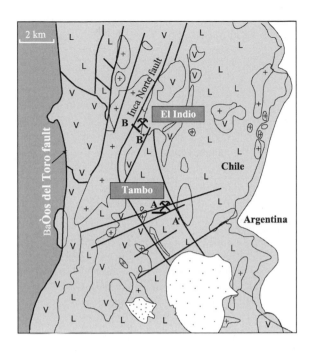

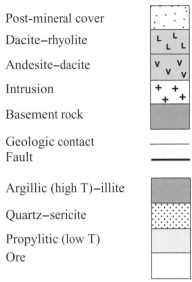

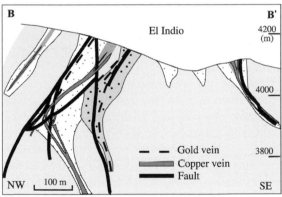

**Fig. 12.20.** (a) Geologic map showing the distribution of volcanic units, major structures and location of the El Indio and Tambo HS Au deposits. (b) Geologic cross-section showing structurally controlled vein Au mineralization, faults, and alteration assemblages at El Indio. (Reprinted with permission of The Society of Resource Geology, Special Publication No. 1, (a) Fig. 10.1, (b) Fig. 10.3, Hedenquist, J. W., Izawa, E., Arribas, A., and White, N. C., 1996.)

residual fluid, and in the near-surface environment condenses within meteoric water to form a hydrochloric–sulfuric acid water (pH 1) that leaches most cations, including Al, but not Si, to produce a siliceous residue referred to as vuggy quartz (Fig. 12.15) (Hayba et al., 1985; Head et al., 1987; Hedenquist et al., 1996, 1998; Hedenquist, 2000).

The actual mechanisms of Au emplacement within the second stage are uncertain but are generally attributed to the ingress of the residual hypersaline magmatic hydrothermal fluid or, more likely, a hybrid of this residual fluid that was derived through mixing with meteoric water, into the vuggy quartz alteration zone where Au is precipitated (Hedenquist et al., 1996, 1998;

Hedenquist, 2000). As in the LS environment, temperature and pH changes resulting from the mixing of residual magmatic hydrothermal fluid with meteoric water, deeper within the system, may have resulted in the precipitation of most base metals at depth, and the transport of Au to surface as a bisulfide complex in a low-salinity, lower-temperature (<300 °C), mixed magmatic–meteoric hydrothermal fluid (Fig. 12.24). However, many uncertainties exist and there is fluid inclusion evidence from some deposits to indicate that Au, and other metals, were preferentially partitioned into the magmatic vapor phase, opening the possibility for Au precipitation in the near-surface environment directly from magmatic volatiles (Heinrich et al., 1992, 1999).

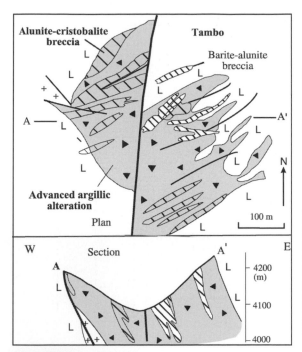

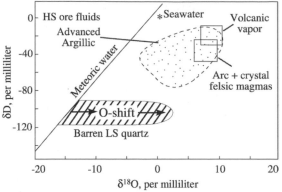

**Fig. 12.22.** Diagram of relationship of $\partial D$ and $\partial^{18}O$ illustrating the involvement of evolved ($\partial^{18}O$ shifted) meteoric or geothermal waters and magmatic waters, and meteoric and magmatic waters in the formation of LS and HS Au deposits respectively. (Reprinted with permission of The Society of Resource Geology, Special Publication No. 1, Fig. 14.4, Hedenquist, J. W., Izawa, E., Arribas, A., and White, N. C., 1996.)

**Fig. 12.21.** Alteration and mineralization plan map and section through the Tambo HS Au deposit illustrating the advanced argillic alteration zone that envelopes the Au zones shown by hatched areas. (Reprinted with permission of The Society of Resource Geology, Special Publication No. 1, Fig. 10.2, Hedenquist, J. W., Izawa, E., Arribas, A., and White, N. C., 1996.)

In the LS environment, precipitation of Au is a product of two main processes, boiling and/or mixing with cooler, oxidizing near-surface meteoric water. Boiling and phase separation (open system and removal of steam) (Seward, 1989) causes destabilization of the $AuHS_2$ complex by the preferential partitioning of S, as $SO_2$, into the volatile phase and the precipitation of Au, illustrated by the simple reaction where:

$$Au(HS)_2 + 4H_2O = Au_{(ppt)} + 2SO_2 + 5H_2$$

$$(12.6)$$

Volume and temperature decrease of the hydrothermal fluid that accompanies boiling also results in the rapid precipitation of other near-saturated components such as $Si(OH)_2$ to form quartz or chalcedony veins. Precipitation of quartz within the fluid conduit results in self-sealing, fluid overpressure, hydraulic fracturing, pressure-release boiling and Au precipitation.

Repetition of this process accounts for the complex, multigeneration and brecciated quartz or chalcedony veins that are typical hosts to many LS Au deposits.

In the HS environment the predominant mechanism for Au precipitation is uncertain, but may be mixing with cool, near-surface, oxidizing groundwater within permeable, acid leached vuggy quartz zones. Mixing destabilizes the $AuHS_2$ complex by oxidizing the reduced sulfur as illustrated Fig. 12.23 and in the simplified reaction where:

$$Au(HS)_2 + 8H_2O = Au_{(ppt)} + 2SO_4 + 17H^+$$

$$(12.7)$$

The schematic cross-section in Fig. 12.24 illustrates a continuum between near-surface epithermal Au deposits and deeper, intrusion-related porphyry Cu (Au, Mo) mineralization. Examples illustrating this spatial and temporal relationship include deposits of the Western Pacific island arcs such as Lepanto, Philippines (Garcia, 1991; Arribas *et al.*, 1995) and LaPepa and El Hueso, central Andes (Sillitoe, 1991). Sillitoe has also documented the occurrence of base-metal-rich LS Au deposits surrounding porphyry Cu deposits in the Philippines, and Setterfield *et al.* (1992)

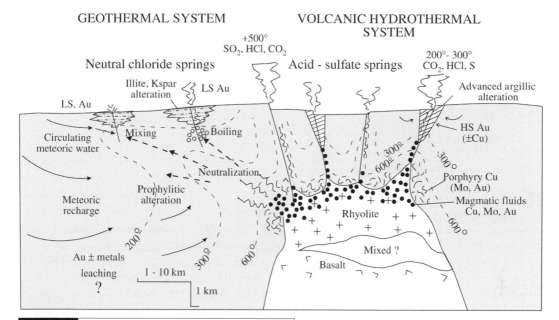

**Fig. 12.24.** Diagrammatic illustration of a genetic model for the formation of LS and HS Au deposits. HS deposits, contemporaneous with volcanism, are interpreted to be a product of magmatic fluids, whereas meteoric water is dominant in LS deposits, although the Au may have had an original magmatic source. Modified from Hedenquist *et al.* (1996) and Hannington *et al.* (1999).

have shown that the sulfide-poor Emperor LS Au deposit is associated with porphyry Cu–Au mineralization. At Lepanto, in particular, the genetic link between spatially associated HS Cu–Au and porphyry Cu–Au deposits is indicated by fluid inclusion and stable isotopic studies, and by the similar 1.5 to 1.2 Ma (K–Ar dating) ages for alunite (HS) and hydrothermal biotite and illite (porphyry Cu–Au) which indicate that the HS epithermal and porphyry systems were active contemporaneously for about 300 Ma (Arribas *et al.*, 1995; Hedenquist *et al.*, 1998).

## Volcanic-associated massive sulfide deposits

Volcanic-associated massive sulfide (VMS) deposits, also referred to as volcanic-hosted (VHMS) or volcanogenic massive sulfide deposits, are major sources of Cu and Zn and, to a lesser extent Pb, Ag, Au, Cd, Se, Sn, Bi, and minor

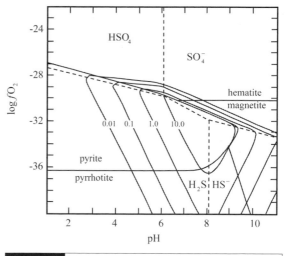

**Fig. 12.23.** Calculated Au solubility as a function of pH and log $fO_2$. Au is carried as a bisulfide complex – $Au(HS)_2$ – and contours are in ppm with temperature fixed at 300 °C and total sulfur at 0.05 M. Note the marked decrease in the solubility of Au decreases during oxidation. (Reprinted with permission of *Ore Deposit Models*, Geoscience Canada Reprint Series no. 3, Geological Association of Canada, Fig. 11, p. 14, Roberts, R. G., 1988.)

amounts of other metals (Franklin, 1990, 1996; Large *et al.*, 2001). For example, in 1988, VMS deposits accounted for 32.8% of the Cu, 56% of the Zn, 29.5% of the Pb, 30% of the Ag, and 3.6%

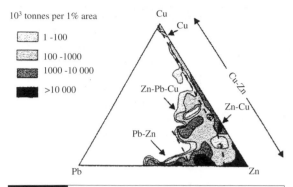

**Fig. 12.25.** Classification of VMS deposits based on Cu–Zn–Pb content. (Reprinted with permission of Economic Geology 75 Anniversary Vol. 1905–80, p. 489, Fig. 1, Franklin, J. M., Lydon, J. W., and Sangster, D. F., 1981.)

of the Au produced from Canadian mines (Lydon, 1988). VMS deposits occur throughout the geologic column and in virtually every major tectonic domain where subaqueous volcanic rocks are an important constituent (Franklin *et al.*, 1975, 1981; Franklin, 1990).

Genetic and empirical models of VMS deposits were initially based on detailed deposit and more regional studies of well-preserved ancient examples particularly those in well-exposed and established mining districts in Japan, Canada, Australia, and Scandinavia. However, the discovery of actively forming "black smoker" deposits on the East Pacific Rise at 21° N in 1977, and subsequent discoveries, validated and refined the existing syngenetic model. Studies of actively forming deposits along mid-ocean ridges and in back arc and incipient spreading environments have made, and will continue to make, significant contributions to our understanding of the primary mineralogy and mineral chemistry, the temperature, and the composition of the mineralizing fluid, as well as the surface growth mechanisms for sulfide mounds. Studies of ancient deposits, on the other hand, are better suited to unraveling the volcanological and structural history of VMS deposits and the distribution and types of alteration assemblages as the hanging wall and footwall strata are available for examination and study.

This chapter reviews the characteristics of VMS deposits, provides examples of VMS deposits

that may have formed in shallower and deeper water environments, and discusses current genetic models. Although most of the examples are from ancient economic deposits, reference is made to actively forming seafloor deposits to illustrate particular processes or characteristics where appropriate.

## Classification

Volcanic-associated (VMS) and sedimentary-associated (SEDEX) deposits are subdivisions of the more general "massive sulfide" class of deposit. Sangster and Scott (1976) define massive sulfide deposits as stratabound and in part stratiform accumulations containing at least 60% sulfide minerals in their stratiform parts. In some deposits the stratiform part may comprise the entire deposit whereas in other deposits appreciable sulfides occur within a discordant stockwork or "stringer" zone located primarily below, but also above the stratiform sulfides in some deposits.

VMS deposits have been classified on the basis of host rock lithology (Sangster and Scott, 1976; Sawkins, 1976; Klau and Large, 1980), ore composition (Hutchinson, 1973; Solomon, 1976; Franklin *et al.*, 1981; Poulson and Hannington, 1996), tectonic environment (Solomon, 1976; Hutchinson, 1980; Barrie and Hannington, 1999; Franklin *et al.*, 1999) and characteristics of the host rocks, i.e., volcaniclastic or flow hosted (Gibson *et al.*, 1999). Lydon's classification in Franklin *et al.* (1981), where massive sulfide deposits are classified on the basis of their average Cu, Zn, and Pb content is probably the most widely used and is useful in identifying processes responsible for compositional variations and anomalies, i.e., the distribution of gold and lead. In this classification scheme VMS deposits are subdivided into Cu–Zn and Zn–Cu–Pb groups (Figs. 12.25 and 12.26). The Cu–Zn group includes deposits of the Canadian and Fennoscandian Shields, the Scandinavian Caledonides, and ophiolite-hosted deposits. The Zn–Cu–Pb group includes deposits of the Bathurst Camp, the Kuroko district, the Iberian (Spanish) Pyrite Belt, Buchans Camp, and the Mount Reed Volcanics (Franklin *et al.*, 1981; Large *et al.*, 2001). Poulson and Hannington (1996) have further

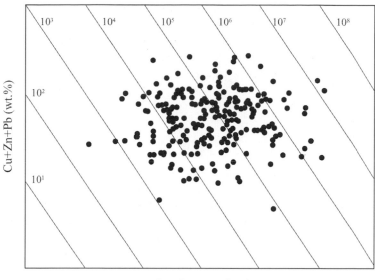

**Fig. 12.26.** Tonnes versus combined Cu–Zn–Pb content of VMS deposits. (Reprinted from the *Geology of Canadian Mineral Deposit Types, Geology of Canada,* No. 8, Fig. 6.3–3, p. 162, Franklin, J. M., 1996, with permission from the Minister of Public works and Government Services Canada and courtesy of the Geological Survey of Canada.)

subdivided VMS deposits into those that are Au-rich and those that are not (Fig. 12.27).

Classification of VMS deposits with respect to tectonic or geologic setting, although more subjective, may be more useful. However, criteria used for establishing the tectonic settings of VMS deposits in ancient successions typically rely on host rock composition and not paleogeographic reconstruction as the latter requires comprehensive volcanological, sedimentological, and structural studies. Unfortunately the tectonic settings of many Archean VMS deposits are, at best, inferred because of uncertainties resulting from the scale of observation, and the quality of geological and geochemical data available. Recognizing these shortcomings, the tectonic classification of Franklin *et al.* (1999) is summarized in Table 12.2. The classification of VMS deposits into volcaniclastic versus flow hosted is done in the field and characteristics of deposits found within these two classes are summarized in Table 12.3.

## General characteristics

Before discussing the different volcanic environments that host VMS deposits and genetic models it is important to review characteristics that are common to VMS deposits regardless of age and tectonic setting.

## Tonnage and grade characteristics

A quantitative summary of the tonnage and grade of Archean, Proterozoic, and Phanerozoic VMS deposits by Boldy (1977) and Sangster (1980) is outlined below.

(1) VMS deposits occur in a natural geometric progression in size. Approximately 80% of the VMS deposits fall in the size range of 0.1 to 10 Mt and about 50% of these are <1 Mt.

(2) Regardless of age, 88% of deposits have combined Cu + Pb + Zn grades of <10% (Fig. 12.26). The most likely combined grade is 6%, roughly in the ratio of 4 : 1 : 1 for Zn : Cu : Pb.

(3) Larger deposits tend to be rich in Zn relative to Cu, and Pb grade varies inversely with respect to Cu and directly with respect to Zn.

(4) VMS deposits tend to occur in clusters or districts containing from 4 to 20 deposits and each district has an average diameter of 33 km. The average total base metal content per district is 4.6 Mt with a coefficient of variation of 32%. Ranked in order of size the largest deposit in each district contains, on average, 67% of the total metal and the second largest about 13%. The remaining deposits decrease in both size and frequency of occurrence (Sangster, 1980).

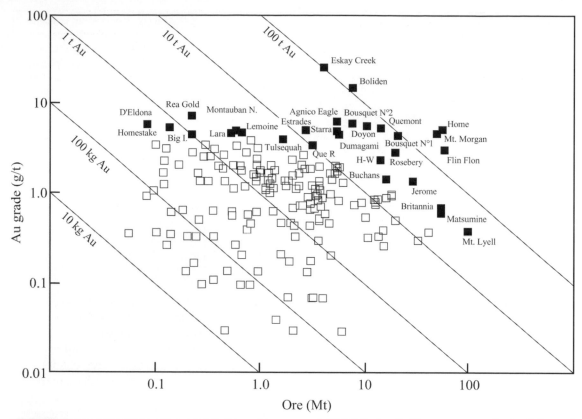

**Fig. 12.27.** Plot of tonnes versus Au content of VMS deposits. (Reprinted from the *Geology of Canadian Mineral Deposit Types, Geology of Canada*, No. 8, Fig. 6.4–1, p. 184, Poulson, K. H. and Hannington, M. D., 1996, with permission from the Minister of Public Works and Government Services Canada and courtesy of the Geological Survey of Canada.)

### Deposit characteristics

Most VMS deposits consist of more than one massive sulfide lens. The morphology of the massive sulfide lens ranges from that of a cone or mound (Fig. 12.28) to a tabular sheet (Fig. 12.29), with the latter morphology more typical of sulfide deposits that have accumulated at and below the seafloor.

Idealized, proximal, VMS deposits consist of a concordant lens of massive sulfide content (>60% sulfides) that is stratigraphically underlain by a discordant stockwork of "stringer sulfide" contained within a pipe-like or irregular envelope of hydrothermally altered rock (Figs. 12.28–12.32

and 12.36 (below)). The upper contact of the sulfide lens is typically sharp, except where stringers extend into the hanging wall, and the lower contact, with the stringer zone, is gradational. The stringer zone represents the near-surface discharge conduit of a subseafloor hydrothermal system. The massive sulfide lens forms by several processes including chimney collapse, hydraulic fracturing, and precipitation of sulfides within the lens as illustrated in Figs. 12.33 and 12.34. Research conducted on ancient massive sulfide deposits indicates that sulfide lenses also grow by the subsurface precipitation of sulfides within void spaces in permeable volcaniclastic or sedimentary units immediately below the seafloor, by hydraulic jacking or displacement of beds, and by replacement of these units (Kerr and Mason, 1990; Gibson and Kerr, 1990, 1993; Kerr and Gibson, 1993; Hannington *et al.*, 1999; Large *et al.*, 2001) (Fig. 12.35). Some deposits consist of fragments of massive sulfide that have been transported as high concentration mass flows from their original site of deposition, such as the

| Table 12.2 | Tectonic classification of VMS deposits | | |
|---|---|---|---|
| Setting | Associated volcanism[a] | Modern examples | Ancient examples |
| Intraocean ocean arc | 1. Volcanic front: A, B, +/− D | Izu-Bonin Northern Tonga-Kermadec Solomon-New Hebrides Mariana Arc | Absent |
| | 2. Arc-related rifts nascent back arc: A, B, +/−D | Harve Trough Eastern Manus Basin Southern Lau Basin | Noranda Sturgeon Lake Flin Flon |
| | 3. Mature back arc spreading center: MORB, sediments | Central Manus Basin Northern Lau Basin Mariana Trough North Fiji Basin | Cyprus Besshi Windy Craggy |
| | 4. Volcanic front | Kurile Arc Southern Tonga-Kermadec Ryukyu Arc | Absent |
| Epi continental ocean–continent–continental margin arc | 5. Arc-related rifts nascent back arc: A, D, R, +/− B | Kagoshima Graben, Kyushu Taupo Volcanic Zone Southern Harve Trough | Eskay Creek Skellefte Tasmania Kuroko |
| | 6. Mature back arc spreading center: A, D, R, +/− B | Okinawa Trough | Bathurst Iberian Pyrite Belt Urals Kazhakstan |
| Ocean rifts | 7. B, +/− R (icelandite) | Mid-Atlantic, East Pacific, Juan de Fuca (includes sedimented rifts) | Semail, Bay of Islands, Josephine |
| Others | 8. Fore-arc rift alkaline | Tabar-Feni Archipelago | Samatosen, BC MM (Yukon) |
| | 9. Rifted margin MORB | Western Woodlark | |

[a] A, andesite; B, basalt; D, dacite; MORB, midocean ridge basalt; R, Rhyolite.
*Source:* After Franklin *et al.* (1999).

Ordovician–Silurian Buchans VMS deposits, Newfoundland (Hutchinson, 1981; Walker and Barbour, 1981). These "distal deposits" are not underlain by a stringer or alteration zone as they have been removed from their hydrothermal vent.

The most common sulfide mineral is pyrite. Pyrrhotite, chalcopyrite, sphalerite, galena, and, less commonly, sulfosalts and bornite occur in subordinate amounts. Magnetite, hematite, and cassiterite are the most common non-sulfide metallic minerals. Quartz, carbonate, sericite, chlorite, barite, and gypsum are the most common gangue minerals (Lydon, 1984). Barite and gypsum are more common in Phanerozoic VMS deposits such as the Miocene Kuroko VMS deposits illustrated in Fig. 12.36.

The textures and structures of the ores are variable and in many instances have been modified during metamorphism and deformation. In cone- or lens-shaped deposits the central base of the sulfide lens is typically massive and structureless (Fig. 12.28). *In situ* breccias and rubbly breccia ore surround and may

| Table 12.3 | Characteristics of flow and volcaniclastic hosted VMS deposits | |
|---|---|---|
| | Flow hosted VMS | Volcaniclastic hosted VMS |
| Deposit composition | Cu-rich, Cu–Zn rich pyrrhotite, pyrite, chalcopyrite, sphalerite, galena | Zn(Pb)-rich (Ag- ± Au-rich) pyrite, pyrrhotite, sphalerite, chalcopyrite, galena |
| Discordant "pipe" alteration | Vertically extensive<br>Well defined and zoned<br>Sericitization: sericite, quartz, pyrite, sphalerite − +K, −Ca, ±Si, +Zn (Fe)<br>Chloritization: Fe- and Mg-chlorite, pyrrhotite, pyrite, chalcopyrite, +Fe, +Mg, −K, +Cu | Irregular and zoned Fe-carbonates: ankerite, siderite: +Fe, +Mg, +$CO_2$, −Na, −Ca<br>Sericitization: sericite, quartz, pyrite, sphalerite (Mg-chlorite), +K, +Zn, ±Si, Fe, −Na, −Ca<br>Chlorite: Fe-chlorite; +Fe, +Mn, +Cu, +Zn, −Mg<br>Argillic/aluminum silicate: andalusite, kyanite, sillmenite, pyrophyllite (paragonite); +Al, +Ca<br>Chloritoid: chloritoid, Fe-chlorite, ±Fe-carbonate; +Fe, +Mn, +Si, ±K, ±Mg, −Na, −Ca |
| Regional semiconformable alteration | Well-developed and cross-cut by "pipe" alteration<br><br>Felsic rocks<br>Spilitization: albite, quartz, sericite<br>Chloritization: Mg-chlorite, +Na, −Ca<br>Silicification: quartz, albite, +Si, +Na, −Fe, −Mg, −Zn<br>Sericitization: sericite, quartz Mg-chlorite, +Mg, ±Si<br><br>Mafic rocks<br>Spilitization: albite, quartz, sericite, Mg-chlorite, +Na, −Ca<br>Silicification: quartz, albite, +Si, +Na, −Fe, −Mg, −Zn<br>Epidote–quartz: epidote, quartz (actinolite, carbonate), +Ca, ±Si, −Fe, −Mg (−Zn, Cu?) | Well-developed, extensive "Transitional" contact with "pipe" alteration<br><br>Felsic rocks<br>Carbonitization: calcite, dolomite, +Ca, +Mg, +$CO_2$, ±Fe, −Na, −Si<br>Sericitization: sericite, quartz (Mg-chlorite), +K, ±Na<br>Chloritoid: chloritoid, Fe-chlorite, ±Fe-carbonate: +Fe, −Mn, +Si, ±K, ±Mg, −Na, −Ca<br><br>Mafic rocks<br>Spilitization: albite, quartz, sericite, Mg-chlorite; +Na, −Ca<br>Epidote–quartz: epidote, quartz (actinolite, carbonate), +Ca, ±Si, −Fe, −Mg (−Zn, Cu?) |
| Examples | Ansil, Norbec, Waite deposits, Millenbach, Corbet, Amulet deposits, Quemont, Delbridge, Kam Kotia, Canadian Jamieson, Izok Lake, Hellyer, Que River, most ophiolite and mid-ocean ridge deposits | Horne, Gallen, Mobrun, Selbaie, Bousquet, Dumagami, Estrades, Louvicout, Mattagami, Orchan, Norite, Coniagas, South Bay, Hacket River, Yava, Garpenberg, Errington and Vermilion, Rosebery, Woodlawn |

*Source:* Modified after Gibson *et al.* (1999).

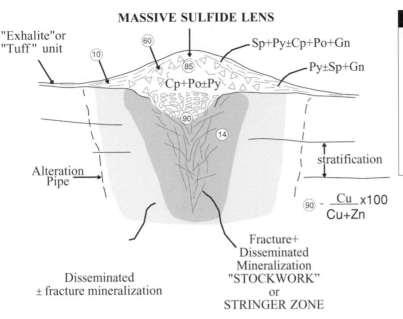

**MASSIVE SULFIDE LENS**

"Exhalite"or "Tuff" unit

Sp+Py±Cp+Po+Gn

Py±Sp+Gn

Cp+Po±Py

stratification

Alteration Pipe

$\frac{Cu}{Cu+Zn} \times 100$

Disseminated ± fracture mineralization

Fracture+ Disseminated Mineralization "STOCKWORK" or STRINGER ZONE

**Fig. 12.28.** Typical sulfide textures, structures, and metal zoning within an idealized massive sulfide "mound" and stringer zone hosted within a flow-dominated succession. (Reprinted with permission from *Giant Ore Deposits*, Society of Economic Geologists Special Publication No. 2, p. 324, Fig. 2, Gibson, H. L. and Kerr. D. J., 1993.)

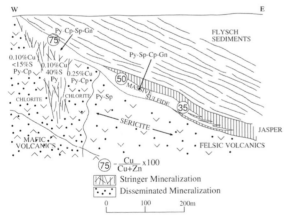

LONGITUDINAL SECTION THROUGH PLANES SANANTONIO, RIO TINTO

$\frac{Cu}{Cu+Zn} \times 100$

Stringer Mineralization

Disseminated Mineralization

**Fig. 12.29.** Typical sulfide textures, structures, and metal zoning within a sheet-like, massive sulfide lens and stringer zone hosted within a volcaniclastic dominated succession. (Reprinted with permission from *Giant Ore Deposits*, Society of Economic Geologists Special Publication No. 2, p. 325, Fig. 3, Gibson, H. L. and Kerr. D. J., 1993.)

overlie the core (Figs. 12.28 and 12.34). The fringe is bedded and consists of finely laminated sulfides (Fig. 12.28). Tabular or sheet-like lenses are often characterized by extremities that are banded or consist of laminated sulfides (Fig. 12.29).

In some deposits a thin, bedded, sulfide- or oxide-bearing siliceous sediment (clastic component is often a fine tuff) referred to as an exhalite, tuff, or tuffaceous exhalite occurs laterally to the deposit and in some instances mantles the deposit (Figs. 12.28, 12.36, and 12.37). In Japan these rocks are called "tetsuskiei" where they are associated with the Miocene Kuroko VMS deposits (Kalogeropoulos and Scott, 1983) (Fig. 12.36). This sedimentary/tuffaceous unit is thought to be a product of suspension sedimentation (waterlain tuff and/or epiclastic sediments) and chemical precipitation that formed during a period of quiescence which marked a hiatus in volcanic activity.

VMS deposits exhibit a pronounced zonation in their mineralogy and chemical composition. The most common zoning pattern is a systematic decrease in the chalcopyrite/sphalerite ratio or the Cu/Cu+Zn ratio upward and outward from the base of the massive sulfide lens and outward from the core of the stringer zone. This zoning pattern is illustrated in Figs. 12.28, 12.29, 12.36, and 12.38.

The pronounced base-metal zoning of massive sulfide deposits is interpreted as a successive replacement of a lower-temperature pyrite–sphalerite sulfide assemblage by a higher-temperature chalcopyrite–pyrrhotite assemblage (Figs. 12.38 and 12.39). Thermal evolution of the

**Fig. 12.30.** Chalcopyrite–pyrrhotite stringers (S) in a chloritized basalt flow (C) within a discordant, chlorite–sericite alteration zone at the Archean, Corbet Deposit, Noranda, Quebec, Canada.

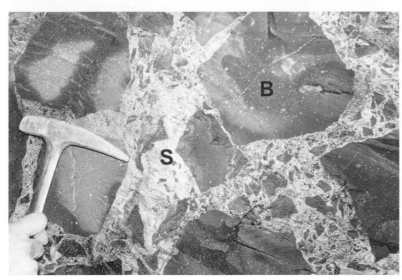

**Fig. 12.31.** *In situ* brecciated basalt flow (B) cemented by pyrite (minor sphalerite and quartz) (S) within the margin of the Corbet stringer zone in Fig. 12.30. The breccia, interpreted to be a product of hydraulic fracturing and weak hydrothermal explosions, was subsequently cemented by sulfides.

sulfide lens is a function of near-surface self-sealing processes that result from sulfide precipitation and alteration of the footwall rocks which restricts the downward movement of entrained seawater. Self-sealing limits the cooling of ascending metal-bearing hydrothermal fluids whose metal content is strongly temperature-dependent (Fig. 12.40). For example, a pronounced drop in the temperature of a hydrothermal fluid with a pH in the 4–5 range results in a drastic decrease in the solubility of $CuCl_2$, but less so for $ZnCl_2$ such that Cu may precipitate initially in the subseafloor environment whereas Zn, being less temperature-dependent,

will precipitate at the seafloor due to rapid cooling and mixing with seawater (increase in pH) (Fig. 12.38). In this model the initial seafloor sulfide lens is sphalerite- or Zn-rich. With progressive sealing of the footwall rocks inhibiting mixing with entrained seawater, ascending fluids would reach the seafloor without significant cooling to precipitate chalcopyrite and resolubilize sphalerite which would be soluble in the higher-temperature fluid. Thus, as Cu sulfides are precipitated within the stringer zone and base of the sulfide lens, Zn is redissolved and moved upward (as either a bisulfide or chloride complex) to precipitate, due to cooling or drop in pH,

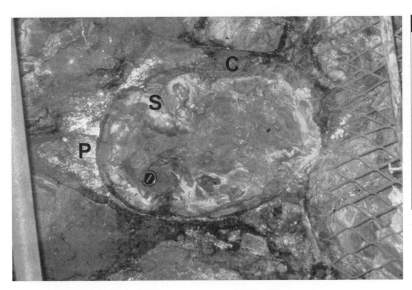

**Fig. 12.32.** Pyrrhotite and chalcopyrite stringers (P) surrounding altered basaltic pillows at the margin of the Corbet Stringer zone in Figs. 12.30 and 12.31. The pillow margins are chloritized (C) and overprint fracture controlled sericite (S) associated with pyrite and sphalerite within the pillow interior. Here the hydrothermal fluids have clearly utilized the primary permeability of the pillowed flows on their ascent to the seafloor.

above and lateral to the Cu-rich core. This successive precipitation–dissolution or zone-refining process illustrated in Fig. 12.41, and first proposed by Eldridge *et al.* (1983) for the Miocene Kuroko deposits, results in a sulfide lens with a Cu-rich core and more Zn-rich fringe and top.

In ancient VMS deposits, Au occurs in two distinct spatial and mineralogical associations: (1) Au–Zn association, associated with pyrite, sphalerite, galena with or without barite toward the top and margin of the sulfide lens; and (2) Au–Cu association, associated with chalcopyrite–pyrrhotite (pyrite) at the base of, and within the core of, the sulfide lens and underlying stringer zone (Knuckey *et al.*, 1982; Large *et al.*, 1989). The "geologic thermodynamic model" proposed by Huston and Large (1989) and Large *et al.* (1989) explains Au distribution within a sulfide lens as it is directly related to the zone-refining process for base metals illustrated in Fig. 12.41.

In their model Au, with other metals, is carried as a chloride complex in the high temperature (>300 °C), hydrothermal fluid. Evidence from active seafloor hydrothermal vents indicates that high-temperature hydrothermal fluids (>350 °C) may contain up to 0.1–0.2 mg/kg Au and are capable of transporting up to 1000 g Au/year in a single deposit (Hannington and Scott, 1989). Cooling of the ascending hydrothermal fluid, primarily a result of mixing with downdrawn seawater, results in the subseafloor precipitation of

Cu, and formation of a Zn-rich seafloor sulfide lens. Some of the Au may be precipitated with the Cu, but as in epithermal deposits, most of the Au is carried as a bisulfide complex in the lower-temperature hydrothermal fluid to be precipitated, along with sphalerite, at the seafloor (Fig. 12.41). During the high-temperature stage (>300 °C) of the zone refining processes where hydrothermal fluids carrying Au and base metals as chloride complexes reach the seafloor they deposit Cu and Au within the underlying stringer zone and base of the sulfide lens, primarily due to a decrease in temperature. These fluids are then capable of leaching Au as a bisulfide complex, along with Zn and if present Pb, from the base of the sulfide mound. This results in a Cu-rich and Zn-, Pb-, and Au-depleted base to the massive sulfide lens. Growth of the Cu-rich base during zone refining (Fig. 12.41) results in continued remobilization of Au, Zn, and Pb and their concentration outward, toward the margins and top of the sulfide lens. In younger VMS deposits which formed in oxygenated oceans Au precipitation is largely a function of oxidation (Eq. 12.7) through mixing with seawater in the Zn- and barite-rich upper part, and margins of the sulfide lens. In Archean VMS deposits, which formed within more "reduced" oceans, oxidation by mixing with seawater may not be a viable precipitation mechanism. Destabilization of the Au bisulfide complex and Au precipitation in this case

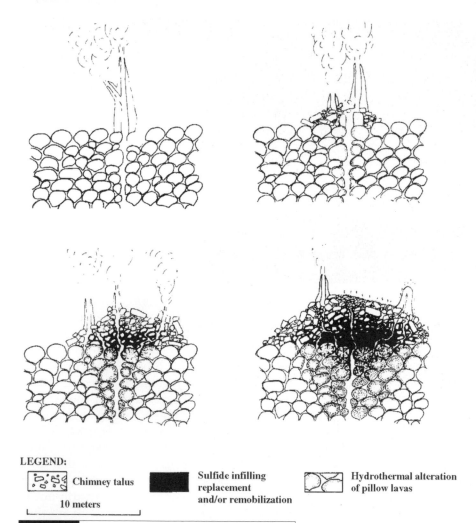

**Fig. 12.33.** Growth of a seafloor sulfide mound through collapse of sulfide chimneys and cementation and replacement of the sulfide chimney debris by sulfide precipitated from ascending hydrothermal fluid. (Reprinted with permission from *Ore Deposit Models*, Geoscience Canada Reprint Series No. 3, Geological Association of Canada, Fig. 2, p. 157, Lydon, J. W., 1988.)

may be a function of: (1) a rapid decrease in temperature and increase in pH during mixing with seawater; and (2) destabilization of the Au bisulfide complex due to consumption of sulfur during precipitation of base-metal sulfide minerals.

The variable Au content of VMS deposits and the occurrence of Au-rich VMS deposits may be a product of the efficiency of the above processes, or variations in the chemical and physical (boiling) characteristics of the hydrothermal

fluids and their Au concentration at the time of Au saturation and precipitation (Hannington and Scott, 1989). As in epithermal deposits, one possible mechanism to increase the Au and metal content of a hydrothermal fluid, and result in rapid precipitation of these metals, is boiling, which decreases temperature, lowers the pH, and precipitates sulfides (Drummond and Ohmoto, 1985). If boiling occurs below surface sulfides will precipitate below the seafloor, unless the fluids are overpressured by self-sealing (Gibson *et al.*, 1999). Boiling, therefore, is restricted to VMS deposits that form in shallower water environments (<1000 m) where the hydrostatic pressure is less. For example at 400 °C the minimum pressure required to prevent boiling is approximately 150 bars or a 1500-m water depth (pure water) and at 300 °C the minimum pressure is approximately

**Fig. 12.34.** Pyrite chimney debris (S) cemented by fine-grained sphalerite (whitish oxidation) as illustrated in Fig. 12.33, from the Cretaceous Lahanos deposit, Turkey.

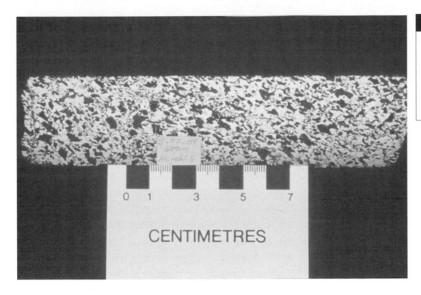

**Fig. 12.35.** Subseafloor sulfides, principally pyrrhotite and chalcopyrite, that cement and replace basalt lapilli comprising volcaniclastic units hosting the Archean, Potter deposit, Munro Township, Ontario, Canada.

90 bars or 900 m of water (Ridge, 1973). Direct evidence for boiling, such as that provided by fluid inclusion studies, is not forthcoming in most VMS deposits but indirect evidence such as the occurrence of advanced argillic alteration may, by analogy with high-sulfidation epithermal Au–Cu deposits, be a product of acid fluids derived from magmatic volatiles including $SO_2$ which can be present at any water depth, or through boiling and phase separation in shallower water environments (Sillitoe *et al.*, 1996; Hannington *et al.*, 1999) (Eq. 12.4).

## Volcanic environment characteristics

VMS deposits occur within submarine volcanic complexes or centers characterized by volcaniclastic rocks and lavas, of both mafic and felsic composition (Cas, 1992). Sedimentary rocks such as shales, graywackes, and carbonates are common and may dominate some VMS successions, as in the Iberian pyrite belt (De Carvalho *et al.*, 1999). The term volcaniclastic is used herein as proposed by Fisher (1966) to refer to primary pyroclastic and autoclastic rocks and their redeposited, syneruptive equivalents. VMS deposits

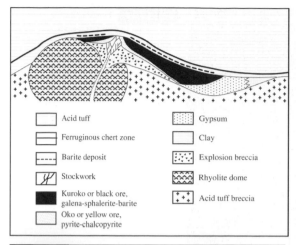

Acid tuff

Ferruginous chert zone

Barite deposit

Stockwork

Kuroko or black ore, galena-sphalerite-barite

Oko or yellow ore, pyrite-chalcopyrite

Gypsum

Clay

Explosion breccia

Rhyolite dome

Acid tuff breccia

**Fig. 12.36.** Cross-section through an idealized Kuroko VMS deposit, Japan showing the distribution of ore types as well as gypsum and barite. (Reprinted from *Ore Geology and Industrial Minerals: An Introduction*, Fig. 15.5, p. 208, Evans, A. M., © 1993, with permission of Blackwell Publishing.)

occur in both volcaniclastic and flow-dominated successions and they may occur in both deeper (>1000 m) and shallower (<1000 m) water, to emergent environments (Gibson *et al.*, 1999).

The characteristic "clustering" of VMS deposits reflects their preferential development within volcanic centers as illustrated for VMS deposits of the Archean Noranda Volcanic Complex in Fig. 12.42. Within each volcanic center the VMS deposits typically occupy a single stratigraphic interval which represents only a small fraction of the total stratigraphic thickness of the host volcanic complex. For example, 80% of the deposits in the Noranda Camp occur within the 3000-m-thick Mine Sequence and of these deposits 90% occur within a stratigraphic interval represented by <500 m of strata (Gibson, 1990) (Fig. 12.43).

Within a volcanic center the structural control on the location of VMS deposits is pronounced. At the deposit scale, VMS deposits are spatially associated with synvolcanic faults and often occur within volcanic vents (both mafic and felsic) that are localized along synvolcanic structures. It is common in some camps, such as the Noranda Camp, for deposits to be "stacked" over vertical distances of up to 700 m along synvolcanic structures. On a larger scale these synvolcanic structures are related to even larger structural features. For example, VMS deposits of the Archean Noranda and Sturgeon Lake Camps and of the Miocene Hokuroko District are localized along synvolcanic faults within interpreted cauldron subsidence structures (Hodgson and Lydon, 1977; Ohmoto and Takahashi, 1983, Gibson and Watkinson, 1990; Morton *et al.*, 1991). Large, multiphase, sill-like, synvolcanic subvolcanic plutons often intrude central volcanic complexes hosting VMS deposits. These subvolcanic intrusions are interpreted to represent the intrusive equivalent of the magma chamber that fed the volcanic complex, and their occurrence often coincides with the location of caldera subsidence structures such as at Noranda and Sturgeon Lake (Gibson and Watkinson, 1990; Morton *et al.*, 1991). Subvolcanic intrusions are interpreted to have, in part, provided the heat or metal required to generate and sustain VMS-forming convective hydrothermal systems (Galley *et al.*, 2000).

## Tectonic characteristics

The tectonic classification in Table 12.2 indicates that approximately 80% of the world's VMS deposits occur in arc-related successions and that the remainder occur in ophiolitic successions that generally are back arc rifts or rifted marginal basins (Franklin *et al.*, 1999). What Table 12.2 does not stress is that the formation of VMS deposits is intimately related in space and time to bimodal volcanism in extensional, rift environments within arc or back arc environments in oceanic or continental crust. Rifting is characterized by high heat flow related to the shallow emplacement of mantle-derived magmas within continental or oceanic crust, and structural permeability: key ingredients in the generation of long-lived, high-temperature hydrothermal systems responsible for the formation of VMS deposits.

## Petrochemical characteristics

Lesher and Groves (1986), Barrie *et al.* (1993), and Lentz (1998) have shown that VMS deposits regardless of their age or tectonic setting are associated with rhyolites of particular trace element compositions referred to by Lesher and Groves (1986) as FII and FIII rhyolites (Fig. 12.44).

**Fig. 12.37.** Thin-bedded to laminated tuff containing sulfide (white) and chert beds typical of mineralized exhalative sediments associated with VMS deposits in the Noranda Camp, Canada. Note the convolute bedding indicative of soft sediment deformation during slumping.

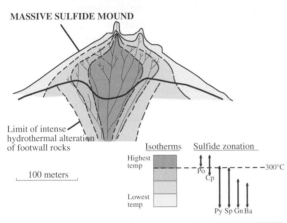

**Fig. 12.38.** Idealized metal and mineral zoning and zone-refining within a seafloor massive sulfide lens. (Reprinted with permission from *Ore Deposit Models*, Geoscience Canada Reprint Series No. 3, Geological Association of Canada, Fig. 7, p. 165, Lydon, J. W., 1988.)

FIII and FII rhyolites are characterized by chondrite normalized REE profiles that are relatively flat with respect to HREE but show slight LREE enrichment ((La/Yb)$n$ ratios of 2/6 and 1/4 respectively) whereas non-VMS-associated FI rhyolites have sloping REE profiles with pronounced LREE enrichment ((La/Yb)$n$ ratios of 6/34) (Fig. 12.43). The distinctive composition of FII and FIII rhyolites is interpreted to reflect their derivation through low pressure, partial melting of hydrated mafic (oceanic) in felsic (continental) crust within rift environments where crustal melting is in response to the emplacement, at high crustal levels (<15 km), of mantle-derived magmas (Lesher *et al.*, 1986; Barrie *et al.*, 1993; Lertz, 1998).

The composition of basaltic rocks associated with VMS deposits is more diverse and therefore less distinctive. For example, mafic flows and intrusions associated with Besshi deposits although dominantly mid-oceanic ridge basalt (MORB) in composition include some compositions that are more alkaline, whereas some andesitic flows in the hanging wall to the Kidd Creek VMS deposit have unusual $P_2O_5$ contents that allow their classification as icelandites (Wyman *et al.*, 1999). Similarly basalts with more alkaline compositions occur as the youngest basalts in volcanic successions hosting VMS deposits in both the Iberian Pyrite Belt and Ordovician Bathurst Camp (McCutcheon *et al.*, 1993; De Carvalho *et al.*, 1999). Distinctive Fe- and Ti-rich and andesites (FeTi basalts) are associated with some seafloor VMS deposits, such as those of the Galapagos spreading centre, where they have been interpreted to be a product of contamination of MORB-type magma by the assimilation of older hydrated basalt (Perfit *et al.*, 1999).

## Alteration characteristics

Alteration associated with VMS deposits is of two types: (1) discordant alteration zones (pipes) that mark the conduit for ascending hydrothermal fluids that formed the deposit; and

**Fig. 12.39.** Cross-section through a sulfide chimney collected in 1993 from the Juan de Fuca Ridge. The chimney is largely composed of sphalerite (wurtzite) (S) except for an inner rind of chalcopyrite (C) that lines the inner chimney wall and progressively replaces sphalerite. This dissolution of lower-temperature sphalerite by higher-temperature fluids (>300 °C) depositing chalcopyrite and the repreciptitation of sphalerite outward, toward the outer chimney walls, is analogous to the zone refining which occurs at the deposit scale as illustrated in Figs. 12.38 and 12.41. (Sample courtesy of Ian R. Jonasson, Geological Survey of Canada, Ottawa, Canada.)

(2) large, regionally extensive, stratabound semi-conformable alteration zones within the stratigraphic succession hosting the deposits. The mineralogical and compositional changes that characterize discordant and semiconformable alteration zones are dependent upon the composition and character of the footwall rock, the permeability of the footwall rocks (volcaniclastic or flow hosted), and, to some extent, the water depth (Gibson *et al.* 1999; Large *et al.*, 2001).

## Volcanic controls on VMS deposit morphology and alteration

### Flow versus volcaniclastic footwall successions

Regardless of tectonic environment, age, or host rock composition VMS deposits can be subdivided into two end members based on whether the footwall succession is dominated by flows – flow hosted – or volcaniclastic rocks – volcaniclastic hosted (Gibson *et al.*, 1999). This subdivision is

an outgrowth of the two-fold subdivision of VMS deposits proposed by Morton and Franklin (1987) and Gibson and Severin (1991) and refers to strata for at least 1 km below the VMS deposit. Gibson *et al.* (1999) have suggested that the subdivision into flow and volcaniclastic environments may correspond, in general, with deeper water versus shallower water volcanic environments. Shallower water environments, that is above 1000 m (and more likely <500 m), are more likely to be dominated by pyroclastic, syneruptive, redeposited pyroclastic and epiclastic deposits, although lava flows and domes also occur. Deeper water environments (>1000 m), because of higher hydrostatic pressure, are dominated by lava flows, although lesser volumes of shallow water pyroclastic and syneruptive pyroclastic deposits may be transported and deposited in this deeper water environment (Cas, 1992). Sedimentary rocks may be a minor or a major component to each class and sediments may be the dominant hanging wall lithology or immediately underlie the deposit. Both flow and volcaniclastic environments are further subdivided into mafic and felsic subtypes based on the dominant composition of the footwall succession. Differences in deposit morphology, alteration morphology and mineralogy of these two deposits types outlined in Table 12.3, and described below reflect, in part, differences in the permeability and glass-rich versus crystalline character of the immediate, and larger footwall strata (Gibson *et al.*, 1999; Large *et al.*, 2001).

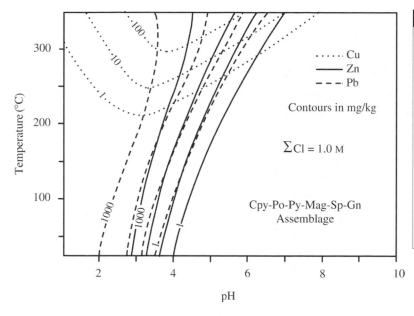

**Fig. 12.40.** The solubility of Cu, Zn, and Pb as chloride complexes with respect to temperature and pH: calculated total metal concentrations in a 1-M NaCl solution in equilibrium with a chalcopyrite–pyrite–pyrrhotite–magnetite–sphalerite–galena assemblage. Note the pronounced temperature control on the solubility of Cu as opposed to Zn. (Reprinted with permission from *Ore Deposit Models*, Geoscience Canada Reprint Series No. 3, Geological Association of Canada, Fig. 6, p. 164, Lydon, J. W., 1988.)

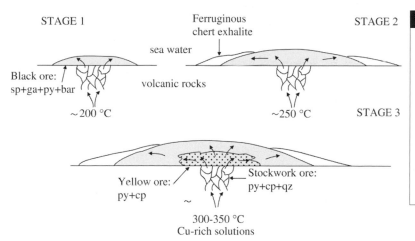

**Fig. 12.41.** Sequence in the development of a massive sulfide lens and zone-refining during thermal evolution of the sulfide deposit (200–350 °C) whereby lower-temperature pyrite and sphalerite are replaced by higher-temperature chalcopyrite and pyrrhotite (pyrite). (Reprinted from *Ore Geology and Industrial Minerals: An Introduction*, Fig. 4.13, p. 74, Evans, A. M., © 1993, with permission of Blackwell Publishing.)

## Volcanic controls on deposit morphology

The morphology of massive sulfide deposits and the mechanisms responsible for their growth on the seafloor differ between deposits in flow and volcaniclastic successions. These differences are primarily attributed to differences in the permeability of the immediate footwall strata. In flow-dominated successions the sulfide lens grows on the seafloor by processes of chimney growth, collapse, replacement of chimney debris, and finally by renewed chimney growth (Figs. 12.33, 12.34, and 12.39) (Lydon, 1984; Hannington et al., 1995). In volcaniclastic rocks, sulfide lenses on the seafloor grow in an analogous manner, but there is abundant evidence derived from studies

of ancient deposits to indicate that the sulfide lenses also grow by subseafloor sulfide precipitation and replacement until the system is essentially self-sealed or was sealed, "capped" by overlying flows (Fig. 12.45) (Gibson and Kerr, 1990, 1993; Hannington et al., 1999; Doyle, 2001; Large et al., 2001).

The Cu-rich, or Cu- and Zn-rich polymetallic core to seafloor and subseafloor massive sulfide deposits is attributed to replacement during zone refining as first described by Eldridge et al. (1983) for Kuroko deposits. The zone refining processes in volcaniclastic-hosted, subseafloor VMS deposits, however, may be restricted to one, or several, localized conduits (stringer zones) below

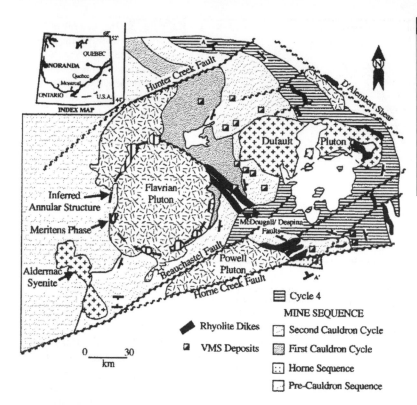

**Fig. 12.42.** Geologic plan of the Archean Central Noranda Volcanic Complex showing the location of a central area of subsidence, the Noranda Cauldron, between the Hunter Creek and Horne Faults, the location of VMS deposits, and subvolcanic Flavrian and Powel plutons which occur in the core of the subsidence structure and volcanic edifice. Modified from Gibson (1990).

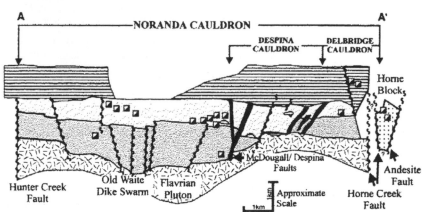

**Fig. 12.43.** Reconstructed geologic cross-section through the Noranda Cauldron showing the stratigraphic control on most of the VMS deposits which occur in both rhyolitic and basaltic flows at the base of the second cauldron subsidence cycle. Location of section and legend as in Fig. 12.42. Modified from Gibson (1990).

the larger sulfide (pyritic) lens or lenses. The result may be a large, primarily pyritic deposit containing localized areas of higher-grade Zn- and Cu-rich ore as at, for example, the Horne,

Roseberry, and Neves Corvo deposits (Gibson and Kerr, 1990; Allen, 1994; Gibson, 1994).

Although subseafloor replacement may favor the formation of larger deposits, the size of the sulfide lens, be it on or below the seafloor, is a function of the longevity of the larger hydrothermal system, and continued long-lived and uninterrupted discharge at one site. The latter is in part dependent on structural reactivation of the hydrothermal vent site as self-sealing due to sulfide precipitation and alteration, or burial by

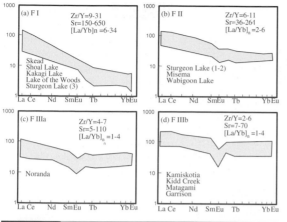

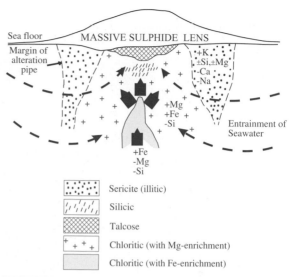

**Fig. 12.44.** Rare earth enrichment (REE) patterns (chondrite normalized) for FI, FII, and FIII rhyolites, modified from Lesher and Groves (1986). FII and FIII rhyolites are the dominant rhyolitic host to Phanerozoic and Archean VMS deposits respectively, whereas FI rhyolites are not known to host these deposits. (Reprinted with permission from *Giant Ore Deposits*, Society of Economic Geologists Special Publication No. 2, p. 322, Fig. 1, Gibson, H. L. and Kerr, D. J., 1993.)

**Fig. 12.46.** Distribution of alteration assemblages and chemical gains and losses associated with VMS deposits hosted by flows as in Fig. 12.28. (Reprinted with permission from *Giant Ore Deposits*, Society of Economic Geologists Special Publication No. 2, p. 327, Fig. 4, Gibson, H. L. and Kerr, D. J., 1993.)

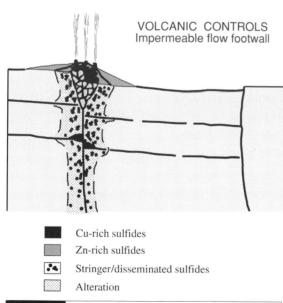

**Fig. 12.45.** Areally restricted but vertically extensive morphology typical of a discordant footwall alteration zone within competent, relatively impermeable flows.

younger deposits or flows will eventually seal the hydrothermal vent site and arrest discharge.

## Volcanic controls on proximal, discordant alteration

Differences in the morphology and mineralogy of proximal, discordant footwall alteration zones, for the most part, are a product of permeability and compositional differences in the footwall strata in flow and volcaniclastic hosted deposits, and perhaps water depth. Discordant, footwall alteration zones below flow-hosted VMS deposits are generally restricted to structures controlling hydrothermal discharge (Fig. 12.45). This results in an alteration zone that is vertically extensive, but laterally restricted, and mineralized footwall rocks which, in many examples, are characterized by both Mg- and Fe-chlorite and sericite-quartz (with or without carbonate) assemblages as illustrated in Figs. 12.30, 12.32, 12.46, and 12.47 and listed in Table 12.3. In this environment ascending hydrothermal fluids are focused to synvolcanic structures and the rocks undergo intense pervasive alteration that because of the high water-to-rock ratio (>500 : 1) results in an

**Fig. 12.47.** Sericitized rhyolite containing amygdules filled by quartz, pyrite, and minor sphalerite (S) overprinted by a pyrrhotite–chalcopyrite stringer mantled by a chloritic envelope (C). The rhyolite occurs in the immediate hanging wall to the Corbet deposit and its alteration attests to the continuation of hydrothermal activity after the Corbet deposit had formed and was buried by thick rhyolite flows which may have capped and eventually arrested hydrothermal discharge.

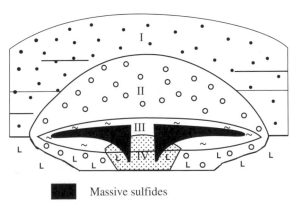

Massive sulfides

Sediments and volcanics

Rhyolite volcanics

Zone I

Zone II

Zone III

Zone IV

**Fig. 12.48.** Mineralogically defined alteration zones within the hanging wall and to a lesser extent the footwall of an idealized Kuroko VMS deposit (Zone I, Zeolite; Zone II, montmorillonite; Zone III, sericite and chlorite; Zone IV, montmorillonite, chloritoid, and sericite). Alteration of hanging wall strata is a product of both ascending hydrothermal fluids and downdrawn seawater. Modified from Shirozo (1974).

essentially monomineralic chlorite core to the alteration pipe. The addition of K and Mg to form sericite and Mg-chlorite at the pipe margins may have been derived from seawater that was downdrawn along the discharge conduits (Fig. 12.46). Downdrawn seawater may also be responsible for the formation of clay minerals, plus sericite and chlorite, within the hanging wall to the Miocene Kuroko deposits of Japan, as illustrated in Fig. 12.48.

VMS deposits in volcaniclastic-dominated successions, on the other hand, are characterized by broad, diffuse zones of hydrothermally altered and mineralized rock that are many times larger, in volume and areal extent, than the massive sulfide deposit itself (Fig. 12.49). Alteration assemblages that characterize these broad alteration zones are typically sericite (chlorite)-quartz or carbonate minerals (Table 12.3) that are commonly overprinted by more localized, discordant zones of Fe-chlorite, argillic alteration, and silicification. Unlike "impermeable" flows the more permeable volcaniclastic rocks do not restrict ascending hydrothermal fluids to synvolcanic structures, and the ascending fluid can move laterally away from the main conduits. Lateral fluid movement is largely a function of the permeability of the footwall rocks and in most cases this permeability increases upward as the units are less compacted, resulting in an upward broadening of the alteration zone (Fig. 12.49). The development of alteration assemblages and precipitation of minor sulfides within broad lateral alteration

## VOLCANIC CONTROLS
### Permeable volcaniclastic footwall

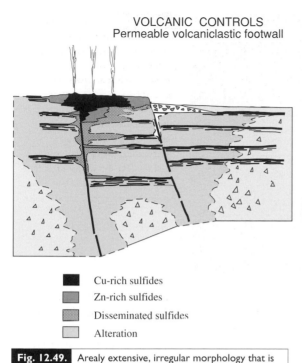

- ■ Cu-rich sulfides
- ▨ Zn-rich sulfides
- ▨ Disseminated sulfides
- ☐ Alteration

**Fig. 12.49.** Arealy extensive, irregular morphology that is typical of discordant footwall alteration within a relatively permeable volcaniclastic succession.

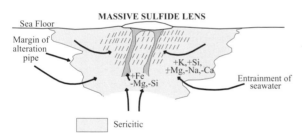

**MASSIVE SULFIDE LENS**

- ☐ Sericitic
- ▨ Silicic
- ▨ Chloritic (with Fe-enrichment)

**Fig. 12.50.** Distribution of alteration assemblages and chemical gains and losses associated with VMS deposits hosted by volcaniclastic rocks as in Fig. 12.29. (Reprinted with permission from *Giant Ore Deposits*, Society of Economic Geologists Special Publication No. 2, p. 328, Fig. 5, Gibson, H. L. and Kerr, D. J., 1993.)

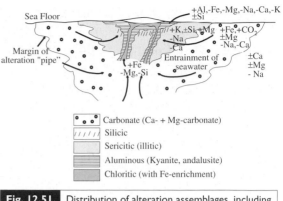

- ⬭ Carbonate (Ca- + Mg-carbonate)
- ▨ Silicic
- ☐ Sericitic (illitic)
- ▤ Aluminous (Kyanite, andalusite)
- ▨ Chloritic (with Fe-enrichment)

**Fig. 12.51.** Distribution of alteration assemblages, including advanced argillic alteration, associated with some volcaniclastic hosted VMS deposits. (Reprinted with permission from *Giant Ore Deposits*, Society of Economic Geologists Special Publication No. 2, p. 329, Fig. 6, Gibson, H. L. and Kerr, D. J., 1993.)

zones eventually decreases their permeability due to hydrothermal lithification or self-sealing and ascending fluid flow is progressively more confined until it is focused along the main structural conduits resulting in more restricted, clearly discordant zones of alteration that are typically Fe-chlorite or argillic/sericitic and superimposed on broader zones of sericite–quartz and/or carbon-ate alteration (Figs. 12.50 and 12.51). The K and Mg added to the rock during sericitization and local chloritization is interpreted to have been derived from seawater which is drawn down into the permeable footwall rocks. The development of argillic alteration, if a product of boiling or magmatic volatiles, may be indicative of shallower water environments.

### Volcanic controls on regional semiconformable alteration zones

Semiconformable alteration occurs as large, regionally extensive areas, of stratabound hydrothermal alteration that may extend for tens to hundreds of kilometers along strike. Semiconformable alteration has been recognized in many VMS hosting successions and has been interpreted as a product of hydrothermal alteration within regional, subseafloor "geothermal systems."

Four types of semiconformable alteration zones have been recognized: spilitization, epidote–quartz alteration, silicification, and carbonate alteration (Table 12.3).

Spilitization was first recognized as a type of broad, semiconformable alteration in ophiolitic successions of East Liguria, Italy, and Cyprus (Spooner and Fyfe, 1973; Heaton and Sheppard, 1977; Spooner, 1977). Spilitic rocks are now recognized within most volcanic successions hosting VMS deposit (Gibson and Kerr, 1993). It is characterized by Na enrichment and Ca depletion and

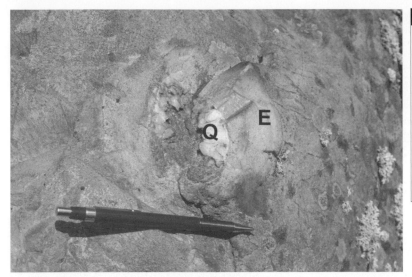

**Fig. 12.52.** Epidote–quartz alteration (patch) (E) developed around a larger amygdule, itself filled by coarse quartz and epidote (Q), within a basaltic andesite flow at Noranda, Quebec, Canada. Epidote–quartz alteration, either as isolated patches or surrounding amygdules, is mappable, and attributed to the reaction between basaltic rocks and evolved seawater at high temperature and low water-to-rock ratio.

**Fig. 12.53.** Silicification patch (S) (quartz–albite assemblage) mantling amygdules within basaltic andesite flows at Noranda, Quebec. Silicification, like epidote–quartz alteration, is a patchy, mappable alteration attributed to the reaction of basaltic or andesitic flows with evolved, silica-saturated seawater.

is typified by a pervasive "greenschist hydrothermal metamorphic assemblage" of albite, chlorite, and quartz in mafic volcanics and albite, sericite, and quartz in felsic volcanics (keratophyres).

Epidote–quartz alteration occurs in subaqueous mafic volcanic rocks. It is characterized by distinct, centimeter- to meter-scale alteration patches composed of epidote and quartz (minor actinolite and carbonate), and enrichment in Ca and depletion in major and trace elements (Fe, Mg, Na, Mn, Cu, and Zn) (Gibson, 1990; Harper, 1999) (Fig. 12.52). Epidote–quartz alteration may represent a high temperature assemblage as alteration of similar mineralogy and morphology is

forming in the Salton Sea hydrothermal system (McKibben *et al.*, 1988), and it also occurs in young fossil hydrothermal systems in Iceland, the modern ocean floor, and in ophiolitic successions (Spooner, 1977; Kristmannsdottir, 1982; Mehegan *et al.*, 1982; Alt, 1999; Harper, 1999).

Silicification is a patchy to pervasive alteration characterized by the mineral assemblage albite–quartz and by the addition of Si and Na and depletion in Fe, Mg, Mn, Zn, and Cu (Fig. 12.53). Silicification, which can extend for kilometers and is often restricted to one or several stratigraphic units, is recognized in flow-dominated, Archean volcanic successions hosting massive

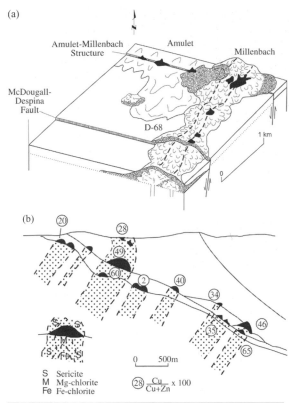

**Fig. 12.54.** (a) Volcanic reconstruction of the Millenbach rhyolite dome–ridge complex showing the pronounced structural control on the location of VMS deposits (black). The Amulet Lower A deposit is localized along a fissure for basaltic–andesite volcanism whereas the Millenbach VMS deposits occur along the top of the rhyolitic dome–ridge and directly above and along the feeding fissure. (Reprinted with permission of *Reviews in Economic Geology*, vol. 8, p. 28, Fig. 15, Gibson, H. L., Morton, R. M., and Hudak, G. J., 1999.) (b) Cross-section through the Amulet and Millenbach VMS deposits showing the stratigraphic stacking of deposits within basaltic and rhyolitic volcanic vent areas and metal zoning within and between individual VMS deposits as illustrated by the ratio (Cu : Cu+Zn) × 100. The progressively more Cu-rich character of VMS deposits at depth within the Amulet system from 20 at the Upper A, 49 at Lower A, and 60 at the #11 shaft deposits is interpreted to reflect subseafloor zone refining due to the continued passage of ascending hydrothermal fluid that strips Zn from the lowermost deposits (replaced by chalcopyrite) and forms a more Zn-rich deposit at the seafloor (Upper A). (Reprinted with permission from *Economic Geology*, vol. 88:6, p. 1424, Fig. 4, Kerr, D. J. and Gibson, H. L., 1993.)

sulfide deposits at Noranda, Matagami, and Snow Lake (MacGechan, 1978; Gibson *et al.*, 1983; Skirrow and Franklin, 1994).

Semiconformable carbonate alteration zones characterized by pervasive groundmass replacement by a carbonate (calcite, dolomite) and/or sericite (with or without quartz, chlorite) assemblage are characterized by Na depletion and enrichment in K, Si, Ca, or Mg and $CO_2$. These semiconformable zones grade into disconformable zones of Fe-carbonate and sericite that occur below the VMS deposits (Franklin, 1986; Gibson and Kerr, 1993).

In flow-dominated successions, regional semiconformable alteration is focused to areas of high permeability such as flow contacts, flow breccias, amygdules, and synvolcanic faults resulting in recognizable differences in the degree of alteration that are commonly mappable (Gibson, 1990). Epidote–quartz, silicification, and carbonate alteration are more common within flow-dominated successions; the latter alteration is particularly well developed in successions thought to be representative of shallower-water environments.

Semiconformable sericite and carbonate alteration are more typical of volcaniclastic successions where they are more pervasive, uniform, and widespread. Variations in alteration intensity are usually not mappable, except by using lithogeochemistry or X-ray diffraction techniques.

The differences in morphology of semiconformable alteration zones in flow- and volcaniclastic-dominated successions, notwithstanding differences imposed by host rock composition, are interpreted to be, in part, related to permeability differences. This permeability contrast results in differences in water : rock ratio, as well as fluid temperature and compositional changes brought about through mixing with seawater.

## Examples of flow-hosted VMS deposits in flow-dominated successions

VMS deposits of the Noranda area occur within the 7-km-thick, Archean (2700 Ma) Noranda Volcanic Complex, the youngest central volcanic complex within the Archean Blake River Group of the Abitibi Subprovince of the Canadian Shield.

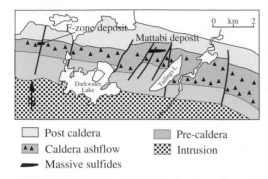

Post caldera      Pre-caldera
Caldera ashflow      Intrusion
Massive sulfides

**Fig. 12.55.** Geologic map of the Sturgeon Lake caldera showing the location of VMS deposits, subaqueous ashflow tuffs, underlying Beidleman Bay subvolcanic intrusion and significant caldera-forming faults. (Reprinted with permission from *Economic Geology*, vol. 86:5, p. 1004, Fig. 2, Morton, R. L., Walker, J. S., Hudak, G. J., and Franklin, J. M., 1991.)

The Noranda Cauldron, a 15 × 20-km subsidence area within the center of the volcanic complex, is host to 17 VMS deposits that occur within a 3000-m-thick, bimodal volcanic succession (Mine Sequence) dominated by rhyolite and basaltic andesite flows and < 1% pyroclastic rocks (Fig. 12.42). All but two of the VMS deposits within the cauldron occur at one stratigraphic interval that marks the onset of a cauldron subsidence cycle (Fig. 12.43) (Gibson and Watkinson, 1990).

VMS deposits in the Millenbach-Amulet area typify flow-hosted VMS deposits and their volcanic environment. At the Millenbach Mine, 15 massive sulfide deposits, ranging in size from approximately 40 000 to 1 000 000 t, occur along a 2-km-long, 240-m-high, blocky rhyolite dome–ridge that directly overlies its feeding fissure defined by rhyolite dikes (Fig. 12.54a). The Amulet VMS deposits are localized along a basalt fissure (Fig. 12.54b). The deposits localized within the basaltic fissure are "stacked" over a 700-m stratigraphic interval indicating that hydrothermal activity was periodically interrupted by volcanism (Fig. 12.54b). Analogous to modern seafloor deposits, these massive sulfide deposits formed on the seafloor as hard, encrusting, sinter-like replacements of sulfide chimney debris. The massive sulfide deposits are underlain by stringer sulfide zones enveloped within a discordant pipe-like volume of chlorite and sericite altered rock. At the Amulet deposits the discordant

chlorite and sericite alteration "pipe" extends into the hanging wall of the Lower A deposit, "connecting" it to the overlying Upper A deposit (Fig. 12.54b). This stratigraphic stacking of VMS deposits attests to the longevity of hydrothermal discharge, discharge that was interrupted and finally arrested by volcanic activity.

## Examples of volcaniclastic-hosted VMS deposits

VMS deposits of the Archean Sturgeon Lake Camp, northwestern Ontario, Canada are well-documented examples of volcaniclastic-hosted VMS deposits. The Mattabi, F-Group, and Sturgeon Lake deposits occur within a shallower-water (<1000 m) volcanic subsidence structure, the Sturgeon Lake caldera (Morton *et al.*, 1991) (Fig. 12.55). The Sturgeon Lake caldera has a minimum strike length of 30 km and, because of its subvertical dip, exposes through surface outcrops and >600 000 m of drill core, some 4500 m of caldera fill material which includes five separate, subaqueously deposited, crystal-rich, pumiceous pyroclastic flow units. Each of the pyroclastic flow units represents a caldera-forming eruption as indicated by the presence of several smaller, nested subsidence structures within the larger inferred caldera.

The Mattabi pyroclastic flow, with a thickness in excess of 800 m and a strike length of at least 30 km, is the third and most voluminous eruptive event associated with the caldera complex. It hosts the 12-Mt Mattabi massive sulfide deposit which is interpreted to have formed on and below the seafloor, the latter through the processes of pore-space filling and replacement. The Mattabi deposit is underlain, and enveloped by, a discordant alteration zone which contains andalusite, kyanite, and pyrophyllite that reflect a metamorphosed advanced argillic alteration which is locally overprinted by Fe–chlorite alteration, and surrounded by a broader more diffuse area of Fe–carbonate alteration (Figs. 12.40 and 12.56). Analogous to HS precious-metal deposits, argillic alteration of the volcaniclastic rocks may have enhanced their primary permeability promoting subseafloor sulfide precipitation, replacement, and growth of the Mattabi sulfide deposit.

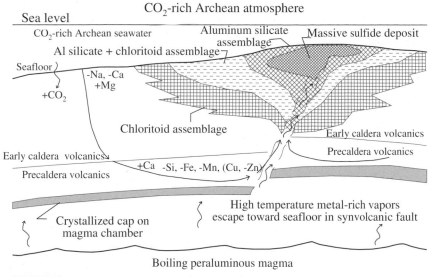

**Fig. 12.56.** Schematic model illustrating the distribution of alteration assemblages surrounding the Mattabi VMS deposit, Sturgeon Lake Camp. (Reprinted with permission from *Reviews in Economic Geology*, vol. 8, p. 46, Fig. 29; Gibson, H. L., Morton, R. M., and Hudak, G. J., 1999.)

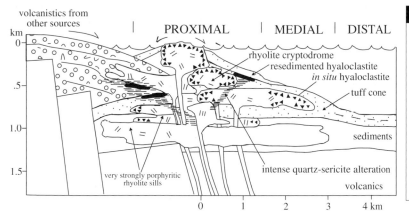

**Fig. 12.57.** Facies model for a subaqueous rhyolite–porphyry cryptodome–tuff cone volcano characterized by post-tuff cone cryptodomes and illustrating the location of VMS deposits and footwall alteration zones, Paleoproterozoic Skellefte district, Sweden. (Reprinted with permission from *Economic Geology*, vol. 91:6, p. 1047, Fig. 14b, Allen, R. L., Weihed, P., and Svenson, S.-A., 1996.)

Another example of a shallower-water VMS deposit, this time associated with monogenic rhyolite volcanoes, is that described by Allen *et al.* (1996) for VMS deposits of the Proterozoic Skellefte district of northern Sweden. Subaqueous rhyolite–porphyry, cryptodome–tuff cone volcanoes that range from 2–10 km in diameter and 250 to 1200 m in thickness host most of the VMS deposits (Fig. 12.57). Most of the rhyolite–porphyry cryptodome–tuff cones were emplaced in shallow water, below stormwave base; a few may have been emergent. The VMS deposits occur within the proximal vent facies, close to the top of these volcanoes, or in overlying units, primarily as subseafloor replacement and cementation of the tuff and pumice units. The sulfide deposits are underlain by irregular footwall alteration zones dominated by sericitic alteration.

## Genetic models

The natural laboratory afforded by actively forming seafloor sulfide deposits and the vast amount of research conducted on ancient deposits and their actively forming analogues has resulted in

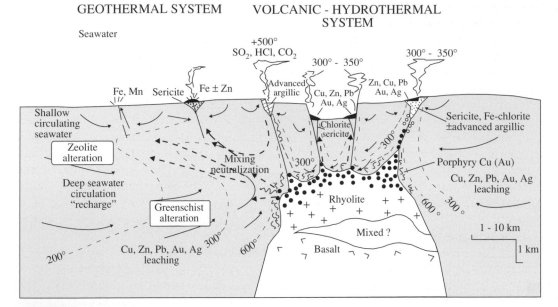

**Fig. 12.58.** Diagrammatic reconstruction of a genetic model for VMS deposits (data from Gibson, 1990; Gibson and Kerr, 1993; Hannington *et al.*, 1999). In this model, VMS deposits form within a volcanic and hydrothermal center defined by proximal volcanic facies, seafloor subsidence structures, an underlying subvolcanic intrusion that represents the intrusive equivalent of the magma chamber that constructed the volcanic edifice, and large-scale semiconformable alteration zones that are a product of evolved seawater interaction with volcanic strata in a high heat flow environment. The VMS deposits are localized along synvolcanic faults that act as the main fluid conduits, or discharge zones, for the underlying hydrothermal system. The metals may have been derived from leaching of volcanic strata by evolved seawater within high temperature semiconformable alteration zones, such as those represented by epidote–quartz alteration and silicification (Figs. 12.52 and 12.53) and/or be derived from magmatic fluids from a deeper source.

a comprehensive genetic model for VMS deposits. VMS deposits are arguably one of the best-understood ore deposit types. We know that VMS deposits are products of submarine hydrothermal fluid discharge localized along synvolcanic structures that are associated, both temporally and spatially, with volcanic vent areas. The deposits occur in a wide variety of rifted, volcanic and tectonic environments (Cas, 1992). The common association with extensional tectonics, rifting, bimodal volcanism, and high heat flow accounts for the location of many VMS deposits in volcano-tectonic depressions such as grabens and calderas and may also account for their association with subvolcanic intrusions and rhyolites of specific trace and REE composition (Hodgson and Lydon, 1977; Gibson and Watkinson, 1990; Morton *et al.*, 1991). VMS deposits occur in both shallower-water and deeper-water environments and in successions dominated by flows or volcaniclastic rocks and sediments.

VMS deposits grow by a combination of sulfide precipitation and open-space filling within the lens and pore-space filling and replacement at and below the seafloor. The characteristic Cu–Zn zoning of the massive sulfide lens and stringer zone is interpreted to be a product of a zone-refining process which in some cases continues after burial of the deposit as revealed by hanging wall alteration and stacked deposits in the Mine Sequence at Noranda (Lydon, 1984; Gibson and Kerr, 1990) (Fig. 12.54b).

What don't we know about the formation of VMS deposits? Obviously many aspects of their formation and location are unknown or uncertain, otherwise our success in discovering this deposit type would be 100%. The greatest uncertainty with respect to the origin of VMS deposits is the source of the metals, the role of seawater versus magmatic fluid in their formation, and environment of formation, particularly water depth. In this regard two end-member hydrodynamic models, the convection model and magmatic model, have been proposed to explain the origin of the hydrothermal fluids and the source of the metals that ultimately form VMS deposits.

In the convection model (Fig. 12.58), metals are interpreted to have been derived through leaching of the underlying volcanic (sedimentary) rocks by convecting, heated and chemically evolved seawater. Extensive areas of semiconformable alteration are interpreted to represent high-temperature reactions zones where evolved, high temperature (>300 °C) seawater leached metals from volcanic aquifer rock (Galley, 1993). Certainly it has been demonstrated that spilitization, silicification, and epidote–quartz alteration, essentially the "greenschist facies" alteration in Fig. 12.58, results in metal depletion and as such may represent high-temperature reaction zones where metals were leached (MacGeehan, 1978; Gibson et al., 1983; Skirrow and Franklin, 1994; Alt, 1999).

However, the connection between semiconformable alteration zones and VMS deposits is not unequivocal. For example, where these alteration types are well exposed, such as in the Noranda District, field relationships indicate that semiconformable spilitization, epidote–quartz, and silicification are cross-cut by discordant chlorite and sericite alteration zones that underlie the VMS deposits. This temporal relationship indicates that alteration genetically tied to the mineralization postdates footwall semiconformable alteration, and that the fluids responsible for the alteration, and VMS deposits, are not derived within known zones of semiconformable alteration. The occurrence of semiconformable alteration within successions that do not host VMS deposits also casts doubt on their genetic relationship to VMS deposits. However, this does not detract from the convection model as a source of the fluids and metal for VMS deposits. It does, however, indicate that we may not have yet found, or perhaps recognized, the high-temperature reaction zones from which the metals were derived.

In the magmatic model the metals are derived from an underlying magma chamber (Fig. 12.58).

In one scenario, convective cooling of a high-level (1–3 km depth) subseafloor subvolcanic intrusion promotes rapid fractionation and exsolution of an aqueous, saline magmatic hydrothermal fluid that preferentially partitioned chloride and metal ions into solution (Holland, 1972; Urabe, 1985; Yang and Scott, 1996). The model in Fig. 12.58 resembles that for LS and HS epithermal Au deposits illustrated in Fig. 12.24, and similar processes may apply. Base metals and Au, carried as chloride complexes within the exsolved, magmatic hydrothermal fluid are transferred to a shallow (1–3 km) subseafloor, high-temperature (350–400 °C) seawater-dominated convective system where they mix. Because of the high temperature and chlorine-rich composition of the seawater the metals are ultimately transported to the seafloor by a seawater-dominated hydrothermal fluid and deposited during cooling, pH increase, and oxidation as a result of near-surface mixing with cold seawater. Boiling, in shallow-water environments, may also have played a role in sulfide precipitation and perhaps the development of argillic alteration zones. In this scenario the magmatic component is overwhelmed by the sheer volume of evolved seawater, and evidence for magmatic fluids, such as high-salinity fluid inclusions, may be difficult to recognize in this seawater-dominated system. However, unusually high metal contents of Hg, As, Sn, Bi, Se, and In in some VMS deposits have been suggested as evidence for a magmatic fluid contribution to VMS deposition (Hannington et al., 1999).

A second scenario, illustrated in Fig. 12.58, involves a direct contribution of magmatic hydrothermal fluids and metals in the formation of VMS deposits, as in HS Au deposits. In this case magmatic hydrothermal fluid exsolved from a high-level subvolcanic intrusion (1–2 km depth) is tapped by a deep penetrating synvolcanic structure and rapidly ascends along this structure to surface. During ascent the magmatic hydrothermal fluid mixes with a minimal amount of evolved, high-temperature convecting seawater and ultimately forms a VMS deposit on the seafloor. Base metals and Au precipitate via the same processes envisaged for the first scenario. However, as in HS Au deposits, separation of the magmatic hydrothermal fluid into a dense, saline, and metal-rich fluid and buoyant, low-density acid vapor phase, and rapid ascent of the acid vapor, may result in the development of advanced argillic alteration within broader zones of sericite-quartz and carbonate alteration produced by ascending evolved seawater and down-drawn, cooler seawater. Regardless, the volume of seawater moving through this system before and after sulfide deposition makes recognition of an original magmatic hydrothermal component difficult. The periodic tapping and expulsion of a magmatic hydrothermal fluid by deep-seated synvolcanic structures ties the formation of VMS deposits to a particular stage or stages in the magmatic, volcanic, and structural evolution of central volcanic complexes, perhaps caldera development, which may better explain the clustering of VMS deposits at one stratigraphic interval, the occurrence of apparently isolated VMS deposits, and the total absence of VMS deposits within some formations and, on a larger scale, the absence of massive sulfide deposits in most submarine successions.

From a broader perspective, the difference between epithermal precious metal and VMS deposits begins to fade. Sillitoe *et al.* (1996), Poulson and Hannington (1996), and Hannington *et al.* (1999) point to similarities between Au-rich VMS deposits and HS and LS epithermal Cu–Au deposits such as acid alteration, in some examples the separation of Au and base metals, argillic alteration, and subsurface or subseafloor replacement. Thus, there may be a continuum between deep and shallower water VMS deposits and subaerial LS and HS epithermal deposits. Indeed this continuum or spectrum may extend to intrusion-hosted porphyry Cu, Cu–Au, and Mo mineralization found within subaerial and subaqueous volcanic successions. The difference between all these deposits types may be a function of subaerial versus subaqueous volcanic environment and volcanism, magmatic versus seawater versus meteoric contributions, and the temperature, composition, and acidity of the ore fluids (Sillitoe, 1973; Hedenquist *et al.*, 1998; Gibson *et al.*, 1999; Hannington *et al.*, 1999; Large *et al.*, 2001).

# Final remarks

An understanding of the volcanology of komatiitic Ni–Cu massive sulfide deposits, epithermal precious metal, and VMS deposits is essential to understanding the processes responsible for their formation, and in the formulation of their genetic models. Volcanic reconstruction through facies analysis is a powerful tool in determining the paleogeographic and tectonic environment of older volcanic successions. In exploration, volcanic reconstruction is a practical method for selecting volcanic successions, units, or facies that represent environments favorable for hosting these deposits types.

# Acknowledgments

I am indebted to many friends, colleagues, and students who have contributed to my understanding of the deposit types discussed. Lecture notes for senior undergraduate and graduate courses in ore deposits geology form the basis of this chapter. As in most review articles this chapter contains summaries of current and past research conducted by numerous individuals, and in this regard I would like to recognize the excellent work of R. A. Allen, J. M. Franklin, M. D. Hannington, J. Hedenquist, R. R. Large, C. M. Lesher, A. J. Naldrett, S. D. Scott, and R. H. Sillitoe which, in some cases, has been simplified and paraphrased. Reviews by D. H. Watkinson, R. E. S. Whitehead, R. R. Keays and B. Lafrance improved the content, but the author is solely responsible for any errors or omissions. Lorraine Dupuis kindly drafted many of the diagrams, often on short notice, and her patience and talent is appreciated.

# References

Allen, R. L. 1994. Synvolcanic subseafloor replacement model for Rosebery and other massive sulfide ores. *Geological Society of Australia Abstracts*, **39**, 107–108.

Allen, R. L., Weihed, P., and Svenson, S.-A. 1996. Setting of Zn–Cu–Au–Ag massive sulfide deposits in the evolution and facies architecture of a 1.9 Ga marine volcanic arc, Skellefte District, Sweden. *Economic Geology*, **91**, 1022–1053.

Alt, J. C. 1999. Hydrothermal alteration and mineralization of oceanic crust: mineralogy, geochemistry and processes. *Reviews in Economic Geology*, **8**, 133–155.

Arndt, N. T. 1976. Melting relations of ultramafic lava (komatiites) at 1 atm and high pressure. *Carnegie Institute of Washington Yearbook*, **75**, 555–562.

Arribas, A., Hedenquist, J. W., Itaya, T., *et al.* 1995. Contemporaneous formation of adjacent porphyry and epithermal Cu–Au deposits over 300 ka in northern Luzon, Philippines. *Geology*, **23**, 337–340.

Ballhaus, C., Ryan, C. G., Mernagh, T. P., *et al.* 1994. The partitioning of Fe, Ni, Cu, Pt, and Au between sulfide, metal and fluid phases. *Geochimica et Cosmochimica Acta*, **58**, 811–826.

Barrie, C. T. and Hannington, M. D. 1999. Introduction: classification of VMS deposits based on host rock compositions. *Reviews in Economic Geology*, **8**, 1–11.

Barrie, C. T., Ludden, J. N., and Green, T. H. 1993. Geochemistry of volcanic rocks associated with Cu–Zn and Ni–Cu deposits in the Abitibi Subprovince. *Economic Geology*, **88**, 1341–1358.

Berger, B. R. and Eimon, P. I. 1983. Conceptual models of epithermal precious metal deposits. In W. C. Shanks III (ed.) *Unconventional Mineral Deposits*. New York, Society of Mining Engineers of AIME, pp. 191–205.

Berger, B. R. and Henley, R. W. 1989. Advances in the understanding of epithermal gold-silver deposits, with special reference to the Western United States. *Economic Geology Monograph*, **6**, 405–423.

Boldy, J. 1977. (Un)certain exploration facts from figures. *Canadian Institute of Mining and Metallurgy Bulletin*, May 1977, 86–95.

Brugmann, G. E., Naldrett, A. J., and MacDonald, S. J. 1989. Platinum Group Element abundances in mafic and ultramafic intrusions of the Lac des Iles Complex, Ontario: implications for the genesis of Cu–Ni mineralization. *Economic Geology*, **84**, 1557–1573.

Buchanan, D. L., Nolan, J., Wilkinson, N., *et al.* 1983. An experimental investigation of sulfur solubility as a function of temperatures in synthetic silicate melts. *Geological Society of South Africa Special Publication*, **7**, 383–391.

Buchanan, L. J. 1981. Precious metal deposits associated with volcanic environments in the Southwest. *Arizona Geological Society Digest*, **14**, 237–262.

Campbell, I. H. and Naldrett, A. J. 1979. The influence of silicate : sulfide ratios on the geochemistry of magmatic sulfides. *Economic Geology*, **74**, 1503–1505.

Candella, P. A. 1989. Felsic magmas, volatiles and metallogenesis. *Reviews in Economic Geology*, **4**, 223–233.

Cas, R. A. F. 1992. Submarine volcanism: eruption styles, products, and relevance to understanding the host-rock successions to volcanic-hosted massive sulfide deposits. *Economic Geology*, **87**, 511–541.

Cas, R. A. F. and Beresford, S. W. 2001. Field characteristics and erosional processes associated with komatiitic lavas: implications for flow behaviour. *Canadian Mineralogist*, **39**, 505–524.

Chauvel, C., Dupre, B., and Jenner, G. A. 1985. The Sm–Nd age of Kambalda volcanics is 500 Ma too old! *Earth and Planetary Science Letters*, **74**, 314–324.

De Carvalho, D., Barriga, F., and Munha, J. 1999. The Iberian Pyrite belt of Portugal and Spain: examples of bimodal–siliciclastic systems. *Reviews in Economic Geology*, **8**, 375–408.

Doyle, M. G. 2001. Volcanic influences on hydrothermal and diagenetic alteration: evidence from Highway-Reward, Mount Windsor Subprovince, Australia. *Economic Geology*, **96**, 1133–1148.

Drummond, J. A. and Ohmoto, H. 1985. Chemical evolution and mineral deposition in boiling hydrothermal systems. *Economic Geology*, **80**, 126–147.

Eldridge, C. S., Birton, P. B., Jr., and Ohmoto, H. 1983. Mineral textures and their bearing on formation of the Kuroke oretodies. *Economic Geology, Monographs*, **5**, 241–281.

Evans, A. M. 1993. *Ore Geology and Industrial Minerals: An Introduction*. Oxford, UK, Blackwell Science.

Fisher, R. V. 1966. Rocks composed of volcanic fragments. *Earth Science Reviews*, **1**, 387–398.

Fournier, R. O. 1985. The behaviour of silica in hydrothermal fluids. *Reviews in Economic Geology*, **2**, 45–62.

Franklin, J. M. 1986. Volcanogenic massive sulfide deposits: an update. In C. J. Andrew *et al.* (eds.) *Geology and Genesis of Mineral Deposits in Ireland*.

Dublin, Irish Association for Economic Geology, pp. 49–69.

1990. Volcanic-associated massive sulfide deposits. *University of Western Australia Publication*, **24**, 211–243.

1996. Volcanic-associated massive sulfide base metals. *Geological Survey of Canada, Geology of Canada*, **8**, 158–183.

Franklin, J. M., Hannington, M. D., Barrei, C. T., *et al.* 1999. *Tectonic Classification of VMS deposits.* Cordilleran Round-Up, Mineral Deposit Research Unit Short Course Notes, University of British Columbia, Vancouver, BC.

Franklin, J. M., Kasarda, J., and Poulson, K. H. 1975. Petrology and chemistry of the Alteration zone at the Mattabi massive sulfide deposit. *Economic Geology*, **70**, 63–79.

Franklin, J. M., Lydon, J. W., and Sangster, D. F. 1981. Volcanic-associated massive sulfide deposits. *Economic Geology, 75th Anniversary Volume*, 485–627.

Galley, A. G. 1993. Characteristics of semi-conformable alteration zones associated with volcanogenic massive sulfide deposits. *Journal of Geochemical Exploration*, **48**, 175–199.

Galley, A. G., Van Breeman, O., and Franklin, J. M. 2000. The relationship between intrusion-hosted Cu–Mo immunization and the VMS deposits of the Archean Sturgeon Lake mining camp. *Economic Geology*, **105**, 1543–1550.

Garcia, J. S. 1991. Geology and mineralization characteristics of the Mankayan mineral district, Benguet, Philippines. In Y. Matsuhisa, M. Aoki, and J. W. Hedenquist (eds.) *High-Temperature Acid Fluids and Associated Alteration and Mineralization*, 3rd Symposium on Deep-Crust Fluids, Tsukuba, Japan, 1990, Report 277, pp. 21–30.

Gibson, H. L. 1990. The mine sequence of the Central Noranda volcanic complex: geology, alteration, massive sulfide deposits and volcanological reconstruction. Ph.D. thesis, Carleton University, Ottawa, Ontario.

1994. Volcanic-associated massive sulfide deposits: the large and the small. In *Proceedings of the Association of Professional Ecologists and Geophysicists of Quebec*, Val d'Or, Quebec, Abstracts.

Gibson, H. L. and Kerr, D. J. 1990. A comparison between the Horne Massive Sulfide deposit and smaller intracauldron deposits of the Mine Sequence, Noranda, Quebec. *Geological Society of America Annual Meeting, Dallas, TX*, vol. 22, Abstract 7.

1993. Giant volcanic-associated massive sulfide deposits: with emphasis on Arhean deposits. *Society of Economic Geologists Special Paper*, **2**, 319–348

Gibson, H. L. and Severin, P. W. S. 1991. Exploration strategies and criteria for volcanic-associated massive sulfide deposits. In *Program with Abstract, Annual Meeting of the Geological Association of Canada/Mineralogical Association of Canada*, Toronto, vol. 16.

Gibson, H. L. and Watkinson, D. H. 1990. Volcanogenic massive sulfide deposits of the Noranda Cauldron and Shield Volcano, Quebec. *Canadian Institute of Mining and Metallurgy Special Volume*, **43**, 119–132.

Gibson, H. L., Morton, R. L., and Hudak, G. 1999. Submarine volcanic processes, deposits and environments favorable for the location of volcanic-associated massive sulfide deposits. *Reviews in Economic Geology*, **8**, 13–51.

Gibson, H. L., Watkinson, D. H., and Comba, C. D. A. 1983. Silicification: hydrothermal alteration in an Archean geothermal system within the Amulet Rhyolite Formation, Noranda, Quebec. *Economic Geology*, **78**, 954–971.

Giggenbach, W. F. 1992a. Magma degassing and mineral deposition in hydrothermal systems along convergent plate boundaries. *Economic Geology*, **87**, 1927–1944.

1992b. Isotopic shifts in waters from geothermal and volcanic systems along convergent plate boundaries and their origin. *Earth and Planetary Science Letters*, **113**, 495–510.

Gresham, J. J. 1986. Depositional environment of volcanic peridotite-associated nickel–copper sulfide deposits with special reference to the Kambalda dome. *Society for Geology Applied to Mineral Deposits Special Publications*, **4**, 63–90.

Gresham, J. J. and Loftus-Hills, G. D. 1981. The geology of the Kambalda Nickel Field Western Australia. *Economic Geology*, **76**, 1373–1416.

Hannington, M. D. and Scott, S. D. 1989. Gold mineralization in volcanogenic massive sulfides: implications of data from active hydrothermal vents on the modern seafloor. *Economic Geology Monograph*, **6**, 491–507.

Hannington, M. D., Jonasson, I. R., Herzig, P. M., *et al.* 1995. Physical and chemical processes of seafloor mineralization at mid-ocean ridges. *Geophysical Monographs*, **91**, 115–157.

Hannington, M. D., Poulson, H., Thompson, J., *et al.* 1999. Volcanogenic gold in the massive sulfide environment. *Reviews in Economic Geology*, **8**, 325–356.

Harper, G. D. 1999. Structural styles of hydrothermal discharge in ophiolite/seafloor systems. *Reviews in Economic Geology*, **8**, 53–73.

Haughton, D. R., Roeder, P. L., and Skinner, B. J. 1974. Solubility of sulfur in mafic magmas. *Economic Geology*, **69**, 451–467.

Hayba, D. O., Bethke, P. M., Head, P., *et al.* 1985. Geologic, mineralogic and geothermal characteristics of volcanic-hosted epithermal precious metal deposits. *Reviews in Economic Geology*, **2**, 129–167.

Head, P., Foley, N. K., and Hayba, D. O. 1987. Comparative anatomy of volcanic-hosted epithermal deposits: acid-sulphate and adularia-sericite types. *Economic Geology*, **82**, 1–26.

Heaton, T. H. E. and Sheppard, S. M. F. 1977. Hydrogen and oxygen isotope evidence for seawater hydrothermal alteration and ore deposition, Troodos Complex, Cyprus. *Geological Society of London Special Paper*, **7**, 42–57.

Hedenquist, J. W. 1987. Mineralization associated with volcanic-related hydrothermal systems in the circum-Pacific basin. *Transactions of the 4th Circum-Pacific Energy and Mineral Resource Conference*, Singapore, 1986, pp. 513–524.

2000. Exploration for epithermal gold deposits. *Reviews in Economic Geology*, **13**, 221–244.

Hedenquist, J. W. and Henley, R. W. 1985. Hydrothermal eruptions in the Waiotapu geothermal system, New Zealand: their origin, associated breccias and relation to precious metal mineralization. *Economic Geology*, **80**, 1640–1668.

Hedenquist, J. W. and Lowenstern, J. B. 1994. Magmatic hydrothermal ore deposits. *Nature*, **370**, 519–527.

Hedenquist, J. W., Arribas, A., and Reynolds, T. J. 1998. Evolution of an intrusion-centered hydrothermal system: Far-Southeast-Lepanto porphyry and epithermal Cu–Au deposits, Philippines. *Economic Geology*, **93**, 373–404.

Hedenquist, J. W., Izawa, E., Arribas, A., *et al.* 1996. *Epithermal Gold Deposits: Styles, Characteristics, and Exploration*, Resource Geology Special Publication No. 1. Tokyo, Japan, Society of Resource Geology.

Heinrich, C. A., Gunther, D., Audetat, A., *et al.* 1999. Metal fractionation between magmatic brine and vapor, determined by microanalysis of fluid inclusions. *Geology*, **27**, 755–758.

Heinrich, C. A., Ryan, C. G., Mernagh, T. P., *et al.* 1992. Segregation of ore metals between magmatic brine and vapor: a fluid inclusion study using PIXE microanalysis. *Economic Geology*, **87**, 1566–1583.

Henley, R. W. 1991. Epithermal gold deposits in volcanic terranes. In R. P. Foster (ed.) *Gold Metallogeny and Exploration*, London, Blackie, pp. 133–164.

Hill, R. E. T., Barnes, S. J., Gole, M. J., *et al.* 1995. The volcanology of komatiites as deduced from field relationships in the Norseman-Wiluna greenstone belt, Western Australia. *Lithos*, **34**, 159–188.

Hill, R. E. T., Gole, M. J., and Barnes, S. J. 1989. Olivine adcumulates in the Norseman-Wiluna greenstone belt, Western Australia: implications for the volcanology of komatiites. In M. D. Prendergast and M. J. Jones (eds.) *Magmatic Sulfides: The Zimbabwe Volume*. London, Institute of Mining and Metallurgy, pp. 189–206.

Hodgson, C. J. and Lydon, J. W. 1977. Geologic setting of volcanogenic massive sulfide deposits and active hydrothermal systems: some implications for exploration. *Canadian Institute of Mining and Metallurgy Bulletin*, **70**, 95–106.

Holland, H. D. 1972. Granites, solutions and base metal deposits. *Economic Geology*, **67**, 281–301.

Huang, W.-L. and Williams, R. J. 1980. Melting relations of portions of the system Fe–S–Si–O to 32 kb with implication to the nature of the mantle core boundary. *Lunar and Planetary Science*, **11**, 468–488.

Huppert, H. E. and Sparks, R. S. J. 1985. Komatiites. I. Eruption and flow. *Journal of Petrology*, **26**, 694–725.

Huston, D. L. and Large, R. R. 1989. A chemical model for the concentration of gold in volcanogenic massive sulfide deposits. *Ore Geology Reviews*, **4**, 171–200.

Hutchinson, R. W. 1973. Volcanogenic sulfide deposits and their metallogenic significance. *Economic Geology*, **68**, 1223–1246.

1980. Massive base-metal sulfide deposits as guides to tectonic evolution. *Geological Association of Canada Special Paper*, **20**, 659–684.

1981. A synthesis and overview of Buchans Geology. *Geological Association of Canada Special Paper*, **12**, 325–350.

Jackson, S. L. and Fyon, J. A. 1992. The Western Abitibi Subprovince in Ontario. In *Geology of Ontario*, Special Volume no. 4, Part 1. Ottawa, Ontario, Ontario Geological Survey, pp. 405–484.

Kalogeropoulos, S. I. and Scott, S. D. 1983. Mineralogy and chemistry of tuffaceous exhalites (tetsuskiei) of the Fukazawa Mine, Hokuroko District, Japan. *Economic Geology Monograph*, **5**, 412–432.

Keays, R. R., Nickel, E. H., Groves, D. I., *et al.* 1982. Iridium and palladium as discriminants of volcanic-exhalative, hydrothermal and magmatic nickel sulfide mineralization. *Economic Geology*, **77**, 1535–1547.

Kerr, D. J. and Gibson, H. L. 1993. A comparison between the Horne massive sulfide deposit and intracauldron deposits of the Mine Sequence, Noranda, Quebec. *Economic Geology*, **88**, 1419–1442.

Kerr, D. J. and Mason, R. 1990. A re-appraisal of the geology and ore deposits of the Horne Mine Complex at Rouyn-Noranda, Quebec. *Canadian Institute of Mining and Metallurgy Special Issue*, **43**, 153–165.

Klau, W. and Large, D. E. 1980. Submarine exhalite Cu–Pb–Zn deposits: a discussion of their classification and metallogenesis. *Geologische Jahrbuch*, **40**, 13–58.

Knuckey, M. J., Comba, C. D. A., and Riverin, G. 1982. The Millenbach deposit, Noranda District, Quebec: an update on structure, metal zoning and wall rock alteration. *Geological Association of Canada Special Paper*, **25**, 297–318.

Kristmannsdottir, H. 1982. Alteration of IRDP drill hole compared with other drill holes in Iceland. *Journal of Geophysical Research*, **87**, 6525–6531.

Lahaye, T., Barnes, S.-J., Frick, L. R., *et al.* 2001. Re–Os isotopic study of komatiitic volcanism and magmatic sulfide formation in the southern Abitibi greenstone belt, Ontario, Canada. *Canadian Mineralogist*, **39**, 473–490.

Large, R. R., Huston, D. L., McGoldrick, P. J., *et al.* 1989. Gold distribution and genesis in Australian volcanogenic massive sulfide deposits and their significance for gold-transport models. *Economic Geology Monograph*, **6**, 520–536.

Large, R. R., McPhie, J., Gemmell, B., *et al.* 2001. The spectrum of ore deposit types, volcanic environments, alteration halos and related exploration vectors in submarine volcanic successions: some examples from Australia. *Economic Geology*, **96**, 913–938.

Lehrman, N. J. 1986. The McLaughlin mine, Napa and Yolo counties, California. *Nevada Bureau of Mines and Geology Report*, **41**, 85–89.

Lentz, D. R. 1998. Petrogenetic evolution of felsic volcanic sequences associated with Phanerozoic volcanic-hosted massive sulfide systems: the role of extensional geodynamics. *Ore Geology Reviews*, **12**, 289–327.

Lesher, C. M. 1989. Komatiite-associated nickel sulfide deposits. *Reviews in Economic Geology*, **4**, 45–101.

Lesher, C. M. and Arndt, N. T. 1995. REE and Nd geochemistry, petrogenesis, and volcanic evolution of contaminated komatiites at Kambalda, Western Australia. *Lithos*, **34**, 127–158.

Lesher, C. M. and Burnham, O. M. 2002. Multicomponent elemental and isotopic mixing in Ni–Cu–(PGE) ores at Kambalda, Western Australia. *Canadian Mineralogist*, **39**, 421–446.

Lesher, C. M. and Campbell, I. H. 1993. Geochemical and fluid-dynamic modeling of compositional varaitions in Archean komatiite-hosted sulfide ores in Western Australia. *Economic Geology*, **88**, 804–816.

Lesher, C. M. and Groves, D. I. 1986. Controls on the formation of komatiite-associated nickel-copper sulfide deposits. In G. H. Freidrich, A. D. Genkin, A. J. Naldrett, *et al.* (eds.) *Geology and Metallogeny of Copper Deposits*. Heidelberg, Germany, Springer-Verlag, pp. 43–62.

Lesher, C. M., Burnham, O. M., Keays, R. R., *et al.* 2001. Trace-element geochemistry and petrogenesis of barren and ore-associated komatiites. *Canadian Mineralogist*, **39**, 673–696.

Lhotka, P. G. and Nesbitt, B. E. 1989. Geology of unmineralized and gold-bearing iron formation, Contwoyoto Lake–Point Lake region, Northwest Territories, Canada. *Canadian Journal of Earth Sciences*, **26**, 46–64.

Lydon, J. W. 1984. Volcanogenic massive sulfide deposits. I. A descriptive model. *Geoscience Canada*, **11**, 195–202.

1988. Volcanogenic massive sulfide deposits. II. Genetic models. *Geoscience Canada*, **15**, 43–65.

MacGeehan, P. J. 1978. The geochemistry of altered volcanic rocks at Matagami, Quebec: a geothermal model for massive sulfide genesis. *Canadian Journal of Earth Sciences*, **15**, 551–570.

MacLean, W. H. 1969. Liquidous phase relations in the $FeS–FeO–Fe_3O_4–SiO_2$ systems and their application in geology. *Economic Geology*, **64**, 865–884.

McCutcheon, S. 1992. Base-metal deposits of the Bathurst–Newcastle district: characteristics and depositional models. *Exploration and Mining Geology*, **1**, 105–120.

McCutcheon, S. R., Langton, J. P., van Staal, C. R., *et al.* 1993. Stratigraphy, tectonic setting and massive sulphide deposits of the Bathurst Mining Camp, northern New Brunswick. In *3rd Annual Field Conference, Geological Society, Canadian Institute of Mining and Metallurgy*, Bathurst, NB, Canada, pp. 1–39.

McKibben, M. A., Andes, J. P., and Williams, A. E. 1988. Active ore formation at a brine interface in metamorphosed deltaic lacustrine sediments: the Salton Sea Geothermal system, California. *Economic Geology*, **83**, 511–523.

Mehegan, J. M., Robinson, P. T., and Delaney, J. R. 1982. Secondary mineralization and hydrothermal alteration in the Reydarifjordur drill core, eastern Iceland. *Journal of Geophysical Research*, **87**, 6511–6524.

Mortenson, J. K. 1987. Preliminary U–Pb zircon ages for volcanic and plutonic rocks of the Noranda–Lac Abitibi area, Abitibi Subprovince, Quebec. *Current Research, Part A, Geological Survey of Canada Paper*, **87-1A**, 581–590.

Morton, R. L. and Franklin, J. M. 1987. Two-fold classification of Archean volcanic-associated massive sulfide deposits. *Economic Geology*, **82**, 1057–1063.

Morton, R. L., Walker, J. S., Hudak, G. J., *et al.* 1991. The early development of an Archean Submarine Caldera Complex with emphasis on the Mattabi Ash-Flow Tuff and its relationship to the Mattabi Massive Sulfide Deposit. *Economic Geology*, **86**, 1002–1011.

Muffler, L. J., White, D. E., and Truesdell, A. H. 1971. Hydrothermal explosion craters in Yellowstone National Park. *Geological Society of America Bulletin*, **82**, 723–740.

Naldrett, A. J. 1973. Nickel sulfide deposits: their classification and genesis with special emphasis on deposits of volcanic association. *Transactions of the Canadian Institute of Mining and Metallurgy*, **76**, 183–201.

1981. Pt group element deposits. *Canadian Institute of Mining and Metallurgy Special Volume*, **23**, 197–232.

1989. *Magmatic Sufide Deposits*. Oxford Monographs on Geology and Geophysics no. 14. Oxford, UK, Oxford University Press.

Naldrett, A. J. and Cabri, L. J. 1976. Ultramafic and related mafic rocks: their classification and genesis with special reference to the concentration of nickel sulfides and platinum-group elements. *Economic Geology*, **71**, 1131–1158.

Ohmoto, H. and Takahashi, T. 1983. Geologic setting of the Kuroko Deposits, Japan. III. Submarine calderas and kuroko genesis. *Economic Geology Monograph*, **5**, 39–54.

Osterberg, S. A., Morton, R. L., and Franklin, J. M. 1987. Hydrothermal alteration and physical volcanology of Archean rocks in the vicinity of the Headway-Coulee massive sulfide occurrence, Onaman Area, northwestern Ontario. *Economic Geology*, **82**, 1501–1520.

Peach, C. L. and Mathez, E. A. 1993. Sulfide melt-silicate distribution coefficients for nickel and iron, and implications for the distribution of other chalcophile elements. *Geochimica et Cosmochimica Acta*, **57**, 3013–3021.

Perfit, M. R., Ridley, W. I., and Jonasson, I. R. 1999. Geologic, petrologic, and geochemical relationships between magmatism and massive sulfide mineralization along the eastern Galapagos spreading center. *Reviews in Economic Geology*, **8**, 75–100.

Poulson, K. H. and Hannington, M. D. 1996. Volcanic-associated massive sulfide gold. *Geological Survey of Canada, Geology of Canada*, **8**, 183–196.

Pyke, D. R., Naldrett, A. J., and Eckstrand, O. R. 1973. Archean ultramafic flows in Munro Township, Ontario. *Geological Society of America Bulletin*, **84**, 955–978.

Rice, A. and Moore, J. 2001. Physical modeling of the formation of komatiite-hosted nickel deposits and a review of the thermal erosion paradigm. *Canadian Mineralogist*, **39**, 491–504.

Ridge, J. D. 1973. Volcanic exhalations and ore deposition in the vicinity of the seafloor, *Mineralium Deposita*, **8**, 332–378.

Roberts, R. G. 1988. Archean lode gold deposits. In R. G. Roberts and P. A. Sheahan (eds.) *Ore Deposits Models*, Geoscience Canada Reprint Series no. 3. pp. 1–19.

Ross, J. R., and Hopkins, G. M. F. 1975. Kambalda nickel-sulfide deposits. In C. L. Knight (ed.) *Economic Geology of Australia and Papua New Guinea*, vol. 1, *Metals*. Australian Institute of Mining and Metallurgy, Monograph 5, pp. 100–121.

Sangster, D. F. 1980. Quantitative characteristics of volcanogenic massive sulfide deposits. *Canadian Institute of Mining and Metallurgy Bulletin*, February 1980, 74–81.

Sangster, D. F. and Scott, S. D. 1976. Precambrian stratabound massive Cu–Zn–Pb sulfide ores of North America. In K. H. Wolf (ed.) *Handbook of Stratabound and Stratiform Ore Deposits*. Amsterdam, Elsevier, pp. 129–222.

Sawkins, F. J. 1976. Massive sulfide deposits in relation to geotectonics. *Geological Association of Canada Special Paper*, **14**, 221–240.

1990. Integrated tectonic-genetic model for volcanic-hosted massive sulfide deposits. *Geology*, **18**, 1061–1064.

Setterfield, T. N., Mussett, A. E., and Oglethorpe, R. D. J. 1992. Magmatism and associated hydrothermal activity during the evolution of the Tavua caldera: $^{40}$Ar – $^{39}$Ar dating of the volcanic, intrusive and hydrothermal events. *Economic Geology*, **87**, 1130–1140.

Seward, T. M. 1989. The hydrothermal chemistry of gold and its implications for ore formation: boiling and conductive cooling as examples. *Economic Geology Monograph*, **6**, 398–404.

Seyfried, W. E., Berndt, M. E., and Seewald, J. S. 1988. Hydrothermal alteration processes at mid-ocean ridges: constraints from diabase alteration experiments, hot spring fluids and composition of the oceanic crust. *Canadian Mineralogist*, **26**, 787–804.

Shima, H. and Naldrett, A. J. 1975. Solubility of sulfur in an ultramafic melt and the relevance of the system Fe–S–O. *Economic Geology*, **70**, 960–967.

Shirozo, H. 1974. Clay minerals in altered wall rocks of the Kuroko-type deposits. *Society of Mining Geologists of Japan Special Issue*, **6**, 303–311.

Sibson, R. H. 1987. Earthquake rupturing as a mineralizing agent in hydrothermal systems. *Geology*, **15**, 701–704.

Sillitoe, R. H. 1973. The tops and bottoms of porphyry copper deposits. *Economic Geology*, **68**, 799–815.

1991. Gold metallogeny of Chile: an introduction. *Economic Geology*, **86**, 1187–1205.

1993. Epithermal Models: Genetic types. Geometrical controls and shallow features. *Geological Association of Canada Special Paper*, **40**, 403–417.

Sillitoe, R. H. and Bonham, H. F., Jr. 1984. Volcanic landforms and ore deposits. *Economic Geology*, **79**, 1286–1298.

Sillitoe, R. H., Hammington, M. D., and Thompson, J. F. M. 1996. High sulfidation deposits in the volcanogenic massive sulfide environment. *Economic Geology*, **91**, 204–211.

Skirrow, R. G. and Franklin, J. M. 1994. Silicification and metal leaching in semiconformable alteration beneath the Chisel Lake massive sulfide deposit, Snow Lake, Manitoba, *Economic Geology*, **89**, 31–50.

Solomon, M. 1976. "Volcanic" massive sulfide deposits and their host rocks: a review and an explanation. In K. H. Wolf (ed.) *Handbook of Stratabound and Stratiform Ore Deposits*. Amsterdam, Elsevier, pp. 21–50.

Spooner, E. T. C. 1977. Hydrodynamic model for the origin of the ophiolitic cupriferous pyrite ore deposits of Cyprus. *Geological Society of London Special Publication*, **7**, 58–71.

Spooner, E. T. C. and Fyfe, W. S. 1973. Sub-seafloor metamorphism, heat and mass transfer. *Contributions to Mineralogy and Petrology*, **42**, 287–304.

Urabe, T. 1985. Aluminous granite as a source magma of hydrothermal ore deposits: an experimental study. *Economic Geology*, **80**, 148–157.

Walker, P. N. and Barbour, D. M. 1981. Geology of the Buchans Ore Horizon Breccias. *Geological Association of Canada Special Paper*, **22**, 161–186.

Wendlandt, R. F. 1982. Sulfide saturation of basalt and andesite melts at high pressures and temperatures, *American Mineralogist*, **67**, 877–885.

Williams, D., Stanton, R. L., and Rambaud, F. 1975. The Planes–San Antonio pyritic deposit of Rio Tinto, Spain: its nature, environment and genesis. *Transactions of the Institute of Mining and Metallurgy, Section B*, **84**, 73–82.

Williams, D. A., Kerr, R. C., and Lesher, C. M. 1998. Emplacement and erosion by Archean komatiite lava flows at Kambalda: revisited. *Journal of Geophysical Research*, **103**, 27533–27549.

Wyman, D. A., Bleeker, W., and Kemich, R. 1999. A 2.7 Ga komatiite, low Ti tholeiite, arc tholeiite transition and inferred proto-arc geodynamic setting of the Kidd Creek Deposit: evidence from precise trace element data. *Economic Geology Monograph*, **10**, 511–528.

Yang, K. and Scott, S. D. 1996. Possible contribution of a metal-rich magmatic fluid to a seafloor hydrothermal system. *Nature*, **383**, 420–423.

# Chapter 13

# Industrial uses of volcanic materials

## Grant Heiken

## Introduction

Wherever there is a natural occurrence of volcanic rocks, humans have developed multiple uses for these materials. These uses range from cut slabs of tuff for construction to very fine ash for polishing automobiles. Some volcanic materials have no immediate benefit apparent to the public, such as the bentonites used in drilling muds. However, others provide esthetic benefits such as the cut ignimbrite blocks that were used to create the Spanish colonial cathedral in Guadalajara, Mexico.

This chapter is a summary of uses for volcanic materials, which have, for clarity, been grouped as (1) pumice and ash, (2) pozzolan, (3) quarried tuff, (4) perlite, (5) basaltic scoria and lava, and (6) volcanic clays and zeolites. For complete coverage of resources, mining and processing technologies, and environmental and public-health problems associated with that mining, the reader should begin by consulting the series *Industrial Minerals and Rocks*, which is updated on a regular basis by the Society of Mining, Metallurgy, and Exploration, Inc. In addition to an extensive literature, there are many World Wide Web pages on specific rock types and their industrial uses.

Most of what is covered in this chapter are the industrial uses of volcanic materials by, for example, corporations that mine bentonite to make drilling muds; there are many publications available on these topics. More difficult to characterize are the small-scale uses of volcanic materials by artisans, farmers, and road builders in developing countries; there are few publications on this topic except in the newsletters and publications of the Association of Geoscientists for International Development.

## Pumice and ash

Pumice and ash resources include thick layers deposited by fallout from explosive eruption plumes (pumice or ash fallout), zones of pumiceous lava in silicic (rhyolite and dacite) lava flows and domes, and deposits left by horizontally moving hot density currents of ash and pumice called pyroclastic flows and pyroclastic surges.

One of the oldest known construction materials, pumice, was used along with pozzolana (fine-grained or zeolitized ash) by the Romans as lightweight aggregate in concrete. Pumice continues to serve as a source of lightweight aggregate for construction, either in poured concrete or in "pumice" or "cinder" blocks. In 1995, 763 000 tons of pumice were consumed in the USA; 63% of that (488 320 tons) was used to make building-blocks, 37% was used for abrasives, concrete, laundry, and "other uses" (US Geological Survey, 1997). All domestic mining in 1995 was in open pits, in relatively remote areas. Domestic production of pumice in the USA in 1995 was 544 000 tons. The remaining 219 000 tons were imported from Greece (67%), Zaire (8%), Turkey (7%), Ecuador (6%), and "other" (12%). Global production estimates are given in Table 13.1.

*Volcanoes and the Environment*, eds. J. Martí and G. G. J. Ernst. Published by Cambridge University Press. © Cambridge University Press 2005.

| Table 13.1 | World mine production, reserves, and reserve base of pumice | | | |
|---|---|---|---|---|
| | Mine production (tonnes × 10³) | | Reserves[a] | Reserve base[a] |
| | 1994 | 1995 | | |
| United States of America | 490 | 544 | 'Large' | 'Large' |
| Chile | 450 | 450 | ND | ND |
| France | 500 | 525 | ND | ND |
| Germany | 650 | 680 | ND | ND |
| Greece | 900 | 1000 | ND | ND |
| Italy | 5200 | 5200 | ND | ND |
| Spain | 700 | 700 | ND | ND |
| Turkey | 1000 | 1100 | ND | ND |
| Other countries | 1500 | 1500 | ND | ND |
| World total | 11 400 | 11 700 | ND | ND |

[a] ND, not determined.

*Source:* US Geological Survey (1997).

The uses for pumice are mostly as construction materials, abrasives, and absorbents (Table 13.2). The broad spectrum of uses ranges from coarse pumice used for cosmetic purposes for thousands of years to the same pumice being used today to "stone-wash" denim fabric.

## Construction

Pumice concrete is appropriate in construction where building weight must be minimized. If pumice sources are reasonably close, the savings in cost of structural steel justify the use of pumice concrete. Within poured concrete, pumice aggregates are uniformly distributed because they do not sink before the concrete hardens.

Because of its elasticity, pumice concrete resists shock; this was demonstrated in Germany during World War II where houses built with pumice concrete resisted bombings much better than those made with conventional concrete (Schmidt, 1956).

Pumice aggregates also increase the insulating value of either poured concrete or masonry. For example, the "K" factor for concrete made with sand and gravel aggregate is 12.5, whereas that for pumice aggregate is 2.4 ("K" factors are BTU/ft$^2$/hr/inch thickness/$°$F); the lower the "K" factor, the better the insulating properties (Schmidt, 1956).

Fine pumice aggregates (<2.5 mm and >0.6 mm) are also used in plaster, which weighs one-third less than conventional plaster. The weight savings for standard gypsum hardwall plaster is 15 kg/m$^2$ (Schmidt, 1956) (Table 13.3).

## Resources

PUMICE-FALL BEDS

The most common and easily mined pyroclastic aggregates are pumice-fall deposits near their source vents. Most quarries are located in tuff deposits that surround rhyolitic or dacitic domes (tuff rings) or in fallout deposits associated with caldera-forming eruptions (Plinian deposits). For example, thick pumice-fall deposits are quarried near Glass Mountain, in northeastern California, where thicknesses range from 2 to 4 m (Chesterman, 1956; Heiken, 1978) (Fig. 13.1). Wall and slope stability is high, allowing open-pit quarrying with little or no structure required to maintain a vertical face. Pumice-fall deposits are easily quarried using front-end loaders and bulldozers. Pumice products are naturally well sorted by size, with median grain sizes of 1 to 10 mm; sizing for aggregates requires little other than coarse screening to eliminate coarser lapilli, bombs, and blocks (a definition by size is: ash <2 mm; lapilli 2–64 mm; blocks and bombs >64 mm). Pumice fragments (pyroclasts) preferred for concrete

| Table 13.2 | Uses for pumice | |
|---|---|---|
| Uses | Processing | Properties |
| *Lightweight aggregate*<br>Structural concrete blocks<br>Panels, floor decking<br>Pozzolan in cement<br>Plaster mix | Crushing, screening | Low density, good crushing strength, fire resistance, acoustical insulator, moisture resistant |
| *Abrasives*<br>Grill cleaner, scouring, cosmetic skin buffing<br>Stonewashing fabric<br>Hand soaps, scouring compound, wood and metal finishing, striking surfaces for matches | Sawn and irregular blocks, coarse granular, crushing and screening | Vesicle edges (bubble walls) form sharp surfaces, continually exposing new surfaces with usage |
| *Absorbents*<br>Potting soils, hydroponic media, cat litter, floor sweeping compound<br>Acid washing, absorbent for fat and grease drippings<br>Carrier for pesticides, herbicides, and fungicides | Granular; crushing and screening; some drying | High porosity, large surface area, low chemical reactivity |
| *Architectural*<br>Loose fill insulation, roofing granules, textured coatings<br>Decorative veneer<br>Landscaping | Granular and sawn blocks | Low density, thermal and acoustical insulator, fire resistance, easily shaped |
| *Fillers*<br>In rubber, paints and plastics; mold release compounds; brake linings | Granular; crushed, dried, and screened | Particle shape, cost |
| *Filter media*<br>Filter animal, vegetable, and mineral oils | Granular; crushed, dried, screened, milled | Particle shape, expandability |

*Source:* Geitgey (1994).

aggregate should be equant to slightly elongate and have a vesicularity (i.e., porosity) between about 25% and 50%. Pumices containing few or no minerals (phenocrysts) are preferred, minimizing pyroclast densities. The pumice-fall deposits should also have low contents of rock fragments (lithic clasts).

RHYOLITE AND DACITE FLOWS AND DOMES

There is a demand for large blocks of pumice to be used as decorative stones, mostly for landscaping. Larger blocks quarried from silicic lava domes have bands of coarse pumice interlayered with layers of fine pumice, obsidian, and crystalline rhyolite. These low-density pumiceous zones form by gas exsolution within slow-moving rhyolite lava flows; the less dense zones may rise buoyantly to the top of the flow, deforming the flow surface (Fink, 1983; Fink and Manley, 1987). The banded appearance of alternating pumice and obsidian is desirable for decorative stone. Quarrying may involve cutting with rock

**Table 13.3**  Comparison of Concrete made with Pumice, Cinders, Expanded Shale, and Sand and Gravel Aggregates

| Concrete made with | Density g/cm$^3$ | Compressive strength (MPa) | Modulus of elasticity (MPa) | Fireproof ratings | Thermal conductivity ($10^{-3}$ (cal/cm)/s °C) |
| --- | --- | --- | --- | --- | --- |
| Pumice | 1.04–1.60 | 7.03–21.09 | 4219–12306 | 5 cm = 4 hr<br>10 cm = 4 hr | 1.5–3.5 |
| Cinders (scoria) | 1.50–1.92 | 7.03–21.09 | 10548–17570 | 15 cm = 4 hr | 3.5–5.0 |
| Expanded shale | 1.50–1.92 | 7.03–35.17 | 10548–17570 | 13 cm = 4 hr | 3.0–5.0 |
| Sand and gravel | 2.32–2.48 | 7.03–49.23 | 24613–35161 | 10 cm = 4 hr | 12.0 |

*Source:* after Schmidt (1956).

**Fig. 13.1.** Pumice quarry in 1360-year-old rhyolitic pumice fallout surrounding the Glass Mountain rhyolite flow, Medicine Lake Highlands, California. Excavation is by bulldozer, front-end loader, and the occasional volcanologist with a shovel.

saws or moving large blocks from the brittle flow carapace.

### PYROCLASTIC FLOW AND SURGE DEPOSITS

Although not as easily quarried as pumice-fall deposits, non-welded portions of pyroclastic flow deposits most distant from the vent may be quarried for ash and pumice. Considerable screening is required to separate lapilli and blocks from the ash fractions. In most cases the main product will be ash (pozzolan), with the pumice lapilli and blocks screened out and disposed of. Quarrying operations around Laacher See, Germany are located in flow and surge deposits; the thickest deposits are found in paleovalley and overbank facies and are tens of meters thick (Schumacher and Schmincke, 1990).

The Minoan Tuff, on the island of Thera, Greece (Santorini), has been quarried for 150 years, mostly for the fine ash that makes up most of the volume of the deposits (pozzolan) (Heiken and McCoy, 1984) but also for pumice fragments. Much of the pyroclastic sequence on Thera is hydrovolcanic, erupted in the seventeenth century BC during a caldera-forming event. The deposits that were quarried until a few years ago are a complex mixture of loosely consolidated fallout, surge, and mudflow.

Sizing of pumice fragments for concrete varies according to use. For monolithic (poured) concrete, all pumice pyroclasts must be less than 2.5 cm in diameter and 90% coarser than 150 μm. For masonry concrete ("cinder" block), the grain size must be smaller, with median fragment sizes ranging from 0.6 to 1.2 mm.

## Abrasives

Pumice, ground pumice, and crystal-free or crystal-poor silicic ash have been used for scouring and polishing over thousands of years. Cut or smoothed pumice stones have been used for everything from scouring griddles to cosmetic removal of warts. In nearly all instances, the preferred starting material has been crystal-free pumice (phenocrysts scratch and gouge the material being polished or can cut skin).

Hatmaker (1932) listed ground pumice uses by size (Table 13.4).

Since Hatmaker summarized these uses in 1932, not a lot has changed. We are still polishing objects and washing our hands with pumice products. Pumice paste and bar hand soap (the most famous brand being "Lava" soap) are still in demand for mechanics and anyone who works with grease, chemicals, or paint. Pumice soaps are recently in demand because they are free of chemicals like acetone and are "environmentally friendly." Pumice soaps are also used to prepare patients for medical examinations where electronic sensors are attached to the body.

The most recent use of pumice is to prewash fabrics for use in clothing manufacture. Hoffer (1996) has studied the "stone-washing" process to understand the characteristics of pumice that are most useful for this process. The stone-washing method was initiated during 1982 and the rapidly expanding market has increased the demand for pumice. Below is a summary of Hoffer's paper:

Garment wet-processing of cloth uses pumice with or without chemicals during laundering of denim fabrics to soften and bleach the garments. Two processes are used:

(1) Pumice is impregnated with oxidizing chemicals and/or bleach, then tumbled with damp denim fabric, or
(2) pumice is tumbled with fabric and water – this reduces the need to dispose of the chemicals.

Pumice suitable for laundry use must have the following qualities:

(1) A "hard" pumice that abrades slowly.
(2) Minimal surface fines produced by the tumbling; the resulting fine shards reduce the amount of chemical absorbed by the pumice pyroclasts.
(3) The pumice purchased must be dry; wet pumice affects the weight paid for and dilutes the amount of chemical absorbed by the pumice pyroclasts before their use in processing.
(4) Absorption capacity is an average of 23.2 wt.%.
(5) Ideal apparent densities are 700–850 $kg/m^3$.
(6) Abrasion loss or the rate of disintegration during tumbling must be minimal.

Abrasion loss is determined by placing several kilograms of pumice in a rifle machine and tumbling for 15 minutes. The tests result in categories of pumice as "hard," "medium," or "soft" (<25% abrasion loss = "hard"; 26–33% = "medium"; and >34% = "soft").

| Table 13.4 | Uses of ground pumice | |
|---|---|---|
| Mesh size | Size (μm) | Uses (in 1932) |
| <200 | <74 | Very fine polishing |
| −200 | 74 | Finishing automobile bodies and certain kinds of soap |
| −160 | 80 | Finishing automobile bodies and certain kinds of soap |
| −140 | 105 | Finishing automobile bodies and certain kinds of soap |
| −100 | 149 | Glass cutting and silverplate finishing |
| −80 | 177 | Glass beveling and piano finishing |
| −60 | 250 | Glass beveling and piano finishing |
| −50 | 300 | Rough rubbing in piano factories; manufacture of combs, pearl buttons, and mechanic's soap |
| −48 | 325 | Rough rubbing in piano factories; manufacture of combs, pearl buttons, and mechanic's soap |
| −40 | 420 | Finishing pearl buttons |
| −30 | 590 | Finishing pearl buttons |
| −000 018 to −5 | 1000 to 4000 | Coarse rubbing of hard materials; use in tumbling barrels |

*Source:* Hatmaker (1932).

## Agricultural uses for pumice

Pumice is sold in garden shops and agricultural distributors as a conditioner to increase soil porosity. Pumice mixed into soil also retains moisture and acts as a feeding regulator for liquid fertilizers. The agricultural applications of pumice range from small flowerpots to commercial farms and fruit groves. Pumice is also used in tanks for the hydroponic cultivation of vegetables in greenhouses. On a regional scale, the pumice lands of North Island, New Zealand, which mantle thick soils, are ideal for the cultivation of pine trees.

## Pozzolan

Pozzolans are natural volcanic silicates that react with lime in water to produce high-strength concrete and are especially useful for constructing marine structures such as breakwaters and piers. Pozzolan hydraulic cement was used by the Romans for the construction of public buildings such as the Pantheon in Rome, roads, and aqueducts; the name "pozzolan" comes from the town of Pozzuoli in the Bay of Naples (Sersale, 1958).

The main components for pozzolan cement are either fine-grained silicic volcanic ash or crushed zeolitized tuffs (pumice and ash cemented with zeolite and clay minerals). Pozzolans must be fine-grained (with a lot of reactive surface area), thus making most hydrovolcanic tuffs excellent resources, whether or not they are glassy or zeolitized. In countries without access to hydrovolcanic tuffs, pulverized fly-ash is often used for pozzolanic cement.

The Roman Vitruvius specified a ratio of 1 part lime to 3 parts of pozzolan for cement used in buildings and 1 part lime to 2 parts pozzolan for underwater structures. The ratio for modern structures using pozzolan cement is more or less the same (Lechtman and Hobbs, 1987). The reaction between the ash or zeolitized tuffs used for pozzolan and lime is slow, but the cement becomes progessively stronger and more durable with time (Mielenz, 1950). Pozzolan cements are resistant to chloride-ion penetration.

Hydrovolcanic tuffs (volcanic ashes formed by rapid quenching of magma by water and cemented with zeolites) from the Campi Flegrei (Phlegrean Fields), near Naples, were the main pozzolan resource for the Romans, although similar resources were available in and near Rome. The fine-grained hydrovolcanic tuffs erupted during the seventeenth-century BC caldera-forming eruption on the island of Thera (Santorini), Greece were quarried as pozzolan during the mid nineteenth century AD and shipped to Egypt for cement lining the Suez Canal. Quarrying of these tuffs for pozzolan cement and lightweight concrete continued until recently, when quarrying ceased because of the greater value of the island for tourism.

Vitric hydrovolcanic ashes can be used as pozzolan after disaggregation. Zeolitized tuffs must be crushed, then heated to 700–800 °C before being used in pozzolan cement.

Natural pozzolan resources are not limited to fine-grained silicic or trachytic vitric tuffs. A zeolitized ignimbrite (the main zeolite phase being clinoptilolite) was mined near Tehachapi, California, and used for pozzolanic cement in construction of the Los Angeles aqueduct (Mielenz *et al.*, 1951). Kitsopoulos and Dunham (1996) and Fragoulis *et al.* (1997) have proposed quarrying clinoptilolite-, heulandite-, and mordenite-bearing tuff deposits on the islands of Polyegos and Kimolos, Aegean Sea, Greece for industrial use in pozzolan cement.

Mining tuffs for use in pozzolan cement requires little more than a bulldozer with a ripper and front-end loaders for quarrying vitric tuffs and the same, plus crushers for zeolitized tuffs. When the quarries on Thera, Greece were active, cargo ships anchored next to the caldera walls to transport the ash. Front-end loaders carried fine-grained hydrovolcanic ash from the quarry face to the cliff edge, dumped the ash through a screen to remove the lithic clasts, and the ash would then go down a chute directly into the ship. Quarrying wasn't always that simple; over much of the last 100 years, quarrying of the thick deposits was accomplished by tunneling under the deposit until it collapsed into the quarry; then the loose ash was carried to the cliff edge to be loaded into ships. Collapsing quarry faces and constant fine ash dust made pozzolan quarrying on Thera a hazardous profession.

## Quarried tuff

Tuff is consolidated pumice and ash. "Solidification" occurs with time by alteration of glassy ash through interaction with water or hydrothermal steam to cement the rock with secondary minerals such as clay or zeolites or by sintering of the glass fragments while the deposit is still hot to form a "welded tuff."

Non-welded or poorly welded massive tuffs are commonly used throughout the world for construction of buildings (Fig. 13.2). The rock is easily cut and dressed into blocks of almost any shape, which are resistant to weathering. The blocks are lighter than most stone of equivalent size, have some insulation value, and are attractive (at least to volcanologists).

In Latin America, tuffs are used frequently for construction, especially for large public buildings (Fig. 13.3). Large buildings in the Peruvian city of Arequipa, known as La Ciudad Blanca ("the White City") are constructed of non-welded white ignimbrite altered by vapor-phase activity (Fenner, 1948). Blocks are cut with saws and dressed with hammer, chisel, and ax; the cooling joints in these ignimbrites are used as block faces when possible. These ignimbrite blocks are also resistant to weathering – an example is the minimal degradation of tuff blocks used in colonial buildings over a period of 300 years (Jenks and Goldich, 1956).

In Naples, Italy, the large-volume, partly hydrovolcanic, fine-grained tuffs (mostly massive pyroclastic flow deposits) deposited during caldera-forming events in the Campi Flegrei have been a constant source of building stones for thousands of years (since Bronze Age Greek times). The Neapolitan Yellow Tuff (12 000 years old) and Campanian Ignimbrite (37 000 years old), and younger tuff rings of the Campi Flegrei cover a large area within and around the city of Naples (Orsi *et al.*, 1996). These tuff deposits are quarried today as a basic building stone of the region. Massive, fine-grained zeolitized tuffs are quarried with large gas-driven, hydraulic rotary saws mounted on light rails. In the past, these tuffs were cut with large handsaws and axes.

Not all quarried tuffs are used for public buildings such as churches and museums. Tuffs are also used globally for the construction of private homes and farm buildings. Meilan's (1984) study of tuff resources, quarrying technology, and cost effectiveness of cut tuff blocks vs.

**Fig. 13.2.** Church of San Pietro, a Romanesque church near Tuscánia Italy, constructed mostly of tuff blocks quarried from the non-welded, zeolitized Tufo Rosi e Scoria Neri. Sculpted facings and rose window are marble. Public buildings and city walls in many of the hill towns of Tuscany and Umbria have been constructed using blocks of tuff.

conventional concrete cinder blocks for home construction in western Argentina is unique. He studied the Toba Rosada, from the Department of Picunches, Neuquén Province, Argentina, where the 0.4 × 0.2 × 0.2 m tuff blocks tested were cut in open-pit quarries. The quarried stone is a pumiceous, non-welded, consolidated vitric tuff that contains phenocrysts of plagioclase, quartz and biotite. The tuff is partly devitrified and zeolitized. It is easily scratched, and is easy to saw. It is abrasive, a characteristic that must be considered when selecting cutting tools.

Meilan (1984) measured physical properties of the Toba Rosada, with the results given in Table 13.5. Meilan concluded that the tuffs tested are good for house construction. There is good insulating quality and the tuff blocks are practical for use in a harsh climate. Water absorption

can be reduced by the application of waterproofing compounds or paints. After analyzing the market and transportation system of the region, Meilan determined that a quarry in the Toba Rosada had to produce 2000 blocks per day to cost-effectively meet the demand. This assumption was for 8 working hours per day and for 250 days per year. The tuff has to be texturally uniform, massive, more or less horizontal, exposed or covered with thin overburden, and with easy access. The tuff quarrying machines are similar to the rotary saws used in Italian tuff quarries.

Even the dust produced by quarrying has some value. Sersale and Frigione (1983) determined that the dust generated during quarrying makes up 25–30% of the quarried material. The dust can be used in pozzolanic cements, as mixtures of between 10% and 40% dust mixed with Portland cement.

Non-welded tuffs (mostly massive ignimbrites) are used as a medium for sculpting. Tuffs are easily carved with basic hand tools and are the basis for a market through North America of carved door and window frames, garden statues, and birdbaths. Most of the statuary is

**Fig. 13.3.** Moldings and facings cut from Tertiary ignimbrites near Guadalajara, Mexico. Non-welded to poorly welded, massive ignimbrite tuffs are used throughout Guadalajara, which is located on a thick ignimbrite sequence.

carved in central-western Mexico and the tuffs are from quarries in the ignimbrites of the Sierra Madre Occidental and the Trans-Mexican Volcanic Province. Monuments composed of tuff are also found throughout Armenia and Georgia.

## Perlite

Most volcanologists first encounter "perlite" in classes on the origins of volcanic rocks. This intriguing rock type is found mostly as glassy cores of rhyolite flows or as densely welded vitrophyres in ignimbrites. Arcuate to spherical microfractures cross all textural components of this glassy, hydrous volcanic rock and give its weathered surfaces the appearance of an onion skin. It is the rock's hydrous nature and structure that give it an industrial value; perlite expands or "pops" to become a lightweight glass foam when heated quickly to plasticity, while evolving steam (the processed product can have densities as low as 32 kg/m$^3$). In 1994 in the USA, there were 68 plants in 34 states for expanding perlite (Bolen,

**Table 13.5** Physical properties of the Toba Rosada, Neuquén Province, Argentina

| Sample[a] | Shear strength (kgf) | Compression strength (kgf/cm$^2$) |
|---|---|---|
| A (dry) | 2750 | 49 |
| B (dry) | 2800 | 50 |
| C (dry) | 2800 | 54 |
| D (dry) | 2600 | 47 |
| E (dry) | 2800 | 45 |
| F (dry) | 2800 | 48 |
| G (wet) | 2520 | 41 |
| H (wet) | 2200 | 37 |
| I (wet) | 2400 | 38 |

[a] Tests were made on nine samples, six dry and three wet (submerged for 72 hours).

1995); 71% of total domestic US perlite sales were used in construction (insulation and acoustical ceiling tile and cryogenic insulation of liquid-gas storage tanks), 10% used for filters, and 19% for agricultural markets (Barker *et al.*, 1996).

Commercial perlite is a hydrated, high-silica (72–77.5% $SiO_2$) rhyolitic volcanic glass, which contains 2–5% water. The rock is characterized by a spherical hydration geometry, which produces the concentric fractures that are visible at many scales. Perlite occurrences are associated globally with Tertiary through middle Quaternary continental volcanic fields (Howell, 1974; Barker *et al.*, 1996). In the USA, New Mexico accounts for 80% of the 710 000 short tons mined in 1994. Greece accounts for most of the imports to Canada and the USA. Most perlite is shipped to expansion plants near the end-users.

The world's largest perlite deposit is a 3.91 Ma rhyolitic dome complex at Cerro No Agua, New Mexico (an odd name for a volcano, considering the hydrated nature of the rock). Mining is by ripping with a bulldozer or by blasting. The rock is then crushed and dried at a temperature of 105 °C, then crushed again and sized for shipping to expansion plants.

## Uses of perlite

### Expanded perlite

Expansion occurs when crushed perlite is heated in a furnace to temperatures of 870–1100 °C; when softened, perlite can expand to as much as 20 times its original volume (Breese and Barker, 1994). Most expanded perlite (56% in 1990) is used in "formed products," which include acoustic ceiling tiles, pipe insulation, and roof insulation board. The second most common uses are as a filter aid for filtering industrial effluent, oils, and fruit juices and as a soil conditioner in agriculture (addition of perlite aggregate reduces soil compaction and retains moisture) (Bolen, 1995). Other uses include lightweight aggregates for concrete and plaster and as low-temperature insulation.

### Unexpanded, ground perlite

Finely ground, untreated perlite is used as an extender in the manufacture of plastics, paint,

polyvinyl chloride (PVC), and nylon. Coarser unexpanded perlite is used as aggregate for concrete and plaster and as foundry slag.

## Resources

Globally, perlite quarries are found on rhyolitic domes of Tertiary through Quaternary age (Bolen, 1995). Perlites occur as chilled dyke margins and as vitrophyres associated with ignimbrites, but these occurrences are too limited in size to be quarried economically.

Rhyolite flows are often constructed of structurally complex, foliated bands of obsidian, coarse pumice, and fine pumice (Fink, 1983; Fink and Manley, 1987). Rhyolite flows overlie tephra deposits and are mantled by breccia. The flow "core" consists of crystalline lava, which grades outward into mixed glassy and finely pumiceous zones – it is the glassy and finely vesicular parts of the flow where perlitic textures are located.

All perlites are hydrous (Table 13.6). The hydration of glassy obsidian has been interpreted by Jezek and Noble (1978) as a secondary process, one of slow, post-emplacement interaction with meteoric water. Chamberlain and Barker (1996), however, favor high-temperature hydration of silica-rich lava prior to or during emplacement.

The glassy, hydrated cores of perlitic rhyolite flows are developed most economically by open-pit quarrying on the summit and slopes of the flows. The outermost breccias, talus, and pyroclastic materials are removed by blasting and/or ripping with bulldozers. In many perlite quarries, crushing and sizing is done nearby and the crushed perlite is shipped to expansion plants near the markets, usually by rail.

## Basaltic scoria and lava

### Scoria

Strombolian and Vulcanian eruptions form scoria (cinder) cones of all sizes (e.g., Riedel *et al.*, 2003). The cones, composed of interbedded ash, lapilli (the volcanic equivalent in size to pebbles), and bombs (equivalent in size to cobbles and boulders) of basaltic or basaltic–andesite composition, are a resource of considerable value, especially if

| Table 13.6 | Some typical perlite compositions (wt.%) | | | | |
|---|---|---|---|---|---|
| | Cerro No Agua, New Mexico | Superior, Arizona | Pioche, Nevada | Big Pine, California | Milos, Greece |
| $SiO_2$ | 72.1 | 73.6 | 73.1 | 73.6 | 74.2 |
| $Al_2O_3$ | 13.5 | 12.7 | 12.8 | 13.8 | 12.3 |
| $Fe_2O_3$ | 0.8 | 0.7 | 0.7 | 0.8 | 0.95 |
| $TiO_2$ | 0.06 | 0.1 | 0.08 | 0.07 | 0.08 |
| CaO | 0.89 | 0.6 | 0.9 | 0.6 | 0.85 |
| MgO | 0.5 | 0.2 | 0.2 | 0.1 | 0.13 |
| $Na_2O$ | 4.6 | 3.2 | 3.0 | 4.1 | 4.0 |
| $K_2O$ | 4.4 | 5.0 | 4.7 | 4.1 | 4.4 |
| $H_2O$ (total) | 3.0 | 3.8 | 3.9 | 3.3 | 2.8 |

Source: Kadey (1983).

the cones are located near a city or railroad. The popularity of scoria as an aggregate is such that, for example, many of Auckland, New Zealand's scoria cones have disappeared; only the inclusion of cones into city parks has preserved some of the volcanic heritage of this city.

Scoria is used mainly to build roads; both gravel and asphalt highways and railroad beds. Major quarries in scoria deposits near Flagstaff, Arizona, are there because of the coincidence of a railroad town with a volcanic field containing hundreds of scoria cones. Graded scoria is used mainly as a sub-base and base for roads; 46% of the 970 000 tons of scoria used in the state of Victoria, Australia, has been used for roadbeds (Guerin, 1992). However, when used as asphalt aggregate, there is excessive tire wear and scoria should be used only for paved highways where high skid resistance is desirable. Scoria is not ideal for stock tracks, because it cuts the hooves of cattle (Guerin, 1992).

It is the skid resistance that makes very fine, ash-size scoria desirable for sanding icy roads. The sharp edges of vesicle walls in individual pyroclasts provide the needed traction on ice. After a major snowstorm in northern New Mexico, roads have a red coating from the application of oxidized ash-size scoria; the scoria is quarried in scoria cones near the city of Santa Fe. Careful crushing and sieving is needed, for lapilli-size particles are kicked up by tires and crack or pit auto windows.

### Scoria resources

Scoria cones are formed during alternating Strombolian and Vulcanian activity; a process of ballistic deposition and subsequent slumping or avalanching provides the sloping beds of naturally sorted scoria (McGetchin et al., 1974; Riedel et al., 2003). These processes give the cones their value; the bouncing and sliding of scoria pyroclasts down cone slopes prevents most welding and actually causes some natural sorting. The most complete study of a scoria cone is that of the Rothenberg cone, east Eifel, Germany (Houghton and Schmincke, 1989), where the outer slopes have the most valuable scoria resource in the form of avalanche deposits and ashfall. The vent area consists of solid dykes and welded and partly welded scoria and is not easily quarried.

### Quarrying

Quarries or "scoria pits" are usually located on the lower slopes of cones, in the avalanche deposits, where screening of large bombs and blocks is unnecessary (Fig. 13.4). Much of the scoria in the lower slopes is not welded. The quarry headwall can be pushed toward the center of the cone and stopped where material is welded and cannot be easily removed. The quarrying can be done with a bulldozer, front-end loader, screens, and dump trucks. In many cases, "pit run" scoria can be used for roadbed construction with no screening required.

**Fig. 13.4.** "Cinder" pit in a scoria cone, near Old Station, north of Lassen Volcanic National Park, California. Well-sorted scoria in avalanche beds on the outer slopes of cinder cones are used for roadbeds in this area (base for asphalt highways and gravel-surfaced logging roads).

If the scoria cone is located near a town, there will be concern about the visual effect of a quarry. Geologists love a good outcrop, but that desire is not shared by the public. Guerin (1992) recommends that the quarries be kept near the base of a cone, preferably on the backside (away from the town) or hidden in irregular topography.

### Other uses of scoria

Scoria is occasionally used as aggregate in concrete blocks, when it is the only aggregate available. It is also used as "shading" or backfill around buried water pipes. Large bombs and scoria are used as decorative rocks in gardens, especially those that are a bright red color. Scoria is sold as a base in barbecues, with the dual purpose of retaining heat and absorbing grease from the meat being grilled.

## Basaltic lava

For scoria quarry owners, basaltic lava is a nuisance. For the Roman engineers building roads in Italy, basaltic and trachytic lavas were a key construction component. The roads were constructed of sand, masonry, and fitted lava blocks cut into hexagonal shapes (the *summa crusta*) (Legget, 1973). Lava blocks were the hardest stones available to the engineers for the road surface. Many of these roads are still being used.

Because of the natural cooling joints and brittle properties, most basaltic lava flows are useless unless crushed to be used as aggregate. The exceptions are when there is well-developed columnar jointing within the lava flows. Lava columns can be broken or cut and used in construction of rock walls, buildings, and as dykes in Holland (Fisher *et al.*, 1997).

# Volcanic clays and zeolites

## Bentonites used for drilling muds, iron ore pelletizing and foundry sands

The name "bentonite" was coined in 1898 for the clay-like material that comprised the Cretaceous-age Benton Formation of the Rock Creek District, Wyoming by Knight (1898). Hewitt (1917)

and Wherry (1917) proposed that bentonites were formed during alteration of volcanic ash in an aqueous environment. The clay making up bentonite was determined to be mostly montmorillonite, and less commonly, beidellite (Ross and Shannon, 1926); this early work presaged the many studies of bentonites over the past 70 years. One of the most comprehensive of these studies was that of the Mowry bentonites of Wyoming. Most bentonites were once ash fallout, evidenced by relict textures of what had been silicic shards and pumice clasts (Khoury and Eberl, 1979; Wise and Weaver, 1979; Heiken and Wohletz, 1985) (Fig. 13.5).

Ross and Shannon (1926) defined bentonite as:

a rock composed of a crystalline clay-like mineral formed by devitrification and accompanying chemical alteration of a glassy igneous material, usually a tuff or volcanic ash; and it often contains variable proportions of accessory crystal grains that were originally phenocrysts . . . the clay-like mineral has . . . a texture inherited from volcanic tuff or ash.

However, within the minerals industry, most, but not all "bentonites" are of volcanic origin.

The major mineral constituents of bentonite are the smectite clays (beidellite, montmorillonite, nontronite, saponite, and hectorite); montmorillonite is the most common in bentonite deposits. The industrial use of bentonite depends upon its swelling capacity when added to water. Bentonite having sodium as the dominant exchangeable ion usually has high swelling capacities and forms gel-like masses when added to water (Elzea and Murray, 1994). Bentonite in which exchangeable calcium is abundant has a lower swelling capacity. The best-quality bentonites, with high swelling capacities, are commonly found in Wyoming and adjacent states.

Bentonites of Cretaceous age are mined or quarried mostly in Wyoming, USA and in Saskatchewan, Canada. These "Western" bentonites supply nearly 70% of the US bentonite production. Reworking of ash deposits in subaerial, sublacustrine, and marine environments result in deposit thicknesses of over 3 m. There are bentonitic ashfall beds interbedded with shale and sandstone that compose the 330-m-thick Mowry

**Fig. 13.5.** Scanning electron micrograph of Cretaceous Mowry Formation bentonite, drillhole in the Casper Arch, Wyoming. The tuffaceous origins of this rock are evident in the relict textures left from pumice and shards. In this image, a relict pumice pyroclast is most evident in what is mostly smectitic clay. Scale bar = 60 μm.

Shale deposits (Wyoming Mining Association, 1997). Although there are bentonite resources on nearly every continent, most of the production remains in the USA.

Exploration for bentonite deposits requires a complete understanding of the geological environment that favors bentonite formation (Elzea and Murray, 1994). A likely target for bentonite is where widespread, large-volume ashfalls were deposited in an ocean or lake. Once identified, the thickness and extent of deposits must be determined, along with mineralogical characterization by X-ray diffraction (mineralogy), scanning electron microscopy (SEM) and thin-section petrography (relict shard identification), and measurement of cation-exchange capacity.

## Uses of bentonite

### DRILLING MUD

Drilling muds must be able to carry cuttings from bit to the surface and allow separation at the surface. They must also cool and clean the drill bit and reduce friction between the drill string and sides of the holes (Castelli, 1996). Sodium bentonite is the preferred drilling mud, with a yield specification (number of barrels of 15 centipoise viscosity mud made from 1 ton of bentonite) of about 90 barrels per ton (American Petroleum Institute).

### OTHER INDUSTRIAL USES

Bentonite is also used for bonding foundry sand and for iron ore pelletizing. It is also utilized as absorbent granules for oil and water.

### CAT LITTER

In 1962, there were 63 million cats in the USA; this number is rising, since cats have replaced dogs as America's favorite pet. Most cats are kept inside in private homes and apartments, requiring the use of a lot of cat litter. At about US$2.00 for a 10-pound (4.5-kg) bag of cat litter, it has a retail value of about US$880 per metric ton. It is shipped to all continents and is expected to increase. In 1994, in the UK alone, the cat litter market was £43–45 million (Austin and Mojtabai, 1996).

Mineral cat-litter materials mined or quarried from volcanic rocks include:

(1) Fuller's earth. This is a common term for Ca-bentonite, acid-activated illite, and palygorskite/attapulgite. Because of its high absorbency it has been popular as a cat litter. However, its use is decreasing because of its weight (higher shipping costs).

(2) Na-bentonite (also known as Na-smectite, Western bentonite or Wyoming bentonite) (smectite clay minerals, most commonly montmorillonite). This is used mostly for scoopable litter (Austin and Mojtabai, 1996).

(3) Zeolites, particularly clinoptilolite, are still a small part of the market, but this is increasing. They are added to other litter materials to absorb odor. For example, 73% of the clinoptilolite mined from the St. Cloud zeolite mine in New Mexico is sold to be blended with clays to make up a cat litter blend (White et al., 1996).

The most desirable material characteristics for cat litter are:

(1) Absorption. Smectites absorb fluid and swell, which is acceptable as long as physical properties of the litter keep it in the cat box. Zeolites absorb fluid on surfaces and in the mineral structure, but do not swell.

(2) Appropriate density (500–900 kg/m$^3$) The bulk density must be high enough to keep granules within the cat box, but light enough for economical shipping.

(3) Granule size. Granules must be large enough not to stick to cat's paws (preferred size range is 1–6 mm) (O'Driscoll, 1992). Rounded grains are preferred.

(4) Odor control. Smectite clays are good for odor control, retarding fermentation of urea to ammonia.

(5) Dust formation. Minimal dust generation is desirable, for both cats and their humans. The processed litter must not break down easily to smaller particles during shipping.

(6) Color (marketing). Customers (not the cats) prefer light-colored litters, which give the illusion of being more antiseptic.

With the increasing popularity of scoopable litter, the producers of Wyoming bentonites (Cretaceous Mowry Bentonite – Slaughter and Earley, 1965) will be producing more for the litter market.

# References

Austin, G. S. and Mojtabai, C. 1996. Cat litter: a growing market for industrial minerals. *New Mexico Bureau of Mines and Mineral Resources Bulletin*, **154**, 267–273.

Barker, J. M., Chamberlain, R. M., Austin, G. S., *et al.* 1996. Economic geology of perlite in New Mexico. *New Mexico Bureau of Mines and Mineral Resources Bulletin*, **154**, 165–185.

Bolen, W. P. 1995. *Perlite: Mineral Industry Surveys, 1990 and 1991*. US Bureau of Mines.

Breese, R. O. Y. and Barker, J. M. 1994. Perlite. In D. D. Carr (ed.) *Industrial Minerals and Rocks,* 6th edn. Society for Mining, Metallurgy, and Exploration, Inc., pp. 735–749.

Chamberlain, R. M. and Barker, J. M. 1996. Genetic aspects of commercial perlite deposits in New Mexico. *New Mexico Bureau of Mines and Mineral Resources Bulletin*, **154**, 171–185.

Chesterman, C. W. 1956. Pumice, pumicite (volcanic ash), and volcanic cinders in California. *California Division of Mines Bulletin*, **174**, 3–98.

Elzea, J. and Murray, H. H. 1994. Bentonite. In D. D. Carr (ed.) *Industrial Minerals and Rocks*, 6th edn. Society for Mining, Metalllurgy and Exploration, Inc., pp. 223–246.

Fenner, C. N. 1948. Incandescent tuff flows in southern Peru. *Bulletin of the Geological Society of America*, **59**, 879–893.

Fink, J. H. 1983. Structure and emplacement of a rhyolitic obsidian flow: Little Glass Mountain, Medicine Lake Highland, northern California. *Geological Society of America Bulletin*, **94**, 362–380.

Fink, J. H. and Manley, C. R. 1987. Origin of pumiceous and glassy textures in rhyolite flows and domes. *Geological Society of America Special Paper*, **212**, 77–88.

Fisher, R. V., Heiken, G. H., and Hulen, J. B. 1997. *Volcanoes: Crucibles of Change.* Princeton, NJ, Princeton University Press.

Fragoulis, D., Chaniotakis, E., and Stamatakis, M. G. 1997. Zeolitic tuffs of Kimolos Island, Aegean Sea, Greece and their industrial potential. *Cement and Concrete Research*, **27**, 889–905.

Geitgey, R. 1994. Pumice and volcanic cinders. In D. D. Carr (ed.) *Industrial Minerals and Rocks*, 6th edn. Society for Mining, Metallurgy, and Exploration, Inc., Golden, pp. 803–813.

Guerin, B. 1992. Review of scoria and tuff quarrying. *Victoria Geological Survey Report*, **96**.

Hatmaker, P. 1932. *Pumice and Pumicite (Volcanic Ash).* US Bureau of Mines Information Circular No. 6560.

Heiken, G. and Wohletz, K. 1985. *Volcanic Ash.* Berkeley, CA, University of California Press.

Hewitt, D. 1917. The origin of bentonite. *Journal of the Washington Academy of Sciences*, **7**, 196–198.

Hoffer, J. M. 1996. Domestic pumice and garment wet-processing. *New Mexico Bureau of Mines and Mineral Resources Bulletin*, **154**, 275–282.

Houghton, B. F. and Schmincke, H.-U. 1989. Rothenberg scoria cone, East Eifel: a complex Strombolian and phreatomagmatic volcano. *Bulletin of Volcanology*, **52**, 28–48.

Howell, W. R. 1974. *The early history of the perlite industry.* Unpublished address to the Perlite Institute, Inc., Colorado Springs, CO, April 21.

Jenks, W. F., and Goldich, S. S. 1956. Rhyolitic tuff flows in southern Peru. *Journal of Geology*, **64**, 156–172.

Jezek, P. A. and Noble, D. C. 1978. Natural hydration and ion exchange of obsidian: an electron microprobe study. *American Mineralogist*, **63**, 266–273.

Kadey, F. L., Jr. 1983. Perlite. In S. J. Lefond (ed.) *Industrial Minerals and Rocks*, 5th edn. New York, pp. 997–1015.

Khoury, H. and Eberl, D. 1979. Bubble-wall shards altered to montmorillonite. *Clays and Clay Minerals*, **27**, 291–292.

Kitsopoulos, K. P. and Dunham, A. C. 1996. Heulandite and mordenite-rich tuffs from Greece: a potential source for pozzolanic materials. *Mineral Deposita*, **31**, 576–583.

Knight, W. C. 1898. Bentonite. *Engineering and Mining Journal*, **66**, 491.

Lechtman, H. N. and Hobbs, L. W. 1987. Roman concrete and the Roman architectural revolution. In W. D. Kingery and E. Lense (eds.) *Ceramics and Civilization*, vol. 3, *High-Technology Ceramics: Past, Present, and Future.* Westerville, OH, American Ceramic Society, pp. 81–125.

Legget, R. F. 1973. *Cities and Geology.* New York, McGraw-Hill.

McGetchin, T. R., Settle, M., and Chouet, B. H. 1974. Cinder cone growth modeled after Northeast Crater, Mt. Etna, Sicily. *Journal of Geophysical Research*, **74**, 3257–3272.

Meilan, D. 1984. Convencia de la utilization de las tobas volcánicas en la construcción de viviendas económicas. M.S. thesis, University of Guanajuato, Mexico.

Mielenz, R. C. 1950. *Materials for Pozzolans: A Report for the Engineering Geologist*, Petrographic Laboratory Report Pet-90B.

Mielenz, R. C., Greene, K. T., and Cyrill Scieltz, N. 1951. Natural pozzolans for concrete. *Economic Geology*, **46**, 311–328.

O'Driscoll, M. 1992. European cat litter: absorbing market growth. *Industrial Minerals*, **299**, 46–65.

Orsi, G., de Vita, S., and di Vito, M. 1996. The restless, resurgent Campi Flegrei nested caldera (Italy): constraints on its evolution and configuration. *Journal of Volcanology and Geothermal Research*, **74**, 179–214.

Ross, C. S. and Shannon, E. V. 1926. The minerals of bentonite and related clays and their physical properties. *American Ceramic Society Journal*, **9**, 77–96.

Schmidt, F. S. 1956. Technology of pumice, pumicite (volcanic ash), and volcanic cinders. *California Division of Mines Bulletin*, **174**, 99–117.

Sersale, R. 1958. Genesi e constituzione del tufo giallo Napolitano. *Rend. Accad. Sci. Fis. Mat.*, **25**, 181–207.

Sersale, R. and Firgione, G. 1983. Utilization of the quarry dust of the Neapolitan yellow tuff for the manufacture of blended cements. *La Chimica e L'Industria*, **65**, 479–481.

Slaughter, M. and Earley, J. W. 1965. *Mineralogy and Geological Significance of the Mowry Bentonites, Wyoming*, Geological Society of America Special Paper no. 83.

US Geological Survey 1997. Mineral commodity summaries, January, 1997. Available online at: http://minerals.er.usgs/minerals/pubs/commod

Wherry, E. 1917. Clay derived from volcanic dust in the Pierre of South Dakota. *Journal of the Washington Academy of Sciences*, **7**, 576–583.

White, J. L., Chavez, W. X., Jr., and Barker, J. M. 1996. Economic geology of the St. Cloud Mining Company (Cuchillo Negro) clinoptilolite deposit, Sierra County, New Mexico. *New Mexico Bureau of Mines and Mineral Resources Bulletin*, **154**, 113–125.

Wise, S., Jr. and Weaver, F. 1979. Volcanic ash: examples of devitrification and early diagenesis. *Scanning Electron Microscopy*, 511–518.

Wyoming Mining Association 1997. Bentonite industry. Available online at: http: www.tcd.net/~wma.bent/benthome.html

# Chapter 14

# Volcanoes, society, and culture

## David K. Chester

## Introduction: relationships between volcanoes, society, and culture in time and space

A volcano is not made on purpose to frighten superstitious people into fits of piety, nor to overwhelm devoted cities with destruction.

James Hutton (1788)

Throughout history volcanoes have fascinated humanity. Even before they were observed by literate observers, eruptions were depicted in art, remembered in legend, and often became incorporated into religious rituals: volcanoes being perceived as agents of benevolence, fear, or vengeance depending on their state of activity and the society involved (see Blong, 1982, 1984; De Boer and Sanders, 2002). In the ancient Near East, the earliest known record of a volcanic eruption is a wall painting from the Neolithic town of Çatal Hüyük in Anatolia (Mellaart, 1967). It shows a Strombolian eruption with an ash cloud and the spasmodic eruption of bombs and blocks (Polinger-Foster and Ritner, 1996), but it is only much later, in Mesopotamia from the 3rd millennium BCE, that volcanoes became part of the written record (Foster, 1996).

As far as written records of eruptions are concerned, even today these are incomplete. It is sobering to recall that:

if a list of . . . volcanoes had been continually kept, it would, at the time of Christ, have contained only the names of 9 Mediterranean volcanoes and West Africa's Mount Cameroon. In the next 10 centuries the list would have grown by only 17 names, 14 of them Japanese. The first historic eruptions of Indonesia were in 1000 and 1006, and newly settled Iceland soon added 9 volcanoes to help swell the list to 48 by 1380 AD . . . The list has continued to grow, with several important volcanic regions such as Hawaii and New Zealand being completely unrepresented until the last 200 years. Only in the present century has the *rate* of growth declined significantly. (Simkin *et al.*, 1981, p. 23.)

Although this statement now requires modification, especially in the light of improved knowledge of eruptions in antiquity (see Stothers and Rampino, 1983; Stothers, 1984; Pang, 1985; Forsyth, 1988; Burgess, 1989; Simkin, 1993; Sigurdsson, 1999 – but see the cautionary note by Simkin, 1994), the fact remains that active volcanoes continue to be discovered. The late Haroun Tazieff, former doyen of European volcanologists, estimated that around 500 volcanoes have erupted in historical time, but admits that this is an underestimate of potential activity because in many parts of the world records are short – often covering less than 100 years – in comparison with the repose periods of many volcanoes (Tazieff, 1983). For example, in 1951 Mt. Lamington in Papua New Guinea erupted killing about 5000 people, yet both written and oral histories led to the belief that this volcano was either dormant or extinct (Taylor, 1958). In fact, subsequent geological evidence and radiometric dating suggest that an eruption had occurred around 13 000 years ago (Baker, 1979). More recent examples

*Volcanoes and the Environment*, eds. J. Martí and G. G. J. Ernst. Published by Cambridge University Press. © Cambridge University Press 2005.

Table 14.1 | Estimated number of people affected, but not killed, by natural disasters 1980–90

| Disaster type | Approximate numbers affected (thousands) | Numbers affected (~% of total) |
|---|---|---|
| Droughts | 952 200 | 57 |
| Floods | 524 600 | 32 |
| Windstorms | 150 300 | 9 |
| Earthquakes | 28 400 | 2 |
| Landslides | 3100 | 0.2 |
| Volcanic eruptions | 620 | 0.04 |
| Wildfires | 610 | 0.04 |
| Tsunamis | ~1 | 0.0001 |
| Total | 1 659 831 | |

*Source:* After UNESCO (1993), with modifications and amendments.

of what were thought to be inactive volcanoes suddenly awakening include: Mt. Arenal (Costa Rica) in 1968; Heimaey (Iceland) in 1973, and Nyos (Cameroon) in 1986. It is a sobering observation that long repose is usually associated with the silicic volcanism of subduction zones (Chester, 1993) and these belts often coincide with major population clusters.

According to the Chinese Government the Tangshan earthquake in 1976 caused the deaths of ~250 000 people but unofficial estimates place the death toll at up to 800 000, with a further 500 000 being injured (Tilling, 1989; Bolt, 1999). Even if the official estimate of mortality is accepted, then this is only slightly fewer that the estimated 270 000 who have died from the effects of all volcanic eruptions during the past 300 years (IAVCEI, IDNDR Task Group, 1990). More recently Tanguy *et al.* (1998) have estimated that about 220 000 people have been killed by eruptions since 1783. As Table 14.1 shows, if figures for the number of people seriously affected, but not killed, by natural disasters for one decade are examined then the impact of eruptions appears to be minimal, when compared with the distress caused by droughts, floods, windstorms, and earthquakes. From these figures it would be easy to conclude that volcanic eruptions pose less of a threat than is commonly supposed and, indeed, Wijkman and Timberlake (1984) go so far as to

suggest that volcanic eruptions are not serious hazards at all. Such a judgment is only true in part, as a more detailed examination of losses makes clear. It should be noted that the annual death toll from volcanic eruptions over the past 300 years still represents an average of nearly 1000 fatalities per annum (McGuire, 1995a), but it is other factors that are of greater significance and which require the conclusions of Wijkman and Timberlake to be revised.

Over the past 300 years, it is only a matter of chance that casualties from eruptions have not been much higher. Several major eruptions have occurred in regions which were at the time and in some cases remain largely uninhabited. Examples include the pyroclastic flows generated by the 1912 eruption of Katmai volcano (Alaska), which produced the so-called "Valley of the Ten Thousand Smokes" and the 1955–6 eruption at Bezimianny (Kamchatka, Russia). If events of this magnitude had occurred in populated areas of global economic importance, such as California or Japan, the effects would have been serious both for these countries and for the world as a whole. In reviewing major volcanic disasters during the period 1980 and 1993, Tilling and Lipman (1993) note that there were three episodes of "restlessness" at major caldera complexes – Long Valley (USA) 1980–4, Campi Flegrei (Italy) 1982–4, and Rabaul (Papua New Guinea)

1983–5 – which did not culminate in eruptions. Even so, these episodes caused major social and economic disruption and, in the case of Campi Flegrei, necessitated the temporary evacuation of some 40 000 people. It is estimated that even a modest eruption at Campi Flegrei could put between five and ten times this number of people at risk (Barberi *et al.*, 1984; Barberi and Carapezza, 1996). When compared with the estimated recurrence intervals of certain types of volcanic activity, historical records are short even in countries like Japan, Italy, and Greece, where knowledge of eruptions stretches back many thousands of years. The Long Valley and Yellowstone volcanic centers in the western USA are still capable of producing large ignimbrites (Cas and Wright, 1987; Wood and Kienle, 1990). If either did so, then as much as 30 000 km$^2$ of land would be devastated and the resulting disaster would rank as one of the world's greatest-ever catastrophes. In passing, it should be noted that many silica-rich volcanoes, which are capable of producing large ignimbrites, have very long repose periods (of up to 10$^5$, or even 10$^6$ years) (Cas and Wright, 1987), when compared with the short span of human history.

It is not only the danger of death which is significant but, as Peter Baxter (Baxter *et al.*, 1982; Baxter, 1990, 1994, 2000) and others (e.g., Sapir, 1993) have shown, there is also increasing evidence that both acute and chronic ill-health may be caused by volcanic eruptions. Furthermore ill-health is not confined to the obvious physical injuries produced by eruption products. For instance on Furnas Volcano (São Miguel, the Azores), the danger to human health of long-term exposure to carbon dioxide, radon, and other pollutants has been highlighted (Baxter *et al.*, 1999).

Hazard is defined as "the interface between an extreme physical event and a vulnerable human population" (Susman *et al.*, 1983, p. 264) and a second reason why volcanic eruptions should not be dismissed as minor threats to life and health is that the severity of volcano hazards may be increasing, though there is no firm evidence of any significant increase in global volcanic activity at least over human timescales. The magnitude, frequency, and recurrence intervals of global volcanism, combined with a varying spatial pattern

of activity – with eruptions sometimes coinciding with population clusters and at other times occurring in sparsely populated regions – means that, in certain decades, losses are inevitably much higher than in others. For instance, and as Fig. 14.1 shows, the 1980s saw more volcano-related deaths than any decade since 1900–9. In the 1990s and the early years of the present century further damaging eruptions have occurred. In 1990–1 major eruptions occurred at Kelud and Unzen Volcanoes in Indonesia and Japan, respectively, and at Mt. Pinatubo in the Philippines. In 1993, The International Federation of Red Cross and Red Crescent Societies – defining a *disaster* as one causing more than ten deaths and/or affecting more than 100 people and/or calling forth assistance from outside the immediate region (Anonymous, 1994a) – recorded four major volcano-related emergencies. Rabaul Volcano (Papua New Guinea) erupted in 1994 destroying a town with the same name (Blong and McKee, 1995) and, more recently, major eruptions have occurred beneath the Vatnajökull ice-cap in Iceland (GVNB, 1996), at Manam Volcano in Papua New Guinea (Tanguy *et al.*, 1998), in the Soufrière Hills (Montserrat, Caribbean), beginning in 1995 (Wadge, 1996; Possekel, 1999), on Piton de la Fournaise (Réunion), and Mt. Etna, Sicily.

"Volcanoes are nature's forges and stills where elements of the Earth, both rare and common, are moved and sorted" (Decker and Decker, 1981, p. 168) and "many of the world's resources of fluorine, sulphur, zinc, copper, lead, arsenic, tin, molybdenum, uranium, tungsten, silver, mercury and gold, were formed by magmatic processes both direct and indirect" (Chester, 1993; see also Fisher *et al.*, 1997) and, since the permanent occupation of towns and villages began with the onset of the agricultural revolution of the Eurasian Neolithic, volcanoes have attracted human settlement. In classical times around the Mediterranean, breached calderas open to the sea – in the Aegean, the Campi Flegrei region of southern Italy and in the Aeolian Islands north of Sicily – were used as natural harbors. In some cases, they are still harbors today, and the 1994 eruption of Rabaul in Papua New Guinea is but the latest example of a port being closed by a volcanic eruption (Blong and McKee, 1995).

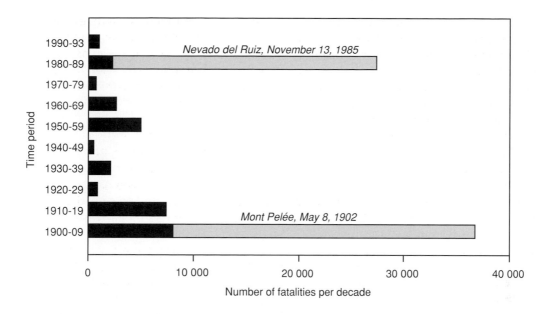

**Fig. 14.1.** The number of deaths caused by volcanic eruptions, 1900–93. (From Tilling and Lipman (1993, Fig. 1, p. 277), based on information in Decker (1986, Fig. 1); Peterson and Tilling (1993, Fig. 14.1); McClelland et al. (1989, Fig. 6), and Tilling (1989).)

It has been claimed that high rural population densities are supported by the intrinsic fertility of volcanic soils (Macdonald, 1972), but an evolving body of research is increasingly challenging this simplistic assertion and further discussion of this point occurs later in the chapter. Volcanoes are proving increasingly attractive for human settlement for reasons besides their agricultural and resource potential and increasing global exposure to eruptions is not primarily a function of the direct allure of volcanic regions. Underlying processes of attraction are rather indirect, subtle, and complex, often being independent of volcanic activity. Attraction represents an interplay of often deep-seated economic, social, and cultural forces which are reflected in population growth, the in-migration of people and changes in the distribution of wealth and poverty both within individual countries and globally (Burton et al., 1978; Dicken, 1986; Hamilton, 1991; Chester et al., 2001). In the densely population region around the Bay of Naples in southern Italy, government-sponsored schemes of economic development and rural-to-urban

migration (King, 1985) have, for example, resulted in at least 700 000 people now being at risk in the event of a future eruption of Vesuvius (Dobran, 1995). Similar rapid growth is occurring in many cities in Central and South America (Anonymous, 1994b) and the Pacific Rim of Asia is rapidly emerging as the world's major economic heartland. In all these cases, the increasing hazardousness in these regions has little to do with volcanism per se (Chester et al., 2001).

Later in this chapter, I will argue that a division of the Earth into a rich "north" – or *first* world – and a poor "south" – or *third* world – is over-simplistic, serving to mask more complex geographies of disaster (Hewitt, 1997, pp. 12–17). If this subdivision is accepted for the present, then it is high rates of population growth in developing countries and particularly in their urban areas that is the principal factor increasing global hazardousness. A disproportionately high number of active volcanoes occur in *economically less developed* countries and many of these, especially those around the Pacific Ocean and in the Caribbean, lie astride belts of active subduction (Fig. 14.2). In the 1980s it was estimated by Don Peterson (1986) that some 10% of the world's population lived either on or in the vicinity of the 600 or so volcanoes which are known to be capable of renewed activity (Peterson, 1986). "By the year 2000 the population at risk . . . is

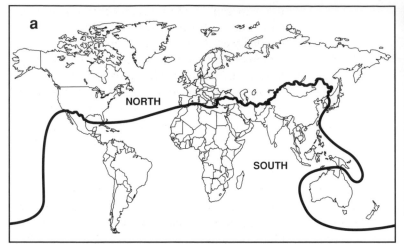

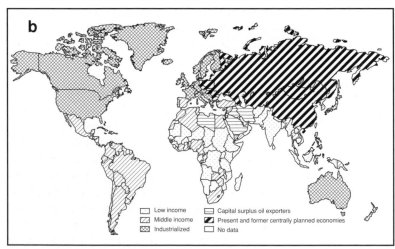

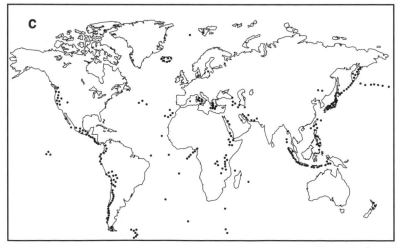

**Fig. 14.2.** World maps showing: (a) Classification of countries into a rich *north* and a poor *south* (after Dickenson *et al.*, 1983, 1996, Fig. 1.3, p. 6). The terms *north* and *south* came into regular use following the publication of the Brandt Report (1980). (b) Economic and political groupings of countries (after Dickenson *et al.*, 1996, Fig. 1.4, p. 7). These represent grouping of countries proposed by the World Bank (1995). (c) Active volcanoes of the world (various sources).

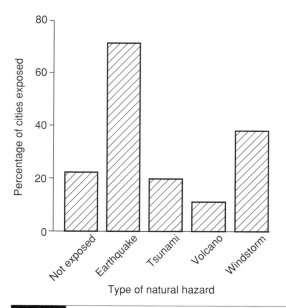

**Fig. 14.3.** Current exposure of the 100 largest cities to natural hazards. Simplified from Degg (1992, Fig. 3, p. 206).

likely to increase to at least 500 million, comparable to the estimated world population at the beginning of the seventeenth century" (Tilling and Lipman, 1993, p. 279). Diagrams produced by Martin Degg (1992) are particularly instructive and potential exposure to future eruptions currently affects ten of the world's 100 largest cities (Fig. 14.3). This is likely to increase because the United Nations estimates that 88% of the 50 fastest-growing cities are both exposed to natural hazards and are located in so-called "Third World" countries (United Nations, 1989; Degg, 1992; see also Anonymous, 1995, 1996). Of these 50 cities, nine are found in areas of active volcanism (Degg, 1992, Fig. 2). More recently it has been argued that volcanoes are the principal hazard facing a growing number of both large and small cities around the world and the contemporary switch in focus of world urbanization has probably impacted more on volcanic risk than any other class of natural hazard (Chester *et al.*, 2001).

A third reason why volcanic eruptions are not a relatively minor hazard is that economic losses are not only high, but are also increasing rapidly. As with mortality and morbidity, so economic losses from eruptions are spiraling due

to demographic pressures and changing national and global patterns of wealth and poverty: much greater wealth is now concentrated in volcanic regions than was the case even a few decades ago. In terms of absolute value, greater financial losses occur in developed countries, but the World Bank calculates that, when losses are related to a country's wealth expressed in terms of its gross national product (GNP), then for events of roughly similar magnitude, costs may be up to 20 times higher in *economically less* than in *economically more* developed countries (Anonymous, 1994b). For poor countries, major natural disasters may serve to wipe out development gains for many years (Harriss *et al.*, 1985; Kates, 1987). This global pattern of losses is reflected in individual volcano-related disasters. The Mount St. Helens eruption of 1980 cost the economy of the USA between US$860 million (Blong, 1984) and US$1 billion (Tilling, 1989), but this only represented around 0.03% of GNP, whereas figures for the 1985 disaster at Nevado del Ruiz (Colombia) show property losses alone of US$300 million: ~1% of GNP (Sigurdsson and Carey, 1986). It has been calculated that property losses in major eruptions commonly comprise only about one-tenth of the total (Chester, 1993, p. 241), so it reasonable to assume that the real cost to the Colombian economy was *c.* 10% of GNP, more than 300 times greater than that inflicted on the economy of the USA four years earlier. In more recent eruptions in developing countries this trend of high GNP losses has continued. Detailed assessments of losses have been compiled for the Philippines, following the eruption of Mt. Pinatubo in 1991 (Tayag and Punongbayan, 1994) and Papua New Guinea, in the aftermath of events at Rabaul Volcano in 1994 (Blong and McKee, 1995) (Table 14.2). In both cases, direct costs were only part of the total financial burden and the long-term drain on the budgets of these developing countries continued for several years. In Rabaul this was particularly serious because the cost of the eruption was high at around 4% of GNP (Table 14.2). The high cost of this eruption occurred mainly in Rabaul township, just one of many settlements of a similar size in Papua New Guinea, representing at the time of eruption only ~0.4% of the country's population. Even the estimated 100 000 people

**Table 14.2** Minimum costs of the 1994 Rabaul eruption (Papua New Guinea); note these figures should be taken as approximations, due to problems of currency conversion and in costing subsistence production and uninsured losses

| Nature of losses | Value of losses (US$ million) |
|---|---|
| Public infrastructure (i.e., roads, air transport, water supply and sanitation, electricity supply, telecommunications, education, health services, law and order) | 50 |
| Public housing | 25 |
| Private sector | |
| Insured losses (~92% commercial losses) | 45 |
| Uninsured losses[a] | 90 |
| Total | 210 |
| Gross national product, 1994[b] | 5085 |
| Minimum GNP cost of eruption | 4% |

[a] Note uninsured losses could have been as high as US$135±30 millions.
[b] GNP figures in this and the rest of the chapter are taken from the World Bank (1995) and later sources.
*Source:* Calculations are based on information in Blong and McKee (1995, pp. 34–39) and Blong (1997, personal communication).

who were affected in some way by the eruption only account for ~2.4% of the country's population. To put the issue into perspective, in Papua New Guinea there are 37 volcanoes which have been active in the Holocene (Chester, 1993) and there is a high probability of a future disaster much greater than that which befell this relatively small town.

It would be a mistake to assume that in terms of monetary losses the world may be split neatly into a rich *north*, where the absolute value of losses is high but GNP costs are low, and a poor *south* (Fig. 14.2a), in which absolute losses are low but the GNP costs are high. Iceland is one of the richest countries in the world (GNP per capita of nearly US$25 000 in 1995), but because it has a population of only 263 000, 1% of GNP only represents around US$65 million. In comparison, 1% of GNP in the USA at the time of the Mount St. Helens eruption represented US$32 billion, whereas 1% of GNP in Japan at the time of the Unzen eruption in 1991 equalled US$39 billion. With some 30% of Iceland at risk from the effects of volcanic activity and a high proportion of the population concentrated in such areas (Sigvalda-

son, 1983) (Fig. 14.4), coping with eruptions is not only extremely taxing for the authorities but also places a strain on the economy. Recovery often requires substantial outside assistance, as was the case following both the Heimaey eruption of 1973 and to a much lesser extent the eruption beneath the Vatnajökull ice-cap in 1996 (Chester, 1993; GVNB, 1996). Japan despite its wealth also illustrates complexity in the nature of its economic losses. As well as having 89 volcanoes that have been active in the Holocene and more than 100 eruptions that are known to have affected people and their activities (Chester, 1993), much of country is also threatened by earthquakes, such as the one which devastated the port city of Kobe in 1995, plus typhoons, landslides, and floods. Indeed, most parts of Japan have been affected by natural disasters, many within living memory. In 1989 the Tokai Bank forecast that a large earthquake in Tokyo could cost the country up to one quarter of its GNP and, because of the pivotal economic position of Japan, this would have significant consequences on the global economy (quoted by UNESCO, 1993). If this were to coincide with, or even occur close in time to, a

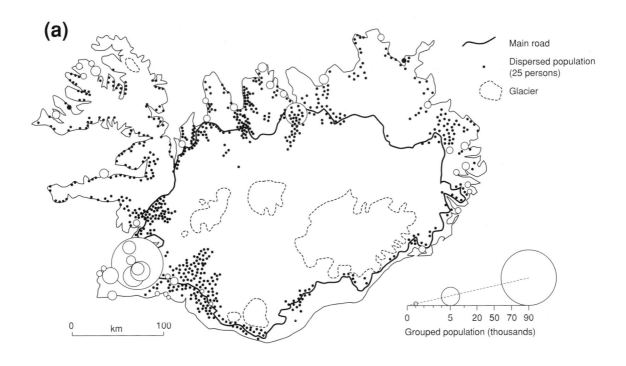

**(a)**

Main road

Dispersed population
(25 persons)

Glacier

0        km        100

Grouped population (thousands)
0    5   20 50 70 90

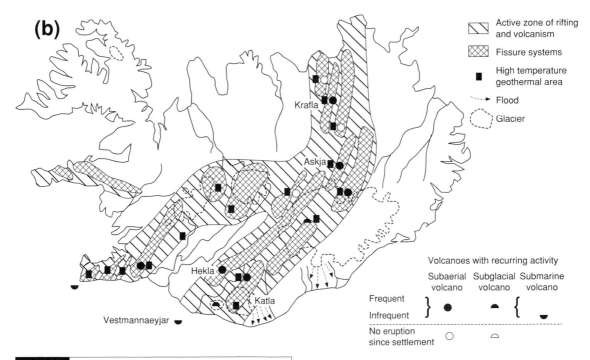

**(b)**

Active zone of rifting
and volcanism

Fissure systems

High temperature
geothermal area

Flood

Glacier

Krafla

Askja

Hekla

Katla

Vestmannaeyjar

Volcanoes with recurring activity

Subaerial    Subglacial  Submarine
volcano      volcano     volcano

Frequent

Infrequent

No eruption
since settlement

**Fig. 14.4.** Volcanic hazards in Iceland (a) Distribution and
density of population. (b) Active volcanic areas in zones of
rifting and fissuring (amended from Sigvaldason (1983)
Figs. 15.1 and 15.2, p. 194).

high-magnitude eruption of one of the country's many volcanoes – such as Mt. Fuji which overlooks Tokyo (Shimozuru, 1983; Hadfield, 1991a, 1991b) – the effect on the global economy would be catastrophic.

As Japan shows, the possible economic effects of eruptions are not confined to one country or region and are becoming increasingly global. In addition to these possible impacts on financial markets, there are three further areas in which potential future losses are currently causing concern, with eruptions even in remote areas being perceived as posing significant dangers. These relate to the impact of eruptions on air transport, upon weather and climate, and the possible effects of volcano-related tsunamis.

Air transport has increased rapidly in the past few decades and currently the fastest rate of growth is in Asia (Seager, 1995), much of it in the region of the circum-Pacific where many of the world's most active explosive volcanoes are to be found (Fig. 14.2c). More than 60 aircraft have been damaged by volcanic activity, the most notable cases being during the eruptions of Galunggung (Indonesia) in 1982, Redoubt (Alaska) in 1989, and Mt. Pinatubo (Philippines) in 1991 (Steenblik, 1990; Casadevall, 1991, 1994). Damage between 1980 and 1998 is estimated to have cost at least US$250 million and there were 12 significant incidents in the 1990s (Miller and Casadevall, 2000). The Redoubt incident alone cost US$80 million and estimates of the costs, including those of litigation, in the event of the total loss of a Boeing 747–700 carrying 300 passengers are as high as US$500 million at 1995 prices (quoted by Tilling, 1995, p. 376).

The nature of volcano-induced changes on weather and climate are reviewed elsewhere in this volume, but the effects of such changes on societies may be equally profound. Although the climatic impact of volcanic eruptions have been studied since the time of Benjamin Franklin (Franklin, 1789; Chester, 1988), it was only with publication of the seminal volume *Volcano Weather: The Story of 1816, the Year without a Summer* (Stommel and Stommel, 1983), that the full impact of eruptions on societies far removed from sites of activity became a major field of study. The evidence presented in *Volcano Weather* showed that the eruption in 1815 of Tambora (Indonesia) had a profound bearing on a wide range of political, economic, and social events in the United States and other countries. It is this recognition of the importance of eruptions, climate change, and humanity – locally and globally, historically and today – that makes this field so important. Volcanoes, climate, and human affairs constitute a present-day research frontier which is discussed at greater length later in the chapter.

The vast majority of tsunamis are generated by submarine tectonic earthquakes or seismically triggered submarine landslides and during the past 50 years few deaths and little devastation have been caused by volcano-induced events, though the harbor of Rabaul town was badly damaged during the 1994 eruption (Blong and McKee, 1995). There are, nevertheless, many examples in the historical record of major damage and many deaths being caused by volcano-induced tsunamis, such as the 5000–6000 estimated to have been killed when Unzen Volcano, Japan erupted in 1792 and the more than 32 000 deaths caused by the 1883 Krakatau eruption in Indonesia (Tanguy *et al.*, 1998). These figures are, however, dwarfed by estimates of current risk that have been made on the basis of past collapses on the Hawaiian Islands and in the Canaries (Moore and Moore, 1984; McGuire 1998; Carracedo, 1999; Ward and Day, 2001). In the case of the Canaries, modeling indicates run-up heights near to the source of up to 375 m and destruction spreading over thousands of kilometers. According to the most extreme and, hence, most unlikely scenario, large sections of the eastern seaboard of the USA would be threatened.

It is clear that the relationship between volcanoes and humanity is closer today than at any time in the past and that eruptions pose a much greater threat to people than is commonly supposed. Because of population growth, ongoing economic development, and global shifts in wealth and poverty, the need for an understanding of the relationship between volcanoes on the one hand and society on the other has never been greater or more pressing.

# Social theory and volcanic eruptions: defining a framework

The concept of hazards as external events impinging on unsuspecting people has been shed in favour of the interpretation that they emerge from interactions between people and environments.

J. K. Mitchell *et al.* (1989)

The related questions of why the innocent suffer during natural disasters and what can be done to reduce their impact are ones that have exercised the minds of people living in disaster-prone regions and those of scientists, philosophers, and theologians for thousands of years. They are issues of universal import which transcend time, place, and culture. In the present century and especially over the past 50 years, two paradigms have been prominent in the "western" tradition of hazard research: the *dominant* approach and its several *radical* alternatives. In contrast to the situation that obtained in the early 1980s (Hewitt, 1983a, 1983b), at present these approaches are not so polarized as they once were, with several research groups now combining a number of strategies (see Chester *et al.* (1999) for an example). There is, however, one way of considering the connections between volcanoes and society – including the hazards they pose – that is as old as the earliest written records and which still persists, albeit with decreasing force – to affect the ways in which people react to disasters. This is theodicy.

## Perspectives from theodicy

*Theodicy* is defined as any attempt to reconcile deistic belief with human suffering. The word theodicy was first coined by the philosopher Gottfried Wilhelm Leibniz in 1710 (Leibniz, 1710), though societies have always attempted both to blame and to propitiate deities when confronted with natural disasters, including volcanic eruptions. In his classic study, *Problems of Suffering in the Religions of the World*, John Bowker (1970) has shown that suffering – caused by both extreme events of nature and by direct human action (e.g., wars, terrorism, civil commotion, and genocide) – has been a concern of first-order importance not

only to the major monotheistic faiths of Judaism, Islam, and Christianity, but has also been a significant strand in Hinduism, Buddhism, Zoroastrianism, Manichaeism, and Jainism. Bowker (1970) goes on to show how suffering is central to much Marxist thought, being a consequence of the operation of capitalism. Because these and other faiths exercise such a profound influence on many aspects of society and culture, theodicy is vital to any understanding of the ways in which people react to volcanic eruptions.

The literature is rich with examples of societies which have viewed eruptions as being expressions of deistic wrath, the only appropriate response being appeasement. Some selected examples are listed in Table 14.3, though it should be noted that accounts that are based on the writings of outside observers should be viewed with caution because cultural presuppositions and prejudices may have been read into the reactions of indigenous peoples. Patricia Plunket (1998, personal communication) notes that many views of pre-Columbian cultures in South America may have been strongly colored by informants who had already been conditioned by processes of conquest and Christian evangelization. Although punishment and vengeance seem to be the most common theodicies, it should not be forgotten that certain cultures have viewed things differently. Vengeance and wrath have not been, for instance, the only theodicies within the Christian tradition, though they were until recently the most prominent (see Chester, 1998; Chester *et al.*, in press), while other traditions of faith sometimes viewed volcanoes as agents bestowing divine favor. Russell Blong (1982, 1984, p. 348) describes the case of a group in Papua New Guinea who so successfully adapted their agriculture to frequent falls of ash that they developed elaborate religious ceremonies to encourage recurrence. Similarly, Mt. Fuji is not only the symbol of Japan, but has also been a focus of worship for many centuries (Fisher *et al.*, 1997).

It cannot be stressed too strongly that attitudes to eruptions involving punishment, divine vengeance, and propitiation are neither confined to historical examples, nor are they merely facets of societies that have been relatively untouched by the forces of modernism and postmodernism.

**Table 14.3** | Selected historical and pre-historic examples of societies which have perceived eruptions as being expressions of divine wrath and vengeance, the only appropriate response being divine appeasement

| Example | Nature |
| --- | --- |
| Mediterranean Europe in the "classical" period | The Etruscan god Velkhan was primarily a god of destruction, but also one of productive fire and the hearth. Hephaestos (Latin equivalent Vulcan) was normally viewed as a constructive craftsman and this aspect of volcanism is stressed in much Latin and Greek literature, but volcanoes are also personified in the form of the destructive and angry Titan. Hephaestos, cripple and craftsman, represented both human brilliance and a reminder of human mortality and reliance on divine benevolence. Vulcan was a long-standing source of worship in archaic and classical Rome. Lucilius Junior (first century AD) notes that on Etna people offered incense to propitiate the gods who were thought to control the mountain (Hyde, 1916; Doonan, 1997; Chester et al., 2000, 2003). |
| Vesuvius (Italy) and Mount Etna (Sicily) from the "classical" period until ~AD 1900 | The skull and two vials of the blood of St. Januarius (San Gennaro), who was martyred in the reign of Diocletian (AD 285–305), were often appealed to by the citizens of Naples during eruptions of Vesuvius. These appeals to God included intercession and displays of the saint's relics at flow fronts in 685, 1631, 1707 and 1767 (Fisher et al., 1997). In AD 252 the people of Catania used the veil of St. Agatha, who had been martyred the previous year, in order to halt a lava flow. This and other relics were used on many occasions (Chester et al., 2000). So efficacious was the veil thought to be, that following the Lisbon earthquake in 1755, many in Portugal believed that St. Agatha should be adopted as their saint to prevent a recurrence. In the event the Iberian St. Francis Borgia was afforded this honor (Kendrick, 1956, p. 72). |
| Slopes of Popocatépetl (Mexico) | Settlements on the slopes of the volcano were destroyed by a Plinian eruption around 50 BC. Excavations show shrines with effigies dedicated to the "volcano" god, suggestive of attempts at divine appeasement (Plunket and Uruñuela, 1997). |
| Hekla, Iceland | In the twelfth century AD and according to Cistercian monks, Hekla was the gateway to hell: a symbol to deter heresy. Such views persisted until at least the sixteenth century (Blong, 1984). |

*Source:* Based on the sources cited; further examples may be obtained from Fisher et al. (1997).

The veil of St. Agatha and other saintly relics (Table 14.3) have been used during many of the eruptions of Etna which have occurred during the present century (Duncan et al., 1996). In the Puna District of Hawaii, many older native Hawaiians have been observed to take little action to protect themselves from losses caused by lava flows, because it is the goddess Pele who controls human fate (Murton and Shimabukuro, 1974; Warrick, 1979). Moreover, even following the eruption of Mount St. Helens (USA) in 1980, some radio evangelists portrayed certain deaths as a warning from God about the evils of drinking strong liquor (quoted by Blong, 1984). Notwithstanding these examples, simple theodicies of vengeance and divine anger have become less prominent features of responses to volcanoes over time. Part of the reason is that the "western" world-view, or *Weltanschauung*, has become more scientifically and rationally focused since the Enlightenment, with eruptions being viewed as expressions of purely natural, rather than supernatural, forces. In the present century, these views have spread not only through mass communications but also by processes of modernization. Today they constitute the *Zeitgeist*. A second reason is that over the past 50 years more sophisticated models of theodicy have become prominent (see Chester (1998) for a review). In the Christian tradition and with reference to disasters in developing countries, one example of this new attitude is that it is now widely recognized that, whilst human sinfulness is expressed in natural disasters, this is quite different from the traditional view which sees individual sin as being punished by the wrath of God in the form of a disaster. Human sin is rather the structural sinfulness of global disparities in wealth, poverty, and power which are expressed in the high number of deaths and major economic losses caused by disasters in developing countries. According to John Schild, an Anglican (i.e., Episcopalian) priest and writer, the proper Christian response to, say, the Montserrat eruption is to "help (people) to hold out while the volcano rages, and in the job of reconstruction when (the eruption) finally (ends)" (Schild, 1997, p. 2). Such a theodicy may also involve aligning Christian opinion with politically and economically progressive forces within developing countries (Chester, 1998). The fact remains, however, that perceived divine vengeance persists as a factor of prime importance that should be taken into account by policy-makers and scientists seeking to reduce the impact of volcano-related disasters (see Fisher *et al.* (1997) and Homan (2001) for an Islamic perspective on earthquake losses in Egypt). It is often the fine detail of culture and place, as much as the broad-brush factors of economic development and the magnitude of physical processes, that determine whether responses to a volcano-related emergency are successful or not.

## The *dominant* perspective

Before the 1980s, there was one way of studying volcanoes and people including the hazards posed by eruptions which eclipsed all others. This became known as the *dominant* approach or paradigm. It was developed in the late 1930s and early 1940s by the American scholar Gilbert White as he considered the range of possible policies which could be introduced to reduce flood losses in the USA. White was inspired by, socialized into, supported and worked within an ethos defined by the interventionist environmental policies initiated by President Roosevelt's administration and, as James Wescoat (1992) has shown, there are broad affinities between the *dominant* approach and the pragmatic tradition of American social thought, especially that espoused by the philosopher John Dewey, whose ideas were particularly influential in the first half of the twentieth century. Dewey's pragmatism stresses four themes:

> the precariousness of human existence;
> a pragmatic – that is practical – focus to scientific investigation;
> learning from experience; and
> open public discourse within a democratic framework.

Although the direct influence of Dewey on White is difficult to establish with certainty, their agendas are remarkably similar (Wescoat, 1992, Fig. 1). Greater confidence may be placed, however, in the assertion that White's pacifist beliefs – so prominent in his writings – were inspired by a personal commitment to Quakerism. "Each of us should ask what in his teaching and research is helping our fellow men [sic] strengthen their capacity to survive in a peaceful world" (White, 1972, p. 322; see also Wescoat, 1992). Between the 1950s and late 1970s, Gilbert White was joined by like-minded scholars, most notably by Robert Kates and Ian Burton, and their approach evolved into a set of techniques that were used not only to study flooding in the USA, but also the totality of natural hazards occurring throughout the world. Because the paths of these pioneer workers crossed at the University of Chicago, an alternative title for the dominant approach is the Chicago School. These scholars quickly achieved a commanding influence on the research literature, which in turn controlled the ways in which policies of hazard reduction were framed by national governments, the United Nations Disaster Relief Office and, more latterly, many academics and policy makers working within the context of the International Decade for Natural Disaster Reduction (IDNDR).

The characteristics of the dominant approach are complex in detail and further information may be obtained from Burton *et al.* (1978), Warrick (1979), Whittow (1987), Chester (1993), Hewitt (1997), and especially Alexander (2000), but simplification is possible. White and his co-workers accepted that factors such as material wealth, experience of hazardous events, systems of belief, and psychological considerations (e.g., Simon, 1957, 1959) are all important in controlling how individuals, social groups, and, indeed, whole societies respond to disasters; but their paradigm is strongly focused around the idea that there exists a range of *adjustments* to natural hazards that are available to individuals and/or societies to deal with extreme events of nature. Table 14.4 is an example of adjustments currently available to deal with hazards from lava flows, and similar tables have been constructed for virtually all categories of volcanic hazard (see Chester, 1993, Ch. 8).

"Bearing the loss" has been the involuntary adjustment which societies have adopted through most of their histories. It is still the situation which obtains in developing countries today. In many so called *pre-industrial* or *folk* societies strategies of harmonizing with nature, combined with often complex sociological adjustments to hazards, have featured (White, 1973). This may be illustrated by reactions to eruptions of Mt. Etna in Sicily over the long period from the close of the Classical Era to the beginning of this century (Table 14.5). With the onset of economic development all changes. When disaster strikes a modern *technological* or *industrial* society, emphasis is placed on shifting the burden of losses from the individual, family, and isolated community to the society, nation, and international agency. Loss-sharing becomes important through aid transfers and insurance, and technology assumes a central role in reducing the threat from subsequent events. Responses focus on control over nature, a narrowing of the range of adjustment, high costs and inflexibility. Finally, there is the *post-industrial* response, which incorporates the best of the *pre-industrial* and *industrial* responses; this response represents a future ideal and it is doubtful whether any society has yet reached this stage in its disaster planning, though Iceland, the USA, and Japan approach it (Chester, 1993). For a given society the three stages – pre-industrial, industrial, and post-industrial – may be sequential but this is not necessarily the case, because features of more than one type of response may be found at the same time amongst different cultural groups and/or regions within a given country or society (White, 1973).

Alternatives to loss-bearing also involve measures to: modify the hazardousness of an extreme event, modify its loss potential, and adjust to losses (see Table 14.4). Such initiatives have been introduced for a wide range of volcanic phenomena in most developed countries of the world. These are discussed in other chapters of this volume and in standard works written or edited by: Walker (1974), Tazieff and Sabroux (1983), Crandell *et al.* (1984), Latter (1989), McGuire *et al.* (1995, 1996), and Sigurdsson *et al.* (2000), to which further reference should be made. Regardless of the type of volcanic hazard, policies conceived within the dominant framework have been successful in reducing deaths and injuries in developed countries and, as discussed earlier, although total economic losses have increased as a result of processes of development and wealth accumulation, the relative toll – expressed as a percentage of national wealth – has typically fallen. There is no doubt that for developed countries the introduction of policies conceived within the dominant paradigm has been a great success, and this is clear from statements issued by official agencies: *Living with Volcanoes,* by Thomas Wright and Thomas Pierson (1992) and published by the US Geological Survey, being an excellent example of this genre.

**Table 14.4** The theoretical range of adjustments to hazards from lava flows. It should be noted that similar tables have been constructed for most types of volcanic hazard. Under the dominant paradigm the column headers remain the same, but the adjustments vary

| | | | | Adjust to losses | | |
| --- | --- | --- | --- | --- | --- | --- |
| Affect the cause | Modify the hazard | Modify the loss potential | Spread the losses | Plan for losses | Bear the losses |
| No known way of altering the eruptive mechanism | (1) Protect high-value installations (2) Alter lava flow direction (3) Arrest forward motion | (1) Introduce warning systems (2) Prepare for a disaster through civil-defence measures (3) Introduce land-planning measures to control future development in particularly hazard-prone areas | (1) Public relief from national and local government (2) Government-sponsored and supported insurance schemes (3) International relief from agencies such as the United Nations Disaster Relief Office | Individual family or company insurance | Individual family, company, or community loss-sharing |
| *Examples and notes* | Use of explosives and bombing to divert flows; has been tried in Hawaii and Etna | Warning systems only available on certain well-monitored volcanoes, in technologically advanced countries, e.g., USA (Hawaii and volcanoes showing signs of activity in the continental USA), Japan, Iceland, and Italy | Public relief available in most countries; the most comprehensive schemes are in the technologically most developed countries, e.g., Canada, USA, Japan, and New Zealand | Possible to a certain extent in more developed countries, but even in the USA it is limited by the discretion of individual companies | This is the traditional form of adjustment and is still widely practiced in many volcanic areas |

(cont.)

**Table 14.4** (cont.)

| Affect the cause | Modify the hazard | Modify the loss potential | Adjust to losses | | |
| --- | --- | --- | --- | --- | --- |
| | | | Spread the losses | Plan for losses | Bear the losses |
| | Emergency barriers tried in Hawaii, Japan and Etna<br><br>Barriers to divert future flows from inhabited areas; have been suggested for the town of Hilo (Hawaii)<br><br>Control forward advance by watering the flow margin; limited success in Hawaii and Heimaey (Iceland) | Emergency evacuation plans have been formulated in several countries, e.g., USA, Japan, Soufrière de Guadeloupe, Italy, and the Azores<br><br>Land-planning policies are in operation in some areas where "general prediction" and hazard mapping have been carried out | Government sponsored insurance schemes available in several countries, e.g., the Russian Federation and New Zealand<br><br>UN Disaster Relief Office established only in 1972; initiatives during the International Decade for National Disaster Reduction (IDNDR) may be significant<br><br>May be of great benefit to developing countries in the face of major losses in the future; international relief given by many developed countries in the past | | |

*Source:* Modified from Chester (1993, Table 8.2), based on Burton *et al.* (1978); Chester *et al.* (1985), and numerous other sources.

**Table 14.5** | Pre-industrial responses to eruptions of Mt. Etna, Sicily; this pre-industrial society developed complex mechanisms to heighten its resilience to frequent lava incursions

| Features of pre-industrial society | Responses on Etna from the Classical period to ~AD 1900 |
|---|---|
| *As identified by White (1973)*<br>1. A wide range of adjustments<br>2. Action by individuals or small groups<br>3. Emphasis on harmonization with, rather than technological control over, nature<br>4. Low capital requirements<br>5. Responses vary over short distances<br>6. Responses are flexible and easily abandoned if unsuccessful | With the exception of maintaining law and order, distributing bread and, in some nineteenth-century eruptions, providing limited financial aid, the role of the state was minimal. Regardless of whether eruptions were interpreted in terms of religious, mythological, or "scientific" world-views, families had to face the reality of loss bearing. There is plentiful evidence that communities were able quickly to recover. Sicilian society was in the past and is today based on extended families and client relationships, and in coping with disasters these networks were vital. Following many eruptions people left their villages to live with relations, leaving those without alternative accommodation to "live in the fields" but many farmers possessed permanent shelters located on their family plots and these were frequently used. Peasant agriculture involves maximizing security rather than profit, and landholding was and is fragmented. Rarely did farmers lose all their land in an eruption and incomes were bolstered by transhumance pastoralism. It is a popular belief that people panic in disasters, but on Etna people usually stayed calm. During many eruptions villagers removed and stored all that could be salvaged from their homes and formed self-help relief committees. |
| *As identified by other writers*<br>7. Losses are perceived as inevitable<br>8. Responses continue over time periods ranging from hundreds to thousands of years | Given time lava will be recolonized by natural processes, but there is evidence that peasant agriculturalists harmonized with nature and both assisted and profited from these processes. Once grasses were established so also was pastoralism and later hardy tree crops — almonds, figs, and pistachios — were deliberately planted. Interpretation of land-use maps and aerial photographs indicate that farmers had an accurate knowledge of recolonization potential. Recolonization was not without innovation and prickly pear cactus (*Opuntia ficus-indica*) was introduced from South America to give farmers a new technique in their quest to assist natural processes of recolonization, the plant's powerful roots being used to help break lava flows. |

*Source*: Based on Chester *et al.* (in press).

Until very recently and in order to spread the benefits of such policies more widely, national governments and international agencies have been wedded to agendas of loss reduction which are strongly grounded within the dominant paradigm. Although no hazard can exist unless there is a human population to be affected, the primary emphasis has been on understanding, controlling, and modifying physical processes by technological means (e.g., hazard mapping and volcano monitoring). Physical processes are considered to be the first-order determinants of a disaster and differences between societies are relegated to a lower, albeit still significant, level of importance (Chester, 1993). Focus on the physically determined nature of natural disasters is reinforced further because conventionally research into hazard reduction has been under the control of scientists with backgrounds in geology, geophysics, engineering, and emergency medicine. In many countries, a physically deterministic locus is present in the mission statements of the agencies that are charged with reducing vulnerability to the effects of volcanic eruptions, and this has tended to apply even when states are at markedly different levels of economic development. The aims and objectives of the US Volcano Hazards Program (Filson, 1987; Wright and Pierson, 1992) are, for instance, remarkably similar to those of Argentina (Zupka, 1996) and many other countries of Latin America (Anonymous, 1994b). The policy aims of national governments and international agencies are defined largely in terms of the transfer of technology and administrative experience from areas where responses are observed to have been successful (i.e., developed countries) to those where they are either non-existent or perceived to have failed (i.e., developing countries) and this thinking is both implicit and explicit in many of the statements which emerged early in the International Decade of Natural Disaster Reduction (IDNDR) and in the years leading up to its inception. Typically, Professor Michel Lechet, a member of the United Nations IDNDR Scientific and Technical Committee (Lechet, 1990), argued that during the decade technology transfer should be one of the principal objectives and for many – but by no means all – participants at international conferences (e.g., Anonymous, 1992, 1993), it was a factor of overwhelming importance.

## Volcanic hazards: radical alternatives

In 1983 Kenneth Hewitt edited a volume, *Interpretations of Calamity*, which has become a classic statement on the complexities inherent in natural-disaster planning (Hewitt, 1983a, 1983b). The book draws on ideas that were inchoate in the 1970s (e.g., Hewitt, 1976, 1980; O'Keefe *et al.*, 1976; Wisner *et al.*, 1976, 1977) and represented a conflation of many disparate strands to produce a critique of the dominant approach. Since 1983, further contributions have been published (e.g., Whittow, 1987; Chester, 1993; Hewitt, 1997; Alexander, 2000). Today radical critiques are exerting a major influence upon the ways in which natural hazards are being studied by scientists, social scientists, and policy-makers.

Focused on developing countries and natural disasters – such as droughts which have a long onset time, are of long duration, and cause damage to large areas – scholars of a radical inclination have disputed the success of measures conceived under the dominant paradigm. Although the arguments they use are involved (reviews Hewitt, 1983a; Chester, 1993), the crux is that most disasters in developing countries have more to do with poverty and deprivation than with extreme meteorological and geophysical events. Kenneth Hewitt (1983b, p. 26) poses a question: "What is more characteristic (for the inhabitants of the frequently drought ridden areas of the Sahel) . . . and to be expected by its long term inhabitants: recurrent droughts or the history of political, economic and social change?" Taking the analysis further, Susman *et al.* (1983) make use of the Marxist concept of *marginalization*, by arguing that the people who suffer most in natural disasters are those who are either economically marginalized (i.e., they are poor) and/or geographically marginalized (i.e., they live in areas that are prone to disaster losses). Relief aid and technological transfers tend to benefit those people who are already well off and can lead to further marginalization of the poorest sections of a community.

The most radical of the radical critics, like Paul Susman *et al.* (1983), propose an openly

Marxist agenda and it comes as no surprise to find that this has not proved popular with international agencies such as the United Nations and the World Bank. What is important about the radical critique is that it emphasizes the uniqueness of place: Hawaii is not Etna, Iceland is not the Azores, and the volcanoes of the Andes and those of the Cascades occur in quite different environments and societies. Even when discussion is restricted to developed countries, successful hazard reduction depends critically not only on understanding volcanological processes per se, but also the impacts these will have on (a) the wider physical environment and (b) the fine detail of the socio-economic conditions and cultural milieu of the society in question. In short, the thrust of the radical critique is that responses and adjustments to hazards must be sensitive to the local environment and "incultured" if they are to be successful.

## Volcanoes and people: an evolving framework of study

A superficial reading of the analysis undertaken by scholars of a radical persuasion might lead to the conclusion that their ideas are of little interest not only in trying to understand the relationships between volcanoes and human society, but also, and more specifically, in proposing strategies for hazard reduction. In contrast to hazards such as drought and desertification, which are the foci of the radicals' concern, volcanic eruptions are usually of short onset, and apart from possible global effects on weather and climate, volcanoes usually affect spatially limited areas and events are, with the exception of longer episodes of "restlessness" which may or may not culminate in a major eruption (Tilling and Lipman, 1993), typically of short duration. Furthermore, virtually all measures to forecast volcanic eruptions require the application of science and technology and high levels of administrative expertise, all of which are firmly grounded within the dominant paradigm. It comes as no surprise to find that applied volcanologists have conventionally been only too eager to anchor their research within this approach. In recent years, however, many elements of the radical critique have been incorporated into day-to-day

practice, so that research is of a different hue from that which was conducted only a few years ago. There are three reasons for this change.

First, some elements of the radical critique have proved too trenchant and persuasive to ignore. Even at the start of the IDNDR in 1990, it was notable that the volcanological community eschewed the dominant paradigm in its purest form, with the International Association of Volcanology and Chemistry of the Earth's Interior (IAVCEI, IDNDR Task Group, 1990) defining an affordable and culturally and socially sensitive program for hazard reduction in developing countries. Volcanology has always had international outlook. There are few researchers who have not visited or worked in countries and cultures other than their own, and the proposed IAVCEI program showed a perspicacious awareness of the cultural, political, and social differences between places, yet at the same time sought to innovate good scientific and administrative practice (Table 14.6). It is perhaps not coincidental that the committee that drew up the program had six out of its 11 members drawn from developing countries (Chester, 1993, p. 308). This strongly "incultured" tone did not become the common wisdom of hazard analysis until much later in the 1990s, as is clear in the Yokohama Strategy published following the World Conference on Natural Disaster Reduction held in 1994 (United Nations, 1995).

Second, important lessons have been learnt from studying the many ways in which eruptions have been dealt with in the different societies and cultures that are to be found on active volcanoes. In 1993 and following a review of major eruptive events which had occurred in the previous half century I argued that many writers, myself included, had either explicitly or implicitly adopted a framework that was based on the recognition of just two causal variables: the *physical characteristics* of the activity and the *development level* of the society in which the eruption took place (Chester, 1993) (Fig. 14.5). I further contended that such a classification was sorely deficient. With regards to the physical characteristics of activity, although the magnitude and frequency of extreme events correlate broadly

**Table 14.6** | Initiatives proposed at the beginning of the International Decade for Natural Disaster Reduction (IDNDR) by the International Association of Volcanology and Chemistry of the Earth's Interior

| Initiative | Details |
|---|---|
| Hazard and risk mapping | Reconnaissance mapping of hazards (i.e., *general prediction*) and risks at previously unmapped volcanoes |
| Volcano surveillance | Baseline monitoring and minimum surveillance (*specific prediction*) at dangerous volcanoes that are not presently monitored; emphasis on affordable procedures and involvement of local people |
| Public education | Improved education about volcanic hazards (e.g., community talks, films, videos, community field trips, observatory open days, workshops in schools, involvement of local volunteers in surveillance, symposia for public officials and decision-makers, simulation of evacuations and establishment of relationships between cities with similar threats) |
| Dialog with public officials | Scientists should discuss emergencies and land-use planning |
| "Decade Volcano" demonstration projects | Proposed that about ten volcanoes be chosen for integrated, multinational, and multidisciplinary study, to demonstrate the range of activities required for mitigation |
| IAVNET (internet) | Electronic mail to be used to allow co-operation and advice before and during an eruption |
| Reference materials | Development of archives on dangerous volcanoes |
| Training | For both new and experienced scientists, civil-defence officials, and planners |
| Low-cost equipment | For developing countries, equipment should be developed which is affordable, reliable, and easy to repair |
| Satellite monitoring | To collect images of geological features and $SO_2$ plumes, etc.; satellites to be used to secure communications during eruptions |
| Crisis assistance | (a) Supplement local personnel during a crisis; (b) build up local expertise; (c) help local scientists and officials prepare for eruptions during periods of calm; and (d) help national scientists become self-sufficient |
| Seed money | Encourage local initiatives and the generation of finance |
| Publication | Wide dissemination of lessons learn from eruptions |

*Source:* IAVCEI, IDNDR Task Group (1990), modified from Chester (1993, p. 309).

**DEVELOPMENT LEVEL**

|  |  | DEVELOPED RESPONSE | UNDERDEVELOPED RESPONSE |
|---|---|---|---|
| **PHYSICAL CHARACTERISTICS** | High magnitude/ low frequency | Examples include: Mount St. Helens, USA (1980) <br><br> Ruapehu volcano, New Zealand (1953) | Examples include: El Chichón, Mexico (1982) <br><br> Agung, Indonesia (1963-64) <br><br> Arenal, Costa Rica (1968) |
|  | Low magnitude/ high frequency | Heimaey, Iceland (1973) <br><br> Mount Etna, Sicily (1983) <br><br> Kilauea, Hawaii (many) | Nyiragongo, Zaire (1977) <br><br> Karthala, Comores, Indian Ocean (1972) |

**Fig. 14.5.** A classification of responses to historical eruptions based on the physical characteristics of the eruption and the level of economic development attained by the country in question. From Chester (1993, Fig. 8.5, p. 247).

with both casualty figures and damage, many eruptions are more individualistic in the threat they pose than is commonly admitted. To quote but one of many examples, the 1985 Nevado del Ruiz eruption in Colombia was small scale and very similar to many other such events in South America in terms of its magnitude and frequency, but it was the interaction with snow and ice at the summit of the volcano that generated the lahars which killed ~25 000 people. More recently, this argument has been given additional force because it is clear that volcanoes may present many and varied dangers even when they are not in eruption, due to complex processes leading to instability, including landslides and the generation of tsunamis (see McGuire *et al.*, 1996; McGuire, 1998). There is also little doubt that the level of economic "development" achieved by a society is related to disaster outcomes, but it has become increasingly clear that "development" is a much more complex concept than many hazard analysts have admitted in the past. Terms such as *north* and *south*, *first*, *second*, and *third worlds*, *developed* and *developing* are but

"euphemisms for a broad classification of the world into rich and poor countries . . . no one set of terms is entirely satisfactory and some are pejorative, implying that a low economic status equates with a position of social and/or cultural inferiority" (Chester, 1993, p. 244). Development in its broadest sense implies, *inter alia*:

i. increases in human capital, through education, training, knowledge and information;
ii. increases in non-human capital, through investment in industry, commerce and communications and
iii. improvement in economic and social organisation, in government, the civil service and local agencies.
(O'Riordan, 1976)

Hence, the classification of countries shown in Fig. 14.2a is oversimplified and misleading. Currently the United Nations classifies countries by means of a Human Development Index. This is reproduced as Fig. 14.6a and it is useful to compare this figure with Figs. 14.2a and b and 14.6b. There is no one geography of volcanic risk and benefit, but many (Hewitt, 1997).

In a pioneering study, Russell Blong reviewed the social impact of four eruptions: Mt. Lamington, Papua New Guinea, 1951; Parícutin, Mexico, 1943–52; Tristan da Cunha, South Atlantic, 1961–2, and Niuafo'ou, Tonga, 1946, and

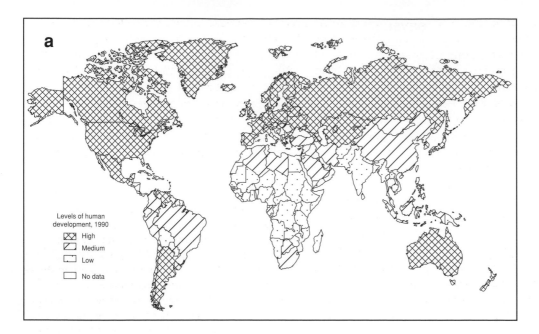

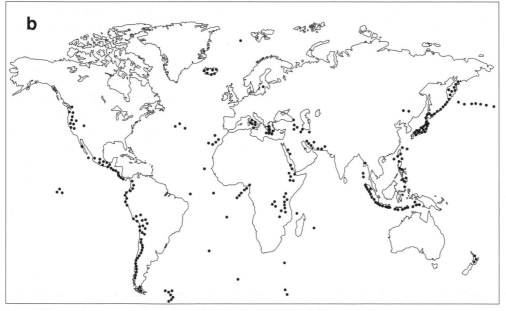

**Fig. 14.6.** World maps showing: (a) Levels of human development 1990. This is based in the UN Human Development Index, a combined measure related to income, literacy rates, education, and life expectancy (based on several sources). The rating of countries according to the Human Development Index underlines the point that income is not the sole factor determining the level of development of a country and that other factors may influence the impact of a disaster. There are many factors that need to be taken into account in determining vulnerability. (b) Principal active volcanoes of the world (various sources).

concluded that outcomes were determined by the broad socio-economic evolution of these countries and that eruptions hastened, but did not alter, existing social trends (Blong, 1984). Later studies have supported Blong's conclusions and have emphasized that it is the fine detail of a society and culture that largely determines both the nature of its relationships to volcanoes in general and to eruptions in particular. Until the mid-1980s, for instance, Italy had an unenviable record of disaster response and recovery despite being a wealthy country, with a GNP per capita of ~US$8570 in 1986, rising to over US$19 000 in the late 1990s (Duncan *et al.*, 1996). Conversely some very poor countries – the best example probably being Indonesia (GNP per capita US$980 in 1996) – managed to cope with a range of natural hazards in a manner that was more redolent of a "rich" country (Zen, 1983; Kusamadinata, 1984; Chester, 1993). Disaster reduction has been a priority within Indonesia's development planning and a mixture of technological – "western" – and lower-cost – "indigenous" – techniques has been used. Italy, in contrast, has had deep-seated historical, cultural, economic, and political difficulties with natural disasters which until recently have caused responses to be less effective than they should have been given the country's wealth. Until major organizational changes occurred in the mid-1980s, Italy had to cope with eruptions and other natural disasters within the context of a system of civil defense that was inefficient and even corrupt. Disaster planning was often subverted and funds destined for relief were sometimes expropriated (Chester *et al.*, 1985).

Since the above comments were first aired, further lessons have been learnt and are now widely disseminated within the volcanological community. In a corpus stretching back to the mid-1980s several American scholars, including Richard Fiske, Peter Lipman, Donald Peterson, Barry Voight, and especially Robert Tilling, have studied human responses to eruptions, by identifying those factors that influence success or failure. In particular they have focused attention on the often critical factor of effective communications between scientists, officials, and the public at large. They argue that this can make all the difference between a successful and an unsuccessful response (Fiske, 1984; Peterson, 1988, 1996; Tilling, 1989, 1995; Peterson and Tilling, 1993; Tilling and Lipman, 1993; Newhall and Punongbayan, 1996; Scarpa and Tilling, 1996; Voight, 1996). Table 14.7 summarizes the work of these scholars. In particular problems are seen to occur when there are significant departures from what is termed an *idealized scenario* which has the following attributes:

(a) scientists have the personnel, equipment and other resources enabling them to progress effectively towards their major goals of understanding the past through geological studies, the present through monitoring and the future utilizing the interpretations built from the past and present; (b) *the scientists maintain full communications with civil authorities and other members of the community and the region*, all of whom are actively interested in and responsive to the information; (c) the authorities of the community and the region respond to the scientific information by developing a land-use plan that is cognizant of the hazards, and they also develop and regularly rehearse emergency plans.

(Peterson and Tilling, 1993, p. 347; my emphasis)

The third reason for change is that there have been further developments in social theory (see Alexander, 2000). In 1994 Piers Blaikie and his colleagues proposed an integrated model – called *vulnerability analysis* – which served to synthesize existing approaches. This model is not only concerned with the physical processes which lead to losses, but also studies the deep-seated root causes of disaster, better to understand how these are channeled by social and economic mechanisms into unsafe conditions for a population (Blaikie *et al.*, 1994). Developing vulnerability analysis for volcanic regions involves conflating conventional hazard analysis with the study of those aspects of the wider physical environment, culture, and society, which either exacerbate or diminish its susceptibility to losses and potential for recovery. Volcanic activity may interact with the physical environment (e.g., the effects of an eruption on river flow, mass movement, and debris flow generation), but the key to understanding vulnerability is to recognize that there are aspects of risk which are independent of a region being volcanic and relate

| Table 14.7 | Factors identified as being critical to successful responses during volcanic crises |
|---|---|

*Effective communications*

Small frequent eruptions induce regular and effective contact between scientists, the authorities, and the populace (e.g., Sakurajima and Oshima (Japan) and Kilauea (Hawaii). Conversely, uncertainty about the outcome of volcanic unrest induces poor relations when the eruption eventually occurs (e.g., Mount St. Helens, USA and Nevado del Ruiz, Colombia). Poor relations have been features of volcanoes which are predicted to erupt, but in the event did not do so (e.g., Long Valley, USA, Rabaul, Papua New Guinea, and Campi Flegrei, Italy). Producing accurate information should be considered an ethical obligation for scientists. Effective communications occur when information is clear and unambiguous (e.g., Galunggung and Colo, Indonesia). False alarms cause strained relationships and scientists must help in the development of the public understanding of "aborted eruptions" (see Banks *et al.*, 1989). The scientist–journalist relationship is crucial.

*Forewarned is forearmed*

Adequate geological and hazard reduction studies before an eruption are vital to success; in this way relationships with authorities and people may be developed in advance and this will aid communication during a crisis.

*Knowledge of society and culture*

Scientific advances though "satisfying and important, are worthless in saving lives or minimizing . . . damage unless they are well integrated with effective communications and interaction between volcanologists, civil authorities and the affected populous" (Scarpa and Tilling, 1996, p. vii). Sociological perspectives are crucially important and "volcanologists must be keenly aware of the characteristics of the society . . . they are seeking to serve" (Peterson, 1988, pp. 41–68).

*Administrative effectiveness*

The Nevado del Ruiz (Colombia) eruption of 1985 shows that tragedy may arise out of cumulative human error over a long period "by misjudgment, indecision and bureaucratic short-sightedness" (Voight, 1988, p. 30).

*Education*

Education must be comprehensive, understandable, and involve local people, decision-makers, and government authorities. Volcanologists must be acutely aware of the society in which they are working.

*Source:* Based on Fiske (1984), Peterson (1988, 1996), Voight (1988), Tilling (1989), Peterson and Tilling (1993), Tilling and Lipman (1993), and Scarpa and Tilling (1996).

to dynamic changes in population, history, culture, and politics. An eruption may be the trigger, but is often not the underlying cause of a disaster. The examples of Indonesia and Italy fit easily into this framework. The inhabitants of the Bay of Naples are vulnerable because of the area's history and geography and vulnerability is expressed in unemployment, poor housing, and crime, as well as in disaster losses. The structural vulnerability of Naples, moreover, continues unabated (see Masood, 1995), though there are some signs of improvement (Chester *et al.*, 2002).

In the early years of the twenty-first century, hazard analysts have placed renewed stress on the importance of involving local communities in hazard reduction programs (e.g., Homan, 2001; Wisner, 2001). Arguing that much work to date has been over directed by government, scientists, and bureaucrats – what is termed "top–down" – these authors argue that to be successful plans should be more "bottom–up." This means that local people should be consulted at all stages in the planning process, so that they feel a sense of "ownership" and involvement.

| DEVELOPED RESPONSE | UNDERDEVELOPED RESPONSE | |
|---|---|---|
| High-magnitude/low frequency events | High-magnitude/low frequency events | Particular responses |
| Examples include: Mount St. Helens, USA (1980); Unzen, Japan (1991) | Examples include: El Chichón, Mexico (1982); Pinatubo, Philippines (1991) | Examples include: |
| Low-magnitude/high frequency events | Low-magnitude/high frequency events | Italy, many eruptions; La Soufrière, Guadeloupe (1976) and Soufrière, St. Vincent (1979) |
| Examples include: Heimaey, Iceland (1973) | Examples include: Island of Njazidja, Comoro Islands (1972, 1977, 1991) | |
| Volcano-related events | Volcano-related events | |
| Examples include: Numerous examples from Iceland, Japan, and the USA | Examples include: Nevado del Ruiz, Colombia (1985) | |

**Fig. 14.7.** The author's 1993 classification of responses and adjustments to volcanic hazards (Chester, 1993, Fig. 8.6, p. 248).

The greater knowledge of volcanic processes, the trenchancy of much of the radical critique, the lessons learnt from the first-hand study of volcanoes and eruptions, and developments in social theory imply that a new framework for the study of volcanoes and society is required, which is more sophisticated than that shown in Fig. 14.5. Because of the rapid progress of research during the early and middle years of the IDNDR, a framework I proposed as recently as 1993 is now outdated (Fig. 14.7). Using the terminology of Fig. 14.7, current research is emphasizing that human responses to volcanoes are all "particular" to some degree. Indeed the implication of the strongly "incultured" studies which have been published in recent years has been to stress the uniqueness of place. A new framework must allow volcanic regions to be studied during eruptions and in periods of quiescence, when societies are enjoying the many benefits that may accrue from being located on the flanks of a volcano. In addition the new framework must take into account technological developments, including techniques of satellite data recovery (Wadge, 1994) and geographical information systems (GIS) (Newhall, 2000). What is proposed is an approach which is similar to the methodology used in environmental impact analysis (EIA) (Jones and Hollier, 1997, pp. 338–340). EIA was developed under legislation

passed in many countries from the late 1960s to assess the impact of large potentially environmentally damaging projects (Mitchell, 1997). In their effects on physical and social environments, volcanoes are in many ways analogous to such projects and there are close parallels between EIA and the frameworks currently being developed to study volcanoes, society, and culture (Fig. 14.8). Indeed, such approaches are diffusing through the volcanological community by means of international conferences. For all volcanoes the myriad of social and physical factors which need to be studied may be expressed in terms of "checklists," while the "overlay" approach is increasingly being used to compare data with spatial dimensions. The latter approach is currently being used on São Miguel Island in the Azores by volcanologists working at the Departimento de Geociencias, Universidade dos Açores. This builds on research carried out in the 1990s as part of the Furnas European Laboratory Project and will cross-reference hazard maps, with other cartography showing aspects of the population, economy, and society of the island (Chester *et al.*, 1995, 1999; McGuire, 1995b; Dibben and Chester, 1999).

Today, applied volcanological research emphasizes linkages across the sciences and social sciences, the complex character of human society, and the uniqueness of place.

| Technique | Features | Comments on and examples of applications in volcanic regions |
|---|---|---|
| Checklist | Lists all the factors - physical, economic, cultural and societal - which need to be considered. Cause/effect relationships are implied but not specified in detail. | The US Geological Survey's programe, Living with Volcanoes (Wright and Pierson, 1992, p. 6) and many initiatives in other countries involve, either implicitly or explicitly, a checklist approach (see also Anonymous,1994b). It is the evolving norm of the IDNDR. |
| Overlays | Traditionally this has relied on overlay maps showing physical, social, historical aspects of the region. Today geographical information systems (GIS) are increasingly being used. | There is much scope for this approach to be used in volcanic regions, because many of the variables are spatial and capable of being either mapped or incorporated into a GIS. The impact of satellite-based systems, significant at present, is likely to be much more prominent in the future (Wadge, 1994). GIS based studies have been used in the Azores (Baxter et al., 1994) |
| Matrices | Matrices are used to identify first-order cause/effect relationships. | At the present time variables are not sufficiently well specified to enable matrices and network based studies to be carried out. There may be much scope in the future. |
| Networks | Used to identify "chains" of complex interactions. Ideally this approach requires mathematical modeling. | |

*Increasing complexity* (vertical label, left side of table)

**Fig. 14.8.** The evolving framework for the study of volcanoes and society. Close parallels with environmental impact analysis (EIA) should be noted. Based on several sources (see Mitchell, 1997, pp. 116–123 for further details).

# Volcanoes and society: contemporary research agendas

> Our Earth survives recurring furies
> of her stomach pains and quakes
> From the bleeding anger of her wounds
> volcanic ash becomes the hope
> that gives rebirth to abundance of seedtimes.
>
> Kofi Anyidoho, *The Homing Call of Earth*, quoted by
> Fisher *et al.* (1997)

As in the past, so present studies of hazard reduction are central to research on societies living on active volcanoes. In the 1990s the Decade Volcanoes were the locus of interest and research took on board many of the points discussed above, being concerned to implement the policy first enunciated by IAVCEI in 1990 (IAVCEI, IDNDR Task Group, 1990) (Table 14.6). The Decade Volcanoes represent different styles of activity, are located in a variety of countries (some developed, others developing), and have been the focus of "intensive, integrated and multidisciplinary research involving international co-operation" (McGuire, 1995b, p. 404). In 1993, Bob Tilling and Peter Lipman (Tilling and Lipman,

| Table 14.8 | Progress being made on the Decade Volcanoes by the middle 1990s |
| --- | --- |
| Volcano | Progress report |
| Colima (Mexico)<br>Galeras (Colombia)<br>Mauna Loa (Hawaii, USA)<br>Merapi (Indonesia)<br>Mt. Rainier (Washington, USA)<br>Santa Maria (Guatemala)<br>Taal (Philippines) | Planning workshops have been held and a variety of research projects are taking place. |
| Etna (Italy)<br>Nyiragongo (Zaire)<br>Sakurajima (Japan)<br>Santorini (Greece)<br>Teide (Spain)<br>Unzen (Japan)<br>Vesuvius (Italy) | In addition to the above additional research has been carried out at these volcanoes. With the exception of Vesuvius, the European volcanoes are also EU Laboratory Volcanoes and have attracted funding from the European Science Foundation. |
| Ulawun (Papua New Guinea)<br>Avachinsky/Koriasksky (Russia) | Little progress was made for some years because of the Rabaul eruption in 1994. Research has resumed. Work started in Russia in late 1995/early 1996. |

For all 16 Decade Volcanoes workshops were convened to review the effectiveness of hazard reduction and to strengthening areas of weakness. Two models were used: (a) integrated meetings of scientists and decision-makers/planners and (b) separate workshops – one for geoscientists – the other for focusing on the use of information and on measures of risk reduction. In model (b) geoscientists have usually been selected for their ability to communicate. Funding remains a major problem; most has come from national governments and local authority sources, but often this represents a redirection of funds rather than *new* money. There has been relatively little involvement of local people.

Much hazard mapping, risk assessment, and monitoring has been carried out on the Decade Volcanoes, and electronic mail – the Smithsonian Institution's Global Volcanism Network – allows the international interchange of information and publication of eruption reports. There has been some volcano training under the auspices of organizations such as European Commission (1996) – now the European Union. Satellite monitoring of eruptions is now routine (see Wadge, 1994; McGuire *et al.*, 1995b; Scarpa and Tilling 1996).

*Source:* Based on the references cited and numerous internet sources.

1993), in reviewing research on the Decade Volcanoes, bemoaned the slow rate of progress. Looking back at the second half of 1990s, it is clear that there were grounds for optimism but no place for complacency (Table 14.8). In July 1997, an editorial in the journal *Nature* argued forcefully that the Achilles' heel of scientific and social scientific research on hazardous volcanoes and the principal reason for its lack of success during the eruption on Montserrat was the lack of an effective and well-funded framework of support (Anonymous, 1997). It is possible to dissent from this particular example (Clay *et al.*, 1999), but the substantive case is well made and funding difficulties have been highlighted by several authors (see Table 14.8). Running alongside the Decade Volcanoes was a further initiative sponsored by the European Science Foundation. In 1992 and 1993, the European Union allocated limited research funds with aims similar to those of the Decade Volcanoes initiative. Known as the Laboratory Volcanoes, those chosen for detailed study were: Mt. Etna (Sicily), Furnas (Azores), Piton da la Fournaise (Réunion, Indian Ocean), Teide (Tenerife), and Santorini (i.e., ancient Thera) in Greece; later Krafla in Iceland was added. Many other countries, most notably the USA, have had major volcano research programs (see Wright and Pierson, 1992) in the 1990s.

Although the IDNDR has now drawn to a close, at the present time (November 2004) the international community is planning successor arrangements (see United Nations, 1999). Termed the International Strategy for Disaster Reduction (ISDR), this was adopted by the UN General Assembly in 1999. Key words and phases in this new strategy are: public awareness, moving from cultures of reaction to cultures of prevention, vulnerable groups, and maintaining the sustainability of hazard-prone areas. It is difficult to know how the new strategy will work in practice, but it seems likely that the strongly "incultured" agenda that has been developed on some of the Decade and Laboratory Volcanoes will be continued and enhanced. In fact, despite the fact that the IDNDR is over, much valuable research is still being carried out on the Decade Volcanoes.

Some idea of the work undertaken during the IDNDR may be illustrated by taking Furnas as an example (Fig. 14.9). Here two international teams have been involved and, as well as "conventional" studies reconstructing the volcano's history, mapping its threat, and monitoring its current activity, special care has been taken to be aware of the cultural milieu of this unique island community (Chester et al., 1999; Dibben and Chester, 1999). In a study of evacuation planning, it became apparent that many of the factors listed in Table 14.9 are critical and will decide whether, or not, people can be successfully removed from their homes in the event of an emergency.

Following the 1783 Laki eruption (Iceland), agricultural yields were depressed in many European countries (Grattan and Charman, 1994; Grattan and Brayshaw, 1995; Stothers, 1996). Putative links between volcanoes, climate/weather, and a range of social, economic, and political issues is a second area of investigation which is forming an increasingly important strand in studies of volcanoes and people. Examples include study of the impact of eruptions on: monsoon reliability and strength in India (Mukherjee et al., 1987); historical crop yields in the USA (Handler, 1985); European grape harvests (Stommel and Swallow, 1983); and the effects of hitherto large and previously unresearched eruptions. With regards to the latter, research on the global climatic impact of the eruption of Huaynaputina volcano in Peru in February 1600 is particularly noteworthy and has been summarized by David Pyle (Pyle, 1998). Models relating eruptions to global weather, climate, and economic circumstances have been proposed by several authors, most notably by Handler (1989). At present neocatastrophism is an important focus in interdisciplinary research, linking the climatic effects of volcanoes to many and varied archeological, historical, and social events. The rise and fall of ancient civilizations – particularly Thera – is under active discussion (Baillie, 1995; Dalfes et al., 1996; Polinger-Foster and Ritner, 1996; Keys, 1999). Thera and many other archeological and historical examples have been subjects of international conferences. Perhaps the most intriguing current suggestion is that the Tambora eruption of 1815 was responsible for poor weather in North America and Europe, and even Napoleon's late arrival and subsequent defeat at the battle of Waterloo (Fresco, 1996; McWilliam, 1996). The Tambora and other examples highlight a potential danger inherent in this research, of uncritically accepting simple one-way environmentally deterministic links between eruptions and human reactions, whereas in reality responses may be far more complex and multi-causal. This lesson, most painfully learnt by academic geography in the early twentieth century (see Peet, 1998), may well have to be relearnt by the applied volcanologist and environmental archeologist/historian. In a contemporary world in which population is growing and the problems involved in feeding it have become so severe, the future impact of eruptions on climate and economy may be very serious indeed. The issues faced by past civilizations (Self, 1994) could pale into relative insignificance.

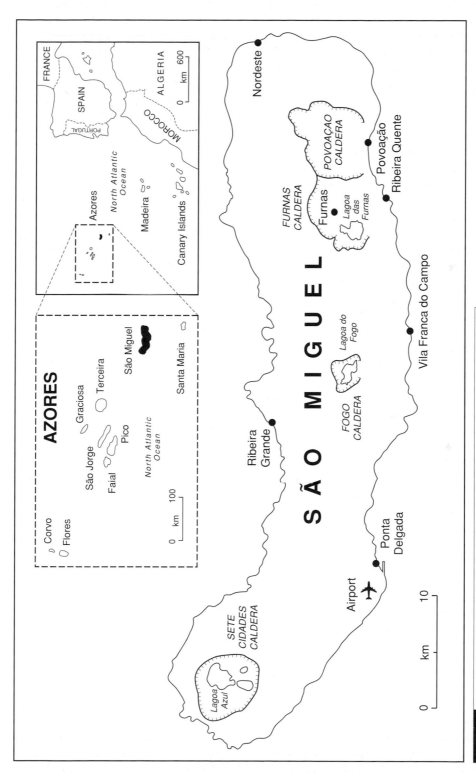

**Fig. 14.9.** The Azores, the island of São Miguel, and Furnas Volcano. Based on Chester *et al.* (1995).

Table 14.9 | A summary of the principal demographic, socio-economic, and cultural–behavioral factors that could complicate an evacuation at Furnas Volcano, São Miguel, Azores

| Factors | Details |
| --- | --- |
| Demographic | The Furnas District had a resident population of 22 644 people in 1991, accommodated in 5693 houses. These figures do not, however, capture: (a) the large number of visitors, especially in summer and (b) the under-occupancy of many houses for most of the year. These are inhabited for some of the time by weekend residents, return migrants from mainland Portugal and abroad, and tourists. Since the Portuguese census date is April 15, a month of relatively low resident population, far more people would potentially require transporting to safety than the published figures suggest. A further demographic factor that could complicate evacuation is the degree to which population is either clustered within the principal settlement of the local authority area, or widely dispersed over its area. |
| Socio-economic | Traditionally an agricultural/fishing area, alternative employment opportunities are limited. This is reflected in recent decades by the permanent or temporary out-migration of many people in the economically active cohort, with the result that dependency ratios (i.e., % under 15 + % over 65 in the population) range from 38% to 46% across the district. The percentage of the population classed as economically active is never greater than 36% of the total in any administrative area. |
| Cultural–behavioral | As a result of an aged population and the fact that before the 1974 Portuguese Revolution many people did not receive even an elementary education, illiteracy ranges from 8% to 23% of the total population across the Furnas District. Behavioral factors are important and include close links between many of the inhabitants and the land, as a result of both active and traditional family-based agricultural ties. At its most simple – and bearing in mind that cattle-rearing and fattening dominate contemporary agriculture – large numbers of livestock, both living and dead, could block roads in the event of an eruption, whilst at another close attachment of people to village, land, and farm could cause some inhabitants to resist evacuation. |

Source: Adapted from Chester et al. (1999), Table 14.4, and based on numerous additional sources and social surveys carried out in the area.

A third research focus concerns what Richard Fisher and his colleagues have recently termed "volcanoes for consumers" (Fisher et al., 1997). Under this portmanteau term are a number of research themes which seek further to explore the positive benefits of volcanoes to society. At its most simple it is well known that volcanoes provide reservoirs for the storage of water in the Azores and Italy, are scenically attractive supporting tourism in the Cascade Mountains and Hawaii, and in restricted areas – such as Iceland and central Italy – are sources of geothermal power (see Fisher et al., 1997; Arnórsson, 2000; Sigurdsson and Lopes-Gautier, 2000), but perhaps the most important current research focus is concerned with volcanic soils.

As mentioned in the introduction, in many parts of the world it has been claimed that high rural population densities are supported by the fertility of volcanic soils (Macdonald, 1972). It has been asserted with confidence that, on the island of Java (Indonesia), exceptionally high rural population densities of 400/km$^2$ in 1955 – rising to a projected figure of 1700/km$^2$ by the year 2000 (Zen, 1983) – are sustained by volcanic soils. Indeed, Mohr (1945) went so far as to suggest that within rural Java there was a strong and simple relationship between population density and volcanic soil type. This untested assertion is sadly widely repeated even in advanced texts. It is now accepted that this relationship is neither as simple nor so strongly physically determined. Reporting research on the tephra-derived soils of Indonesia, much of which was originally carried out by Tan (1964), Ugolini and Zasoski (1979) maintained that volcanic Andosols are not intrinsically but only potentially fertile and that over the centuries farmers have had to manage soils with great care through judicious applications of phosphorus and organic matter. Highly relevant to this discussion is research from the Zapotitán Valley in El Salvador. Here Olson (1983, p. 56) concludes that the common contention that falls of tephra improve soils is a myth, which cannot be supported either by detailed field work or laboratory data. On the slopes of Mt. Etna (Sicily), similarly high rural population densities of up to 400/km$^2$ are found, in comparison with only around 130/km$^2$ for the island as a whole. Once again, the reason is not simple environmental determinism as is often stated, but rather a close symbiosis between a Mediterranean rural peasant society and a potentially fertile soil resource, which has been developing over more than 3000 years of human occupation. The lava soils of Etna weather slowly, require additions of organic matter to maintain fertility, can usually only yield their full potential under irrigation, and have to be conserved on steep slopes by the construction of elaborate systems of terraces (Chester et al., 1985, in press).

Whereas much has been published on tephra soils (e.g., Ping, 2000), partly because of their importance in geochronology and archeology, a search of the literature on lava soils produces very little (see James et al., 2000, for a review). This is despite the high agricultural value placed on both tephra and lava soils in many parts of the world and the fact that careful management and conservation measures are often required to prevent erosion. A pioneer study of erosion in antiquity in the Methana Region of Greece (James et al., 1994) showed the great potential of this area of integrated earth science and further examples of the use, management, and abuse of lava and ash soils are to be found in Fisher et al. (1997, Ch. 13 and references).

Until comparatively recently, volcanology was dominated by geology in general and igneous petrology in particular. Research was carried out by scientists from the "First World" working both in their own and as expatriates in other countries. Today under the twin stimuli of the IDNDR and the ISDR, and the growing interest in environmental history, environmental archeology, and climate change, applied volcanology is changing and is now truly international and multidisciplinary, and addresses a much wider research agenda than hitherto. Volcanology is concerned not only with volcanoes per se, but also with the people living on and affected by them.

Ne plus haustae aut obrutae urbes – "no more shall the cities be destroyed" – has been the unofficial motto of the Hawaiian Volcano Observatory since its founding in 1912. At the present time and for the foreseeable future, this is an unachievable goal, but the study of volcanoes and society seeks to capture the spirit of this maxim: to minimize the effects of damaging eruptions; to understand why throughout history people have been attracted to volcanic regions; and to demonstrate the means by which the positive endowments provided by volcanoes for humanity may be maximized and at the same time the risks minimized.

# References

Alexander, D. 2000. Confronting Catastrophe. Harpenden, UK, Terra Publishing.

Anonymous, 1992. Opportunities for British Involvement in the International Decade for Natural Disaster Reduction

(*IDNDR*), Proceedings of a workshop held on March 27, 1992. London, The Royal Society and The Royal Society of Engineering.

1993. *Medicine in the International Decade for Natural Disaster Reduction (IDNDR)*, Proceedings of a workshop at The Royal Society, London. London, The Royal Society and The Royal Society of Engineering.

1994a. *World Disasters Report 1994*. Geneva, Switzerland, International Federation of Red Cross and Red Crescent Societies.

1994b. *A World Safe from Natural Disasters: The Journey of Latin America and the Caribbean*. Washington, DC, Pan-American Health Organization.

1995. *Megacities: Reducing Vulnerability to Natural Disasters*. London, Institution of Civil Engineers.

1996. *Cities at Risk: Making Cities Safer before Disaster Strikes*. Geneva, Switzerland, United Nation IDNDR Contact Group.

1997. Co-operation can help to get the message across. *Nature*, **388**, 1.

Arnórsson, S. 2000. Exploitation of geothermal resources. In H. Sigurdsson, B. Houghton, S. R. McNutt, *et al.* (eds.) *Encyclopedia of Volcanoes*. San Diego, CA, Academic Press, pp. 1243–1258.

Baillie, M. G. L. 1995. *A Slice through Time*. London, Batsford.

Baker, P. E. 1979. Geological aspects of volcano prediction. *Journal of the Geological Society*, **136**, 341–345.

Banks, N. G., Tilling, R. I., Harlow, D. H., *et al.* 1989. Volcano monitoring and short-term forecasts. In R. I. Tilling (ed.) *Short Course in Geology*, vol. 1, *Volcanic Hazards*. Washington, DC, American Geophysical Union, pp. 51–80.

Barberi, F. and Carapezza, M. L. 1996. The problem of volcanic unrest: the Campi Flegrei case history. In R. Scarpa and R. I. Tilling (eds.) *Monitoring and Mitigation of Volcanic Hazards*. Heidelberg, Germany, Springer-Verlag, pp. 771–785.

Barberi, F., Corrado, G., Innocenti, F., *et al.* 1984. Phlegraean Fields 1982–1984: brief chronicle of a volcanic emergency in a densely populated area. *Bulletin Volcanologique*, **47**(2), 175–185.

Baxter, P. J. 1990. Medical aspects of volcanic eruptions. I. Main causes of death and injury. *Bulletin of Volcanology*, **52**, 532–544.

1994. Vulnerability in volcanic eruptions. In *Medicine in the International Decade for Natural Disaster Reduction (IDNDR): Research, Preparedness and Response for Sudden Impact Disasters in the 1990s*. London, The Royal Society of Engineering, pp. 15–17.

2000. Impacts of eruptions on human health. In H. Sigurdsson, B. Houghton, S. R. McNutt, *et al.* (eds.) *Encyclopedia of Volcanoes*. San Diego, CA, Academic Press, pp. 1035–1043.

Baxter, P. J., Baubron, J.-C., Chester, D. K., *et al.* 1994. Assessing vulnerability at Furnas Volcano, São Miguel, Azores. *International Workshop on European Laboratory Volcanoes*, Commission of the European Communities – Director General for Science, Catania, Sicily, June 18–21, 1994, conference abstract.

Baxter, P. J., Baubron, J.-C., and Coutinho, R. 1999. Health hazards and disaster potential of ground gas emissions at Furnas volcano, São Miguel, Azores. *Journal of Volcanology and Geothermal Research*, **92**, 95–106.

Baxter, P. J., Bertstein, R. S., Falk, H., *et al.* 1982. Medical aspects of volcanic disasters: an outline of the hazards and emergency response measures. *Disasters*, **6**, 268–276.

Blaikie, P., Cannon, T., Davis, I., *et al.* 1994. *At Risk: Natural Hazards, People's Vulnerability and Disasters*. London, Routledge.

Blong, R. A. 1982. *The Time of Darkness: Local Legends and Volcanic Reality in Papua New Guinea*. Seattle, WA, University of Washington Press.

1984. *Volcanic Hazards*. Sydney, Australia, Academic Press.

Blong, R. A. and McKee, C. 1995. *The Rabaul Eruption 1994*. Sydney, Australia, Natural Hazards Research Centre, Macquarie University.

Bolt, B. A. 1999. *Earthquakes*, 4th edn. New York, W. H. Freeman.

Bowker, J. 1970. *Problems of Suffering in the Religions of the World*. Cambridge, UK, Cambridge University Press.

Brandt Report 1980. *North–South: A Programme for Survival. Report of the Independent Commission on International Development Issues*. Cambridge, MA, MIT Press.

Burgess, C. 1989. Volcanoes, catastrophe and the global crisis of the second millennium BC. *Current Archaeology*, **117**, 325–329.

Burton, I., Kates, R. W., and White, G. 1978. *The Environment as Hazard*. New York, Oxford University Press.

Carracedo, J. C. 1999. Growth, structure, instability and collapse of Canarian volcanoes and comparisons with Hawaiian volcanoes. *Journal of Volcanology and Geothermal Research*, **94**, 1–19.

Cas, R. A. F. and Wright, J. V. 1987. *Volcanic Successions: Modern and Ancient*. London, Allen and Unwin.

Casadevall, T. J. (ed.) 1991. *Volcanic Ash and Aviation Safety*, *Program and Abstracts*, 1st International Symposium, Seattle, WA, July 8–12, 1991. Washington, DC, US Government Printing Office.

1994. *Volcanic Ash and Aviation Safety, Proceedings Volume*, 1st International Symposium, Seattle, WA, July 8–12, 1991. Washington, DC, US Government Printing Office.

Chester, D. K. 1988. Volcanoes and climate. *Progress in Physical Geography*, **12**, 1–36.

1993. *Volcanoes and Society*. London, Edward Arnold.

1998. The theodicy of natural disasters. *Scottish Journal of Theology*, **51**, 485–505.

Chester, D. K., Degg, M., Duncan, A. M., *et al.* 2001. The increasing exposure of cities to the effects of volcanic eruptions: a global survey. *Environmental Hazards*, **2**, 89–103.

Chester, D. K., Dibben, C., and Coutinho, R. 1995. *Report of the Evacuation of the Furnas District, São Miguel, Azores in the Event of a Future Eruption*, Volcano, Eruptive History and Hazard Open File Report no. 4. London, University College London, Commission for the European Communities, European Science Foundation Laboratory.

Chester, D. K., Dibben, C., Coutinho, R. J., *et al.* 1999. Human adjustments and social vulnerability to volcanic hazards: the case of Furnas Volcano, São Miguel, Azores. *Geological Society of London Special Publication*, **161**, 189–207.

Chester, D. K., Dibben, C. J. L., and Duncan, A. M. 2002. Volcanic hazard assessment in western Europe. *Journal of Volcanology and Geothermal Research*, **115**, 411–435.

Chester, D. K., Duncan, A. M., and Guest, J. E. (in press). Loss-bearing on Etna from the Classical Period to 1900 CE. In P. Johnston (ed.) *The Cultural Response to Volcanic Landscape*.

Chester, D. K., Duncan, A. M., Guest, J. E., *et al.* 1985. *Mount Etna: The Anatomy of a Volcano*. London, Chapman and Hall.

2000. Human responses to Etna Volcano during the Classical Period. *Geological Society of London Special Publication*, **171**, 179–188.

Clay, E., Barrow, C., Benson, C., *et al.* 1999. *An Evaluation of HMG's Response to the Montserrat Volcanic Emergency*. London, UK Department for International Development.

Crandell, D. R., Booth, B., Kusamadinata, K., *et al.* 1984. *Source-Book for Volcanic Hazards Zonation*. Paris, UNESCO.

Dalfes, H. M., Kukla, G., and Weiss, H. 1996. *Third Millennium BC Climate Change and Old World Collapse*. New York, Springer-Verlag.

Decker, R. W. 1986. Forecasting volcanic eruptions. *Annual Reviews in Earth and Planetary Sciences*, **14**, 267–291.

Decker, R. W. and Decker, B. 1981. *Volcanoes*. San Francisco, CA, W. H. Freeman.

Degg, M. 1992. Natural disasters: recent trends and future prospects. *Geography*, **77**, 198–209.

Dibben, C. and Chester, D. K. 1999. Human vulnerability in volcanic environments: the case of Furnas, São Miguel, Azores. *Journal of Volcanology and Geothermal Research*, **92**, 133–150.

Dicken, P. 1986. *Global Shifts: Industrial Change in a Turbulent World*. London, Harper and Row.

Dickenson, J. P., Clark, C. G., Gould, W. T. S., *et al.* 1983. *A Geography of the Third World*. London, Methuen.

Dickenson, J. P., Gould, W. T. S., Clarke, C. G., *et al.* 1996. *A Geography of the Third World*, 2nd edn. London, Routledge.

Dobran, F. 1995. A risk assessment methodology at Vesuvius based on the global volcanic simulation. In T. Horlick-Jones, A. Amendola, and R. Casale (eds.) *Natural Risk and Civil Protection*. London, E. and F. N. Spon, pp. 131–136.

Doonan, R. C. P. 1997. Vulcanism and the furnace: the social production of the technological metaphor. In *Volcanoes, Earthquakes and Archaeology* (conference abstracts). London, The Geological Society.

Duncan, A. M., Dibben, C., Chester, D. K., *et al.* 1996. The 1928 eruption of Mount Etna Volcano, Sicily, and the destruction of the town of Mascali. *Disasters*, **20**, 1–20.

European Commission (Union) 1996. *The Mitigation of Volcanic Hazards*, Proceedings of the Course. Brussels, The European Union.

Filson, J. R. 1987. Geological hazards: programs and research in the USA. *Episodes*, **10**, 292–295.

Fisher, R. V., Heiken, G., and Hulen, J. B. 1997. *Volcanoes: Crucibles of Change*. Princeton, NJ, Princeton University Press.

Fiske, R. S. 1984. Volcanologists, journalists, and the concerned public. In *Explosive Volcanism: Inception, Evolution and Hazards*. Washington, DC, National Academy Press, pp. 170–176.

Forsyth, P. Y. 1988. In the wake of Etna, 44 BC. *Classical Antiquity*, **7**, 49–57.

Foster, B. R. 1996. Volcanic phenomena in Mesopotamian sources. *Journal of Near Eastern Studies*, **55**, 1–14.

Franklin, B. 1789. Meteorological imaginations and conjectures. *Memoirs of the Manchester Literary and Philosophical Society*, **2**, 373–377.

Fresco, A. 1996. Explosive theory casts new light on the Battle of Waterloo. *The Times*, May 21.

Grattan, J. and Brayshaw, M. 1995. An amazing and portentous summer: environmental and social responses in Britain to the 1783 eruption of an Iceland volcano. *Geographical Journal*, **161**, 125–134.

Grattan, J. and Charman, D. J. 1994. Non-climatic factors and the environmental impact of volcanic volatiles: implications of the Laki fissure eruption of AD 1783. *The Holocene*, **4**, 101–106.

GVNB 1996. *Grimsvotn*. Global Volcanism Network Bulletin no. 21. Washington, DC, Smithsonian Institution.

Hadfield, P. 1991a. The global earthquake. *The Sunday Times*, April 28, 20–32.

1991b. Telltale 'swarms' warned of Japanese eruption. *New Scientist*, June 8, 13.

Hamilton, F. E. I. 1991. Global economic change. In R. J. Bennett and R. C. Estall (eds.) *Global Change and Challenge*. London, Routledge, pp. 80–102.

Handler, P. 1985. Possible association between the climatic effects of stratospheric aerosols and corn yields in the United States. *Agricultural Meteorology*, **35**, 205–328.

1989. The effects of volcanic aerosols on global climate. *Journal of Volcanology and Geothermal Research*, **37**, 233–249.

Harriss, R. W., Hohenemser, C., and Kates, R. W. 1985. Human and non-human mortality. In R. W. Kates, C. Hohenemser, and J. X. Kasperson (eds.) *Perilous Progress: Managing the Hazards of Technology*. Boulder, CO, Westview Press, pp. 129–155.

Hewitt, K. 1976. Earthquake hazards in the mountains. *Natural Hazards*, **85**, 30–37.

1980. Review: The environment as hazard. *Annals of the Association of American Geographers*, **70**, 306–311.

(ed.) 1983a. *Interpretations of Calamity*. London, Allen and Unwin.

1983b. The idea of calamity in a technocratic age. In K. Hewitt (ed.) *Interpretations of Calamity*, London, Allen and Unwin, pp. 3–30.

1997. *Regions of Risk: A Geographical Introduction to Disasters*. Harlow, UK, Longman.

Homan, J. 2001. A culturally sensitive approach to risk? "Natural" hazard perception in Egypt and the UK. *Australian Journal of Emergency Management*, **16**(2), 14–18.

Hutton, J. 1788. Theory of the Earth: or an investigation of the laws observable in the composition, dissolution and restoration of land upon the globe. *Transactions of the Royal Society of Edinburgh*, **1**, 209–304.

Hyde, W. W. 1916. The volcanic history of Etna. *Geographical Review*, **1**, 401–418.

IAVCEI IDNDR Task Group 1990. Reducing volcanic disasters in the 1990s. *Bulletin of the Volcanological Society of Japan*, **35**, 80–95.

James, P. A., Mee, C. B., and Taylor, G. J. 1994. Soil erosion and the archaeological landscape of Methana, Greece. *Journal of Field Archaeology*, **21**, 295–416.

James, P., Chester, D., and Duncan, A. 2000. Volcanic soils: their nature and significance for archaeology. *Geological Society of London Special Publication*, **171**, 317–338.

Jones, G. and Hollier, G. 1997. *Resources, Society and Environmental Management*. London, Paul Chapman.

Kates, R. W. 1987. Hazard assessment and management. In D. J. Mclaren and B. J. Skinner (eds.) *Resources and World Development*. Chichester, UK, John Wiley, pp. 741–753.

Kendrick, T. D. 1956. *The Lisbon Earthquake*. London, Methuen.

Keys, D. 1999. *Catastrophe: An Investigation into the Origins of the Modern World*. London, Century.

King, R. 1985. *The Industrial Geography of Italy*. Beckenham, UK, Croom Helm.

Kusamadinata, K. 1984. Indonesia. In D. R. Crandell, B. Booth, K. Kusamadinata, *et al.* (eds.) *Source-Book for Volcanic-Hazards Zonation*. Paris, UNESCO, pp. 55–60.

Latter, J. H. (ed.) 1989. *Volcanic Hazards Assessment and Monitoring*. Berlin, Springer-Verlag.

Lechet, M. F. 1990. The International Decade for Natural Disaster Reduction: background and objectives. *Disasters*, **14**, 1–6.

Leibniz, G. W. von 1710. *Essais de Théodicée sur la bonté de Dieu, à la liberté de l'homme et l'origine du mal*. Amsterdam. [Facsimile microfilm: Microfilm Research Publications, New Haven, CT (1973); English translation: ed. A. Farrer, transl. E. M. Huggard. London, Routledge and Kegan Paul (1952).]

Macdonald, G. A. 1972. *Volcanoes*. Englewood Cliffs, NJ, Prentice Hall.

Masood, E. 1995. Rows erupt over evacuation plans for Mount Vesuvius. *Nature*, **377**, 471.

McClelland, L., Simkin, T., Summers, M., *et al.* 1989. *Global Volcanism 1975–1985*. Englewood Cliffs, NJ, Prentice Hall.

McGuire, W. J. 1995a. Monitoring active volcanoes: an introduction. In W. J. McGuire, C. Kilburn, and J. Murray (eds.) *Monitoring Active Volcanoes*. London, University College London Press, pp. 1–31.

1995b. Prospects for volcano surveillance. In W. J. McGuire, C. Kilburn, and J. Murray (eds.) *Monitoring Active Volcanoes*. London, University College London Press, pp. 403–410.

1998. Volcanic hazards and their mitigation. Geological Society of London Engineering Geology Special Publication, **15**, 79–95.

McGuire, W. J., Kilburn, C., and Murray, J. (eds.) 1995. *Monitoring Active Volcanoes*. London, University College London Press.

McGuire, W. J., Jones, A. P., and Neuberg, J. (eds.) 1996. *Volcano Instability on the Earth and Other Planets*, Special Publication no. 110. London, Geological Society.

McWilliam, F. 1996. Late arrival at Waterloo. *Geographical Magazine* **68**, 7.

Mellaart, J. 1967. *Çatal Hüyük: A Neolithic Town in Anatolia*. London, Thames and Hudson.

Miller, T. P. and Casadevall, T. J. 2000. Volcanic ash hazards to aviation. In H. Sigurdsson, B. Houghton, S. R. McNutt, *et al.* (eds.) *Encyclopedia of Volcanoes*. San Diego, CA, Academic Press, pp. 915–930.

Mitchell, B. 1997. *Geography and Resource Analysis*. Harlow, UK, Longman.

Mitchell, J. K., Devine, N., and Jaggar, K. 1989. A contextual model of natural hazard. *Geographical Review*, **79**, 391–409.

Mohr, E. C. J. 1945. The relationship between soil and population density in the Netherlands Indies. In P. Honig and F. Verdoorn (eds.) *Science and Scientists in the Netherlands Indies*. New York, Board for the Netherlands Indies, Surinam and Curaçae, pp. 254–262.

Moore, J. G. and Moore, G. W. 1984. Deposit from a giant wave on the island of Lanai, Hawaii. *Science*, **226**, 1312–1315.

Mukherjee, B. K., Indira, K., and Dani, K. K. 1987. Low latitude volcanic eruptions and their effects on Sri Lankan rainfall during the north east monsoon. *Journal of Climatology*, **7**, 145–155.

Murton, B. J. and Shimabukuro, S. 1974. Human adjustment and volcanic hazard in the Puna District, Hawaii. In G. F. White (ed.) *Natural Hazards: Local, National and Global*. New York, Oxford University Press, pp. 151–161.

Newhall, C. G. 2000. Volcano warnings. In H. Sigurdsson, B. Houghton, S. R. McNutt, *et al.* (eds.) *Encyclopedia of Volcanoes*. San Diego, CA, Academic Press, pp. 1185–1197.

Newhall, C. G. and Punongbayan, R. S. 1996. The narrow margin of successful volcanic-risk mitigation. In R. Scarpa and R. I. Tilling (eds.) *Monitoring and Mitigation of Volcanic Hazards*. Berlin, Springer-Verlag, pp. 807–838.

O'Keefe, P., Westgate, K., and Wisner, B. 1976. Taking the naturalness out of natural disaster. *Nature*, **260**, 566–567.

Olson, G. W. 1983. An evaluation of soil properties and potentials in different volcanic deposits. In P. D. Sheets (ed.) *Archaeology and Volcanism in Central America: The Zapotitán Valley of El Salvador*. Austin, TX, University of Texas Press, pp. 52–56.

O'Riordan, T. 1976. *Environmentalism*. London, Pion Books.

Pang, K. D. 1985. Three very large volcanic eruptions in antiquity and their effects on the climate of the ancient world. *Eos*, **66**, 816.

Peet, R. 1998. *Modern Geographical Thought*. Oxford, UK, Blackwell.

Peterson, D. W. 1986. Volcanoes: tectonic setting and impact on society. In *Active Tectonics*. Washington, DC, National Academy Press, pp. 231–246.

1988. Volcanic hazards and public response. *Journal of Geophysical Research*, **93**(B5), 4161–4170.

1996. Mitigation measures and preparedness plans for Volcanic emergencies. In R. Scarpa and R. I. Tilling (eds.) *Monitoring and Mitigation of Volcanic Hazards*. Berlin, Springer-Verlag, pp. 701–718.

Peterson, D. W. and Tilling, R. I. 1993. Interactions between scientists, civil authorities and the public at hazardous volcanoes. In C. R. J. Kilburn and G. Luongo (eds.) *Active Lavas*. London, University College London Press, pp. 339–365.

Ping, C.-L. 2000. Volcanic soils. In H. Sigurdsson, B. Houghton, S. R. McNutt, *et al.* (eds.) *Encyclopedia of Volcanoes*. San Diego, CA, Academic Press, pp. 1259–1270.

Plunket, P. and Uruñuela, G. 1997. Revelations of a Plinian eruption of the Popocatépetl Volcano in Central Mexixo. In *Volcanoes, Earthquakes and Archaeology* (conference abstracts). London, The Geological Society, pp. 33–34.

Polinger-Foster, K. and Ritner, R. K. 1996. Texts, storms and the Thera eruption. *Journal of Near Eastern Studies*, **55**, 1–14.

Possekel, A. J. 1999. *Living with the Unexpected*. Berlin, Springer-Verlag.

Pyle, D. M. 1998. How did the summer go? *Nature*, **393**, 415–416.

Sapir, D. G. 1993. Health effects of earthquakes and volcanoes: epidemiological and policy issues. *Disasters*, **17**, 255–262.

Scarpa, R. and Tilling, R. I. (eds.) 1996. *Monitoring and Mitigation of Volcanic Hazards*. Berlin, Springer-Verlag.

Schild, J. 1997. Danger and depression lurk under the volcano. *Church Times* (London), July 4, 2.

Seager, J. 1995. *The State of the Environment Atlas*. Harmondsworth, UK, Penguin Books.

Self, S. 1994. Volcanic activity and climatic catastrophes. *Volcanic Studies Group – Geological Society (London)*, 4th Thematic and Research in Progress Meeting, University of Liverpool, January 5–6, 1994 (Abstract), pp. 33–34.

Shimozuru, D. 1983. Volcanic hazard assessment of Mount Fuji. *Natural Disaster Science*, **5**(2), 15–31.

Sigurdsson, H. 1999. *Melting the Earth: The History of Ideas on Volcanic Eruptions*. Oxford, UK, Oxford University Press.

Sigurdsson, H. and Carey, S. 1986. Volcanic disasters in Latin America and the 13th November 1985 eruption of Nevado del Ruiz volcano in Colombia. *Disasters*, **10**, 205–216.

Sigurdsson, H. and Lopes-Gautier, R. 2000. Volcanoes and tourism. In H. Sigurdsson, B. Houghton, S. R. McNutt, *et al.* (eds.) *Encyclopedia of Volcanoes*. San Diego, CA, Academic Press, pp. 1283–1299.

Sigurdsson, H., Houghton, B., McNutt, S. R., *et al.* (eds.) 2000. *Encyclopedia of Volcanoes*. San Diego, CA, Academic Press.

Sigvaldason, G. E. 1983. Volcanic prediction in Iceland. In H. Tazieff and J.-C. Sabroux (eds.) *Developments in Volcanology*, vol. 1, *Forecasting Volcanic Events*. Amsterdam, Elsevier, pp. 193–215.

Simkin, T. 1993. Terrestrial volcanism in time and space. *Annual Reviews in Earth and Planetary Sciences*, **21**, 427–452.

1994. Distant effects of volcanism – how big and how often? *Science*, **264**, 913–914.

Simkin, T., Siebert, L., McClelland, L., *et al.* 1981. *Volcanoes of the World*. Washington, DC, Smithsonian Institution.

Simon, H. A. 1957. *Administrative Behaviour*. New York, Macmillan.

1959. Theories of decision making in economic and behavioral science. *American Economic Review*, **49**, 253–283.

Steenblik, J. W. 1990. Volcanic ash: a rain of terror. *Airline Pilot*, **59**, 10–15, 56.

Stommel, H. and Stommel, E. 1983. *Volcano Weather: The Story of 1816, the Year without a Summer*. Newport, RI, Seven Seas Press.

Stommel, H. and Swallow, J. C. 1983. Do late grape harvests follow large volcanic eruptions? *Bulletin of the American Meteorological Society*, **64**, 794–795.

Stothers, R. B. 1984. Mystery cloud of AD 536. *Nature*, **307**, 344–345.

1996. The great dry fog of 1783. *Climatic Change*, **32**, 79–89.

Stothers, R. B. and Rampino, M. R. 1983. Volcanic eruptions in the Mediterranean region before AD 630 from written and archaeological sources. *Journal of Geophysical Research*, **88**(B), 6357–6371.

Susman, P., O'Keefe, P., and Wisner, B. 1983. Global disasters, a radical interpretation. In K. Hewitt (ed.) *Interpretations of Calamity*. London, Allen and Unwin, pp. 263–283.

Tan, K. H. 1964. The andosols in Indonesia. *World Soil Resources Report*, **14**, 101–110.

Tanguy, J.-C., Ribière, C., Scarth, A., *et al.* 1998. Victims from volcanic eruptions: a revised database. *Bulletin of Volcanology*, **60**, 137–144.

Tayag, J. C. and Punongbayan, R. S. 1994. Volcanic disaster mitigation in the Philippines: experience from Mount Pinatubo. *Disasters*, **18**, 1–15.

Taylor, G. A. 1958. The 1951 eruption of Mount Lamington, Papua. *Australian Bureau of Mineral Resources Geological and Geophysical Bulletin*, **38**, 1–117.

Tazieff, H. 1983. Some general points about volcanism. In H. Tazieff and J. C. Sabroux (eds.) *Developments in Volcanology*, vol. 1, *Forecasting Volcanic Events*. Amsterdam, Elsevier, pp. 9–25.

Tazieff, H. and Sabroux, J. C. (eds.) 1983. *Developments in Volcanology*, vol. 1, *Forecasting Volcanic Events*. Amsterdam, Elsevier.

Tilling, R. I. 1989. Volcanic hazards and their mitigation: progress and problems. *Reviews of Geophysics*, **27**, 237–267.

1995. The role of monitoring in forecasting volcanic events. In W. J. McGuire, C. Kilburn, and J. Murray (eds.) *Monitoring Active Volcanoes*. London, University College London Press, pp. 369–402.

Tilling R. I. and Lipman, R. W. 1993. Lessons in reducing volcanic risk. *Nature*, **364**, 277–280.

Ugolini, F. G. and Zasoski, R. J. 1979. Soils derived from tephra. In P. D. Sheets and D. K. Grayson (eds.) *Volcanic Activity and Human Ecology*. New York, Academic Press, pp. 83–124.

UNESCO 1993. *Disaster Reduction*, Environment and Development Briefs no. 5. London, Banson.

United Nations 1989. *Prospects of World Urbanization, 1988*, Population Studies no. 112 (ST/ESA/SER.A), New York, United Nations, Department of International Economic and Social Affairs.

1995. *Yokohama Strategy and Plan of Action for a Safer World: Guidelines for Natural Disaster Prevention, Preparedness and Mitigation*. New York, United Nations.

1999. *International Decade for Natural Disaster Reduction: Successor Arrangements*. New York, United Nations.

Voight, B. 1988. Countdown to catastrophe. *Earth and Mineral Sciences* (Pennsylvania State University), **57**(2), 17–35.

1996. The management of volcano emergencies: Nevado del Ruiz. In R. Scarpa and R. I. Tilling (eds.) *Monitoring and Mitigation of Volcanic Hazards*. Berlin, Springer-Verlag, pp. 719–769.

Wadge, G. (ed.) 1994. *Natural Hazards and Remote Sensing*. London, The Royal Society and The Royal Society of Engineering.

(ed.) 1996. *The Soufrière Hills Eruption*, abstracts of papers presented at a discussion meeting, November 27, 1996. London, The Geological Society.

Walker, G. P. L. 1974. Volcanic hazards and the prediction of volcanic eruptions. *Geological Society (London) Miscellaneous Paper*, **3**, 23–41.

Ward, S. N. and Day, S. 2001. Cumbre Vieja Volcano: Potential collapse and tsunami at La Palma, Canary Islands. *Geophysical Research Letters*, **28**, 3397–3400.

Warrick, R. A., 1979. Volcanoes as hazard: an overview. In P. D. Sheets and D. K. Grayson (eds.) *Volcanic Activity and Human Ecology*. New York, Academic Press, pp. 161–189.

Wescoat, J. L., Jr. 1992. Common themes in the work of Gilbert White and John Dewey: a pragmatic appraisal. *Annals of the Association of American Geographers*, **82**, 587–607.

White, G. F. 1972. Geography and public policy. *Professional Geographer*, **24**, 302–309.

1973. Natural hazards research. In R. J. Chorley (ed.), *Directions in Geography*. Methuen, London, 193–212.

Whittow, J. 1987. Hazards: adjustment and mitigation. In M. J. Clark, K. J. Gregory, and A. M. Gurnell (eds.) *Horizons in Physical Geography*. London, Macmillan, pp. 307–319.

Wijkman, A. and Timberlake, L. 1984. *Natural Disaster: Acts of God or Acts of Man?* London, Earthscan and International Institute for Environment and Development.

Wisner, B. 2001. Capitalism and the shifting spatial and social distribution of hazard and vulnerability. *Australian Journal of Emergency Management*, **16**(2), 44–50.

Wisner, B., Westgate, K., and O' Keefe, P. 1976. Poverty and disaster. *New Society* **9**, 547–548.

Wisner, B., O' Keefe, P., and Westgate, K. 1977. Global systems and local disasters: the untapped power of people's science. *Disasters*, **1**(1), 47–57.

Wood, C. A. and Kienle, J. 1990. *Volcanoes of North America: United States and Canada*. Cambridge, UK, Cambridge University Press.

World Bank 1995. *World Tables* (CD-ROM). Baltimore, MD, Johns Hopkins University Press.

Wright, T. L. and Pierson, T. C. 1992. *Living with Volcanoes*, US Geological Survey Circular no. 1073: Washington, DC, US Geological Survey Volcano Hazards Program.

Zen, M. T. 1983. Mitigating volcanic disasters in Indonesia. In H. Tazieff and J.-C. Sabroux (eds.) *Developments in Volcanology*, vol. 1, *Forecasting Volcanic Events*. Amsterdam, Elsevier, pp. 219–236.

Zupka, D. 1996. Working to reduce volcanic risks in Argentina. *Stop Disasters*, **29**, 26–27.

# Chapter 15

# Volcanoes and the economy

Charlotte Benson

## Introduction

Volcanic eruptions can have significant economic implications, both nationally and for individual households and communities. However, the nature and magnitude of such impacts are dependent on a number of factors. These include the precise nature of the volcanic activity; the timing of various eruptive phases relative to climatic seasons and agricultural and other cyclical economic activities (e.g., tourism); population density and types of economic activity in the affected area; and its economic significance for the rest of the country. Economic performance in the period immediately prior to an eruption, both nationally and internationally, can also play some role.

As a preliminary exercise in examining the economic impacts of volcanic activity, it is useful to consider the distinction commonly made between sudden-impact and slow-onset natural disasters. The former involve sudden, short-lived, destructive events, causing potentially severe damage to infrastructure, factories, and other productive capital, effectively destroying the means of production. They also affect social infrastructure, including housing. In the short term they are therefore potentially more disruptive to an economy than slow-onset disasters, which basically comprise droughts. However, the reverse is true in the longer term as the typically lengthier duration of slow-impact disasters can affect underlying economic variables such as levels of savings, investment, and domestic demand,

as well as having a direct severe impact on water-intensive sectors.

Volcanic hazards are conventionally classified as sudden-impact disasters. However, at least from an economic perspective, they can display certain characteristics of slow-onset as well as sudden-impact disasters. More specifically, eruptive phases can continue for a considerable period of time, creating considerable uncertainty and fear amongst local communities and dampening economic performance. Meanwhile major eruptions can be followed be several years of destructive lahars. These most commonly occur during the months of highest precipitation, rapidly destroying infrastructure and creating economic difficulties of a nature associated with sudden-impact disasters. The threat of further activity can continue to perpetuate a climate of extreme uncertainty, undermining public and private confidence and delaying the replacement of destroyed assets or any new investment. Both lava flows and lahars can also change the depth and course of rivers, potentially resulting in future flooding and thus creating further uncertainty.

These various factors complicate efforts to examine the economic impacts of volcanic hazards. There might be no major eruption and relatively little physical damage yet the disruption to normal economic life could still be relatively high. Alternatively there could be a major eruption, causing considerable physical damage to property and infrastructure. However, rather than prompting a post-disaster construction boom as might be expected in the aftermath of an earthquake, the threat of further activity could depress the local economy as both public and private investors hold back. Thus, the economic costs of volcanic activity could extend considerably beyond the cost of total physical damage to homes, factories, roads, and other infrastructure.

Both the potential physical and other economic impacts of volcanic hazards are explored in more detail below. However, the chapter begins with a broad overview of the methodological framework commonly used in measuring the economic impacts of natural disasters. A brief overview of the economy-wide impacts of volcanic eruptions is then provided. This is followed by an examination of some of the main direct and indirect impacts of volcanic eruptions and of their government budgetary consequences. The chapter concludes with a discussion of the role that economic considerations play in determining the nature and scale of various forms of disaster management. The chapter excludes any explicit examination of the impact of volcanic-related tsunamis, earthquakes, or global climatic variability, all of which can also have potentially substantial additional economic implications.

The chapter draws together existing documentation on the economic impacts of volcanic activity, to the extent that such information exists. However, it is also limited by the very scope and depth of that work: the economic impacts of all natural hazards is an underexplored area of research and there is a clear need for further work in this area to help ensure the adoption of adequate and appropriate risk reduction measures.

## Measuring the "cost" of disasters: current practices

The economic costs of disasters are commonly categorized as direct costs, indirect costs, and secondary effects (for example, United Nations, 1979; Anderson, 1991; Bull, 1992; OECD, 1994; Otero and Marti, 1995). *Direct costs* relate to the physical damage to capital assets, including buildings, infrastructure, industrial plants, and inventories of finished, intermediate, and raw

materials, destroyed or damaged by the actual impact of a disaster. Crop production losses are often included in estimates of direct costs. *Indirect costs* refer to damage to the flow of goods and services, including lower output from damaged or destroyed assets and infrastructure; loss of earnings due to damage to marketing infrastructure such as roads and ports and to lower effective demand; and the costs associated with the use of more expensive inputs following the destruction of cheaper usual sources of supply. They also include the costs in terms of both medical expenses and lost productivity arising from increased incidence of disease, injury, and death. *Secondary effects* concern both the short- and long-term impacts of a disaster on overall economic performance, such as deterioration in trade and government budget balances and increased indebtedness as well as the impact on the distribution of income or the scale and incidence of poverty. They can also include shifts in government monetary and fiscal policy to, for example, contain the effects of increased disaster-induced inflation or to finance additional government expenditure. Direct losses can therefore be roughly equated with stock losses whilst indirect costs and secondary effects both constitute flow losses.

According to existing studies, the relative balance of direct, indirect, and secondary costs varies between disasters depending on the nature and extent of damage, although it is generally agreed that indirect costs are often considerably higher (e.g., Otero and Marti, 1995). However, most estimates of the costs of disasters are based on post-disaster damage assessments, which are undertaken with the aim of providing information upon which appropriate relief and rehabilitation responses can be based. These focus particularly on damage to buildings, infrastructure (such as roads and bridges, electricity and power cables, and irrigation networks), capital equipment, and standing crops but provide little evidence on the wider indirect and secondary economic impacts of disasters. Additional estimates provided by the insurance industry are even more biased, focusing largely on insured physical damage. Thus, the various damage assessments typically available underestimate the true cost of disasters. They also offer little insight on adaptive behavior which may be undertaken to minimize the impacts of disasters or on the precise nature and scale of a particular economy's, or community's, hazard vulnerability. In consequence, they may provide insufficient information to facilitate the making of appropriate decisions on disaster mitigation and preparedness strategies and investments.[1]

Moreover, even the accuracy of the estimated cost of damage produced as part of this assessment process can be uncertain. Governments sometimes lack any formal methodology, showing little concern for the quality of the data collated by leaving the design of damage assessments to people on the ground rather than trying to standardize practices. Assessments may also be incomplete, as, for example, in the Philippines where they only take account of damage to income groups and subsectors that may be eligible for government assistance (Chardin, 1996). In addition, some assessments report the replacement cost of destroyed assets while others are based on the present value of the net income that an asset would have produced over the remainder of its life, thus taking into account the age of the asset and any depreciation through wear and tear. Further problems exist in establishing a counterfactual – that is, in determining how an economy would have performed in the absence of a disaster. Economies are dynamic rather than static entities, constantly changing under the influence of various factors. Thus, it cannot simply be assumed that all economic indicators – for example, the rate of growth, the rate of inflation, interest rates, levels of savings, the external trade balance, the government budgetary balance – would have held constant in the absence of a particular disaster event.

Further problems are created by the speed with which damage assessments are finalized. By definition, such assessments should begin in the immediate aftermath of a disaster and indeed, the first assessments are typically completed very rapidly. However, although respective governments in affected countries may prepare a subsequent report detailing relief and rehabilitation efforts, assessments of the longer-term impacts of a disaster are undertaken only very rarely.

Reflecting these various difficulties, estimates of the economic impacts of a particular disaster can vary considerably depending on the methodology involved, the impacts examined, and, most fundamentally, the objective of the assessment, as illustrated in the case of the 1985 eruption of Nevado del Ruiz, Colombia (Box 15.1). Bryant (1991) provides a further example of the sometimes substantial gaps between different

---

**Box 15.1** | Discrepancies in estimating the "cost" of volcanic eruptions: an example from the 1985 Nevado del Ruiz eruption, Colombia

The 1985 eruption of Nevado del Ruiz killed 21 800 people and destroyed over 5000 buildings and 2500 vehicles in the town of Armero alone. The eruption also disrupted cash crop and livestock production in an important agricultural region of the country.

The Colombian Government initially estimated total losses at Peso 34.9 billion (US$218 million) (Mileti et al., 1991). However, another source later estimated total damage to the Department of Northern Tolima alone at US$400 million (Praez, 1986) while OFDA (1995) estimated total damage at US$1 billion. True costs may have been even higher again, in part reflecting Armero's role as a major regional agricultural service centre and thus the knock-on effects of the eruption on the wider region.

---

[1] Disasters themselves can inhibit efforts to estimate their flow impacts by disrupting the collection and reporting of economic data as, for example, in Montserrat in 1996 and 1997.

estimates, claiming that Washington State's calculation of the economic cost of the Mount St. Helens' 1980 eruption of US$2.7 billion was an overestimate whilst even the US Federal Congress appropriations of US$1 billion were too high, in part reflecting media interest in the eruption. Bryant estimated the total cost at US$844 million, comprising clean-up operations totaling US$270 million, agricultural losses of US$39 million, property damage, largely to roads and bridges, of US$85 million, and commercial timber losses amounting to US$450 million. However, this is a very narrow definition of economic losses, based largely on direct losses and ignoring a number of indirect and secondary costs.

More fundamentally, from an economic perspective even the desire to estimate the total cost of a disaster with a single figure needs to be questioned. Such ambitions reflect a naive conception of the economic consequences of disasters. Impacts such as a rise in the external trade deficit or the rate of inflation cannot be aggregated into a single figure yet may have serious implications.

Reflecting these difficulties and the considerable question marks over virtually all estimates of the "total cost" of particular disasters, this chapter therefore steers away from attempts to place total figures on the economic impact of individual eruptions. Furthermore, no attempt is made to construct a "league table" of the world's "most costly" eruptions. Instead, the chapter takes a more disaggregated, qualitative approach, considering both the physical damage and the indirect and secondary impacts of volcanic activity. In so doing, it aims to provide an overview of the complex economic consequences of volcanic hazards – a hopefully more fruitful exercise in terms of identifying the lessons for preparedness, mitigation, and relief of future eruptions.

## The benefits of volcanic activity

Before examining the various adverse impacts of volcanic hazards, it is also important to stress that any efforts to assess the economic impacts of a disaster need to take account of any potential benefits. With regard to volcanic hazards, volcanic soils are potentially highly fertile as the ash can contain trace elements beneficial to the growth of plants. This can encourage a relatively high population density and intense agriculture in the vicinity of a volcano. Indeed, volcanoes are often located in major agricultural-producing regions of a country. Volcanoes can also constitute important tourist attractions, providing sometimes stunning natural beauty. Low-level volcanic activity can even attract additional tourists, as, for example, demonstrated by the 1963–5 eruption of Volcán Irazú in Costa Rica which was reported to have been largely responsible for a 4.3% increase in the number of tourists between 1963 and 1964. Lahars generated as a consequence of the 1991 eruption of Mt. Pinatubo in the Philippines have also benefited the province of Pampanga by making it less susceptible to flooding.

Disaster-preparedness activities initiated in response to indications of renewed volcanic activity and the threat of a relatively imminent eruption can also entail certain indirect economic benefits. For example, volcanic activity at Rabaul, Papua New Guinea, between 1983 and 1985 resulted in a number of improvements to infrastructure in the region, including to airstrips, wharves, roads, bridges, water and power supplies and sewerage, storage, health facilities, and communications. Moreover, food production increased as a number of people cultivated food gardens outside the danger zone.

In addition, volcanic eruptions can imply certain redistributional impacts, to the advantage of certain geographical areas or individuals. For example, new investments may be redirected to neighboring towns or regions, tourism increasingly focused on another region or airplanes rerouted, implying that one region's or town's loss could be another's gain. Similar redistributional impacts can also occur at an individual level. For instance, following the destruction of the town of Rabaul, Papua New Guinea, as a consequence of the 1994 volcanic eruption, housing rental prices in the area increased substantially to the advantage of landlords if not tenants. As a further example, a survey of the impact of the 1980 eruption of Mount St. Helens on two towns, Toutle and Lexington, questioned respondents about changes in their individual standard of living as a consequence of the eruption. Although only 38% of Toutle and 11% of Lexington respondents reported any change, a third of Toutle and

9% of Lexington respondents reported a positive change, reflecting the fact that although some jobs were eliminated, others were created. In particular, the massive log salvaging effort provided steady work for many (Perry and Lindell, 1990).

## Overall impacts

The economic impacts of a volcanic eruption are typically localized, affecting only a small part of a country except in the case of small island economies (Box 15.2) or particularly major eruptions.[2] The localized impact of an eruption is significant from an economic perspective. It implies that the impacts may not be captured in broad economic aggregates such as the performance of a country's gross domestic product (GDP) or export sector or even the rate of inflation. For example, despite a volcanic disaster which almost totally destroyed the town of Rabaul in 1994 and had a significant impact on the provincial economy, Papua New Guinea experienced a modest real increase in national GDP year on year.

Nevertheless, at a more localized level, natural disasters still constitute potentially severe exogenous shocks and can also have particularly profound implications in reinforcing economic

---

**Box 15.2** | **The economic consequences of the eruption of La Soufrière, Montserrat, 1995–1997**

Activity at the volcano of La Soufrière on the southern tip of the small Caribbean island of Montserrat began in June 1995, continuing until the present time (December 2004). The island's economy had been performing well prior to the eruption, in part boosted by an important tourist industry. However, the onset of volcanic activity seriously disrupted the island's economy. This box focuses on events up to the end of 1996.

At the beginning of 1995, economic prospects for the island had been considered good, with projected improvements in performance in the tourism, construction, agricultural, and manufacturing sectors. Data for the first half of 1995 confirm these patterns. However, the onset of volcanic activity in June of that year had an immediate adverse impact on economic performance. Although manufacturing activity improved marginally due to the commissioning of a new rice mill in May 1995, a number of construction projects were halted whilst numbers of tourist arrivals fell 27% year on year in the third quarter of the year. The situation deteriorated further during the final months of 1995 as intensified volcanic activity resulted in the evacuation of the island's capital, Plymouth, which was located in the south of the island. Construction activity contracted again, in part as home-builders were no longer able to obtain insurance or therefore building loans. Some public-sector projects were also suspended as government priorities were reassessed, while new investments and developments were delayed by the continuing uncertainty. Manufacturing output from most subsectors was also poor. However, the country's new rice mill continued to operate despite the fact that it was located in

---

[2] Exceptional major eruptions, occurring on a scale experienced only a few times each century, have been linked to the large-scale emission of greenhouse gases, with significant consequences for global climatic conditions. For example, the 1991 eruption of Mt. Pinatubo is estimated to have caused an average 0.1 °C fall in global temperature (Rymer, 1993). The eruption of Tambora in Indonesia in 1815 resulted in global temperature declines of around 1 °C for several months and reportedly led to crop failure, famine, and disease as far away as Europe (Rymer, 1993).

the danger zone, boosting both total exports and manufacturing output. Indeed, rice and electronic components dominated the island's export earnings, with overall export earnings more than three-fold those for the same period in 1994 as a direct consequence of the new rice exports.

Overall economic performance remained poor in 1996. Crops were damaged by increased emissions of gas and ash. Agricultural operations were also seriously curtailed by another evacuation of the south of the island in April 1996, although farmers were latterly permitted to return to harvest crops. Meanwhile, the tourism sector performed very weakly while the production of electronics components was disrupted by the industry's relocation out of the danger zone. Other manufacturing output declined following the closure of the industrial estate, which was also located within the danger zone, in the second quarter of the year. In consequence, although no official data are available, non-rice exports are believed to have declined and imports to have risen, the latter reflecting stronger demand for food and other relief supplies. More positively, construction activity picked up from the second quarter of 1996 as the construction of new permanent residences on the north of the island was begun and some temporary commercial buildings were replaced by more permanent ones. This activity was partly motivated by an agreement by the insurance industry to provide 40% cover of insured items against volcanic risk, in part reversing its earlier decision to provide no cover at all. The rice mill also continued in operation, dominating manufacturing output and export earnings.

Recent events in Montserrat also offer some evidence of the impact of extended volcanic activity on the monetary sector. Between the last quarters of 1994 and 1996, demand and time deposits, which together formed 71% of total money supply (M2) in the last quarter of 1996, declined by 12% and 11% respectively, in part reflecting the drawdown of personal savings to supplement household incomes which had been disrupted by the eruption. However, this decline was approximately offset by a fall in lending as the uncertainty generated by the eruption dampened both commercial and, particularly, household investment and thus demand for loans, and presumably as banks assigned higher levels of risk to lending.

*Source: Eastern Caribbean Central Bank (1994–6).*

poverty. They can bring productive activities to a temporary halt as power, communications, and transport systems are damaged. In addition, they can dampen local commerce, rates of investment, levels of employment, and many other aspects of a healthy economic life. For example, in the three months following the 1985 eruption of Nevado del Ruiz volcano, Colombia, the nearby town of Honda, which had suffered some damage from lahars, experienced a more than 50% decline in economic activity (Mileti *et al.*, 1991). The local fishing industry was particularly adversely affected by the contamination of the river whilst tourist trade also declined significantly. Regional GDP was estimated to have declined by 8% in 1985 as a direct consequence of the eruption (Martinez, 1986).

Such impacts are further illustrated by evidence from the Philippines following the eruption of Mt. Pinatubo in June 1991. The provinces of Zambales, Pampanga, and Tarlac as well as part of Bataan, all in Region III, were most severely affected by the initial eruption. Some 80 000 ha of agricultural land and fishponds as well as whole villages and small towns were completely buried by ash and the initial lahars, with sometimes just a church steeple or the roofs of buildings remaining visible above the mud. Transport, communications, power, irrigation, and other infrastructure as well as houses and public buildings were also damaged; drains and other water conduits were blocked, increasing the risk of flooding; commercial and industrial establishments in the cities of Angeles and Olongapo were forced to suspend operations; some 600 000 jobs were lost, equivalent to around a quarter of total employment in Central Luzon; and, at the height of the eruption, 200 000 people were evacuated. Total damage to industry, including agri-based industries, was estimated at US$32 million. Ash deposits further afield created additional disruption, forcing, for example, the closure of both schools in Manila and the international airport for several days (PHIVOLCS, 1991). Subsequent lahars have occurred in every year since the eruption. Those in 1991 and 1992 alone affected almost 260 000 people, destroyed 4190 houses, and caused total damage exceeding Peso 1 billion (US$39 million at the 1991 rate of exchange).

In terms of the implications for overall economic activity, GDP for Region III fell by 3.2% year on year in 1991. Manufacturing GDP alone declined by 6.8% and construction by 6.2%, although agricultural GDP increased by 2.4%. Despite the continuing threat of lahars, construction activity picked up considerably again in 1992, increasing by a massive 76.7% year on year. This no doubt partly reflected infrastructural rehabilitation efforts begun in late 1991. Positive growth was also restored to other sectors of the regional economy, although at relatively low rates compared to those in the years leading up to the eruption. This presumably reflected the impact of continued uncertainty in undermining confidence in the local economy.

More positively, major eruptions can also generate large-scale reconstruction programs, potentially providing a major stimulus to the local economy if building materials and labor are sourced locally. For example, following the 1994 Rabaul eruption in Papua New Guinea which, as already noted, totally destroyed the important commercial town of Rabaul, many businesses relocated to the nearby town of Kokopo, the new administrative and business center developed by the government. This reconstruction program was expected to provide a strong stimulus to the local economy (Fairbairn, 1996).

Meanwhile, even volcanic crises which ultimately result in little direct damage can lead to considerable economic losses, as illustrated by the earlier 1983–5 eruptive phase at Rabaul (Box 15.3).

## Direct impacts

As already indicated, the direct impacts of a volcanic eruption relate to damage to homes, hospitals, schools, factories, machinery, crops, livestock, inventories of inputs and finished goods, transport, telecommunications, and power supply infrastructure. The precise scale and nature of these impacts depends on the eruption itself. More violent manifestations, involving pyroclastic flows, lava, or lahars, can cause extensive physical damage. Slower-evolving eruptions can permit sufficient warning to be given for the timely evacuation of the immediately affected population, reducing loss of life and damage to non-fixed assets. However, heavy deposits of ash can cause buildings to collapse, prevent the use of agricultural lands for many years, contaminate water supplies, block and damage both water and sewage systems, cause power outages and telecommunications blackouts, damage transport and other public infrastructure, and obstruct rivers, subsequently causing flooding. Flooding can also occur as a consequence of the eruption of volcanoes capped by glaciers as, for example, in the case of the 1980 eruption of Mount St. Helens in Washington state, USA. The direction of prevailing winds plays a potentially significant role, in determining the pattern and scale of damage.

Loss of human life, injury, and illness pose further potential economic costs. Pyroclastic flows and surges have caused the most deaths and injuries in volcanic eruptions (Baxter, 1994) and although only an estimated 4% of volcanic eruptions entail loss of life (Rymer, 1993), some eruptions have had tragic consequences leaving many dead. For example, during the violent 1902 eruption of Mt. Pelée on the Caribbean island of

---

**Box 15.3** | The 1983–5 Rabaul crisis: the cost of uncertainty

Between September 1983 and July 1985, the Rabaul caldera in Papua New Guinea experienced a period of considerable unrest. During the crisis, a Stage-2 alert, which implied that an eruption was expected within a few weeks to months, was maintained for 13 months, with significant adverse implications for both the private and public sectors. Fortunately, no major eruption actually occurred. Nevertheless considerable economic costs were incurred, estimated at over Kina 20 million (US$22.2 million at the 1984 rate of exchange), including over Kina 6 million in government disaster expenditure (Lowenstein, 1988).

The volcanic activity resulted in a substantial fall in property values whilst some schools were closed. Shortages of some imported items developed as shops allowed stocks to run down while people purchased additional supplies as part of their disaster-preparedness activities. Seismic risk levies on insurance premiums also rose substantially, from 0.5% in late 1983 to 3.5% at the end of 1985. Some companies even ceased renewing existing cover against volcanic or seismic risk or issuing new such policies (Lowenstein, 1988). As of June 1988, none of the decisions taken by insurance companies during the 1983–5 crisis had been reported to have been reversed, resulting in continued lack of finance for new investments and limiting the level of new development in Rabaul (Lowenstein, 1988).

A non-random survey of 273 people affected by the crisis undertaken in May 1984 indicated that about 40% of those living in the Red Zone, the highest area of risk, had voluntarily evacuated and resettled outside the danger zone for 4–5 months whilst, in total, an estimated 10 000 people, including 7000 from Rabaul town, had moved. The same survey indicated that 39% of the sample group had moved belongings out of the danger zone whilst 27% had built houses and planted food gardens outside Rabaul and a further 13% had done one or other of the latter two activities. Some 17% of respondents believed there would be an eruption in 1984, 18% that there would not, and 60% were uncertain (Kuester and Forsyth, 1985).

---

Martinique 28 000 people perished. Depending on its precise chemical composition, volcanic ash can also be highly toxic, contaminating water, crops, and air (Baxter, 1994). Moreover, volcanic activity can have psychological impacts, whether as a consequence of major eruptions, the continuing threat of renewed activity, or disruptions to employment (Baxter *et al.*, 1982). However, there are particular difficulties in valuing loss of human life or the cost of ill-health. Indeed, this is one of the more contentious areas of economics. This chapter therefore goes no further than to highlight the potential additional economic costs associated with such losses, both to individual households and to the wider economy.

## Agriculture and livestock

Volcanic areas often form some of a country's most productive agricultural regions as already noted, reflecting the benefits of volcanic soils. In areas with relatively frequent eruptions, some plants may have genetically adapted to cope with repeated volcanism. For example, in the case of Mount St. Helens, there is some evidence that fir trees located on the northeast flank of the volcano have undergone such adaptation over the centuries (J. H. Means and J. K. Winjum, unpublished data). Nevertheless, agricultural losses often form a significant part of the total direct costs of volcanic activity and can also imply considerable medium-term loss of livelihoods to farming communities.

Crops, trees, and other vegetation can be damaged or destroyed in a number of ways. In more extreme cases, volcanic activity can even result in nutritional stress. For example, the Laki fissure eruption in 1783 in Iceland triggered a famine which killed an estimated 10 000 people.

Vast tracts of agricultural land can be buried under lava or lahars, perhaps preventing production for a number of years. For example, during the 1928 eruption of Mt. Etna, "1.6 km$^2$ of agricultural land was inundated by lava, effectively putting it out of action for more than a lifetime" (Chester *et al.*, 1996, p. 58). Similarly, lahars associated with the 1991 eruption of Mt. Pinatubo in the Philippines caused extensive damage (Box 15.4). Gas emissions can also destroy crops, as demonstrated by the 1946 eruption of Masaya volcano, Nicaragua, during which fumes damaged some 130 km$^2$ of coffee crops and killed an estimated 60 million coffee trees (Sigurdsson and Carey, 1986). In addition, crops can be seriously affected by deposits of volcanic ash, which can both lower the efficiency of photosynthesis through the leaves of plants and can be highly acidic. The toxicity can also result in the slow death of plants and trees. For example, coconut trees left standing after the 1984 eruption of Mayon Volcano in the Philippines gradually died over the following year (Umbal, 1987).

The overall impact on agricultural production is partly dependent on the timing of eruptions relative to the agricultural year, as already indicated, and the relative balance of permanent and annual cropping. For example, Costa Rican coffee

---

**Box 15.4** | Agricultural impacts of the 1991 eruption of Mt. Pinatubo

Prior to the eruption of Mt. Pinatubo in July 1991, Central Luzon had been the Philippines' prime rice growing region, accounting for 17.5% of national rice production in 1988–92 and 14.7% of gross acreage. Sugar cultivation and aquaculture were also important. For example, the region produced 45% of the country's total fishpond production in 1990. Agricultural activities were supported by an extensive irrigation and river control network.

However, the volcanic eruption and subsequent lahars have had severe implications for agricultural production. Ash to a depth of 5 cm or more was deposited over an area of some 550 000 ha as a consequence of the initial eruption in June 1991, severely damaging some 385 500 ha of agricultural land as well as forestry and aquaculture (Rantucci, 1994). Lighter ashfalls were deposited over a much wider area whilst subsequent lahars have caused further severe damage. As a result, brackish-water fishponds were severely disrupted and much of the irrigation and river control network damaged or destroyed. The eruption completely altered eight major river systems, raising river beds some 3–7 m and altering river courses (US Army Corps of Engineers, 1994). Five national, 163 communal, and 14 smaller irrigation systems were reported to have been partly or totally damaged. Livestock were also killed, both as a direct consequence of the eruption and the related collapse of buildings and as the result of subsequent shortages of fodder.

In the second half of 1991, in the immediate wake of the initial eruption, rice production in Central Luzon declined by 21.2% year on year whilst acreage fell by 14%. The first crop of 1992 was also 15% lower year on year, reflecting lower yields and acreage; and by 1994 production and yields had still not recovered to their 1990 levels, although gross acreage had. As of 1994, aquaculture output was still only some 60% of previous levels due to the obstruction of water flows and tidal exchange and the destruction of breeding grounds. These problems were expected to continue as more ash was washed downstream (US Army Corps of Engineers, 1994).

Some 87 000 ha of agricultural lands with less than 15 cm of ash deposits were reported to be recoverable. As the chemical properties of the ash reduced fertility but did not render the land unusable, farmers were able to till the land to mix the ash with soil and restart planting operations (Fernandez and Gordon, 1993). However, in areas with deeper ash deposits it was expected to take some 20 years or more before adapted vegetation types could grow (Rantucci, 1994). The International Rice Research Institute has had some success in planting rice in volcano-damaged fields using heavy inputs of fertilizer. However, there are questions about the economic viability of such methods, as rice production in the Philippines is already comparatively expensive (McBeth, 1991).

The Asian Development Bank estimated that the eruption resulted in US$220 million direct damage to agricultural production, forestry, fisheries, and livestock, equivalent to 2.3% of national agricultural GDP in 1991. Damage to agricultural facilities totaled US$15.2 million, whilst a further US$ 179.6 million was lost in foregone revenues. As of August 1991, the eruption was also expected to cause some US$890 million losses in agricultural production over the following five years, with considerable multiplier effects through the economy.

production declined by 24%, or 15 000 tonnes, in 1963/4 after ash emitted from Volcan Irazú settled on coffee bushes while they were in flower, effectively destroying the crop (Benson, 1995). Fortunately, the coffee bushes themselves were not destroyed but the dispersion of ash led to an increase in insects which severely damaged their leaves. Total damage to the coffee industry was estimated at over US$20 million. The reduced coffee harvest also resulted in the loss of some 22 000 jobs in 1964 (Gobierna de Costa Rica, 1964) and lowered export earnings.

In the case of damage to heavily forested areas there may be some opportunity for salvage operations, as for example following the 1980 eruption of Mount St. Helens. This volcano is located in a relatively sparsely populated, forested area, also popular for recreational activities such as climbing, backpacking, hunting, and fishing. Within 15 minutes of the initial blast – one of the most violent this century – some 61 000 ha of forest were devastated. However, a considerable amount of timber was retrievable. In particular, trees killed by the eruption but only charred or pitted were undamaged beneath the bark. Salvage operations on some 19 000 ha were planned to be completed by the end of 1984, generating an estimated

2.7 billion board feet of timber. Moreover, overall logging costs were only slightly more expensive than normal. Replanting on 2400 ha of land with under 15 cm of tephra began the first winter after the eruption. A further 5600 ha were reforested in 1982 whilst it was envisaged that a total 28 000 ha would be reforested by 1985. However, in some areas planting costs were considerably higher than normal, because of more costly site preparation costs and planting difficulties (Means and Winjum, unpublished data).

Pyroclastic flows, lava, lahars, and ashfall can also damage or destroy agricultural infrastructure, including irrigation and drainage systems, machinery and equipment, and crop-storage facilities. For example, the 1985 eruption of Nevado del Ruiz, located in an important agricultural region of Colombia, blocked hundreds of irrigation channels and severely damaged storage facilities for cotton, rice, and coffee as well as destroying some 3400 ha of agricultural land (Praez, 1986). The quality and quantity of both surface water and groundwater can also be affected, in part due to contamination from ash and other deposits and to hydrological changes resulting from pyroclastic flows, ashfalls, lava, and lahars, including alterations to the drainage and pattern of river flow. Meanwhile, the destruction of upstream trees can result in increased runoff, again reducing groundwater levels and resulting in water shortages as, for example, occurred following the 1982 eruption of Mt. Galunggung, Indonesia (Bakornas PBA, 1982).

Livestock can be killed from ingestion of large quantities of ash, poisoning from toxic constituents of ash or gas emissions or, indirectly, from shortages of availability of uncontaminated pasture (Seaman *et al.*, 1984). For example, several thousand cattle were reported to have died after eating toxic ash-covered vegetation during the 1943 eruption of Paricutín in Mexico (Sigurdsson and Carey, 1986). Inland fisheries can be similarly disrupted, as illustrated in the case of the 1991 eruption of Mt. Pinatubo (see Box 15.4). For example, ash and lava can pollute water sources as well as resulting in the shallowing or destruction of fishponds and other water systems. Meanwhile livelihoods can be adversely affected from reduced milk, egg, and fish yields and the loss of draft animals. Shortages of fodder can also result in the widespread sale of livestock at much reduced prices as, for example, was reported in the aftermath of the 1982 eruption of Mt. Galunggung (Bakornas PBA, 1982).

Finally, agricultural activities can be disrupted by labor shortages during and in the aftermath of eruptions. For example, the repatriation of labor following the 1994 Rabaul eruption in Papua New Guinea was reported to be an important factor resulting in lower cocoa and copra production. Prolonged evacuations can also disrupt farming activities. For example, agricultural operations were curtailed by the evacuation of the south of the island of Montserrat in April 1996, although farmers were latterly permitted to return to harvest crops.

Agricultural recovery can take a number of years, depending on the depth and chemical composition of deposits of ash, larger debris, lava, or lahars. Some cultivation may be possible if lava or lahar deposits are not too deep. For example, Chang (1992) reports that papayas, macadamia nuts, and various horticultural crops grow extremely well in the volcanic East Rift Zone of Hawaii, even on lava flows that have been broken up and planted. In the case of the eruption of Mount St. Helens, some plants even sprouted up through the tephra during the first spring following the eruption, having survived in buried soils (Means and Winjum, unpublished data; see Dale *et al.*, Chapter 8, this volume). However, despite some indications of relatively rapid recovery, there is little documented evidence on the precise longer-term agricultural impact of volcanic eruptions or the speed of recovery, again reflecting the speed with which post-disaster damage assessments are taken. Such documents often report the area of damaged agricultural land and include statements to the effect that reclamation is expected to take perhaps several or perhaps even many years but the actual rate of recovery, the particular problems encountered, the solutions adopted, or the implications for agricultural communities are seldom documented.

## Damage to buildings and infrastructure

Lava, pyroclastic flows, and lahars can cause extensive physical damage to buildings, factories, machinery, and other infrastructure, including clean water, sewage disposal, transportation, power and telecommunication facilities, in their path, with various indirect implications for economic performance. For example, the 1928 eruption of Mt. Etna totally devastated the town of Mascali, with an estimated population of over 2000. During the same eruption, some smaller settlements as well as a narrow-gauge and main-line railroad track were also destroyed, seriously disrupting communications in the whole of eastern Sicily. Total damage was estimated at US$1.3 million in 1928 prices (Chester *et al.*, 1996). As a further example, almost 41 000 homes were totally destroyed and 68 000 damaged by the initial eruption of Mt. Pinatubo in 1991 (Tayag and Punongbayan, 1994). A further 6500 houses were damaged or destroyed by lahars in 1992, with additional destruction as a consequence of lahars in subsequent years. Total losses were estimated at Peso 10.6 billion (US$385 million) in 1991 and Peso 1.2 billion (US$30.6 million) in 1992 alone, including Peso 3.8 billion damage to public infrastructure. Meanwhile, the 1980 eruption of Mount St. Helens, although occurring in a sparsely populated area, buried or destroyed over 500 miles of road, 16 miles of railroad, and 12 bridges together with three non-resident logging camps and dozens of machines and vehicles. As a final example, the 1994 eruption of Rabaul was estimated to cause total Kina 100 million (US$91 million) damage to public infrastructure and buildings, based on estimates for the replacement and, in some cases, upgrading of destroyed assets, equivalent to approximately twice the annual provincial government budget and 6% of total central government expenditure (including net lending) for 1994 (Fairbairn, 1996). These costs included a Kina 10 million upgrading of an airport to replace the larger one in the immediate vicinity of Rabaul which had been destroyed during the eruption. An estimated additional Kina 180 million in losses were incurred by the private sector (Blong and McKee, 1995), with further unassessed losses to villages. Meanwhile, the rehabilitation program, including the development of Kokopo as the new administrative and subregional center, was estimated to cost US$100 million, much of which was expected to be financed by international donors.

Large explosive eruptions can also cause severe damage, as in the case of the 1902 eruption of Mt. Pelée which destroyed masonry walls 1 m thick in buildings located over 6 km from the eruption (Schmalz, 1992). Meanwhile, ash deposits can cause the collapse of roofs. For example, the 1994 Rabaul eruption resulted in the deposit of up to 900 mm of wet ash on roofs in the town of Rabaul, equivalent to over 1.5 t/m$^2$, resulting in their widescale collapse (Blong and McKee, 1995). Mudflows to depths of 200–300 mm also passed under doorways, causing considerable damage to building interiors and contents.

In addition to the threat to transport infrastructure, vehicles and airplanes can also be damaged. Ash can corrode metal and clog engines, fuel filters, and the like. In addition, aircraft can be exposed to danger in mid-air should they fly directly through an ash cloud. For example, a Boeing 747–400 plane traveled directly through an ash cloud emitted by Redoubt Volcano, Alaska, in December 1989, resulting in the loss of power to all four of its engines. Although the engines were subsequently restarted and the plane landed safely, repair costs were estimated at US$80 million (Tuck et al., 1992). Two Boeing 747 jet passenger planes were similarly endangered during the 1982 eruption of Mt. Galunggung (Bryant, 1991), resulting in the subsequent suspension of flights in the region until the volcanic activity ceased. Lava or lahar flows in coastal areas can also potentially reduce the navigational capacity of rivers and canals and block ports. For example, the 1883 eruption of Krakatau filled one Sumatran trading bay with sheets of pumice, forcing the transport of food and other supplies some 35 km overland (Simkin and Fiske, 1983). International trading routes were also disrupted as the volcano is located in the Sunda Straits, one of the busiest and oldest shipping routes in the world. Similarly, debris deposits and lahars also created navigational and other problems in the river systems draining Mount St. Helens in the aftermath of the 1980 eruption.

However, once again, the timing of various eruptive phases relative to local climatic seasons plays an important role in determining the extent of damage. Emissions of ash during the rainy season are likely to result in less damage and require smaller-scale cleaning operations as ash clouds are effectively washed out before traveling very far. Rain also washes ash off roofs and roads, further reducing the necessity for clean-up operations.

There are also some opportunities for mitigating the extent of damage to buildings and infrastructure during periods of low-level volcanic activity involving relatively little ashfall via regular clean-up operations. For example, during the 1963–5 eruption of Volcán Irazú, Costa Rica, a major nightly cleaning operation of streets, roofs, water supply, and drains was implemented in the nearby capital of San José. The government was reported to employ up to 500 people in these operations. With adequate warning, the local population can also remove non-fixed assets at risk from pyroclastic flows, lava, lahars, or ash to safer locations, as in the case of the 1928 eruption of Mt. Etna when the population of Mascali were assisted by soldiers and given access to military transport to remove furniture, doors, ceramic and roof tiles, and other household objects (Duncan et al., 1996). Similarly, during the eruption of Pu'u O'o-Kupaianaha, Hawaii, in 1983–91 some smaller dwellings as well as a church were raised whole from their foundations and moved to safer locations while some larger dwellings were dismantled (Chang, 1992).

Basic transportation networks and essential services such as power, water supply, and telecommunications can also often be restored relatively quickly, minimizing related economic losses. For example, within two weeks of the 1985 eruption of Nevado del Ruiz many roads in the affected area had been re-opened and temporary bridges erected whilst a number of other restoration activities had also been achieved. However, this assumes that there is no further risk of an immediate eruption, potentially placing reconstruction efforts at risk.

## Indirect impacts

### Productive sectors

Volcanic activity can have significant indirect implications for productive sectors, as already indicated. However, there have been relatively few attempts to quantify such impacts.

One notable exception has been an investigation of the economic costs of the 1989–90 Redoubt eruptions, Alaska, to the aviation and local oil industries, providing a flavor of the types of costs involved (Tuck et al., 1992). This study used a social accounting framework, comparing the economic activity which would have taken place in the absence of any eruption with that which actually occurred. Total losses to the aviation industry were estimated at US$101 million, including losses to airport support industries, the cost of passenger waiting time, and reduced landing and other fees as well as US$80 million in direct equipment damage costs. Activities at an oil terminal located on a river at the base of Mt. Redoubt to which oil is delivered from a nearby oilfield for loading into tankers was also affected. Parts of the facility were damaged by flooding as mud and debris deposits altered the flow of the river whilst the amount of oil that could be stored at the facility was reduced to avoid the threat of a major oil spill should a major explosion occur. The latter together with tanker loading difficulties forced a reduction and then a total halt in production from oilfields shipping

through the terminal, with below-normal production sustained for a total of nine months. Total costs were estimated at US$47.3 million, including modifications and repairs to the terminal as well as lost revenues.

### Employment

Volcanic eruptions can result in the temporary or even permanent loss of jobs if places of work – be they fields, factories, or offices – fall within the danger zone, effectively prohibiting access, or are damaged or destroyed. Such losses can create considerable hardship for individual households as well as reducing aggregate expenditure in the local economy, in turn contributing to any recessionary implications from an eruption.

For example, around 624 000 persons or 28% of the total locally employed population were reported to lose their sources of livelihood either temporarily or permanently as a consequence of the eruption of Mt. Pinatubo in April 1991 (Tayag and Punongbayan, 1994). Even more minor eruptions can result in some disruption. For example, the Kenai Peninsula governmental sector, Alaska, lost at least 1010 days of labor because of closures during the Mt. Redoubt eruptions in 1989–90 (Tuck et al., 1992). Around 80 worker days were also lost as workers in the cities of Soldotna and Homer were sent home whilst 30 days were lost when a hospital sent non-essential workers home for half a day.

However, volcanic activities can also create new job opportunities, as already indicated. In particular, they may generate openings in the construction industry as reconstruction efforts get under way. Rehabilitation programs can also play an important role in helping to restore livelihoods. For example, following the 1928 eruption of Mt. Etna, the authorities secured new agricultural work for farm-workers left unemployed following the eruption. Similar efforts were also made to provide new jobs at some resettlement centers following the 1991 Mt. Pinatubo eruption. For example, at one such site, a series of warehouses were constructed in the hope of encouraging industry, and thus new jobs, into the area. However, these have unfortunately remained empty as most of those in the resettlement camps lack the necessary skills to attract high technology industries, such as computer technology (Pye-Smith, 1997).[3]

## Inflation

In theory, natural disasters are likely to have a net inflationary impact. Prices can rise as a consequence of supply shortages, reflecting damage both to goods (especially foodstuffs) and to transport and marketing infrastructure. Demand for certain items, such as building materials, can also increase, again forcing up prices. Such rises may be partly offset by lower demand for luxury goods or non-essentials, in turn reflecting the generally recessionary nature of a disaster and reduced levels of employment and income. Nevertheless, the net impact of a disaster is likely to remain inflationary.

In practice, localized temporary price hikes have undoubtedly occurred as a consequence of volcanic activity. For example, the 1994 Rabaul disaster was reported to result in a marked temporary rise in prices for many types of local produce (Fairbairn, 1996). Similarly, the 1963–5 eruption of Volcán Irazú had some impact on food prices in the San José region, with a 5.6% rise in 1964 despite the receipt of some food aid imports from the USA to boost supply (Benson, 1995; Young et al., 1998).[4]

However, prices typically fall again as supplies are brought in from non-affected areas whilst localized rises are rarely reflected in national inflation indices. Moreover, governments often undertake considerable efforts to minimize rises in the price of basic consumer goods, undertaking deliberate efforts to ensure that price hikes are avoided through, for example, the implementation of various anti-inflationary laws.

Property rental prices can also increase if there is widespread damage to or destruction of houses and commercial properties. In contrast, property and land prices in higher risk areas may decline. However, such falls are largely hypothetical as few property sales are likely to take place

during or in the immediate aftermath of periods of more intensive volcanic activity.

## Investment

Once again, disasters often have little discernible impact on national rates of investment other than in small-island economies, such as Montserrat (see Box 15.2), but can affect more localized rates and patterns of investment. In more extreme cases, eruptions even influence the regional and provincial distribution of investment resources, with possible implications for widening regional inequalities. Volcanic eruptions can destroy capital stock and infrastructure, conceivably boosting overall investment in the local economy as assets are replaced. However, they can also act as a disincentive to new investment, particularly in their immediate aftermath when people's perceptions of hazard risks are heightened whilst the economy appears less stable.

For example, the three Philippine provinces most severely affected by the 1991 eruption of Mt. Pinatubo – Zambales, Pampanga, and Tarlac (all in Region III) – received 2% of the total cost of projects approved under various investment laws in 1990, the year prior to the eruption (Benson, 1997). This figure fell to 1.1% in 1991 and averaged 0.5% in 1992, demonstrating the extent to which the eruption and the subsequent threat of lahars acted as a major disincentive to would-be investors. Indeed, despite various efforts, it has proved extremely difficult to attract investors into the Mt. Pinatubo region (Box 15.5), although overall levels of investment in the area have been partly boosted by considerable structural investments to provide protection against lahars.

Even low-level volcanic activity which does not present any immediate danger to human life or property can bring rates of new investment to a virtual halt by creating considerable confusion and uncertainty and undermining

---

[3] Employment opportunities received less consideration in deciding the location of certain other resettlement camps established in the wake of the 1991 Mt. Pinatubo eruption. For example, one camp was established 29 km from the nearest probable place of work, with the cost of public transport to and from this town equivalent to over half the average rate of pay for a day laborer (Pye-Smith, 1997).

[4] The role of the eruption in fueling this rise in prices is supported by a subsequent 1.4% decline in the San José food price index in 1965, after the cessation of the volcanic activity. The city's overall consumer price index declined by 0.2%.

## Box 15.5 | The 1991 Mt. Pinatubo eruption and investment: the Clark Base

Investment in the Clark mini-industrial zone in the aftermath of the Mt. Pinatubo eruption provides an interesting example of investor behavior and risk perception. The former US airbase was covered in a thin layer of volcanic ash and rubble as a consequence of the eruption. This precipitated the final US withdrawal from the base although a pull-out had already been likely following the expiry of the 1947 US–Philippines Military Bases Agreement and strong domestic opposition to its extension. After the US withdrawal, the decision was taken to redevelop both Clark and another former US base, the Subic Bay naval base, into agro-industrial, civil aviation, and tourism centers. In the case of Clark, these plans included the development of a holiday resort and a new international airport.

However, the threat of both lahars and possible further eruptions has effectively hampered efforts to promote Clark's redevelopment. Another major eruption of Mt. Pinatubo is considered highly unlikely for centuries but the base remains potentially threatened by lahars in the short to medium term. For example, the city of Angeles, which is adjacent to the base, was damaged by a lahar in 1991. The base also experiences occasional light ashfalls due to secondary explosions.

Thus, by late 1995 Clark had yet to attract any major investors, although its land leases were particularly cheap and tax incentives were also available to companies locating within both Clark and Subic Bay (Benson, 1997). Instead, although some 50 projects up to a total cost of Peso 6.2 billion (US$0.23 billion) had been approved by the end of 1994, most of the enterprises were relatively small, with under 50 employees, reflecting the high perceived risk of damage from lahars and further eruptions (Tiglao and Tasker, 1995). Transport difficulties in the rainy season had proved a further disincentive to potential investors as heavy rains generated mudflows, sometimes entailing extensive road detours and thus additional transportation costs as well as delays (Tiglao and Tasker, 1995).

Even those companies which have invested in Clark have sometimes faced difficulties relating to others' perceptions of risks. For example, one company which has constructed a 22-hole golf course, a hotel, and a casino at the former base reported certain difficulties in promoting its enterprise because of the general public's poor perception, in turn, of the safety of Clark (Benson, 1997). The company has therefore had to work hard to assure potential customers of its safety.

The Clark Base also includes a resettlement camp for victims of the Mt. Pinatubo eruption. With few job opportunities to date in the camp, prostitution was reported to be becoming increasingly common (Watkins, 1995).

confidence in the local economy. For example, the 1963–5 eruption of Volcán Irazú, Costa Rica, led to a reduction in the level of new capital investment as some projects were delayed, reflecting uncertainty about how serious the eruption would be. In particular, the Costa Rican government stopped virtually all building works in 1964. Levels of investment in residential and non-residential buildings also fell by 27% and 23% in 1964, although picking up again in 1965 following the cessation of activity in the February of that year. Meanwhile, although investment in machinery, transport, and other equipment continued unabated, there was some shift in the latter towards projects stimulated by the eruption such as the construction of dams and bridges.

Political factors can play some role in the scale and nature of government response to disasters and thus in their impact on post-eruptive levels of investment. As an extreme example, the substantial and rapid investment in the reconstruction of the Italian town of Mascali on a new site following the 1928 eruption of Mt. Etna has been interpreted as "a propaganda exercise to illustrate the effectiveness of the (Mussolini) regime" (Chester *et al.*, 1996, p. 58).

## The external sector

In theory, natural disasters would be expected to create balance-of-payments difficulties to the extent that they reduce the production of goods for export, whether by the direct destruction of agricultural and other sectors or by disrupting the transport of inputs and finished products, power supply, and so forth. They could also result in an increase in the level of imports to meet disaster-related domestic food deficits and repair damage although, to the extent that natural disasters have a domestic recessionary effect, demand for non-essential imports could decline, partly offsetting some of the pressure on the balance of payments if not on public finance. Depending on levels of foreign-exchange reserves, these various changes could result in an increase in external borrowing, with implications for future levels of debt servicing and, ultimately, economic growth. Any worsening of the balance-of-payments position could also exert pressure on the exchange rate and thus on a country's international competitiveness, again with potentially serious consequences. The impact of natural disasters on exports and imports also has certain budgetary implications in so far as government revenue is derived from various export and import duties and tariffs.

Once again, in practice it is difficult to discern much impact of volcanic activity on national trade statistics, other than in cases where the area immediately surrounding a volcano has been the main center of production for a particular export. Moreover, even then a decline in output may be offset by a rise in world prices. Indeed, contemporaneous fluctuations in world commodity prices can play a significant role in working to either offset or reinforce the impact of volcanic eruptions and other disasters in a number of developing countries for whom a handful of primary commodities form a significant share of export earnings yet which have little influence on international price movements, as illustrated by the 1994 Rabaul eruption, Papua New Guinea. The province normally accounts for around half of national cocoa and two-thirds of coconut oil exports but the eruption resulted in a Kina 12 million (US$10.9 million at the 1994 rate of exchange) reduction in regional exports of these items, reflecting both lower output and the dislocation of shipping, which was halted for almost three months. However, the loss in earnings was partly compensated by higher international prices while national export earnings from these two products actually rose in 1994 and 1995. Indeed, coupled with substantial increases in other agricultural export earnings, the country actually recorded a significant reduction in its overall balance-of-payments deficit in 1994 and a surplus was achieved in 1995.

Volcanic eruptions can also have important implications for tourism earnings, as already indicated in the case of Montserrat (see Box 15.2). Again, the precise impact on tourism is partly time-dependent, with highly publicized and more active behavior during the peak tourist season likely to have more serious implications. However, minor activity can encourage tourism whilst the number of visitors to volcanic regions can also increase in the aftermath of an eruption as, for example, occurred following the 1980 eruption of Mount St. Helens. In this particular case the growth in tourism was viewed as a mixed blessing by local residents, providing a boost to the local economy but also threatening to crowd facilities and to destroy "the pristine beauty of the countryside" (Perry and Lindell, 1990, p. 151).

As a further example, the hot-spring resort town of Toyako-Onsen, Japan, sought to exploit the potential tourism benefits of the 1977 eruption of nearby Mt. Usu to its advantage. Its efforts included the opening of a Museum of Volcano Science and the construction of a promenade ground which, since it would also serve as an emergency bypass route, was co-funded by the national and prefectural governments rather than by the town. Roads were also improved to secure safe evacuation routes and a new sewage works constructed as part of the disaster recovery assistance, again largely with government funding and, at the time, making Toyako-Onsen one of only very few Japanese lakeside tourist areas to have sewage treatment equipment. However, despite such investments, some four years after the 1977 eruption the annual number of tourists to the town had still yet to reach its pre-eruption level whilst those who had come often chose to make a day trip to the town and to stay elsewhere (Hirose, 1982).

Finally, despite perhaps playing a crucial role in meeting emergency and rehabilitation needs, international assistance provided in response to a volcanic event can also entail long-term debt legacies if provided in the form of soft loans rather than grants. For example, much of the US$100 million 1994 Rabaul disaster rehabilitation program was expected to be financed by international donors, in part through soft loans (Fairbairn, 1996). This assistance included a US$25 million loan from the World Bank. In the case of the international response to the 1991 eruption of Mt. Pinatubo, part of the infrastructure financed under an Asian Development Bank loan was even subsequently destroyed by lahars. Nevertheless, the loan still had to be repaid.

## Government budgetary aspects of volcanic eruptions

Natural disasters can have several important implications for both provincial and national government budgetary resources. Public-financed mitigation and prevention measures exert continual and potentially significant demands on limited budgetary resources whilst relief and rehabilitation operations can imply additional unplanned expenditure or the partial redeployment of previously allocated funds. The precise cost of relief efforts is partly determined by the social structure of the affected population. For example, following the 1985 eruption of Nevado del Ruiz much of the local population found temporary housing with relatives or friends or rented or purchased alternative accommodation rather than seeking temporary shelter in the refugee camps established for survivors. However, other associated costs, particularly infrastructural rehabilitation as well as the construction of structural measures to reduce the level of damage incurred from subsequent lahars can be substantial. For example, such costs accounted for 75% of total expenditure on the Mt. Pinatubo relief and rehabilitation program between 1991 and 1994.

Government revenue may also be adversely affected in the event of a disaster as depressed economic activity, possibly including a decline in imports and exports, translates into lower direct and indirect tax revenues. Certain tax exemptions or credits may also be granted to disaster victims, further reducing revenue. For example, the Colombian government established some credit, taxation, and import incentives in the wake of the 1985 Nevado del Ruiz eruption, in part to support the creation of new employment opportunities (Praez, 1986). Meanwhile, under US federal income tax law, a casualty loss can be deducted in the event of qualifying disasters for losses which are not compensated for by insurance or other forms of relief (Murphy, 1980). For example, the whole of Washington state as well as parts of Idaho were declared eligible for casualty loss claims following the 1980 eruption of Mount St. Helens. Claims can be made against loss of trade and business and damage to property, although declines in the market value of a property are not covered.

Thus, a government may face increasing budgetary pressures as a consequence of both increased expenditure and lower revenue which it will be obliged to meet by increasing the money supply, running down foreign-exchange reserves, or increasing levels of domestic and/or external borrowing. These various financing options, in turn, have potentially significant knock-on effects. The creation of base money is inflationary. Domestic borrowing exerts upward pressure on interest rates and can result in a credit squeeze. Foreign borrowing can result in an appreciation of the exchange rate, reducing prices of imports and increasing those of exports.

It can also place future strains on the economy via higher debt-servicing costs. A rundown of foreign-exchange reserves is limited by the very size of those reserves and, again, entails an appreciation in the exchange rate, with possible associated risks of capital flight and a balance-of-payments crisis (Fischer and Easterly, 1990).

In practice, however, it is often difficult to discern much impact of natural disasters, or volcanic activity more specifically, on either government expenditure or revenue at the aggregate, national level. Efforts to measure their public finance implications are often frustrated by the fact that the precise sourcing of relief and rehabilitation expenditure, including any reallocation of planned expenditure, is often poorly documented. There are also certain measurement difficulties in isolating the impacts of disasters from broader trends. Various budgetary reforms, in particular as governments have sought to reduce their budget deficits, can also obscure the impacts of disasters, with increased disaster-related expenditure perhaps offset by additional cutbacks elsewhere. Indeed, in the current prevailing climate which typically emphasizes careful budgetary management, natural disasters tend to force a reallocation of resources rather than a substantial increase in expenditure. For example, the total reconstruction program implemented by the Colombian government in the wake of the 1985 eruption of Nevado del Ruiz cost Col$51 million (US$0.4 million). Although part of this cost was met from donations from private and international donations, some previously budgeted sources were reassigned to the rehabilitation program, including Col$2.4 million previously committed to other investments. The Ministry of Works provided Col$11.3 million, again presumably involving a reallocation of resources, with an additional Col$8.8 million from other ministries and Col$8.5 million raised through a tax on gasoline (Praez, 1986).

Nevertheless, although the impact may be obscured, volcanic activity can place considerable strains on government finance. For example, data provided by the Philippine Department of Budget and Management indicate that the Philippine government spent Peso 24.3 billion (at real 1994 prices) (US$922 million) on the Mt. Pinatubo relief and rehabilitation program between 1991 and 1994, equivalent to between 0.9% and 3.1% of total annual government expenditure and between 2.3% and 7.4% of discretionary expenditure over the same period (Benson, 1997).[5,6]

The 1994 Rabaul crisis provides further evidence on the budgetary impacts of volcanic eruptions. As a direct consequence of the crisis, provincial government tax revenue fell from Kina 8 million (US$7.3 million at the 1994 rate of exchange) to Kina 2 million, largely reflecting lower revenues from sales taxes and export levies (Fairbairn, 1996). Meanwhile, the provincial government spent Kina 5 million on relief and rehabilitation efforts, with further expenditure anticipated over subsequent years relating to the purchase of land for the relocation of up to 18 villages. By mid 1996, the central government had also spent some Kina 12 million on emergency and restoration work, in part financed by delaying a number of planned capital investment projects and by making some current expenditure cutbacks. Total government expenditure was eventually expected to reach as much as Kina 30 million.

Even volcanic events which do not involve major eruptions can still impose considerable budgetary costs. For example, during the 1983–5 Rabaul crisis, the Papua New Guinea National Government allocated around Kina 4.5 million (US$4.1 million at the 1984 rate of exchange) to previously unbudgeted disaster preparations,

---

[5] A further Peso 2 billion had been allocated for 1995 as of early December 1995. Peso 50 million had been earmarked for the Mt. Pinatubo Fund under the 1996 budget together with Peso 40 million of the total Peso 540 million proceeds generated from the sale of military camps. These allocations for 1996 were substantially lower than the Fund's request of Peso 5 billion. Certain government departments were effectively expected to meet part of the shortfall.

[6] Total non-discretionary payments, primarily consisting of substantial debt servicing and repayment costs and the public sector wage bill, totaled around 77% of total government revenue in 1990, 70% in 1991 and 71% in each of 1992, 1993, and 1994. Thus, discretionary budgetary resources are very limited in the Philippines.

including funding for the construction and upgrade of various transport and port infrastructure. An additional Kina 1.5 million was indirectly spent on related activities but absorbed into the budgets of various government departments (Blong and Aislabie, 1988).

Finally, the budgetary implications of volcanic eruptions are potentially particularly serious for small island economies, as illustrated by recent evidence from Montserrat. In the 18 months preceding the eruption, the country had run a current account surplus on its fiscal operations. Moreover, government revenue had been expected to increase during the remainder of 1995, reflecting a forecast improvement in economic activity (see Box 15.2). However, in the event, government fiscal operations actually moved into an immediate deficit in the third quarter of 1995 as a direct consequence of the adverse economic effects of the eruption. Total current revenue declined by 39% and tax revenue by 66% year on year while current expenditure fell by a more modest 17% The evacuation of the capital, Plymouth, in the last quarter of 1995 further disrupted government revenue collection efforts, resulting in a 6.9% fall in total government revenue and a 27.4% increase in expenditure for the year as a whole. Tax revenue remained very depressed in 1996, totaling only 71% of the 1994 level. Moreover total government expenditure continued to rise to a level 27% higher than in 1994, reflecting increased spending relating to the emergency. However, considerable external budgetary support of EC$7.6 million (US$2.8 million), equivalent to 15% of total expenditure for the year, was provided by the British government in the third quarter of 1996 to offset the decline in revenue as a consequence of the crisis. In consequence, the final deficit for 1996 stood at EC$5.5 million, equivalent to 11% of total annual expenditure.

## Disaster management

Disaster prevention and mitigation measures can play potentially important roles in reducing the broader economic consequences of natural hazards as well as minimizing loss of human life.

Increased expenditure on disaster prevention and mitigation also implies lower relief and rehabilitation spending in the longer term.

Such measures can be divided into structural and non-structural measures. Various structural measures have been devised to mitigate the impacts of volcanic eruptions, before, during, and, where lahars remain a threat, after eruptive phases. These include the construction of barriers to control the flow of sedimentation and, during eruption, the use of explosives to block or divert flows and of water to cool and solidify flows. Non-structural measures comprise an array of essentially non-engineered activities. For example, the construction of buildings with steeper roof pitches and the engineering of structures to withstand the weight of considerable ash deposit can reduce damage from ashfall. Afforestation of the slopes of volcanoes can play an important role in increasing the water retention capacity of the deposit and minimizing surface area (Umbal, 1987) as well as reducing the speed of flow of lava or lahars. Other non-structural measures include public education and training; vulnerability assessments; land-use control; insurance; the operation of monitoring, forecasting, and warning systems; and the management of structural features.

This section provides some flavor of the role that economic considerations – as one of a number of factors – can play in determining the nature and scale of certain forms of disaster management. More generally, Chester (1993) makes a distinction between the responses of more and less developed countries, reflecting various factors including differences in wealth, scientific expertise, the availability of insurance, and the efficiency of civil administrations.

### Structural measures

Structural measures are typically more expensive to implement than non-structural ones. They may therefore need to be partly justified on economic grounds, in turn requiring a careful assessment of their potential economic benefits.

However, they can entail certain indirect costs which should be taken into account in any such analysis. For example, the construction of diversion channels may entail the loss of agricultural land as, for instance, in the Philippines in the case of a mega-dyke constructed to protect Angeles City and San Fernando from possible lahars subsequent to the 1991 eruption of Mt. Pinatubo (Pye-Smith, 1997).[7]

The construction of diversion channels also implies certain trade-offs, with some areas protected at the expense of others. Thus, for example, Chang (1992) reports that potential legal liabilities relating to the destruction of properties which might remain undamaged in the absence of any diversion channels has halted the construction of such channels in Hawaii. Meanwhile, Blong (1984) points to potential insurance complications as the diversion of potential lahar flows could result in alterations to insurance provision.

## Hazard maps

One of the most effective ways of reducing potential human, financial, and economic losses as a consequence of volcanic eruptions is through detailed hazard mapping, land-use planning, and the related restriction of habitation and economic activity, including the construction of infrastructure, in high-risk zones. For example, following the 1980 eruption of Mount St. Helens, 44 000 ha of land in the vicinity of the volcano was established as the Mount St. Helens National Volcanic Monument, available for public enjoyment and scientific study (Means and Winjum, unpublished data). Some controlled access zones were also created. The Costa Rican and Hawaiian authorities acted similarly, designating certain volcanic areas as national parks, thus aiming to maximize their tourist appeal while minimizing potential human and financial losses.

Various mechanisms, such as conditional insurance and credit policies as well as legislative procedures, can be used to enforce hazard maps. For example, since the 1991 eruption of Mt. Pinatubo, local governments and banks in the affected provinces have issued clearance and finance for land development only on the presen-

tation of a certificate from the Philippine Institute for Volcanology and Seismology (PHIVOLCS) to the effect that the proposed site falls outside the hazard zone.

However, there have apparently been relatively few attempts to use hazard maps to analyze and attempt to reduce vulnerability to potential volcanic activity. For example, there has been little, if any, long-term land-use planning in the vicinity of Vesuvius, one of the world's most dangerous volcanoes, and buildings are now spreading up the flanks of the volcano (Baxter, 1994). Moreover, even where such maps do exist, a number of difficulties can be encountered in establishing and enforcing strict land-use codes. High population density can place considerable pressure on land resources. Volcanic soils also offer potentially important agricultural benefits, as already noted. Meanwhile, the introduction of designated hazard zones can meet with opposition from the local community because of its implications for many aspects of economic life ranging from property prices and the cost of insurance premiums to the number of overnight tourists and livelihood opportunities. For example, human settlements are still located in areas designated as lahar risk areas around Mt. Mayon in the Philippines, in part because local livelihoods are partly dependent on the land (Umbal, 1987). Indeed, in a survey undertaken by PHIVOLCS following the 1984 eruption of Mt. Mayon – in which over 8000 ha of land was rendered partially or totally useless, 158 houses damaged or destroyed, and total damage to an estimated Peso 45 million (US$2.7 million) inflicted to infrastructure – 94% of respondents nevertheless indicated that they planned to reoccupy the same land and resume their pre-eruption activities (Tayag and Rimando, 1987). This partly reflected the fact that in the five years prior to the eruption, the sample barangays (local districts) had experienced an average annual growth rate of 3.4% compared to a national one of 2.7%. Moreover, many of the

---

[7] Difficulties in securing land can even place the structures themselves at risk, potentially resulting in heavy losses. For example, again in the Philippines, a 10-km dyke costing over US$0.5 million was left with a 200-m gap because one landlord refused to allow access to his land, in turn leading to the subsequent destruction of the area by a lahar (Watkins, 1995). Local infighting over the location of other dykes in the Mt. Pinatubo area has resulted in some delays in construction, with demands from many groups to have their own particular area protected.

respondents had limited formal education and therefore few transferable skills for urban employment, further explaining their desire to remain in the vicinity of Mt. Mayon.[8]

Perceptions of risk also play an important role in determining the scale and nature of human activity, as illustrated by evidence from Hawaii (Box 15.6). Various factors, including systems of belief, can influence such perceptions. For example, indigenous Aeta communities living in the locality of Mt. Pinatubo believed that the volcano was the dwelling place of the spirits, including "Apo Namalyari," the Supreme Being and Creator. Some Aeta families were therefore initially

---

### Box 15.6 | Perceptions of risk: the East Rift Zone, Hawaii

A 150-year respite of activity in the volcanic East Rift Zone of the island of Hawaii led many local inhabitants to overlook volcanologists' warnings about hazard risks (Chang, 1992). Instead, the area became a focus of real-estate development from the early 1950s. Although two volcanoes became active again in 1955 and a subsequent eruption in 1960 totally destroyed one village, the damage incurred was largely confined to forest areas. Moreover, damage from eruptions over the following 20 years was generally confined to areas within the Hawaii Volcanoes National Park. Thus, the rate of agricultural expansion and real-estate development in the area continued to increase, fueled by the declaration of statehood in 1959. Almost 50 000 lots were created in the Puna District, on or near the East Rift Zone, while the value of many lots of land were reported to increase by over US$20 000 before finally becoming totally valueless, at least temporarily, following lava flows in the 1980s.

A 1972 survey of local residents revealed that most people were aware of the volcanic risks but believed the threat lay to property rather than lives and thus that, although other parts of the island were safer, the advantages of remaining in the region outweighed the disadvantages (Murton and Shimabukuro, 1972). A later survey also revealed that many believed that the state and federal governments would provide financial assistance in the event of an eruption, apparently unaware that such assistance could be partly conditional upon evidence of certain mitigative behavior.

However, outsiders purchasing land in the East Rift Zone area were often entirely unaware of the risk of volcanic hazards, tending to live on the US mainland. Realtors and developers also played down the volcanic risk, as demonstrated when some researchers posed as prospective buyers and casually enquired about the risk of volcanic eruptions (Sorensen and Gersmehl, 1980). The researchers were shown no volcanic risk maps or maps of recent lava flows. Instead, two of the six realtors approached said that their properties were far from the danger zone, two stated that there had been no damage from lava flows in the area in recent decades, a fifth simply told the researchers not to worry, and a sixth

---

[8] Similarly, PHIVOLCS considers the volcanic island of Taal a danger zone which should be prohibited from permanent human settlement. Yet, despite 33 recorded eruptions since 1572, including an eruption in 1911 which devastated most of the island and killed 1335 people, the island has experienced rapid demographic growth, with its population increasing from 1830 inhabitants in 1977 to 3247 by 1986 (Tayag and Rimando, 1987).

admitted some damage in the area but accompanied this information with the old adage that "lightning does not strike twice in the same place."

In contrast, the insurance industry recognized the volcanic hazard risk in the area and therefore sought to minimize its exposure, in turn effectively limiting the extent of property development: in part as a result of its actions, houses were constructed on fewer than 5% of the lots created in the Puna District. Insurance companies began to refuse policies on constructions in the highest risk zones from around 1971, implying that people were unable to secure building loans. Such constraints became more common with the publication of a volcanic hazards map in 1974 and a revised map in 1987. Finally, following further significant damage in 1990, hazard maps were begun to be used more explicitly as a basis for recommendations which would limit development in many areas around the East Rift Zone, rather than relying on insurance company actions (Germanoski and Malinconico, 1992). Nevertheless, Chang (1992, p. 481) comments that "various agencies and businesses have also tended to consider the volcanic hazard zones as being absolute, with well-defined boundary lines separating each one . . . (whilst ignoring) the complexities of approximations and scientific probabilities."

unwilling to evacuate the area immediately prior to the June 1991 eruption because they believed it impossible that their God would want to harm them (Seitz, 1998). In a somewhat different vein, Sorensen and Gersmehl (1980, p. 134) point to the particular difficulties in "establishing the credibility of (a) hazard" to incoming populations because such groups, in being attracted into the area, have "already been subjected to a barrage of risk-denial, that is information that either denigrates lava flows to harmless events or refutes their likelihood of occurrence."

Additional problems are created by sometimes poor historical eruptive records, in some cases preventing appropriate land-use planning. Volcanoes for which there is no record of activity are considered dormant or extinct whilst some are even thought to be mountains (Sigurdsson and Carey, 1986). Ironically, longer intervals between eruptions typically imply more violent manifestations when they do occur.

Finally, the expected frequency of an eruption is also relevant in determining appropriate land usage. Thus, for example, Taupo Volcano in New Zealand has had great explosive eruptions in the past but the average recurrence interval is estimated in the order of 2000 years. The volcano is therefore kept under surveillance

but considerable investment in forestry, farming, hydroelectric power generation, and tourism has taken place in the region despite potentially high losses in the event of an eruption (Crandell *et al.*, 1984).

## Evacuation

Adequate warning systems combined with effective emergency planning can play a crucial role in reducing the physical and human costs of volcanic eruptions. They can permit the removal of livestock, food stores, vehicles, agricultural machinery, household equipment, and all other non-fixed assets from areas at highest risk, as already noted. However, the precise timing of a major eruption or even whether one will occur cannot be forecast with any certainty, even in a volcano which is already displaying some activity. Thus, with hindsight, an evacuation may prove to have been unnecessary and, moreover, to have resulted in significant economic losses as well as social displacement. For example, a small eruption of La Soufrière volcano in the West Indian island of Guadeloupe in 1976 precipitated the evacuation of most of the 73 000 inhabitants of the lower flanks of the volcano as well as of livestock and fishing boats for up to three and a half months, at considerable economic and social costs. The evacuation was subsequently proved to have been unnecessary (Blong, 1984).[9]

The economic opportunity cost as well as the social and psychological impacts of evacuation can influence both the decision of public authorities to implement an evacuation program and of local communities to comply willingly. Perhaps the most notorious example of a situation where economic considerations overrode scientific concerns was the 1985 eruption of Nevado del Ruiz, Colombia (see Box 15.1). A hazard map had been drawn up only a month before the eruption and the extremely dangerous situation of the town of Armero highlighted. Indeed, Armero had been constructed on lahars generated by the 1595 eruption of the same volcano. However, prior to the 1985 eruption, the Colombian Congress criticized both the Colombian National Institute of Geology and Mines and the Colombian Civil Defense for frightening the local population with their efforts to increase hazard awareness (Mileti *et al.*, 1991). Moreover, Armero's authorities asked the volcanologists to stop issuing bulletins because fears of an eruption were disrupting normal economic activity.

A similar, if far less tragic, example of the influence of economic considerations is provided by the 1977 eruption of Mt. Usu on the northern Japanese island of Hokkaido. The eruption seriously disrupted tourist trade at the hot-spring resort of Toyako-Onsen, located directly below the volcano (Hirose, 1982). In early August, most of the town's residents were evacuated but the evacuation order was rescinded by the mayor during daytime hours several weeks later following claims by the tourist industry that the evacuation would ruin a number of businesses. However, both the police and prefectural authorities were opposed to this action, claiming that it

was premature. The police therefore continued to restrict traffic in the area until late September, effectively curtailing any recovery of the tourist industry.

Even the drawing up of emergency plans can invoke opposition from local municipalities. For example, civic leaders were unwilling to collaborate with volcanologists in the creation of an emergency plan for Galeres volcano, in Colombia, which threatens the nearby city of Pasto, because they did not want to arouse anxiety or create any economic disruption (Baxter, 1994).

## Disaster insurance

The insurance industry is often portrayed by those working in the disaster sphere as a panacea. This reflects an apparently widely held belief that if disaster insurance coverage was widespread then both the public and private burden of post-disaster relief and rehabilitation programs and potentially the very vulnerability of a country's building stock and infrastructure would be considerably reduced, the latter by making the issue of insurance coverage conditional upon the completion of disaster-proofing measures.

In reality, however, insurance does not offer a "solution" to the economic problems posed by potential disaster losses but merely a mechanism for the transfer of risk, effectively altering the implications of a disaster – in particular, by spreading its costs over perhaps many years – but not necessarily reducing them. The application of deductibles can also reduce the level of insurance pay-outs, especially where damage is widespread but relatively minor as, for example, might be the situation in the case of ashfall over a wide area.

Most countries have some form of insurance industry, often offering various forms of disaster insurance. However, coverage is typically far from universal, particularly in developing countries. For example, Blong and McKee (1995) estimated

---

[9] Seismic disturbances and other indications of volcanic activity had begun in July 1975, precipitating a number of spontaneous voluntary evacuations following more frightening episodes. Following a period of intensified activity, the entire area at risk from an eruption was evacuated in mid August as "given the immediate circumstances, there was no prudent alternative to a prompt transfer of the population at risk" (Zelinsky and Kosinski, 1991, p. 71). However, as the abnormal activity gradually subsided whilst economic pressures to return increased, the population were gradually permitted to return. In the first instance, some workers were permitted into the evacuated area during the day, returning to their temporary accommodation at night. By mid September, normal economic activities were reported to have recommenced in some two-thirds of the evacuation zone.

total uninsured private sector losses as a consequence of the 1994 Rabaul eruption, Papua New Guinea, at Kina 120 million (US$109 million) compared to insured private sector losses of Kina 60 million. Volcanic eruptions are also often relatively infrequent events, in some cases occurring centuries apart, implying that there may be little demand for volcanic insurance as a separate item. Instead, coverage may be provided under general household and business policies while certain claims relating to damage caused as an indirect consequence of an eruption may also be made under flood policies.

Volcanic insurance cover may be withdrawn in the aftermath of an eruption if the insurance industry experiences heavy losses, as demonstrated by relatively recent events in Montserrat and the Philippines. A number of local insurance companies finally canceled all volcanic insurance policies in the island of Montserrat in late August 1997, after the most recent scientific evidence indicated that there was an extremely high probability of very widespread losses. Meanwhile, in the Philippines, insurance against volcanic eruptions used to be provided to households and businesses located over 15 km from the nearest volcano at no extra charge. However, this practice was halted following the 1991 Mt. Pinatubo eruption while some insurance policies in lahar-vulnerable areas were canceled. As of late 1995, it was extremely difficult to obtain insurance policies for properties or businesses within a zone of 100 km around Mt. Pinatubo, presumably with knock on implications for investment in the region (Benson, 1997). Similarly, following the 1980 eruption of Mount St. Helens, some insurance companies were expected to reduce the extent of cover offered against volcanic hazards (Johnson and Jarvis, 1980).

### Relief and rehabilitation

Finally, appropriate and carefully targeted post-disaster relief and rehabilitation efforts can play an important role both in meeting immediate humanitarian needs and in ensuring a rapid economic recovery in the aftermath of a disaster. Such efforts should extend beyond the satisfaction of basic subsistence needs to include efforts to restore livelihood opportunities. Although, although there may be an urgent need to reconstruct damaged properties and infrastructure, it is also important to consider the risk of further volcanic hazards and to incorporate mitigation strategies into any redevelopment. Most obviously, careful hazard zoning is required. The reconstruction program following the 1985 eruption of Nevado del Ruiz provides an example of the types of such measures which may be included in rehabilitation and reconstruction programs (Box 15.7).

## Conclusion

Volcanic eruptions can have potentially serious impacts on the local economy, not only relating to the cost of physical damage but also to a wide and complex range of indirect and secondary impacts. Even relatively minor eruptions causing little damage to property or loss of life can entail significant economic costs.

However, to date there have been relatively few detailed studies of the economic impacts of volcanic hazards. Most available evidence is based on post-disaster damage assessments undertaken in the immediate aftermath of eruptions before the full extent of the longer-term economic impacts have become apparent. Meanwhile, there have been few attempts to explore the full short- and longer-term economic impacts of any eruption. Further research, going beyond a mere aggregation of the cost of physical damage, is required. This is not an academic exercise. Instead, such studies, by improving the general understanding of the economic consequences of volcanic eruptions, could play a vital role in ensuring both the adoption of appropriate prevention, mitigation, and preparedness strategies which reduce economic vulnerability to volcanic hazards and the design of relief and rehabilitation responses which address economic as well as short-term humanitarian needs.

The need for such research is particularly pertinent in view of the fact that a number of factors are likely to lead to an increase in the scale of magnitude of the economic impacts of volcanic eruptions in the future. Demographic

**Box 15.7** | Reconstruction following the 1985 Nevado del Ruiz eruption, Colombia

The reconstruction plan implemented in the wake of the 1985 eruption of Nevado del Ruiz listed the organization, recuperation, and development of economic activities amongst its objectives (Rocha-Jaramillo, 1986). The plan was divided into three components: risk prevention (both nationally and as a consequence of further eruptions of Nevado del Ruiz), with, as of 1986, a budget of Col$753 million (US$3.9 million); social rehabilitation, with a budget of Col$5980 million; and socio-economic and material reconstruction, with a budget of Col$44 191 million. The latter included programs for the technological development of the agricultural sector, in part to adapt to the new ecological conditions created as a consequence of the disaster, and employment development as well as reconstruction assistance. It also entailed certain mitigation measures, including the construction of some new roads to avoid the isolation of certain regions and cities in the event of further eruptions. As part of the rehabilitation process, the telecommunications network was also expanded and modernized and several new sub-power stations constructed. Similarly, the agricultural rehabilitation included assistance to modernize production and marketing practices. Thus the program not only rehabilitated the area but also aimed to improve the standard of living and accelerate the pace of economic development in the affected region.

In the wake of the eruption it was also recognized that greater emphasis should be placed on the integration of disaster mitigation and prevention measures into development planning at the local, regional, and national levels – for example, with regard to land use and the location of public utilities and facilities (Praez, 1986). Thus, the Tolima Regional Development Plan, for instance, included some proposals by which regional development could contribute to disaster prevention and post-disaster rehabilitation (Martinez, 1986).

pressures in certain regions of the world are already resulting in the increased occupation, particularly by poorer households, of volcanic areas. Environmental degradation is also playing an aggravating role, in some places increasing the potential scale of damage. For example, deforestation is reported to have played a role in facilitating the mobilization of sedimentation deposited as a consequence of the 1991 eruption of Mt. Pinatubo in the Philippines (Rantucci, 1994). Finally, economic development itself implies increased infrastructural and other investments and increasingly complex economic interdependencies between both different sectors and regions of an economy. If such developments occur without due regard to hazard-related risks then they could imply a rise in the direct, indi-

rect, and secondary economic impacts of volcanic events. However, such patterns are not inevitable. Instead, detailed risk assessments can play an important role in identifying appropriate strategies to mitigate the economic impacts of volcanic eruptions.

# References

Anderson, M. B. 1991. Which costs more: prevention or recovery? In A. Kreimer and M. Munasinghe (eds.) *Managing Natural Disasters and the Environment*, Selected Materials from the Colloquium on the Environment and Natural Disasters Management. Washington, DC, The World Bank.

Bakornas, P. B. A. 1982. *Report and Recommendations of the National Workshop on Mt. Galunggung Volcanic*

*Risk Management*, Workshop held in Bandung, Indonesia, September 20–25, 1981. Bandung, Indonesia, Badan Koordinasis Nasional Penanggulangan Bencana Alam.

Baxter, P. J. 1994. Vulnerability in volcanic hazards. In UK National Coordination Committee for the International Decade for Natural Disaster Reduction, Medicine in the International Decade for Natural Disaster Reduction (IDNDR) *Research, Preparedness and Response for Sudden Impact Disasters in the 1990s*, Proceedings of a Workshop held at the Royal Society, London. London, Royal Academy of Engineering.

Baxter, P. J., Berstein, R. S., Falk, H., *et al.* 1982. Medical aspects of volcanic disasters: an outline of hazards and emergency response measures. *Disasters*, **6**, 268–276.

Benson, C. 1995. *Economic and Social Consequences of the Activity of Volcan Irazú*. London, Overseas Development Institute.

1997. *The Economic Impact of Natural Disasters in the Philippines*, Working Paper Series no. 99. London, Overseas Development Institute.

Blong, R. 1984. *Volcanic Hazards: A Sourcebook on the Effects of Eruptions*. New York, Academic Press.

Blong, R. and Aislabie, C. 1988. *The Impact of Volcanic Hazards at Rabaul, Papua New Guinea*, Institute of National Affairs Discussion Paper no. 33. Port Moresby, Papua New Guinea, Institute of National Affairs.

Blong, R. and McKee, C. 1995. *The Rabaul Eruption 1994: Destruction of a Town*. Sydney, Australia, Natural Hazards Research Centre, Macquarie University.

Bryant, E. 1991. *Natural Hazards*. Cambridge, UK, Cambridge University Press.

Bull, R. 1992. Disaster economics. In *Disaster Management Training Programme*, 1st edn. New York, United Nations Development Programme and United Nations Disaster Relief Office.

Chang, M. 1992. Human decisions and natural hazards: a case of the East Rift Zone of Kilauea volcano on the island of Hawaii. In S. K. Majumdar, G. S. Forbes, E. W. Miller, *et al.* (eds.) *Natural and Technological Disasters: Causes, Effects and Preventive Measures*. Pittsburgh, PA, Pennsylvania Academy of Science, pp. 469–483.

Chardin, J.-P. 1996. *Report on Mission to the Philippines, 10–22 May*. Geneva, Switzerland, International Decade for Natural Disaster Reduction Secretariat.

Chester, D. 1993. *Volcanoes and Society*. London, Edward Arnold.

Chester, D. K., Dibben, C., Duncan, A. M., *et al.* 1996. The 1928 eruption of Mount Etna and the destruction of the town of Mascali: the hazard implications of short-lived, high effusion eruptions. In P. J. Gravestock and W. J. McGuire (eds.) *Etna: Fifteen Years On*. Cheltenham, UK, Cheltenham and Gloucester College of Higher Education, pp. 58–60.

Crandell, D. R., Booth, B., Kusumadinata, K., *et al.* 1984. *Source-Book for Volcanic-Hazards Zonation*. Paris, UNESCO.

Duncan, A. M., Dibben, C., Chester, D. K., *et al.* 1996. The 1928 eruption of Mount Etna volcano, Sicily, and the destruction of the town of Mascali. *Disasters*, **20**, 1–20.

Eastern Caribbean Central Bank 1994–6. *Economic and Financial Review*. Basseterre, St Kitts, and Nevis, Research Department.

Fairbairn, T. I. J. 1996. *The Economic Impact of Natural Disasters in the South Pacific with Special Reference to Papua New Guinea, Fiji, Western Samoa and Niue*, Report prepared for United Nations Department of Humanitarian Affairs, South Pacific Programme Office, Suva. Charlestown, Australia, Fairbairn Pacific Consultants Pty. Ltd.

Fernandez, C. and Gordon, J. 1993. Natural disasters and their human consequences: overcoming the vacuum between humanitarian aid and long term rehabilitation. In P. A. Merriman and C. W. A. Browitt (eds.) *Natural Disasters: Protecting Vulnerable Communities*. London, Thomas Telford, pp. 432–446.

Fischer, S. and Easterly, W. 1990. The economics of the government budget constraint. *The World Bank Research Observer*, **5**, 127–224.

Gobierno de Costa Rica 1964. *Consecuencias Económicas y Sociales de la Actividad del Volcán Irazú*. San José, Costa Rica, Gobierno de Costa Rica, Oficina de Planificación.

Germanoski, D. and Malinconico, L. L. 1992. Hazards associated with volcanoes and volcanic eruption. In S. K. Majumdar, G. S. Forbes, E. W. Miller, *et al.* (eds.) *Natural and Technological Disasters: Causes, Effects and Preventive Measures*. Pittsburgh, PA, Pennsylvania Academy of Science.

Hirose, H. 1982. Volcanic eruption in northern Japan. *Disasters*, **6**, 89–91.

Johnson, R. E. and Jarvis, J. S. 1980. The insurance industry response. In *Natural Disasters: Selected Problems and Implications*, seminar sponsored by the Continuing Legal Education Committee.

Washington, DC, Washington State Bar Association, pp. 3.1–3.42.

Kuester, I. and Forsyth, S. 1985. Rabaul eruption risk: population awareness and preparedness survey. *Disasters*, **9**, 179–182.

Lowenstein, P. L. 1988. *Rabaul Seismo-Deformational Crisis of 1983–85: Monitoring, Emergency Planning and Interaction with the Authorities, the Media and the Public*, Report no. 88/32. Rabaul, Papua New Guinea, Rabaul Volcanological Observatory.

Martinez, G. 1986. *Recent Colombian Experience in Dealing with the Postdisaster Rehabilitation/ Reconstruction Programme following the Nevado del Ruiz Volcanic Eruption*, paper prepared for United Nations Centre for Regional Development International Seminar on Regional Development Planning for Disaster Prevention. Bogota, Colombia, Unidad de Desarrollo Regional y Urabno, Departmento Nacional de Planeación.

McBeth, J. 1991. Misery to come. *Far Eastern Economic Review*, September 19.

Mileti, D. S., Bolton, P. A., Fernandez, G., *et al.* 1991. *Natural Disasters Studies: An Investigative Series of the Committee on Natural Disasters*, vol. 4, *The Eruption of Nevado del Ruiz Volcano, Colombia, South America*. Washington, DC, National Academy Press.

Murphy, T. P. 1980. Federal income tax aspects of casualty losses. In *Natural Disasters: Selected Problems and Implications*, Seminar sponsored by the Continuing Legal Education Committee. Washington, DC, Washington State Bar Association, pp. 7.1–7.16.

Murton, B. J. and Shimabukuro, S. 1972. Human adjustment to volcanic hazard in Puna District, Hawaii. *Paper presented at the 22nd International Geographical Congress*, July 24–30.

OECD 1994. *Guidelines on Aid and Environment*, no. 7, *Guidelines for Aid Agencies on Disaster Mitigation*. Paris, Development Assistance Committee, Organization for Economic Co-operation and Development.

OFDA 1995. *Disaster History: Significant Data on Major Disasters Worldwide, 1990–Present*. Washington, DC, Office of US Foreign Disaster Assistance, Agency for International Development.

Otero, R. C. and Marti, R. Z. 1995. The impacts of natural disasters on developing economies: implications for the international development and disaster community. In M. Munasinghe and C. Clarke (eds.) *Disaster Prevention for Sustainable Development: Economic and Policy Issues*, Report from the Yokohama World Conference on Natural Disaster Reduction, May 23–27, 1994. Washington, DC, The World Bank, pp. 11–40.

Perry, R. W. and Lindell, M. K. 1990. *Living with Mount St. Helens: Human Adjustment to Volcano Hazards*. Pullman, WA, Washington University Press.

PHIVOLCS 1991. *Mt. Pinatubo Wakes from 600 Years Slumber*. Quezon City, Philippines, Philippine Institute of Volcanology and Seismology, Department of Science and Technology.

Praez, G. M. 1986. Recent Colombian experience in dealing with the post-disaster rehabilitation/ reconstruction programme following the Nevado del Ruiz volcanic eruption. In *Planning for Crisis Relief: Towards Comprehensive Resource Management and Planning for Natural Disaster Prevention*. Nagoya, Japan United Nations Centre for Regional Development, pp. 217–228.

Pye-Smith, C. 1997. *The Philippines: In Search of Justice*. Oxford, UK, Oxfam UK and Ireland.

Rantucci, G. 1994. *Geological Disasters in the Philippines: The July 1990 Earthquake and the June 1991 Eruption of Mount Pinatubo – Description, Effects and Lessons Learned*, report in cooperation with the Philippine Institute of Volcanology and Seismology (PHIVOLCS). Rome, Italian Ministry of Foreign Affairs, Directorate General for Development Cooperation.

Rocha-Jaramillo, C. A. 1986. Rehabilitation programme for the area affected by the eruption of the volcano Nevado del Ruiz. In *Planning for Crisis Relief: Towards Comprehensive Resource Management and Planning for Natural Disaster Prevention*. Nagoya, Japan, United Nations Centre for Regional Development, pp. 229–235.

Rymer, H. 1993. Predicting volcanic eruptions using microgravity, and the mitigation of volcanic hazard. In P. A. Merriman and C. W. A. Browitt (eds.) *Natural Disasters: Protecting Vulnerable Communities*. London, Thomas Telford, pp. 252–269.

Schmalz, R. F. 1992. Volcanic eruptions. In S. K. Majumdar, G. S. Forbes, E. W. Miller, *et al.* (eds.) *Natural and Technological Disasters: Causes, Effects and Preventive Measures*. Pittsburg, PA, Pennsylvania Academy of Science.

Seaman, J., Leivesley, S., and Hogg, C. 1984. *Contributions to Epidemiology and Biostatistics*, vol. 5, *Epidemiology of Natural Disasters*. Basel, Switzerland, Karger.

Seitz, S. 1998. Coping strategies in an ethnic minority group: the Aeta of Mount Pinatubo. *Disasters*, **22**, 76–90.

Sigurdsson, H. and Carey, S. 1986. Volcanic disasters in Latin America and the 13th November 1985 eruption of Nevado del Ruiz volcano in Colombia. *Disasters*, **10**, 205–216.

Simkin, T. and Fiske, R. S. 1983. *Krakatau: The Volcanic Eruption and its Effects*. Washington, DC, Smithsonian Institution Press.

Sorensen, J. H. and Gersmehl, P. J. 1980. Volcanic hazard warning system: persistence and transferability. *Environmental Management*, **4**, 125–136.

Tayag, J. C. and Punongbayan, R. S. 1994. Volcanic disaster mitigation in the Philippines: experience from Mt. Pinatubo. *Disasters,* **18**, 1–15.

Tayag, J. C. and Rimando, R. E. 1987. Assessment of public awareness regarding geological hazards. In PHIVOLCS (ed.) *Geological Hazards and Disaster Preparedness Systems*. Manila, Philippines, Philippine Institute of Volcanology and Seismology, Department of Science and Technology.

Tiglao, R. and Tasker, R. 1995. Back in business. Far Eastern Economic Review, December 14, pp. 62–64.

Tuck, B. H., Huskey, L., and Talbot, L. 1992. *The Economic Consequences of the 1989–90 Mt. Redoubt Eruptions*, paper prepared for US Geological Survey, Alaska Volcano Observatory. Anchorage, University of Alaska, Institute of Social and Economic Research.

Umbal, J. V. 1987. Recent lahars of Mayon Volcano. In PHIVOLCS (ed.) *Geological Hazards and Disaster Preparedness Systems*. Manila, Philippines, Philippine Institute of Volcanology and Seismology, Department of Science and Technology, pp. 56–76.

United Nations 1979. *Disaster Prevention and Mitigation: A Compendium of Current Knowledge*, vol. 7, *Economic Aspects*. New York, United Nations.

US Army Corps of Engineers 1994. *Mount Pinatubo Recovery Action Plan Abbreviated Version: Eight River Basins, Republic of the Philippines*. Portland District, OR, US Army Corps of Engineers.

Watkins, K. 1995. Shadowland. *Guardian*, London, December 20, p. 20.

Young, S. R., Baxter, P. J., Pomonis, A., Ernst, G. G. J., and Benson, C. 1998. *Volcanic Hazards and Community Preparedness at Volcán Irazú, Costa Rica*. British Geological Survey Technical Report WC/98/16R. London, British Geological Survey.

Zelinsky, W. and Kosinski, L. A. 1991. *The Emergency Evacuation of Cities: A Cross-National Historical and Geographical Study*. Savage, MD, Rowman and Littlefield.

# Index